D0882682

Texts in Applied Mathematics **17**

Texts in Applied Mathematics

1. *Sirovich:* Introduction to Applied Mathematics.
2. *Wiggins:* Introduction to Applied Nonlinear Dynamical Systems and Chaos.
3. *Hale/Koçak:* Dynamics and Bifurcations.
4. *Chorin/Marsden:* A Mathematical Introduction to Fluid Mechanics, 3rd ed.
5. *Hubbard/West:* Differential Equations: A Dynamical Systems Approach, Part I: Ordinary Differential Equations.
6. *Sontag:* Mathematical Control Theory: Deterministic Finite Dimensional Systems.
7. *Perko:* Differential Equations and Dynamical Systems.
8. *Seaborn:* Hypergeometric Functions and Their Applications.
9. *Pipkin:* A Course on Integral Equations.
10. *Hoppensteadt/Peskin:* Mathematics in Medicine and the Life Sciences.
11. *Braun:* Differential Equations and Their Applications, 4th ed.
12. *Stoer/Bulirsch:* Introduction to Numerical Analysis, 2nd ed.
13. *Renardy/Rogers:* A First Graduate Course in Partial Differential Equations.
14. *Banks:* Growth and Diffusion Phenomena: Mathematical Frameworks and Applications.
15. *Brenner/Scott:* The Mathematical Theory of Finite Element Methods.
16. *Van de Velde:* Concurrent Scientific Computing.
17. *Marsden/Ratiu:* Introduction to Mechanics and Symmetry.

Jerrold E. Marsden Tudor S. Ratiu

Introduction to Mechanics and Symmetry

A Basic Exposition of Classical Mechanical Systems

With 43 Illustrations

Springer-Verlag
New York Berlin Heidelberg London Paris
Tokyo Hong Kong Barcelona Budapest

Jerrold E. Marsden
Department of Mathematics and EECS
University of California
Berkeley, CA 94720, USA

Tudor S. Ratiu
Department of Mathematics
University of California
Santa Cruz, CA 95064, USA

Mathematics Subject Classifications: 57R25, 58F05, 70Hxx

Library of Congress Cataloging-in-Publication Data
Marsden, Jerrold E.
Introduction to mechanics and symmetry / Jerrold E. Marsden, Tudor S. Ratiu
p. cm. — (Texts in applied mathematics ; 17)
Includes bibliographical references and index.
ISBN 0-387-97275-7 (New York). — ISBN 3-540-97275-7 (Berlin)
1. Mechanics, Analytic. 2. Symmetry (Physics) I. Ratiu, Tudor S. II. Title. III. Series.
QA808.M33 1994 94-10793
531—dc20

Printed on acid-free paper.

Production managed by Laura Carlson; manufacturing supervised by Gail Simon.
Photocomposed copy prepared from the authors' LaTeX files.
Printed and bound by R.R. Donnelley and Sons, Harrisonburg, VA.
Printed in the United States of America.

9 8 7 6 5 4 3 2 1

ISBN 0-387-94275-7 Springer-Verlag New York Berlin Heidelberg (hardcover edition)
ISBN 3-540-94275-7 Springer-Verlag Berlin Heidelberg New York

ISBN 0-387-94347-1 Springer-Verlag New York Berlin Heidelberg (softcover edition)
ISBN 3-540-94347-1 Springer-Verlag Berlin Heidelberg New York

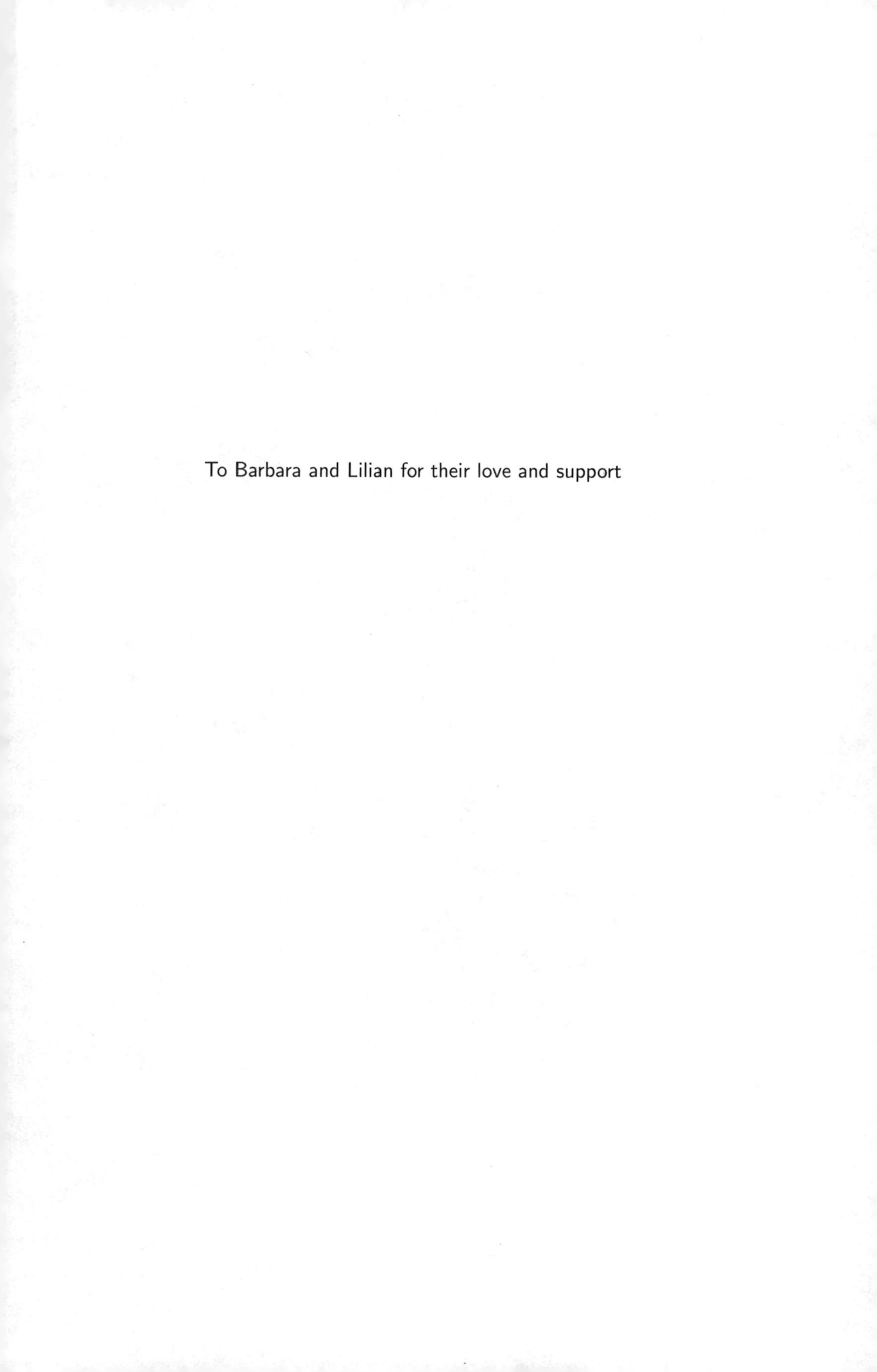

To Barbara and Lilian for their love and support

Series Preface

Mathematics is playing an ever more important role in the physical and biological sciences, provoking a blurring of boundaries between scientific disciplines and a resurgence of interest in the modern as well as the classical techniques of applied mathematics. This renewal of interest, both in research and teaching, has led to the establishment of the series: *Texts in Applied Mathematics (TAM)*.

The development of new courses is a natural consequence of a high level of excitement on the research frontier as newer techniques, such as numerical and symbolic computer systems, dynamical systems, and chaos, mix with and reinforce the traditional methods of applied mathematics. Thus, the purpose of this textbook series is to meet the current and future needs of these advances and encourage the teaching of new courses.

TAM will publish textbooks suitable for use in advanced undergraduate and beginning graduate courses, and will complement the *Applied Mathematical Sciences (AMS)* series, which will focus on advanced textbooks and research level monographs.

Preface

Symmetry and mechanics have been close partners since the time of the founding masters, namely, Newton, Euler, Lagrange, Laplace, Poisson, Jacobi, and Hamilton, and its subsequent developers, including Noether, Riemann, Routh, Kelvin, Poincaré, and Cartan. To this day, symmetry has continued to play a strong role, especially with the modern work of Arnold, Guillemin, Kirillov, Kostant, Moser, Smale, Souriau, Sternberg, and many others. This book is about these developments, with an emphasis on concrete applications that we hope will make it accessible to a wide variety of readers, especially senior undergraduate and graduate students in mathematics, physics, and engineering.

The geometric point of view in mechanics combined with solid analysis has been a phenomenal success in linking various diverse areas, both within and across standard disciplinary lines. It has provided both insight into fundamental issues in mechanics (such as Hamiltonian structures in continuum mechanics, fluid mechanics, and plasma physics) and provided useful tools in specific models such as new stability and bifurcation criteria using the energy-Casimir and energy-momentum methods, new numerical codes based on geometrically exact update procedures, and new reorientation techniques in control theory and robotics.

The role of symmetry in mechanical problems, which was already widely used by the founders of the subject, has been developed considerably in recent times to gain further understanding into such diverse phenomena as reduction, stability, and bifurcation relative to prescribed symmetries (symmetry breaking), methods of finding explicit solutions for integrable systems, and a deeper penetration into special systems, such as the Kowalewski top. We hope this book will provide a reasonable avenue to, and foundation for, these exciting developments.

Because of the extensive and complex set of possible directions in which one can develop the theory, we have provided a fairly lengthy introduction. It is intended to be read lightly at the beginning and then consulted from time to time as the text itself is read. This volume contains much of the basic theory of mechanics and should prove to be a useful foundation for further, as well as more specialized topics. In particular, due to space limitations we warn the reader that many important topics in mechanics are not treated in this volume. We are preparing a second volume on general reduction theory and its applications. With luck and a little support, it will be available in the near future.

A solution manual is available that contains complete solutions to many of the exercises and other supplementary comments. To obtain one, send a mailing label along with $15 to cover printing and postage to J. Marsden, Department of Mathematics, University of California, Berkeley, CA 94720.

We thank Alan Weinstein, Rudolf Schmid, and Rich Spencer for helping with an early set of notes that helped us on our way. Our many colleagues, students, and readers, especially Henry Abarbanel, Vladimir Arnold, Larry Bates, Michael Berry, Tony Bloch, Marty Golubitsky, Mark Gotay, George Haller, Aaron Hershman, Darryl Holm, Phil Holmes, Sameer Jalnapurkar, Edgar Knobloch, P.S. Krishnaprasad, Debra Lewis, Robert Littlejohn, Richard Montgomery, Phil Morrison, Richard Murray, Oliver O'Reilly, George Patrick, Octavian Popp, Matthias Reinsch, Shankar Sastry, Juan Simo, Hans Troger, and Steve Wiggins have our deepest gratitude for their encouragement and suggestions. We also collectively thank all our students and colleagues who have used these notes and have provided valuable advice. We are also indebted to Carol Cook, Anne Kao, Nawoyuki Gregory Kubota, Sue Knapp, Barbara Marsden, Marnie McElhiney, June Meyermann, Teresa Wild, and Ester Zack for their dedicated and patient work on the typesetting and artwork for this book. We want to single out with special thanks, Nawoyuki Gregory Kubota for his special effort with the typesetting and the figures (including the cover illustration and his skillful use of Mathematica). We also thank the staff at Springer-Verlag, especially Laura Carlson, Ken Dreyhaupt, Rüdiger Gebauer, and Karen Kosztolnyik for their skillful editorial work and production of the book.

Berkeley, CA
Santa Cruz, CA
Spring, 1994

Jerry Marsden
Tudor Ratiu

Contents

About the Authors

Jerrold E. Marsden is professor of mathematics and EECS at the University of California at Berkeley. He has done extensive research in the area of geometric mechanics, with applications to rigid body systems, fluid mechanics, elasticity theory, and plasma theory, as well as to general field theory, including relativistic fields. He also works in dynamical systems and control theory, especially how it relates to mechanical systems and systems with symmetry. He was one of the original founders in the early 1970s of reduction theory for mechanical systems with symmetry, which remains an active and much-studied area of research today. He was the recipient of the prestigious Norbert Wiener prize of the American Mathematical Society and the Society for Industrial and Applied Mathematics in 1990. He has been a Carnegie Fellow at Heriot-Watt University (1977), a Killam Fellow at the University of Calgary (1979), a Miller Fellow at the University of California, Berkeley (1981–1982), a Humboldt Fellow in Germany (1991), and a Fairchild Fellow at Caltech (1992). He has served in several administrative capacities, such as director of the Research Group in Nonlinear Systems and Dynamics at Berkeley (1984–1986), the National Science Foundation Advisory Panel for Mathematics, the Advisory Committee of the Mathematical Sciences Institute at Cornell, and as director of the Fields Institute (1990–1994). He has served as an editor for Springer-Verlag's Applied Mathematical Sciences series since 1982 and is an editor for Springer-Verlag's *Journal of Nonlinear Science*.

Tudor S. Ratiu is professor of mathematics at the University of California at Santa Cruz. He has previously taught at the University of Michigan, Ann Arbor, as a T.H. Hildebrandt Research Assistant Professor (1980–1983) and at the University of Arizona, Tucson (1983–1987). His research interests center on geometric mechanics, symplectic geometry, global analysis, and infinite dimensional Lie theory, together with their applications to integrable systems, nonlinear dynamics, continuum mechanics, plasma physics, and bifurcation theory. He has been a National Science Foundation Postdoctoral Fellow (1983–1986), an A.P. Sloan Foundation Fellow (1984–1987), and a Miller Research Professor at the University of California at Berkeley (1994). He was a member of the Mathematical Sciences Research Institute in Berkeley (1983–1984, 1989–1990, 1994), the Center of Nonlinear Science at the Los Alamos National Laboratory (1983), the Max Planck Institute of Mathematics in Bonn, Germany (1984, 1990), the Mathematical Sciences Institute at Cornell University (1990), the University of Montpellier, France (1990), the Fields Institute (1993), the Erwin Schrödinger Institute of Mathematical Physics in Vienna, Austria (1993, 1994), and the Institut des Hautes Etudes Scientifiques in Bures-sur-Yvette, France (1994–1995). Since his arrival at UC Santa Cruz in 1987, he has been on the executive committee of the Nonlinear Sciences Organized Research Unit. He is currently on the editorial board of the *Annals of Global Analysis* and the *Annals of the University of Timisoara*.

1
Introduction and Overview

1.1 Lagrangian and Hamiltonian Formalisms

Classical mechanics deals with the dynamics of particles, rigid bodies, continuous media (fluid, plasma, and solid mechanics), and other fields (such as electromagnetism, gravity, etc.). This theory also plays a crucial role in quantum mechanics, in control theory and other areas of physics, engineering and even chemistry and biology. Clearly classical mechanics is a large subject that plays a fundamental role in science. Throughout history, mechanics has also played a key role in the development of mathematics. Starting with the creation of calculus stimulated by Newton's mechanics, it continues today with exciting developments in group representations, geometry, and topology; these mathematical developments in turn are being applied to interesting problems in physics and engineering.

Symmetry has always played an important role in mechanics, from fundamental formulations of basic principles to concrete applications, such as stability criteria for rotating structures. The theme of this book is to emphasize the role of symmetry in various aspects of mechanics.

Warning This introduction treats a collection of topics fairly rapidly. The student should not expect to understand everything perfectly at this stage. *We will return to many of the topics in subsequent chapters.*

Mechanics has two main branches, ***Lagrangian mechanics*** and ***Hamiltonian mechanics***. In one sense, Lagrangian mechanics is more fundamental since it is based on variational principles and it is what generalizes most directly to the general relativistic context. In another sense, Hamiltonian mechanics is more fundamental, since it is based directly on the energy concept and it is what is more closely tied to quantum mechanics. Fortunately, in many cases these branches are equivalent as we shall see in detail in Chapter 7. Needless to say, the merger of quantum mechanics and general relativity remains one of the main problems of mechanics.

The Lagrangian formulation of mechanics can be based on the observation that there are variational principles behind the fundamental laws of force balance as given by Newton's law in $\mathbf{F} = m\mathbf{a}$. One chooses a configuration space Q with coordinates $q^i, i = 1, \ldots, n$, that describe the ***configuration*** of the system under study. Then one introduces the ***Lagrangian*** $L(q^i, \dot{q}^i, t)$, which is shorthand notation for $L(q^1, \ldots, q^n, \dot{q}^1, \ldots, \dot{q}^n)$. Usually, L is the kinetic *minus* the potential energy of the system and one takes $\dot{q}^i = dq^i/dt$ regarded as the velocity. The ***variational principle of Hamilton*** states

$$\delta \int_a^b L(q^i, \dot{q}^i, t)\, dt = 0. \tag{1.1.1}$$

In (1.1.1), we choose curves $q^i(t)$ joining two fixed points in Q over a fixed time interval $[a, b]$, and calculate the integral regarded as a function of this curve. Then (1.1.1) states that this function has a critical point. If we let δq^i be a variation of the curve (and proceed somewhat formally at first), then by the chain rule, (1.1.1) is equivalent to

$$\sum_{i=1}^n \int_a^b \left(\frac{\partial L}{\partial q^i} \delta q^i + \frac{\partial L}{\partial \dot{q}^i} \delta \dot{q}^i \right) dt = 0 \tag{1.1.2}$$

for all variations δq^i.

Using $\delta \dot{q}^i = \frac{d}{dt} \delta q^i$ (which is essentially the equality of mixed partials), integrating the second term by parts, and using the boundary conditions $\delta q^i = 0$ at $t = a$ and b, (1.1.2) becomes

$$\sum_{i=1}^n \int_a^b \left(\frac{\partial L}{\partial q^i} - \frac{d}{dt} \left(\frac{\partial L}{\partial \dot{q}^i} \right) \right) \delta q^i\, dt = 0. \tag{1.1.3}$$

Since δq^i is arbitrary (apart from being zero at the endpoints), (1.1.2) is equivalent to the ***Euler-Lagrange equations***

$$\frac{d}{dt} \frac{\partial L}{\partial \dot{q}^i} - \frac{\partial L}{\partial q^i} = 0, \quad i = 1, \ldots, n. \tag{1.1.4}$$

(This topic will be discussed at greater length in §7.3 and §8.1). For the case of kinetic minus potential energy for a system of particles, where L has the form

$$L(q^i, \dot{q}^i, t) = \frac{1}{2} \sum_{i=1}^n m_i \|\dot{q}^i\|^2 - V(q^i), \tag{1.1.5}$$

(1.1.4) reduces to

$$\frac{d}{dt}(m_i \dot{q}^i) = -\frac{\partial V}{\partial q^i}, \tag{1.1.6}$$

which is $\mathbf{F} = m\mathbf{a}$ for the motion of a particle in the potential field V.

Already at this stage, interesting links with geometry emerge. If $g_{ij}(q)$ is a given metric tensor (for now, just think of this as a q-dependent positive-definite symmetric $n \times n$ matrix) and we consider the kinetic energy Lagrangian

$$L(q^i, \dot{q}^i) = \frac{1}{2} \sum_{i,j=1}^{n} g_{ij}(q)\dot{q}^i \dot{q}^j, \tag{1.1.7}$$

then *the Euler-Lagrange equations are equivalent to the equations of geodesic motion*, as can be directly verified (see §7.5 for details). Conservation laws that are a result of symmetry in a mechanical context can then be applied to yield interesting geometric facts. For instance, we will see that theorems about geodesics on surfaces of revolution can be readily proved this way.

The Lagrangian formalism can be extended to the infinite dimensional case. Here the q^i are replaced by *fields* $\varphi^1, \ldots, \varphi^m$ which are, for example, functions of spatial points x^i and time. Then L is a function of $\varphi^1, \ldots, \varphi^m, \dot{\varphi}^1, \ldots, \dot{\varphi}^m$ and the spatial derivatives of the fields. We shall deal with various examples of this later, but we emphasize that properly interpreted, the variational principle and the Euler-Lagrange equations remain intact. One simply replaces the partial derivatives in the Euler-Lagrange equations by *functional derivatives* defined below.

To pass to the Hamiltonian formalism, introduce the ***conjugate momenta***

$$p_i = \frac{\partial L}{\partial \dot{q}^i}, \qquad i = 1, \ldots, n, \tag{1.1.8}$$

make the change of variables $(q^i, \dot{q}^i) \mapsto (q^i, p_i)$, and introduce the Hamiltonian

$$H(q^i, p_i, t) = \sum_{j=1}^{n} p_j \dot{q}^j - L(q^i, \dot{q}^i, t). \tag{1.1.9}$$

Remembering the change of variables, we make these computations:

$$\frac{\partial H}{\partial p_i} = \dot{q}^i + \sum_{j=1}^{n} \left(p_j \frac{\partial \dot{q}^j}{\partial p_i} - \frac{\partial L}{\partial \dot{q}^j} \frac{\partial \dot{q}^j}{\partial p_i} \right) = \dot{q}^i \tag{1.1.10}$$

and

$$\frac{\partial H}{\partial q^i} = \sum_{j=1}^{n} p_j \frac{\partial \dot{q}^j}{\partial q^i} - \frac{\partial L}{\partial q^i} - \sum_{j=1}^{n} \frac{\partial L}{\partial \dot{q}^j} \frac{\partial \dot{q}^j}{\partial q^i} = -\frac{\partial L}{\partial q^i}, \tag{1.1.11}$$

where (1.1.8) has been used twice. Using (1.1.4) and (1.1.8), we see that (1.1.11) is equivalent to

$$\frac{\partial H}{\partial q^i} = -\frac{d}{dt} p_i. \tag{1.1.12}$$

Thus, *the Euler-Lagrange equations are equivalent to* ***Hamilton's equations***

$$\frac{dq^i}{dt} = \frac{\partial H}{\partial p_i}, \qquad \frac{dp_i}{dt} = -\frac{\partial H}{\partial q^i}, \qquad i = 1, \ldots, n. \tag{1.1.13}$$

The analogous Hamiltonian partial differential equations for time dependent *fields* $\varphi^1, \ldots, \varphi^m$ and their conjugate momenta $\pi_1, ..., \pi_m$, are

$$\frac{\partial \varphi^a}{\partial t} = \frac{\delta H}{\partial \pi_a}, \qquad \frac{\partial \pi_a}{\partial t} = -\frac{\delta H}{\partial \varphi^a}, \qquad a = 1, \ldots, m, \tag{1.1.14}$$

where H is a functional of the fields φ^a and π_a, and the ***variational*** or ***functional derivatives*** are defined through the equation

$$\begin{aligned} \int_{\mathbb{R}^n} \frac{\delta H}{\delta \varphi^1} \delta\varphi^1 \, d^n x &= \lim_{\varepsilon \to 0} \frac{1}{\varepsilon} [H(\varphi^1 + \varepsilon \delta \varphi^1, \varphi^2, \ldots, \varphi^m, \pi_1, \ldots, \pi_m) \\ &\qquad - H(\varphi^1, \varphi^2, \ldots, \varphi^m, \pi_1, \ldots, \pi_m)], \end{aligned} \tag{1.1.15}$$

and similarly for $\delta H/\delta\varphi^2, \ldots, \delta H/\delta \pi_m$. Both equations (1.1.13) and (1.1.14) can be recast in ***Poisson bracket form***

$$\dot{F} = \{F, H\}, \tag{1.1.16}$$

where the brackets in the respective cases are given by

$$\{F, G\} = \sum_{i=1}^{n} \left(\frac{\partial F}{\partial q^i} \frac{\partial G}{\partial p_i} - \frac{\partial F}{\partial p_i} \frac{\partial G}{\partial q^i} \right) \tag{1.1.17}$$

and

$$\{F, G\} = \sum_{a=1}^{m} \int_{\mathbb{R}^n} \left(\frac{\delta F}{\delta \varphi^a} \frac{\delta G}{\delta \pi_a} - \frac{\delta F}{\delta \pi_a} \frac{\delta G}{\delta \varphi^a} \right) d^n x. \tag{1.1.18}$$

There is also a variational principle valid directly on the Hamiltonian side. For the Euler-Lagrange equations, we deal with curves in q-space, whereas for Hamilton's equations we deal with curves in (q,p)-space. The principle is

$$\delta \int_a^b \sum_{i=1}^{n} [p_i \dot{q}^i - H(q^j, p_j)] \, dt = 0 \tag{1.1.19}$$

as is readily verified; one requires $p_i \delta q^i = 0$ at the endpoints.

This formalism is the basis for the analysis of many important systems in particle dynamics and field theory, as described in standard texts such as Whittaker [1927], Goldstein [1980], Arnold [1989], Thirring [1978], and Abraham and Marsden [1978]. The underlying geometric structures that are important for this formalism are those of *symplectic* and *Poisson geometry*. How these structures are related to the Euler-Lagrange equations and variational principles via the Legendre transformation is an essential ingredient of the story. Furthermore, in the infinite-dimensional case it is fairly well understood how to deal rigorously with many of the functional analytic difficulties that arise; see, for example, Chernoff and Marsden [1974] and Marsden and Hughes [1983].

Exercise

1.1-1 *Show by* ***direct calculation*** *that the classical Poisson bracket satisfies the* ***Jacobi identity****. That is, if F and K are both functions of the $2n$ variables $(q^1, q^2, \ldots, q^n, p_1, p_2, \ldots, p_n)$ and we define*

$$\{F, K\} = \sum_{i=1}^{n} \left(\frac{\partial F}{\partial q^i} \frac{\partial K}{\partial p_i} - \frac{\partial K}{\partial q^i} \frac{\partial F}{\partial p_i} \right),$$

then the identity $\{L, \{F, K\}\} + \{K, \{L, F\}\} + \{F, \{K, L\}\} = 0$ holds.

1.2 The Rigid Body

It was already clear in the last century that certain systems apparently resist the canonical formalism outlined in §1.1. For example, to obtain a Hamiltonian description for fluids, Clebsch [1857, 1859] found it necessary to introduce certain nonphysical potentials. (For a modern account of Clebsch potentials and further references, see Marsden and Weinstein [1983], Marsden, Ratiu, and Weinstein [1984a,b], Cendra and Marsden [1987], and Cendra, Ibort, and Marsden [1987].) We will discuss fluids in §1.4 below.

An analogous situation is exhibited by the Euler equations of rigid body dynamics. In the absence of external forces, the equations are usually written as follows, as we shall derive in detail in Chapter 15:

$$\begin{aligned} I_1 \dot{\Omega}_1 &= (I_2 - I_3)\Omega_2\Omega_3, \\ I_2 \dot{\Omega}_2 &= (I_3 - I_1)\Omega_3\Omega_1, \\ I_3 \dot{\Omega}_3 &= (I_1 - I_2)\Omega_1\Omega_2, \end{aligned} \qquad (1.2.1)$$

where $\Omega = (\Omega_1, \Omega_2, \Omega_3)$ is the body angular velocity vector and I_1, I_2, I_3 are the moments of inertia. Are equations (1.2.1) Lagrangian or Hamiltonian in any sense? Since there are an odd number of equations, they cannot be put in canonical Hamiltonian form in the sense of equations (1.1.13).

One possible answer is that to see the Lagrangian (or Hamiltonian) structure of the rigid body equations, one can use the description in terms of Euler angles θ, φ, ψ and their velocities $\dot{\theta}, \dot{\varphi}, \dot{\psi}$ (or conjugate momenta $p_\theta, p_\varphi, p_\psi$), relative to which the equations *are* in Euler-Lagrange (or canonical Hamiltonian) form. However, this procedure requires using *six equations* while many questions are easier to study using the *three equations* (1.2.1).

To see how (1.2.1) are Lagrangian in some sense, introduce the Lagrangian

$$L(\Omega) = \frac{1}{2}(I_1\Omega_1^2 + I_2\Omega_2^2 + I_3\Omega_3^2) \tag{1.2.2}$$

which, as we will see in detail in Chapter 15, is the kinetic energy of the rigid body. Regarding $I\Omega = (I_1\Omega_1, I_2\Omega_2, I_3\Omega_3)$ as a vector, we write (1.2.1) as

$$\frac{d}{dt}\frac{\partial L}{\partial \Omega} = \frac{\partial L}{\partial \Omega} \times \Omega \tag{1.2.3}$$

which is the rigid body form of the Euler-Lagrange equations. These equations appear explicitly in Lagrange [1788] (Volume 2, p.212) and were generalized to arbitrary Lie algebras by Poincaré [1901]. We will discuss the general case in Chapter 13. We can also write a variational principle for (1.2.3) that is analogous to that for the Euler-Lagrange equations, but is written *directly* in terms of Ω. Namely, (1.2.3) is equivalent to

$$\delta \int_a^b L\, dt = 0, \tag{1.2.4}$$

where variations of Ω are of the form

$$\delta\Omega = \dot{\Sigma} + \Omega \times \Sigma, \tag{1.2.5}$$

and where Σ vanishes at the endpoints. This may be proved in the same way as we proved that the variational principle (1.1.1) is equivalent to the Euler-Lagrange equations (1.1.4); see Exercise **1.2-2**. In fact, later on in Chapter 13, we shall see how to *derive* this variational principle from the more "primitive" one (1.1.1).

If, instead of variational principles, we concentrate on Poisson brackets and drop the requirement that they be in the canonical form (1.1.17), then there is also a simple and beautiful Hamiltonian structure for the rigid body equations. To state it, introduce the ***angular momenta***

$$\Pi_i = I_i\Omega_i = \frac{\partial L}{\partial \Omega_i}, \quad i = 1, 2, 3, \tag{1.2.6}$$

so that the Euler equations become

$$\dot{\Pi}_1 = \frac{I_2 - I_3}{I_2 I_3}\Pi_2\Pi_3, \quad \dot{\Pi}_2 = \frac{I_3 - I_1}{I_3 I_1}\Pi_3\Pi_1, \quad \dot{\Pi}_3 = \frac{I_1 - I_2}{I_1 I_2}\Pi_1\Pi_2, \tag{1.2.7}$$

that is,

$$\dot{\mathbf{\Pi}} = \mathbf{\Pi} \times \mathbf{\Omega}. \tag{1.2.8}$$

Introduce the following rigid body Poisson bracket on functions of the $\mathbf{\Pi}$'s:

$$\{F, G\}(\mathbf{\Pi}) = -\mathbf{\Pi} \cdot (\nabla F \times \nabla G) \tag{1.2.9}$$

and the Hamiltonian

$$H = \frac{1}{2}\left(\frac{\Pi_1^2}{I_1} + \frac{\Pi_2^2}{I_2} + \frac{\Pi_3^2}{I_3}\right). \tag{1.2.10}$$

One checks (Exercise **1.2-3**) that Euler's equations (1.2.7) are equivalent to

$$\dot{F} = \{F, H\}. \tag{1.2.11}$$

This simple result is implicit in many works, such as Arnold [1966, 1969], and is given explicitly in this form for the rigid body in Sudarshan and Mukunda [1974]. (Some preliminary versions were given by Pauli [1953], Martin [1959], and Nambu [1973].) On the other hand, the variational form (1.2.4) appears to be due to Poincaré [1901b] and Hamel [1904], at least implicitly. It is given explicitly in Marsden and Scheurle [1993a,b].

Notice that conservation of total angular momentum holds for *any* equation of the form (1.2.11), regardless of the Hamiltonian; indeed, with

$$C(\mathbf{\Pi}) = \frac{1}{2}(\Pi_1^2 + \Pi_2^2 + \Pi_3^2),$$

we have $\nabla C(\mathbf{\Pi}) = \mathbf{\Pi}$, and so

$$\begin{aligned}
\frac{d}{dt}\frac{1}{2}(\Pi_1^2 + \Pi_2^2 + \Pi_3^2) &= \{C, H\}(\mathbf{\Pi}) \\
&= -\mathbf{\Pi} \cdot (\nabla C \times \nabla H) \\
&= -\mathbf{\Pi} \cdot (\mathbf{\Pi} \times \nabla H) = 0.
\end{aligned}$$

Functions such as these that Poisson commute with *every* function are called ***Casimir functions***; they play an important role in the study of *stability*, as we shall see later.

Historical Note H.B.K. Casimir was a student of P. Ehrenfest and wrote his thesis on the quantum mechanics of the rigid body. Ehrenfest in turn wrote his thesis under Boltzman around 1900 on variational principles in fluid dynamics and was one of the first to study them in material, rather than Clebsch representation. This is a seed for many important ideas in this book. See also the introductory comments in §1.4.

Exercises

1.2-1 *Show by direct calculation that the rigid body Poisson bracket satisfies the Jacobi identity. That is, if F and K are both functions of* (Π_1, Π_2, Π_3) *and we define*

$$\{F, K\}(\mathbf{\Pi}) = -\mathbf{\Pi} \cdot (\nabla F \times \nabla K),$$

then the identity $\{L, \{F, K\}\} + \{K, \{L, F\}\} + \{F, \{K, L\}\} = 0$ *holds.*

1.2-2 *Verify directly that the Euler equations for a rigid body are equivalent to*

$$\delta \int L \, dt = 0$$

for variations $\delta\Omega = \dot{\Sigma} + \Omega \times \Sigma$, *where* Σ *vanishes at the endpoints.*

1.2-3 *Verify directly that the Euler equations for a rigid body are equivalent to the equations*

$$\frac{d}{dt} F = \{F, H\},$$

where $\{\,,\}$ *is the rigid body Poisson bracket and H is the rigid body Hamiltonian.*

1.2-4(a) *Show that the rotation group* $SO(3)$ *can be identified with the* ***Poincaré sphere****: that is, the unit circle bundle of the two sphere* S^2, *that is, the set of unit tangent vectors to the two-sphere in* $\mathbb{R}^3$.

(b) *Using the known fact that any (continuous) vector field on* S^2 *must vanish somewhere, show that* $SO(3)$ *cannot be written as* $S^2 \times S^1$.

1.3 Lie-Poisson Brackets, Poisson Manifolds, and Momentum Maps

The rigid body variational principle and the rigid body Poisson bracket are special cases of general constructions associated to any ***Lie algebra*** $\mathfrak{g}$, that is, a vector space together with a bilinear, antisymmetric bracket $[\xi, \eta]$ satisfying ***Jacobi's identity***:

$$[[\xi, \eta], \zeta] + [[\zeta, \xi], \eta] + [[\eta, \zeta], \xi] = 0 \qquad (1.3.1)$$

for all $\xi, \eta, \zeta \in \mathfrak{g}$. For example, the Lie algebra associated to the rotation group is $\mathfrak{g} = \mathbb{R}^3$ with bracket $[\xi, \eta] = \xi \times \eta$, the ordinary vector cross product.

The construction of a variational principle on $\mathfrak{g}$, as we shall see in Chapter 13, replaces $\delta\Omega = \dot{\Sigma} + \Omega \times \Sigma$ by $\delta\xi = \dot{\eta} + [\eta, \xi]$. The resulting general equations on $\mathfrak{g}$, which we will study in detail in Chapter 13, are called the ***Euler-Poincaré equations***. These equations are valid for either finite or infinite dimensional Lie algebras. For now, we state them in the finite dimensional case. To do so, we use the following notation. We choose a basis $e_1, \ldots, e_r$ of $\mathfrak{g}$ (so dim $\mathfrak{g} = r$), the ***structure constants*** C_{ab}^d are defined by

$$[e_a, e_b] = \sum_{d=1}^{r} C_{ab}^d e_d, \tag{1.3.2}$$

where a, b run from 1 to r. If ξ is an element of the Lie algebra, its components relative to this basis are denoted ξ^a. If $e^1, \ldots, e^n$ is the corresponding dual basis, then the components of the differential of the Lagrangian L are just the partial derivatives $\partial L/\partial \xi^a$. Then the Euler-Poincaré equations are

$$\frac{d}{dt}\frac{\partial L}{\partial \xi^d} = c_{ad}^b \frac{\partial L}{\partial \xi^b} \xi^a.$$

For example, for $L : \mathbb{R}^3 \to \mathbb{R}$, the Euler-Poincaré equations become

$$\frac{d}{dt}\frac{\partial L}{\partial \Omega} = \frac{\partial L}{\partial \Omega} \times \Omega,$$

which generalize the Euler equations for rigid body motion. As we mentioned earlier, these equations were written down for a fairly general class of L by Lagrange [1788, Volume 2, Equation A on p. 212.] It was Poincaré [1901b] who generalized them to any Lie algebra.

We can also generalize the rigid body Poisson bracket as follows: Let F, G be defined on the dual space $\mathfrak{g}^*$. Denoting elements of $\mathfrak{g}^*$ by μ, let the ***functional derivative*** of F at μ be the unique element $\delta F/\delta \mu$ of $\mathfrak{g}$ defined by

$$\lim_{\varepsilon \to 0} \frac{1}{\varepsilon}[F(\mu + \varepsilon \delta\mu) - F(\mu)] = \left\langle \delta\mu, \frac{\delta F}{\delta \mu} \right\rangle, \tag{1.3.3}$$

where $\langle \, , \rangle$ denotes the pairing between $\mathfrak{g}^*$ and $\mathfrak{g}$. This definition (1.3.3) is consistent with the definition of $\delta F/\delta \varphi$ given in (1.1.15) when $\mathfrak{g}$ and $\mathfrak{g}^*$ are chosen to be appropriate spaces of fields. Define the ($\pm$) ***Lie-Poisson brackets*** by

$$\{F, G\}_\pm(\mu) = \pm \left\langle \mu, \left[\frac{\delta F}{\delta \mu}, \frac{\delta G}{\delta \mu}\right] \right\rangle. \tag{1.3.4}$$

Using the coordinate notation introduced above, the ($\pm$) Lie-Poisson brackets become

$$\{F, G\}_\pm = \pm \sum_{a,b,d=1}^{r} C_{ab}^d \mu_d \frac{\partial F}{\partial \mu_a} \frac{\partial G}{\partial \mu_b}, \tag{1.3.5}$$

where $\mu = \mu_a e^a$.

The Lie-Poisson bracket and the canonical brackets from the last section have four simple but crucial properties that are readily verified:

PB1 $\{F, G\}$ is a real bilinear in F and G.
PB2 $\{F, G\} = -\{G, F\}$, antisymmetry.
PB3 $\{\{F, G\}, H\} + \{\{H, F\}, G\} + \{\{G, H\}, F\} = 0$, Jacobi identity.
PB4 $\{FG, H\} = F\{G, H\} + \{F, H\}G$, Leibniz identity.

A manifold (that is, an n-dimensional "smooth surface") P together with a bracket operation on $C^\infty(P)$, the space of smooth functions on P, and satisfying properties **PB1–PB4**, is called a ***Poisson manifold***. In particular, *$\mathfrak{g}^*$ is a Poisson manifold*. In Chapter 10 we will study the general concept of a Poisson manifold.

For example, if we choose $\mathfrak{g} = \mathbb{R}^3$ with the bracket taken to be the cross product $[\mathbf{x}, \mathbf{y}] = \mathbf{x} \times \mathbf{y}$, and identify $\mathfrak{g}^*$ with $\mathfrak{g}$ using the dot product on $\mathbb{R}^3$ (so $\langle \mathbf{\Pi}, \mathbf{x} \rangle = \mathbf{\Pi} \cdot \mathbf{x}$ is the usual dot product), then the $(-)$ Lie-Poisson bracket becomes the rigid body bracket. The theory of Poisson manifolds reduces to that of symplectic manifolds in case the bracket is nondegenerate (that is, the only local casimir functions are constants); this result of Pauli [1953] and Jost [1964] will be proved in Chapter 10.

On a Poisson manifold $(P, \{\cdot\,, \cdot\})$, associated to any function H there is a vector field, denoted by X_H, which has the property that for any smooth function $F : P \to \mathbb{R}$ we have the identity $\langle \mathbf{d}F, X_H \rangle = \mathbf{d}F \cdot X_H = \{F, H\}$. We say that *the vector field X_H is **generated** by the function H* and that X_H is a ***Hamiltonian vector field***. We define the associated ***dynamical system*** whose points z in phase space evolve in time by the differential equation

$$\dot{z} = X_H(z). \tag{1.3.6}$$

This definition is consistent with the equations in Poisson bracket form (1.1.16). The function H may have the interpretation of the energy of the system, but of course the definition (1.3.6) makes sense for *any* function. For canonical systems with the Poisson bracket given by (1.1.17), X_H is given by the formula

$$X_H(q^i, p_i) = \left(\frac{\partial H}{\partial p_i}, -\frac{\partial H}{\partial q^i} \right), \tag{1.3.7}$$

whereas for the rigid body bracket given on $\mathbb{R}^3$ by (1.2.9),

$$X_H(\mathbf{\Pi}) = \mathbf{\Pi} \times \nabla H(\mathbf{\Pi}). \tag{1.3.8}$$

Two features of the rigid body equations carry over to more general Lie algebras. First of all, the Lie-Poisson brackets arise from canonical brackets on the cotangent bundle (phase space) T^*G associated with a Lie group G with $\mathfrak{g}$ as its Lie algebra. That is, there is a general construction underlying the association

$$(\theta, \varphi, \psi, p_\theta, p_\varphi, p_\psi) \mapsto (\Pi_1, \Pi_2, \Pi_3) \tag{1.3.9}$$

given explicitly by:

$$\begin{aligned}
\Pi_1 &= \frac{1}{\sin\theta}[(p_\varphi - p_\psi \cos\theta)\sin\psi + p_\theta \sin\theta\cos\psi], \\
\Pi_2 &= \frac{1}{\sin\theta}[(p_\varphi - p_\psi \cos\theta)\cos\psi - p_\theta \sin\theta\sin\psi], \\
\Pi_3 &= p_\psi.
\end{aligned} \tag{1.3.10}$$

This rigid body map takes the canonical bracket in the variables (θ, φ, ψ) and their conjugate momenta $(p_\theta, p_\varphi, p_\psi)$ to the $(-)$ Lie-Poisson bracket; that is, if F and K are functions of Π_1, Π_2, Π_3, they determine functions of $(\theta, \varphi, \psi, p_\theta, p_\varphi, p_\psi)$ by substituting (1.3.10). Then a (slightly tedious but straightforward) exercise using the chain rule shows that

$$\{F, K\}_{(-)\{\text{Lie-Poisson}\}} = \{F, K\}_{\text{canonical}}. \tag{1.3.11}$$

We say that the map defined by (1.3.10) is a ***canonical map*** or a Poisson map and that the $(-)$ Lie-Poisson bracket has been obtained from the canonical bracket by ***reduction***.

For the rigid body, G is the (proper) rotation group $SO(3)$ and the Euler angles and their conjugate momenta are coordinates for T^*G. The choice of T^*G as the primitive phase space is made according to the classical procedures of mechanics: the configuration space $SO(3)$ is chosen since each element $A \in SO(3)$ describes the position of the rigid body relative to a reference configuration, that is, the rotation maps the reference configuration to the current configuration. For the description using Lagrangian mechanics, one forms the velocity-phase space $TSO(3)$ with coordinates $(\theta, \varphi, \psi, \dot\theta, \dot\varphi, \dot\psi)$. The Hamiltonian description is obtained as in §1.1 by using the Legendre transform which maps TG to T^*G.

The passage from T^*G to the space of $\mathbf{\Pi}$'s (body angular momentum space) given by (1.3.10) turns out to be determined by *left* translation on the group. This mapping is in fact an example of a *momentum map*; that is, a mapping whose components are the "Noether quantities" associated with a symmetry group. The map (1.3.10) being a Poisson (canonical) map (see equation (1.3.11)) *is a general fact about momentum maps* proved in §12.6. To get to *space coordinates* one would use *right* translations and the $(+)$ bracket. This is what is done to get the standard description of fluid dynamics.

The second feature that carries over to general groups is the fact that Lie-Poisson systems on $\mathbb{R}^3$ conserve the total angular momenta; that is, leave the spheres in $\mathbf{\Pi}$-space invariant. The generalization of these objects are called ***coadjoint orbits***.

Coadjoint orbits are smooth immersed submanifolds of $\mathfrak{g}^*$, with the property that any Lie-Poisson system $\dot{F} = \{F, H\}$ leaves them invariant. We shall also see how these spaces are symplectic (and hence Poisson) manifolds in their own right and are related to the right $(+)$ or left $(-)$ invariance of the system regarded on T^*G, and the corresponding conserved Noether quantities (*momentum maps* again).

On a general Poisson manifold $(P, \{\cdot\,, \cdot\})$, the definition of a momentum map is as follows. We assume that a Lie group G with Lie algebra $\mathfrak{g}$ acts on P by canonical transformations. As we shall review later (see Chapter 9), the infinitesimal way of specifying the action is to associate to each Lie algebra element $\xi \in \mathfrak{g}$ a vector field ξ_P on P. A ***momentum map*** is a map $\mathbf{J} : P \to \mathfrak{g}^*$ with the property that for every $\xi \in \mathfrak{g}$, the function $\langle \mathbf{J}, \xi \rangle$ (the pairing of the $\mathfrak{g}^*$ valued function $\mathbf{J}$ with the vector ξ) generates the vector field ξ_P. As we shall see later, this definition generalizes the usual notions of linear and angular momentum. Hopefully the rigid body example above conveys the fact that the notion has in fact much wider interest. A fundamental fact about momentum maps is that if the Hamiltonian H is invariant under the action of the group G, then the vector valued function $\mathbf{J}$ is a constant of the motion for the dynamics of the Hamiltonian vector field X_H associated to H.

One of the important notions related to momentum maps is that of ***infinitesimal equivariance*** or the ***classical commutation*** relations, which state that

$$\{\langle \mathbf{J}, \xi \rangle, \langle \mathbf{J}, \eta \rangle\} = \langle \mathbf{J}, [\xi, \eta] \rangle \tag{1.3.12}$$

for any Lie algebra elements ξ and η. Relations like this are well known for the angular momentum, and can be directly checked using the Lie algebra of the rotation group. Later, in Chapter 12 we shall see that the relations (1.3.12) hold for a large important class of momentum maps that are given by computable formulas. Remarkably, it is the condition (1.3.12) that is exactly what is needed to prove that $\mathbf{J}$ *is, in fact, a Poisson map*. It is via this route that one gets an intellectually satisfying generalization of the fact that the map defined by equations (1.3.10) is a Poisson map, that is, equation (1.3.11) holds.

The Lie-Poisson bracket has an interesting history. It was discovered by Sophus Lie (Lie [1890], Vol. II, p. 237). However, Lie's bracket and his related work was not given much attention until the work of Kirillov, Kostant, and Souriau (and others) revived it in the mid-60s. Meanwhile, it was noticed by Pauli and Martin around 1950 that the rigid body equations are in Hamiltonian form using the rigid body bracket, but they were apparently unaware of the underlying Lie theory. Meanwhile, the generalization of the Euler equations to any Lie algebra $\mathfrak{g}$ by Poincaré [1901b] and Hamel [1904] proceeded as well, but without much contact with Lie's work until recently.

The symplectic structure on coadjoint orbits also has a complicated history and itself goes back to Lie (Lie [1890], Ch. 20).

The general notion of a Poisson manifold also goes back to Lie, but the four defining properties of the Poisson bracket have been isolated by many authors such as Dirac [1964], p. 10. The term "Poisson manifold" was coined by Lichnerowicz [1977]. We shall give more historical information on Poisson manifolds in §10.3.

The notion of the momentum map (the English translation of the French word "application moment") also has roots going back to the work of Lie. We shall give some comments on its history in §11.2.

Momentum maps have found an astounding array of applications beyond those already mentioned. For instance, they are used in the study of the space of all solutions of a relativistic field theory (see Chapter 13) and in the study of singularities in algebraic geometry (see Atiyah [1983] and Kirwan [1984a]). They also enter into convex analysis in many interesting ways, such as the Schur-Horn theorem (Schur [1923], Horn [1954]) and its generalizations (Kostant [1973]) and in the theory of integrable systems (Bloch, Brockett, and Ratiu [1990, 1992] and Bloch, Flaschka, and Ratiu [1990, 1993]). It turns out that the image of the momentum map has remarkable convexity properties: see Atiyah [1982], Guillemin and Sternberg [1982, 1984], Kirwan [1984b], Delzant [1988], and Lu and Ratiu [1991].

Exercises

1.3-1 *A linear operator D on the space of smooth functions on $\mathbb{R}^n$ is called a* ***derivation*** *if it satisfies the Leibniz identity: $D(FG) = (DF)G + F(DG)$. Accept the fact from the theory of manifolds (see Chapter 4) that in local coordinates the expression of DF takes the form*

$$(DF)(x) = \sum_{i=1}^{n} a^i(x) \frac{\partial F}{\partial x^i}(x)$$

for some smooth functions $a^1, \ldots, a^n$.

(a) *Use the fact above to prove that for any Poisson bracket $\{\,,\}$ on $\mathbb{R}^n$, we have*

$$\{F, G\} = \sum_{i,j=1}^{n} \{x^i, x^j\} \frac{\partial F}{\partial x^i} \frac{\partial G}{\partial x^j}.$$

(b) *Show that the Jacobi identity holds for a Poisson bracket $\{\,,\}$ on $\mathbb{R}^n$ if and only if it holds for the coordinate functions.*

1.3-2 **(a)** *Define, for a fixed function $f : \mathbb{R}^3 \to \mathbb{R}$*

$$\{F, K\}_f = \nabla f \cdot (\nabla F \times \nabla K).$$

Show that this is a Poisson bracket.

(b) *What does the bracket in part* **(a)** *have to do with Nambu* [1973].

1.3-3 *Verify directly that* (1.3.10) *defines a Poisson map.*

1.3-4 *Show that a bracket satisfying the Leibniz identity also satisfies* $F\{K, L\} - \{FK, L\} = \{F, K\}L - \{F, KL\}$.

1.4 Incompressible Fluids

Arnold [1966a, 1969] showed that the Euler equations for an incompressible fluid could be given a Lagrangian and Hamiltonian description similar to that for the rigid body. His approach has the appealing feature that one sets things up just the way Lagrange and Hamilton would have done: one begins with Q, forms a Lagrangian L on TQ and then H on T^*Q, just as was outlined in §1.1. Thus, one automatically has variational principles, etc. Here $Q = G$ is the group $\mathrm{Diff}_{\mathrm{vol}}(\Omega)$ of volume preserving transformations of the fluid container (a region Ω in $\mathbb{R}^2$ or $\mathbb{R}^3$, or a Riemannian manifold in general, with smooth boundary). The group multiplication in G is simply composition. The Hamiltonian description of fluids using the group $\mathrm{Diff}_{\mathrm{vol}}(\Omega)$ as configuration space apparently goes back to the thesis of Paul Ehrenfest, written under the direction of Boltzman around 1900 (see Klein [1970]). The reason we select $G = \mathrm{Diff}_{\mathrm{vol}}(\Omega)$ as the configuration space is similar to that for the rigid body; namely, each φ in G is a mapping of Ω to Ω which takes a reference point $X \in \Omega$ to a current point $x = \varphi(X) \in \Omega$; thus, knowing φ tells us where each particle of fluid goes and hence gives us the ***fluid configuration***. We ask that φ be a diffeomorphism to exclude discontinuities, cavitation, and fluid interpenetration, and we ask that φ be volume preserving to correspond to the assumption of incompressibility.

A ***motion*** of a fluid is a family of time-dependent elements of G, which we write as $x = \varphi(X, t)$. The ***material velocity*** field is defined by $\mathbf{V}(X, t) = \partial\varphi(X, t)/\partial t$, and the spatial velocity field is defined by $\mathbf{v}(x, t) = \mathbf{V}(X, t)$ where x and X are related by $x = \varphi(X, t)$. If we suppress "t" and write $\dot{\varphi}$ for $\mathbf{V}$, note that

$$\mathbf{v} = \dot{\varphi} \circ \varphi^{-1} \quad \text{i.e.,} \quad \mathbf{v}_t = \mathbf{V}_t \circ \varphi_t^{-1}, \tag{1.4.1}$$

where $\varphi_t(x) = \varphi(X, t)$. We can regard (1.4.1) as a map from the space of $(\varphi, \dot{\varphi})$ (material or Lagrangian description) to the space of $\mathbf{v}$'s (spatial or Eulerian description). Like the rigid body, the material to spatial map (1.4.1) takes the canonical bracket to a Lie-Poisson bracket; one of our goals is to understand this reduction. Notice that if we replace φ by $\varphi \circ \eta$ for a fixed (time-independent) $\eta \in \mathrm{Diff}_{\mathrm{vol}}(\Omega)$, then $\dot{\varphi} \circ \varphi^{-1}$ is independent of η; this reflects the *right* invariance of the Eulerian description ($\mathbf{v}$ is invariant under composition of φ by η on the right). This is also called the ***particle relabeling symmetry*** of fluid dynamics. The spaces TG and T^*G represent the Lagrangian (material) description and we pass to the Eulerian (spatial) description by right translations and use the (+) Lie-Poisson bracket. One of the things we want to do later is to better understand the reason for the switch between right and left in going from the rigid body to fluids.

The Euler equations for an ideal, incompressible, homogeneous fluid moving in the region Ω expressed in the spatial representation are

$$\frac{\partial \mathbf{v}}{\partial t} + (\mathbf{v} \cdot \nabla)\mathbf{v} = -\nabla p \tag{1.4.2}$$

with the constraint $\operatorname{div} \mathbf{v} = 0$ and boundary conditions: $\mathbf{v}$ is tangent to $\partial\Omega$.

Here the pressure p is determined implicitly by the divergence-free (volume preserving) constraint $\operatorname{div} \mathbf{v} = 0$. (See Chorin and Marsden [1993] for more information.) The associated Lie algebra $\mathfrak{g}$ is taken to be the space of all divergence-free vector fields tangent to the boundary, endowed with the *negative* ***Jacobi-Lie bracket*** of vector fields given by

$$[v, w]_L^i = \sum_{j=1}^{n} \left(w^j \frac{\partial v^i}{\partial x^j} - v^j \frac{\partial w^i}{\partial x^j} \right). \tag{1.4.3}$$

(The sub L on $[\cdot\,,\cdot]$ refers to the fact that it is the *left* Lie algebra bracket on $\mathfrak{g}$. The most common convention for the Lie bracket of vector fields, also the one we adopt, has the opposite sign.) We identify $\mathfrak{g}$ and $\mathfrak{g}^*$ using the pairing

$$\langle \mathbf{v}, \mathbf{w} \rangle = \int_\Omega \mathbf{v} \cdot \mathbf{w} \, d^3x \tag{1.4.4}$$

and introduce the (+) Lie-Poisson bracket, called the ***ideal fluid bracket***, on functions of $\mathbf{v}$ by

$$\{F, G\}(\mathbf{v}) = \int_\Omega \mathbf{v} \cdot \left[\frac{\delta F}{\delta \mathbf{v}}, \frac{\delta G}{\delta \mathbf{v}} \right]_L d^3x, \tag{1.4.5}$$

where $\delta F/\delta \mathbf{v}$ is defined by

$$\lim_{\varepsilon \to 0} \frac{1}{\varepsilon}[F(\mathbf{v} + \varepsilon \delta\mathbf{v}) - F(\mathbf{v})] = \int_\Omega \left(\delta\mathbf{v} \cdot \frac{\delta F}{\delta \mathbf{v}} \right) d^3x. \tag{1.4.6}$$

With the energy function chosen to be the kinetic energy,

$$H(\mathbf{v}) = \frac{1}{2} \int_\Omega \|\mathbf{v}\|^2 \, d^3x, \tag{1.4.7}$$

one can verify that the Euler equations (1.4.2) are equivalent to the Poisson bracket equations

$$\dot{F} = \{F, H\} \tag{1.4.8}$$

for all functions F on $\mathfrak{g}$. For this, one uses the orthogonal decomposition $\mathbf{w} = \mathbb{P}\mathbf{w} + \nabla p$ of a vector field $\mathbf{w}$ into a divergence-free part $\mathbb{P}\mathbf{w}$ in $\mathfrak{g}$ and a gradient. The Euler equations can be written

$$\frac{\partial \mathbf{v}}{\partial t} + \mathbb{P}(\mathbf{v} \cdot \nabla \mathbf{v}) = 0. \tag{1.4.9}$$

One can express the Hamiltonian structure in terms of the vorticity as a basic dynamic variable, and show that the preservation of coadjoint orbits amounts to Kelvin's circulation theorem. Marsden and Weinstein [1983] show that the Hamiltonian structure in terms of Clebsch potentials fits naturally into this Lie-Poisson scheme, and that Kirchhoff's Hamiltonian description of point vortex dynamics, vortex filaments, and vortex patches can be derived in a natural way from the Hamiltonian structure described above.

The general geometric framework of the Euler-Poincaré and the Lie-Poisson equations gives other insights as well. For example, they show that the Euler equations are derivable from the variational principle

$$\delta \int_a^b \int_\Omega \frac{1}{2} \|\mathbf{v}\|^2 \, d^3x = 0$$

for all $\delta \mathbf{v}$ of the form

$$\delta \mathbf{v} = \dot{\mathbf{u}} + [\mathbf{u}, \mathbf{u}]_L$$

where $\mathbf{u}$ is a vector field (representing the infinitesimal particle displacement) vanishing at the temporal endpoints. This principle is due to Newcomb [1962]; see also Bretherton [1970]. For the case of general Lie algebras, it is due to Marsden and Scheurle [1993b]; see also Bloch, Krishnaprasad, Marsden and Ratiu [1994b]. See also the review article of Morrison [1994] for a somewhat different perspective.

There are important functional analytic differences between working in Lagrangian representation (that is, on T^*G) and in Eulerian representation, that is, on $\mathfrak{g}^*$ that are important for proving existence and uniqueness theorems, theorems on the limit of zero viscosity, and the convergence of numerical algorithms (see Ebin and Marsden [1970], Marsden, Ebin, and Fischer [1972], and Chorin, Hughes, Marsden, and McCracken [1978]). Finally, we note that for *two-dimensional flow*, a collection of Casimir functions is given by

$$C(\omega) = \int_\Omega \Phi(\omega(x)) \, d^2x \tag{1.4.10}$$

for $\Phi : \mathbb{R} \to \mathbb{R}$ any (smooth) function where $\omega\mathbf{k} = \nabla \times \mathbf{v}$ is the ***vorticity***. For three-dimensional flow, (1.4.10) is no longer a Casimir.

Exercise

1.4-1 *Show that any divergence-free vector field X on $\mathbb{R}^3$ can be written globally as a curl of another vector field and can locally be written as $X = \nabla f \times \nabla g$ where f and g are real-valued functions on $\mathbb{R}^3$. Assume this (so-called Clebsch-Monge) representation also holds globally. Show that the particles of fluid, which follow trajectories satisfying $\dot{x} = X(x)$, are trajectories of a Hamiltonian system with a bracket in the form of Exercise* **1.3-2**.

1.5 The Maxwell-Vlasov System

The period 1970–1980 saw the development of noncanonical Hamiltonian structures for the Korteweg-de Vries (KdV) equation (Gardner [1971]) and other soliton equations. This quickly became entangled with the attempts to understand integrability of Hamiltonian systems and the development of the algebraic approach; see, for example, Gelfand and Dorfman [1979], Manin [1979] and references therein. More recently these approaches have come together again; see, for instance, Reyman and Semenov-Tian-Shansky [1990]. KdV type models are usually derived from or are approximations to more fundamental fluid models and it seems fair to say that the reasons for their complete integrability are not yet completely understood. For fluid and plasma systems, some of the key early works on Poisson bracket structures were Dashen and Sharp [1968], Goldin [1971], Iwinski and Turski [1976], Dzyaloshinski and Volovick [1980], Morrison and Greene [1980], and Morrison [1980]. In Sudarshan and Mukunda [1974], Guillemin and Sternberg [1982], and Ratiu [1980, 1982], a general theory for Lie-Poisson structures for special kinds of Lie algebras, called semidirect products, was begun. This was quickly recognized (see, for example, Marsden [1982], Marsden, Weinstein, Ratiu, Schmid, and Spencer [1983], Holm and Kuperschmidt [1983], and Marsden, Ratiu and Weinstein [1984a,b]) to be relevant to the brackets for compressible flow; see §1.7 below.

A rational scheme for systematically deriving brackets is needed, since, for one thing, a direct verification of Jacobi's identity can be inefficient and time-consuming. Here we describe a derivation of the Maxwell-Vlasov bracket by Marsden and Weinstein [1982]. The method is similar to Arnold's, namely by performing a reduction starting with:

i canonical brackets in a Lagrangian representation for the plasma; and

ii a potential representation for the electromagnetic field.

Parallel developments can be given for many other brackets, such as the charged fluid bracket by Spencer and Kaufman [1982]. Another method, based primarily on Clebsch potentials, was developed in a series of papers by Holm and Kupershmidt (for example, [1983]) and applied to a number of interesting systems, including superfluids and superconductors. They also pointed out that semidirect products were appropriate for the MHD bracket of Morrison and Greene [1980].

The Maxwell-Vlasov equations for a collisionless plasma are the fundamental equations in plasma physics. See, for example, Clemmow and Dougherty [1959], Van Kampen and Felderhof [1967], Krall and Trivelpiece [1973], Davidson [1972], Ichimaru [1973], and Chen [1974]. In Euclidean space, the basic dynamical variables are:

$f(\mathbf{x}, \mathbf{v}, t)$: the plasma particle number density per phase space volume $d^3x\, d^3v$;
$\mathbf{E}(\mathbf{x}, t)$: the electric field;
$\mathbf{B}(\mathbf{x}, t)$: the magnetic field.

The equations for a collisionless plasma consisting of a single species of particles with mass m and charge e are

$$\frac{\partial f}{\partial t} + \mathbf{v} \cdot \frac{\partial f}{\partial \mathbf{x}} + \frac{e}{m}\left(\mathbf{E} + \frac{1}{c}\mathbf{v} \times \mathbf{B}\right) \cdot \frac{\partial f}{\partial \mathbf{v}} = 0, \tag{1.5.1}$$

$$\frac{1}{c}\frac{\partial \mathbf{B}}{\partial t} = -\operatorname{curl} \mathbf{E}, \tag{1.5.2}$$

$$\frac{1}{c}\frac{\partial \mathbf{E}}{\partial t} = \operatorname{curl} \mathbf{B} - \frac{1}{c}\mathbf{j}_f, \tag{1.5.3}$$

$$\operatorname{div} \mathbf{E} = \rho_f \quad \text{and} \quad \operatorname{div} \mathbf{B} = 0. \tag{1.5.4}$$

The ***current*** defined by f is given by $\mathbf{j}_f = e \int \mathbf{v} f(\mathbf{x}, \mathbf{v}, t)\, d^3v$ and the ***charge density*** by $\rho_f = e \int f(\mathbf{x}, \mathbf{v}, t)\, d^3v$. Also, $\partial f / \partial \mathbf{x}$ and $\partial f / \partial \mathbf{v}$ denote the gradients of f with respect to $\mathbf{x}$ and $\mathbf{v}$, respectively, and c is the speed of light. The evolution equation for f results from the Lorentz force law and standard transport assumptions. The remaining equations are the standard Maxwell equations with charge density ρ_f and current $\mathbf{j}_f$ produced by the plasma.

Two limiting cases will aid our discussions. First, if the plasma is constrained to be static, that is, f is concentrated at $\mathbf{v} = 0$ and t-independent, we get the ***charge-driven Maxwell equations***:

$$\frac{1}{c}\frac{\partial \mathbf{B}}{\partial t} = -\operatorname{curl} \mathbf{E}, \quad \operatorname{div} \mathbf{E} = \rho, \quad \frac{1}{c}\frac{\partial \mathbf{E}}{\partial t} = \operatorname{curl} \mathbf{B}, \quad \operatorname{div} \mathbf{B} = 0. \tag{1.5.5}$$

Second, if we let $c \to \infty$, electrodynamics becomes electrostatics, and we get the ***Poisson-Vlasov equation***:

$$\frac{\partial f}{\partial t} + \mathbf{v} \cdot \frac{\partial f}{\partial \mathbf{x}} - \frac{e}{m}\frac{\partial \varphi_f}{\partial \mathbf{x}} \cdot \frac{\partial f}{\partial \mathbf{v}} = 0, \tag{1.5.6}$$

where $-\nabla^2 \varphi_f = \rho_f$. In this context, the name "Poisson-Vlasov" seems quite appropriate. The equation is, however, formally the same as the earlier

Jeans [1919] equation of stellar dynamics. Henon [1982] has proposed calling it the "collisionless Boltzmann equation."

1.6 The Maxwell and Poisson-Vlasov Brackets

Let us discuss the vacuum Maxwell equations in more detail. For simplicity, we let $m = e = c = 1$. As the basic configuration space, we take the space $\mathcal{A}$ of vector potentials $\mathbf{A}$ on $\mathbb{R}^3$ (for the Yang-Mills equations this is generalized to the space of connections on a principal bundle over space). The corresponding phase space $T^*\mathcal{A}$ is identified with the set of pairs $(\mathbf{A}, \mathbf{Y})$ where $\mathbf{Y}$ is also a vector field on $\mathbb{R}^3$. The canonical Poisson bracket is used on $T^*\mathcal{A}$:

$$\{F, G\} = \int \left(\frac{\delta F}{\delta \mathbf{A}} \frac{\delta G}{\delta \mathbf{Y}} - \frac{\delta F}{\delta \mathbf{Y}} \frac{\delta G}{\delta \mathbf{A}} \right) d^3x. \tag{1.6.1}$$

The electric field will be $\mathbf{E} = -\mathbf{Y}$ and the magnetic field is given by $\mathbf{B} = \operatorname{curl} \mathbf{A}$. With the Hamiltonian

$$H(\mathbf{A}, \mathbf{Y}) = \frac{1}{2} \int (\|\mathbf{E}\|^2 + \|\mathbf{B}\|^2)\, d^3x, \tag{1.6.2}$$

Hamilton's canonical field equations (1.1.14) are readily seen to be equations for $\partial \mathbf{E}/\partial t$ and $\partial \mathbf{A}/\partial t$ which imply the vacuum Maxwell's equations. Alternatively, one can begin with $T\mathcal{A}$ and the Lagrangian

$$L(\mathbf{A}, \dot{\mathbf{A}}) = \frac{1}{2} \int \left(\|\dot{\mathbf{A}}\|^2 - \|\nabla \times \mathbf{A}\|^2 \right) d^3x \tag{1.6.3}$$

and then go the route of the Euler-Lagrange equations and variational principles.

It is of interest to incorporate the equation $\operatorname{div} \mathbf{E} = \rho$ and, correspondingly to use directly the field strengths $\mathbf{E}$ and $\mathbf{B}$, rather than $\mathbf{E}$ and $\mathbf{A}$. To do this, we introduce the gauge group $\mathcal{G}$, the additive group of real-valued functions $\psi : \mathbb{R}^3 \to \mathbb{R}$. Each $\psi \in \mathcal{G}$ transforms the fields according to the rule

$$(\mathbf{A}, \mathbf{E}) \mapsto (\mathbf{A} + \nabla \psi, \mathbf{E}). \tag{1.6.4}$$

Each such transformation leaves the Hamiltonian H invariant and is a canonical transformation, that is, it leaves Poisson brackets intact. In this situation, as above, there will be a corresponding conserved quantity, or *momentum map* in the same sense as in §1.3. As mentioned there, some simple general formulae for computing them will be studied in detail later. These momentum maps transform the phase space to the dual of the Lie algebra and are consistent with its Lie-Poisson structure, that is, they are

Poisson maps. For the action (1.6.4) of $\mathcal{G}$ on $T^*\mathcal{A}$, the associated momentum map is

$$\mathbf{J}(\mathbf{A}, \mathbf{Y}) = \operatorname{div} \mathbf{E}, \tag{1.6.5}$$

so we recover the fact that div $\mathbf{E}$ is preserved by Maxwell's equations (this is easy to verify directly using div curl $= 0$). Thus we see that we can incorporate the equation div $\mathbf{E} = \rho$ by restricting our attention to the set $\mathbf{J}^{-1}(\rho)$. The theory of reduction is a general process whereby one reduces the dimension of a phase space by exploiting conserved quantities and symmetry groups. In the present case, the reduced space is $\mathbf{J}^{-1}(\rho)/\mathcal{G}$ which is identified with Max_ρ, the space of $\mathbf{E}$'s and $\mathbf{B}$'s with div $\mathbf{E} = \rho$ and div $\mathbf{B} = 0$.

The space Max_ρ inherits a Poisson structure as follows. If F and K are functions on Max_ρ, we substitute $\mathbf{E} = -\mathbf{Y}$ and $\mathbf{B} = \nabla \times \mathbf{A}$ to express F and K as functionals of $(\mathbf{A}, \mathbf{Y})$. Then we compute the canonical brackets on $T^*\mathcal{A}$ and express the result in terms of $\mathbf{E}$ and $\mathbf{B}$. Carrying this out using the chain rule gives

$$\{F, K\} = \int \left(\frac{\delta F}{\delta \mathbf{E}} \cdot \operatorname{curl} \frac{\delta K}{\delta \mathbf{B}} - \frac{\delta K}{\delta \mathbf{E}} \cdot \operatorname{curl} \frac{\delta F}{\delta \mathbf{B}} \right) d^3x, \tag{1.6.6}$$

where $\delta F/\delta \mathbf{E}$ and $\delta F/\delta \mathbf{B}$ are *divergence-free* vector fields defined in the usual way; for example,

$$\lim_{\varepsilon \to 0} \frac{1}{\varepsilon}[F(\mathbf{E} + \varepsilon \delta \mathbf{E}, \mathbf{B}) - F(\mathbf{E}, \mathbf{B})] = \int \frac{\delta F}{\delta \mathbf{E}} \cdot \delta \mathbf{E} \, d^3x. \tag{1.6.7}$$

This bracket makes Max_ρ into a Poisson manifold and the map $(\mathbf{A}, \mathbf{Y}) \mapsto (-\operatorname{div} \mathbf{Y}, \nabla \times \mathbf{A})$ into a Poisson map. The bracket (1.6.6) was discovered (by a different procedure) by Pauli [1933] and Born and Infeld [1935]. We refer to (1.6.6) as the ***Pauli-Born-Infeld bracket*** or the Maxwell bracket for Maxwell's equations.

With the energy H given by (1.6.2) regarded as a function of $\mathbf{E}$ and $\mathbf{B}$, Hamilton's equations in bracket form $\dot{F} = \{F, H\}$ on Max_ρ captures the full set of Maxwell's equations (with external charge density ρ).

Morrison [1980] showed that the Poisson-Vlasov equations are Hamiltonian with

$$H(f) = \frac{1}{2} \int \|\mathbf{v}\|^2 f(\mathbf{x}, \mathbf{v}, t) \, d^3x \, d^3v + \frac{1}{2} \int \|\nabla \varphi_f\|^2 \, d^3x \tag{1.6.8}$$

and the Poisson-Vlasov bracket

$$\{F, G\} = \int f \left\{ \frac{\delta F}{\delta f}, \frac{\delta G}{\delta f} \right\}_{\mathrm{xv}} d^3x \, d^3v, \tag{1.6.9}$$

where $\{\, , \}_{\mathrm{xv}}$ is the canonical bracket on $(\mathbf{x}, \mathbf{v})$-space. As was observed in Gibbons [1981] and Marsden and Weinstein [1982], this is the $(+)$ Lie-Poisson bracket associated with the Lie algebra $\mathfrak{g}$ of functions of $(\mathbf{x}, \mathbf{v})$ with Lie bracket the canonical Poisson bracket.

According to the general theory, this Lie-Poisson structure comes from canonical brackets on the cotangent bundle of the group underlying $\mathfrak{g}$, just as was the case for the rigid body and incompressible fluids. This time the group $G = \text{Diff}_{\text{can}}$ is the group of canonical transformations of $(\mathbf{x}, \mathbf{v})$-space. The Poisson-Vlasov equations can equally well be written in canonical form on T^*G. This is the Lagrangian description of a plasma, and the Hamiltonian description here goes back to Low [1958], Katz [1961], and Lundgren [1963]. Thus, one can start with the Lagrangian description with canonical brackets and, through reduction, derive the brackets here.

There are other approaches to the Hamiltonian formulation using analogues of Clebsch potentials; see, for instance, Su [1961], Zakharov [1971], and Gibbons, Holm, and Kupershmidt [1982].

1.7 The Poisson-Vlasov to Fluid Map

Before going on to the Maxwell-Vlasov equations, we point out a remarkable connection between the Poisson-Vlasov bracket (1.6.9) and the bracket for compressible flow.

The Euler equations for compressible flow in a region Ω in $\mathbb{R}^3$ are

$$\rho\left(\frac{\partial \mathbf{v}}{\partial t} + (\mathbf{v} \cdot \nabla)\mathbf{v}\right) = -\nabla p \tag{1.7.1}$$

and

$$\frac{\partial \rho}{\partial t} + \operatorname{div}(\rho \mathbf{v}) = 0, \tag{1.7.2}$$

with the boundary condition

$$\mathbf{v} \quad \text{tangent to} \quad \partial\Omega.$$

Here the pressure p is determined from an internal energy function per unit mass given by $p = \rho^2 w'(\rho)$ where $w = w(\rho)$ is the constitutive relation. (We ignore entropy for the present discussion—its inclusion is easy to deal with.) The ***compressible fluid Hamiltonian*** is

$$H = \frac{1}{2}\int_\Omega \rho\|\mathbf{v}\|^2\, d^3x + \int_\Omega \rho w(\rho)\, d^3x. \tag{1.7.3}$$

The relevant Poisson bracket is most easily expressed if we use the momentum density $\mathbf{M} = \rho\mathbf{v}$ and density ρ as our basic variables. The ***compressible fluid bracket*** is

$$\begin{aligned}\{F, G\} = &\int_\Omega \mathbf{M} \cdot \left[\left(\frac{\delta G}{\delta \mathbf{M}} \cdot \nabla\right)\frac{\delta F}{\delta \mathbf{M}} - \left(\frac{\delta F}{\delta \mathbf{M}} \cdot \nabla\right)\frac{\delta G}{\delta \mathbf{M}}\right] d^3x \\ &+ \int_\Omega \rho \left[\left(\frac{\delta G}{\delta \mathbf{M}} \cdot \nabla\right)\frac{\delta F}{\delta \rho} - \left(\frac{\delta F}{\delta \mathbf{M}} \cdot \nabla\right)\frac{\delta G}{\delta \rho}\right] d^3x.\end{aligned} \tag{1.7.4}$$

The space of $(\mathbf{M}, \rho)$'s can be shown to be the dual of a semidirect product Lie algebra and that (1.7.4) is the associated (+) Lie-Poisson bracket (see Marsden, Weinstein, Ratiu, Schmid, and Spencer [1983], Holm and Kupershmidt [1983], and Marsden, Ratiu, and Weinstein [1984a,b]).

The relationship with the Poisson-Vlasov bracket is this: suppressing the time variable, define the map $f \mapsto (\mathbf{M}, \rho)$ by

$$\mathbf{M}(\mathbf{x}) = \int_\Omega \mathbf{v} f(\mathbf{x}, \mathbf{v}) d^3 v \quad \text{and} \quad \rho(\mathbf{x}) = \int_\Omega f(\mathbf{x}, \mathbf{v})\, d^3 v. \tag{1.7.5}$$

Remarkably, this plasma to fluid map is a Poisson map taking the Poisson-Vlasov bracket (1.6.9) to the compressible fluid bracket (1.7.4). In fact, this map is a momentum map (Marsden, Weinstein, Ratiu, Schmid, and Spencer [1983]). The Poisson-Vlasov Hamiltonian is *not* invariant under the associated group action, however.

1.8 The Maxwell-Vlasov Bracket

The bracket for the Maxwell-Vlasov equations was given by Iwinski and Turski [1976] and Morrison [1980]. Marsden and Weinstein [1982] used systematic procedures involving reduction and momentum maps to derive the bracket from a canonical bracket.

We start with the material description of the plasma as the cotangent bundle of the group $\mathrm{Diff}_{\mathrm{can}}$ of canonical transformations of $(\mathbf{x}, \mathbf{p})$-space and the space $T^*\mathcal{A}$ for Maxwell's equations. We justify this by noticing that the motion of a charged particle in a fixed, but (possibly time-dependent) electromagnetic field via the Lorentz force law defines a (time-dependent) canonical transformation. On $T^*\mathrm{Diff}_{\mathrm{can}} \times T^*\mathcal{A}$ we put the sum of the two canonical brackets, and then we reduce. First we reduce by $\mathrm{Diff}_{\mathrm{can}}$, which acts on $T^*\mathrm{Diff}_{\mathrm{can}}$ by right translation, but does not act on $T^*\mathcal{A}$. Thus we end up with densities $f_{\mathrm{mom}}(\mathbf{x}, \mathbf{p}, t)$ on position-momentum space and with the space $T^*\mathcal{A}$ used for the Maxwell equations. On this space we get the (+) Lie-Poisson bracket, plus the canonical bracket on $T^*\mathcal{A}$. Recalling that $\mathbf{p}$ is related to $\mathbf{v}$ and $\mathbf{A}$ by $\mathbf{p} = \mathbf{v} + \mathbf{A}$, we let the gauge group $\mathcal{G}$ of electromagnetism act on this space by

$$\begin{aligned}(f_{\mathrm{mom}}(\mathbf{x}, \mathbf{p}, t), \mathbf{A}(\mathbf{x}, t), \mathbf{Y}(\mathbf{x}, t)) &\mapsto \\ (f_{\mathrm{mom}}(\mathbf{x}, \mathbf{p} + \nabla\varphi(\mathbf{x}), t), &\mathbf{A}(\mathbf{x}, t) + \nabla\varphi(x), \mathbf{Y}(\mathbf{x}, t)).\end{aligned} \tag{1.8.1}$$

The momentum map associated with this action is computed to be

$$\mathbf{J}(f_{\mathrm{mom}}, \mathbf{A}, \mathbf{Y}) = \operatorname{div} \mathbf{E} - \int f_{\mathrm{mom}}(\mathbf{x}, \mathbf{p})\, d^3 p. \tag{1.8.2}$$

This corresponds to $\operatorname{div} \mathbf{E} - \rho_f$ if we write $f(\mathbf{x}, \mathbf{v}, t) = f_{\mathrm{mom}}(\mathbf{x}, \mathbf{p} - \mathbf{A}, t)$. This reduced space $\mathbf{J}^{-1}(0)/\mathcal{G}$ can be identified with the space $\mathcal{MV}$ of triples

$(f, \mathbf{E}, \mathbf{B})$, satisfying $\operatorname{div} \mathbf{E} = \rho_f$ and $\operatorname{div} \mathbf{B} = 0$. The bracket on $\mathcal{MV}$ is computed by the same procedure as for Maxwell's equations. These computations yield the following **Maxwell-Vlasov bracket**:

$$\begin{aligned} \{F, K\}(f, \mathbf{E}, \mathbf{B}) &= \int f \left\{ \frac{\delta F}{\delta f}, \frac{\delta K}{\delta f} \right\}_{\mathrm{xv}} d^3x\, d^3v \\ &+ \int \left(\frac{\delta F}{\delta \mathbf{E}} \cdot \operatorname{curl} \frac{\delta K}{\delta \mathbf{B}} - \frac{\delta K}{\delta \mathbf{E}} \cdot \operatorname{curl} \frac{\delta F}{\delta \mathbf{B}} \right) d^3x \\ &+ \int \left(\frac{\delta F}{\delta \mathbf{E}} \cdot \frac{\delta f}{\delta \mathbf{v}} \frac{\delta K}{\delta f} - \frac{\delta K}{\delta \mathbf{E}} \cdot \frac{\delta f}{\delta \mathbf{v}} \frac{\delta F}{\delta f} \right) d^3x\, d^3v \\ &+ \int f \mathbf{B} \cdot \left(\frac{\partial}{\partial \mathbf{v}} \frac{\delta F}{\delta f} \times \frac{\partial}{\partial \mathbf{v}} \frac{\delta K}{\delta f} \right) d^3x\, d^3v. \end{aligned} \tag{1.8.3}$$

With the ***Maxwell-Vlasov Hamiltonian***

$$\begin{aligned} H(f, \mathbf{E}, \mathbf{B}) &= \frac{1}{2} \int \|\mathbf{v}\|^2 f(\mathbf{x}, \mathbf{v}, t)\, d^3x\, d^3v \\ &+ \frac{1}{2} \int (\|\mathbf{E}(x, t)\|^2 + \|\mathbf{B}(\mathbf{x}, t)\|^2)\, d^3x, \end{aligned} \tag{1.8.4}$$

the Maxwell-Vlasov equations take the Hamiltonian form

$$\dot{F} = \{F, H\} \tag{1.8.5}$$

on the Poisson manifold $\mathcal{MV}$.

1.9 The Heavy Top

The equations of motion for a rigid body with a fixed point in a gravitational field provide another interesting example of a system which is Hamiltonian relative to a Lie-Poisson bracket. The underlying Lie algebra consists of the algebra of infinitesimal Euclidean motions in $\mathbb{R}^3$. (These do *not* arise as Euclidean motions of the body since the body has a fixed point). As we shall see, there is a close parallel with the Poisson structure for compressible fluids.

The basic phase space we start with is again $T^*SO(3)$, coordinatized by Euler angles and their conjugate momenta. In these variables, the equations are in canonical Hamiltonian form; however, the presence of gravity breaks the symmetry and the system is no longer $SO(3)$ invariant, so it cannot be written entirely in terms of the body angular momentum $\mathbf{\Pi}$. One also needs to keep track of $\mathbf{\Gamma}$, the "direction of gravity" as seen from the body ($\mathbf{\Gamma} = \mathbf{A}^{-1}\mathbf{k}$ where $\mathbf{k}$ points upward and $\mathbf{A}$ is the element of $SO(3)$ describing the current configuration of the body). The equations of motion are

$$\dot{\Pi}_1 = \frac{I_2 - I_3}{I_2 I_3} \Pi_2 \Pi_3 + Mgl(\Gamma^2 \chi^3 - \Gamma^3 \chi^2),$$

$$\begin{aligned}\dot{\Pi}_2 &= \frac{I_3 - I_1}{I_3 I_1}\Pi_3\Pi_1 + Mgl(\Gamma^3\chi^1 - \Gamma^1\chi^3), \\ \dot{\Pi}_3 &= \frac{I_1 - I_2}{I_1 I_2}\Pi_1\Pi_2 + Mgl(\Gamma^1\chi^2 - \Gamma^2\chi^1)\end{aligned} \tag{1.9.1}$$

and

$$\dot{\mathbf{\Gamma}} = \mathbf{\Gamma} \times \Omega \tag{1.9.2}$$

where M is the body's mass, g is the acceleration of gravity, $\boldsymbol{\chi}$ is the unit vector on the line connecting the fixed point with the body's center of mass, and l is the length of this segment.

The Lie algebra of the Euclidean group is $\mathfrak{se}(3) = \mathbb{R}^3 \times \mathbb{R}^3$ with the Lie bracket

$$[(\boldsymbol{\xi}, \mathbf{u}), (\boldsymbol{\eta}, \mathbf{v})] = (\boldsymbol{\xi} \times \boldsymbol{\eta}, \boldsymbol{\xi} \times \mathbf{v} - \boldsymbol{\eta} \times \mathbf{u}). \tag{1.9.3}$$

We identify the dual space with pairs $(\mathbf{\Pi}, \mathbf{\Gamma})$; the corresponding $(-)$ Lie-Poisson bracket called the ***heavy top bracket*** is

$$\begin{aligned}\{F, G\}(\mathbf{\Pi}, \mathbf{\Gamma}) &= -\mathbf{\Pi} \cdot (\nabla_\Pi F \times \nabla_\Pi G) \\ &\quad -\mathbf{\Gamma} \cdot (\nabla_\Pi F \times \nabla_\Gamma G - \nabla_\Pi G \times \nabla_\Gamma F).\end{aligned} \tag{1.9.4}$$

The above equations for $\mathbf{\Pi}, \mathbf{\Gamma}$ can be checked to be equivalent to

$$\dot{F} = \{F, H\}, \tag{1.9.5}$$

where the ***heavy top Hamiltonian***

$$H(\mathbf{\Pi}, \mathbf{\Gamma}) = \frac{1}{2}\left(\frac{\Pi_1^2}{I_1} + \frac{\Pi_2^2}{I_2} + \frac{\Pi_3^2}{I_3}\right) + Mgl\mathbf{\Gamma} \cdot \boldsymbol{\chi} \tag{1.9.6}$$

is the total energy of the body (Sudarshan and Mukunda [1974]).

The Lie algebra of the Euclidean group has a structure which is a special case of what is called a *semidirect product*. Here it is the product of the group of rotations with the translation group. It turns out that semidirect products occur under rather general circumstances when the symmetry in T^*G is broken. In particular, notice the similarities in structure between the Poisson bracket (1.7.4) for compressible flow and (1.9.4). For compressible flow it is the density which prevents a full $\mathrm{Diff}(\Omega)$ invariance; the Hamiltonian is only invariant under those diffeomorphisms that preserve the density. The general theory for semidirect products was developed by Sudarshan and Mukunda [1974], Ratiu [1980, 1981, 1982], Guillemin and Sternberg [1982], Marsden, Weinstein, Ratiu, Schmid, and Spencer [1983], Marsden, Ratiu, and Weinstein [1984a,b], and Holm and Kupershmidt [1983].

Exercise

1.9-1 *Verify that $\dot{F} = \{F, H\}$ are equivalent to the heavy top equations using the heavy top Hamiltonian and bracket.*

1.10 Nonlinear Stability

There are various meanings that can be given to the word "stability." Intuitively, stability means that small disturbances do not grow large as time passes. Being more precise about this notion is not just mathematical nitpicking; indeed, different interpretations of the word stability can lead to *different* stability criteria. Examples like the double spherical pendulum and stratified shear flows that are sometimes used to model oceanographic phenomena, show that one can get *different* criteria if one uses linearized or nonlinear analyses (see Marsden and Scheurle [1993a] and Abarbanel, Holm, Marsden, and Ratiu [1986]).

The history of stability theory in mechanics is very complex, but certainly has its roots in the work of Riemann [1860, 1861], Routh [1877], Thomson and Tait [1879], Poincaré [1885, 1892], and Liapunov [1892, 1897].

Since these early references, the literature has become too vast to even survey roughly. We do mention however, that a guide to the large Soviet literature may be found in Mikhailov and Parton [1990].

The basis of the nonlinear stability method discussed below was originally given by Arnold [1965b, 1966b] and applied to two-dimensional ideal fluid flow, substantially augmenting the pioneering work of Lord Rayleigh [1880]. Related methods were also found in the plasma physics literature, notably by Newcomb [1958], Fowler [1963], and Rosenbluth [1964]. However, these works did not provide a general setting or key convexity estimates needed to deal with the nonlinear nature of the problem. In retrospect, we may view other stability results, such as the stability of solitons in the Korteweg-de Vries (KdV) equations due to Benjamin [1972] and Bona [1975] (see also Maddocks and Sachs [1992]) as being instances of the same method used by Arnold. A crucial part of the method exploits the fact that the basic equations of nondissipative fluid and plasma dynamics are Hamiltonian in character. We shall explain below how the Hamiltonian structures discussed in the previous sections are used in the stability analysis.

Stability is a dynamical concept. To explain it, we shall use some fundamental notions from the theory of dynamical systems (see, for example, Hirsch and Smale [1974] and Wiggins [1988, 1990]). The laws of dynamics are usually presented as equations of motion which we write in the abstract form of a ***dynamical system***:

$$\dot{u} = X(u). \tag{1.10.1}$$

Here, u is a variable describing the state of the system under study, X is a system-specific function of u and $\dot{u} = du/dt$, where t is time. The set of all allowed u's forms the phase space P. For a classical mechanical system, u is often a $2n$-tuple $(q^1, \ldots, q^n, p_1, \ldots, p_n)$ of positions and momenta and, for fluids, u is a velocity field in physical space. As time evolves, the state of the system changes; the state follows a curve $u(t)$ in P. The trajectory $u(t)$ is assumed to be uniquely determined if its initial condition $u_0 = u(0)$ is specified. An ***equilibrium state*** is a state u_e such that $X(u_e) = 0$. The unique trajectory starting at u_e is u_e itself; that is, u_e does not move in time.

The language of dynamics has been an extraordinarily useful tool in the physical and biological sciences, especially during the last few decades. The study of systems which develop spontaneous oscillations through a mechanism called the Poincaré-Andronov-Hopf bifurcation is an example of such a tool (see Marsden and McCracken [1976], Chow and Hale [1982], and Wiggins [1990], for example). More recently, the concept of "chaotic dynamics" has sparked a resurgence of interest in dynamical systems. This occurs when dynamical systems possess trajectories that are so complex that they behave as if they were random. Some believe that the theory of turbulence will use such notions in its future development. We are not concerned with chaos directly, although it plays a role in some of what follows. In particular, we remark that in the definition of stability below, stability does not preclude chaos. In other words, the trajectories near a stable point can still be temporally very complex; stability just prevents them from moving very far from equilibrium.

To define stability, we choose a measure of nearness in P using a "metric" d. For two points u_1 and u_2 in P, d determines a positive number denoted $d(u_1, u_2)$, which is called the ***distance*** from u_1 to u_2. In the course of a stability analysis, it is necessary to specify, or construct, a metric appropriate for the problem at hand. In this setting, one says that an equilibrium state u_e is ***stable*** when trajectories which start near u_e remain near u_e for all $t \geq 0$. In precise terms, given any number $\epsilon > 0$, there is $\delta > 0$ such that if $d(u_0, u_e) < \delta$, then $d(u(t), u_e) < \epsilon$ for all $t > 0$. Figure 1.10.1 shows examples of stable and unstable equilibria for dynamical systems whose state space is the plane.

Fluids can be stable relative to one distance measure and, simultaneously, unstable relative to another. This seeming pathology actually reflects important physical processes; see Wan and Pulvirente [1984].

A well-known physical example illustrating the definition of stability is the motion of a free rigid body. This system can be simulated by tossing a book, held shut with a rubber band, into the air. It rotates stably when spun about its longest and shortest axes, but unstably when spun about the middle axis (Figure 1.10.2). The distance measure defining stability in this example is a metric in body angular momentum space. We shall return to this example in detail in Chapter 15 when we study rigid body stability.

There are two other ways of treating stability. First of all, one can lin-

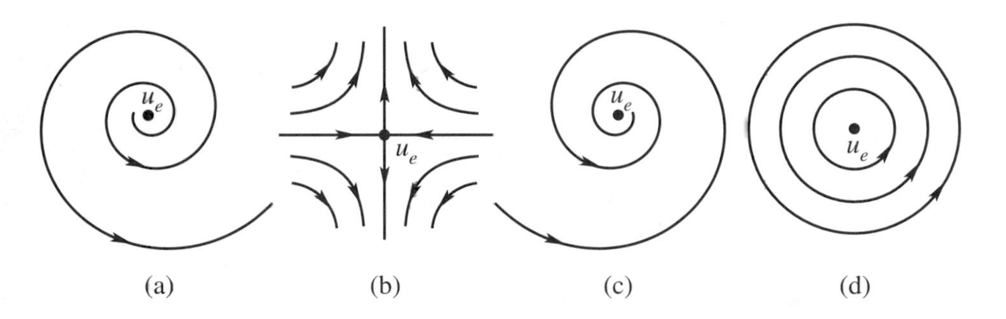

FIGURE 1.10.1. The equilibrium point (a) is unstable because the trajectory $u(t)$ does not remain near u_e. Similarly (b) is unstable since most trajectories (eventually) move away from u_e. The equilibria in (c) and (d) are stable because all trajectories near u_e stay near u_e.

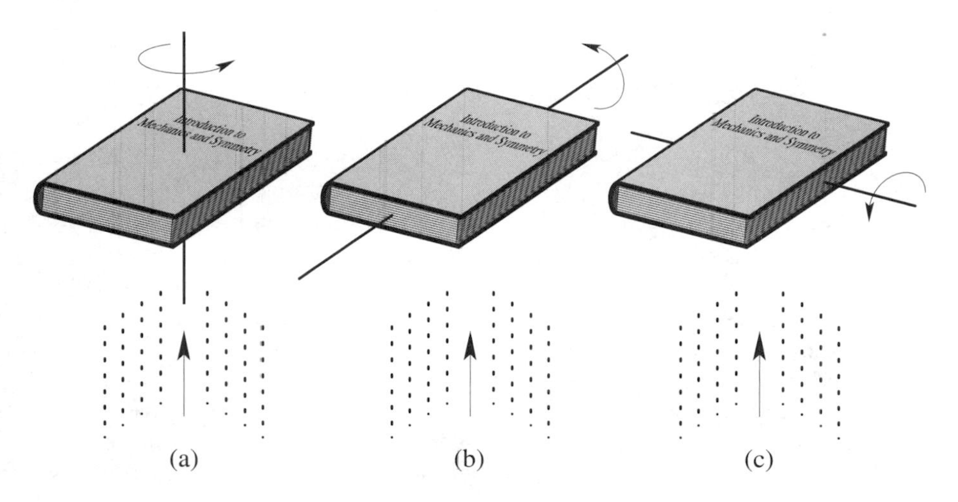

FIGURE 1.10.2. If you toss a book into the air, you can make it spin stably about (a) its shortest axis and (b) its longest axis, but it is unstable when it rotates about (c) its middle axis.

earize equation (1.10.1); if δu denotes a variation in u and $X'(u_e)$ denotes the linearization of X at u_e (the matrix of partial derivatives in the case of finitely many degrees of freedom), the linearized equations describe the time evolution of "infinitesimal" disturbances of u_e:

$$\frac{d}{dt}(\delta u) = X'(u_e) \cdot \delta u. \tag{1.10.2}$$

Equation (1.10.1), on the other hand, describes the nonlinear evolution of *finite* disturbances $\Delta u = u - u_e$. We say u_e is ***linearly stable*** if (1.10.2) is stable at $\delta u = 0$, in the sense defined above. Intuitively, this means that there are no infinitesimal disturbances which are growing in time. If $(\delta u)_0$ is an eigenfunction of $X'(u_e)$, that is, if

$$X'(u_e) \cdot (\delta u)_0 = \lambda(\delta u)_0 \tag{1.10.3}$$

for a complex number λ, then the corresponding solution of (1.10.2) with initial condition $(\delta u)_0$ is

$$\delta u = e^{t\lambda}(\delta u)_0. \tag{1.10.4}$$

This is growing when λ has positive real part. This leads us to the third notion of stability: we say that (1.10.1) or (1.10.2) is ***spectrally stable*** if the eigenvalues (more precisely points in the spectrum) all have non-positive real parts. In finite dimensions and, under appropriate technical conditions in infinite dimensions, one has the following implications:

$$\text{(stability)} \Rightarrow \text{(spectral stability)}$$

and

$$\text{(linear stability)} \Rightarrow \text{(spectral stability)}.$$

If the eigenvalues all lie strictly in the left half-plane, then a classical result of Liapunov guarantees stability. (See, for instance, Hirsch and Smale [1974] and Wiggins [1990] for the finite-dimensional case and Marsden and McCracken [1976], or Abraham, Marsden, and Ratiu [1988] for the infinite-dimensional case.) However, in systems of interest to us, the dissipation is very small; our systems will often be conservative. For such systems the eigenvalues must be symmetrically distributed under reflection in the real and imaginary axis. This implies that the only possibility for spectral stability is when the eigenvalues lie exactly on the imaginary axis. Thus, this version of *the Liapunov theorem is of no help in the Hamiltonian case.*

In general, *spectral stability typically does not imply stability*; instabilities can be generated (even in Hamiltonian systems) through *resonance* and *Arnold diffusion*. (See, for example, Galavotti [1983] and Lichtenberg and Lieberman [1983] for an account of much of what is known, both theoretical and numerical.) Thus, to obtain general stability results, one must use other techniques to augment or replace the linearized theory. We give such a technique below.

Here is a simple planar example of a system which is spectrally stable at the origin, but which is unstable there. In polar coordinates (r, θ), consider the evolution of $u = (r, \theta)$ given by

$$\dot{r} = r^3(1 - r^2) \quad \text{and} \quad \dot{\theta} = 1. \tag{1.10.5}$$

In (x, y) coordinates this system takes the form

$$\dot{x} = x(x^2 + y^2)(1 - x^2 - y^2) - y, \; \dot{y} = y(x^2 + y^2)(1 - x^2 - y^2) + x.$$

The eigenvalues of the linearized system at the origin are readily verified to be $\pm\sqrt{-1}$, so the origin is spectrally stable; however, the phase portrait, shown in Figure 1.10.3 shows that the origin is unstable. (We include the factor $1-r^2$ to give the system an attractive periodic orbit—this is merely to enrich the example and show how a stable periodic orbit can attract the orbits expelled by an unstable equilibrium.) This is not, however, a conservative system; next we give examples of Hamiltonian systems with similar features.

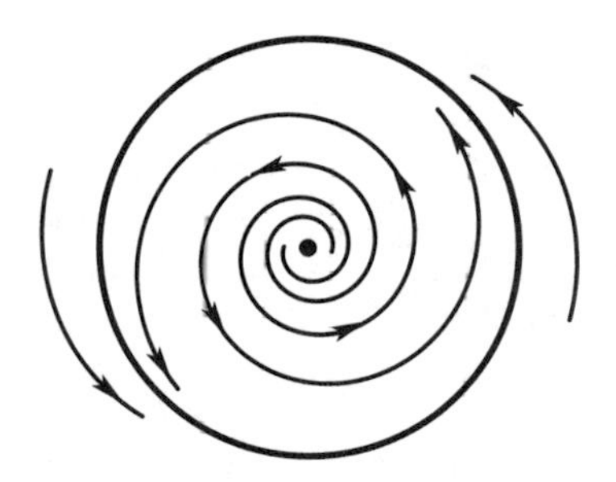

FIGURE 1.10.3. The phase portrait for $\dot{r} = r^3(1-r^2)$; $\dot{\theta} = 1$.

First, we consider an example involving resonance. The linear system in $\mathbb{R}^2$ whose Hamiltonian is given by $H(q,p) = \frac{1}{2}p^2 + \frac{1}{2}q^2 + pq$ has zero as a double eigenvalue so it is spectrally stable. On the other hand, $q(t) = (q_0+p_0)t + q_0$ and $p(t) = -(q_0+p_0)t + p_0$ is the solution of this system with initial condition (q_0, p_0), which clearly leaves any neighborhood of the origin no matter how close to it (q_0, p_0) is. Thus *spectral stability need not imply even linear stability*. An even simpler example of the same phenomenon is given by the free particle Hamiltonian $H(q,p) = \frac{1}{2}p^2$. Another higher-dimensional example with resonance in $\mathbb{R}^8$ is given by the linear system whose Hamiltonian is $H = q_2p_1 - q_1p_2 + q_4p_3 - q_3p_4 + q_2q_3$. The general solution with initial condition $(q_1^0, \ldots, p_4^0)$ is given by

$$
\begin{aligned}
q_1(t) &= q_1^0 \cos t + q_2^0 \sin t,\\
q_2(t) &= -q_1^0 \sin t + q_2^0 \cos t,\\
q_3(t) &= q_3^0 \cos t + q_4^0 \sin t,\\
q_4(t) &= -q_3^0 \sin t + q_4^0 \cos t,\\
p_1(t) &= -\frac{q_3^0}{2} t \sin t + \frac{q_4^0}{2}(t \cos t - \sin t) + p_1^0 \cos t + p_2^0 \sin t,\\
p_2(t) &= -\frac{q_3^0}{2}(t \cos t + \sin t) - \frac{q_4^0}{2} t \sin t - p_1^0 \sin t + p_2^0 \cos t,\\
p_3(t) &= \frac{q_1^0}{2} t \sin t - \frac{q_2^0}{2}(t \cos t + \sin t) + p_3^0 \cos t + p_4^0 \sin t,\\
p_4(t) &= \frac{q_1^0}{2}(t \cos t - \sin t) + \frac{q_2^0}{2} t \sin t - p_3^0 \sin t + p_4^0 \cos t.
\end{aligned}
$$

One sees that $p_i(t)$ leaves any neighborhood of the origin, no matter how close to the origin the initial conditions $(q_1^0, \ldots, p_4^0)$ are, that is, the system is linearly unstable. On the other hand, all eigenvalues of this linear system are $\pm i$, each a quadruple eigenvalue. Thus this linear system is spectrally stable.

Here is Cherry's example, giving a *system that is spectrally stable* ***and*** *linearly stable but is nonlinearly unstable*. Consider the Hamiltonian on $\mathbb{R}^4$ given by

$$H = \frac{1}{2}(q_1^2 + p_1^2) - (q_2^2 + p_2^2) + \frac{1}{2}p_2(p_1^2 - q_1^2) - q_1 q_2 p_1. \tag{1.10.6}$$

This system has an equilibrium at the origin, which is linearly stable since the linearized system consists of two uncoupled oscillators in the $(\delta q_2, \delta p_2)$ and $(\delta q_1, \delta p_1)$ variables, respectively, with frequencies in the ratio $2 : 1$ (the eigenvalues are $\pm i$ and $\pm 2i$, so the frequencies are in resonance). A family of solutions (parametrized by a constant τ) of Hamilton's equations for (1.10.6) is given by

$$\left.\begin{aligned} q_1 &= -\sqrt{2}\frac{\cos(t-\tau)}{t-\tau}, & q_2 &= \frac{\cos 2(t-\tau)}{t-\tau}, \\ p_1 &= \sqrt{2}\frac{\sin(t-\tau)}{t-\tau}, & p_2 &= \frac{\sin 2(t-\tau)}{t-\tau}. \end{aligned}\right\} \tag{1.10.7}$$

The solutions (1.10.7) clearly blow up in finite time; however, they start at time $t = 0$ at a distance $\sqrt{3}/\tau$ from the origin, so by choosing τ large, we can find solutions starting arbitrarily close to the origin, yet going to infinity in a finite time, so *the origin is nonlinearly unstable*.

Despite the above situation relating the linear and nonlinear theories, there has been much effort devoted to the development of spectral stability methods. When *instabilities* are present, spectral estimates do give important information on growth rates. As far as stability goes, spectral stability gives necessary, but not sufficient, conditions for stability. In other words, for the nonlinear problems *spectral instability can predict instability, but not stability*, this is a basic result of Liapunov; see Abraham, Marsden, and Ratiu [1988], for example. Our immediate purpose is the opposite: *to describe sufficient conditions for stability*.

Besides the energy, there are other conserved quantities associated with group symmetries such as linear and angular momentum. Some of these may be termed "reduction remnants" since they are associated with the group that underlies the passages from material to spatial or body coordinates. These are called *Casimir functions*; such a quantity, denoted C, is characterized by the fact that it Poisson commutes with every function, that is

$$\{C, F\} = 0 \tag{1.10.8}$$

for all functions F on phase space P. We shall study such functions and their relation with momentum maps in Chapters 10 and 11. For example, if Φ is any function of one variable, the quantity

$$C(\mathbf{\Pi}) = \Phi(\|\mathbf{\Pi}\|^2) \tag{1.10.9}$$

is a Casimir for the rigid body bracket, as is seen by using the chain rule. Likewise,

$$C(\omega) = \int_\Omega \Phi(\omega)\, dx\, dy \tag{1.10.10}$$

is a Casimir function for the two-dimensional fluid bracket. (This calculation ignores boundary terms that arise in an integration by parts—see Lewis, Marsden, Montgomery, and Ratiu [1986] for a treatment of these boundary terms.)

Casimir functions are conserved by the dynamics associated with any Hamiltonian H since $\dot{C} = \{C, H\} = 0$. Conservation of (1.10.9) corresponds to conservation of total angular momentum for the rigid body, while conservation of (1.10.10) represents Kelvin's circulation theorem for the Euler equations. It provides infinitely many independent constants of the motion that mutually Poisson commute; that is, $\{C_1, C_2\} = 0$, but this does *not* imply that these equations are integrable.

For Hamiltonian systems in canonical form, there is the classical Lagrange-Dirichlet stability criterion. First of all, notice that an equilibrium point (q_e, p_e) is a point at which the partial derivatives of H vanish, that is, it is a critical point of H. If the $2n \times 2n$ *matrix* $\delta^2 H$ *of second partial derivatives evaluated at* (q_e, p_e) *is positive- or negative-definite* (*that is, all the eigenvalues have the same sign*), *then* (q_e, p_e) *is stable*. This follows from conservation of energy and the fact from calculus, that the level sets of H near (q_e, p_e) are approximately ellipsoids. As mentioned earlier, this condition implies, but is not implied by, spectral stability. Apart from KAM (Kolmogorov, Arnold, Moser) theory, which gives stability of periodic solutions for two degree of freedom systems, the Lagrange-Dirichlet theorem is the only known *general* stability theorem for Hamiltonian systems.

For example, let us apply the Lagrange-Dirichlet theorem to a classical mechanical system whose Hamiltonian is the form kinetic plus potential energy. If (q_e, p_e) is an equilibrium, it follows that p_e is zero. Moreover, the matrix $\delta^2 H$ of second-order partial derivatives of H evaluated at (q_e, p_e) block diagonalizes with one of the blocks being the matrix of the quadratic form of the kinetic energy which is always positive-definite. Therefore, if $\delta^2 H$ is definite, it must be positive-definite and this in turn happens if and only if $\delta^2 V$ is positive-definite at q_e, where V is the potential energy of the system. We conclude that *for a classical mechanical system*, $(q_e, 0)$ *is a stable equilibrium, provided the matrix* $\delta^2 V(q_e)$ *of second-order partial derivatives of the potential* V *at* q_e *is positive-definite (or, more generally,*

q_e is a strict local minimum for V). If $\delta^2 V$ at q_e has a negative-definite direction, then q_e is an unstable equilibrium.

The second statement is seen in the following way. The linearized Hamiltonian system at $(q_e, 0)$ is again a Hamiltonian system whose Hamiltonian is of the form kinetic plus potential energy, the potential energy being given by the quadratic form $\delta^2 V(q_e)$. From a standard theorem in linear algebra, which states that two quadratic forms, one of which is positive-definite, can be simultaneously diagonalized, we conclude that the linearized Hamiltonian system decouples into a family of Hamiltonian systems of the form

$$\frac{d}{dt}(\delta p_k) = -c_k \delta q^k, \qquad \frac{d}{dt}(\delta q^k) = \frac{1}{m_k}\delta p_k,$$

where $1/m_k > 0$ are the eigenvalues of the positive-definite quadratic form given by the kinetic energy in the variables δp_j, and c_k are the eigenvalues of $\delta^2 V(q_e)$. Thus the eigenvalues of the linearized system are given by $\pm\sqrt{-c_k/m_k}$. Therefore, if some c_k is negative, the linearized system has at least one positive eigenvalue and thus $(q_e, 0)$ is spectrally and hence linearly and nonlinearly unstable.

The energy-momentum-Casimir method is a generalization of the classical Lagrange-Dirichlet method. Given an equilibrium u_e for $\dot{u} = X(u)$, it proceeds in the following steps:

Energy-Momentum-Casimir Method

To test an equilibrium (satisfying $X_H(z_e) = 0$)) for stability:

1. *Find a conserved function* $C - \langle \mathbf{J}, \xi \rangle$ (C will typically be a Casimir and $\mathbf{J}$ will be the momentum map for an *additional* symmetry group) *such that the first variation vanishes*:

$$\delta(H + C - \langle \mathbf{J}, \xi \rangle)(z_e) = 0.$$

2. *Calculate the second variation*

$$\delta^2(H + C - \langle \mathbf{J}, \xi \rangle)(z_e).$$

3. *If $\delta^2(H + C - \langle \mathbf{J}, \xi \rangle)(z_e)$ is definite (either positive or negative), then z_e is called* ***formally stable***.

With regard to Step 3, we point out that an equilibrium solution need not be a critical point of H alone; in general, $\delta H(z_e) \neq 0$. An example where this occurs is a rigid body spinning about one of its principal axes of inertia. In this case, a critical point of H alone would have zero angular velocity; but a critical point of $H + C$ is a (nontrivial) stationary rotation about one of the principal axes.

The argument used to establish the Lagrange-Dirichlet test formally works in infinite dimensions too. Unfortunately, for systems with infinitely many degrees of freedom (like fluids and plasmas), there is a technical snag. The calculus argument used before runs into problems; one might think these are just technical and that we just need to be more careful with the calculus arguments. In fact, there is widespread belief in this "energy criterion" (see, for instance, the discussion and references in Marsden and Hughes [1983], Ch. 6, and Potier-Ferry [1982]). However, Ball and Marsden [1984] have shown by means of an example from elasticity theory that the difficulty is genuine: they produce a critical point of H at which $\delta^2 H$ is positive-definite, yet this point is *not* a local minimum of H. On the other hand, Potier-Ferry [1982] shows that asymptotic stability is restored if suitable dissipation is added. Another way to overcome this difficulty is to modify Step 3 using a convexity argument of Arnold [1966b].

Convexity Analysis

Modified Step 3
Assume P is a *linear* space.

(a) *Let* $\Delta u = u - u_e$ *denote a finite variation in phase space.*

(b) *Find quadratic functions* Q_1 *and* Q_2 *such that*

$$Q_1(\Delta u) \leq H(u_e + \Delta u) - H(u_e) - \delta H(u_e) \cdot \Delta u$$

and

$$Q_2(\Delta u) \leq C_\xi(u_e + \Delta u) - C_\xi(u_e) - \delta C_\xi(u_e) \cdot \Delta u,$$

where $C_\xi = C - \langle \mathbf{J}, \xi \rangle$.

(c) *Require that* $Q_1(\Delta u) + Q_2(\Delta u) > 0$ *for all* $\Delta u \neq 0$.

(d) *Introduce the norm* $\|\Delta u\|$ *by*

$$\|\Delta u\|^2 = Q_1(\Delta u) + Q_2(\Delta u),$$

so $\|\Delta u\|$ *is a measure of the distance from* u *to* u_e :

$$d(u, u_e) = \|\Delta u\|.$$

(e) *Require that*

$$| \; H(u_e + \Delta u) - H(u_e) \; | \leq C_1 \|\Delta u\|^\alpha$$

and

$$| \; C_\xi(u_e + \Delta u) - C_\xi(u_e) \; | \leq C_2 \|\Delta u\|^\alpha$$

for constants $\alpha, C_1, C_2 > 0$, *and* $\|\Delta u\|$ *sufficiently small.*

These conditions guarantee stability of u_e and provide the distance measure relative to which stability is defined. The key part of the proof is simply the observation that if we add the two inequalities in (b), we get

$$\|\Delta u\|^2 \leq H(u_e + \Delta u) + C_\xi(u_e + \Delta u) - H(u_e) - C_\xi(u_e)$$

using the fact that $\delta H(u_e) \cdot \Delta u$ and $\delta C_\xi(u_e) \cdot \Delta u$ add up to zero by Step 1. But H and C_ξ are constant in time so

$$\|(\Delta u)_{\text{time}=t}\|^2 \leq [H(u_e + \Delta u) + C_\xi(u_e + \Delta u) - H(u_e) - C_\xi(u_e)]\,|_{\text{time}=0}\,.$$

Now employ the inequalities in (e) to get

$$\|(\Delta u)_{\text{time}=t}\|^2 \leq (C_1 + C_2)\|(\Delta u)_{\text{time}=0}\|^\alpha.$$

This estimate bounds the temporal growth of finite perturbations in terms of initial perturbations, which is what is needed for stability. For a survey of this method, additional references and numerous examples, see Holm, Marsden, Ratiu, and Weinstein [1985].

There are some situations (such as the stability of elastic rods) in which the above techniques do not apply. The chief reason is that there may be a lack of sufficiently many Casimir functions to even achieve the first step. For this reason a modified (but more sophisticated) method has been developed called the "energy-momentum method." The key to the method is to avoid the use of Casimir functions by applying the method *before* any reduction has taken place. This method was developed in a series of papers of Simo, Posbergh, and Marsden [1990, 1991] and Simo, Lewis, and Marsden [1991]. A discussion and additional references may be found in Marsden [1992] and in Volume II.

The distinctions between "stability by energy methods, that is, ***energetics***" and "spectral stability," become especially interesting when one adds dissipation. In fact, building on the classical work of Kelvin and Chetaev, one can prove that if $\delta^2 H$ is indefinite, yet the spectrum is on the imaginary axis, then adding dissipation necessarily makes the system *linearly unstable*. That is, at least one pair of eigenvalues of the linearlized equations move into the right half-plane. This is a phenomenon called ***dissipation induced instabilities.*** This result, along with related developments, is proved in Bloch, Krishnaprasad, Marsden, and Ratiu [1991, 1994a,b]. For example, consider the linear ***gyroscopic system***

$$M\ddot{\mathbf{q}} + S\dot{\mathbf{q}} + V\mathbf{q} = 0, \tag{1.10.11}$$

where $\mathbf{q} \in \mathbb{R}^n$, M is a positive-definite symmetric $n \times n$ matrix, S is skew, and V is symmetric. This system is Hamiltonian (Exercise **1.10-2**). If V has negative eigenvalues, then (1.10.11) is *formally unstable*. However, due to S, the system can be spectrally stable. However, if R is positive-definite symmetric and $\epsilon > 0$ is small, the system with friction

$$M\ddot{\mathbf{q}} + S\dot{\mathbf{q}} + \epsilon R\dot{\mathbf{q}} + V\mathbf{q} = 0 \tag{1.10.12}$$

is linearly unstable. A specific example is given in Exercise **1.10-3**.

Exercises

1.10-1 *Work out Cherry's example of the Hamiltonian system in* $\mathbb{R}^4$ *whose energy function is given by* (1.10.6). *Show explicitly that the origin is a linearly and spectrally stable equilibrium but that it is nonlinearly unstable by proving that* (1.10.7) *is a solution for every* $\tau > 0$ *which can be chosen to start arbitrarily close to the origin and which goes to infinity for* $t \to \tau$.

1.10-2 *Show that* (1.10.11) *is Hamiltonian with* $\mathbf{p} = M\dot{\mathbf{q}}$,

$$H(\mathbf{q}, \mathbf{p}) = \frac{1}{2}\mathbf{p} \cdot M^{-1}\mathbf{p} + \frac{1}{2}\mathbf{q} \cdot V\mathbf{q}$$

and

$$\{F, K\} = \frac{\partial F}{\partial q^i}\frac{\partial K}{\partial p_i} - \frac{\partial K}{\partial q^i}\frac{\partial F}{\partial p_i} - S^{ij}\frac{\partial F}{\partial p_i}\frac{\partial K}{\partial p_j}.$$

1.10-3 *Show that the characteristic polynomial for the linear system* (1.10.11) *is*

$$p(\lambda) = \det[\lambda^2 M + \lambda S + V]$$

and that this actually is a polynomial of degree n *in* λ^2.

1.10-4 *Consider the two degree of freedom system*

$$\begin{aligned} \ddot{x} - g\dot{y} + \gamma\dot{x} + \alpha x &= 0, \\ \ddot{y} + g\dot{x} + \delta\dot{y} + \beta y &= 0. \end{aligned}$$

(a) *Write it in the form* (1.10.12).

(b) *For* $\gamma = \delta = 0$ *show:*

- **i** *it is spectrally stable if* $\alpha > 0, \beta > 0$;
- **ii** *for* $\alpha\beta < 0$, *it is spectrally unstable;*
- **iii** *for* $\alpha < 0, \beta < 0$, *it is formally unstable (that is, the energy function, which is a quadratic form, is indefinite); and*
  - **A** *if* $D := (g^2 + \alpha + \beta)^2 - 4\alpha\beta < 0$, *then there are two roots in the right half-plane and two in the left; the system is spectrally unstable;*
  - **B** *if* $D = 0$ *and* $g^2 + \alpha + \beta \geq 0$ *the system is spectrally stable, but if* $g^2 + \alpha + \beta < 0$ *then it is spectrally unstable; and*
  - **C** *if* $D > 0$ *and* $g^2 + \alpha + \beta \geq 0$ *the system is spectrally stable, but if* $g^2 + \alpha + \beta < 0$, *then it is spectrally unstable.*

(c) *For a polynomial* $p(\lambda) = \lambda^4 + \rho_1\lambda^3 + \rho_2\lambda^2 + \rho_3\lambda + \rho_4$, *the Routh-Hurwitz criterion (see Gantmacher* [1959], *Volume 2) says that the number of right half-plane zeros of p is the number of sign changes of the sequence*

$$\left\{1, \rho_1, \frac{\rho_1\rho_2 - \rho_3}{\rho_1}, \frac{\rho_3\rho_1\rho_2 - \rho_3^2 - \rho_4\rho_1^2}{\rho_1\rho_2 - \rho_3}, \rho_4\right\}.$$

Apply this to the case in which $\alpha < 0, \beta < 0, g^2 + \alpha + \beta > 0$, *and at least one of* γ or δ *is positive to show that the system is spectrally unstable.*

1.11 Bifurcation

When the energy-momentum-Casimir method indicates that an instability might be possible, techniques of bifurcation theory can be brought to bear to determine the emerging dynamical complexities such as the development of multiple equilibria and periodic orbits. For example, consider a particle moving with no friction in a rotating hoop (Figure 1.11.1).

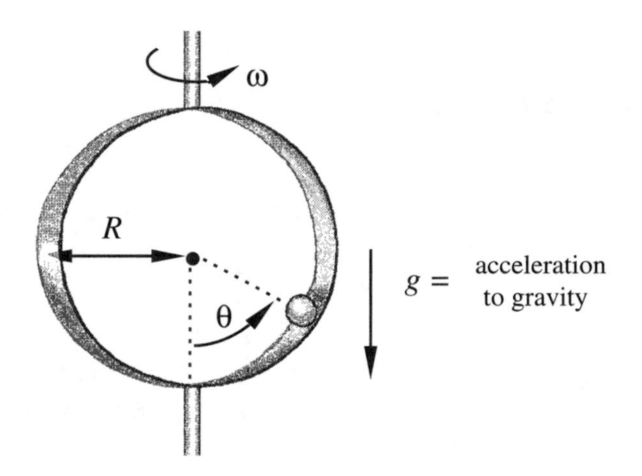

FIGURE 1.11.1. A particle moving in a hoop rotating with angular velocity ω.

In §2.10 we derive the equations and study the phase portraits for this system. One finds that as ω increases past $\sqrt{g/R}$, the stable equilibrium at $\varphi = 0$ becomes unstable through a *Hamiltonian pitchfork bifurcation* and two new solutions are created. These solutions are symmetric in the vertical axis, a reflection of the original $\mathbb{Z}_2$ symmetry of the mechanical system in Figure 1.11.1. Breaking this symmetry by, for example, putting the rotation axis slightly off-center is an interesting topic that we shall discuss in §2.10.

Another example of this type concerns bifurcations of a rotating liquid drop: the system consists of the two-dimensional Euler equations for an ideal fluid with a free boundary. An equilibrium solution consists of a rigidly rotating circular drop. The energy-momentum-Casimir method shows stability provided that

$$\Omega < 2\sqrt{\frac{3\tau}{R^3}}. \tag{1.11.1}$$

In this formula, Ω is the angular velocity of the circular drop, R is its radius, and τ is the surface tension, a constant. As Ω increases and (1.11.1) is violated, the stability of the circular solution is lost and is picked up by elliptical-like solutions with $\mathbb{Z}_2 \times \mathbb{Z}_2$ symmetry. The bifurcation is actually subcritical relative to Ω (that is, the new solutions occur *below* the critical value of Ω) and is supercritical (the new solutions occur *above* criticality) relative to the angular momentum. This is proved in Lewis, Marsden, and Ratiu [1987] and Lewis [1989], where other references may also be found (see Figure 1.11.2).

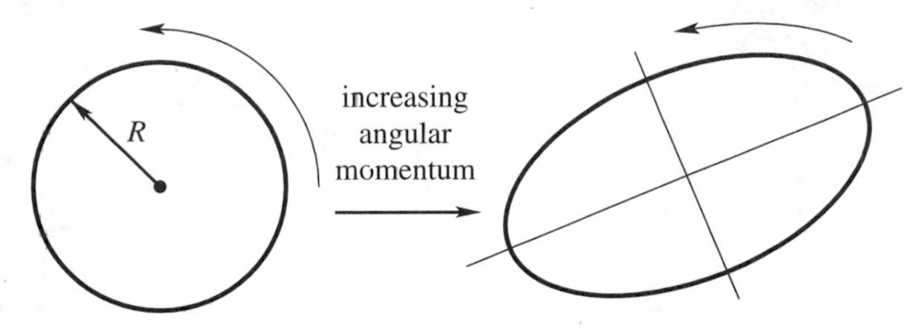

FIGURE 1.11.2. A circular liquid drop losing its stability and its symmetry.

For the ball in the hoop, the eigenvalue evolution for the linearized equations is shown in Figure 1.11.3(a). For the rotating liquid drop the movement of eigenvalues is the same: they are constrained to *stay* on the imaginary axis because of the symmetry of the problem. Without this symmetry, eigenvalues typically split, as in Figure 1.11.3(b). These are examples of a general theory of the movement of such eigenvalues given in Golubitsky and Stewart [1987].

A somewhat more sophisticated example is the heavy top: a rigid body with one point fixed, moving in a gravitational field. When the top makes the transition from a fast top to a slow top,

$$\omega \downarrow \frac{2\sqrt{MglI_1}}{I_3}, \tag{1.11.2}$$

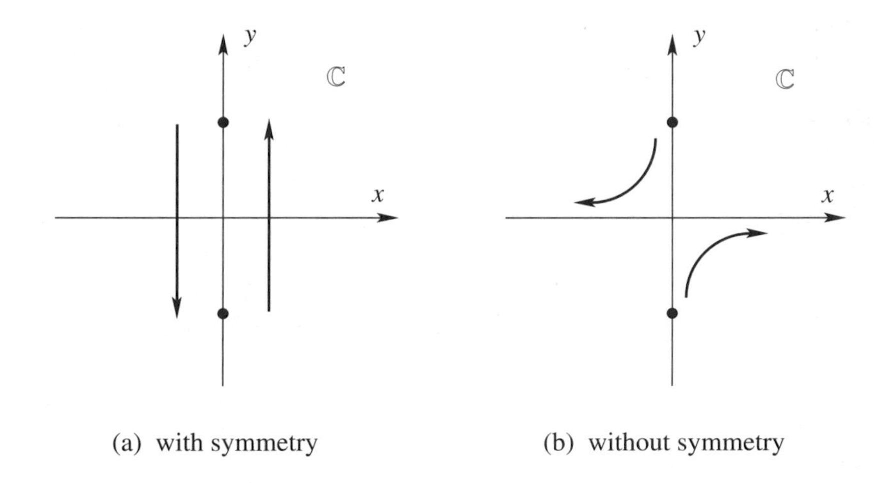

FIGURE 1.11.3. The movement of eigenvalues in bifurcation of equilibria.

stability is lost and a ***resonance bifurcation*** occurs. Here, when the bifurcation occurs, the eigenvalues of the equations linearized at the equilibrium behave as in Figure 1.11.4.

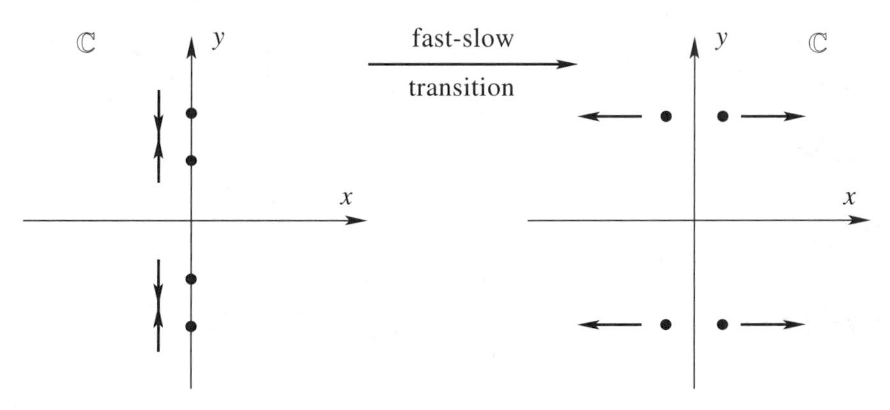

FIGURE 1.11.4. Eigenvalue movement in the Hamiltonian Hopf bifurcation.

For an extensive study of bifurcations and stability in the dynamics of a heavy top, see Lewis, Ratiu, Simo, and Marsden [1992]. Behavior of this sort is sometimes called a ***Hamiltonian Krein-Hopf bifurcation***, or a ***gyroscopic instability*** (see Van der Meer [1985, 1990]). Here more complex dynamic behavior ensues, including periodic and chaotic motions (see Holmes and Marsden [1983]). In some systems with symmetry, the eigenvalues can ***pass*** as well as ***split***, as has been shown by Dellnitz, Melbourne, and Marsden [1992] and references therein.

More sophisticated examples, such as the dynamics of two coupled three-dimensional rigid bodies requires a systematic development of the basic theory of Golubitsky and Schaeffer [1985] and Golubitsky, Stewart, and Schaeffer [1988]. This theory is begun in, for example, Duistermaat [1983], Lewis, Marsden, and Ratiu [1987], Lewis [1989], Patrick [1989], Meyer and Hall [1992], Broer, Chow, Kim, and Vegter [1993], and Golubitsky, Marsden, Stewart, and Dellnitz [1994]. For bifurcations in the double spherical pendulum (which includes a Hamiltonian-Krein-Hopf bifurcation), see Dellnitz, Marsden, Melbourne, and Scheurle [1992] and Marsden and Scheurle [1993a].

Exercises

1.11-1 *Study the bifurcations (changes in the phase portrait) for the equation*

$$\ddot{x} + \mu x + x^2 = 0$$

as μ passes through zero. Use the second derivative test on the potential energy discussed in §1.10.

1.11-2 *Repeat Exercise* **1.11-1** *for*

$$\ddot{x} + \mu x + x^3 = 0$$

as μ passes through zero.

1.12 The Poincaré-Melnikov Method and Chaos

To begin with a simple example, consider the equation of a forced pendulum

$$\ddot{\phi} + \sin \phi = \epsilon \cos \omega t. \tag{1.12.1}$$

Here ω is a constant angular forcing frequency and ϵ is a small parameter. Systems of this or a similar nature arise in many interesting situations. For example, a double planar pendulum and other "executive toys" exhibit chaotic motion that is analogous to the behavior of this equation; see Burov [1986] and Shinbrot, Grebogi, Wisdom, and Yorke [1992].

For $\epsilon = 0$ this has the phase portrait of a simple pendulum (the same as shown later in Figure 2.10.2a). For ϵ small but nonzero, (1.12.1) possesses no analytic integrals of the motion. In fact, it possesses transversal intersecting stable and unstable manifolds (separatrices); that is, the Poincaré maps $P_{t_0} : \mathbb{R}^2 \to \mathbb{R}^2$ that advance solutions by one period $T = 2\pi/\omega$ starting at time t_0 possess transversal homoclinic points. This type of dynamic behavior has several consequences, besides precluding the existence of analytic integrals, that lead one to use the term "chaotic." For example, (1.12.1) has infinitely many periodic solutions of arbitrarily high period.

Also, using the shadowing lemma, one sees that given any bi-infinite sequence of zeros and ones (for example, use the binary expansion of e or π), there exists a corresponding solution of (1.12.1) that successively crosses the plane $\phi = 0$ (the pendulum's vertically downward configuration) with $\phi > 0$ corresponding to a zero and $\phi < 0$ corresponding to a one. The origin of this chaos on an intuitive level lies in the motion of the pendulum near its unperturbed homoclinic orbit, the orbit that does one revolution in infinite time. Near the top of its motion (where $\phi = \pm\pi$) small nudges from the forcing term can cause the pendulum to fall to the left or right in a temporally complex way.

The dynamical systems theory needed to justify the preceding statements is available in Smale [1967], Moser [1973], Guckenheimer and Holmes [1983], and Wiggins [1988, 1990]. Some key people responsible for the development of the basic theory are Poincaré, Birkhoff, Kolmogorov, Melnikov, Arnold, Smale, and Moser. The idea of transversal intersecting separatrices comes from Poincaré's famous paper on the three-body problem (Poincaré [1890]). His goal, not quite achieved for reasons we shall comment on later, was to prove the nonintegrability of the restricted three body problem and that various series expansions used up to that point diverged (he began the theory of asymptotic expansions and dynamical systems in the course of this work).

Although Poincaré had all the essential tools needed to prove that equations like (1.12.1) are not integrable (in the sense of having no analytic integrals), his interests lay with harder problems and so he did not develop the easier basic theory very much. Important contributions were made by Melnikov [1963] and Arnold [1964] which lead to a simple procedure for proving that (1.12.1) is not integrable. The Poincaré-Melnikov method was revived by Chirikov [1979], Holmes [1980b] and Chow, Hale, and Mallet-Paret [1980]. We shall give the method for Hamiltonian systems. We refer to Guckenheimer and Holmes [1983] and to Wiggins [1988, 1990] for generalizations and further references.

The Poincaré-Melnikov Method

1. Write the dynamical equation to be studied in abstract form as

$$\dot{x} = X_0(x) + \epsilon X_1(x,t), \tag{1.12.2}$$

where $x \in \mathbb{R}^2, X_0$ is a Hamiltonian vector field with energy H_0, X_1 is periodic with period T and is Hamiltonian with energy a T-periodic function H_1. Assume that X_0 has a homoclinic orbit $\overline{x}(t)$ so $\overline{x}(t) \to x_0$, a hyperbolic saddle point, as $t \to \pm\infty$.

2. Compute the ***Poincaré-Melnikov function*** defined by

$$M(t_0) = \int_{-\infty}^{\infty} \{H_0, H_1\}(\overline{x}(t - t_0), t)\, dt \tag{1.12.3}$$

where $\{\,,\}$ denotes the Poisson bracket.

If $M(t_0)$ has simple zeros as a function of t_0, then (1.12.2) has, for sufficiently small ϵ, homoclinic chaos in the sense of transversal intersecting separatrices (in the sense of Poincaré maps as mentioned above).

We shall give a proof of this result in §2.11. To apply it to equation (1.12.1) one proceeds as follows. Let $x = (\phi, \dot{\phi})$ so we get

$$\frac{d}{dt}\begin{bmatrix} \phi \\ \dot{\phi} \end{bmatrix} = \begin{bmatrix} \dot{\phi} \\ -\sin\phi \end{bmatrix} + \epsilon \begin{bmatrix} 0 \\ \cos\omega t \end{bmatrix}.$$

The homoclinic orbits for $\epsilon = 0$ are given by (see Exercise **1.12-1**)

$$\overline{x}(t) = \begin{bmatrix} \phi(t) \\ \dot{\phi}(t) \end{bmatrix} = \begin{bmatrix} \pm 2\tan^{-1}(\sinh t) \\ \pm 2\,\mathrm{sech}\, t \end{bmatrix}$$

and one has

$$H_0(\phi, \dot{\phi}) = \frac{1}{2}\dot{\phi}^2 - \cos\phi \quad \text{and} \quad H_1(\phi, \dot{\phi}, t) = \phi\cos\omega t. \tag{1.12.4}$$

Hence (1.12.3) gives

$$M(t_0) = \int_{-\infty}^{\infty} \left(\frac{\partial H_0}{\partial \phi}\frac{\partial H_1}{\partial \dot{\phi}} - \frac{\partial H_0}{\partial \dot{\phi}}\frac{\partial H_1}{\partial \phi} \right) (\overline{x}(t - t_0), t)\, dt$$

$$= -\int_{-\infty}^{\infty} \dot{\phi}(t-t_0)\cos\omega t\,dt$$

$$= \mp\int_{-\infty}^{\infty} [2\,\mathrm{sech}(t-t_0)\cos\omega t]\,dt.$$

Changing variables and using the fact that sech is even and sin is odd, we get

$$M(t_0) = \mp 2\left(\int_{-\infty}^{\infty} \mathrm{sech}\,t\cos\omega t\,dt\right)\cos(\omega t_0).$$

The integral is evaluated by residues (see Excercise **1.12-2**):

$$M(t_0) = \mp 2\pi\,\mathrm{sech}\left(\frac{\pi\omega}{2}\right)\cos(\omega t_0), \tag{1.12.5}$$

which clearly has simple zeros. Thus, this equation has chaos for ϵ small enough.

Exercise

1.12-1 *Verify directly that the homoclinic orbits for the simple pendulum equation* $\ddot{\phi} + \sin\phi = 0$ *are given by* $\phi(t) = \pm 2\tan^{-1}(\sinh t)$.

1.12-2 *Evaluate the integral* $\int_{-\infty}^{\infty} \mathrm{sech}\,t\cos\omega t\,dt$ *to prove* (1.12.5) *as follows. Write* $\mathrm{sech}\,t = 2/(e^t + e^{-t})$ *and note that there is a simple pole of*

$$f(z) = \frac{e^{i\omega z} + e^{-i\omega z}}{e^z + e^{-z}}$$

in the complex plane at $z = \pi i/2$. *Evaluate the residue there and apply Cauchy's theorem.* (Consult a book on complex variables such as Marsden and Hoffman, *Basic Complex Analysis*, Second Edition, Freeman, 1988.)

1.13 Resonances, Geometric Phases, and Control

The work of Smale [1970] shows that topology plays an important role in mechanics. Smale's work employs Morse theory applied to conserved quantities such as the energy-momentum map. In this section we point out other ways in which geometry and topology enter mechanical problems.

We begin with a discussion of the ***one-to-one resonance*** following Cushman and Rod [1982]. When one considers resonant systems one often encounters Hamiltonians of the form

$$H = \frac{1}{2}(q_1^2 + p_1^2) + \frac{\lambda}{2}(q_2^2 + p_2^2) + \text{ higher-order terms.} \tag{1.13.1}$$

The oscillators have the same frequency when $\lambda = 1$, which is why one speaks of a one-to-one resonance. To analyze the dynamics of H, it is important to utilize a good geometric picture for the critical case

$$H_0 = \frac{1}{2}(q_1^2 + p_1^2 + q_2^2 + p_2^2). \tag{1.13.2}$$

The energy level $H_0 =$ constant is the three-sphere $S^3 \subset \mathbb{R}^4$. If we think of H_0 as a function on $\mathbb{C}^2$ by letting

$$z_1 = q_1 + ip_1 \quad \text{and} \quad z_2 = q_2 + ip_2,$$

then $H_0 = (|z_1|^2 + |z_2|^2)/2$ and so H_0 is left-invariant by the action of $SU(2)$, the complex 2×2 unitary matrices of determinant one. The corresponding conserved quantities are

$$\left.\begin{aligned} W_1 &= 2(q_1q_2 + p_1p_2), \\ W_2 &= 2(q_2p_1 - q_1p_2), \\ W_3 &= q_1^2 + p_1^2 - q_2^2 - p_2^2, \end{aligned}\right\} \tag{1.13.3}$$

which comprise the components of a (momentum) map

$$\mathbf{J} : \mathbb{R}^4 \to \mathbb{R}^3. \tag{1.13.4}$$

From the relation $4H_0^2 = W_1^2 + W_2^2 + W_3^2$, one finds that $\mathbf{J}$ restricted to S^3 gives a map

$$j : S^3 \to S^2. \tag{1.13.5}$$

The fibers j^{-1}(point) are circles and the dynamics of H_0 moves along these circles. The map j is the ***Hopf fibration*** which describes S^3 as a topologically nontrivial circle bundle over S^2. The role of the Hopf fibration in mechanics was known to Reeb [1949].

One also finds that the study of systems like (1.13.1) that are close to H_0 can, to a good approximation, be reduced to dynamics on S^2. These dynamics are in fact Lie-Poisson and S^2 sits as a coadjoint orbit in $\mathfrak{so}(3)^*$, so the evolution is of rigid body type, just with a different Hamiltonian. For a computer study of the Hopf fibration in the one-to-one resonance, see Kocak, Bisshopp, Banchoff, and Laidlaw [1986].

The Hopf fibration occurs in a number of other interesting mechanical systems. One of these is the free rigid body. When doing reduction for the rigid body, one studies the reduced space $\mathbf{J}^{-1}(\mu)/G_\mu = \mathbf{J}^{-1}(\mu)/S^1$, which in this case is the sphere S^2. Also, as we shall see in Chapter 15, $\mathbf{J}^{-1}(\mu)$ is topologically the same as the rotation group $SO(3)$, which in turn is the same as $S^3/\mathbb{Z}_2$. Thus, the reduction map is a map of $SO(3)$ to S^2. Such a map is given explicitly by taking an orthogonal matrix A and mapping it to the vector on the sphere given by $A\mathbf{k}$, where $\mathbf{k}$ is the unit vector along the z-axis. This map that does the projection is in fact a restriction of a momentum map and is, when composed with the map of $S^3 \cong SU(2)$ to $SO(3)$, just the Hopf fibration again. Thus, not only does the Hopf fibration occur in the one-to-one resonance, *it occurs in the rigid body in a natural way as the reduction map from material to body representation!*

Next we briefly discuss some ideas in the theory of *geometric phases.* The history of this concept is complex. We refer to Berry [1990] for a discussion of the history, going back to Bortolotti in 1926, Vladimirskii and Rytov in 1938 in the study of polarized light, to Kato in 1950 and Longuet-Higgins and others in 1958 in atomic physics. Some additional historical comments regarding phases in rigid body mechanics are given below. We shall pick up the story with the classical example of the Foucault pendulum. The Foucault pendulum gives an interesting phase shift (a shift in the angle of the plane of the pendulum's swing) when the overall system undergoes a cyclic evolution (the pendulum is carried in a circular motion due to the Earth's rotation). This phase shift is geometric in character: if one parallel transports an orthonormal frame along the same line of latitude, it returns with a phase shift equaling that of the Foucault pendulum. This phase shift $\Delta\theta = 2\pi \cos\alpha$ (where α is the co-latitude) has the geometric meaning shown in Figure 1.13.1.

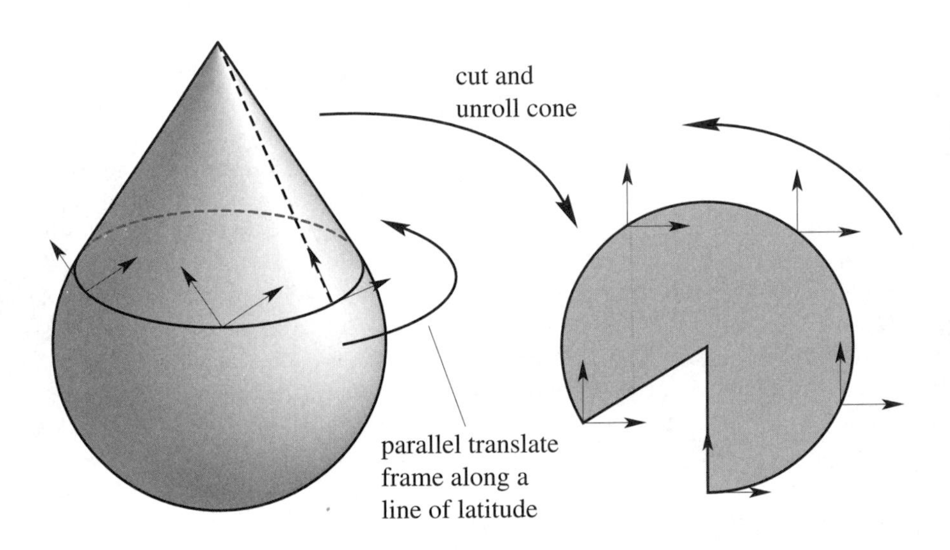

FIGURE 1.13.1. The geometric interpretation of the Foucault pendulum phase shift.

In geometry, when an orthonormal frame returns after traversing a closed path to its original position but rotated, the rotation is referred to as

holonomy (or ***anholonomy***). This is a unifying mathematical concept that underlies many geometric phases in systems such as fiber optics, MRI (magnetic resonance imaging), amoeba propulsion, molecular dynamics, micromotors, and other effects. These applications represent one reason why the subject is of such current interest.

In the quantum case a seminal paper on geometric phases is Kato [1950]. It was Berry [1984, 1985], Simon [1984], Hannay [1985], and Berry and Hannay [1988] who realized that holonomy is the crucial geometric unifying thread. On the other hand, Golin, Knauf, and Marmi [1989], Montgomery [1988], and Marsden, Montgomery, and Ratiu [1989, 1990] demonstrated that averaging connections and reduction of mechanical systems with symmetry also plays an important role, both classically and quantum mechanically. Aharonov and Anandan [1987] have shown that the geometric phase for a closed loop in projectivized complex Hilbert space occurring in quantum mechanics equals the exponential of the symplectic area of a two-dimensional manifold whose boundary is the given loop. The symplectic form in question is naturally induced on the projective space from the canonical symplectic form of complex Hilbert space (minus the imaginary part of the inner product) via reduction. Marsden, Montgomery, and Ratiu [1990] show that this formula is the holonomy of the closed loop relative to a principal S^1-connection on the unit ball of complex Hilbert space and is a particular case of the holonomy formula in principal bundles with abelian structure group.

Another class of examples where geometric phases naturally occur is in families of integrable systems depending on parameters. Consider an integrable system with action-angle variables $(I_1, I_2, \ldots, I_n, \theta_1, \theta_2, \ldots, \theta_n)$; assume the Hamiltonian $H(I_1, I_2, \ldots I_n; m)$ depends on a parameter $m \in M$. This just means that we have a Hamiltonian independent of the angular variables θ and we can identify the configuration space with an n-torus $\mathbb{T}^n$. Let c be a loop based at a point m_0 in M. We want to compare the angular variables in the torus over m_0, once the system is slowly changed as the parameters undergo the circuit c. Since the dynamics in the fiber varies as we move along c, even if the actions vary by a negligible amount, there will be a shift in the angle variables due to the frequencies $\omega^i = \partial H / \partial I^i$ of the integrable system; correspondingly, one defines

$$\textbf{dynamic phase} = \int_0^1 \omega^i \left(I, c(t)\right) dt.$$

Here we assume that the loop is contained in a neighborhood whose standard action coordinates are defined. In completing the circuit c, we return to the same torus, so a comparison between the angles makes sense. The actual shift in the angular variables during the circuit is the ***dynamic phase*** plus a correction term called the ***geometric phase***. One of the key results is that this geometric phase is the holonomy of an appropriately constructed connection called the ***Hannay-Berry connection*** on the torus bundle over M which is constructed from the action-angle variables. The corresponding angular shift, computed by Hannay [1985], is called ***Hannay's angles***, so the actual phase shift is given by

$$\Delta\theta = \text{dynamic phases} + \text{Hannay's angles}.$$

The geometric construction of the Hannay-Berry connection for classical systems is given in terms of momentum maps and averaging in Golin, Knauf, and Marmi [1989] and Montgomery [1988]. Weinstein [1990] makes precise the geometric structures which make possible a definition of the Hannay angles for a cycle in the space of lagrangian submanifolds, even without the presence of an integrable system. Berry's phase is then seen as a "primitive" for the Hannay angles. A summary of this work is given in Woodhouse [1992].

Another class of examples where geometric phases naturally arise is in the dynamics of coupled rigid bodies. The three dimensional single rigid body is discussed below. For several coupled rigid bodies, the dynamics can be quite complex. For instance, even for bodies in the plane, the dynamics is known to be chaotic, despite the presence of stable relative equilibria; see Oh, Sreenath, Krishnaprasad, and Marsden [1989]. Geometric phase phenomena for this type of example are quite interesting and are related to some of the work of Wilczek and Shapere on locomotion in micro-organisms. (See, for example, Shapere and Wilczek [1987, 1989] and Wilczek and Shapere [1989].) In this problem, control of the system's *internal variables* can lead to phase changes in the *external variables*. These choices of variables are related to the variables in the reduced and the unreduced phase spaces. In this setting one can formulate interesting questions of optimal control such as "When a cat falls and turns itself over in mid-flight (all the time with zero angular momentum!) does it do so with optimal efficiency in terms of, say, energy expended?" There are interesting answers to these questions that are related to the dynamics of Yang-Mills particles moving in the associated gauge field of the problem. See Montgomery [1984, 1990] and references therein.

We give two simple examples of how geometric phases for linked rigid bodies works. Additional details can be found in Marsden, Montgomery, and Ratiu [1990]. First, consider three uniform coupled bars (or coupled planar rigid bodies) linked together with pivot (or pin) joints, so the bars are free to rotate relative to each other. Assume the bars are moving freely in the plane with no external forces and that the angular momentum is zero. However, assume that the joint angles can be controlled with, say, motors in the joints. Figure 1.13.2 shows how the joints can be manipulated, each one going through an angle of 2π and yet the overall assemblage rotates through an angle π. Here we assume that the moments of inertia of the two outside bars (about an axis through their centers of mass and perpendicular to the page) are each one-half that of the middle bar. The statement is verified by examining the equation for zero angular momentum (see, for example Sreenath, Oh, Krishnaprasad, and Marsden [1988] and Oh, Sreenath, Krishnaprasad, and Marsden [1989]). General formulas for the reconstruction phase applicable to examples of this type are given in Krishnaprasad [1989].

A second example is the dynamics of linkages. This type of example is considered in Krishnaprasad [1989], Yang and Krishnaprasad [1990], including comments on the relation with the three-manifold theory of Thurston. Here one considers a linkage of rods, say four rods linked by pivot joints as in Figure 1.13.3. The system is free to rotate without external forces or torques, but there are assumed to be torques at the joints. When one turns the small "crank" the whole assemblage turns even though the angular momentum, as in the previous example, stays zero.

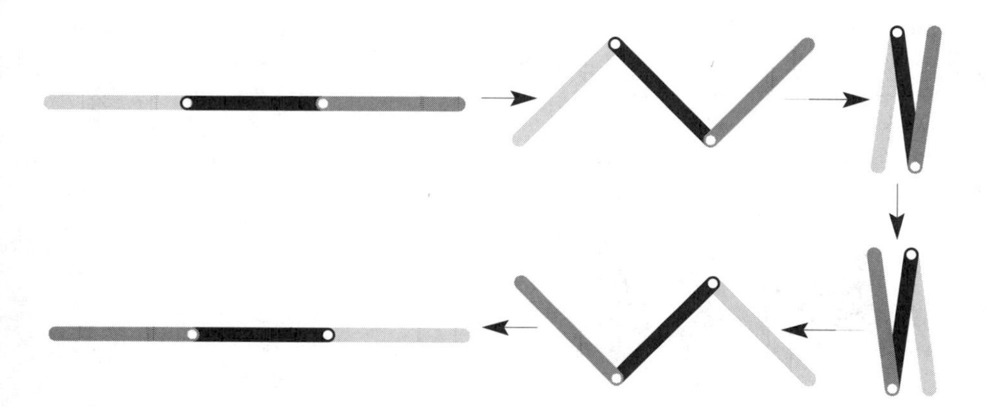

FIGURE 1.13.2. Manipulating the joint angles can lead to an overall rotation of the system.

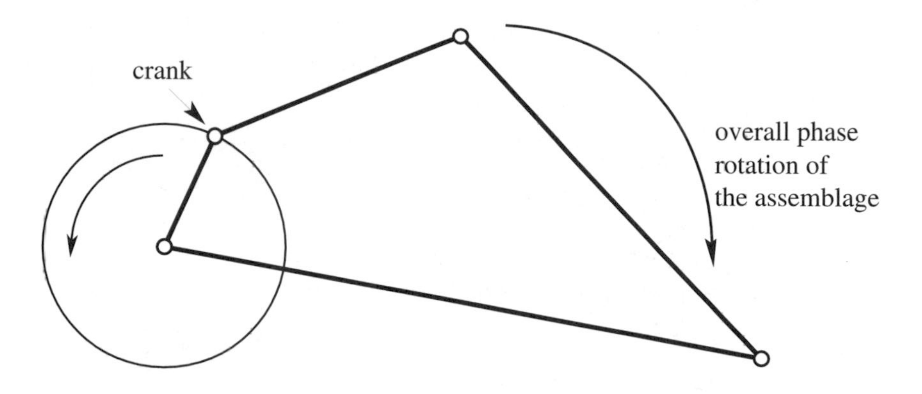

FIGURE 1.13.3. Turning the crank can lead to an overall phase shift.

As we shall see in Chapter 15, the motion of a rigid body is a geodesic with respect to a left-invariant Riemannian metric (the inertia tensor) on $SO(3)$. The corresponding phase space is $P = T^*SO(3)$ and the momentum map $\mathbf{J} : P \to \mathbb{R}^3$ for the *left* $SO(3)$ action is *right* translation to the identity. We identify $\mathfrak{so}(3)^*$ with $\mathfrak{so}(3)$ via the Killing form and identify $\mathbb{R}^3$ with $\mathfrak{so}(3)$ via the map $v \mapsto \hat{v}$ where $\hat{v}(w) = v \times w$, $\times$ being the standard cross product. Points in $\mathfrak{so}(3)^*$ are regarded as the left reduction of $T^*SO(3)$ by $G = SO(3)$ and are the angular momenta as seen from a *body-fixed* frame. The reduced spaces $P_\mu = \mathbf{J}^{-1}(\mu)/G_\mu$ are identified with spheres in $\mathbb{R}^3$ of Euclidean radius $\|\mu\|$, with their symplectic form $\omega_\mu = -dS/\|\mu\|$ where dS is the standard area form on a sphere of radius $\|\mu\|$ and where G_μ consists of rotations about the μ-axis. The trajectories of the reduced dynamics are obtained by intersecting a family of homothetic ellipsoids (the energy ellipsoids) with the angular momentum spheres. In particular, all but at most four of the reduced trajectories are periodic. These four exceptional trajectories are the well-known homoclinic trajectories; we shall determine them explicitly in §15.8.

Suppose a reduced trajectory $\mathbf{\Pi}(t)$ is given on P_μ, with period T. *After time T, by how much has the rigid body rotated in space?* The spatial angular momentum is $\pi = \mu = g\mathbf{\Pi}$, which is the conserved value of $\mathbf{J}$. Here $g \in SO(3)$ is the attitude of the rigid body and $\mathbf{\Pi}$ is the body angular momentum. If $\mathbf{\Pi}(0) = \mathbf{\Pi}(T)$, then $\mu = g(0)\mathbf{\Pi}(0) = g(T)\mathbf{\Pi}(T)$ and so $g(T)^{-1}\mu = g(0)^{-1}\mu$, that is, $g(T)^{-1}\mu$ is a rotation about the axis μ. We want to give the angle of this rotation.

To answer this question, let $c(t)$ be the corresponding trajectory in $\mathbf{J}^{-1}(\mu) \subset P$. Identify $T^*SO(3)$ with $SO(3) \times \mathbb{R}^3$ by left trivialization, so $c(t)$ gets identified with $(g(t), \mathbf{\Pi}(t))$. Since the reduced trajectory $\mathbf{\Pi}(t)$ closes after time T, we recover the fact that $c(T) = gc(0)$ for some $g \in G_\mu$. Here, $g = g(T)g(0)^{-1}$ in the preceding notation. Thus, we can write

$$g = \exp[(\Delta\theta)\zeta], \tag{1.13.6}$$

where $\zeta = \mu/\|\mu\|$ identifies $\mathfrak{g}_\mu$ with $\mathbb{R}$ by $a\zeta \mapsto a$, for $a \in \mathbb{R}$. Let D be one of the two spherical caps on S^2 enclosed by the reduced trajectory, let Λ be the corresponding oriented solid angle, that is, $|\Lambda| = (\text{area } D)/\|\mu\|^2$, and let H_μ be the energy of the reduced trajectory. See Figure 1.13.4. All norms are taken relative to the Euclidean metric of $\mathbb{R}^3$. Montgomery [1991a] and Marsden, Montgomery, and Ratiu [1990] show that modulo 2π, we have the rigid body phase formula

$$\Delta\theta = \frac{1}{\|\mu\|}\left\{\int_D \omega_\mu + 2H_\mu T\right\} = -\Lambda + \frac{2H_\mu T}{\|\mu\|}. \tag{1.13.7}$$

The history of this formula is quite interesting and seems to have proceeded independently of the other historical developments above[1] The first

[1]We thank V. Arnold for valuable help with these comments.

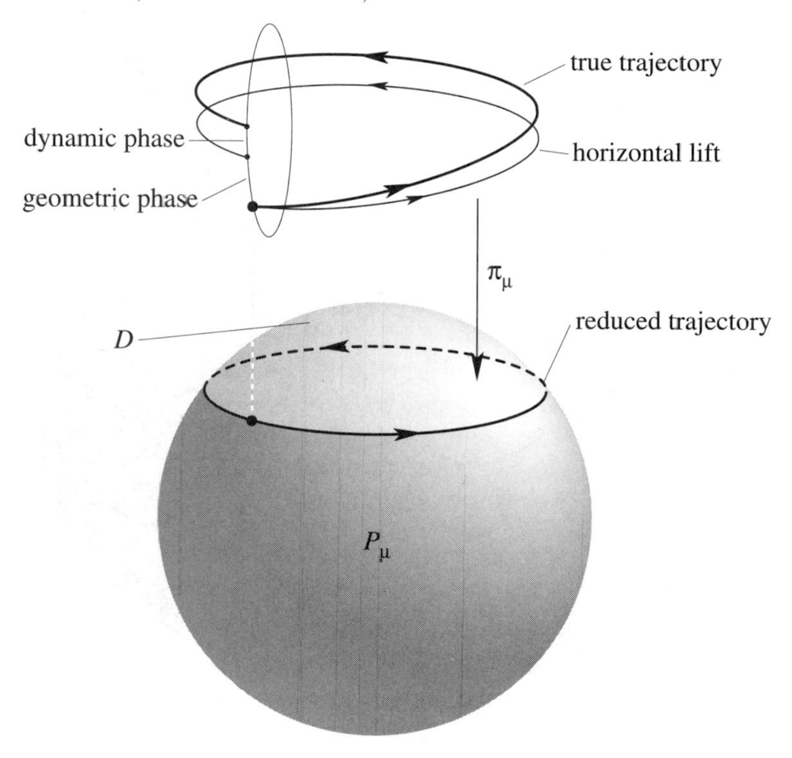

FIGURE 1.13.4. The geometry of the rigid body phase shift formula.

instance of a special case of this formula that we know of is given in the book of Ishlinskii [1952]; see also Ishlinskii [1963]. On page 195 of a later book on mechanics, Ishlinskii [1976] notes that "the formula was found by the author in 1943 and was published in Ishlinskii [1952]." The formula referred to in the works of Ishlinskii actually just covers a special case in which only the geometric phase is present. For example, in certain precessional motions in which, up to a certain order in averaging, one can ignore the dynamic phase and only the geometric phase survives. Even though Ishlinskii only found special cases of the result, he did recognize that it is related to the geometric concept of parallel transport. The general case of the formula presented above was found by Goodman and Robinson [1958]; their proof is based on the Gauss-Bonnet theorem. The special case of this formula for a *symmetric* free rigid body was given by Hannay [1985] and Anandan [1988], formula (20). The proof of the general formula based on the theory of connections and the formula for holonomy in terms of curvature, was given by Montgomery in Marsden, Montgomery and Ratiu [1990]; see also Montgomery [1991]. The approach using the Gauss-Bonnet theorem and its relation to the Poinsot construction along with additional results is taken up by Levi [1993].

Another example which naturally gives rise to geometric phases is the rigid body with one or more internal rotors. Figure 1.13.5 illustrates the system considered.

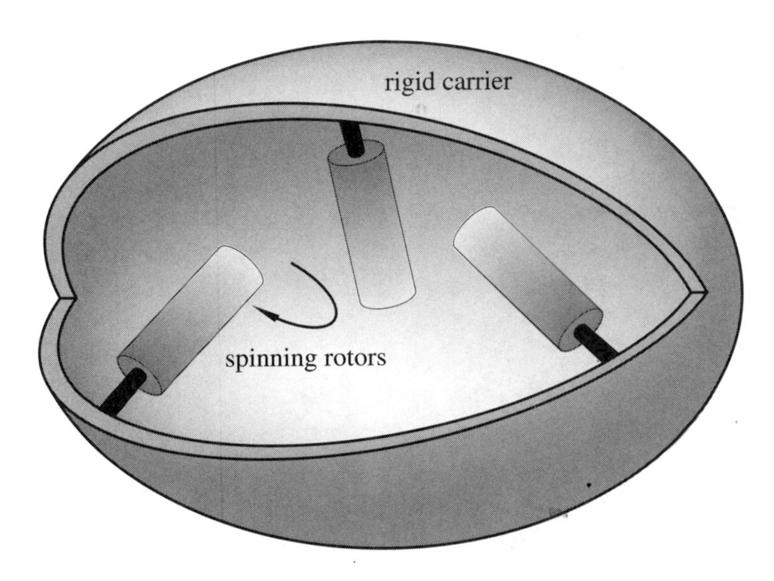

FIGURE 1.13.5. The rigid body with internal rotors.

To specify the position of this system we need an element of the group of rigid motions of $\mathbb{R}^3$ to place the center of mass and the attitude of the carrier, and an angle (element of S^1) to position each rotor. Thus the configuration space is $Q = SE(3) \times S^1 \times S^1 \times S^1$. The equations of motion of this system are an extension of Euler's equations of motion for a free spinning rotor. Just as holding a spinning bicycle wheel while sitting on a swivel chair can affect the carrier's motion, so the spinning rotors can affect the dynamics of the rigid carrier.

In this example, one can analyze equilibria and their stability in much the same way as one can with the rigid body. However, what one often wants to do is to forcibly spin, or control, the rotors so that one can achieve attitude control of the structure in the same spirit that a falling cat has *control* of its attitude by manipulating its body parts while falling. For example, one can attempt to prescribe a relation between the rotor dynamics and the rigid body dynamics by means of a *feedback law*. This has the property that the total system angular momentum is still preserved and that the resulting dynamic equations can be expressed entirely in terms of the free rigid body variable. (A falling cat has zero angular momentum even though it is able to turn over!) In some cases the resulting equations are again Hamiltonian on the invariant momentum sphere. Using this fact, one can compute the geometric phase for the problem generalizing the free rigid body phase formula. (See Bloch, Krishnaprasad, Marsden, and Sanchez

[1992] for details.) One hopes that this type of analysis will be useful in designing and understanding attitude control devices.

There are many continuum mechanical examples to which the techniques of geometric mechanics apply. Some of those we shall study later are free boundary problems (Lewis, Marsden, Montgomery, and Ratiu [1986], Montgomery, Marsden, and Ratiu [1984], Mazer and Ratiu [1989]), spacecraft with flexible attachments (Krishnaprasad and Marsden [1987]), elasticity (Holm and Kupershmidt [1983], Kupershmidt and Ratiu [1983], Marsden, Ratiu, and Weinstein [1984a,b], Simo, Marsden, and Krishnaprasad [1988]), and reduced MHD (Morrison and Hazeltine [1984] and Marsden and Morrison [1984]). We also wish to look at these theories from both the spatial (Eulerian) and body (convective) points of view as reductions of the canonical material picture. These two reductions are, in an appropriate sense, dual to each other.

Reduction also finds use in a number of other diverse areas as well. We mention just a few.

- Integrable systems (Moser [1980], Perelomov [1990], Adams Harnad, and Previato [1988], Fomenko and Trofimov [1989], Fomenko [1989], Reyman and Semenov-Tian-Shansky [1990]).
- Applications of integrable systems to numerical analysis (like the QR algorithm and sorting algorithms); see Deift and Li [1989] and Bloch, Brockett, and Ratiu [1990, 1992].
- Hamiltonian chaos (Arnold [1964], Ziglin [1980a,b, 1981], Holmes and Marsden [1981, 1982a,b, 1983], Wiggins [1988]).
- Averaging (Cushman and Rod [1982], Iwai [1982, 1985], Ercolani, Forest, McLaughlin, and Montgomery [1987]).
- Hamiltonian bifurcations (Van der Meer [1985], Golubitsky and Schaeffer [1985], Golubitsky and Stewart [1987], Golubitsky, Stewart, and Schaeffer [1988], Lewis, Marsden, and Ratiu [1987], Lewis, Ratiu, Simo, and Marsden [1992], Montaldi, Roberts, and Stewart [1988], Golubitsky, Marsden, Stewart, and Dellnitz [1994]).
- Algebraic geometry (Atiyah [1982, 1983], Kirwan [1984, 1985, 1988]).
- Celestial mechanics (Deprit [1983]).
- Vortex dynamics (Ziglin [1980b], Koiller, Soares, and Melo Neto [1985], Wan and Pulvirente [1984], Wan [1986, 1988a,b,c], Szeri and Holmes [1988]).
- Solitons (Flaschka, Newell, and Ratiu [1983a,b], Newell [1985], Kovacic and Wiggins [1992], McLaughlin, Overman, Wiggins, and Xion [1993], Alber and Marsden [1992]).

- Relativity and Yang-Mills theory (Fischer and Marsden [1972, 1979], Arms [1981], Arms, Marsden, and Moncrief [1981, 1982]).
- Fluid variational principles using Clebsch variables and "Lin constraints" (Seliger and Whitham [1968], Cendra and Marsden [1987], Cendra, Ibort, and Marsden [1987]).
- Control and satellite dynamics (Krishnaprasad [1985], van der Shaft and Crough [1987], Aeyels and Szafranski [1988], Bloch, Krishnaprasad, Marsden and Sanchez [1992], Wang, Krishnaprasad and Maddocks [1991]).
- Nonholonomic systems (Naimark and Fufaev [1972], Koiller [1992], Bates and Sniatycki [1993], Bloch, Krishnaprasad, Marsden and Murray [1994]).

Reduction is a natural historical culmination of the works of Liouville (for integrals in involution) and of Jacobi (for angular momentum) for reducing the phase space dimension in the presence of first integrals. It is intimately connected with work on momentum maps and its forerunners appear already in Jacobi [1866], Lie [1890], Cartan [1922], and Whittaker [1927]. It was developed later in Kirillov [1962], Arnold [1966a], Kostant [1970], Souriau [1970], Smale [1970], Nekhoroshev [1977], Meyer [1973], and Marsden and Weinstein [1974]. See also Guillemin and Sternberg [1984] and Marsden and Ratiu [1986] for the Poisson case and Sjamaar and Lerman [1991] for the singular symplectic case.

2

Hamiltonian Systems on Linear Symplectic Spaces

A natural arena for Hamiltonian mechanics is a symplectic or Poisson manifold. The first chapters concentrate on the symplectic case while Chapter 10 introduces the Poisson case. The symplectic context focuses on the symplectic two-form $\sum dq^i \wedge dp_i$ and its infinite-dimensional analogues, while the Poisson context looks at the Poisson bracket as the fundamental object. To facilitate the understanding of a number of points, we begin this chapter with the theory in linear spaces. This linear setting is already adequate for a number of interesting examples such as the wave equation and Schrödinger's equation. Later in Chapter 4 we make the transition to manifolds.

2.1 Introduction

To motivate the introduction of symplectic geometry in mechanics, we briefly recall from §1.1 the classical transition from Newton's second law to the Lagrange and Hamilton equations. ***Newton's Second Law*** for a particle moving in Euclidean three-space $\mathbb{R}^3$, under the influence of a ***potential energy*** $V(\mathbf{q})$, is

$$\mathbf{F} = m\mathbf{a}, \tag{2.1.1}$$

where $\mathbf{q} \in \mathbb{R}^3$, $\mathbf{F}(\mathbf{q}) = -\nabla V(\mathbf{q})$ is the force, m is the mass of the particle, and $\mathbf{a} = d^2\mathbf{q}/dt^2$ is the acceleration (assuming we start in a postulated privileged coordinate frame called an ***inertial frame***). The potential energy V is introduced through the notion of work and the assumption that the force field is conservative. The introduction of the ***kinetic energy*** $K = \frac{1}{2}m\|d\mathbf{q}/dt\|^2$ is through the ***power***, or ***rate of work equation***:

$$\frac{dK}{dt} = m\left\langle \dot{\mathbf{q}}, \ddot{\mathbf{q}} \right\rangle = \left\langle \dot{\mathbf{q}}, \mathbf{F} \right\rangle,$$

where $\langle\, , \rangle$ denotes the inner product on $\mathbb{R}^3$.

The ***Lagrangian*** is defined by

$$L(q^i, \dot{q}^i) = \frac{m}{2}\|\dot{\mathbf{q}}\|^2 - V(\mathbf{q}) \tag{2.1.2}$$

and one checks by direct calculation that Newton's second law is equivalent to the ***Euler-Lagrange equations***:

$$\frac{d}{dt}\frac{\partial L}{\partial \dot{q}^i} - \frac{\partial L}{\partial q^i} = 0, \tag{2.1.3}$$

which are second-order differential equations in q^i; the equations (2.1.3) are worthy of independent study for a general L since they are the equations for stationary values of the ***action integral***:

$$\delta \int_{t_1}^{t_2} L(q^i, \dot{q}^i)\, dt = 0 \tag{2.1.4}$$

as will be detailed later. These ***variational principles*** play a fundamental role throughout mechanics—both in particle mechanics and field theory.

It is easily verified that $dE/dt = 0$, where E is the ***total energy***: $E = \frac{1}{2}m\|\dot{\mathbf{q}}\|^2 + V(\mathbf{q})$. Lagrange and Hamilton observed that it is convenient to introduce the momentum $p_i = m\dot{q}^i$ and rewrite E as a function of p_i and q^i by letting

$$H(\mathbf{q}, \mathbf{p}) = \frac{\|\mathbf{p}\|^2}{2m} + V(\mathbf{q}), \tag{2.1.5}$$

for then Newton's second law is equivalent to ***Hamilton's canonical equations***

$$\dot{q}^i = \frac{\partial H}{\partial p_i}, \quad \dot{p}_i = -\frac{\partial H}{\partial q^i}, \tag{2.1.6}$$

which is a *first-order* system in $(\mathbf{q}, \mathbf{p})$-space, or ***phase space***.

For a deeper understanding of Hamilton's equations, we recall some matrix notation (see Abraham, Marsden, and Ratiu [1988], §5.1 for more details). Let E be a vector space with a basis $e_1, \ldots, e_n$ and the associated dual basis for E^* denoted $e^1, \ldots, e^n$; that is, e^i is defined by $\left\langle e^i, e_j \right\rangle := e^i(e_j) = \delta^i_j$, which equals 1 if $i = j$ and 0 if $i \neq j$. Vectors $v \in E$ are written $v = v^i e_i$ (sum on i) and covectors $\alpha \in E^*$ as $\alpha = \alpha_i e^i$. If $A : E \to F$

is a linear transformation, its ***matrix*** relative to bases $e_1, \ldots, e_n$ of E and $f_1, \ldots, f_m$ of F is denoted $A^j_{\ i}$ and is defined by

$$A(e_i) = A^j_{\ i} f_j, \quad \text{i.e.,} \quad [A(v)]^j = A^j_{\ i} v^i. \tag{2.1.7}$$

Thus, the columns of the matrix of A are $A(e_1), \ldots, A(e_n)$; the upper index is the row index and the lower index is the column index. For other linear transformations, we place the indices in their corresponding places. For example, if $A : E^* \to F$ is a linear transformation, its matrix A^{ij} satisfies $A(e^j) = A^{ij} f_i$, that is, $[A(\alpha)]^i = A^{ij} \alpha_j$.

If $B : E \times F \to \mathbb{R}$ is a bilinear form, its ***matrix*** B_{ij} is defined by

$$B_{ij} = B(e_i, f_j); \quad \text{i.e.,} \quad B(v, w) = v^i B_{ij} w^j. \tag{2.1.8}$$

Define the ***associated*** linear map $B^\flat : E \to F^*$ by $B^\flat(v)(w) = B(v, w)$ and observe that $B^\flat(e_i) = B_{ij} f^j$. Since $B^\flat(e_i)$ is the ith column of the matrix representing the linear map $B^\flat$, it follows that *the matrix of $B^\flat$ in the bases $e_1, \ldots, e_n, f^1, \ldots, f^n$ is the transpose of B_{ij}*; that is,

$$[B^\flat]_{ji} = B_{ij}. \tag{2.1.9}$$

Let Z denote the vector space of (q, p)'s and write $z = (q, p)$. Let the indexes on q, p be collectively denoted by $z^I, I = 1, \ldots, 2n$. One reason for the notation z is that if one thinks of z as a *complex variable* $z = q + ip$, then as will be explained in Exercise **2.3-2**, Hamilton's equations are equivalent to the following complex form of Hamilton's equations:

$$\dot{z} = -2i \frac{\partial H}{\partial \bar{z}}. \tag{2.1.10}$$

We can interpret Hamilton's equations (2.1.6) as follows. Think of the operation

$$\mathbf{d}H(z) = \left(\frac{\partial H}{\partial q^i}, \frac{\partial H}{\partial p_i} \right) \mapsto \left(\frac{\partial H}{\partial p_i}, -\frac{\partial H}{\partial q^i} \right) =: X_H(z), \tag{2.1.11}$$

which forms a vector field X_H, called the ***Hamiltonian vector field***, from the differential of H as a linear map

$$R : Z^* \to Z.$$

The matrix of R is

$$[R^{AB}] = \begin{bmatrix} \mathbf{0} & \mathbf{I} \\ -\mathbf{I} & \mathbf{0} \end{bmatrix} =: \mathbb{J}, \tag{2.1.12}$$

where we write $\mathbb{J}$ for the specific matrix (2.1.12) sometimes called the ***symplectic matrix***. Thus,

$$X_H(z) = R \cdot \mathbf{d}H(z) \tag{2.1.13}$$

or, if the components of X_H are denoted X^I, $I = 1, \ldots, 2n$,

$$X^I = R^{IJ}\frac{\partial H}{\partial z^J}, \quad \text{i.e.,} \quad X_H = \mathbb{J}\nabla H \tag{2.1.14}$$

where ∇H is the naive gradient of H; that is, the row vector $\mathbf{d}H$ regarded as a column vector.

Let $B(\alpha, \beta) = \langle \alpha, R(\beta) \rangle$ be the bilinear form associated to R where $\langle\, , \rangle$ denotes the canonical pairing between Z^* and Z. One calls either the bilinear form B or its associated linear map R, the ***Poisson structure***. The classical ***Poisson bracket*** (consistent with what we defined in Chapter 1) is defined by

$$\{F, G\} = B(\mathbf{d}F, \mathbf{d}G) = \mathbf{d}F \cdot \mathbb{J}\nabla G. \tag{2.1.15}$$

The ***symplectic structure*** Ω is the bilinear form associated to $R^{-1} : Z \to Z^*$, that is, $\Omega(v, w) = \left\langle R^{-1}(v), w \right\rangle$ or, equivalently, $\Omega^\flat = R^{-1}$. The matrix of Ω is $\mathbb{J}$ in the sense that

$$\Omega(v, w) = v^T \mathbb{J} w. \tag{2.1.16}$$

To unify notation we shall sometimes write

Ω	for the symplectic form,	$Z \times Z \to \mathbb{R}$	with matrix $\mathbb{J}$,
$\Omega^\flat$	for the associated linear map,	$Z \to Z^*$	with matrix $\mathbb{J}^T$,
$\Omega^\sharp$	for the inverse map $(\Omega^\flat)^{-1} = R$,	$Z^* \to Z$	with matrix $\mathbb{J}$,
B	for the Poisson form,	$Z^* \times Z^* \to \mathbb{R}$	with matrix $\mathbb{J}$.

Hamilton's equations are

$$\dot{z} = X_H(z) = \Omega^\sharp\, \mathbf{d}H(z). \tag{2.1.17}$$

Multiplying both sides by $\Omega^\flat$, we get

$$\Omega^\flat X_H(z) = \mathbf{d}H(z). \tag{2.1.18}$$

In terms of the symplectic form, (2.1.18) reads

$$\Omega(X_H(z), v) = \mathbf{d}H(z) \cdot v \tag{2.1.19}$$

for all $z, v \in Z$.

Problems like rigid body dynamics, quantum mechanics as a Hamiltonian system, and the motion of a particle in a rotating reference frame motivate the need to generalize these concepts. We shall do this in subsequent chapters and deal with both symplectic and Poisson structures in due course.

2.2 Symplectic Forms on Vector Spaces

Let Z be a real Banach space, possibly infinite dimensional, and let $\Omega : Z \times Z \to \mathbb{R}$ be a continuous bilinear form on Z. The form Ω is said to be ***nondegenerate*** (***or weakly nondegenerate***) if $\Omega(z_1, z_2) = 0$ for all $z_2 \in Z$ implies $z_1 = 0$. As in §2.1, the induced continuous linear mapping $\Omega^\flat : Z \to Z^*$ is defined by

$$\Omega^\flat(z_1)(z_2) = \Omega(z_1, z_2). \tag{2.2.1}$$

Nondegeneracy of Ω is equivalent to injectivity of $\Omega^\flat$; that is, to the condition "$\Omega^\flat(z) = 0$ implies $z = 0$." The form Ω is said to be ***strongly nondegenerate*** if $\Omega^\flat$ is an isomorphism, that is, $\Omega^\flat$ is onto as well as being injective. The open mapping theorem guarantees that if Z is a Banach space and $\Omega^\flat$ is one-to-one and onto, then its inverse is continuous. In most of the infinite-dimensional examples discussed in this book Ω will be only (weakly) nondegenerate.

A linear map between finite-dimensional spaces of the same dimension is one-to-one if and only if it is onto. Hence, *when Z is finite dimensional, weak nondegeneracy and strong nondegeneracy are equivalent*. If Z is finite dimensional, the matrix elements of Ω relative to a basis $\{e_I\}$ are defined by

$$\Omega_{IJ} = \Omega(e_I, e_J).$$

If $\{e^J\}$ denotes the basis for Z^* that is dual to $\{e_I\}$, that is, $\langle e^J, e_I \rangle = \delta_I^J$ and if we write $z = z^I e_I$ and $w = w^I e_I$, then

$$\Omega(z, w) = z^I \Omega_{IJ} w^J \quad \text{(sum over } I, J\text{)}.$$

Since the matrix of $\Omega^\flat$ relative to the bases $\{e_I\}$ and $\{e^J\}$ equals the transpose of the matrix of Ω relative to $\{e_I\}$; that is $(\Omega^\flat)_{JI} = \Omega_{IJ}$, nondegeneracy is equivalent to $\det[\Omega_{IJ}] \neq 0$. In particular, Z is even dimensional, since the determinant of a skew-symmetric matrix with an odd number of rows (and columns) is zero.

Definition 2.2.1 *A **symplectic form** Ω on a vector space Z is a nondegenerate skew-symmetric bilinear form on Z. The pair (Z, Ω) is called a **symplectic vector space**. If Ω is strongly nondegenerate, (Z, Ω) is called a **strong symplectic vector space**.*

Exercise

2.2-1 *Let (Z, Ω) be a finite-dimensional symplectic vector space and let $V \subset Z$ be a linear subspace. Assume that V is symplectic*; that is, Ω *restricted to $V \times V$ is nondegenerate. Let*

$$V^\Omega = \{z \in Z \mid \Omega(z, v) = 0 \quad \text{for all} \quad v \in V\}.$$

Show that V^Ω is symplectic and $Z = V \oplus V^\Omega$.

2.3 Examples

(a) Canonical Forms Let W be a vector space, and let $Z = W \times W^*$. Define the ***canonical symplectic form*** Ω on Z by

$$\Omega((w_1, \alpha_1), (w_2, \alpha_2)) = \alpha_2(w_1) - \alpha_1(w_2), \tag{2.3.1}$$

where $w_1, w_2 \in W$ and $\alpha_1, \alpha_2 \in W^*$.

More generally, let W and W' be two vector spaces in duality, that is, there is a weakly nondegenerate pairing $\langle \, , \rangle : W' \times W \to \mathbb{R}$. Then on $W \times W'$,

$$\Omega((w_1, \alpha_1), (w_2, \alpha_2)) = \langle \alpha_2, w_1 \rangle - \langle \alpha_1, w_2 \rangle \tag{2.3.2}$$

is a weak symplectic form. ♦

(b) The Space of Functions Let $\mathcal{F}(\mathbb{R}^3)$ be the space of smooth functions $\varphi : \mathbb{R}^3 \to \mathbb{R}$, and let $\mathrm{Den}_c(\mathbb{R}^3)$ be the space of smooth densities on $\mathbb{R}^3$ with compact support. We write a density $\pi \in \mathrm{Den}_c(\mathbb{R}^3)$ as a function $\pi' \in \mathcal{F}(\mathbb{R}^3)$ with compact support times the volume element d^3x on $\mathbb{R}^3$ as $\pi = \pi' \, d^3x$. The spaces $\mathcal{F}$ and Den_c are in weak nondegenerate duality by the pairing $\langle \varphi, \pi \rangle = \int \varphi \pi' \, d^3x$. Therefore, from (2.3.2), we get the symplectic form Ω on the vector space $Z = \mathcal{F}(\mathbb{R}^3) \times \mathrm{Den}_c(\mathbb{R}^3)$:

$$\Omega((\varphi_1, \pi_1), (\varphi_2, \pi_2)) = \int_{\mathbb{R}^3} \varphi_1 \pi_2 - \int_{\mathbb{R}^3} \varphi_2 \pi_1. \tag{2.3.3}$$

We choose densities with compact support so that the integrals in this formula will be finite. Other choices of spaces could be used as well. ♦

(c) Finite-Dimensional Canonical Form Suppose that W is a real vector space of dimension n. Let $\{e_i\}$ be a basis of W, and let $\{e^i\}$ be the dual basis of W^*. With $Z = W \times W^*$ and defining $\Omega : Z \times Z \to \mathbb{R}$ as in (2.3.1), one computes that the matrix of Ω in the basis $\{(e_1, 0), \ldots, (e_n, 0), (0, e^1), \ldots, (0, e^n)\}$ is

$$\mathbb{J} = \begin{bmatrix} \mathbf{0} & \mathbf{I} \\ -\mathbf{I} & \mathbf{0} \end{bmatrix}, \tag{2.3.4}$$

where $\mathbf{I}$ and $\mathbf{0}$ are the $n \times n$ identity and zero matrices. ♦

(d) Symplectic Form Associated to an Inner Product Space If $(W, \langle \, , \rangle)$ is a real inner product space, W is in duality with itself, so we obtain a symplectic form on $Z = W \times W$ from (2.3.2):

$$\Omega((w_1, w_2), (z_1, z_2)) = \langle z_2, w_1 \rangle - \langle z_1, w_2 \rangle . \tag{2.3.5}$$

As a special case of (2.3.5), let $W = \mathbb{R}^3$ with the usual inner product

$$\langle \mathbf{q}, \mathbf{v} \rangle = \mathbf{q} \cdot \mathbf{v} = \sum_{i=1}^{3} q^i v^i.$$

The corresponding symplectic form on $\mathbb{R}^6$ is given by

$$\Omega((\mathbf{q}_1, \mathbf{v}_1), (\mathbf{q}_2, \mathbf{v}_2)) = \mathbf{v}_2 \cdot \mathbf{q}_1 - \mathbf{v}_1 \cdot \mathbf{q}_2, \tag{2.3.6}$$

where $\mathbf{q}_1, \mathbf{q}_2, \mathbf{v}_1, \mathbf{v}_2 \in \mathbb{R}^3$. This coincides with Ω defined in Example **(c)** for $W = \mathbb{R}^3$, provided $\mathbb{R}^3$ is identified with $(\mathbb{R}^3)^*$. ♦

Bringing Ω to canonical form using elementary linear algebra results in the following statement. *If (Z, Ω) is a p-dimensional symplectic vector space, then p is even. Furthermore, Z is isomorphic to $W \times W^*$ and there is a basis of W in which the matrix of Ω is $\mathbb{J}$.* Such a basis is called ***canonical***, as are the corresponding coordinates. See Exercise **2.3-3**.

(e) Symplectic Form on $\mathbb{C}^n$ Write elements of complex n-space $\mathbb{C}^n$ as n-tuples $z = (z_1, \ldots, z_n)$ of complex numbers. The ***Hermitian inner product*** is

$$\langle z, w \rangle = \sum_{j=1}^{n} z_j \overline{w}_j = \sum_{j=1}^{n} (x_j u_j + y_j v_j) + i \sum_{j=1}^{n} (u_j y_j - v_j x_j),$$

where $z_j = x_j + i y_j$ and $w_j = u_j + i v_j$. Thus, $\operatorname{Re} \langle z, w \rangle$ is the real inner product and $-\operatorname{Im} \langle z, w \rangle$ is the symplectic form if $\mathbb{C}^n$ is identified with $\mathbb{R}^n \times \mathbb{R}^n$. ♦

(f) Quantum Mechanical Symplectic Form The following symplectic vector space arises in quantum mechanics, as we shall explain in Chapter 3. Recall that a ***Hermitian inner product*** $\langle \, , \rangle : \mathcal{H} \times \mathcal{H} \to \mathbb{C}$ on a complex Hilbert space $\mathcal{H}$ is linear in its first argument, antilinear in its second, and $\langle \psi_1, \psi_2 \rangle$ is the complex conjugate of $\langle \psi_2, \psi_1 \rangle$, where $\psi_1, \psi_2 \in \mathcal{H}$.

Set

$$\Omega(\psi_1, \psi_2) = -2\hbar \operatorname{Im} \langle \psi_1, \psi_2 \rangle,$$

where $\hbar$ is Planck's constant. One checks that Ω is a strong symplectic form on $\mathcal{H}$. Let $\mathcal{H}$ be the complexification of a real Hilbert space H, so it is identified with $H \times H$, and the inner product is given by

$$\langle (u_1, u_2), (v_1, v_2) \rangle = \langle u_1, v_1 \rangle + \langle u_2, v_2 \rangle + i(\langle u_2, v_1 \rangle - \langle u_1, v_2 \rangle).$$

This form coincides with $2\hbar$ times that in (2.3.5). On the other hand, if we embed $\mathcal{H}$ into $\mathcal{H} \times \mathcal{H}^*$ via $\psi \mapsto (i\psi, \psi)$ then the restriction of $\hbar$ times the canonical symplectic form (2.3.5) on $\mathcal{H} \times \mathcal{H}^*$, namely,

$$((\psi_1, \varphi_1), (\psi_2, \varphi_2)) \mapsto \hbar \operatorname{Re}[\langle \varphi_2, \psi_1 \rangle - \langle \varphi_1, \psi_2 \rangle],$$

coincides with Ω. ♦

Exercises

2.3-1 *Verify that the formula for the symplectic form for* $\mathbb{R}^{2n}$ *as a matrix* $\mathbb{J} = \begin{bmatrix} \mathbf{0} & \mathbf{I} \\ -\mathbf{I} & \mathbf{0} \end{bmatrix}$ *coincides with the definition of the symplectic form as the canonical form on* $\mathbb{R}^{2n}$ *regarded as the product* $\mathbb{R}^n \times (\mathbb{R}^n)^*$.

2.3-2 *Write* $z = q + ip$ *and show that Hamilton's equations are equivalent to*

$$\dot{z} = -2i \frac{\partial H}{\partial \overline{z}}.$$

(Give a plausible definition of the right-hand side as part of your answer.)

2.3-3 *Find a canonical basis for a symplectic form* Ω *on* Z *as follows. Let* $e_1 \in Z, e_1 \neq 0$. *Find* $e_2 \in Z$ *with* $\Omega(e_1, e_2) \neq 0$. *By rescaling* e_2, *assume* $\Omega(e_1, e_2) = 1$. *Let* V *be the span of* e_1 *and* e_2. *Apply Exercise* **2.2-1** *and repeat this construction on* V^Ω.

2.4 Canonical Transformations or Symplectic Maps

To motivate the definition of symplectic maps (synonymous with canonical transformations), start with Hamilton's equations:

$$\dot{q}^i = \frac{\partial H}{\partial p_i}, \quad \dot{p}_i = -\frac{\partial H}{\partial q^i}, \tag{2.4.1}$$

and a transformation $\varphi : Z \to Z$ of phase space to itself. Write

$$(\overline{q}, \overline{p}) = \varphi(q, p) \quad \text{i.e.,} \quad \overline{z} = \varphi(z). \tag{2.4.2}$$

Assume $z(t) = (q(t), p(t))$ satisfies Hamilton's equations, that is,

$$\dot{z}(t) = X_H(z(t)) = \Omega^\sharp \, \mathbf{d}H(z(t)), \tag{2.4.3}$$

where $\Omega^\sharp$ is the linear map of $Z^* \to Z$ with matrix $\mathbb{J}$. By the chain rule, $\overline{z} = \varphi(z)$ satisfies

$$\dot{\overline{z}}^I = \frac{\partial \varphi^I}{\partial z^J} \dot{z}^J =: A^I{}_J \dot{z}^J \tag{2.4.4}$$

(sum on J). Substituting (2.4.3) into (2.4.4) employing coordinate notation, and the chain rule implies

$$\dot{\overline{z}}^I = A^I{}_J B^{JK} \frac{\partial H}{\partial z^K} = A^I{}_J B^{JK} A^L{}_K \frac{\partial H}{\partial \overline{z}^L}, \tag{2.4.5}$$

where the matrix of $\Omega^\sharp$ is B^{JK}. Thus, the equations (2.4.5) are Hamiltonian if and only if

$$A^I{}_J B^{JK} A^L{}_K = B^{IL}. \tag{2.4.6}$$

In terms of composition of linear maps, (2.4.6) means

$$A \circ \Omega^\sharp \circ A^T = \Omega^\sharp, \tag{2.4.7}$$

or in matrix notation

$$A \mathbb{J} A^T = \mathbb{J}, \tag{2.4.8}$$

since the matrix of $\Omega^\sharp$ in canonical coordinates is $\mathbb{J}$ (see §2.1). A transformation satisfying (2.4.6) is called a ***canonical transformation, a symplectic transformation***, or a ***Poisson transformation***. In Chapter 10, where Poisson structures can be different from symplectic ones, we will see that (2.4.7) generalizes to the Poisson context.

Taking determinants of (2.4.8), shows that $\det A = \pm 1$ and in particular that A is invertible; taking the inverse of (2.4.7) gives

$$(A^T)^{-1} \circ \Omega^\flat \circ A^{-1} = \Omega^\flat \quad \text{i.e.,} \quad A^T \circ \Omega^\flat \circ A = \Omega^\flat, \tag{2.4.9}$$

which has the matrix form

$$A^T \mathbb{J} A = \mathbb{J}$$

since the matrix of $\Omega^\flat$ in canonical coordinates is $-\mathbb{J}$ (see §2.1). Note that (2.4.8) and (2.4.10) are equivalent (the inverse of one gives the other). As bilinear forms, (2.4.9) reads

$$\Omega(\mathbf{D}\varphi(z) \cdot z_1, \mathbf{D}\varphi(z) \cdot z_2) = \Omega(z_1, z_2). \tag{2.4.10}$$

With (2.4.10) as a guideline, we write the general condition for a symplectic map.

Definition 2.4.1 *If (Z, Ω) and (Y, Ξ) are symplectic vector spaces, a smooth map $f : Z \to Y$ is called* ***symplectic*** *or* ***canonical*** *if it preserves the symplectic forms, that is, if*

$$\Xi(\mathbf{D}f(z) \cdot z_1, \mathbf{D}f(z) \cdot z_2) = \Omega(z_1, z_2) \tag{2.4.11}$$

for all $z, z_1, z_2 \in Z$, where $\mathbf{D}f$ is the derivative of f (the Jacobian matrix in finite dimensions).

In the following box we introduce a convenient notation for these sorts of transformations.

Pull Back Notation

$\varphi^* f$ to ***pull back a function***: $\varphi^* f = f \circ \varphi$

$\varphi_* g$ to ***push forward a function***: $\varphi_* g = g \circ \varphi^{-1}$

$\varphi_* X$ to ***push forward a vector field*** X by φ:

$$(\varphi_* X)(\varphi(z)) = \mathbf{D}\varphi(z) \cdot X(z);$$

in components,

$$(\varphi_* X)^I = \frac{\partial \varphi^I}{\partial z^J} X^J.$$

$\varphi^* Y$ to ***pull back a vector field*** Y by $\varphi : \varphi^* Y = (\varphi^{-1})_* Y$

$\varphi^* \Omega$ to ***pull back a bilinear form*** Ω on Z, construct the following bilinear form $\varphi^* \Omega$ depending on the point $z \in Z$:

$$(\varphi^* \Omega)_z(z_1, z_2) = \Omega(\mathbf{D}\varphi(z) \cdot z_1, \mathbf{D}\varphi(z) \cdot z_2);$$

in components,

$$(\varphi^* \Omega)_{IJ} = \frac{\partial \varphi^K}{\partial z^I} \frac{\partial \varphi^L}{\partial z^J} \Omega_{KL};$$

$\varphi_* \Xi$ to ***push forward a bilinear form*** Ξ by φ, pull back by the inverse:

$$\varphi_* \Xi = (\varphi^{-1})^* \Xi.$$

In this pull back notation, (2.4.11) reads: $(f^* \Xi)_z = \Omega$, or $f^* \Xi = \Omega$ for short.

It is simple to verify that if (Z, Ω) is a finite-dimensional symplectic vector space, the set of all linear symplectic mappings $T : Z \to Z$ forms a group under composition. It is called the ***symplectic group*** and is denoted by $Sp(Z, \Omega)$. As we have seen, in a canonical basis, a matrix A is symplectic if and only if

$$A^T \mathbb{J} A = \mathbb{J}, \tag{2.4.12}$$

where A^T is the transpose of A. For $Z = W \times W^*$ and a canonical basis, if A has the matrix

$$A = \begin{bmatrix} A_{qq} & A_{qp} \\ A_{pq} & A_{pp} \end{bmatrix}, \tag{2.4.13}$$

then one checks (Exercise **2.4-2**) that (2.4.12) *is equivalent to either of the two conditions*:

1. $A_{qq} A_{qp}^T$ *and* $A_{pp} A_{pq}^T$ *are symmetric and* $A_{qq} A_{pp}^T - A_{qp} A_{pq}^T = \mathbf{I}$; or
2. $A_{pq}^T A_{qq}$ *and* $A_{qp}^T A_{pp}$ *are symmetric and* $A_{qq}^T A_{pp} - A_{pq}^T A_{pq} = \mathbf{I}$.

In infinite dimensions $Sp(Z, \Omega)$ is, by definition, the set of elements of $GL(Z)$ (invertible bounded linear operators of Z to Z) that leave Ω fixed.

If (Z, Ω) is a (weak) symplectic space and E and F are subspaces of Z, we define $E^{\Omega} = \{z \in Z \mid \Omega(z, e) = 0 \text{ for all } e \in E\}$, called the ***symplectic orthogonal complement*** of E. We leave it to the reader to check that

i E^{Ω} is closed;

ii $E \subset F$ implies $F^{\Omega} \subset E^{\Omega}$;

iii $E^{\Omega} \cap F^{\Omega} = (E + F)^{\Omega}$;

iv if Z is finite dimensional, then $\dim E + \dim E^{\Omega} = \dim Z$ (to show this, use the fact that $E^{\Omega} = \ker(i^* \circ \Omega^\flat)$, where $i : E \to Z$ is the inclusion and $i^* : Z^* \to E^*$ is its dual, $i^*(\alpha) = \alpha \circ i$, which is surjective);

v if Z is finite dimensional, $E^{\Omega\Omega} = E$ (this is also true in infinite dimensions if E is closed); and

vi if E and F are closed, then $(E \cap F)^{\Omega} = E^{\Omega} + F^{\Omega}$ (to prove this use **iii** and **v**).

Exercises

2.4-1 *Show that a transformation $\varphi : \mathbb{R}^{2n} \to \mathbb{R}^{2n}$ is symplectic in the sense that its derivative matrix $A = \mathbf{D}\varphi(z)$ satisfies the condition $A^T \mathbb{J} A = \mathbb{J}$ if and only if the condition*

$$\Omega(Az_1, Az_2) = \Omega(z_1, z_2)$$

holds for all $z_1, z_2 \in \mathbb{R}^{2n}$.

2.4-2 *Let $Z = W \times W^*$, let $A : Z \to Z$ and, using canonical coordinates, write the matrix of A as*

$$A = \begin{bmatrix} A_{qq} & A_{qp} \\ A_{pq} & A_{pp} \end{bmatrix}.$$

Show that A being symplectic is equivalent to either of the two conditions:

i *$A_{qq}A_{qp}^T$ and $A_{pp}A_{pq}^T$ are symmetric and $A_{qq}A_{pp}^T - A_{qp}A_{pq}^T = \mathbf{I}$; or*

ii *$A_{pq}^T A_{qq}$ and $A_{qp}^T A_{pp}$ are symmetric and $A_{qq}^T A_{pp} - A_{pq}^T A_{qp} = \mathbf{I}$. (Here, $\mathbf{I}$ is the $n \times n$ identity.)*

2.4-3 *Let f be a given function of $\mathbf{q} = (q^1, q^2, \ldots, q^n)$. Define the map $\varphi : \mathbb{R}^{2n} \to \mathbb{R}^{2n}$ by $\varphi(\mathbf{q}, \mathbf{p}) = (\mathbf{q}, \mathbf{p} + \mathbf{d}f(\mathbf{q}))$. Show that φ is a canonical (symplectic) transformation.*

2.4-4 **(a)** *Let $A \in GL(n, \mathbb{R})$ be an invertible linear transformation. Show that the map $\varphi : \mathbb{R}^{2n} \to \mathbb{R}^{2n}$ given by $(\mathbf{q}, \mathbf{p}) \mapsto (A\mathbf{q}, (A^{-1})^T \mathbf{p})$ is a canonical transformation.*

(b) If $\mathbf{R}$ *is a rotation in $\mathbb{R}^3$, show that the map $(\mathbf{q}, \mathbf{p}) \mapsto (\mathbf{R}\mathbf{q}, \mathbf{R}\mathbf{p})$ is a canonical transformation.*

2.5 The Abstract Hamilton Equations

The concrete form of Hamilton's equations we have already encountered is a special case of a construction on symplectic spaces. Here, we discuss this formulation for systems whose phase space is linear; in subsequent sections we will generalize the setting to phase spaces which are symplectic manifolds and in Chapter 10 to spaces where only a Poisson bracket is given. These generalizations will all be important in our study of specific examples.

Definition 2.5.1 *Let (Z, Ω) be a symplectic vector space. A vector field $X : Z \to Z$ is called **Hamiltonian** if*

$$\Omega^\flat(X(z)) = \mathbf{d}H(z), \tag{2.5.1}$$

*for all $z \in Z$, for some C^1 function $H : Z \to \mathbb{R}$. Here $\mathbf{d}H(z) = \mathbf{D}H(z)$ is alternative notation for the derivative of H. If such an H exists, we write $X = X_H$ and call H a **Hamiltonian function**, or **energy function** for X.*

In a number of important examples, especially infinite-dimensional ones, H need not be defined on all of Z. We shall briefly discuss some of the technicalities involved in §3.3.

If Z is finite dimensional, nondegeneracy of Ω implies that $\Omega^\flat : Z \to Z^*$ is an isomorphism, which guarantees that X_H exists for any given function H. However, if Z is infinite dimensional and Ω is only weakly nondegenerate, we do not know *a priori* that X_H exists for a given H. If it does exist, it is unique since $\Omega^\flat$ is one-to-one.

The set of Hamiltonian vector fields on Z is denoted $\mathfrak{X}_{\mathrm{Ham}}(Z)$, or simply $\mathfrak{X}_{\mathrm{Ham}}$. Thus $X_H \in \mathfrak{X}_{\mathrm{Ham}}$ is the vector field determined by the condition

$$\Omega(X_H(z), \delta z) = \mathbf{d}H(z) \cdot \delta z \quad \text{for all} \quad z, \delta z \in Z. \tag{2.5.2}$$

If X is a vector field, the ***interior product*** $\mathbf{i}_X \Omega$ is defined to be the dual vector (that is, a one form) given at a point $z \in Z$ as follows:

$$(\mathbf{i}_X \Omega)_z \in Z^*; \quad (\mathbf{i}_X \Omega)_z(v) := \Omega(X(z), v),$$

for all $v \in Z$. Then condition (2.5.1) or (2.5.2) may be written as

$$\mathbf{i}_X \Omega = \mathbf{d}H. \tag{2.5.3}$$

To express H in terms of X_H and Ω, we integrate the identity $\mathbf{d}H(tz)\cdot z = \Omega(X_H(tz), z)$ from $t = 0$ to $t = 1$. The fundamental theorem of calculus gives

$$\begin{aligned} H(z) - H(0) &= \int_0^1 \frac{dH(tz)}{dt} dt = \int_0^1 \mathbf{d}H(tz) \cdot z \, dt \\ &= \int_0^1 \Omega(X_H(tz), z) \, dt. \end{aligned} \tag{2.5.4}$$

Let us now abstract the calculation we did in arriving at (2.4.8).

Proposition 2.5.2 *Let (Z, Ω) and (Y, Ξ) be symplectic vector spaces and $f : Z \to Y$ a diffeomorphism. Then f is a symplectic transformation if and only if for all Hamiltonian vector fields X_H on Y, we have $f_* X_{H\circ f} = X_H$; that is,*

$$\mathbf{D}f(z) \cdot X_{H\circ f}(z) = X_H(f(z)). \tag{2.5.5}$$

Proof Note that for $v \in Z$,

$$\begin{aligned} \Omega(X_{H\circ f}(z), v) &= \mathbf{d}(H \circ f)(z) \cdot v \\ &= \mathbf{d}H(f(z)) \cdot \mathbf{D}f(z) \cdot v \\ &= \Xi(X_H(f(z)), \mathbf{D}f(z) \cdot v). \end{aligned} \tag{2.5.6}$$

If f is symplectic, then $\Xi(\mathbf{D}f(z) \cdot X_{H\circ f}(z), \mathbf{D}f(z) \cdot v) = \Omega(X_{H\circ f}(z), v)$ and thus by nondegeneracy of Ξ and the fact that $\mathbf{D}f(z) \cdot v$ is an arbitrary element of Y (because f is a diffeomorphism and hence $\mathbf{D}f(z)$ is an ismorphism), (2.5.5) holds. Conversely, if (2.5.5) holds, then (2.5.6) implies that $\Xi(\mathbf{D}f(z) \cdot X_{H\circ f}(z), \mathbf{D}f(z) \cdot v) = \Omega(X_{H\circ f}(z), v)$ for any $v \in Z$ and any C^1 map $H : Y \to \mathbb{R}$. However, $X_{H\circ f}(z)$ equals an arbitrary element $w \in Z$ for a correct choice of the Hamiltonian function H, namely, $(H \circ f)(z) = \Omega(w, z)$. Thus f is symplectic. ■

Definition 2.5.3 ***Hamilton's equations*** *for H is the system of differential equations defined by X_H. Letting $c : \mathbb{R} \to Z$ be a curve, they are the equations*

$$\frac{dc(t)}{dt} = X_H(c(t)). \tag{2.5.7}$$

2.6 The Classical Hamilton Equations

We now relate the abstract form (2.5.7) to the classical form of Hamilton's equations. In the following, an n-tuple $(q^1, \ldots, q^n)$ will be denoted simply by (q^i), etc.

Proposition 2.6.1 *Let (Z, Ω) be a $2n$-dimensional symplectic vector space, and let $(q^i, p_i) = (q^1, \dots, q^n, p_1, \dots, p_n)$ denote canonical coordinates, with respect to which Ω has matrix $\mathbb{J}$. Then in this coordinate system, $X_H : Z \to Z$ is given by*

$$X_H = \left(\frac{\partial H}{\partial p_i}, -\frac{\partial H}{\partial q^i} \right) = \mathbb{J} \cdot \nabla H. \tag{2.6.1}$$

Thus, Hamilton's equations in canonical coordinates are

$$\frac{dq^i}{dt} = \frac{\partial H}{\partial p_i}, \qquad \frac{dp_i}{dt} = -\frac{\partial H}{\partial q^i}. \tag{2.6.2}$$

More generally, if $Z = V \times V'$, $\langle \cdot , \cdot \rangle : V \times V' \to \mathbb{R}$ is a weakly nondegenerate pairing, and $\Omega((e_1, \alpha_1), (e_2, \alpha_2)) = \langle \alpha_2, e_1 \rangle - \langle \alpha_1, e_2 \rangle$, then

$$X_H(e, \alpha) = \left(\frac{\delta H}{\delta \alpha}, -\frac{\delta H}{\delta e} \right), \tag{2.6.3}$$

where $\delta H / \delta \alpha \in V$ and $\delta H / \delta e \in V'$ are the ***partial functional derivatives*** *defined by*

$$\mathbf{D}_2 H(e, \alpha) \cdot \beta = \left\langle \beta, \frac{\delta H}{\delta \alpha} \right\rangle \tag{2.6.4}$$

for any $\beta \in V'$ and similarly for $\delta H / \delta e$; in (2.6.3) it is assumed that the functional derivatives exist.

Proof If $(f, \beta) \in V \times V'$, then

$$\begin{aligned} \Omega\left(\left(\frac{\delta H}{\delta \alpha}, -\frac{\delta H}{\delta e} \right), (f, \beta) \right) &= \left\langle \beta, \frac{\delta H}{\delta \alpha} \right\rangle + \left\langle \frac{\delta H}{\delta e}, f \right\rangle \\ &= \mathbf{D}_2 H(e, \alpha) \cdot \beta + \mathbf{D}_1 H(e, \alpha) \cdot f \\ &= \langle \mathbf{d} H(e, \alpha), (f, \beta) \rangle . \quad \blacksquare \end{aligned}$$

Proposition 2.6.2 (Conservation of Energy) *Let $c(t)$ be an integral curve of X_H. Then $H(c(t))$ is constant in t. If φ_t denotes the flow of X_H, that is, $\varphi_t(z)$ is the solution of (2.5.7) with initial conditions $z \in Z$, then $H \circ \varphi_t = H$.*

Proof By the chain rule,

$$\begin{aligned} \frac{d}{dt} H(c(t)) &= \mathbf{d} H(c(t)) \cdot \frac{d}{dt} c(t) = \Omega\left(X_H(c(t)), \frac{d}{dt} c(t) \right) \\ &= \Omega\left(X_H(c(t)), X_H(c(t)) \right) = 0, \end{aligned}$$

where the final equality follows from the skew-symmetry of Ω. $\blacksquare$

Exercise

2.6-1 *Let the skew-symmetric bilinear form Ω on $\mathbb{R}^{2n}$ have the matrix*

$$\begin{bmatrix} \mathbf{B} & \mathbf{I} \\ -\mathbf{I} & \mathbf{0} \end{bmatrix},$$

where $\mathbf{B} = [B_{ij}]$ is a skew-symmetric $n \times n$ matrix.

(a) *Show that Ω is nondegenerate and hence a symplectic form on $\mathbb{R}^{2n}$.*

(b) *In canonical coordinates $(q^1, \ldots, q^n, p_1, \ldots, p_n)$ for $\mathbb{J}$ show that Hamilton's equations relative to Ω are*

$$\frac{dq^i}{dt} = \frac{\partial H}{\partial p_i}, \quad \frac{dp_i}{dt} = -\frac{\partial H}{\partial q^i} - B_{ij}\frac{\partial H}{\partial p_j}.$$

2.7 When Are Equations Hamiltonian?

Having seen how to derive Hamilton's equations on (Z, Ω) given H, it is natural to consider the converse: when are a given set of equations

$$\frac{dz}{dt} = X(z), \quad \text{where} \quad X : Z \to Z \quad \text{is a vector field,} \tag{2.7.1}$$

Hamilton's equations for some H? If X is linear, the answer is given by the following.

Proposition 2.7.1 *Let the vector field $A : Z \to Z$ be linear. Then A is Hamiltonian if and only if A is Ω-skew; that is,*

$$\Omega(Az_1, z_2) = -\Omega(z_1, Az_2)$$

for all $z_1, z_2 \in Z$. Furthermore, in this case one can take $H(z) = \frac{1}{2}\Omega(Az, z)$.

Proof Differentiating the defining relation

$$\Omega(X_H(z), v) = \mathbf{d}H(z) \cdot v \tag{2.7.2}$$

with respect to z in the direction u and using bilinearity of Ω, one gets

$$\Omega(\mathbf{D}X_H(z) \cdot u, v) = \mathbf{D}^2 H(z)(v, u). \tag{2.7.3}$$

From this and the symmetry of the second partial derivatives, we get

$$\begin{aligned} \Omega(\mathbf{D}X_H(z) \cdot u, v) &= \mathbf{D}^2 H(z)(v, u) = \Omega(\mathbf{D}X_H(z) \cdot v, u) \\ &= -\Omega(u, \mathbf{D}X_H(z) \cdot v). \end{aligned} \tag{2.7.4}$$

If $A = X_H$ for some H, then $\mathbf{D}X_H(z) = A$, and (2.7.4) becomes $\Omega(Au, v) = -\Omega(u, Av)$; hence A is Ω-skew. Conversely, suppose that A is Ω-skew. Defining $H(z) = \frac{1}{2}\Omega(Az, z)$, we claim that $A = X_H$. Indeed,

$$\begin{aligned} \mathbf{d}H(z) \cdot u &= \frac{1}{2}\Omega(Au, z) + \frac{1}{2}\Omega(Az, u) \\ &= -\frac{1}{2}\Omega(u, Az) + \frac{1}{2}\Omega(Az, u) \\ &= \frac{1}{2}\Omega(Az, u) + \frac{1}{2}\Omega(Az, u) = \Omega(Az, u). \quad \blacksquare \end{aligned}$$

In canonical coordinates, where Ω has matrix $\mathbb{J}$, Ω-skewness of A is equivalent to symmetry of the marix $\mathbb{J}A$; that is, $\mathbb{J}A + A^T\mathbb{J} = 0$. The vector space of all linear transformations of Z satisfying this condition is denoted by $\mathfrak{sp}(Z, \Omega)$ and its elements are called ***infinitesimal symplectic transformations***. In canonical coordinates, if $Z = W \times W^*$ and if A has the matrix

$$A = \begin{bmatrix} A_{qq} & A_{qp} \\ A_{pq} & A_{pp} \end{bmatrix}, \tag{2.7.5}$$

then one checks that *A is infinitesimally symplectic if and only if A_{qp} and A_{pq} are both symmetric and $A_{qq}^T + A_{pp} = \mathbf{0}$*. Compare with Exercise **2.7-1**.

In the complex linear case, we use Example **(f)** in §2.3 ($2\hbar$ times the negative imaginary part of a Hermitian inner product $\langle\,,\rangle$ is the symplectic form) to arrive at the following.

Corollary 2.7.2 *Let $\mathcal{H}$ be a complex Hilbert space with Hermitian inner product $\langle\,,\rangle$ and let $\Omega(\psi_1, \psi_2) = -2\hbar\,\mathrm{Im}\,\langle\psi_1, \psi_2\rangle$. Let $A : \mathcal{H} \to \mathcal{H}$ be a complex linear operator. There exists an $H : \mathcal{H} \to \mathbb{R}$ such that $A = X_H$ if and only if iA is symmetric or, equivalently, satisfies*

$$\langle iA\psi_1, \psi_2\rangle = \langle \psi_1, iA\psi_2\rangle\,. \tag{2.7.6}$$

*In this case, H may be taken to be $H(\psi) = \hbar\langle iA\psi, \psi\rangle$. We let $H_{op} = i\hbar A$ and thus Hamilton's equations $\dot{\psi} = A\psi$ becomes the **Schrödinger equation**:*

$$i\hbar\frac{\partial \psi}{\partial t} = H_{op}\psi. \tag{2.7.7}$$

(This example is continued in §2.8 and in §3.1.)

Proof A is Ω-skew if and only if $\mathrm{Im}\,\langle A\psi_1, \psi_2\rangle = -\mathrm{Im}\,\langle\psi_1, A\psi_2\rangle$ for all $\psi_1, \psi_2 \in \mathcal{H}$. Replacing everywhere ψ_1 by $i\psi_1$ and using the relation $\mathrm{Im}(iz) = \mathrm{Re}\,z$, this is equivalent to $\mathrm{Re}\,\langle A\psi_1, \psi_2\rangle = -\mathrm{Re}\,\langle\psi_1, A\psi_2\rangle$. Since

$$\langle iA\psi_1, \psi_2\rangle = -\mathrm{Im}\,\langle A\psi_1, \psi_2\rangle + i\,\mathrm{Re}\,\langle A\psi_1, \psi_2\rangle$$

and

$$\langle\psi_1, iA\psi_2\rangle = \mathrm{Im}\,\langle\psi_1, A\psi_2\rangle - i\,\mathrm{Re}\,\langle\psi_1, A\psi_2\rangle\,,$$

we see that Ω-skewness of A is equivalent to iA being symmetric. Finally

$$\begin{aligned} \hbar \langle iA\psi, \psi\rangle &= \hbar \operatorname{Re} i \langle A\psi, \psi\rangle = -\hbar \operatorname{Im} \langle A\psi, \psi\rangle \\ &= \frac{1}{2}\Omega(A\psi, \psi) \end{aligned}$$

and the corollary follows from Proposition **2.7.1**. ■

For nonlinear differential equations, the analogue of Proposition **2.7.1** is the following.

Proposition 2.7.3 *Let $X : Z \to Z$ be a (smooth) vector field on a symplectic vector space (Z, Ω). Then $X = X_H$ for some $H : Z \to \mathbb{R}$ if and only if $\mathbf{D}X(z)$ is Ω-skew for all z.*

Proof We have seen the "only if" part in the proof of Proposition **2.7.1**. Conversely, if $\mathbf{D}X(z)$ is Ω-skew, define

$$H(z) = \int_0^1 \Omega(X(tz), z)\, dt + \text{ constant}; \tag{2.7.8}$$

we claim that $X = X_H$. Indeed,

$$\begin{aligned} \mathbf{d}H(z) \cdot v &= \int_0^1 [\Omega(\mathbf{D}X(tz) \cdot tv, z) + \Omega(X(tz), v)]\, dt \\ &= \int_0^1 [\Omega(t\mathbf{D}X(tz) \cdot v, z) + \Omega(X(tz), v)]\, dt \\ &= \Omega\left(\int_0^1 [t\mathbf{D}X(tz) \cdot z + X(tz)]\, dt, v\right) \\ &= \Omega\left(\int_0^1 \frac{d}{dt}[tX(tz)]\, dt, v\right) = \Omega(X(z), v). \quad ■ \end{aligned}$$

Aside Looking ahead to Chapter 4 on differential forms, one can check that (2.7.8) for H is reproduced by the proof of the Poincaré lemma applied to the one-form $\mathbf{i}_X\Omega$. That $\mathbf{D}X(z)$ is Ω-skew is equivalent to $\mathbf{d}(\mathbf{i}_X\Omega) = 0$. ♦

Using the straightening out theorem (see, for example, Abraham, Marsden, and Ratiu ([1988], p. 194) it is easy to see that on an even-dimensional manifold *any* vector field is locally Hamiltonian near points where it is nonzero, relative to *some* symplectic form. However, it is not so simple to get a general criterion of this sort that is global, covering singular points as well.

An interesting characterization of Hamiltonian vector fields involves the Cayley transform. Let (Z, Ω) be a symplectic vector space and $A : Z \to Z$ a linear transformation such that $I - A$ is invertible. Then *A is Hamiltonian* *iff its **Cayley transform** $C = (I + A)(I - A)^{-1}$ is symplectic.* See Exercise **2.7-2**. For applications, see Laub and Meyer [1974], Paneitz [1981], Feng [1986], and Austin and Krishnaprasad [1993]. The Cayley transform is useful in some Hamiltonian numerical algorithms, as this last reference and Marsden [1992] shows.

Exercises

2.7-1 *Let $Z = W \times W^*$ and use a canonical basis to write the matrix of the linear map $A : Z \to Z$ as*

$$A = \begin{bmatrix} A_{qq} & A_{qp} \\ A_{pq} & A_{pp} \end{bmatrix}.$$

Show that A is infinitesimally symplectic, that is, $\mathbb{J}A + A^T\mathbb{J} = 0$ if and only if A_{qp} and A_{pq} are both symmetric and $A_{qq}^T + A_{pp} = \mathbf{0}$.

2.7-2 *Let (Z, Ω) be a symplectic vector space. Let $A : Z \to Z$ be a linear map and assume that $(I - A)$ is invertible. Show that A is Hamiltonian iff its Cayley transform*

$$(I + A)(I - A)^{-1}$$

is symplectic. Give an example of a linear Hamiltonian vector field such that $(I - A)$ is not invertible.

2.7-3 *Let (Z, Ω) be a finite-dimensional symplectic vector space and let $\varphi : Z \to Z$ be symplectic. If λ is an eigenvalue of multiplicity k, then so is $1/\lambda$. Prove this using the characteristic polynomial of φ.*

2.7-4 *Let (Z, Ω) be a finite-dimensional symplectic vector space and let $A : Z \to Z$ be Hamiltonian. Show that the generalized kernel of A* (that is, *$\{z \in Z \mid A^k z = 0$ for some integer $k \geq 1\}$) is a symplectic subspace.*

2.8 Hamiltonian Flows

This subsection discusses flows of Hamiltonian vector fields a little further. The next subsection gives the abstract definition of the Poisson bracket, relates it to the classical definitions, and then shows how it may be used in describing the dynamics. Later on, Poisson brackets will play an increasingly important role.

Let X_H be a Hamiltonian vector field on a symplectic vector space (Z, Ω) with Hamiltonian $H : Z \to \mathbb{R}$. The ***flow*** of X_H is the collection of maps

$\varphi_t : Z \to Z$ satisfying

$$\frac{d}{dt}\varphi_t(z) = X_H(\varphi_t(z)) \tag{2.8.1}$$

for each $z \in Z$ and real t. Here and in the following, all statements concerning the map $\varphi_t : Z \to Z$ are to be considered only for those z and t such that $\varphi_t(z)$ is defined, as determined by differential equations theory.

First consider the case in which A is a (bounded) *linear* vector field. The flow of A may be written as $\varphi_t = e^{tA}$; that is, the solution of $dz/dt = Az$ with initial condition z_0 is given by $z(t) = \varphi_t(z_0) = e^{tA}z_0$.

Proposition 2.8.1 *The flow φ_t of a linear vector field $A : Z \to Z$ consists of (linear) canonical transformations if and only if A is Hamiltonian.*

Proof For all $u, v \in Z$ we have

$$\begin{aligned}
\frac{d}{dt}(\varphi_t^*\Omega)(u,v) &= \frac{d}{dt}\Omega(\varphi_t(u), \varphi_t(v)) \\
&= \Omega\left(\frac{d}{dt}\varphi_t(u), \varphi_t(v)\right) + \Omega\left(\varphi_t(u), \frac{d}{dt}\varphi_t(v)\right) \\
&= \Omega(A\varphi_t(u), \varphi_t(v)) + \Omega(\varphi_t(u), A\varphi_t(v)).
\end{aligned}$$

Therefore, A is Ω-skew, that is, A is Hamiltonian, if and only if each φ_t is a linear canonical transformation. ■

For nonlinear flows, there is a corresponding result.

Proposition 2.8.2 *The flow φ_t of a (nonlinear) Hamiltonian vector field X_H consists of canonical transformations. Conversely, if the flow of a vector field X consists of canonical transformations, then it is Hamiltonian.*

Proof Let φ_t be the flow of a vector field X. Using (2.8.1) and the chain rule:

$$\frac{d}{dt}[\mathbf{D}\varphi_t(z) \cdot v] = \mathbf{D}\left[\frac{d}{dt}\varphi_t(z)\right] \cdot v = \mathbf{D}X(\varphi_t(z)) \cdot (\mathbf{D}\varphi_t(z) \cdot v).$$

Using this, we get

$$\begin{aligned}
&\frac{d}{dt}\Omega(\mathbf{D}\varphi_t(z) \cdot u, \mathbf{D}\varphi_t(z) \cdot v) \\
&\quad = \Omega(\mathbf{D}X(\varphi_t(z)) \cdot [\mathbf{D}\varphi_t(z) \cdot u], \mathbf{D}\varphi_t(z) \cdot v) \\
&\qquad + \Omega(\mathbf{D}\varphi_t(z) \cdot u, \mathbf{D}X(\varphi_t(z)) \cdot [\mathbf{D}\varphi_t(z) \cdot v]).
\end{aligned}$$

If $X = X_H$, then $\mathbf{D}X_H(\varphi_t(z))$ is Ω-skew by Proposition **2.7.3**, so, $\Omega(\mathbf{D}\varphi_t(z) \cdot u, \mathbf{D}\varphi_t\ (z) \cdot v) =$ constant. At $t = 0$ this equals $\Omega(u, v)$, so $\varphi_t^*\Omega = \Omega$. Conversely, if φ_t is canonical, this calculation shows that $\mathbf{D}X(\varphi_t(z))$ is Ω-skew, whence by Proposition **2.7.3**, $X = X_H$ for some H. ■

Later on we give another proof of Proposition **2.8.2** using the notation of differential forms.

Proposition 2.8.3 (Schrödinger's Equation) *Let $A : \mathcal{H} \to \mathcal{H}$ be a complex linear map on a complex Hilbert space $\mathcal{H}$. The flow φ_t of A is canonical, that is, consists of canonical transformations with respect to the symplectic form Ω defined in Example* **(f)** *of §2.3, if and only if φ_t is unitary.*

Proof By definition,

$$\Omega(\psi_1, \psi_2) = -2\hbar \operatorname{Im} \langle \psi_1, \psi_2 \rangle ,$$

so

$$\Omega(\varphi_t \psi_1, \varphi_t \psi_2) = -2\hbar \operatorname{Im} \langle \varphi_t \psi_1, \varphi_t \psi_2 \rangle$$

for $\psi_1, \psi_2 \in \mathcal{H}$. Thus φ_t is canonical if and only if $\operatorname{Im} \langle \varphi_t \psi_1, \varphi_t \psi_2 \rangle = \operatorname{Im} \langle \psi_1, \psi_2 \rangle$ and this in turn is equivalent to unitarity by complex linearity of φ_t since $\langle \psi_1, \psi_2 \rangle = -\operatorname{Im} \langle i\psi_1, \psi_2 \rangle + i \operatorname{Im} \langle \psi_1, \psi_2 \rangle$. ■

This shows that the flow of the Schrödinger equation is canonical and unitary, and so preserves the probability amplitude of any wave function that is a solution:

$$\langle \varphi_t \psi, \varphi_t \psi \rangle = \langle \psi, \psi \rangle ,$$

where φ_t is the flow of A. Later we shall see how this conservation of the norm also results from a symmetry-induced conservation law.

2.9 Poisson Brackets

Definition 2.9.1 *Given a symplectic vector space (Z, Ω) and two functions $F, G : Z \to \mathbb{R}$, the* ***Poisson bracket*** *$\{F, G\} : Z \to \mathbb{R}$ of F and G is defined by*

$$\{F, G\}(z) = \Omega(X_F(z), X_G(z)). \tag{2.9.1}$$

Using the definition of a Hamiltonian vector field, we find that equivalent expressions are

$$\{F, G\}(z) = \mathbf{d}F(z) \cdot X_G(z) = -\mathbf{d}G(z) \cdot X_F(z). \tag{2.9.2}$$

In (2.9.2) we write $\pounds_{X_G} F = \mathbf{d}F \cdot X_G$, for the derivative of F in the direction X_G.

Lie Derivative Notation

$\pounds_X f = \mathbf{d}f \cdot X$ is the ***directional derivative*** of f in the direction X. In coordinates it is given by

$$\pounds_X f = \frac{\partial f}{\partial z^I} X^I \quad \text{(sum on } I).$$

Functions F, G which are such that $\{F, G\} = 0$ are said to be in ***involution*** or to ***Poisson commute***.

Now we turn to some examples of Poisson brackets.

(a) Canonical Bracket Suppose that Z is $2n$-dimensional. Then in canonical coordinates $(q^1, \ldots, q^n, p_1, \ldots, p_n)$ we have

$$\begin{aligned} \{F, G\} &= \left[\frac{\partial F}{\partial p_i}, -\frac{\partial F}{\partial q^i}\right] \mathbb{J} \begin{bmatrix} \dfrac{\partial G}{\partial p_i} \\ -\dfrac{\partial G}{\partial q^i} \end{bmatrix} \\ &= \frac{\partial F}{\partial q^i}\frac{\partial G}{\partial p_i} - \frac{\partial F}{\partial p_i}\frac{\partial G}{\partial q^i} \quad \text{(sum on } i). \end{aligned} \tag{2.9.3}$$

From this, we get the ***fundamental Poisson brackets***:

$$\{q^i, q^j\} = 0, \quad \{p_i, p_j\} = 0, \quad \text{and} \quad \{q^i, p_j\} = \delta^i_j. \tag{2.9.4}$$

In terms of the Poisson structure, that is, the bilinear form B from §2.1, the Poisson bracket takes the form

$$\{F, G\} = B(\mathbf{d}F, \mathbf{d}G). \quad \blacklozenge \tag{2.9.5}$$

(b) The Space of Functions Let (Z, Ω) be defined as in Example **(b)** of §2.3 and let $F, G : Z \to \mathbb{R}$. Using (2.6.3) and (2.9.1) above, we get

$$\begin{aligned} \{F, G\} = \Omega(X_F, X_G) &= \Omega\left(\left(\frac{\delta F}{\delta \pi}, -\frac{\delta F}{\delta \varphi}\right), \left(\frac{\delta G}{\delta \pi}, -\frac{\delta G}{\delta \varphi}\right)\right) \\ &= \int_{\mathbb{R}^3} \left(\frac{\delta G}{\delta \pi}\frac{\delta F}{\delta \varphi} - \frac{\delta F}{\delta \pi}\frac{\delta G}{\delta \varphi}\right). \end{aligned} \tag{2.9.6}$$

This example will be used in the next chapter when we study classical field theory. $\blacklozenge$

The ***Jacobi-Lie bracket*** $[X, Y]$ of two vector fields X and Y on a vector space Z is defined by demanding that

$$\mathbf{d}f \cdot [X, Y] = \mathbf{d}(\mathbf{d}f \cdot Y) \cdot X - \mathbf{d}(\mathbf{d}f \cdot X) \cdot Y$$

for all real-valued functions f. In Lie derivative notation, this reads

$$\pounds_{[X,Y]} f = \pounds_X \pounds_Y f - \pounds_Y \pounds_X f.$$

One checks that this condition becomes, in vector analysis notation,

$$[X, Y] = (X \cdot \nabla)Y - (Y \cdot \nabla)X,$$

and in coordinates,

$$[X, Y]^J = X^I \frac{\partial}{\partial z^I} Y^J - Y^I \frac{\partial}{\partial z^I} X^J.$$

Proposition 2.9.2 *Let $[\,,]$ denote the Jacobi-Lie bracket of vector fields, and let $F, G \in \mathcal{F}(Z)$. Then*

$$X_{\{F,G\}} = -[X_F, X_G]. \tag{2.9.7}$$

Proof

$$\begin{aligned}
\Omega(X_{\{F,G\}}(z), u) &= \mathbf{d}\{F, G\}(z) \cdot u = \mathbf{d}(\Omega(X_F(z), X_G(z))) \cdot u \\
&= \Omega(\mathbf{D}X_F(z) \cdot u, X_G(z)) + \Omega(X_F(z), \mathbf{D}X_G(z) \cdot u) \\
&= \Omega(\mathbf{D}X_F(z) \cdot X_G(z), u) - \Omega(\mathbf{D}X_G(z) \cdot X_F(z), u) \\
&= \Omega(\mathbf{D}X_F(z) \cdot X_G(z) - \mathbf{D}X_G(z) \cdot X_F(z), u) \\
&= \Omega(-[X_F, X_G](z), u).
\end{aligned}$$

Weak nondegeneracy of Ω implies the result. ■

Proposition 2.9.3 *The Poisson bracket $\{\,,\} : \mathcal{F}(Z) \times \mathcal{F}(Z) \to \mathcal{F}(Z)$ makes $\mathcal{F}(Z)$ into a **Lie algebra**. That is, this bracket is real bilinear, skew-symmetric, and satisfies Jacobi's identity.*

Proof To verify Jacobi's identity note that for $F, G, H : Z \to \mathbb{R}$, we have

$$\begin{aligned}
\{F, \{G, H\}\} &= -\pounds_{X_F}\{G, H\} = \pounds_{X_F}\pounds_{X_G} H, \\
\{G, \{H, F\}\} &= -\pounds_{X_G}\{H, F\} = \pounds_{X_G}\pounds_{X_F} H
\end{aligned}$$

and

$$\{H, \{F, G\}\} = \pounds_{X_{\{F,G\}}} H,$$

so that

$$\{F, \{G, H\}\} + \{G, \{H, F\}\} + \{H, \{F, G\}\} = \pounds_{X_{\{F,G\}}} H + \pounds_{[X_F, X_G]} H.$$

The result thus follows by (2.9.7). ■

Corollary 2.9.4 *The set of Hamiltonian vector fields* $\mathfrak{X}_{\mathrm{Ham}}(Z)$ *forms a Lie subalgebra of* $\mathfrak{X}(Z)$.

Proposition 2.9.5 *Let* $\varphi : Z \to Z$ *be a diffeomorphism. Then* φ *is symplectic iff it preserves Poisson brackets, that is,*

$$\{\varphi^* F, \varphi^* G\} = \varphi^*\{F, G\}, \tag{2.9.8}$$

for all $F, G : Z \to \mathbb{R}$.

Proof We use the identity

$$\varphi^*(\pounds_X f) = \pounds_{\varphi^* X}(\varphi^* f),$$

which follows from the chain rule. Thus,

$$\varphi^*\{F, G\} = \varphi^* \pounds_{X_G} F = \pounds_{\varphi^* X_G}(\varphi^* F)$$

and

$$\{\varphi^* F, \varphi^* G\} = \pounds_{X_{G\circ\varphi}}(\varphi^* F).$$

Thus φ preserves Poisson brackets iff $\varphi^* X_G = X_{G\circ\varphi}$ for every $G : Z \to \mathbb{R}$, that is, iff φ is symplectic by (2.5.5). ■

Proposition 2.9.6 *Let* X_H *be a Hamiltonian vector field on* Z, *with Hamiltonian* H *and flow* φ_t. *Then for* $F : Z \to \mathbb{R}$,

$$\frac{d}{dt}(F \circ \varphi_t) = \{F \circ \varphi_t, H\} = \{F, H\} \circ \varphi_t. \tag{2.9.9}$$

Proof By the chain rule and the definition of X_F,

$$\begin{aligned} \frac{d}{dt}[(F \circ \varphi_t)(z)] &= \mathbf{d}F(\varphi_t(z)) \cdot X_H(\varphi_t(z)) \\ &= \Omega(X_F(\varphi_t(z)), X_H(\varphi_t(z))) \\ &= \{F, H\}(\varphi_t(z)). \end{aligned}$$

By (2.9.8), this equals $\{F \circ \varphi_t, H \circ \varphi_t\}(z) = \{F \circ \varphi_t, H\}(z)$ by conservation of energy. ■

Corollary 2.9.7 *Let* $F, G : Z \to \mathbb{R}$. *Then* F *is constant along integral curves of* X_G *if and only if* G *is constant along integral curves of* X_F *and this is true if and only if* $\{F, G\} = 0$.

Proposition 2.9.8 *Let* $A, B : Z \to Z$ *be linear Hamiltonian vector fields with corresponding energy functions* $H_A(z) = \frac{1}{2}\Omega(Az, z)$ *and* $H_B(z) = \frac{1}{2}\Omega(Bz, z)$. *Letting* $[A, B] = A \circ B - B \circ A$ *be the operator commutator, we have*

$$\{H_A, H_B\} = H_{[A,B]}. \tag{2.9.10}$$

Proof By definition, $X_{H_A} = A$ and so

$$\{H_A, H_B\}(z) = \Omega(Az, Bz).$$

Since A and B are Ω-skew, we get

$$\begin{aligned} \{H_A, H_B\}(z) &= \frac{1}{2}\Omega(ABz, z) - \frac{1}{2}\Omega(BAz, z) \\ &= \frac{1}{2}\Omega([A, B]z, z) = H_{[A,B]}(z). \quad \blacksquare \end{aligned}$$

As we shall see in §3.3, (2.9.10) in the case of quantum mechanics leads one to statements like: "Commutators in quantum mechanics are not only *analogous* to Poisson brackets, they *are* Poisson brackets." Even more striking are *true statements* like this "Don't tell me that quantum mechanics is right and classical mechanics is wrong—after all quantum mechanics is a *special case* of classical mechanics."

2.10 A Particle in a Rotating Hoop

In this subsection we take a break from the abstract theory to do an example the "old-fashioned" way. This and other examples will also serve as excellent illustrations of the theory we are developing.

Consider a particle constrained to move on a circular hoop; for example a bead sliding in a hula-hoop. The particle is assumed to have mass m and to be acted on by gravitational and frictional forces, as well as constraint forces that keep it on the hoop. The hoop itself is spun about a vertical axis with constant angular velocity ω, as in Figure 2.10.1.

The position of the particle in space is specified by the angles θ and φ, as shown in Figure 2.10.1. We can take $\varphi = \omega t$, so the position of the particle becomes determined by θ alone. Let the orthonormal frame along the coordinate directions $\mathbf{e}_\theta, \mathbf{e}_\varphi$, and $\mathbf{e}_r$ be as shown.

The forces acting on the particle are:

1. Friction, proportional to the velocity of the particle relative to the hoop: $-\nu R\dot{\theta}\mathbf{e}_\theta$, where $\nu \geq 0$ is a constant.

2. Gravity: $-mg\mathbf{k}$.

3. Constraint forces in the directions $\mathbf{e}_r$ and $\mathbf{e}_\varphi$ to keep the particle in the hoop.

The equations of motion are derived from Newton's second law $\mathbf{F} = m\mathbf{a}$. To get them, we need to calculate the acceleration $\mathbf{a}$; here $\mathbf{a}$ means the acceleration relative to the *fixed inertial frame xyz* in space; it does not

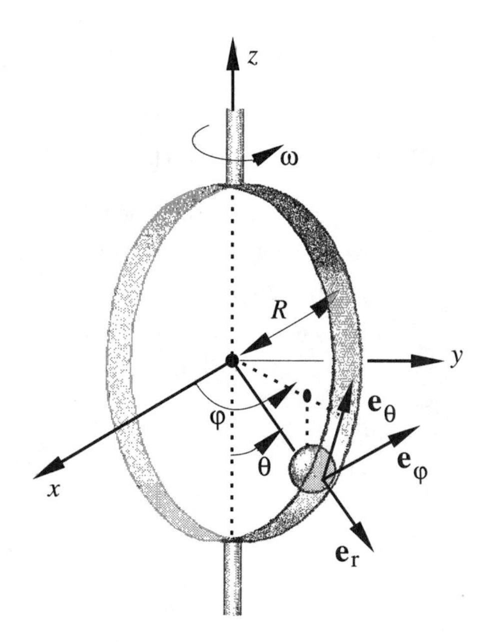

FIGURE 2.10.1. A particle moving in a hoop rotating with angular velocity ω.

mean $\ddot{\theta}$. Relative to this xyz coordinate system, we have

$$\left.\begin{aligned} x &= R\sin\theta\cos\varphi, \\ y &= R\sin\theta\sin\varphi, \\ z &= -R\cos\theta. \end{aligned}\right\} \tag{2.10.1}$$

Calculating the second derivatives using $\varphi = \omega t$ and the chain rule gives

$$\left.\begin{aligned} \ddot{x} &= -\omega^2 x - \dot{\theta}^2 x + (R\cos\theta\cos\varphi)\ddot{\theta} - 2R\omega\dot{\theta}\cos\theta\sin\varphi, \\ \ddot{y} &= -\omega^2 y - \dot{\theta}^2 y + (R\cos\theta\sin\varphi)\ddot{\theta} + 2R\omega\dot{\theta}\cos\theta\cos\varphi, \\ \ddot{z} &= -z\dot{\theta}^2 + (R\sin\theta)\ddot{\theta}. \end{aligned}\right\} \tag{2.10.2}$$

If $\mathbf{i}$, $\mathbf{j}$, $\mathbf{k}$, denote unit vectors along the x, y, and z axes, respectively, we have the easily verified relation

$$\mathbf{e}_\theta = (\cos\theta\cos\varphi)\mathbf{i} + (\cos\theta\sin\varphi)\mathbf{j} + \sin\theta\mathbf{k}. \tag{2.10.3}$$

Now consider the vector equation $\mathbf{F} = m\mathbf{a}$, where $\mathbf{F}$ is the sum of the three forces described earlier and

$$\mathbf{a} = \ddot{x}\mathbf{i} + \ddot{y}\mathbf{j} + \ddot{z}\mathbf{k}. \tag{2.10.4}$$

The $\mathbf{e}_\varphi$ and $\mathbf{e}_r$ components of $\mathbf{F} = m\mathbf{a}$ only tell us what the constraint forces must be; the equation of motion comes from the $\mathbf{e}_\theta$ component:

$$\mathbf{F} \cdot \mathbf{e}_\theta = m\mathbf{a} \cdot \mathbf{e}_\theta. \tag{2.10.5}$$

Using (2.10.3), the left side of (2.10.5) is

$$\mathbf{F} \cdot \mathbf{e}_\theta = -\nu R\dot{\theta} - mg\sin\theta \tag{2.10.6}$$

while from (2.10.2), (2.10.3), and (2.10.4), the right side of (2.10.5) is

$$\begin{aligned} m\mathbf{a} \cdot \mathbf{e}_\theta &= m\{\ddot{x}\cos\theta\cos\varphi + \ddot{y}\cos\theta\sin\varphi + \ddot{z}\sin\theta\} \\ &= m\{\cos\theta\cos\varphi[-\omega^2 x - \dot{\theta}^2 x + (R\cos\theta\cos\varphi)\ddot{\theta} \\ &\quad - 2R\omega\dot{\theta}\cos\theta\sin\varphi] + \cos\theta\sin\varphi[-\omega^2 y - \dot{\theta}^2 y \\ &\quad + (R\cos\theta\sin\varphi)\ddot{\theta} + 2R\omega\dot{\theta}\cos\theta\cos\varphi] \\ &\quad + \sin\theta[-z\dot{\theta}^2 + (R\sin\theta)\ddot{\theta}\}. \end{aligned}$$

Using (2.10.1), this simplifies to

$$m\mathbf{a} \cdot \mathbf{e}_\theta = mR\{\ddot{\theta} - \omega^2\sin\theta\cos\theta\}. \tag{2.10.7}$$

Comparing (2.10.5), (2.10.6), and (2.10.7), we get

$$\ddot{\theta} = \omega^2\sin\theta\cos\theta - \frac{\nu}{m}\dot{\theta} - \frac{g}{R}\sin\theta \tag{2.10.8}$$

as our final equation of motion. Several remarks concerning it are in order:

i If $\omega = 0$ and $\nu = 0$, (2.10.8) reduces to the ***pendulum equation***

$$R\ddot{\theta} + g\sin\theta = 0.$$

In fact, our system can be viewed just as well as a ***whirling pendulum***.

ii For $\nu = 0$, (2.10.8) is Hamiltonian with respect to $q = \theta, p = mR^2\dot{\theta}$, canonical bracket structure

$$\{F, K\} = \frac{\partial F}{\partial q}\frac{\partial K}{\partial p} - \frac{\partial K}{\partial q}\frac{\partial F}{\partial p}, \tag{2.10.9}$$

and the energy

$$H = \frac{p^2}{2mR^2} - mgR\cos\theta - \frac{mR^2\omega^2}{2}\sin^2\theta. \tag{2.10.10}$$

We now use Lagrangian methods to derive (2.10.8). In Figure 2.10.1, the velocity is $\mathbf{v} = R\dot{\theta}\mathbf{e}_\theta + (\omega R\sin\theta)\mathbf{e}_\varphi$, so the kinetic energy is

$$T = \frac{1}{2}m\|\mathbf{v}\|^2 = \frac{1}{2}m(R^2\dot{\theta}^2 + [\omega R\sin\theta]^2), \tag{2.10.11}$$

while the potential energy is

$$V = -mgR\cos\theta. \tag{2.10.12}$$

Thus the Lagrangian is given by

$$L = T - V = \frac{1}{2}mR^2\dot{\theta}^2 + \frac{mR^2\omega^2}{2}\sin^2\theta + mgR\cos\theta \qquad (2.10.13)$$

and the Euler-Lagrange equations, namely,

$$\frac{d}{dt}\frac{\partial L}{\partial \dot{\theta}} = \frac{\partial L}{\partial \theta},$$

(see §1.1 or §2.1) become

$$mR^2\ddot{\theta} = mR^2\omega^2 \sin\theta\cos\theta - mgR\sin\theta,$$

which are the same equations we derived by hand in (2.10.8) for $\nu = 0$. The Legendre transform gives $p = mR^2\dot{\theta}$ and the Hamiltonian (2.10.10).

Consider ***equilibrium solutions***; that is, solutions satisfying $\dot{\theta} = 0, \ddot{\theta} = 0$; (2.10.8) gives

$$R\omega^2 \sin\theta\cos\theta = g\sin\theta. \qquad (2.10.14)$$

Certainly $\theta = 0$ and $\theta = \pi$ solve (2.10.14) corresponding to the particle at the bottom or top of the hoop. If $\theta \neq 0$ or π, (2.10.14) becomes

$$R\omega^2 \cos\theta = g \qquad (2.10.15)$$

which has two solutions when $g/R\omega^2 < 1$. The value

$$\omega_c = \sqrt{\frac{g}{R}} \qquad (2.10.16)$$

is the ***critical rotation rate***. (Notice that ω_c is the frequency of linearized oscillations for the simple pendulum, that is, for $R\ddot{\theta} + g\theta = 0$.) For $\omega < \omega_c$ there are only *two* solutions $\theta = 0,\ \pi$, while for $\omega > \omega_c$ there are *four solutions*,

$$\theta = 0,\ \pi,\ \pm\cos^{-1}\left(\frac{g}{R\omega^2}\right). \qquad (2.10.17)$$

We say that a ***bifurcation*** (or a ***Hamiltonian pitchfork bifurcation*** to be accurate) has occurred as ω crosses ω_c. We can see this graphically in computer generated solutions of (2.10.8). Set $x = \theta, y = \dot{\theta}$ and rewrite (2.10.8) as

$$\left.\begin{aligned} \dot{x} &= y, \\ \dot{y} &= \frac{g}{R}(\alpha\cos x - 1)\sin x - \beta y, \end{aligned}\right\} \qquad (2.10.18)$$

where $\alpha = R\omega^2/g$ and $\beta = \nu/m$. Taking $g = R$ for illustration, Figure 2.10.2 shows representative orbits in the phase portraits of (2.10.18) for various α, β.

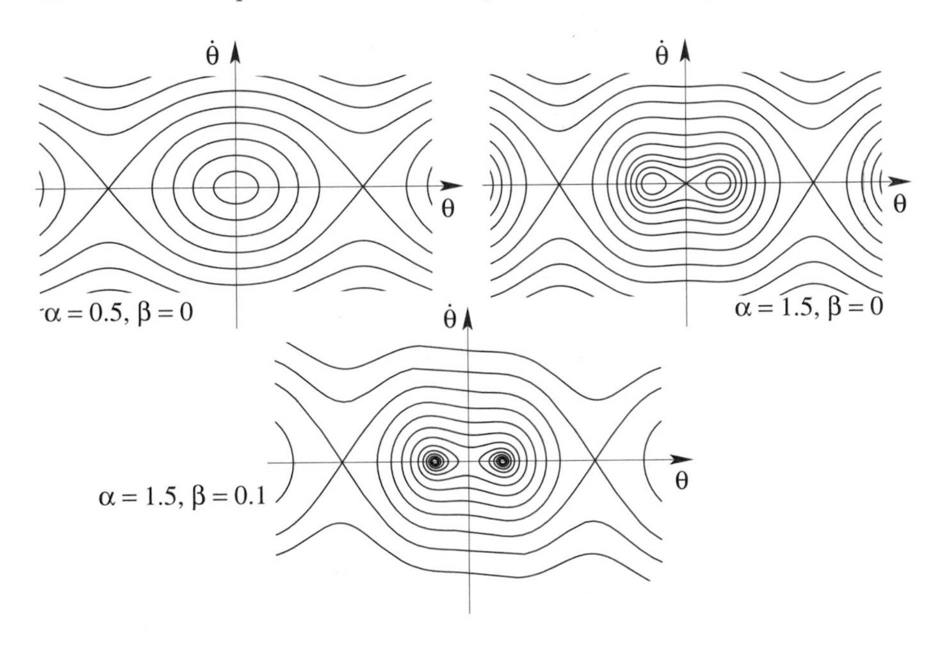

FIGURE 2.10.2. Phase portraits of the ball in the rotating hoop.

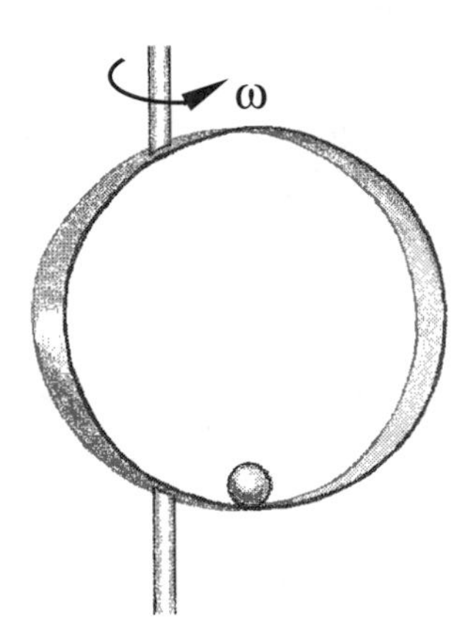

FIGURE 2.10.3. A ball in an off-center rotating hoop.

This system with $\nu = 0$; that is, $\beta = 0$, is symmetric in the sense that the $\mathbb{Z}_2$-action given by $\theta \mapsto -\theta$ and $\dot{\theta} \mapsto -\dot{\theta}$ leaves the phase portrait invariant. If this $\mathbb{Z}_2$ symmetry is broken, by setting the rotation axis a little off center, for example, then one side gets preferred, as in Figure 2.10.3.

The evolution of the phase portrait for $\nu = 0$ is shown in Figure 2.10.4.

Near $\theta = 0$, the potential function has changed from the symmetric bifur-

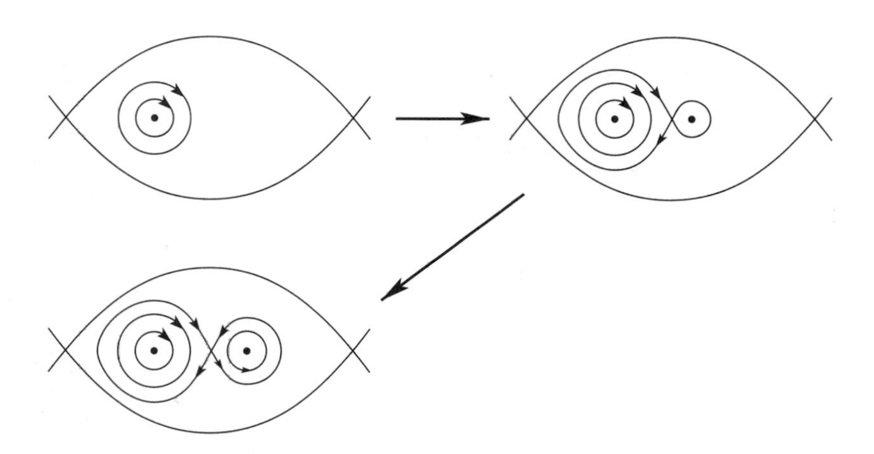

FIGURE 2.10.4. The phase portraits for the ball in the off-centered hoop as the angular velocity increases.

cation in Figure 2.10.5(a) to the unsymmetric one in Figure 2.10.5(b). This is what is known as the ***cusp catastrophe***; see Golubitsky and Schaeffer [1985] and Arnold [1968, 1984] for more information.

In (2.10.8), imagine that the hoop is subject to small periodic pulses; say $\omega = \omega_0 + \rho\cos(\eta t)$. Using the Melnikov method described in the introduction and in the following section, it is presumably true (but a messy calculation to prove) that the resulting time-periodic system has horseshoe chaos if ϵ and ν are small, but ρ/ν exceeds a critical value. See Exercise **2.10-3** and §2.11.

Exercises

2.10-1 *Derive the equations of motion for a particle in a hoop spinning about a line a distance ϵ off center. What can you say about the equilibria as functions of ϵ and ω?*

2.10-2 *Derive the formula for the homoclinic orbit (the orbit tending to the saddle point as $t \to \pm\infty$) of a pendulum $\ddot{\psi} + \sin\psi = 0$. Do this using conservation of energy, determining the value of the energy on the homoclinic orbit, solving for $\dot{\psi}$ and then integrating.*

2.10-3 *Using the method of the preceding exercise, derive an integral formula for the homoclinic orbit of the frictionless particle in a rotating hoop.*

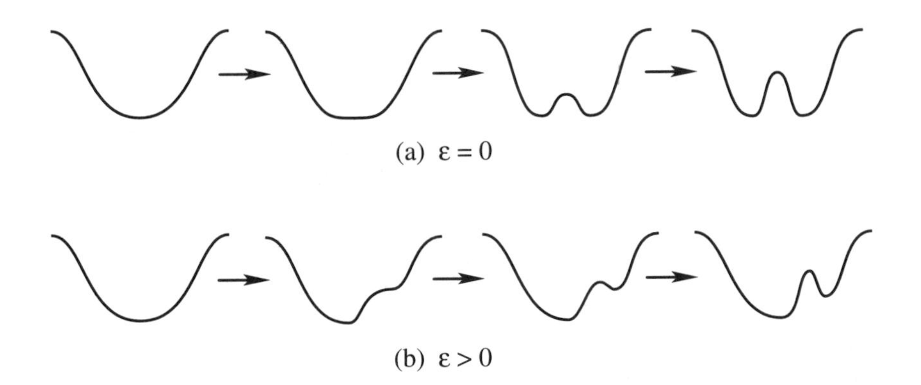

FIGURE 2.10.5. The evolution of the potential for the ball in the centered and the off-centered hoop.

2.10-4 *Determine all equilibria of Duffing's equation*

$$\ddot{x} - \beta x + \alpha x^3 = 0,$$

where α and β are positive constants and study their stability. Derive a formula for the two homoclinic orbits.

2.11 The Poincaré-Melnikov Method and Chaos

Recall from the introduction that in the simplest version of the Poincaré-Melnikov method we are concerned with dynamical equations that perturb a planar Hamiltonian system

$$\dot{x} = X_0(x) \tag{2.11.1}$$

to one of the form

$$\dot{x} = X_0(x) + \epsilon X_1(x, t), \tag{2.11.2}$$

where ϵ is a small parameter, $x \in \mathbb{R}^2$, X_0 is a Hamiltonian vector field with energy H_0, X_1 is periodic with period T, and is Hamiltonian with energy a T-periodic function H_1. We assume that X_0 has a homoclinic orbit $\overline{x}(t)$ so $\overline{x}(t) \to x_0$, a hyperbolic saddle point, as $t \to \pm\infty$. Define the ***Poincaré-Melnikov function*** by

$$M(t_0) = \int_{-\infty}^{\infty} \{H_0, H_1\}(\overline{x}(t - t_0), t)\, dt \tag{2.11.3}$$

where $\{\,,\}$ denotes the Poisson bracket. A major contribution of Arnold [1964] is to the study of higher-dimensional cases and the attendant introduction of an instability mechanism now called ***Arnold diffusion***. We shall comment on this and other extensions below.

There are two convenient ways of visualizing the dynamics of (2.11.2). Introduce the ***Poincaré map*** $P_\epsilon^s : \mathbb{R}^2 \to \mathbb{R}^2$, which is the time T map for (2.11.2) starting at time s. For $\epsilon = 0$, the point x_0 and the homoclinic orbit are invariant under P_0^s, which is independent of s. The hyperbolic saddle x_0 persists as a nearby family of saddles x_ϵ for $\epsilon > 0$, small, and we are interested in whether or not the stable and unstable manifolds of the point x_ϵ for the map P_ϵ^s intersect transversally (if this holds for one s, it holds for all s). If so, we say (2.11.2) has ***horseshoes*** *for* $\epsilon > 0$.

The second way to study (2.11.2) is to look directly at the suspended system on $\mathbb{R}^2 \times S^1$, where S^1 is the circle; (2.11.2) becomes the autonomous ***suspended system***

$$\left.\begin{aligned} \dot{x} &= X_0(x) + \epsilon X_1(x, \theta), \\ \dot{\theta} &= 1. \end{aligned}\right\} \tag{2.11.4}$$

From this point of view the curve

$$\gamma_0(t) = (x_0, t)$$

is a periodic orbit for (2.11.2). This orbit has ***stable manifolds*** and ***unstable manifolds*** denoted $W_0^s(\gamma_0)$ and $W_0^u(\gamma_0)$ defined as the set of points tending exponentially to γ_0 as $t \to \infty$ and $t \to -\infty$, respectively. (See Abraham, Marsden, and Ratiu [1988], Guckenheimer and Holmes [1983], or Wiggins [1988, 1990, 1992] for more details.) In this example, they coincide: $W_0^s(\gamma_0) = W_0^u(\gamma_0)$.

For $\epsilon > 0$ the (hyperbolic) closed orbit γ_0 perturbs to a nearby (hyperbolic) closed orbit which has stable and unstable manifolds $W_\epsilon^s(\gamma_\epsilon)$ and $W_\epsilon^u(\gamma_\epsilon)$. If $W_\epsilon^s(\gamma_\epsilon)$ and $W_\epsilon^u(\gamma_\epsilon)$ intersect transversally, we again say that (2.11.2) has ***horseshoes***. These two definitions of admitting horseshoes are readily seen to be equivalent.

Theorem 2.11.1 (Poincaré-Melnikov Theorem) *Let the Poincaré-Melnikov function be defined by* (2.11.3). *Assume* $M(t_0)$ *has simple zeros as a* T*-periodic function of* t_0. *Then, for sufficiently small* ϵ, (2.11.2) *has horseshoes; that is, homoclinic chaos in the sense of transversal intersecting separatrices.*

Proof In the suspended picture, we use the energy function H_0 to measure the first-order movement of $W_\epsilon^s(\gamma_\epsilon)$ at $\overline{x}(0)$ at time t_0 as ϵ is varied. Note that points of $\overline{x}(t)$ are regular points for H_0 since H_0 is constant on $\overline{x}(t)$ and $\overline{x}(0)$ is not a fixed point. That is, the differential of H_0 does not vanish at $\overline{x}_0(0)$. Thus, the values of H_0 give an accurate measure of the distance from the homoclinic orbit. If $(x_\epsilon^s(t, t_0), t)$ is the curve on $W_\epsilon^s(\gamma_\epsilon)$ that is an integral curve of the suspended system and has an initial condition $x_\epsilon^s(t_0, t_0)$ that is the perturbation of $W_0^s(\gamma_0) = \{\text{the plane } t = t_0\}$ in the normal direction to the homoclinic orbit, then $H_0(x_\epsilon^s(t_0, t_0))$ measures the normal distance. But

$$H_0(x_\epsilon^s(T,t_0)) - H_0(x_\epsilon^s(t_0,t_0)) = \int_{t_0}^{T} \frac{d}{dt} H_0(x_\epsilon^s(t,t_0))\,dt$$

$$= \int_{t_0}^{T} \{H_0, H_0 + \epsilon H_1\}(x_\epsilon^s(t,t_0),t)\,dt. \tag{2.11.5}$$

Since $x_\epsilon^s(T,t_0)$ is ϵ-close to $\overline{x}(t-t_0)$ (uniformly as $T \to +\infty$), we see that $\mathbf{d}(H_0 + \epsilon H^1)(x_\epsilon^s(t,t_0),t) \to 0$ exponentially as $t \to +\infty$, and since $\{H_0, H_0\} = 0$, (2.11.5) becomes

$$H_0(x_\epsilon^s(T,t_0)) - H_0(x_\epsilon^s(t_0,t_0))$$

$$= \epsilon \int_{t_0}^{T} \{H_0, H_1\}(\overline{x}(t-t_0,t))\,dt + O(\epsilon^2). \tag{2.11.6}$$

Similarly,

$$H_0(x_\epsilon^u(t_0,t_0)) - H_0(x_\epsilon^u(-S,t_0))$$

$$= \epsilon \int_{-S}^{t_0} \{H_0, H_1\}(\overline{x}(t-t_0,t))\,dt + O(\epsilon^2). \tag{2.11.7}$$

Since $x_\epsilon^s(T,t_0) \to \gamma_\epsilon$, a periodic orbit for the perturbed system as $T \to +\infty$, we can choose T and S such that $H_0(x_\epsilon^s(T,t_0)) - H_0(x_\epsilon^u(-S,t_0)) \to 0$ as $T, S \to \infty$. Thus, adding (2.11.6) and (2.11.7), and letting $T, S \to \infty$, we get

$$H_0(x_\epsilon^u(t_0,t_0)) - H_0(x_\epsilon^s(t_0,t_0))$$

$$= \epsilon \int_{-\infty}^{\infty} \{H_0, H_1\}(\overline{x}(t-t_0,t))\,dt + O(\epsilon^2). \tag{2.11.8}$$

It follows that if $M(t_0)$ has a simple zero at time t_0, then $x_\epsilon^u(t_0,t_0)$ and $x_\epsilon^s(t_0,t_0)$ intersect transversally near the point $\overline{x}(0)$ at time t_0. (Since $\mathbf{d}H_0 \to 0$ exponentially at the saddle points, the integrals involved in this criterion are automatically convergent.) ■

We now describe a few of the extensions and applications of this technique. The literature in this area is growing very quickly and we make no claim to be comprehensive (the reader can track down many additional references by consulting the references cited).

If in (2.11.2), only X_0 is Hamiltonian, the same conclusion holds if (2.11.3) is replaced by

$$M(t_0) = \int_{-\infty}^{\infty} (X_0 \times X_1)(\overline{x}(t-t_0),t)\,dt \tag{2.11.9}$$

where $X_0 \times X_1$ is the (scalar) cross product for planar vector fields. In fact, X_0 need not even be Hamiltonian if an area expansion factor is inserted.

Example A Equation (2.11.9) applies to the forced damped Duffing equation

$$\ddot{u} - \beta u + \alpha u^3 = \epsilon(\gamma \cos \omega t - \delta \dot{u}). \tag{2.11.10}$$

Here the homoclinic orbits are given by (see Exercise **2.10-4**)

$$u(t) = \pm\sqrt{\frac{2\beta}{\alpha}} \operatorname{sech}(\sqrt{\beta} t) \tag{2.11.11}$$

and (2.11.9) becomes, after a residue calculation,

$$M(t_0) = \gamma\pi\omega\sqrt{\frac{2}{\alpha}} \operatorname{sech}\left(\frac{\pi\omega}{2\sqrt{\beta}}\right) \sin(\omega t_0) - \frac{4\delta\beta^{3/2}}{3\alpha}, \tag{2.11.12}$$

so one has simple zeros and hence chaos of the horseshoe type if

$$\frac{\gamma}{\delta} > \frac{2\sqrt{2}\beta^{3/2}}{3\omega\sqrt{\alpha}} \cosh\left(\frac{\pi\omega}{2\sqrt{\beta}}\right) \tag{2.11.13}$$

and ϵ is small. ♦

Example B Another interesting example, due to Montgomery [1985], concerns the equations for superfluid ^{3}He. These are the Leggett equations and we shall confine ourselves to what is called the A phase for simplicity (see Montgomery's paper for additional results). The equations are

$$\dot{s} = -\frac{1}{2}\left(\frac{\chi\Omega^2}{\gamma^2}\right) \sin 2\theta$$

and

$$\dot{\theta} = \left(\frac{\gamma^2}{\chi}\right) s - \epsilon\left(\gamma B \sin \omega t + \frac{1}{2}\Gamma \sin 2\theta\right). \tag{2.11.14}$$

Here s is the spin, θ an angle (describing the "order parameter"), and $\gamma, \chi, \ldots$ are physical constants. The homoclinic orbits for $\epsilon = 0$ are given by

$$\overline{\theta}_{\pm} = 2\tan^{-1}(e^{\pm\Omega t}) - \pi/2 \quad \text{and} \quad \overline{s}_{\pm} = \pm 2\frac{\Omega e^{\pm 2\Omega t}}{1 + e^{\pm 2\Omega t}}. \tag{2.11.15}$$

One calculates the Poincaré-Melnikov function to be

$$M_{\pm}(t_0) = \mp\frac{\pi\chi\omega B}{8\gamma} \operatorname{sech}\left(\frac{\omega\pi}{2\Omega}\right) \cos \omega t - \frac{2}{3}\frac{\chi}{\gamma^2}\Omega\Gamma, \tag{2.11.16}$$

so that (2.11.14) has chaos in the sense of horseshoes if

$$\frac{\gamma B}{\Gamma} > \frac{16}{3\pi}\frac{\Omega}{\omega} \cosh\left(\frac{\pi\omega}{2\Omega}\right) \tag{2.11.17}$$

and if ϵ is small. ♦

For references and information on higher-dimensional versions of the method and applications, see Wiggins [1988]. We shall comment on some aspects of this shortly. There is even a version of the Poincaré-Melnikov method applicable to PDEs (due to Holmes and Marsden [1981]). One basically still uses formula (2.11.9) where $X_0 \times X_1$ is replaced by the symplectic pairing between X_0 and X_1. However, there are two new difficulties in addition to standard technical analytic problems that arise with PDEs. The first is that there is a serious problem with resonances. This can be dealt with using the aid of damping. Second, the problem seems to be *not* reducible to two dimensions; the horseshoe involves all the modes. Indeed, the higher modes do seem to be involved in the physical buckling processes for the beam model discussed next.

Example C A PDE model for a buckled forced beam is

$$\ddot{w} + w'''' + \Gamma w' - \kappa \left(\int_0^1 [w']^2 \, dz \right) w'' = \epsilon (f \cos \omega t - \delta \dot{w}), \tag{2.11.18}$$

where $w(z,t), 0 \leq z \leq 1$, describes the deflection of the beam, $\dot{} = \partial/\partial t$, $' = \partial/\partial z$, and $\Gamma, \kappa, \ldots$ are physical constants. For this case, one finds that if

i $\pi^2 < \Gamma < 4\rho^3$ (first mode is buckled);

ii $j^2\pi^2(j^2\pi^2 - \Gamma) \neq \omega^2, j = 2, 3, \ldots$ (resonance condition);

iii $\dfrac{f}{\delta} > \dfrac{\pi(\Gamma - \pi^2)}{2\omega\sqrt{\kappa}} \cosh \left(\dfrac{\omega}{2\sqrt{\Gamma - \omega^2}} \right)$ transversal zeros for $M(t_0)$);

iv $\delta > 0$;

and ϵ is small, then (2.11.18) has horseshoes. Experiments (see Moon [1988]) showing chaos in a forced buckled beam provided the motivation which lead to the study of (2.11.18). ♦

This kind of result can also be used for a study of chaos in a van der Waals fluid (Slemrod and Marsden [1985]) and for soliton equations (see Birnir [1986], Ercolani, Forest, and McLaughlin [1990], and Birnir and Grauer [1994]). For example, in the damped, forced sine-Gordon equation one has chaotic transitions between breathers and kink-antikink pairs and in the Benjamin-Ono equation one can have chaotic transitions between solutions with different numbers of poles.

For Hamiltonian systems with two degrees of freedom, Holmes and Marsden [1982a] show how the Melnikov method may be used to prove the existence of horseshoes on energy surfaces in nearly integrable systems. The class of systems studied have a Hamiltonian of the form

$$H(q,p,\theta,I) = F(q,p) + G(I) + \epsilon H_1(q,p,\theta,I) + O(\epsilon^2), \tag{2.11.19}$$

where (θ, I) are action-angle coordinates for the oscillator $G; G(0) = 0, G' > 0$. It is assumed that F has a homoclinic orbit $\overline{x}(t) = (\overline{q}(t), \overline{p}(t))$ and that

$$M(t_0) = \int_{-\infty}^{\infty} \{F, H_1\}\, dt, \tag{2.11.20}$$

the integral taken along $(\overline{x}(t - t_0), \Omega t, I)$ has simple zeros. Then (2.11.19) has horseshoes on energy surfaces near the surface corresponding to the homoclinic orbit and small I; the horseshoes are taken relative to a Poincaré map strobed to the oscillator G. The paper by Holmes and Marsden [1982a] also studies the effect of positive and negative damping. These results are related to those for forced one degree of freedom systems since one can often reduce a two degrees of freedom Hamiltonian system to a one degree of freedom forced system.

For some systems in which the variables do not split as in (2.11.19), such as a nearly symmetric heavy top, one needs to exploit a symmetry of the system and this complicates the situation to some extent. The general theory for this is given in Holmes and Marsden [1983] and was applied to show the existence of horseshoes in the nearly symmetric heavy top; see also some closely related results of Ziglin [1980a].

This theory has been used by Ziglin [1980b] and Koiller [1985] in vortex dynamics, for example, to give a proof of the non-integrability of the restricted four vortex problem. Koiller, Soares and Melo Neto [1985] gives applications to the dynamics of general relativity showing the existence of horseshoes in Bianchi IX models. See Oh, Sreenath, Krishnaprasad, and Marsden [1989] for applications to the dynamics of coupled rigid bodies.

Arnold [1964] extended the Poincaré-Melnikov theory to systems with several degrees of freedom. In this case the transverse homoclinic manifolds are based on KAM tori and allow the possibility of chaotic drift from one torus to another. This drift, now known as ***Arnold diffusion*** is a basic ingredient in the study of chaos in Hamiltonian systems (see for instance, Chirikov [1979] and Lichtenberg and Lieberman [1983] and references therein). Instead of a single Melnikov function, one now has a ***Melnikov vector*** given schematically by

$$\mathbf{M} = \begin{pmatrix} \int_{-\infty}^{\infty} \{H_0, H_1\}\, dt \\ \\ \int_{-\infty}^{\infty} \{I_1, H_1\}\, dt \\ \cdots \\ \int_{-\infty}^{\infty} \{I_n, H_1\}\, dt \end{pmatrix}, \tag{2.11.21}$$

where $I_1, \ldots, I_n$ are integrals for the unperturbed (completely integrable) system and where $\mathbf{M}$ depends on t_0 and on angles conjugate to $I_1, \ldots, I_n$.

One requires $\mathbf{M}$ to have transversal zeros in the vector sense. This result was given by Arnold for forced systems and was extended to the autonomous case by Holmes and Marsden [1982b, 1983]; see also Robinson [1988]. These results apply to systems such as a pendulum coupled to several oscillators and the many vortex problems. It has also been used in power systems by Salam, Marsden, and Varaiya [1983], building on the horseshoe case treated by Kopell and Washburn [1982]. See also Salam and Sastry [1985]. There have been a number of other directions of research on these techniques. For example, Grundler [1985] developed a multidimensional version applicable to the spherical pendulum and Greenspan and Holmes [1983] showed how it can be used to study subharmonic bifurcations. See Wiggins [1988] for more information.

In Poincaré's celebrated memoir [1890] on the three-body problem, he introduced the mechanism of transversal intersection of separatrices which obstructs the integrability of the equations and the attendant convergence of series expansions for the solutions. This idea has been developed by Birkhoff and Smale using the horseshoe construction to describe the resulting chaotic dynamics. However, in the region of phase space studied by Poincaré, it has never been proved (except in some generic sense that is not easy to interpret in specific cases) that the equations really are nonintegrable. In fact, Poincaré himself traced the difficulty to the presence of terms in the separatrix splitting which are exponentially small. A crucial component of the measure of the splitting is given by the following formula of Poincaré [1890, p. 223]:

$$J = \frac{-8\pi i}{\exp\left(\frac{\pi}{\sqrt{2\mu}}\right) + \exp\left(-\frac{\pi}{\sqrt{2\mu}}\right)},$$

which is exponentially small (or beyond all orders) in μ. Poincaré was aware of the difficulties that this exponentially small behavior causes; on page 224 of his article, he states: "En d'autres termes, si on regarde μ comme un infiniment petit du premier ordre, la distance BB', sans être nulle, est un infiniment petit d'ordre infini. C'est ainsi que la fonction $e^{-1/\mu}$ est un infiniment petit d'ordre infini sans être nulle ... Dans l'example particulier que nous avons traité plus haut, la distance BB' est du mème ordre de grandeur que l'integral J, c'est à dire que $\exp(-\pi/\sqrt{2\mu})$."

This is a serious difficulty that arises when one uses the Melnikov method near an elliptic fixed point in a Hamiltonian system or in bifurcation problems giving birth to homoclinic orbits. The difficulty is related to those described by Poincaré. Near elliptic points, one sees homoclinic orbits in normal forms and after a temporal rescaling this leads to a rapidly oscillatory perturbation that is modeled by the following variation of the pendulum equation:

$$\ddot{\phi} + \sin\phi = \epsilon\cos\left(\frac{\omega t}{\epsilon}\right). \tag{2.11.22}$$

If one formally computes $M(t_0)$ one finds:

$$M(t_0, \epsilon) = \pm 2\pi \operatorname{sech}\left(\frac{\pi\omega}{2\epsilon}\right) \cos\left(\frac{\omega t_0}{\epsilon}\right). \tag{2.11.23}$$

While this has simple zeros, the proof of the Poincaré-Melnikov theorem is no longer valid since $M(t_0, \epsilon)$ is now of order $\epsilon^{-\pi/2\epsilon}$ and the error analysis in the proof only gives errors of order ϵ^2. In fact, no expansion in powers of ϵ can detect exponentially small terms like $\epsilon^{-\pi/2\epsilon}$.

Holmes, Marsden, and Scheurle [1988] and Delshams and Seara [1991] show that (2.11.22) has chaos that is, in a suitable sense, *exponentially small* in ϵ. The idea is to expand expressions for the stable and unstable manifolds in a Perron type series whose terms are of order $\epsilon^\kappa \epsilon^{-\pi/2\epsilon}$. To do so, the extension of the system to complex time plays a crucial role. One can hope that since such results for (2.11.22) can be proved, it may be possible to return to Poincaré's 1890 work and complete the arguments he left unfinished.

To illustrate how exponentially small phenomena enter bifurcation problems, consider the problem of a Hamiltonian saddle node bifurcation

$$\ddot{x} + \mu x + x^2 = 0 \tag{2.11.24}$$

with the addition of higher-order terms and forcing:

$$\ddot{x} + \mu x + x^2 + \text{h.o.t.} = \delta f(t). \tag{2.11.25}$$

The phase portrait of (2.11.24) is shown in Figure 2.11.1.

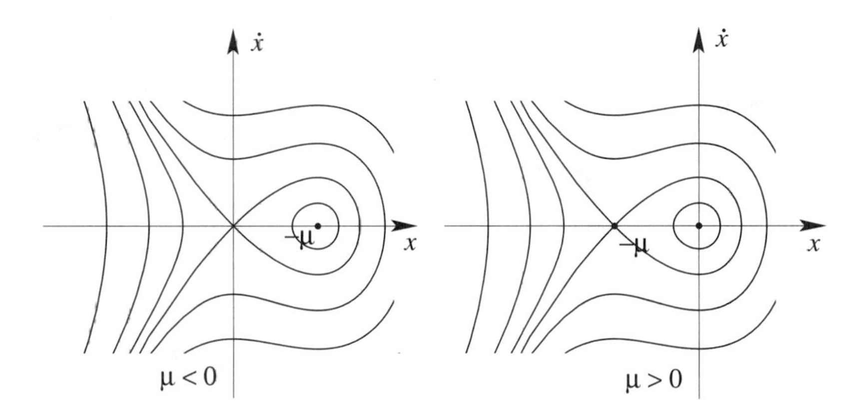

FIGURE 2.11.1. Phase portraits of $\ddot{x} + \mu x + x^2 = 0$.

The system (2.11.24) is Hamiltonian with

$$H(x, \dot{x}) = \frac{1}{2}\dot{x}^2 + \frac{1}{2}\mu x^2 + \frac{1}{3}x^3. \tag{2.11.26}$$

Let us first consider the system without higher-order terms:

$$\ddot{x} + \mu x + x^2 = \delta f(t). \tag{2.11.27}$$

To study it, we rescale to blow up the singularity:

$$x(t) = \lambda \xi(\tau), \tag{2.11.28}$$

where $\lambda = \mid \mu \mid$ and $\tau = t\sqrt{\lambda}$. Letting $' = d/d\tau$, we get

$$\left.\begin{aligned} \xi'' - \xi + \xi^2 &= \frac{\delta}{\mu^2} f\left(\frac{\tau}{\sqrt{-\mu}}\right), \quad \mu < 0, \\ \xi'' + \xi + \xi^2 &= \frac{\delta}{\mu^2} f\left(\frac{\tau}{\sqrt{\mu}}\right), \quad \mu > 0, \end{aligned}\right\} \tag{2.11.29}$$

The exponentially small estimates of Holmes, Marsden, and Scheurle [1988] apply to (2.11.29). One gets exponentially small upper and lower estimates in certain algebraic sectors of the (δ, μ) plane that depend on the nature of f. The estimates for the splitting have the form $C(\delta/\mu^2)\exp(-\pi/\sqrt{|\mu|})$. Now consider

$$\ddot{x} + \mu x + x^2 + x^3 = \delta f(t). \tag{2.11.30}$$

With $\delta = 0$, there are equilibria at

$$x = 0, \ -r, \quad \text{or} \quad -\frac{\mu}{r} \quad \text{and} \quad \dot{x} = 0, \tag{2.11.31}$$

where

$$r = \frac{1 + \sqrt{1 - 4\mu}}{2}, \tag{2.11.32}$$

which is approximately 1 when $\mu \approx 0$. The phase portrait of (2.11.30) with $\delta = 0$ and $\mu = -\frac{1}{2}$ is shown in Figure 2.11.2. As μ passes through 0, the small lobe in Figure 2.11.2 undergoes the same bifurcation as in Figure 2.11.1, with the large lobe changing only slightly.

Again we rescale to give

$$\left.\begin{aligned} \ddot{\xi} - \xi + \xi^2 - \mu\xi^3 &= \frac{\delta}{\mu^2} f\left(\frac{\tau}{\sqrt{-\mu}}\right), \quad \mu < 0, \\ \ddot{\xi} + \xi + \xi^2 + \mu\xi^3 &= \frac{\delta}{\mu^2} f\left(\frac{\tau}{\sqrt{\mu}}\right), \quad \mu > 0. \end{aligned}\right\} \tag{2.11.33}$$

Notice that for $\delta = 0$, the phase portrait is μ-dependent. The homoclinic orbit surrounding the small lobe for $\mu < 0$ is given explicitly in terms of ξ by

$$\xi(\tau) = \frac{4e^\tau}{\left(e^\tau + \frac{2}{3}\right)^2 - 2\mu}, \tag{2.11.34}$$

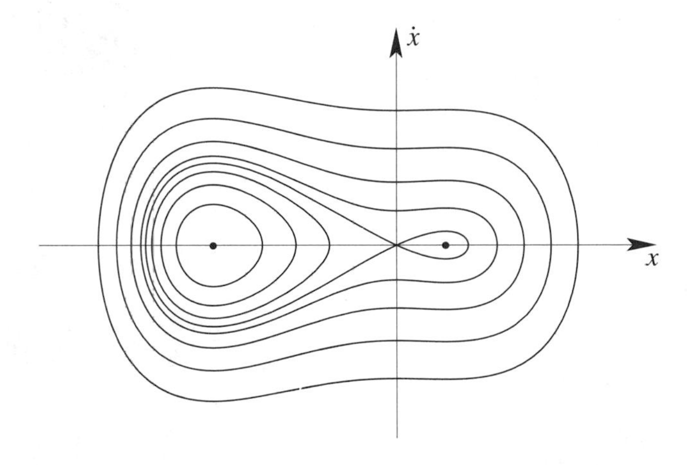

FIGURE 2.11.2. The phase portrait of $\ddot{x} - \frac{1}{2}x + x^2 + x^3 = 0$.

which is μ-dependent. An interesting technicality is that without the cubic term, we get μ-independent *double* poles at $t = \pm i\pi + \log 2 - \log 3$ in the complex τ-plane, while (2.11.34) has a pair of simple poles that splits these double poles to the pairs of simple poles at

$$\tau = \pm i\pi + \log\left(\frac{2}{3} \pm i\sqrt{2\lambda}\right), \tag{2.11.35}$$

where again $\lambda = |\mu|$. (There is no particular significance to the real part, such as $\log 2 - \log 3$ in the case of no cubic term; this can always be gotten rid of by a shift in the base point $\xi(0)$.)

If a quartic term x^4 is added, these pairs of simple poles will split into quartets of branch points and so on. Thus, while the analysis of higher-order terms has this interesting μ-dependence, it seems that the basic exponential part of the estimates, namely

$$\exp\left(-\frac{\pi}{\sqrt{|\mu|}}\right), \tag{2.11.36}$$

remains intact.

3

An Introduction to Infinite-Dimensional Systems

A common choice of configuration space for classical field theory is an infinite-dimensional vector space of functions or tensor fields on space or spacetime, the elements of which are called ***fields***. Here we relate our treatment of infinite-dimensional Hamiltonian systems discussed in §2.1 to classical Lagrangian and Hamiltonian field theory and then give examples. Classical field theory is a large subject with many aspects not covered here; we treat only a few topics that are basic to subsequent developments; see Chapters 6 and 7 for further information and references.

3.1 Lagrange's and Hamilton's Equations for Field Theory

As with finite-dimensional systems, one can begin with a Lagrangian and a variational principle, and then pass to the Hamiltonian via the Legendre transformation. At least formally, all the constructions we did in the finite-dimensional case go over to the infinite-dimensional one.

For instance, suppose we choose our configuration space $Q = \mathcal{F}(\mathbb{R}^3)$ to be the space of fields φ on $\mathbb{R}^3$. Our Lagrangian will be a function $L(\varphi, \dot{\varphi})$ from $Q \times Q$ to $\mathbb{R}$. The variational principle is

$$\delta \int_a^b L(\varphi, \dot{\varphi})\, dt = 0, \tag{3.1.1}$$

which is equivalent to the Euler-Lagrange equations

$$\frac{d}{dt}\frac{\delta L}{\delta \dot{\varphi}} = \frac{\delta L}{\delta \varphi} \tag{3.1.2}$$

in the usual way. Here,

$$\pi = \frac{\delta L}{\delta \dot{\varphi}} \tag{3.1.3}$$

is the conjugate momentum which we regard as a density on $\mathbb{R}^3$, as in Chapter 2. The corresponding Hamiltonian is

$$H(\varphi, \pi) = \int \pi \dot{\varphi} - L(\varphi, \dot{\varphi}) \tag{3.1.4}$$

in accordance with our general theory. We also know that the Hamiltonian should generate the canonical Hamilton equations. We verify this now.

Proposition 3.1.1 *Let $Z = \mathcal{F}(\mathbb{R}^3) \times \mathcal{D}en(\mathbb{R}^3)$, with Ω defined as in Example* **(b)** *of §2.3. Then the Hamiltonian vector field $X_H : Z \to Z$ corresponding to a given energy function $H : Z \to \mathbb{R}$ is given by*

$$X_H = \left(\frac{\delta H}{\delta \pi}, -\frac{\delta H}{\delta \varphi} \right). \tag{3.1.5}$$

Hamilton's equations on Z are

$$\frac{\partial \varphi}{\partial t} = \frac{\delta H}{\delta \pi}, \quad \frac{\partial \pi}{\partial t} = -\frac{\delta H}{\delta \varphi}. \tag{3.1.6}$$

Remarks

1. Here $\mathcal{F}$ and $\mathcal{D}en$ stand for function spaces included in the space of all functions and densities, chosen appropriate to the functional analysis needs of the particular problem. In practice this often means, among other things, that appropriate conditions at infinity are imposed to permit integration by parts.

2. The equations of motion for a curve $z(t) = (\varphi(t), \pi(t))$ in the form $\Omega(dz/dt, \delta z) = \mathbf{d}H(z(t)) \cdot \delta z$ for all $\delta z \in Z$ with compact support, are called the ***weak form of the equations of motion***. They can still be valid when there is not enough smoothness or decay at infinity to justify the literal equality $dz/dt = X_H(z)$; this situation can occur, for example, if one is considering shock waves.

Proof of Proposition 3.1.1 To derive the partial functional derivatives, we use the natural pairing

$$\langle\, , \rangle : \mathcal{F}(\mathbb{R}^3) \times \mathcal{D}en(\mathbb{R}^3) \to \mathbb{R}, \quad \text{where} \quad \langle \varphi, \pi \rangle = \int \varphi \pi' \, d^3x, \tag{3.1.7}$$

where we write $\pi = \pi' d^3x \in \mathcal{D}en$. Recalling that $\delta H/\delta\varphi$ is a density, let

$$X = \left(\frac{\delta H}{\delta \pi}, -\frac{\delta H}{\delta \varphi}\right).$$

We need to verify that $\Omega(X(\varphi,\pi),(\delta\varphi,\delta\pi)) = \mathbf{d}H(\varphi,\pi)\cdot(\delta\varphi,\delta\pi)$. Indeed,

$$\begin{aligned}
\Omega(X(\varphi,\pi),(\delta\varphi,\delta\pi)) &= \Omega\left(\left(\frac{\delta H}{\delta \pi}, -\frac{\delta H}{\delta \varphi}\right),(\delta\varphi,\delta\pi)\right) \\
&= \int \frac{\delta H}{\delta \pi}(\delta\pi)' d^3x + \int \delta\varphi \left[\frac{\delta H}{\delta\varphi}\right]' d^3x \\
&= \left\langle \frac{\delta H}{\delta \pi}, \delta\pi \right\rangle + \left\langle \delta\varphi, \frac{\delta H}{\delta\varphi}\right\rangle \\
&= \mathbf{D}_\pi H(\varphi,\pi)\cdot\delta\pi + \mathbf{D}_\varphi H(\varphi,\pi)\cdot\delta\varphi \\
&= \mathbf{d}H(\varphi,\pi)\cdot(\delta\varphi,\delta\pi). \quad \blacksquare
\end{aligned}$$

3.2 Examples: Hamilton's Equations

(a) The Wave Equation Consider $Z = \mathcal{F}(\mathbb{R}^3)\times\mathcal{D}en(\mathbb{R}^3)$ as above. Let φ denote the configuration variable, that is, the first component in the phase space $\mathcal{F}(\mathbb{R}^3)\times\mathcal{D}en(\mathbb{R}^3)$, and interpret φ as a measure of the displacement from equilibrium of a homogeneous elastic medium. Writing $\pi' = \rho\, d\varphi/dt$, where ρ is the mass density, the ***kinetic energy*** is

$$T = \frac{1}{2}\int \frac{1}{\rho}[\pi']^2\, d^3x.$$

For small displacements φ, one assumes a linear restoring force such as the one given by the ***potential energy***

$$\frac{k}{2}\int \|\nabla\varphi\|^2\, d^3x,$$

for an (elastic) constant k. Because we are considering a homogeneous medium, ρ and k are constants, so let us work in units in which they are unity. Nonlinear effects can be modeled in a naive way by introducing a nonlinear term, $U(\varphi)$ into the potential. However, for an elastic medium one really should use constitutive relations based on the principles of continuum mechanics; see Marsden and Hughes [1983]. For the naive model, the Hamiltonian $H : Z \to \mathbb{R}$ is the ***total energy***

$$H(\varphi,\pi) = \int\left[\frac{1}{2}(\pi')^2 + \frac{1}{2}\|\nabla\varphi\|^2 + U(\varphi)\right] d^3x. \tag{3.2.1}$$

Using the definition of the functional derivative, we find that

$$\frac{\delta H}{\delta \pi} = \pi', \quad \frac{\delta H}{\delta \varphi} = (-\nabla^2 \varphi + U'(\varphi)) d^3 x. \tag{3.2.2}$$

Therefore, the equations of motion are

$$\frac{\partial \varphi}{\partial t} = \pi', \quad \frac{\partial \pi'}{\partial t} = \nabla^2 \varphi - U'(\varphi), \tag{3.2.3}$$

or, in second-order form,

$$\frac{\partial^2 \varphi}{\partial t^2} = \nabla^2 \varphi - U'(\varphi). \tag{3.2.4}$$

Various choices of U correspond to various physical applications. When $U' = 0$, we get the linear wave equation, with unit propagation velocity. Another choice, $U(\varphi) = \frac{1}{2} m^2 \varphi^2 + \lambda \varphi^4$, occurs in the quantum theory of self-interacting mesons; the parameter m is related to the meson mass, and φ^4 governs the nonlinear part of the interaction. When $\lambda = 0$, we get

$$\nabla^2 \varphi - \frac{\partial^2 \varphi}{\partial t^2} = m^2 \varphi, \tag{3.2.5}$$

which is called the ***linear Klein-Gordon equation***. ♦

Technical Aside For the wave equation, one appropriate choice of function space is $Z = H^1(\mathbb{R}^3) \times L^2_{\mathrm{Den}}(\mathbb{R}^3)$, where $H^1(\mathbb{R}^3)$ denotes the H^1-functions on $\mathbb{R}^3$, that is, functions which, along with their first derivatives, are square integrable, and $L^2_{\mathrm{Den}}(\mathbb{R}^3)$ denotes the space of densities $\pi = \pi' \, d^3 x$ where the function π' on $\mathbb{R}^3$ is square integrable. Note that the Hamiltonian vector field

$$X_H(\varphi, \pi) = (\pi', (\nabla^2 \varphi - U'(\varphi)) d^3 x)$$

is defined only on the dense subspace $H^2(\mathbb{R}^3) \times H^1_{\mathrm{Den}}(\mathbb{R}^3)$ of Z. This is a common occurrence in the study of Hamiltonian partial differential equations; we return to this in §3.3. ♦

In the preceding example, Ω was given by the canonical form with the result that the equations of motion were in the standard form (3.1.5). In addition, the Hamiltonian function was given by the actual energy of the system under consideration. We now give examples in which these statements require reinterpretation but which nevertheless fall into the framework of the general theory developed so far.

(b) The Schrödinger Equation Let $\mathcal{H}$ be a complex Hilbert space, for example, the space of complex-valued functions ψ on $\mathbb{R}^3$ with the inner product

$$\langle \psi_1, \psi_2 \rangle = \int \psi_1(x)\overline{\psi}_2(x)\, d^3x,$$

where the overbar denotes complex conjugation. For a self-adjoint, complex-linear operator $H_{\text{op}} : \mathcal{H} \to \mathcal{H}$, the Schrödinger equation is

$$i\hbar \frac{\partial \psi}{\partial t} = H_{\text{op}}\psi, \tag{3.2.6}$$

where $\hbar$ is Planck's constant. Define

$$A = \frac{-i}{\hbar} H_{\text{op}}$$

so that the Schrödinger equation becomes

$$\frac{\partial \psi}{\partial t} = A\psi. \tag{3.2.7}$$

The symplectic form on $\mathcal{H}$ is given by $\Omega(\psi_1, \psi_2) = -2\hbar \operatorname{Im} \langle \psi_1, \psi_2 \rangle$. Self-adjointness of H_{op} is a condition stronger than symmetry and is essential for proving well-posedness of the initial-value problem for (3.2.6); for an exposition, see, for instance, Abraham, Marsden, and Ratiu [1988]. Historically, it was Kato [1950] who established this for important problems such as the hydrogen atom.

From §2.7, we know that since H_{op} is symmetric, A is Hamiltonian. The Hamiltonian is

$$H(\psi) = \hbar \langle iA\psi, \psi \rangle = \langle H_{\text{op}}\psi, \psi \rangle \tag{3.2.8}$$

which is the ***expectation value*** of H_{op} at ψ, defined by $\langle H_{\text{op}} \rangle (\psi) = \langle H_{\text{op}}\psi, \psi \rangle$. ♦

(c) The Korteweg-de Vries (KdV) Equation Denote by Z the vector subspace $\mathcal{F}(\mathbb{R})$ consisting of those functions u with $|u(x)|$ decreasing sufficiently fast as $x \to \pm\infty$ so that the integrals we will write are defined and integration by parts is justified. As we shall see later, the Poisson brackets for the KdV equation are quite simple, and historically they were found first (see Gardner [1971] and Zakharov [1971, 1974]). To be consistent with our exposition, we begin with the somewhat more complicated symplectic structure. Pair Z with itself using the L^2 inner product. Let the KdV symplectic structure Ω be defined by

$$\Omega(u_1, u_2) = \frac{1}{2} \left(\int_{-\infty}^{\infty} [\hat{u}_1(x)u_2(x) - \hat{u}_2(x)u_1(x)]\, dx \right), \tag{3.2.9}$$

where $\hat{u}$ denotes a primitive of u, that is,

$$\hat{u} = \int_{-\infty}^{x} u(y)\, dy.$$

In §8.5 we shall see a way to *construct* this form. The form Ω is clearly skew-symmetric. Note that if $u_1 = \partial v/\partial x$ for some $v \in Z$, then

$$\begin{aligned}
\int_{-\infty}^{\infty} \hat{u}_2(x) u_1(x)\, dx &= \int_{-\infty}^{\infty} \hat{u}_2(x) \frac{\partial \hat{u}_1(x)}{\partial x}\, dx \\
&= \hat{u}_1(x)\hat{u}_2(x)\Big|_{-\infty}^{\infty} - \int_{-\infty}^{\infty} \hat{u}_1(x) u_2(x)\, dx \\
&= \left(\int_{-\infty}^{\infty} \frac{\partial v(x)}{\partial x}\, dx\right)\left(\int_{-\infty}^{\infty} u_2(x)\, dx\right) - \int_{-\infty}^{\infty} \hat{u}_1(x) u_2(x)\, dx \\
&= \left(v(x)\Big|_{-\infty}^{\infty}\right)\left(\int_{-\infty}^{\infty} u_2(x)\, dx\right) - \int_{-\infty}^{\infty} \hat{u}_1(x) u_2(x)\, dx \\
&= -\int_{-\infty}^{\infty} \hat{u}_1(x) u_2(x)\, dx.
\end{aligned}$$

Thus, if $u_1(x) = \partial v(x)/\partial x$, then Ω can be written as

$$\Omega(u_1, u_2) = \int_{-\infty}^{\infty} \hat{u}_1(x) u_2(x)\, dx = \int_{-\infty}^{\infty} v(x) u_2(x)\, dx. \tag{3.2.10}$$

To prove weak nondegeneracy of Ω, we check that if $v \neq 0$, there is a w such that $\Omega(w, v) \neq 0$. Indeed, if $v \neq 0$ and we let $w = \partial v/\partial x$, then $w \neq 0$ because $v(x) \to 0$ as $|x| \to \infty$. Hence by (3.2.10),

$$\Omega(w, v) = \Omega\left(\frac{\partial v}{\partial x}, v\right) = \int_{-\infty}^{\infty} (v(x))^2\, dx \neq 0.$$

Suppose that a Hamiltonian $H : Z \to \mathbb{R}$ is given. We claim that the corresponding Hamiltonian vector field X_H is given by

$$X_H(u) = \frac{\partial}{\partial x}\left(\frac{\delta H}{\delta u}\right). \tag{3.2.11}$$

Indeed, by (3.2.10)

$$\Omega(X_H(v), w) = \int_{-\infty}^{\infty} \frac{\delta H}{\delta v}(x) w(x)\, dx = \mathbf{d}H(v) \cdot w.$$

It follows from (3.2.11) that the corresponding Hamilton equations are

$$u_t = \frac{\partial}{\partial x}\left(\frac{\delta H}{\delta u}\right), \tag{3.2.12}$$

where, in (3.2.12) and in the following, subscripts denote derivatives. As a special case, consider the function

$$H_1(u) = -\frac{1}{6}\int_{-\infty}^{\infty} u^3\, dx.$$

Then

$$\frac{\partial}{\partial x}\frac{\delta H}{\delta u} = -uu_x,$$

and so (3.2.12) becomes the ***one-dimensional transport equation***

$$u_t + uu_x = 0. \tag{3.2.13}$$

Next, let

$$H_2(u) = \int_{-\infty}^{\infty}\left(\frac{1}{2}u_x^2 - u^3\right) dx; \tag{3.2.14}$$

then (3.2.12) becomes

$$u_t + 6uu_x + u_{xxx} = 0. \tag{3.2.15}$$

This is the ***Korteweg-de Vries (KdV) equation***, describing shallow water waves. For a concise presentation of its famous complete set of integrals, see Abraham and Marsden [1978], §6.5, and for more information, see Newell [1985].

If we look for traveling wave solutions of (3.2.15), that is, $u(x,t) = \varphi(x - ct)$, for a constant $c > 0$ and a positive function φ, we see that u satisfies the KdV equation iff φ satisfies

$$c\varphi' - 6\varphi\varphi' - \varphi''' = 0. \tag{3.2.16}$$

Integrating once gives

$$c\varphi - 3\varphi^2 - \varphi'' = C, \tag{3.2.17}$$

where C is a constant. This equation is Hamiltonian in the canonical variables (φ, φ') with Hamitonian function

$$h(\varphi, \varphi') = \frac{1}{2}(\varphi')^2 - \frac{c}{2}\varphi^2 + \varphi^3 + C\varphi. \tag{3.2.18}$$

From conservation of energy, $H(\varphi, \varphi') = D$, it follows that

$$\varphi' = \pm\sqrt{c\varphi^2 - 2\varphi^3 - 2C\varphi + 2D}, \tag{3.2.19}$$

or, writing $s = x - ct$, we get

$$s = \pm\int \frac{d\varphi}{\sqrt{c\varphi^2 - 2\varphi^3 - 2C\varphi + 2D}}. \tag{3.2.20}$$

We seek solutions which, together with their derivatives vanish at $\pm\infty$. Then (3.2.17) and (3.2.19) give $C = D = 0$, so

$$s = \pm \int \frac{d\varphi}{\sqrt{c\varphi^2 - 2\varphi^3}} = \pm \frac{1}{\sqrt{c}} \log \left| \frac{\sqrt{c - 2\varphi} - \sqrt{c}}{\sqrt{c - 2\varphi} + \sqrt{c}} \right| + K \tag{3.2.21}$$

for some constant K that will be determined below.

For $C = D = 0$, the Hamiltonian (3.2.18) becomes

$$h(\varphi, \varphi') = \frac{1}{2}(\varphi')^2 - \frac{c}{2}\varphi^2 + \varphi^3 \tag{3.2.22}$$

and thus the two equilibria given by $\partial h/\partial\varphi = 0, \partial h/\partial\varphi' = 0$, are $(0, 0)$ and $(c/3, 0)$. The matrix of the linearized Hamiltonian system at these equilibria is $\begin{bmatrix} 0 & 1 \\ \pm c & 0 \end{bmatrix}$ which shows that $(0, 0)$ is a saddle and $(c/3, 0)$ is spectrally stable. The second variation criterion on the potential energy (see §1.10) $-\frac{c}{2}\varphi^2 + \varphi^3$ at $(c/3, 0)$ shows that this equilibrium is stable. Thus, if $(\varphi(s), \varphi'(s))$ is a homoclinic orbit emanating and ending at $(0, 0)$, the value of the Hamiltonian function (3.2.22) on it is $H(0, 0) = 0$. From (3.2.22) it follows that $(c/2, 0)$ is a point on this homoclinic orbit and thus (3.2.20) for $C = D = 0$ is its expression. Taking the initial condition of this orbit at $s = 0$ to be $\varphi(0) = c/2, \varphi'(0) = 0$, (3.2.21) forces $K = 0$ and so

$$\left| \frac{\sqrt{c - 2\varphi} - \sqrt{c}}{\sqrt{c - 2\varphi} + \sqrt{c}} \right| = e^{\pm\sqrt{c}s}.$$

Since $\varphi \geq 0$ by hypothesis, the expression in the absolute value is negative and thus

$$\frac{\sqrt{c - 2\varphi} - \sqrt{c}}{\sqrt{c - 2\varphi} + \sqrt{c}} = -e^{\pm\sqrt{c}s},$$

whose solution is

$$\varphi(s) = \frac{2ce^{\pm\sqrt{c}s}}{(1 + e^{\pm\sqrt{c}s})^2} = \frac{c}{2\cosh^2(\sqrt{c}s/2)}.$$

This produces the ***soliton solution***

$$u(x, t) = \frac{c}{2}\operatorname{sech}^2\left[\frac{\sqrt{c}}{2}(x - ct)\right]. \quad \blacklozenge \tag{3.2.23}$$

(d) Sine-Gordon Equation For functions $u(x, t)$, where x and t are real variables, the ***sine-Gordon equation*** is $u_{tt} = u_{xx} + \sin u$. Equation (3.2.4) shows that it is Hamiltonian with $\pi = u_t\, d^3x$, so $\pi' = u_t$,

$$H(u) = \int_{-\infty}^{\infty} \left(\frac{1}{2}u_t^2 + \frac{1}{2}u_x^2 + \cos u\right) dx, \tag{3.2.24}$$

and the canonical bracket structure, as in the wave equation. This equation also has a complete set of integrals; see again Newell [1985]. ♦

(e) Abstract Wave Equation Let $\mathcal{H}$ be a real Hilbert space and $B : \mathcal{H} \to \mathcal{H}$ a linear operator. On $\mathcal{H} \times \mathcal{H}$, put the symplectic structure Ω given by (2.3.5). One can check that:

i $A = \begin{bmatrix} 0 & I \\ -B & 0 \end{bmatrix}$ is Ω-skew if and only if B is a symmetric operator on $\mathcal{H}$; and

ii if B is symmetric, then a Hamiltonian for A is

$$H(x,y) = \frac{1}{2}(\|y\|^2 + \langle Bx, x \rangle). \tag{3.2.25}$$

The equations of motion (2.6.3) give the ***abstract wave equation***:

$$\ddot{x} + Bx = 0. \quad ♦ \tag{3.2.26}$$

(f) Linear Elastodynamics On $\mathbb{R}^3$ consider the equations

$$\rho \mathbf{u}_{tt} = \operatorname{div}(\mathbf{c} \cdot \nabla \mathbf{u}), \quad \text{i.e.,} \quad \rho u^i_{tt} = \frac{\partial}{\partial x^j}\left[c^{ijkl}\frac{\partial u^k}{\partial x^l}\right], \tag{3.2.27}$$

where ρ is a positive function, and $\mathbf{c}$ is a fourth-order tensor field (the ***elasticity tensor***) on $\mathbb{R}^3$ with the symmetries $c^{ijkl} = c^{klij} = c^{jikl}$.

On $\mathcal{F}(\mathbb{R}^3;\mathbb{R}^3) \times \mathcal{F}(\mathbb{R}^3;\mathbb{R}^3)$ (or more precisely on

$$H^1(\mathbb{R}^3;\mathbb{R}^3) \times L^2(\mathbb{R}^3;\mathbb{R}^3)$$

with suitable decay properties at infinity), define

$$\Omega((\mathbf{u},\dot{\mathbf{u}}),(\mathbf{v},\dot{\mathbf{v}})) = \int_{\mathbb{R}^3} \rho(\dot{\mathbf{v}} \cdot \mathbf{u} - \dot{\mathbf{u}} \cdot \mathbf{v})\, d^3x. \tag{3.2.28}$$

The form Ω is the canonical symplectic form (2.3.2) for fields $\mathbf{u}$ and their conjugate momenta $\pi = \rho\dot{\mathbf{u}}$.

On the space of functions $\mathbf{u} : \mathbb{R}^3 \to \mathbb{R}^3$, consider the ρ-weighted L^2-inner product

$$\langle \mathbf{u}, \mathbf{v} \rangle_\rho = \int_{\mathbb{R}^3} \rho \mathbf{u} \cdot \mathbf{v}\, d^3x. \tag{3.2.29}$$

Then the operator $B\mathbf{u} = -(1/\rho)\operatorname{div}(\mathbf{c}\cdot\nabla\mathbf{u})$ is symmetric with respect to this inner product and thus by Example **(e)** above, the operator $A(\mathbf{u},\dot{\mathbf{u}}) = (\dot{\mathbf{u}}, (1/\rho)\operatorname{div}(\mathbf{c}\cdot\nabla\mathbf{u}))$ is Ω-skew.

The equations (3.2.27) of linear elastodynamics are Hamiltonian with respect to Ω given by (3.2.28) with energy

$$H(\mathbf{u}, \dot{\mathbf{u}}) = \frac{1}{2} \int \rho \|\dot{\mathbf{u}}\|^2 \, d^3x + \frac{1}{2} \int c^{ijkl} e_{ij} e_{kl} \, d^3x, \tag{3.2.30}$$

where

$$e_{ij} = \frac{1}{2} \left(\frac{\partial u^i}{\partial x^j} + \frac{\partial u^j}{\partial x^i} \right). \quad \blacklozenge \tag{3.2.31}$$

Exercises

3.2-1 **(a)** *Let* $\varphi : \mathbb{R}^{n+1} \to \mathbb{R}$. *Show directly that the sine-Gordon equation*

$$\frac{\partial^2 \varphi}{\partial t^2} - \nabla^2 \varphi + \sin \varphi = 0$$

are the Euler-Lagrange equations of a suitable Lagrangian.

(b) *Let* $\varphi : \mathbb{R}^{n+1} \to \mathbb{C}$. *Write the nonlinear Schrödinger equation*

$$i \frac{\partial \varphi}{\partial t} + \nabla^2 \varphi + \beta \varphi |\varphi|^2 = 0$$

as a Hamiltonian system.

3.2-2 *Find a "soliton" solution for the sine-Gordon equation*

$$\frac{\partial^2 \varphi}{\partial t^2} - \frac{\partial^2 \varphi}{\partial x^2} + \sin \varphi = 0$$

in one-space dimension.

3.2-3 *Consider the complex nonlinear Schrödinger equation in one-space dimension*

$$i \frac{\partial \varphi}{\partial t} + \frac{\partial^2 \varphi}{\partial x^2} + \beta \varphi |\varphi|^2 = 0, \quad \beta \neq 0.$$

(a) *Show that the function* $\psi : \mathbb{R} \to \mathbb{C}$ *defining the traveling wave solution* $\varphi(x, t) = \psi(x - ct)$ *for* $c > 0$ *satisfies a second-order differential equation equivalent to a Hamiltonian system in* $\mathbb{R}^4$ *relative to the non-canonical symplectic form whose matrix is given by*

$$\mathbb{J}_c = \begin{bmatrix} 0 & c & 1 & 0 \\ -c & 0 & 0 & 1 \\ -1 & 0 & 0 & 0 \\ 0 & -1 & 0 & 0 \end{bmatrix}.$$

(See Exercise **2.6-1***.)*

(b) *Analyze the equilibria of the resulting Hamiltonian system in $\mathbb{R}^4$ and determine their linear stability properties.*

(c) *Let $\psi(s) = e^{ics/2}a(s)$ for a real function $a(s)$ and determine a second-order equation for $a(s)$. Show that the resulting equation is Hamiltonian and has heteroclinic orbits for $\beta < 0$. Find them.*

(d) *Find "soliton" solutions for the complex nonlinear Schrödinger equation.*

3.3 Examples: Poisson Brackets and Conserved Quantities

(a) Schrödinger Bracket In Example **(b)** of §3.2, we saw that if H_{op} is a self-adjoint complex linear operator on a Hilbert space $\mathcal{H}$, then $A = H_{\mathrm{op}}/i\hbar$ is Hamiltonian and the corresponding energy function H_A is the expectation value $\langle H_{\mathrm{op}}\rangle$ of H_{op}. Letting H_{op} and K_{op} be two such operators, and applying the Poisson bracket-commutator correspondence (2.9.10), or a direct calculation, we get

$$\{\langle H_{\mathrm{op}}\rangle, \langle K_{\mathrm{op}}\rangle\} = \langle [H_{\mathrm{op}}, K_{\mathrm{op}}]\rangle . \tag{3.3.1}$$

In other words, *the expectation value of the commutator is the Poisson bracket of the expectation values.*

Notice that if we take $K_{\mathrm{op}}\psi = \psi$, the identity operator, the corresponding Hamiltonian function is $p(\psi) = \|\psi\|^2$ and from (3.3.1) we see that p is a conserved quantity for any choice of H_{op}, a fact that is central to the probabilistic interpretation of quantum mechanics. Later we shall see that p is the conserved quantity associated to the ***phase symmetry*** $\psi \mapsto e^{i\theta}\psi$.

More generally, if F and G are two functions on $\mathcal{H}$ with $\delta F/\delta\psi = \nabla F$, the gradient of F taken relative to the real inner product $\mathrm{Re}\langle\, ,\rangle$ on H, one finds that

$$X_F = \frac{1}{2i\hbar}\nabla F \tag{3.3.2}$$

and

$$\{F, G\} = -\frac{1}{2\hbar}\,\mathrm{Im}\,\langle\nabla F, \nabla G\rangle . \tag{3.3.3}$$

Notice that (3.3.2), (3.3.3), and $\mathrm{Im}\, z = -\mathrm{Re}(iz)$ give

$$\begin{aligned}
\mathbf{d}F \cdot X_G &= \mathrm{Re}\,\langle\nabla F, X_G\rangle = \frac{1}{2\hbar}\mathrm{Re}\,\langle\nabla F, -i\nabla G\rangle \\
&= \frac{1}{2\hbar}\mathrm{Re}\,\langle i\nabla F, \nabla G\rangle = -\frac{1}{2\hbar}\mathrm{Im}\,\langle\nabla F, \nabla G\rangle \\
&= \{F, G\}
\end{aligned}$$

as expected. ♦

(b) KdV Bracket Using the definition of the bracket (2.9.1), the symplectic structure, and the Hamiltonian vector field formula from Example **(c)** of §3.2, one finds that

$$\{F, G\} = \int_{-\infty}^{\infty} \frac{\delta F}{\delta u} \frac{\partial}{\partial x} \left(\frac{\delta G}{\delta u} \right) dx \tag{3.3.4}$$

for functions F, G of u having functional derivatives that vanish at $\pm\infty$. ♦

(c) Linear and Angular Momentum for the Wave Equation The wave equation on $\mathbb{R}^3$ discussed in Example **(a)** of §3.2 has the Hamiltonian

$$H(\varphi, \pi) = \int_{\mathbb{R}^3} \left[\frac{1}{2}(\pi')^2 + \frac{1}{2}\|\nabla\varphi\|^2 + U(\varphi) \right] d^3x. \tag{3.3.5}$$

Define the ***linear momentum*** in the x-direction by

$$P_x(\varphi) = \int \pi' \frac{\partial \varphi}{\partial x} \, d^3x. \tag{3.3.6}$$

By (3.3.6), $\delta P_x/\delta\pi = \partial\varphi/\partial x$, and $\delta P_x/\delta\varphi = (-\partial\pi'/\partial x)\, d^3x$, so we get from (3.2.2)

$$\begin{aligned} \{H, P_x\}(\varphi, \pi) &= \int_{\mathbb{R}^3} \left(\frac{\delta P_x}{\delta \pi} \frac{\delta H}{\delta \varphi} - \frac{\delta H}{\delta \pi} \frac{\delta P_x}{\delta \varphi} \right) \\ &= \int_{\mathbb{R}^3} \left[\frac{\partial \varphi}{\partial x}(-\nabla^2\varphi + U'(\varphi)) + \pi' \frac{\partial \pi'}{\partial x} \right] d^3x \\ &= \int_{\mathbb{R}^3} \frac{\partial}{\partial x} \left[\frac{1}{2}\|\nabla\varphi\|^2 + U(\varphi) + \frac{1}{2}(\pi')^2 \right] d^3x = 0 \end{aligned} \tag{3.3.7}$$

assuming the fields and U vanish appropriately at ∞. Thus, P_x is conserved. The conservation of P_x is connected with invariance of H under translations in the x-direction. Deeper insights into this connection are explored later. Of course, similar conservation laws hold in the y- and z-directions.

Likewise, the angular momenta $\mathbf{J} = (J_x, J_y, J_z)$, where, for example,

$$J_z(\varphi) = \int_{\mathbb{R}^3} \pi' \left(x\frac{\partial}{\partial y} - y\frac{\partial}{\partial x} \right) \varphi \, d^3x \tag{3.3.8}$$

are constants of the motion. This is proved in an analogous way. (For precise function spaces in which these operations can be justified, see Chernoff and Marsden [1974].) ♦

(d) Linear and Angular Momentum for the Schrödinger Equation

i In Example **(b)** of §3.3, assume that $\mathcal{H}$ is the space of complex-valued L^2-functions on $\mathbb{R}^3$ and that the self-adjoint linear operator $H_{\rm op}\colon \mathcal{H} \to \mathcal{H}$ commutes with infinitesimal translations of the argument by a fixed vector $\xi \in \mathbb{R}^3$, that is, $H_{\rm op}(\mathbf{D}\psi(\cdot)\cdot\xi) = \mathbf{D}(H_{\rm op}\psi(\cdot))\cdot\xi$ for any ψ whose derivative is in $\mathcal{H}$. One checks, using (3.3.1) that

$$P_\xi(\psi) = \left\langle \frac{i}{\hbar}\mathbf{D}\psi\cdot\xi, \psi \right\rangle. \tag{3.3.9}$$

Poisson commutes with $\langle H_{\rm op}\rangle$. If ξ is the unit vector along the x-axis, the corresponding conserved quantity is

$$P_x(\psi) = \left\langle \frac{i}{\hbar}\frac{\partial\psi}{\partial x}, \psi \right\rangle.$$

ii Assume that $H_{\rm op}\colon \mathcal{H} \to \mathcal{H}$ commutes with infinitesimal rotations by a fixed skew-symmetric 3×3 matrix $\hat{\omega}$, that is,

$$H_{\rm op}(\mathbf{D}\psi(x)\cdot\hat{\omega}x) = \mathbf{D}((H_{\rm op}\psi)(x))\cdot\hat{\omega}x \tag{3.3.10}$$

for every ψ whose derivative is in $\mathcal{H}$, where, on the left-hand side, $H_{\rm op}$ is thought of as acting on the function $x \mapsto \mathbf{D}\psi(x)\cdot\hat{\omega}x$. Then the angular momentum function

$$\mathbf{J}(\hat{\omega}) : y \mapsto \langle i\mathbf{D}\psi(x)\cdot\hat{\omega}(x)/\hbar, \psi(x)\rangle. \tag{3.3.11}$$

Poisson commutes with $\mathcal{H}$ so is a conserved quantity. If we choose $\omega = (0,0,1)$; that is,

$$\hat{\omega} = \begin{bmatrix} 0 & -1 & 0 \\ 1 & 0 & 0 \\ 0 & 0 & 0 \end{bmatrix},$$

this corresponds to an infinitesimal rotation around the z-axis. Explicitly, the angular momentum around the x^l-axis is given by

$$J_l(\psi) = \left\langle \frac{i}{\hbar}\left(x^j\frac{\partial\psi}{\partial x^k} - x^k\frac{\partial\psi}{\partial x^j}\right), \psi \right\rangle,$$

where (j,k,l) is a cyclic permutation of $(1,2,3)$. ♦

(e) Linear and Angular Momentum for Linear Elastodynamics Consider again the equations of linear elastodynamics; see Example **(f)** of §3.2. Observe that the Hamiltonian is invariant under translations if the elasticity tensor $\mathbf{c}$ is homogeneous (independent of (x, y, z)); the corresponding conserved linear momentum in the x-direction is

$$P_x = \int_{\mathbb{R}^3} \rho \dot{\mathbf{u}} \cdot \frac{\partial \mathbf{u}}{\partial x} d^3x. \tag{3.3.12}$$

Likewise the Hamiltonian is invariant under rotations if $\mathbf{c}$ is isotropic; that is, invariant under rotations, which is equivalent to $\mathbf{c}$ having the form

$$c^{ijkl} = \mu(\delta^{ik}\delta^{jl} + \delta^{il}\delta^{jk}) + \lambda\delta^{ij}\delta^{kl},$$

where μ and λ are constants (see Marsden and Hughes [1983], §4.3, for the proof). The conserved angular momentum about the z-axis is

$$J = \int_{\mathbb{R}^3} \rho \dot{\mathbf{u}} \cdot \left(x\frac{\partial \mathbf{u}}{\partial y} - y\frac{\partial \mathbf{u}}{\partial x} \right) d^3x. \quad \blacklozenge \tag{3.3.13}$$

In Chapter 11, we will gain a deeper insight into the significance and construction of these conserved quantities.

Some Technicalities for Infinite-Dimensional Systems In general, unless the symplectic form on the Banach space Z is strong, the Hamiltonian vector field X_H is *not* defined on the whole of Z but only on a dense subspace. For example, in the case of the wave equation $\partial^2\varphi/\partial t^2 = \nabla^2\varphi - U'(\varphi)$, a possible choice of phase space is $H^1(\mathbb{R}^3) \times L^2(\mathbb{R}^3)$, but X_H is defined only on the dense subspace $H^2(\mathbb{R}^3) \times H^1(\mathbb{R}^3)$. It can also happen that the Hamiltonian H is not even defined on the whole of Z. For example, if $H_{\mathrm{op}} = \nabla^2 + V$ for the Schrödinger equation on $L^2(\mathbb{R}^3)$, then H could have domain containing $H^2(\mathbb{R}^3)$ which coincides with the domain of the Hamiltonian vector field iH_{op}. If V is singular, the domain need not be exactly $H^2(\mathbb{R}^3)$. As a quadratic form, H might be extendable to $H^1(\mathbb{R}^3)$. See Reed and Simon [1974, Volume II] or Kato [1984] for details.

The problem of existence and even uniqueness of solutions can be quite delicate. For linear systems one often appeals to Stone's theorem for the Schrödinger and wave equations, and to the Hille-Yosida theorem in the case of more general linear systems. We refer to Marsden and Hughes [1983], Chapter 6, for the theory and examples. In the case of nonlinear Hamiltonian systems, the theorems of Segal [1962], Kato [1975], and Hughes, Kato, and Marsden [1977] are relevant.

For infinite-dimensional nonlinear Hamiltonian systems technical differentiability conditions on its flow φ_t are needed to ensure that each φ_t is a symplectic map; see Chernoff and Marsden [1974], and especially Marsden and Hughes [1983], Chapter 6. These technicalities really are needed in many interesting examples. In the next section, we prove versions of these theorems with simplified hypotheses. $\blacklozenge$

Exercises

3.3-1 *Show that* $\{F_i, F_j\} = 0,\ i, j = 0, 1, 2, 3,$ *where the Poisson bracket is the KdV bracket and where:*

$$\begin{aligned}
F_0(u) &= \int_{-\infty}^{\infty} u\, dx \\
F_1(u) &= \int_{-\infty}^{\infty} \frac{1}{2} u^3\, dx \\
F_2(u) &= \int_{-\infty}^{\infty} \left(-u^3 + \frac{1}{2}(u_x)^2 \right) dx \qquad \textit{(the KdV Hamiltonian)} \\
F_3(u) &= \int_{-\infty}^{\infty} \left(\frac{5}{2} u^4 + 5uu_x^2 + (u_{xx})^2 \right) dx.
\end{aligned}$$

4
Interlude: Manifolds, Vector Fields, and Differential Forms

In preparation for later chapters, it will be necessary for the reader to learn a little bit about manifold theory. We recall a few basic facts here, beginning with the finite-dimensional case. (See Abraham, Marsden, and Ratiu [1988] for a full account.) The reader should not attempt to master all of this material now, but rather read through it for general sense and come back to it repeatedly as our development of mechanics proceeds.

4.1 Manifolds

Given a set M, a ***chart*** on M is an open set U in Euclidean space $\mathbb{R}^n$ with coordinates $(x^1, \ldots, x^n)$ (more generally U can be open in a Banach space) together with a one-to-one map φ of U onto some subset of M,

$$\varphi : U \to \varphi(U) \subset M.$$

We call M a ***differentiable manifold*** if the following hold:

M1 *It is covered by a collection of charts, that is, every point is represented in at least one chart.*

M2 *If two charts U, U' have an overlapping image in M, then $V = \varphi^{-1}(\varphi(U) \cap \varphi'(U'))$ and $V' = (\varphi')^{-1}(\varphi(U) \cap \varphi'(U'))$ are open sets in $\mathbb{R}^n$. Hence the mapping $\varphi'^{-1} \circ \varphi : V \to V'$ from an open subset of $\mathbb{R}^n$ to a subset of $\mathbb{R}^n$ is defined (Figure 4.1.1). The charts U, U' are called* ***compatible*** *if these n functions of n variables $\varphi'^{-1} \circ \varphi$ are C^∞.*

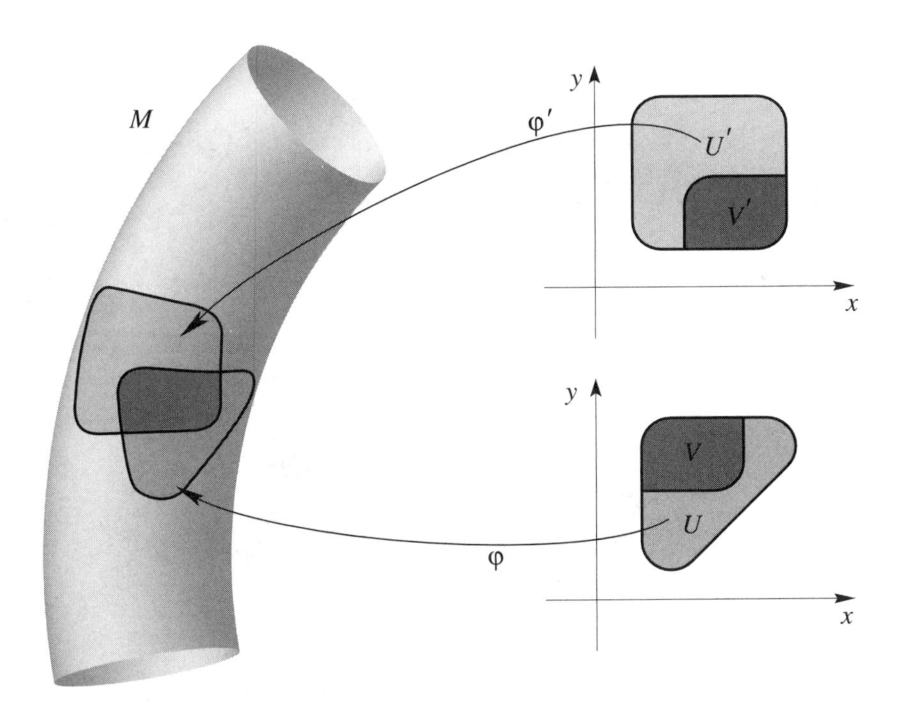

FIGURE 4.1.1. Overlapping charts on a manifold.

M3 *M has an **atlas**; that is, M can be written as a union of compatible charts.*

Two atlases are called ***equivalent*** if their union is also an atlas. One often rephrases the definition by saying that a differentiable structure on a manifold is an equivalence class of atlases.

A ***neighborhood*** of a point x in a manifold M is the image under a map $\varphi : U \to M$ of a neighborhood of the representation of x in a chart U. Neighborhoods define open sets and one checks that the open sets in M define a topology. *Usually we assume without explicit mention that the topology is Hausdorff*: two different points x, x' in M have nonintersecting neighborhoods. A differentiable manifold M is called an n-***manifold*** if every chart has domain in an n-dimensional vector space.

Another useful viewpoint is to think of M as a set covered by a collection of coordinate charts with local coordinates $(x^1, \ldots, x^n)$ with the property that all mutual changes of coordinates are smooth maps.

Two curves $t \mapsto c_1(t)$ and $t \mapsto c_2(t)$ in an n-manifold M are called ***equivalent at*** x if

$$c_1(0) = c_2(0) = x \quad \text{and} \quad (\varphi^{-1} \circ c_1)'(0) = (\varphi^{-1} \circ c_2)'(0)$$

in some chart φ. It is easy to check that this definition is chart independent.

A ***tangent vector*** v to a manifold M at a point $x \in M$ is an equivalence class of curves at x. One proves that the set of tangent vectors to M at x forms a vector space. It is denoted T_xM and is called the ***tangent space*** to M at $x \in M$. Given a curve $c(t)$, we denote by $c'(s)$ the tangent vector at $c(s)$ defined by the equivalence class of $t \mapsto c(s+t)$ at $t = 0$.

Let U be a chart of an atlas for the manifold M with coordinates $(x^1, \ldots, x^n)$. The ***components*** of the tangent vector v to the curve $t \mapsto (\varphi^{-1} \circ c)(t)$ are the numbers $v^1, \ldots, v^n$ defined by

$$v^i = \frac{d}{dt}(\varphi^{-1} \circ c)^i \bigg|_{t=0},$$

where $i = 1, \ldots, n$. The ***tangent bundle*** of M, denoted by TM, is the differentiable manifold whose underlying set is the disjoint union of the tangent spaces to M at the points $x \in M$, that is,

$$TM = \bigcup_{x \in M} T_xM.$$

Thus, a point of TM is a vector v that is tangent to M at some point $x \in M$. To define the differentiable structure on TM, we need to specify how to construct local coordinates on TM. To do this, let $x^1, \ldots, x^n$ be local coordinates on M and let $v^1, \ldots, v^n$ be components of a tangent vector in this coordinate system. Then the $2n$ numbers $x^1, \ldots, x^n, v^1, \ldots, v^n$ give a local coordinate system on TM. Notice that $\dim TM = 2 \dim M$.

The ***natural projection*** is the map $\tau_M : TM \to M$ that takes a tangent vector v to the point $x \in M$ at which the vector v is attached (that is, $v \in T_xM$). The inverse image $\tau_M^{-1}(x)$ of a point $x \in M$ under the natural projection τ_M is the tangent space T_xM. This space is called the ***fiber*** of the tangent bundle over the point $x \in M$.

Let $f : M \to N$ be a map of a manifold M to a manifold N. We call f ***differentiable*** (or C^k) if in local coordinates on M and N it is given by differentiable (or C^k) functions. The ***derivative*** of a differentiable map $f : M \to N$ at a point $x \in M$ is defined to be the linear map

$$T_xf : T_xM \to T_{f(x)}N$$

constructed in the following way. For $v \in T_xM$, choose a curve $c :]-\epsilon, \epsilon[\to M$ with $c(0) = x$, and velocity vector $dc/dt\,|_{t=0} = v$. Then $T_xf \cdot v$ is the velocity vector at $t = 0$ of the curve $f \circ c : \mathbb{R} \to N$, that is,

$$T_xf \cdot v = \frac{d}{dt} f(c(t)) \bigg|_{t=0}.$$

The vector $T_xf \cdot v$ does not depend on the curve c but only on the vector v. If M and N are manifolds and $f : M \to N$ is of class C^{r+1}, then

$Tf : TM \to TN$ is a mapping of class C^r. Note that

$$\left.\frac{dc}{dt}\right|_{t=0} = T_0 c \cdot 1.$$

A ***vector field*** X on a manifold M is a map $X : M \to TM$ that assigns a vector $X(x)$ at the point $x \in M$; that is, $\tau_M \circ X = \text{identity}$. An ***integral curve*** of X with initial condition x_0 at $t = 0$ is a (differentiable) map $c :]a, b[\to M$ such that $]a, b[$ is an open interval containing $0, c(0) = x_0$ and

$$c'(t) = X(c(t))$$

for all $t \in]a, b[$. In formal presentations we usually suppress the domain of definition, even though this is technically important. The ***flow*** of X is the collection of maps

$$\varphi_t : M \to M$$

such that $t \mapsto \varphi_t(x)$ is the integral curve of X with initial condition x. Existence and uniqueness theorems from ordinary differential equations guarantee φ is smooth in x and t (where defined) if X is. From uniqueness, we get the ***flow property***

$$\varphi_{t+s} = \varphi_t \circ \varphi_s$$

along with the initial conditions $\varphi_0 = \text{identity}$. The flow property generalizes the situation where $M = V$ is a *linear* space, $X(x) = Ax$ for a (bounded) *linear* operator A, and where

$$\varphi_t(x) = e^{tA}x$$

to the *nonlinear* case.

If $f : M \to \mathbb{R}$ is a smooth function, we can differentiate it at any point $x \in M$ to obtain a map $T_x f : T_x M \to T_{f(x)}\mathbb{R}$. Identifying the tangent space of $\mathbb{R}$ at any point with itself (a process we usually do in any vector space), we get a linear map $\mathbf{d}f(x) : T_x M \to \mathbb{R}$. That is, $\mathbf{d}f(x) \in T_x^* M$, the dual of the vector space $T_x M$.

In coordinates, the ***directional derivatives*** defined by $\mathbf{d}f(x) \cdot v$, where $v \in T_x M$, are given by

$$\mathbf{d}f(x) \cdot v = \sum_{i=1}^{n} \frac{\partial f}{\partial x^i} v^i.$$

We will employ the ***summation convention*** and drop the summation sign when there are repeated indices. We also call $\mathbf{d}f$ the ***differential*** of f.

One can show that specifying the directional derivatives completely determines a vector, and so we can identify a basis of T_xM using the operators $\partial/\partial x^i$. We write

$$(e_1, \ldots, e_n) = \left(\frac{\partial}{\partial x^1}, \ldots, \frac{\partial}{\partial x^n} \right)$$

for this basis so that $v = v^i \partial/\partial x^i$.

If we replace each vector space T_xM with its dual T_x^*M, we obtain a new $2n$-manifold called the ***cotangent bundle*** and denoted T^*M. The dual basis to $\partial/\partial x^i$ is denoted dx^i. Thus, relative to a choice of local coordinates we get the basic formula

$$\mathbf{d}f(x) = \frac{\partial f}{\partial x^i} dx^i$$

for any smooth function $f : M \to \mathbb{R}$.

Exercises

4.1-1 *Show that the two-sphere $S^2 \subset \mathbb{R}^3$ is a 2-manifold.*

4.1-2 *If $\varphi_t : S^2 \to S^2$ rotates points on S^2 through an angle t, show that φ_t is the flow of a certain vector field on S^2.*

4.1-3 *Let $f : S^2 \to \mathbb{R}$ be defined by $f(x, y, z) = z$. Compute $\mathbf{d}f$ relative to spherical coordinates (θ, φ).*

4.2 Differential Forms

We next review some of the basic definitions, properties, and operations on differential forms, without proofs (see Abraham, Marsden, and Ratiu [1988] and references therein). *The main idea of differential forms is to provide a generalization of the basic operations of vector calculus*, div, grad, *and* curl, *and the integral theorems of Green, Gauss, and Stokes to manifolds of arbitrary dimension.*

A ***2-form*** Ω on a manifold M is a function $\Omega(x) : T_xM \times T_xM \to \mathbb{R}$ that assigns to each point $x \in M$ a skew-symmetric bilinear form on the tangent space T_xM to M at x. More generally, a ***k-form*** α (sometimes called a ***differential form of degree*** k) on a manifold M is a function $\alpha(x) : T_xM \times \ldots \times T_xM$ (there are k factors) $\to \mathbb{R}$ that assigns to each point $x \in M$ a skew-symmetric k-multilinear map on the tangent space T_xM to M at x. Without the skew-symmetry assumption, α would be called a $(0, k)$-***tensor***. A map $\alpha : V \times \ldots \times V$ (there are k factors) $\to \mathbb{R}$ is

multilinear when it is linear in each of its factors, that is,

$$\alpha(v_1, \ldots, av_j + bv_j', \ldots, v_k) = a\alpha(v_1, \ldots, v_j, \ldots, v_k) + b\alpha(v_1, \ldots, v_j', \ldots, v_k)$$

for all j with $1 \leq j \leq k$. A k-multilinear map $\alpha : V \times \ldots \times V \to \mathbb{R}$ is ***skew*** (or ***alternating***) when it changes sign whenever two of its arguments are interchanged, that is, for all $v_1, \ldots, v_k \in V$,

$$\alpha(v_1, \ldots, v_i, \ldots, v_j, \ldots, v_k) = -\alpha(v_1, \ldots, v_j, \ldots, v_i, \ldots, v_k).$$

Let $x^1, \ldots, x^n$ denote coordinates on M, let

$$\{e_1, \ldots, e_n\} = \{\partial/\partial x^1, \ldots, \partial/\partial x^n\}$$

be the corresponding basis for T_xM, and let $\{e^1, \ldots, e^n\} = \{dx^1, \ldots, dx^n\}$ be the dual basis for T_x^*M. Then at each $x \in M$, we can write a 2-form as

$$\Omega_x(v, w) = \Omega_{ij}(x)v^i w^j, \quad \text{where} \quad \Omega_{ij}(x) = \Omega_x\left(\frac{\partial}{\partial x^i}, \frac{\partial}{\partial x^j}\right),$$

and more generally a k-form can be written

$$\alpha_x(v_1, \ldots, v_k) = \alpha_{i_1 \ldots i_k}(x) v_1^{i_1} \ldots v_k^{i_k},$$

where there is a sum on $i_1, \ldots, i_k$ and where

$$\alpha_{i_1 \ldots i_k}(x) = \alpha_x\left(\frac{\partial}{\partial x^{i_1}}, \ldots, \frac{\partial}{\partial x^{i_k}}\right),$$

and where $v_i = v_i^j \partial/\partial x^j$, with a sum on j.

If α is a $(0,k)$-tensor on a manifold M, and β is a $(0,l)$-tensor, their ***tensor product*** $\alpha \otimes \beta$ is the $(0, k+l)$-tensor on M defined by

$$(\alpha \otimes \beta)_x(v_1, \ldots, v_{k+l}) = \alpha_x(v_1, \ldots, v_k)\beta_x(v_{k+1}, \ldots, v_{k+l}) \tag{4.2.1}$$

at each point $x \in M$.

If t is a $(0,p)$-tensor, define the ***alternation operator*** $\mathbf{A}$ acting on t by

$$\mathbf{A}(t)(v_1, \ldots, v_p) = \frac{1}{p!} \sum_{\pi \in S_p} \operatorname{sgn}(\pi) t(v_{\pi(1)}, \ldots, v_{\pi(p)}), \tag{4.2.2}$$

where $\operatorname{sgn}(\pi)$ is the ***sign*** of the permutation π:

$$\operatorname{sgn}(\pi) = \begin{cases} +1 & \text{if } \pi \text{ is even}, \\ -1 & \text{if } \pi \text{ is odd}, \end{cases} \tag{4.2.3}$$

and S_p is the group of all permutations of the numbers $1, 2, \ldots, p$. The operator $\mathbf{A}$ therefore skew-symmetrizes p-multilinear maps.

If α is a k-form and β is an l-form on M, their ***wedge product*** $\alpha \wedge \beta$ is the $(k+l)$-form on M defined by

$$\alpha \wedge \beta = \frac{(k+l)!}{k!\, l!} \mathbf{A}(\alpha \otimes \beta). \tag{4.2.4}$$

[**Note**: The numerical factor in (4.2.4) agrees with the convention of Abraham and Marsden [1978], Abraham, Marsden, and Ratiu [1988], and Spivak [1976], but *not* that of Arnold [1989], Guillemin and Pollack [1974], or Kobayashi and Nomizu [1963]; it is the Bourbaki [1971] convention.]

For example, if α and β are one-forms,

$$(\alpha \wedge \beta)(v_1, v_2) = \alpha(v_1)\beta(v_2) - \alpha(v_2)\beta(v_1)$$

while if α is a 2-form and β is a 1-form,

$$(\alpha \wedge \beta)(v_1, v_2, v_3) = \alpha(v_1, v_2)\beta(v_3) + \alpha(v_3, v_1)\beta(v_2) + \alpha(v_2, v_3)\beta(v_1).$$

We state the following without proof:

Proposition 4.2.1 *The wedge product has the following properties:*

i *$\alpha \wedge \beta$ is **associative**: $\alpha \wedge (\beta \wedge \gamma) = (\alpha \wedge \beta) \wedge \gamma$.*

ii *$\alpha \wedge \beta$ is **bilinear** in α, β:*

$$\begin{aligned} (a\alpha_1 + b\alpha_2) \wedge \beta &= a(\alpha_1 \wedge \beta) + b(\alpha_2 \wedge \beta), \\ \alpha \wedge (c\beta_1 + d\beta_2) &= c(\alpha \wedge \beta_1) + d(\alpha \wedge \beta_2). \end{aligned}$$

iii *$\alpha \wedge \beta$ is **anticommutative**: $\alpha \wedge \beta = (-1)^{kl} \beta \wedge \alpha$, where α is a k-form and β is an l-form.*

In terms of the dual basis dx^i, any k-form can be written locally as

$$\alpha = \alpha_{i_1 \ldots i_k} dx^{i_1} \wedge \cdots \wedge dx^{i_k}$$

where the sum is over all i_j satisfying $i_1 < \cdots < i_k$.

Let $\varphi : M \to N$ be a C^∞ map from the manifold M to the manifold N and α be a k-form on N. Define the ***pull back*** $\varphi^*\alpha$ of α by φ to be the k-form on M given by

$$(\varphi^*\alpha)_x(v_1, \ldots, v_k) = \alpha_{\varphi(x)}(T_x\varphi \cdot v_1, \ldots, T_x\varphi \cdot v_k). \tag{4.2.5}$$

If φ is a diffeomorphism, the ***push forward*** φ_* is defined by $\varphi_* = (\varphi^{-1})^*$.

Here is another basic property.

Proposition 4.2.2 *The pull back of a wedge product is the wedge product of the pull backs:*

$$\varphi^*(\alpha \wedge \beta) = \varphi^*\alpha \wedge \varphi^*\beta. \tag{4.2.6}$$

Let α be a k-form on a manifold M and X a vector field. The ***interior product*** $\mathbf{i}_X\alpha$ (sometimes called the contraction of X and α, and written $\mathbf{i}(X)\alpha$) is defined by

$$(\mathbf{i}_X\alpha)_x(v_2, \ldots, v_k) = \alpha_x(X(x), v_2, \ldots, v_k). \tag{4.2.7}$$

Proposition 4.2.3 *Let α be a k-form and β an l-form on a manifold M. Then*

$$\mathbf{i}_X(\alpha \wedge \beta) = (\mathbf{i}_X\alpha) \wedge \beta + (-1)^k \alpha \wedge (\mathbf{i}_X\beta). \tag{4.2.8}$$

The ***exterior derivative*** $\mathbf{d}\alpha$ of a k-form α on a manifold M is the $(k+1)$-form on M determined by the following proposition:

Proposition 4.2.4 *There is a unique mapping* $\mathbf{d}$ *from k-forms on M to $(k+1)$-forms on M such that:*

i *If α is a 0-form ($k = 0$), that is, $\alpha = f \in C^\infty(M)$, then $\mathbf{d}f$ is the one-form which is the differential of f.*

ii $\mathbf{d}\alpha$ *is **linear** in α, that is,*

$$\mathbf{d}(c_1\alpha_1 + c_2\alpha_2) = c_1\mathbf{d}\alpha_1 + c_2\mathbf{d}\alpha_2 \quad \textit{for} \quad c_1, c_2, \in \mathbb{R}.$$

iii $\mathbf{d}\alpha$ *satisfies the **product rule**, that is,*

$$\mathbf{d}(\alpha \wedge \beta) = \mathbf{d}\alpha \wedge \beta + (-1)^k \alpha \wedge \mathbf{d}\beta,$$

where α is a k-form and, β is an l-form.

iv $\mathbf{d}^2 = 0$, *that is,* $\mathbf{d}(\mathbf{d}\alpha) = 0$ *for any k-form α.*

v $\mathbf{d}$ *is a **local operator**, that is, $\mathbf{d}\alpha(x)$ only depends on α restricted to any open neighborhood of x; in fact, if U is open in M, then*

$$\mathbf{d}(\alpha|U) = (\mathbf{d}\alpha)|U.$$

If α is a k-form given in coordinates by

$$\alpha = \alpha_{i_1\ldots i_k} dx^{i_1} \wedge \cdots \wedge dx^{i_k} \quad (\text{sum on } i_1 < \cdots < i_k),$$

then the coordinate expression for the exterior derivative is

$$\mathbf{d}\alpha = \frac{\partial \alpha_{i_1\ldots i_k}}{\partial x^j} dx^j \wedge dx^{i_1} \wedge \cdots \wedge dx^{i_k} \quad (\text{sum on all } j \text{ and } i_1 < \cdots < i_k). \tag{4.2.9}$$

Formula (4.2.9) can be taken as the definition of the exterior derivative, provided one shows that (4.2.9) has the above-described properties and, correspondingly, is independent of the choice of coordinates.

Next is a useful proposition that, in essence, rests on the chain rule:

Proposition 4.2.5 *Exterior differentiation commutes with pull back, that is,*

$$\mathbf{d}(\varphi^*\alpha) = \varphi^*(\mathbf{d}\alpha), \tag{4.2.10}$$

where α is a k-form on a manifold N and φ is a smooth map from a manifold M to N.

A k-form α is called ***closed*** if $\mathbf{d}\alpha = 0$ and *exact* if there is a $(k-1)$-form β such that $\alpha = \mathbf{d}\beta$. By Proposition **4.2.4 ii** every exact form is closed. Exercise **4.4-2** gives an example of a closed nonexact one-form.

Proposition 4.2.6 (Poincaré Lemma) *A closed form is locally exact, that is, if $\mathbf{d}\alpha = 0$ there is a neighborhood about each point on which $\alpha = \mathbf{d}\beta$.*

See Exercise **4.2-5** for the proof.

The table below on "Vector calculus and differential forms" summarizes how forms are related to the usual operations of vector calculus. We now elaborate on a few items in this table. In item 4, note that

$$\mathbf{d}f = \frac{\partial f}{\partial x}dx + \frac{\partial f}{\partial y}dy + \frac{\partial f}{\partial z}dz = (\operatorname{grad} f)^\flat = (\nabla f)^\flat$$

which is equivalent to $\nabla f = (\mathbf{d}f)^\sharp$.

The Hodge star operator on $\mathbb{R}^3$ maps k-forms to $(3-k)$-forms and is uniquely determined by linearity and the properties in item 2. (This operator can be defined on general Riemannian manifolds; see Abraham, Marsden, and Ratiu [1988].)

In item 5, if we let $F = F_1\mathbf{e}_1 + F_2\mathbf{e}_2 + F_3\mathbf{e}_3$, so $F^\flat = F_1\,dx + F_2\,dy + F_3\,dz$, then,

$$\begin{aligned}
\mathbf{d}(F^\flat) &= \mathbf{d}F_1 \wedge dx + F_1\mathbf{d}(dx) + \mathbf{d}F_2 \wedge dy + F_2\mathbf{d}(dy) + \mathbf{d}F_3 \wedge dz \\
&\quad + F_3\mathbf{d}(dz) \\
&= \left(\frac{\partial F_1}{\partial x}dx + \frac{\partial F_1}{\partial y}dy + \frac{\partial F_1}{\partial z}dz\right) \wedge dx \\
&\quad + \left(\frac{\partial F_2}{\partial x}dx + \frac{\partial F_2}{\partial y}dy + \frac{\partial F_2}{\partial z}dz\right) \wedge dy \\
&\quad + \left(\frac{\partial F_3}{\partial x}dx + \frac{\partial F_3}{\partial y}dy + \frac{\partial F_3}{\partial z}dz\right) \wedge dz \\
&= -\frac{\partial F_1}{\partial y}dx \wedge dy + \frac{\partial F_1}{\partial z}dz \wedge dx + \frac{\partial F_2}{\partial x}dx \wedge dy - \frac{\partial F_2}{\partial z}dy \wedge dz \\
&\quad - \frac{\partial F_3}{\partial x}dz \wedge dx + \frac{\partial F_3}{\partial y}dy \wedge dz \\
&= \left(\frac{\partial F_2}{\partial x} - \frac{\partial F_1}{\partial y}\right) dx \wedge dy + \left(\frac{\partial F_1}{\partial z} - \frac{\partial F_3}{\partial x}\right) dz \wedge dx \\
&\quad + \left(\frac{\partial F_3}{\partial y} - \frac{\partial F_2}{\partial z}\right) dy \wedge dz.
\end{aligned}$$

Hence, using item 2,

$$\begin{aligned} *(\mathbf{d}(F^\flat)) &= \left(\frac{\partial F_2}{\partial x} - \frac{\partial F_1}{\partial y}\right) dz + \left(\frac{\partial F_1}{\partial z} - \frac{\partial F_3}{\partial x}\right) dy \\ &\quad + \left(\frac{\partial F_3}{\partial y} - \frac{\partial F_2}{\partial z}\right) dx, \\ (*(\mathbf{d}(F^\flat)))^\sharp &= \left(\frac{\partial F_3}{\partial y} - \frac{\partial F_2}{\partial z}\right) \mathbf{e}_1 + \left(\frac{\partial F_1}{\partial z} - \frac{\partial F_3}{\partial x}\right) \mathbf{e}_2 \\ &\quad + \left(\frac{\partial F_2}{\partial x} - \frac{\partial F_1}{\partial y}\right) \mathbf{e}_3 \\ &= \operatorname{curl} F = \nabla \times F. \end{aligned}$$

With reference to item 6, let $F = F_1\mathbf{e}_1 + F_2\mathbf{e}_2 + F_3\mathbf{e}_3$, so $F^\flat = F_1\, dx + F_2\, dy + F_3\, dz$. Thus $*(F^\flat) = F_1\, dy \wedge dz + F_2(-dx \wedge dz) + F_3\, dx \wedge dy$, and so

$$\begin{aligned} \mathbf{d}(*(F^\flat)) &= \mathbf{d}F_1 \wedge dy \wedge dz - \mathbf{d}F_2 \wedge dx \wedge dz + \mathbf{d}F_3 \wedge dx \wedge dy \\ &= \left(\frac{\partial F_1}{\partial x} dx + \frac{\partial F_1}{\partial y} dy + \frac{\partial F_1}{\partial z} dz\right) \wedge dy \wedge dz \\ &\quad - \left(\frac{\partial F_2}{\partial x} dx + \frac{\partial F_2}{\partial y} dy + \frac{\partial F_2}{\partial z} dz\right) \wedge dx \wedge dz \\ &\quad + \left(\frac{\partial F_3}{\partial x} dx + \frac{\partial F_3}{\partial y} dy + \frac{\partial F_3}{\partial z} dz\right) \wedge dx \wedge dy \\ &= \frac{\partial F_1}{\partial x} dx \wedge dy \wedge dz + \frac{\partial F_2}{\partial y} dx \wedge dy \wedge dz + \frac{\partial F_3}{\partial z} dx \wedge dy \wedge dz \\ &= \left(\frac{\partial F_1}{\partial x} + \frac{\partial F_2}{\partial y} + \frac{\partial F_3}{\partial z}\right) dx \wedge dy \wedge dz = (\operatorname{div} F) dx \wedge dy \wedge dz. \end{aligned}$$

Therefore $*(\mathbf{d}(*(F^\flat))) = \operatorname{div} F = \nabla \cdot F$.

The definition and properties of vector-valued forms are direct extensions of these for usual forms on vector spaces and manifolds. One can think of a vector-valued form as an array of usual forms (see Abraham, Marsden, and Ratiu [1988]).

Vector Calculus and Differential Forms

1. **Sharp and Flat** (Using standard coordinates in $\mathbb{R}^3$)
   (a) $v^\flat = v^1\,dx + v^2\,dy + v^3\,dz =$ one-form corresponding to the vector $v = v^1\mathbf{e}_1 + v^2\mathbf{e}_2 + v^3\mathbf{e}^3$.
   (b) $\alpha^\sharp = \alpha_1\mathbf{e}_1 + \alpha_2\mathbf{e}_2 + \alpha_3\mathbf{e}_3 =$ vector corresponding to the one-form $\alpha = \alpha_1\,dx + \alpha_2\,dy + \alpha_3\,dz$.
2. **Hodge Star Operator**
   (a) $*1 = dx \wedge dy \wedge dz$.
   (b) $*dx = dy \wedge dz$, $*dy = -dx \wedge dz$, $*dz = dx \wedge dy$, $*(dy \wedge dz) = dx$, $*(dx \wedge dz) = -dy$, $*(dx \wedge dy) = dz$.
   (c) $*(dx \wedge dy \wedge dz) = 1$.
3. **Cross Product and Dot Product**
   (a) $v \times w = [*(v^\flat \wedge w^\flat)]^\sharp$.
   (b) $(v \cdot w)dx \wedge dy \wedge dz = v^\flat \wedge *(w^\flat)$.
4. **Gradient** $\nabla f = \mathrm{grad} f = (\mathbf{d}f)^\sharp$.
5. **Curl** $\nabla \times F = \mathrm{curl}\, F = [*(\mathbf{d}F^\flat)]^\sharp$.
6. **Divergence** $\nabla \cdot F = \mathrm{div}\, F = *\mathbf{d}(*F^\flat)$.

Exercises

4.2-1 *Let* $\varphi : \mathbb{R}^3 \to \mathbb{R}^2$ *be given by* $\varphi(x, y, z) = (x + z, xy)$. *For* $\alpha = e^v\,du + u\,dv \in \Omega^1(\mathbb{R}^2)$ *and* $\beta = u\,du \wedge dv$ *compute* $\alpha \wedge \beta, \varphi^*\alpha, \varphi^*\beta$, *and* $\varphi^*\alpha \wedge \varphi^*\beta$.

4.2-2 *Given*

$$\alpha = y^2\,dx \wedge dz + \sin(xy)\,dx \wedge dy + e^x\,dy \wedge dz \in \Omega^2(\mathbb{R}^3)$$

and

$$X = 3\partial/\partial x + \cos z\partial/\partial y - x^2\partial/\partial z \in \mathfrak{X}(\mathbb{R}^3),$$

compute $\mathbf{d}\alpha$ *and* $\mathbf{i}_X\alpha$.

4.2-3 **(a)** *Denote by $\Lambda^k(\mathbb{R}^n)$ the vector space of all skew-symmetric k-linear maps on $\mathbb{R}^n$. Prove that this space has dimension $n!/k!\,(n-k)!$ by showing that a basis is given by $\{e^{i_1} \wedge \cdots \wedge e^{i_k} \mid i_1 < \ldots < i_k\}$ where $\{e_1, \ldots, e_n\}$ is a basis of $\mathbb{R}^n$ and $\{e^1, \ldots, e^n\}$ is its dual basis, that is, $e^i(e_j) = \delta^i_j$.*

(b) *If $\mu \in \Lambda^n(\mathbb{R}^n)$ is nonzero, prove that the map $v \in \mathbb{R}^n \mapsto \mathbf{i}_v\mu \in \Lambda^{n-1}(\mathbb{R}^n)$ is an isomorphism.*

(c) *If M is a smooth n-manifold and $\mu \in \Omega^n(M)$ is nowhere vanishing (in which case it is called a volume form), show that the map $X \in \mathfrak{X}(M) \mapsto \mathbf{i}_X\mu \in \Omega^{n-1}(M)$ is a module isomorphism over $\mathcal{F}(M)$.*

4.2-4 *Let $\alpha = \alpha_i\, dx^i$ be a closed one-form in a ball around the origin in $\mathbb{R}^n$. Show that $\alpha = \mathbf{d}f$ for*

$$f(x^1, \ldots, x^n) = \int_0^1 \alpha_j(tx^1, \ldots, tx^n)x^j\, dt.$$

4.2-5 **(a)** *Let U be an open ball around the origin in $\mathbb{R}^n$ and $\alpha \in \Omega^k(U)$ a closed form. Verify that $\alpha = \mathbf{d}\beta$ for*

$$\beta(x^1, \ldots, x^n) = \left(\int_0^1 t^{k-1}\alpha_{ji_1\ldots i_{k-1}}(tx^1, \ldots, tx^n)x^j\, dt\right) dx^{i_1} \wedge \ldots \wedge dx^{i_{k-1}},$$

where the sum is over $i_1 < \cdots < i_{k-1}$, $\alpha = \alpha_{j_1\ldots j_k}\, dx^{j_1} \wedge \ldots \wedge dx^{j_k}$, $j_1 < \cdots < j_k$ and where α is extended to be skew-symmetric in its lower indices.

(b) *Deduce the Poincaré lemma from* **(a)**.

4.3 The Lie Derivative

The *dynamic definition* of the Lie derivative is as follows. Let α be a k-form and let X be a vector field with flow φ_t. The ***Lie derivative*** of α along X is given by

$$\pounds_X\alpha = \lim_{t\to 0}\frac{1}{t}[(\varphi_t^*\alpha) - \alpha] = \left.\frac{d}{dt}\varphi_t^*\alpha\right|_{t=0}. \tag{4.3.1}$$

Theorem 4.3.1 (Lie Derivative Theorem)

$$\frac{d}{dt}\varphi_t^*\alpha = \varphi_t^*\pounds_X\alpha. \tag{4.3.2}$$

This formula holds also for *time-dependent* vector fields.

If f is a real-valued function on a manifold M and X is a vector field on M, the ***Lie derivative of f along X*** is the ***directional derivative***

$$\pounds_X f = X[f] := \mathbf{d}f \cdot X. \tag{4.3.3}$$

If M is finite-dimensional,

$$\pounds_X f = X^i \frac{\partial f}{\partial x^i}. \tag{4.3.4}$$

If Y is a vector field on a manifold N and $\varphi : M \to N$ is a diffeomorphism, the ***pull back*** $\varphi^* Y$ is a vector field on M defined by

$$(\varphi^* Y)(x) = T_x \varphi^{-1} \circ Y \circ \varphi. \tag{4.3.5}$$

Two vector fields X on M and Y on N are said to be ***φ-related*** if

$$T\varphi \circ X = Y \circ \varphi. \tag{4.3.6}$$

Clearly, if $\varphi : M \to N$ is a diffeomorphism and Y is a vector field on N, $\varphi^* Y$ and Y are φ-related. For a diffeomorphism φ, the ***push forward*** is defined, as for forms, by $\varphi_* = (\varphi^{-1})^*$.

If M is finite dimensional and C^∞ then the set of vector fields on M coincides with the set of derivatives on $\mathcal{F}(M)$. The same result is true for C^k manifolds and vector fields if $k \geq 2$. This property is false for infinite-dimensional manifolds; see Abraham, Marsden, Ratiu [1988]. If M is C^∞ and smooth, then the derivation $f \mapsto X[Y[f]] - Y[X[f]]$, where $X[f] = \mathbf{d}f \cdot X$, determines a unique vector field denoted by $[X, Y]$ and called the ***Jacobi-Lie bracket*** of X and Y. Defining $\pounds_X Y = [X, Y]$ gives the ***Lie derivative*** of Y along X. Then the Lie derivative theorem (4.3.2) holds with α replaced by Y and the pull back operation given by (4.3.5).

If M is infinite-dimensional, then one defines the Lie derivative of Y along X by

$$\left.\frac{d}{dt}\right|_{t=0} \varphi_t^* Y = \pounds_X Y, \tag{4.3.7}$$

where φ_t is the flow of X. Then formula (4.3.2) with α replaced by Y holds and the action of the vector field $\pounds_X Y$ on a function f is given by $X[Y[f]] - Y[X[f]]$ which is denoted, as in the finite-dimensional case, $[X, Y][f]$. As before $[X, Y] = \pounds_X Y$ is also called the Jacobi-Lie bracket of vector fields.

If M is finite-dimensional,

$$(\pounds_X Y)^j = X^i \frac{\partial Y^j}{\partial x^i} - Y^i \frac{\partial X^j}{\partial x^i} = (X \cdot \nabla) Y^i - (Y \cdot \nabla) X^j, \tag{4.3.8}$$

and in general, where we identify X, Y with their local representatives

$$[X, Y] = \mathbf{D}Y \cdot X - \mathbf{D}X \cdot Y. \tag{4.3.9}$$

The *algebraic approach* to the Lie derivative on forms or tensors proceeds as follows. Extend the definition of the Lie derivative from functions and vector fields to differential forms, by requiring that the Lie derivative is a derivation; for example, for one-forms α, write

$$\pounds_X\langle\alpha, Y\rangle = \langle\pounds_X\alpha, Y\rangle + \langle\alpha, \pounds_X Y\rangle, \tag{4.3.10}$$

where X, Y are vector fields and $\langle\alpha, Y\rangle = \alpha(Y)$. More generally,

$$\pounds_X(\alpha(Y_1, \ldots, Y_k)) = (\pounds_X\alpha)(Y_1, \ldots, Y_k) + \sum_{i=1}^{k}\alpha(Y_1, \ldots, \pounds_X Y_i, \ldots, Y_k), \tag{4.3.11}$$

where $X, Y_1, \ldots, Y_k$ are vector fields and α is a k-form.

Proposition 4.3.2 *The dynamic and algebraic definitions of the Lie derivative of a differential k-form are equivalent.*

Cartan's Magic Formula

$$\pounds_X\alpha = \mathbf{d}\mathbf{i}_X\alpha + \mathbf{i}_X\mathbf{d}\alpha. \tag{4.3.12}$$

Another property of the Lie derivative is the following: if $\varphi : M \to N$ is a diffeomorphism,

$$\varphi^*\pounds_Y\beta = \pounds_{\varphi^*Y}\varphi^*\beta$$

for $Y \in \mathfrak{X}(N), \beta \in \Omega^k(M)$. More generally, if $X \in \mathfrak{X}(M)$ and $Y \in \mathfrak{X}(N)$ are ψ related, that is, $T\psi \circ X = Y \circ \psi$ for $\psi : M \to N$ a smooth map, then $\pounds_X\psi^*\beta = \psi^*\pounds_Y\beta$ for all $\beta \in \Omega^k(N)$.

An n-manifold M is said to be ***orientable*** if there is a nowhere vanishing n-form μ on it; μ is called a ***volume form*** and it is a basis of $\Omega^n(M)$ over $\mathcal{F}(M)$. Two volume forms μ_1 and μ_2 on M are said to define the same ***orientation*** if there is an $f \in \mathcal{F}(M)$, with $f > 0$ and such that $\mu_2 = f\mu_1$. Connected orientable manifolds admit precisely two orientations. A basis $\{v_1, \ldots v_n\}$ of T_mM is said to be ***positively oriented*** relative to the volume form μ on M if $\mu(m)(v_1, \ldots, v_n) > 0$. Note that the volume forms defining the same orientation form a convex cone in $\Omega^n(M)$, that is, if $a > 0$ and μ is a volume form, then $a\mu$ is again a volume form and if $t \in [0, 1]$ and μ_1, μ_2 are volume forms, then $t\mu_1 + (1-t)\mu_2$ is again a volume form. The first property is obvious. To prove the second, let $m \in M$ and let $\{\sigma_1, \ldots \sigma_n\}$ be a positively oriented basis of T_mM relative to the orientation defined by μ_1, or equivalently (by hypothesis) by μ. Then $\mu_1(m)(v_1, \ldots, v_n) > Q, \mu_2(m)(v_1, \ldots, v_n) > 0$ so that their convex combination is again strictly positive.

If $\mu \in \Omega^n(M)$ is a volume form, since $\pounds_X\mu \in \Omega^n(M)$ there is a function, called the ***divergence*** of X relative to μ and denoted $\operatorname{div}_\mu(X)$ or simply $\operatorname{div}(X)$, such that

$$\pounds_X\mu = \operatorname{div}_\mu(X)\mu. \tag{4.3.13}$$

From the dynamic approach to Lie derivatives it follows that $\operatorname{div}_\mu(X) = 0$ iff $F_t^*\mu = \mu$, where F_t is the flow of X. This condition says that F_t is ***volume preserving***. If $\varphi : M \to M$, since $\varphi^*\mu \in \Omega^n(M)$ there is a function, called the ***Jacobian*** of φ and denoted $J_\mu(\varphi)$ or simply $J(\varphi)$, such that

$$\varphi^*\mu = J_\mu(\varphi)\mu. \tag{4.3.14}$$

Thus, φ is *volume preserving iff* $J_\mu(\varphi) = 1$. From the inverse function theorem, we see that φ *is a local diffeomorphism iff* $J_\mu(\varphi) \neq 0$ *on* M.

There are a number of valuable identities relating the Lie derivative, the exterior derivative and the interior product. For example, if Θ is a one form and X and Y are vector fields, identity 6 in the following table gives

$$\mathbf{d}\Theta(X, Y) = X[\Theta(Y)] - Y[\Theta(X)] - \Theta([X, Y]). \tag{4.3.15}$$

Exercises

4.3-1 *Let M be an n-manifold, $\omega \in \Omega^n(M)$ a volume form, $X, Y \in \mathfrak{X}(M)$, and $f, g : M \to \mathbb{R}$ smooth functions such that $f(m) \neq 0$ for all m. Prove the following identities:*

(a) $\operatorname{div}_{f\omega}(X) = \operatorname{div}_\omega(X) + X[f]/f$;

(b) $\operatorname{div}_\omega(gX) = g \operatorname{div}_\omega(X) + X[g]$; *and*

(c) $\operatorname{div}_\omega([X, Y]) = X[\operatorname{div}_\omega(Y)] - Y[\operatorname{div}_\omega(X)]$.

4.3-2 *Show that the partial differential equation*

$$\frac{\partial f}{\partial t} = \sum_{i=1}^{n} X^i(x^1, \ldots, x^n) \frac{\partial f}{\partial x^i}$$

with initial condition $f(x, 0) = g(x)$ has the solution $f(x, t) = g(F_t(x))$, where F_t is the flow of the vector field $(X^1, \ldots, X^n)$ in $\mathbb{R}^n$ whose flow is assumed to exist for all time. Show that the solution is unique. Generalize this exercise to the equation

$$\frac{\partial f}{\partial t} = X[f]$$

for X a vector field on a manifold M.

4.3-3 *Show that if M and N are orientable manifolds, so is $M \times N$.*

4.4 Stokes' Theorem

Basically, one integrates an n-form ω on an oriented n-manifold by summing up the ordinary integrals of $f(x^1, \ldots, x^n)\, dx^1 \cdots dx^n$ in charts, where

$$\Omega = f(x^1, \ldots, x^n)\, dx^1 \wedge \cdots \wedge dx^n$$

is the local representative of ω, being careful not to count overlaps twice. The change of variables formula guarantees that the result is well defined.

Theorem 4.4.1 (Stokes' Theorem) *Suppose that M is a compact, oriented k-dimensional manifold with boundary ∂M. Let α be a smooth $(k-1)$-form on M. Then*

$$\int_M \mathbf{d}\alpha = \int_{\partial M} \alpha. \tag{4.4.1}$$

Special cases of Stokes' theorem are as follows:

The Integral Theorems of Calculus

(a) Fundamental Theorem of Calculus

$$\int_b^a f'(x)\, dx = f(b) - f(a). \tag{4.4.2}$$

(b) Green's Theorem For a region $\Omega \subset \mathbb{R}^2$:

$$\iint_\Omega \left(\frac{\partial Q}{\partial x} - \frac{\partial P}{\partial y} \right) dx\, dy = \int_{\partial\Omega} P\, dx + Q\, dy. \tag{4.4.3}$$

(c) Divergence Theorem For a region $\Omega \subset \mathbb{R}^3$:

$$\iiint_\Omega \operatorname{div} \mathbf{F}\, dV = \iint_{\partial\Omega} \mathbf{F} \cdot n\, dA. \tag{4.4.4}$$

(d) Classical Stokes' Theorem For a surface $S \subset \mathbb{R}^3$:

$$\begin{aligned}
\iint_S \bigg\{ & \left(\frac{\partial R}{\partial y} - \frac{\partial Q}{\partial z} \right) dy \wedge dz \\
& + \left(\frac{\partial P}{\partial z} - \frac{\partial R}{\partial x} \right) dz \wedge dx + \left(\frac{\partial Q}{\partial x} - \frac{\partial P}{\partial y} \right) dx \wedge dy \bigg\} \\
= & \iint_S \mathbf{n} \cdot \operatorname{curl} \mathbf{F}\, dA \\
= & \int_{\partial S} P\, dx + Q\, dy + R\, dz,
\end{aligned} \tag{4.4.5}$$

where $\mathbf{F} = (P, Q, R)$.

Notice that the Poincaré lemma generalizes the vector calculus theorems in $\mathbb{R}^3$ saying that if $\operatorname{curl} \mathbf{F} = 0$, then $\mathbf{F} = \nabla f$ and if $\operatorname{div} \mathbf{F} = 0$, then $\mathbf{F} = \nabla \times \mathbf{G}$. Recall that it states: *If α is closed, then locally α is exact; that is, if $\mathbf{d}\alpha = 0$, then locally $\alpha = \mathbf{d}\beta$ for some β.*

The failure of closed forms to be globally exact leads to the study of a very important topological invariant of M, the ***de Rham cohomology***. The kth de Rham cohomology group, denoted $H^k(M)$ is defined by

$$H^k(M) := \frac{\ker(\mathbf{d} : \Omega^k(M) \to \Omega^{k+1}(M))}{\operatorname{range}(\mathbf{d} : \Omega^{k-1}(M) \to \Omega^k(M))}.$$

The de Rham theorem states that these abelian groups are isomorphic to the so-called singular cohomology groups of M defined in algebraic topology in terms of simplexes and that depend only on the topological structure of M and not on its differentiable structure. The isomorphism is provided by integration and the fact that the integration map drops to the preceding quotient is guaranteed by Stokes' theorem. A useful particular case of this theorem is the following: if M is an orientable compact boundaryless n-manifold, then $\int_M \mu = 0$ if and only if the n-form μ is exact. This statement is equivalent to $H^n(M) = \mathbb{R}$.

Another basic result in integration theory is the global change of variables formula.

Theorem 4.4.2 (Change of Variables) *Let M and N be oriented n-manifolds and let $F : M \to N$ be an orientation-preserving diffeomorphism. If α is an n-form on N (with, say, compact support), then*

$$\int_M F^*\alpha = \int_N \alpha.$$

We also mention a basic result called ***Frobenius' theorem***. If $E \subset TM$ is a vector subbundle, it is said to be ***involutive*** if for any two vector fields X, Y on M with values in E, $[X, Y]$ is also a vector field with values in E. The subbundle E is said to be ***integrable*** if for each point $m \in M$ there is a local submanifold of M containing m such that its tangent bundle equals E restricted to this submanifold. If E is integrable, the local integral manifolds can be extended to get, through each $m \in M$, a maximal integral manifold, which is an immersed submanifold of M. The collection of all maximal integral manifolds through all points of M forms a foliation.

The Frobenius theorem states that the involutivity of E is equivalent to the integrability of E, which in turn is equivalent to the existence of a foliation on M whose tangent bundle equals E.

We summarize some additional properties in the following table:

Identities for Vector Fields and Forms

1. Vector fields on M with the bracket $[X,Y]$ form a ***Lie algebra***; that is, $[X,Y]$ is real bilinear, skew-symmetric, and ***Jacobi's identity*** holds:

$$[[X,Y],Z] + [[Z,X],Y] + [[Y,Z],X] = 0.$$

Locally,

$$[X,Y] = \mathbf{D}Y \cdot X - \mathbf{D}X \cdot Y = (X \cdot \nabla)Y - (Y \cdot \nabla)X$$

and on functions,

$$[X,Y][f] = X[Y[f]] - Y[X[f]].$$

2. For diffeomorphisms φ and ψ,

$$\varphi_*[X,Y] = [\varphi_* X, \varphi_* Y] \quad \text{and} \quad (\varphi \circ \psi)_* X = \varphi_* \psi_* X.$$

3. The forms on a manifold comprise a real associative algebra with $\wedge$ as multiplication. Furthermore, $\alpha \wedge \beta = (-1)^{kl} \beta \wedge \alpha$ for k and l-forms α and β, respectively.

4. For maps φ and ψ,

$$\varphi^*(\alpha \wedge \beta) = \varphi^*\alpha \wedge \varphi^*\beta \quad \text{and} \quad (\varphi \circ \psi)^*\alpha = \psi^*\varphi^*\alpha.$$

5. $\mathbf{d}$ is a real linear map on forms, $\mathbf{dd}\alpha = 0$, and

$$\mathbf{d}(\alpha \wedge \beta) = \mathbf{d}\alpha \wedge \beta + (-1)^k \alpha \wedge \mathbf{d}\beta$$

for α a k-form.

6. For α a k-form and $X_0, \ldots, X_k$ vector fields,

$$(\mathbf{d}\alpha)(X_0, \ldots, X_k) = \sum_{i=0}^{k} (-1)^i X_i[\alpha(X_0, \ldots, \hat{X}_i, \ldots, X_k)]$$

$$+ \sum_{0 \le i < j \le k} (-1)^{i+j} \alpha([X_i, X_j], X_0, \ldots, \hat{X}_i, \ldots, \hat{X}_j, \ldots, X_k)$$

where $\hat{X}_i$ means that X_i is omitted. Locally,

$$\mathbf{d}\alpha(x)(v_0, \ldots, v_k) = \sum_{i=0}^{k} (-1)^i \mathbf{D}\alpha(x) \cdot v_i(v_0, \ldots, \hat{v}_i, \ldots, v_k).$$

7. For a map φ, $\varphi^* \mathbf{d}\alpha = \mathbf{d}\varphi^*\alpha$.

8. **Poincaré Lemma** If $\mathbf{d}\alpha = 0$, then α is locally exact; that is, there is a neighborhood U about each point on which $\alpha = \mathbf{d}\beta$. The same result holds globally on a contractible manifold.

9. $\mathbf{i}_X\alpha$ is real bilinear in X, α and for $h : M \to \mathbb{R}$,

$$\mathbf{i}_{hX}\alpha = h\mathbf{i}_X\alpha = \mathbf{i}_X h\alpha.$$

Also, $\mathbf{i}_X\mathbf{i}_X\alpha = 0$ and

$$\mathbf{i}_X(\alpha \wedge \beta) = \mathbf{i}_X\alpha \wedge \beta + (-1)^k \alpha \wedge \mathbf{i}_X\beta$$

for α a k-form.

10. For a diffeomorphism φ,

$$\varphi^*(\mathbf{i}_X\alpha) = \mathbf{i}_{\varphi^* X}(\varphi^*\alpha);$$

if $f : M \to N$ is a mapping and Y is f-related to X, that is, $Tf \circ X = Y \circ f$, then

$$\mathbf{i}_Y f^*\alpha = f^*\mathbf{i}_X\alpha.$$

11. $\pounds_X\alpha$ is real bilinear in X, α and

$$\pounds_X(\alpha \wedge \beta) = \pounds_X\alpha \wedge \beta + \alpha \wedge \pounds_X\beta.$$

12. **Cartan's Magic Formula**

$$\pounds_X\alpha = \mathbf{d}\mathbf{i}_X\alpha + \mathbf{i}_X\mathbf{d}\alpha.$$

13. For a diffeomorphism φ,

$$\varphi^*\pounds_X\alpha = \pounds_{\varphi^* X}\varphi^*\alpha;$$

if $f : M \to N$ is a mapping and Y is f-related to X, then

$$\pounds_Y f^*\alpha = f^*\pounds_X\alpha.$$

14. $$(\pounds_X \alpha)(X_1, \ldots, X_k) = X[\alpha(X_1, \ldots, X_k)] - \sum_{i=0}^{k} \alpha(X_1, \ldots, [X, X_i], \ldots, X_k).$$

Locally,

$$\begin{aligned}(\pounds_X \alpha)(x) \cdot (v_1, \ldots, v_k) &= (\mathbf{D}\alpha_x \cdot X(x))(v_1, \ldots, v_k) \\ &\quad + \sum_{i=0}^{k} \alpha_x(v_1, \ldots, \mathbf{D}X_x \cdot v_i, \ldots, v_k).\end{aligned}$$

15. The following identities hold:

    1. $\pounds_{fX}\alpha = f\pounds_X\alpha$, $\pounds_X f\alpha = f\pounds_X\alpha + \mathbf{d}f \wedge \mathbf{i}_X\alpha$;

    2. $\pounds_{[X,Y]}\alpha = \pounds_X\pounds_Y\alpha - \pounds_Y\pounds_X\alpha$;

    3. $\mathbf{i}_{[X,Y]}\alpha = \pounds_X\mathbf{i}_Y\alpha - \mathbf{i}_Y\pounds_X\alpha$;

    4. $\pounds_X\mathbf{d}\alpha = \mathbf{d}\pounds_X\alpha$; and

    5. $\pounds_X\mathbf{i}_X\alpha = \mathbf{i}_X\pounds_X\alpha$.

16. If M is a finite-dimensional manifold, $X = X^l\partial/\partial x^l$, and $\alpha = \alpha_{i_1\ldots i_k} dx^{i_1} \wedge \cdots \wedge dx^{i_k}$, where $i_1 < \cdots < i_k$, then the following formulas hold:

$$\begin{aligned}\mathbf{d}\alpha &= \left(\frac{\partial \alpha_{i_1\ldots i_k}}{\partial x^l}\right) dx^l \wedge dx^{i_1} \wedge \cdots \wedge dx^{i_k}, \\ \mathbf{i}_X\alpha &= X^l \alpha_{l i_2 \ldots i_k} dx^{i_2} \wedge \cdots \wedge dx^{i_k}, \\ \pounds_X\alpha &= X^l \left(\frac{\partial \alpha_{i_1\ldots i_k}}{\partial x^l}\right) dx^{i_1} \wedge \cdots \wedge dx^{i_k} \\ &\quad + \alpha_{l i_2 \ldots i_k} \left(\frac{\partial X^l}{\partial x^{i_1}}\right) dx^{i_1} \wedge dx^{i_2} \wedge \ldots \wedge dx^{i_k} + \ldots.\end{aligned}$$

Exercises

4.4-1 *Let Ω be a closed bounded region in $\mathbb{R}^2$. Use Green's theorem to show that the area of Ω equals the line integral*

$$\frac{1}{2}\int_{\partial\Omega}(x\,dy - y\,dx).$$

4.4-2 *On $\mathbb{R}^2\backslash\{(0,0)\}$ consider the one-form $\alpha = (x\,dy - y\,dx)/(x^2+y^2)$.*

(a) *Show that this form is closed.*

(b) *Using the angle θ as a variable on S^1, compute $i^*\alpha$, where $i : S^1 \to \mathbb{R}^2$ is the standard embedding.*

(c) *Show that α is not exact.*

4.4-3 The magnetic monopole *Let $\mathbf{B} = g\mathbf{r}/r^3$ be a vector field on Euclidean three-space minus the origin where $r = \|\mathbf{r}\|$. Show that $\mathbf{B}$ cannot be written as the* curl *of something.*

5

Hamiltonian Systems on Symplectic Manifolds

Now we are ready to geometrize Hamiltonian mechanics to the context of manifolds. First we make phase spaces nonlinear and then we study Hamiltonian systems in this context.

5.1 Symplectic Manifolds

Definition 5.1.1 *A* ***symplectic manifold*** *is a pair* (P, Ω) *where* P *is a manifold and* Ω *is a closed (weakly) nondegenerate two-form on* P*. If* Ω *is strongly nondegenerate, we speak of a* ***strong symplectic manifold****.*

As in the linear case, strong nondegeneracy of the two-form Ω means that at each $z \in P$, the bilinear form $\Omega_z : T_zP \times T_zP \to \mathbb{R}$ is nondegenerate, that is, Ω_z defines an isomorphism $\Omega_z^\flat : T_zP \to T_z^*P$. For a (weak) symplectic form, the induced map $\Omega^\flat : \mathfrak{X}(P) \to \mathfrak{X}^*(P)$ between vector fields and one-forms is one-to-one, but in general is not surjective. We will see later that Ω is required to be closed, that is, $\mathbf{d}\Omega = 0$, where $\mathbf{d}$ is the exterior derivative, so that the induced Poisson bracket satisfies the Jacobi identity and so that the flows of Hamiltonian vector fields will consist of canonical transformations. In coordinates z^I on P in the finite-dimensional case, if $\Omega = \Omega_{IJ}\, dz^I \wedge dz^J$ (sum over all $I < J$), then $\mathbf{d}\Omega = 0$ becomes the condition

$$\frac{\partial \Omega_{IJ}}{\partial z^K} + \frac{\partial \Omega_{KI}}{\partial z^J} + \frac{\partial \Omega_{JK}}{\partial z^I} = 0. \tag{5.1.1}$$

Examples

(a) Symplectic Vector Spaces If (Z, Ω) is a symplectic vector space, then it is also a symplectic manifold. The requirement $\mathbf{d}\Omega = 0$ is satisfied automatically since Ω is a *constant* form (that is, $\Omega(z)$ is independent of $z \in Z$).

(b) The cylinder $S^1 \times \mathbb{R}$ with coordinates (θ, p) is a symplectic manifold with $\Omega = d\theta \wedge dp$.

(c) The torus $\mathbb{T}^2$ with periodic coordinates (θ, φ) is a symplectic manifold with $\Omega = d\theta \wedge d\varphi$.

(d) The two-sphere S^2 of radius r is symplectic with Ω the standard *area element* $\Omega = r^2 \sin\theta \, d\theta \wedge d\varphi$ on the sphere as the symplectic form. ♦

Given a manifold Q, we will show in Chapter 6 that the cotangent bundle T^*Q has a natural symplectic structure. When Q is the ***configuration space*** of a mechanical system, T^*Q is called the ***momentum phase space***. This important example generalizes the linear examples with phase spaces of the form $W \times W^*$ that we studied in Chapter 2.

The next result says that, in principle, every strong symplectic manifold is, in suitable local coordinates, a symplectic vector space. (By contrast, a corresponding result for Riemannian manifolds is not true unless they have zero curvature; i.e., are flat.)

Theorem 5.1.2 (Darboux' Theorem) *Let (P, Ω) be a strong symplectic manifold. Then in a neighborhood of each $z \in P$, there is a local coordinate chart in which Ω is constant.*

Proof We can assume $P = E$ and $z = 0 \in E$ where E is a Banach space. Let Ω_1 be the constant form equaling $\Omega(0)$. Let $\Omega' = \Omega_1 - \Omega$ and $\Omega_t = \Omega + t\Omega'$, for $0 \leq t \leq 1$. For each t, the bilinear form $\Omega_t(0) = \Omega(0)$ is nondegenerate. Hence by openness of the set of linear isomorphisms of E to E^* and compactness of $[0, 1]$, there is a neighborhood of 0 on which Ω_t is strongly nondegenerate for all $0 \leq t \leq 1$. We can assume that this neighborhood is a ball. Thus by the Poincaré lemma, $\Omega' = \mathbf{d}\alpha$ for some one-form α. Replacing α by $\alpha - \alpha(0)$, we can suppose $\alpha(0) = 0$. Define a smooth time-dependent vector field X_t by

$$\mathbf{i}_{X_t}\Omega_t = -\alpha,$$

which is possible since Ω_t is strongly nondegenerate. Since $\alpha(0) = 0$ we get $X_t(0) = 0$, and so from the local existence theory for ordinary differential equations, there is a ball on which the integral curves of X_t are defined for a time at least one; see Abraham, Marsden, and Ratiu [1988], §4.1, for the technical theorem. Let F_t be the flow of X_t starting at $F_0 =$ identity. By the Lie derivative formula for *time-dependent* vector fields, we have

$$\begin{aligned} \frac{d}{dt}(F_t^*\Omega_t) &= F_t^*(\pounds_{X_t}\Omega_t) + F_t^*\frac{d}{dt}\Omega_t \\ &= F_t^*\mathbf{d}\mathbf{i}_{X_t}\Omega_t + F_t^*\Omega' = F_t^*(\mathbf{d}(-\alpha) + \Omega') = 0. \end{aligned}$$

Thus, $F_1^*\Omega_1 = F_0^*\Omega_0 = \Omega$, so F_1 provides a chart transforming Ω to the constant form Ω_1. ■

This proof is due to Moser [1965]. Weinstein [1971] noted that it generalizes to the infinite-dimensional *strong* symplectic case. Unfortunately, many interesting infinite-dimensional symplectic manifolds are *not* strong. In fact, the analogue of Darboux's theorem is not valid for weak symplectic forms. For an example, see Abraham and Marsden [1978], Exercise 3.2H, and for conditions under which it is valid, see Marsden [1981]. See also Olver [1988].

Corollary 5.1.3 *If (P, Ω) is a finite-dimensional symplectic manifold, then P is even dimensional, and in a neighborhood of $z \in P$ there are local coordinates $(q^1, \ldots, q^n, p_1, \ldots, p_n)$ (where* $\dim P = 2n$*) such that*

$$\Omega = \sum_{i=1}^{n} dq^i \wedge dp_i. \tag{5.1.2}$$

This follows from Darboux's theorem and the canonical form for linear symplectic forms. As in the vector space case, coordinates in which Ω takes the above form are called ***canonical coordinates***.

Corollary 5.1.4 *If (P, Ω) is a $2n$-dimensional symplectic manifold, then P is oriented by the* ***Liouville volume***

$$\Lambda = \frac{(-1)^{[n/2]}}{n!}\Omega \wedge \cdots \wedge \Omega \quad (n \text{ times}). \tag{5.1.3}$$

In canonical coordinates $(q^1, \ldots, q^n, p_1, \ldots, p_n)$, Λ has the expression

$$\Lambda = dq^1 \wedge \cdots \wedge dq^n \wedge dp_1 \wedge \cdots \wedge dp_n. \tag{5.1.4}$$

Thus, if (P, Ω) is a $2n$-dimensional symplectic manifold, then (P, Λ) is a ***volume manifold*** (that is, a manifold with a volume element). The measure associated to Λ is called the ***Liouville measure***. The factor $(-1)^{[n/2]}/n!$ is chosen so that in canonical coordinates, Λ has the expression (5.1.4).

Exercises

5.1-1 *Show how to construct (explicitly) canonical coordinates for the symplectic form $\Omega = f\mu$ on S^2, where μ is the standard area element and where $f : S^2 \to \mathbb{R}$ is a positive function.*

5.1-2 (Moser [1965]) *Let μ_0 and μ_1 be two volume elements (nowhere vanishing n-forms) on the compact boundaryless n-manifold M giving M the same orientation. Assume that $\int_M \mu_0 = \int_M \mu_1$. Show that there is a diffeomorphism $\varphi : M \to M$ such that $\varphi^* \mu_1 = \mu_0$.*

5.1-3 (Requires some functional analysis) *Prove that Darboux' theorem fails for the following weak symplectic form. Let H be a real Hilbert space and $S : H \to H$ a compact, self-adjoint, and positive operator whose range is dense in H, but not equal to H. Let $A_x = S + \|x\|^2 I$ and*

$$g_x(e, f) = \langle A_x e, f \rangle.$$

Let Ω be the weak symplectic form on $H \times H$ associated to g. Show that there is no coordinate chart about $(0,0) \in H \times H$ on which Ω is constant.

5.2 Symplectic Transformations

Definition 5.2.1 *Let (P_1, Ω_1) and (P_2, Ω_2) be symplectic manifolds. A C^∞-mapping $\varphi : P_1 \to P_2$ is called* ***symplectic*** *or* ***canonical*** *if*

$$\varphi^* \Omega_2 = \Omega_1. \tag{5.2.1}$$

Recall that $\Omega_1 = \varphi^* \Omega_2$ means that for each $z \in P$, and all $v, w \in T_z P_1$, we have the following identity:

$$\Omega_{1z}(v, w) = \Omega_{2\varphi(z)}(T_z\varphi \cdot v, T_z\varphi \cdot w),$$

where Ω_{1z} means Ω_1 evaluated at the point z and where $T_z\varphi$ is the tangent (derivative) of φ at z.

If $\varphi : (P_1, \Omega_1) \to (P_2, \Omega_2)$ is canonical, the property $\varphi^*(\alpha \wedge \beta) = \varphi^*\alpha \wedge \varphi^*\beta$ implies that $\varphi^*\Lambda = \Lambda$; that is, φ also preserves the Liouville measure. Thus we get the following:

Proposition 5.2.2 *A smooth canonical transformation between symplectic manifolds of the same dimension is volume preserving and is a local diffeomorphism.*

The last statement comes from the inverse function theorem: if φ is volume preserving, its Jacobian determinant is 1, so φ is locally invertible. It is clear that the set of canonical diffeomorphisms of P form a subgroup of Diff(P), the group of all diffeomorphisms of P. This group, denoted $\mathrm{Diff}_{\mathrm{can}}(P)$, plays a key role in the study of plasma dynamics.

If Ω_1 and Ω_2 are exact, say $\Omega_1 = -\mathbf{d}\Theta_1$ and $\Omega_2 = -\mathbf{d}\Theta_2$, then (5.2.1) is equivalent to

$$\mathbf{d}(\varphi^*\Theta_2 - \Theta_1) = 0. \tag{5.2.2}$$

Let $M \subset P_1$ be an oriented two manifold with boundary ∂M. Then if (5.2.2) holds, we get

$$0 = \int_M \mathbf{d}(\varphi^*\Theta_2 - \Theta_1) = \int_{\partial M} (\varphi^*\Theta_2 - \Theta_1),$$

that is,

$$\int_{\partial M} \varphi^*\Theta_2 = \int_{\partial M} \Theta_1. \tag{5.2.3}$$

Proposition 5.2.3 *The map $\varphi : P_1 \to P_2$ is canonical iff* (5.2.3) *holds for every oriented two manifold $M \subset P_1$ with boundary ∂M.*

The converse is proved by choosing M to be a small disk in P_1 and using the statement: if the integral of a two-form over any small disk vanishes, then the form is zero. The latter assertion is proved by contradiction, constructing a two-form on a two-disk whose coefficient is a bump function. Equation (5.2.3) is an example of an ***integral invariant***. For more information, see Arnold [1989] and Abraham and Marsden [1978].

Exercises

5.2-1 *Let $\varphi : \mathbb{R}^{2n} \to \mathbb{R}^{2n}$ be a map of the form $\varphi(q, p) = (q, p + \alpha(q))$. Use the canonical one-form to determine when φ is symplectic.*

5.2-2 *Let $\mathbb{T}^6$ be the six torus with symplectic form*

$$\Omega = d\theta_1 \wedge d\theta_2 + d\theta_3 \wedge d\theta_4 + d\theta_5 \wedge d\theta_6.$$

Show that if $\varphi : \mathbb{T}^6 \to \mathbb{T}^6$ is symplectic and $M \subset \mathbb{T}^6$ is a compact oriented four-manifold with boundary, then

$$\int_{\partial M} \varphi^*(\Omega \wedge \Theta) = \int_{\partial M} \Omega \wedge \Theta,$$

where $\Theta = \theta_1\, d\theta_2 + \theta_3\, d\theta_4 + \theta_5\, d\theta_6$.

5.3 Complex Structures and Kähler Manifolds

This section develops the relation between complex and symplectic geometry a little further. We shall return to this topic in more depth in Volume II.

We begin with the case of vector spaces. By a ***complex structure*** on a real vector space Z, we mean a linear map $\mathbb{J} : Z \to Z$ such that $\mathbb{J}^2 = -\text{Identity}$. Setting $iz = \mathbb{J}(z)$ gives Z the structure of a complex vector space.

Note that if Z is finite dimensional, the hypothesis on $\mathbb{J}$ implies that $(\det \mathbb{J})^2 = (-1)^{\dim Z}$, so $\dim Z$ must be an even number since $\det \mathbb{J} \in \mathbb{R}$. The complex dimension of Z is half the real dimension. Conversely, if Z is a complex vector space, it is also a real vector space by restricting scalar multiplication to the real numbers. In this case, $\mathbb{J}z = iz$ is the complex structure on Z. As before, the real dimension of Z is twice the complex dimension since the vectors z and iz are linearly independent.

We have already seen that the imaginary part of a complex inner product is a symplectic form. Conversely, *if $\mathcal{H}$ is a real Hilbert space and Ω is a skew-symmetric weakly nondegenerate bilinear form on $\mathcal{H}$, then there is a complex structure $\mathbb{J}$ on $\mathcal{H}$ and a real inner product s such that*

$$s(z, w) = -\Omega(\mathbb{J}z, w). \tag{5.3.1}$$

The expression

$$h(z, w) = s(z, w) - i\Omega(z, w) \tag{5.3.2}$$

defines a Hermitian inner product, and h or s is complete on $\mathcal{H}$ iff Ω is strongly nondegenerate. (See Abraham and Marsden [1978], p.173, for the proof.) Moreover, given any two of $(s, \mathbb{J}, \Omega)$, there is at most one third structure such that (5.3.1) holds.

If we identify $\mathbb{C}^n$ with $\mathbb{R}^{2n}$ and write

$$z = (z_1, \ldots, z_n) = (x_1 + iy_1, \ldots, x_n + iy_n) = ((x_1, y_1), \ldots, (x_n, y_n)),$$

then

$$\begin{aligned} -\operatorname{Im} \langle (z_1, \ldots, z_n), (z_1', \ldots, z_n') \rangle &= -\operatorname{Im}(z_1 \overline{z}_1' + \cdots + z_n \overline{z}_n') \\ &= -(x_1' y_1 - x_1 y_1' + \cdots + x_n' y_n - x_n y_n'). \end{aligned}$$

Thus, the canonical symplectic form on $\mathbb{R}^{2n}$ may be written

$$\Omega(z, z') = -\operatorname{Im} \langle z, z' \rangle = \operatorname{Re} \langle iz, z' \rangle, \tag{5.3.3}$$

which, by (5.3.1), agrees with the convention that $\mathbb{J} : \mathbb{R}^{2n} \to \mathbb{R}^{2n}$ is multiplication by i.

An ***almost complex stucture*** $\mathbb{J}$ on a manifold M is a smooth tangent bundle isomorphism $\mathbb{J} : TM \to TM$ covering the identity map on M such that for each point $z \in M, \mathbb{J}_z = \mathbb{J}(z) : T_zM \to T_zM$ is a complex structure on the vector space T_zM. A manifold with an almost complex structure is called an ***almost complex manifold***.

A manifold M is called a ***complex manifold*** if it admits an atlas $\{(U_\alpha, \varphi_\alpha)\}$ whose charts $\varphi_\alpha : U_\alpha \subset M \to E$ map to a complex Banach space E and the transition functions $\varphi_\beta \circ \varphi_\alpha^{-1} : \varphi_\alpha(U_\alpha \cap U_\beta) \to \varphi_\beta(U_\alpha \cap U_\beta)$ are holomorphic maps. The complex structure on E (multiplication by i) induces via the chart maps φ_α an almost complex structure on each chart domain U_α. Since the transition functions are biholomorphic diffeomorphisms, the almost complex structures on $U_\alpha \cap U_\beta$ induced by φ_α and φ_β coincide. This shows that a complex manifold is also almost complex. The converse is not true.

If M is an almost complex manifold, T_zM is endowed with the structrue of a complex vector space. A ***Hermitian metric*** on M is a smooth assignment of a (possibly weak) complex inner product on T_zM for each $z \in M$. As in the case of vector spaces, the imaginary part of the Hermitian metric defines a non-degenerate (real) two-form on M. The real part of a Hermitian metric is a Riemannian metric on M. If the complex inner product on each tangent space is strongly nondegenerate, the metric is ***strong***; in this case both the real and imaginary parts of the Hermitian metric are strongly nondegenerate over $\mathbb{R}$.

An almost complex manifold M with a Hermitian metric $\langle\, , \rangle$ is called a ***Kähler manifold***, if M is a complex manifold and the two-form $-\mathrm{Im}\,\langle\, , \rangle$ is a closed two form on M. There is an equivalent definition that is often useful: A Kähler manifold is a smooth manifold with a Riemannian metric g and an almost complex structure $\mathbb{J}$ such that $\mathbb{J}_z$ is g-skew for each $z \in M$ and such that $\mathbb{J}$ is covariantly constant with respect to g. (One requires some Riemannian geometry to understand this definition—it will not be required in what follows.) The important fact used later on is the following:

Any Kähler manifold is also symplectic, with symplectic form given by

$$\Omega_z(v_z, w_z) = \langle \mathbb{J}_z v_z, w_z \rangle\,. \tag{5.3.4}$$

In this second definition of Kähler manifolds, the condition $\mathbf{d}\Omega = 0$ follows from $\mathbb{J}$ being covariantly constant. (We shall prove this, along with an account of how Kähler manifolds fit into the theory of complex manifolds, in Volume II.) A ***strong Kähler manifold*** is a Kähler manifold whose Hermitian inner product is strong.

Any complex Hilbert space $\mathcal{H}$ is trivially a strong Kähler manifold. As an example of a nontrivial Kähler manifold, we shall consider the projectivization $\mathbb{P}\mathcal{H}$ of a complex Hilbert space $\mathcal{H}$. Recall from Example **(f)** of §2.3 that $\mathcal{H}$ is a symplectic vector space relative to the quantum mechanical symplectic form $\Omega(\psi_1, \psi_2) = -2\hbar\,\mathrm{Im}\,\langle \psi_1, \psi_2 \rangle$ where $\langle\, , \rangle$ is the Hermitian inner product on $\mathcal{H}$, $\hbar$ is Planck's constant, and $\psi_1, \psi_2 \in \mathcal{H}$. Recall also that $\mathbb{P}\mathcal{H}$ is the space of complex lines through the origin in $\mathcal{H}$. Denote by $\pi : \mathcal{H}\backslash\{0\} \to \mathbb{P}\mathcal{H}$ the canonical projection which sends a vector $\psi \in \mathcal{H}\backslash\{0\}$ to the complex line it spans, denoted by $[\psi]$ when thought of as a point in $\mathbb{P}\mathcal{H}$ and by $\mathbb{C}\psi$ when interpreted as a subspace of $\mathcal{H}$. The space $\mathbb{P}\mathcal{H}$ is a smooth complex manifold, π is a smooth map, and the tangent space

$T_{[\psi]}\mathbb{P}\mathcal{H}$ is isomorphic to $\mathcal{H}/\mathbb{C}\psi$. π is a surjective submersion. (See Abraham, Marsden, Ratiu [1988], Chapter 3.) Since the kernel of $T_\psi\pi : \mathcal{H} \to T_{[\psi]}\mathbb{P}\mathcal{H}$ is $\mathbb{C}\psi$, the map $T_\psi\pi \,|\, (\mathbb{C}\psi)^\perp$ is a complex linear isomorphism from $(\mathbb{C}\psi)^\perp$ to $T_\psi\mathbb{P}\mathcal{H}$ that depends on the chosen representative ψ in $[\psi]$.

If $U : \mathcal{H} \to \mathcal{H}$ is a unitary operator (that is, U is invertible and

$$\langle U\psi_1, U\psi_2\rangle = \langle \psi_1, \psi_2\rangle$$

for all $\psi_1, \psi_2 \in \mathcal{H}$), then the rule $[U][\psi] := [U\psi]$ defines a biholomorphic diffeomorphism on $\mathbb{P}\mathcal{H}$.

Proposition 5.3.1

i *If* $[\psi] \in \mathbb{P}\mathcal{H}$, $\|\psi\| = 1$, *and* $\varphi_1, \varphi_2 \in (\mathbb{C}\psi)^\perp$, *the formula*

$$\langle T_\psi\pi(\varphi_1), T_\psi\pi(\varphi_2)\rangle = 2\hbar\,\langle \varphi_1, \varphi_2\rangle \tag{5.3.5}$$

gives a well-defined strong Hermitian inner product on $T_{[\psi]}\mathbb{P}\mathcal{H}$, *that is, the left hand side does not depend on the choice of* ψ *in* $[\psi]$. *The dependence on* $[\psi]$ *is smooth and so (5.3.5) defines a Hermitian metric on* $\mathbb{P}\mathcal{H}$ *called the* ***Fubini-Study metric***. *This metric is invariant under the action of the maps* $[U]$, *for all unitary operators* U *on* $\mathcal{H}$.

ii *For* $[\psi] \in \mathbb{P}\mathcal{H}$, $\|\psi\| = 1$, *and* $\varphi_1, \varphi_2 \in (\mathbb{C}\psi)^\perp$,

$$g_{[\psi]}(T_\psi\pi(\varphi_1), T_\psi\pi(\varphi_2)) = 2\hbar\,\mathrm{Re}\,\langle \varphi_1, \varphi_2\rangle \tag{5.3.6}$$

defines a strong Riemannian metric on $\mathbb{P}\mathcal{H}$ *invariant under all transformations* $[U]$.

iii *For* $[\psi] \in \mathbb{P}\mathcal{H}$, $\|\psi\| = 1$, *and* $\varphi_1, \varphi_2 \in (\mathbb{C}\psi)^\perp$,

$$\Omega_{[\psi]}(T_\psi\pi(\varphi_1), T_\psi\pi(\varphi_2)) = -2\hbar\,\mathrm{Im}\,\langle \varphi_1, \varphi_2\rangle \tag{5.3.7}$$

defines a strong symplectic form on $\mathbb{P}\mathcal{H}$ *invariant under all transformations* $[U]$.

Remark One can give a conceptually cleaner, but more advanced approach to this process using general reduction theory. The proof below is by a direct argument.

Proof We first prove **i**. If $\lambda \in \mathbb{C}\backslash\{0\}$, then $\pi(\lambda(\psi + t\varphi)) = \pi(\psi + t\varphi)$, and since

$$(T_{\lambda\psi}\pi)(\lambda\varphi) = \left.\frac{d}{dt}\pi(\lambda\psi + t\lambda\varphi)\right|_{t=0} = \left.\frac{d}{dt}\pi(\psi + t\varphi)\right|_{t=0} = (T_\psi\pi)(\varphi),$$

we get $(T_{\lambda\psi}\pi)(\lambda\varphi) = (T_\psi\pi)(\varphi)$. Thus, if $\|\lambda\psi\| = \|\psi\| = 1$, it follows that $|\lambda| = 1$. We have by (5.3.5),

$$\begin{aligned}\langle (T_{\lambda\psi}\pi)(\lambda\varphi_1), (T_{\lambda\psi}\pi)(\lambda\varphi_2)\rangle &= 2\hbar\,\langle \lambda\varphi_1, \lambda\varphi_2\rangle = 2\hbar|\lambda|^2\,\langle \varphi_1, \varphi_2\rangle \\ &= 2\hbar\,\langle \varphi_1, \varphi_2\rangle = \langle (T_\psi\pi)(\varphi_1), (T_\psi\pi)(\varphi_2)\rangle\,.\end{aligned}$$

This shows that the definition (5.3.5) of the Hermitian inner product is independent on the normalized representative $\psi \in [\psi]$ chosen in order to define it. This Hermitian inner product is strong since it coincides with the inner product on the complex Hilbert space $(\mathbb{C}\psi)^{\perp}$.

A straightforward computation (see exercise **5.3-3**) shows that for $\psi \in \mathcal{H}\backslash\{0\}$ and $\varphi_1, \varphi_2 \in \mathcal{H}$ arbitrary, the Hermitian metric is given by

$$\langle T_\psi\pi(\varphi_1), T_\psi\pi(\varphi_2)\rangle = 2\hbar\|\psi\|^{-2}(\langle\varphi_1,\varphi_2\rangle - \|\psi\|^{-2}\langle\varphi_1,\psi\rangle\langle\psi,\varphi_2\rangle). \quad (5.3.8)$$

Since the right hand side is smooth in $\psi \in \mathcal{H}\backslash\{0\}$ and this formula drops to $\mathbb{P}\mathcal{H}$, it follows that (5.3.5) is smooth in $[\psi]$.

If U is a unitary map on $\mathcal{H}$ and $[U]$ is the induced map on $\mathbb{P}\mathcal{H}$, we have

$$\begin{aligned} T_{[\psi]}[U]\cdot T_\psi\pi(\varphi) &= T_{[\psi]}[U]\cdot \frac{d}{dt}[\psi + t\varphi]\bigg|_{t=0} = \frac{d}{dt}[U][\psi+t\varphi]\bigg|_{t=0} \\ &= \frac{d}{dt}[U(\psi+t\varphi)]\bigg|_{t=0} = T_{U\psi}\pi(U\varphi). \end{aligned}$$

Therefore, since $\|U\psi\| = \|\psi\| = 1$ and $\langle U\varphi_j, U\psi\rangle = 0$, we get by (5.3.5),

$$\begin{aligned} \left\langle T_{[\psi]}[U]\cdot T_\psi\pi(\varphi_1), T_{[\psi]}[U]\cdot T_\psi\pi(\varphi_2)\right\rangle &= \langle T_{U\psi}\pi(U\varphi_1), T_{U\psi}\pi(U\varphi_2)\rangle \\ &= \langle U\varphi_1, U\varphi_2\rangle = \langle\varphi_1,\varphi_2\rangle \\ &= \langle T_\psi\pi(\varphi_1), T_\psi\pi(\varphi_2)\rangle, \end{aligned}$$

which proves the invariance of the Hermitian metric under the action of the transformation $[U]$.

Part **ii** is obvious as the real part of the Hermitian metric (5.3.5).

Finally we prove **iii**. From the invariance of the metric it follows that the form Ω is also invariant under the action of unitary maps, that is, $[U]^*\Omega = \Omega$. So, also $[U]^*\mathbf{d}\Omega = \mathbf{d}\Omega$. Now consider the unitary map U_0 on $\mathcal{H}$ defined by $U_0\psi = \psi$ and $U_0 = -$Identity on $(\mathbb{C}\psi)^\perp$. Then from $[U_0]^*\Omega = \Omega$ we have for $\varphi_1, \varphi_2, \varphi_3 \in (\mathbb{C}\psi)^\perp$

$$\begin{aligned} &\mathbf{d}\Omega([\psi])(T_\psi\pi(\varphi_1), T_\psi\pi(\varphi_2), T_\psi\pi(\varphi_3)) \\ &\quad = \mathbf{d}\Omega([\psi])(T_{[\psi]}[U_0]\cdot T_\psi\pi(\varphi_1), T_{[\psi]}[U_0]\cdot T_\psi\pi(\varphi_2), T_{[\psi]}[U_0]\cdot T_\psi\pi(\varphi_3)). \end{aligned}$$

But $T_{[\psi]}[U_0]\cdot T_\psi\pi(\varphi) = T_\psi\pi(-\varphi) = -T_\psi\pi(\varphi)$, which implies by trilinearity of $\mathbf{d}\Omega$ that $\mathbf{d}\Omega = 0$.

The symplectic form Ω is strongly nondegenerate since on $T_{[\psi]}\mathbb{P}\mathcal{H}$ it restricts to the corresponding quantum mechanical symplectic form on the Hilbert space $(\mathbb{C}\psi)^\perp$. ■

The results above prove that $\mathbb{P}\mathcal{H}$ is an infinite dimensional Kähler manifold on which the unitary group $U(\mathcal{H})$ acts by isometries. This can be generalized to Grassmannian manifolds of finite (or infinite) dimensional subspaces of $\mathcal{H}$, and even more, to flag manifolds (see Besse [1987], Pressley and Segal [1985]).

Exercises

5.3-1 *On $\mathbb{C}^n$, show that $\Omega = -\mathbf{d}\Theta$, where $\Theta(z) \cdot w = \frac{1}{2} \operatorname{Im} \langle z, w \rangle$.*

5.3-2 *Let P be a manifold that is both symplectic, with symplectic form Ω and is Riemannian, with metric g.*

(a) *Show that P has an almost complex structure $\mathbb{J}$ such that $\Omega(u, v) = g(\mathbb{J}u, v)$ if and only if*

$$\Omega(\nabla F, v) = -g(X_F, v)$$

for all $F \in \mathcal{F}(P)$.

(b) *Under the hypothesis of* (**a**)*, show that a Hamiltonian vector field X_H is locally a gradient iff $\pounds_{\nabla H}\Omega = 0$.*

5.3-3 *Show that for any vectors $\varphi_1, \varphi_2 \in \mathcal{H}$ and $\psi \neq 0$ the Fubini-Study metric can be written:*

$$\langle T_\psi \pi(\varphi_1), T_\psi \pi(\varphi_2) \rangle = 2\hbar \|\psi\|^{-2} (\langle \varphi_1, \varphi_2 \rangle - \|\psi\|^{-2} \langle \varphi_1, \psi \rangle \langle \psi, \varphi_2 \rangle).$$

Conclude that the Riemannian metric and symplectic forms are given by

$$g_{[\psi]}(T_\psi \pi(\varphi_1), T_\psi \pi(\varphi_2)) = \frac{2\hbar}{\|\psi\|^4} \operatorname{Re}(\langle \varphi_1, \varphi_2 \rangle \|\psi\|^2 - \langle \varphi_1, \psi \rangle \langle \psi, \varphi_2 \rangle)$$

and

$$\Omega_{[\psi]}(T_\psi \pi(\varphi_1), T_\psi \pi(\varphi_2)) = -\frac{2\hbar}{\|\psi\|^4} \operatorname{Im}(\langle \varphi_1, \varphi_2 \rangle \|\psi\|^2 - \langle \varphi_1, \psi \rangle \langle \psi, \varphi_2 \rangle).$$

5.3-4 *Prove that $\mathbf{d}\Omega = 0$ on $\mathbb{P}\mathcal{H}$ directly without using the invariance under the maps $[U]$, for U a unitary operator on $\mathcal{H}$.*

5.3-5 *For $\mathbb{C}^{n+1}$, show that in a projective chart of $\mathbb{CP}^n$ the symplectic form Ω is given by: $(1+|z|^2)^{-1}(\mathbf{d}\sigma - (1+|z|^2)^{-1}\sigma \wedge \overline{\sigma})$ where $\mathbf{d}|z|^2 = \sigma + \overline{\sigma}$ (explicitly $\sigma = \sum_{i=1}^n z_i \mathbf{d}\overline{z}_i$). Then show that $\mathbf{d}\Omega = 0$. Note the similarity between this formula and the corresponding one in* **5.3-3**.

5.4 Hamiltonian Systems

Definition 5.4.1 *Let (P, Ω) be a symplectic manifold. A vector field X on P is called* ***Hamiltonian*** *if there is a function $H : P \to \mathbb{R}$ such that*

$$\mathbf{i}_X \Omega = \mathbf{d}H; \tag{5.4.1}$$

that is, for all $v \in T_zP$, we have the identity

$$\Omega_z(X(z), v) = \mathbf{d}H(z) \cdot v.$$

In this case we write X_H for X. The set of all Hamiltonian vector fields on P is denoted $\mathfrak{X}_{\mathrm{Ham}}(P)$. ***Hamilton's equations*** *are the evolution equations*

$$\dot{z} = X_H(z).$$

In finite dimensions, Hamilton's equations in canonical coordinates are

$$\frac{dq^i}{dt} = \frac{\partial H}{\partial p_i}, \quad \frac{dp^i}{dt} = -\frac{\partial H}{\partial q^i}.$$

A vector field X is called ***locally Hamiltonian*** if $\mathbf{i}_X\Omega$ is closed. This is equivalent to $\pounds_X\Omega = 0$ where $\pounds_X\Omega$ denotes Lie differentiation of Ω along X, because

$$\pounds_X\Omega = \mathbf{i}_X\mathbf{d}\Omega + \mathbf{d}\mathbf{i}_X\Omega = \mathbf{d}\mathbf{i}_X\Omega.$$

If X is locally Hamiltonian, it follows from the Poincaré lemma that there locally exists a function H such that $\mathbf{i}_X\Omega = \mathbf{d}H$, so locally $X = X_H$ and thus the terminology is consistent. Moreover, the flow φ_t of a locally Hamiltonian vector field X satisfies $\varphi_t^*\Omega = \Omega$ since

$$\frac{d}{dt}\varphi_t^*\Omega = \varphi_t^*\pounds_X\Omega = 0,$$

and thus one gets the following:

Proposition 5.4.2 *The flow φ_t of a vector field X consists of symplectic transformations (that is, for each $t, \varphi_t^*\Omega = \Omega$ where defined) if and only if X is locally Hamiltonian.*

A constant vector field on the torus $\mathbb{T}^2$ gives an example of a locally Hamiltonian vector field that is not Hamiltonian. (See Exercise **5.4-1**.)

If X_H is Hamiltonian with flow φ_t, then

$$\frac{d}{dt}(H \circ \varphi_t) = \varphi_t^* X_H[H] = \varphi_t^*\Omega(X_H, X_H) = 0, \tag{5.4.2}$$

since Ω is skew. Thus $H \circ \varphi_t$ is constant in t. We have proved the following:

Proposition 5.4.3 (Conservation of Energy) *If φ_t is the flow of X_H on the symplectic manifold P, then $H \circ \varphi_t = H$ (where defined).*

The same argument given in the vector space case proves:

Proposition 5.4.4 *A diffeomorphism $\varphi : P_1 \to P_2$ of symplectic manifolds is symplectic if and only if it satisfies*

$$\varphi^* X_H = X_{H\circ\varphi} \tag{5.4.3}$$

for all functions $H : U \to \mathbb{R}$ (such that X_H is defined) where U is any open subset of P_2.

The same qualifications on technicalities pertinent to the infinite-dimensional case that were discussed for vector spaces apply to the present context as well. For instance, given H, there is no *a priori* guarantee that X_H exists: we usually assume it abstractly and verify it in examples. Also, we may wish to deal with X_H's that have dense domains rather than everywhere defined smooth vector fields. These technicalities are important, but do not affect many of the main goals of this book. We shall, for simplicity, deal only with everywhere defined vector fields and refer the reader to Chernoff and Marsden [1974] and Marsden and Hughes [1983] for the general case. We shall also tacitly restrict our attention to functions *which have* Hamiltonian vector fields. Of course in the finite-dimensional case these technical problems disappear.

Exercises

5.4-1 *Let X be a constant nonzero vector field on the two-torus. Show that X is locally Hamiltonian but is not globally Hamiltonian.*

5.4-2 *Show that the bracket of two locally Hamiltonian vector fields on a symplectic manifold (P, Ω) is globally Hamiltonian.*

5.4-3 *Consider the equations on $\mathbb{C}^2$ given by*

$$\begin{aligned} \dot{z}_1 &= -iw_1 z_1 + ip\overline{z}_2 + iz_1(a|z_1|^2 + b|z_2|^2), \\ \dot{z}_2 &= -iw_2 z_2 + iq\overline{z}_1 + iz_2(c|z_1|^2 + d|z_2|^2). \end{aligned}$$

Show it is Hamiltonian iff $p = q$ and $b = c$ with

$$H = \frac{1}{2}(w_2|z_2|^2 + w_1|z_1|^2) - p\,\mathrm{Re}(z_1 z_2) - \frac{a}{4}|z_1|^4 - \frac{b}{2}|z_1 z_2|^2 - \frac{d}{4}|z_2|^4.$$

5.5 Poisson Brackets on Symplectic Manifolds

Analogous to the vector space treatment, we define the ***Poisson bracket*** of two functions $F, G : P \to \mathbb{R}$ by

$$\{F, G\}(z) = \Omega(z)(X_F(z), X_G(z)). \tag{5.5.1}$$

From Proposition **5.4.4** we get (see the proof of Proposition **2.9.5**):

Proposition 5.5.1 *A diffeomorphism $\varphi : P_1 \to P_2$ is symplectic if and only if*

$$\{F, G\} \circ \varphi = \{F \circ \varphi, G \circ \varphi\} \tag{5.5.2}$$

for all functions $F, G \in \mathcal{F}(U)$, where U is an arbitrary open subset of P_2.

Therefore, Proposition **5.4.2** shows that

if φ_t is the flow of a Hamiltonian vector field X_H (or a locally Hamiltonian vector field), then

$$\varphi_t^*\{F, G\} = \{\varphi_t^* F, \varphi_t^* G\}$$

for all $F, G \in \mathcal{F}(P)$ (or restricted to an open set if the flow is not everywhere defined).

Corollary 5.5.2 *The following derivation identity holds:*

$$X_H[\{F, G\}] = \{X_H[F], G\} + \{F, X_H[G]\} \tag{5.5.3}$$

where we use the notation $X_H[F] = \pounds_{X_H} F$ for the derivative of F in the direction X_H.

Proof We Differentiate the identity

$$\varphi_t^*\{F, G\} = \{\varphi_t^* F, \varphi_t^* G\}$$

in t at $t = 0$ where φ_t is the flow of X_H. The left-hand side clearly gives the left side of (5.5.3). To evaluate the right-hand side, first notice that

$$\begin{aligned} \Omega_z^\flat \left[\left. \frac{d}{dt} \right|_{t=0} X_{\varphi_t^* F}(z) \right] &= \left. \frac{d}{dt} \right|_{t=0} \Omega_z^\flat X_{\varphi_t^* F}(z) \\ &= \left. \frac{d}{dt} \right|_{t=0} \mathbf{d}(\varphi_t^* F)(z) \\ &= (\mathbf{d} X_H[F])(z) = \Omega_z^*(X_{X_H[F]}(z)). \end{aligned}$$

Thus,

$$\left. \frac{d}{dt} \right|_{t=0} X_{\varphi_t^* F} = X_{X_H[F]}.$$

Therefore,

$$\begin{aligned} \left. \frac{d}{dt} \right|_{t=0} \{\varphi_t^* F, \varphi_t^* G\} &= \left. \frac{d}{dt} \right|_{t=0} \Omega_z(X_{\varphi_t^* F}(z), X_{\varphi_t^* G}(z) \\ &= \Omega_z(X_{X_H[F]}, X_G(z)) + \Omega_z(X_F(z), X_{X_H[G]}(z)) \\ &= \{X_H[F], G\}(z) + \{F, X_H[G]\}(z). \quad \blacksquare \end{aligned}$$

Proposition 5.5.3 *The functions $\mathcal{F}(P)$ form a Lie algebra under the Poisson bracket.*

Proof Since $\{F, G\}$ is obviously real bilinear and skew-symmetric, the only thing to check is Jacobi's identity. From $\{F, G\} = \mathbf{i}_{X_F}\Omega(X_G) = \mathbf{d}F(X_G) =$

$X_G[F]$, we have $\{\{F,G\},H\} = X_H[\{F,G\}]$ and so by Corollary **5.5.2** we get

$$\begin{aligned}\{\{F,G\},H\} &= \{X_H[F],G\} + \{F, X_H[G]\} \\ &= \{\{F,H\},G\} + \{F,\{G,H\}\},\end{aligned} \tag{5.5.4}$$

which is Jacobi's identity. ■

This derivation gives us additional insight: *Jacobi's identity is just the infinitesimal statement of* φ_t *being canonical.*

In the same spirit, one can check that *if* Ω *is a nondegenerate two-form with the Poisson bracket defined by* (5.5.1), *then the Poisson bracket satisfies the Jacobi identity if and only if* Ω *is closed* (see Exercise **5.5-1**).

The ***Poisson bracket-Lie derivative identity***

$$\{F,G\} = X_G[F] = -X_F[G] \tag{5.5.5}$$

we derived in this proof will be useful.

Proposition 5.5.4 *The set of Hamiltonian vector fields* $\mathfrak{X}_{\mathrm{Ham}}(P)$ *is a Lie subalgebra of* $\mathfrak{X}(P)$ *and, in fact,*

$$[X_F, X_G] = -X_{\{F,G\}}. \tag{5.5.6}$$

Proof As derivations,

$$\begin{aligned}[X_F, X_G][H] &= X_F X_G[H] - X_G X_F[H] \\ &= X_F[\{H,G\}] - X_G[\{H,F\}] \\ &= \{\{H,G\},F\} - \{\{H,F\},G\} \\ &= -\{H,\{F,G\}\} = -X_{\{F,G\}}[H],\end{aligned}$$

by Jacobi's identity. ■

Proposition 5.5.5 *We have*

$$\frac{d}{dt}(F\circ\varphi_t) = \{F\circ\varphi_t, H\} = \{F,H\}\circ\varphi_t, \tag{5.5.7}$$

where φ_t *is the flow of* X_H *and* $F \in \mathcal{F}(P)$.

Proof By (5.5.5) and the chain rule,

$$\frac{d}{dt}(F\circ\varphi_t)(z) = \mathbf{d}F(\varphi_t(z))\cdot X_H(\varphi_t(z)) = \{F,H\}(\varphi_t(z)).$$

Since φ_t is symplectic, this becomes

$$\{F\circ\varphi_t, H\circ\varphi_t\}(z)$$

which also equals $\{F \circ \varphi_t, H\}(z)$ by conservation of energy. This proves (5.5.7). ■

Equation (5.5.7), often written more compactly as

$$\dot{F} = \{F, H\}, \tag{5.5.8}$$

is called the ***equation of motion in Poisson bracket form***. We indicated in Chapter 1 why the formulation (5.5.8) is important.

Corollary 5.5.6 *$F \in \mathcal{F}(P)$ is a constant of the motion for X_H iff $\{F, H\} = 0$.*

Proposition 5.5.7 *Assume that the functions f, g, and $\{f, g\}$ are integrable relative to the Liouville volume $\Lambda \in \Omega^{2n}(P)$ on a $2n$-dimensional symplectic manifold (P, Ω). Then*

$$\int_P \{f, g\}\Lambda = \int_{\partial P} f\mathbf{i}_{X_g}\Lambda = -\int_{\partial P} g\mathbf{i}_{X_f}\Lambda.$$

Proof Since $\pounds_{X_g}\Omega = 0$, it follows that $\pounds_{X_g}\Lambda = 0$ so that $\operatorname{div}(fX_g) = X_g[f] = \{f, g\}$. Therefore, by Stokes' theorem

$$\int_P \{f, g\}\Lambda = \int_P \operatorname{div}(fX_g)\Lambda = \int_P \pounds_{fX_g}\Lambda = \int_P \mathbf{d}\mathbf{i}_{fX_g}\Lambda = \int_{\partial P} f\mathbf{i}_{X_g}\Lambda,$$

the second equality following by skew-symmetry of the Poisson bracket.■

Corollary 5.5.8 *Assume that $f, g, h \in \mathcal{F}(P)$ have compact support or decay fast enough such that they and their Poisson brackets are L^2 integrable relative to the Liouville volume on a $2n$-dimensional symplectic manifold (P, Ω). Assume also that at least one of f and g vanish on ∂P, if $\partial P \neq \varnothing$. Then the L^2-inner product is bi-invariant on the Lie algebra $(\mathcal{F}(P), \{\,,\})$, that is,*

$$\int_P f\{g, h\}\Lambda = \int_P \{f, g\}h\Lambda.$$

Proof From $\{hf, g\} = h\{f, g\} + f\{h, g\}$ and Proposition **5.5.7**,

$$0 = \int_P \{hf, g\}\Lambda = \int_P h\{f, g\}\Lambda + \int_P f\{h, g\}\Lambda$$

which proves the corollary. ■

Exercises

5.5-1 *Let Ω be a nondegenerate two-form on a manifold P. Form Hamiltonian vector fields and the Poisson bracket using the same definitions as in* *the linear case. Show that Jacobi's identity holds if and only if the two-form* Ω *is closed.*

5.5-2 *Let* P *be a compact boundaryless symplectic manifold. Show that the space of functions* $\mathcal{F}_0(P) = \{f \in \mathcal{F}(P) \mid \int_P f\Lambda = 0\}$ *is a Lie subalgebra of* $(\mathcal{F}(P), \{\,,\})$ *isomorphic to the Lie algebra of Hamiltonian vector fields on* P .

5.5-3 *Let* $J : \mathbb{C}^2 \to \mathbb{R}$ *be defined by*

$$J = \frac{1}{2}(|z_1|^2 - |z_2|^2).$$

Show that

$$\{H, J\} = 0,$$

where H *is given in Exercise* **5.4-2**.

6
Cotangent Bundles

In many mechanics problems, the phase space is the cotangent bundle T^*Q of configuration space Q. There is an "intrinsic" symplectic structure on T^*Q that can be described in various equivalent ways. Assume first that Q is n-dimensional. Pick local coordinates $(q^1, \ldots, q^n)$ on Q. Since $(dq^1, \ldots, dq^n)$ is a basis of T_q^*Q, we can write any $\alpha \in T_q^*Q$ as $\alpha = p_i\, dq^i$. Then $(q^1, \ldots, q^n, p_1, \ldots, p_n)$ are local coordinates on T^*Q. Define the ***canonical symplectic form*** on T^*Q by

$$\Omega = dq^i \wedge dp_i.$$

This defines a two-form Ω that can be checked to be independent of the choice of coordinates $(q^1, \ldots, q^n)$. Observe that Ω is locally constant, that is, is independent of the base point $(q^1, \ldots, q^n, p_1, \ldots, p_n)$ and so $\mathbf{d}\Omega = 0$. In this section we show how to do this construction intrinsically and we will study this canonical symplectic structure in some detail.

6.1 The Linear Case

To motivate a coordinate independent definition of Ω, consider the case in which Q is a vector space W (which may be infinite dimensional), so that $T^*Q = W \times W^*$. We have already described the canonical two-form on $W \times W^*$:

$$\Omega_{(w,\alpha)}((u, \beta), (v, \gamma)) = \langle \gamma, u \rangle - \langle \beta, v \rangle\,, \tag{6.1.1}$$

where $(w, \alpha) \in W \times W^*$ is the base point, $u, v \in W$, and $\beta, \gamma \in W^*$. This canonical two-form will be constructed from the ***canonical one-form*** Θ, defined as follows:

$$\Theta_{(w,\alpha)}(u, \beta) = \langle \alpha, u \rangle . \tag{6.1.2}$$

The next proposition shows that the canonical two-form (6.1.1) is exact:

$$\Omega = -\mathbf{d}\Theta. \tag{6.1.3}$$

We begin with a computation that reconciles these formulae with their coordinate expressions.

Proposition 6.1.1 *In the finite-dimensional case the symplectic form Ω defined by* (6.1.1) *can be written $\Omega = dq^i \wedge dp_i$ in coordinates $q^1, \ldots, q^n$ on W and corresponding dual coordinates $p_1, \ldots, p_n$ on W^*. The associated canonical one-form is given by $\Theta = p_i\, dq^i$ and* (6.1.3) *holds.*

Proof If $(q^1, \ldots, q^n, p_1, \ldots, p_n)$ are coordinates on T^*W then

$$\left(\frac{\partial}{\partial q^1}, \ldots, \frac{\partial}{\partial q^n}, \frac{\partial}{\partial p_1}, \ldots, \frac{\partial}{\partial p_n} \right)$$

denotes the induced basis for $T_{(w,\alpha)}(T^*W)$, and $(dq^1, \ldots, dq^n, dp_1, \ldots, dp_n)$ denotes the associated dual basis of $T^*_{(w,\alpha)}(T^*W)$. Write

$$(u, \beta) = \left(u^j \frac{\partial}{\partial q^j}, \; \beta_j \frac{\partial}{\partial p_j} \right)$$

and similarly for (v, γ). Hence

$$\begin{aligned} (dq^i \wedge dp_i)_{(w,\alpha)}((u, \beta), (v, \gamma)) &= (dq^i \otimes dp_i - dp_i \otimes dq^i)((u, \beta), (v, \gamma)) \\ &= dq^i(u, \beta) dp_i(v, \gamma) - dp_i(u, \beta) dq^i(v, \gamma) \\ &= u^i \gamma_i - \beta_i v^i . \end{aligned}$$

Also, $\Omega_{(w,\alpha)}((u, \beta), (v, \gamma)) = \gamma(u) - \beta(v) = \gamma_i u^i - \beta_i v^i$. Thus,

$$\Omega = dq^i \wedge dp_i .$$

Similarly,

$$(p_i\, dq^i)_{(w,\alpha)}(u, \beta) = \alpha_i\, dq^i(u, \beta) = \alpha_i u^i ,$$

and

$$\Theta_{(w,\alpha)}(u, \beta) = \alpha(u) = \alpha_i u^i .$$

Comparing, we get $\Theta = p_i\, dq^i$. Therefore,

$$-\mathbf{d}\Theta = -\mathbf{d}(p_i\, dq^i) = dq^i \wedge dp_i = \Omega. \quad \blacksquare$$

To verify (6.1.3) for the infinite-dimensional case, use (6.1.2) and the second formula 6 of the table at the end of §4.4 to give

$$\begin{aligned}\mathbf{d}\Theta_{(w,\alpha)}((u_1,\beta_1),(u_2,\beta_2)) &= [\mathbf{D}\Theta_{(w,\alpha)}\cdot(u_1,\beta_1)]\cdot(u_2,\beta_2)\\ &\quad - [\mathbf{D}\Theta_{(w,\alpha)}\cdot(u_2,\beta_2)]\cdot(u_1,\beta_1)\\ &= \langle\beta_1,u_2\rangle - \langle\beta_2,u_1\rangle\,,\end{aligned}$$

since $\mathbf{D}\Theta_{(w,\alpha)}\cdot(u,\beta) = \langle\beta,\cdot\rangle$. But this equals $-\Omega_{(w,\alpha)}((u_1,\beta_1),(u_2,\beta_2))$.

To give an intrinsic interpretation to Θ, let us prove that

$$\Theta_{(w,\alpha)}\cdot(u,\beta) = \left\langle\alpha, T_{(w,\alpha)}\pi_W(u,\beta)\right\rangle, \tag{6.1.4}$$

where $\pi_W : W\times W^*\to W$ is the projection. Indeed (6.1.4) coincides with (6.1.2) since $T_{(w,\alpha)}\pi_W : W\times W^*\to W$ is the projection on the first factor.

Exercise

6.1-1 Jacobi-Haretu Coordinates *Consider the configuration space* $Q=\mathbb{R}^3\times\mathbb{R}^3\times\mathbb{R}^3$ *with elements denoted* $\mathbf{r}_1,\mathbf{r}_2$, *and* $\mathbf{r}_3$. *Call the conjugate momenta* $\mathbf{p}_1,\mathbf{p}_2,\mathbf{p}_3$ *and equip the phase space* T^*Q *with the canonical symplectic structure* Ω. *Let* $\mathbf{j}=\mathbf{p}_1+\mathbf{p}_2+\mathbf{p}_3$. *Let* $\mathbf{r}=\mathbf{r}_2-\mathbf{r}_1$ *and let* $\mathbf{s}=\mathbf{r}_3-\frac{1}{2}(\mathbf{r}_1+\mathbf{r}_2)$. *Show that the form* Ω *pulled back to the level sets of* $\mathbf{j}$ *has the form* $\Omega=\mathbf{d}\mathbf{r}\wedge d\pi+\mathbf{d}\mathbf{s}\wedge d\sigma$, *where the variables* π *and* σ *are defined by* $\pi=\frac{1}{2}(\mathbf{p}_2-\mathbf{p}_1)$ *and* $\sigma=\mathbf{p}_3$.

6.2 The Nonlinear Case

Definition 6.2.1 *Let* Q *be a manifold. We define* $\Omega=-\mathbf{d}\Theta$, *where* Θ *is the one-form on* T^*Q *defined analogous to* (6.1.4), *namely*

$$\Theta_\beta(v) = \langle\beta, T\pi_Q\cdot v\rangle\,, \tag{6.2.1}$$

where $\beta\in T^*Q$, $v\in T_\beta(T^*Q)$, $\pi_Q : T^*Q\to Q$ *is the projection, and* $T\pi_Q : T(T^*Q)\to TQ$ *is the tangent map of* π_Q.

The computations in Proposition **6.1.6** show that $(T^*Q,\Omega=-\mathbf{d}\Theta)$ is a symplectic manifold; indeed, in local coordinates with $(w,\alpha)\in U\times W^*$, where U is open in W, and where $(u,\beta),(v,\gamma)\in W\times W^*$, the two-form $\Omega=-\mathbf{d}\Theta$ is given by

$$\Omega_{(w,\alpha)}((u,\beta),(v,\gamma)) = \gamma(u)-\beta(v). \tag{6.2.2}$$

Darboux's theorem and its corollary can be interpreted as asserting that any (strong) symplectic manifold locally looks like $W\times W^*$ in suitable local coordinates.

For a function $H : T^*Q \to \mathbb{R}$, the Hamiltonian vector field X_H on the cotangent bundle T^*Q is given in canonical cotangent bundle charts $U \times W^*$, where U is open in W, by

$$X_H(w, \alpha) = \left(\frac{\delta H}{\delta \alpha}, -\frac{\delta H}{\delta w} \right). \tag{6.2.3}$$

Indeed, denoting $X_H(w, \alpha) = (w, \alpha, v, \gamma)$, for any $(u, \beta) \in W \times W^*$ we have

$$\begin{aligned} \mathbf{d}H_{(w,\alpha)} \cdot (u, \beta) &= \mathbf{D}_w H_{(w,\alpha)} \cdot u + \mathbf{D}_\alpha H_{(w,\alpha)} \cdot \beta \\ &= \left\langle \frac{\delta H}{\delta w}, u \right\rangle + \left\langle \beta, \frac{\delta H}{\delta \alpha} \right\rangle \end{aligned} \tag{6.2.4}$$

which, by definition and (6.2.2), equals

$$\Omega_{(w,\alpha)}(X_H(w, \alpha), (u, \beta)) = \langle \beta, v \rangle - \langle \gamma, u \rangle . \tag{6.2.5}$$

Comparing (6.2.4) and (6.2.5) gives (6.2.3). In finite dimensions, (6.2.3) is the familiar right-hand side of Hamilton's equations.

Formula (6.2.3) and the definition of the Poisson bracket show that in canonical cotangent bundle charts,

$$\{f, g\}(w, \alpha) = \left\langle \frac{\delta f}{\delta w}, \frac{\delta g}{\delta \alpha} \right\rangle - \left\langle \frac{\delta g}{\delta w}, \frac{\delta f}{\delta \alpha} \right\rangle, \tag{6.2.6}$$

which in finite dimensions becomes

$$\{f, g\}(q^i, p_i) = \sum_{i=1}^{n} \left(\frac{\partial f}{\partial q^i} \frac{\partial g}{\partial p_i} - \frac{\partial f}{\partial p_i} \frac{\partial g}{\partial q^i} \right). \tag{6.2.7}$$

Another characterization of the canonical one-form that is sometimes useful is the following:

Proposition 6.2.2 *Θ is the unique one-form on T^*Q such that*

$$\alpha^* \Theta = \alpha \tag{6.2.8}$$

for any local one-form α on Q, where, on the left-hand side, α is regarded as a map (of some open subset of) Q to T^*Q.

Proof In finite dimensions, if $\alpha = \alpha_i(q^j)\, dq^i$ and $\Theta = p_i\, dq^i$, then to calculate $\alpha^*\Theta$ means that we substitute $p_i = \alpha_i(q^j)$ into Θ, a process which clearly gives back α, so $\alpha^*\Theta = \alpha$. The general argument is as follows. If Θ is the canonical one-form on T^*Q, and $v \in T_qQ$, then

$$\begin{aligned} (\alpha^*\Theta)_q \cdot v &= \Theta_{\alpha(q)} \cdot T_q\alpha(v) = \langle \alpha(q), T_{\alpha(q)}\pi_Q(T_q\alpha(v)) \rangle \\ &= \langle \alpha(q), T_q(\pi_Q \circ \alpha)(v) \rangle = \alpha(q) \cdot v \end{aligned}$$

since $\pi_Q \circ \alpha =$ identity on Q.

For the converse, assume that Θ is a one-form on T^*Q satisfying (6.2.8). We will show that it must then be the canonical one-form (6.2.1). In finite dimensions this is straightforward: if $\Theta = A_i \, dq^i + B^i \, dp_i$ for A_i, B^i functions of (q^j, p_j), then

$$\alpha^*\Theta = (A_i \circ \alpha) \, dq^i + (B^i \circ \alpha) \, d\alpha_i = \left(A_j \circ \alpha + (B^i \circ \alpha) \frac{\partial \alpha_i}{\partial q^j} \right) dq^j$$

which equals $\alpha = \alpha_i \, dq^i$ iff

$$A_j \circ \alpha + (B^i \circ \alpha) \frac{\partial \alpha_i}{\partial q^j} = \alpha_j.$$

Since this must hold for all α_j, putting $\alpha_1, \ldots, \alpha_n$ constant it follows that $A_j \circ \alpha = \alpha_i$, that is, $A_j = p_j$. Therefore, the remaining equation is

$$(B^i \circ \alpha) \partial \alpha_i / \partial q^j = 0$$

for any α_i; choosing $\alpha_i(q^1, \ldots, q^n) = q_0^i + (q^i - q_0^i) p_i^0$ (no sum) implies $0 = (B^i \circ \alpha)(q_0^1, \ldots, q_0^n) = p_i^0$ for all (q_0^j, p_j^0), that is, $B^i = 0$ and thus $\Theta = p_i \, dq^i$.

[Optional. In infinite dimensions, the proof is slightly different. We will show that if (6.2.8) holds then Θ is locally given by (6.1.4) and thus it is the canonical one-form. If $U \subset E$ is the chart domain the Banach space E modeling Q for any $v \in E$ we have

$$(\alpha^*\Theta)_u \cdot (u, v) = \Theta(u, \alpha(u)) \cdot (v, \mathbf{D}\alpha(u) \cdot v),$$

where α is given locally by $u \mapsto (u, \alpha(u))$ for $\alpha : U \to E^*$. Thus (6.2.8) is equivalent to

$$\Theta_{(u, \alpha(u))} \cdot (v, \mathbf{D}\alpha(u) \cdot v) = \langle \alpha(u), v \rangle$$

which would imply (6.1.4) and hence Θ being the canonical one-form, provided we can show that for prescribed $\gamma, \delta \in E^*, u \in U, v \in E$ there is an $\alpha : U \to E^*$ such that $\alpha(u) = \gamma, \mathbf{D}\alpha(u) \cdot v = \delta$. Such a mapping is constructed in the following way. For $v = 0$ choose $\alpha(u)$ to equal γ for all u. For $v \neq 0$, by the Hahn-Banach theorem one can find a $\varphi \in E^*$ such that $\varphi(v) = 1$. Now set $\alpha(x) = \gamma - \varphi(u)\delta + \varphi(x)\delta$.] ■

Exercises

6.2-1 *Let N be a submanifold of M and denote by Θ_N and Θ_M the canonical one-forms on the cotangent bundles $\pi_N : T^*N \to N$ and $\pi_M : T^*M \to M$, respectively. Let $\pi : (T^*M)|N \to T^*N$ be the projection defined by $\pi(\alpha_n) = \alpha_n | T_n N$ where $n \in N$ and $\alpha_n \in T_n^*M$. Show that $\pi^*\Theta_N = i^*\Theta_M$,* *where* $i : (T^*M)|N \to T^*M$ *is the inclusion.*

6.2-2 *Let* $f : Q \to \mathbb{R}$ *and* $X \in \mathfrak{X}(T^*Q)$. *Show that* $\Theta(X) \circ \mathbf{d}f = X[f \circ \pi_Q]$.

6.2-3 *Let* Q *be a given configuration manifold and let the* ***extended phase space*** *be defined by* $(T^*Q) \times \mathbb{R}$. *Given a vector field* X *on* T^*Q, *extend it to a vector field* $\overline{X}$ *on* $(T^*Q) \times \mathbb{R}$ *by* $\overline{X} = (X, 1)$.

Let H *be a (possibly time-dependent) function on* $(T^*Q) \times \mathbb{R}$ *and set*

$$\Omega_H = \Omega + dH \wedge dt,$$

where Ω *is the canonical two-form. Show that* X *is the Hamiltonian vector field for* H *if and only if*

$$\mathbf{i}_{\overline{X}} \Omega_H = 0.$$

6.2-4 *Give an example of a symplectic manifold* (P, Ω) *where* Ω *is exact, but* P *is not a contangent bundle.*

6.3 Cotangent Lifts

We now describe an important way to create symplectic transformations on cotangent bundles.

Definition 6.3.1 *Given two manifolds* Q *and* S *and a diffeomorphism* $f : Q \to S$, *the* ***cotangent lift*** $T^*f : T^*S \to T^*Q$ *of* f *is defined by*

$$\langle T^*f(\alpha_s), v \rangle = \langle \alpha_s, (Tf \cdot v) \rangle, \tag{6.3.1}$$

where

$$\alpha_s \in T_s^*S, \quad v \in T_qQ, \quad \textit{and} \quad s = f(q).$$

The importance of this construction is that T^*f is guaranteed to be symplectic; it is often called a "point transformation" because it arises from a diffeomorphism on points in configuration space. Notice that while Tf covers f, T^*f covers f^{-1}. Denote by $\pi_Q : T^*Q \to Q$ and $\pi_S : T^*S \to S$, the canonical cotangent bundle projections.

Proposition 6.3.2 *A diffeomorphism* $\varphi : T^*S \to T^*Q$ *preserves the canonical one-forms* Θ_Q *and* Θ_S *on* T^*Q *and* T^*S, *respectively, if and only if* φ *is the cotangent lift* T^*f *of some diffeomorphism* $f : Q \to S$.

Proof First assume that $f : Q \to S$ is a diffeomorphism. Then for arbitrary $\beta \in T^*S$ and $v \in T_\beta(T^*S)$, we have

$$\begin{aligned}((T^*f)^*\Theta_Q)_\beta \cdot v &= (\Theta_Q)_{T^*f(\beta)} \cdot TT^*f(v) \\ &= \langle T^*f(\beta), (T\pi_Q \circ TT^*f) \cdot v \rangle \\ &= \langle \beta, T(f \circ \pi_Q \circ T^*f) \cdot v \rangle \\ &= \langle \beta, T\pi_S \cdot v \rangle = \Theta_{S\beta} \cdot v,\end{aligned}$$

since $f \circ \pi_Q \circ T^*f = \pi_S$.

Conversely, assume that $\varphi^*\Theta_Q = \Theta_S$, that is,

$$\langle \varphi(\beta), T(\pi_Q \circ \varphi)(v) \rangle = \langle \beta, T\pi_S(v) \rangle \tag{6.3.2}$$

for all $\beta \in T^*S$ and $v \in T_\beta(T^*S)$. Since φ is a diffeomorphism, the range of $T_\beta(\pi_Q \circ \varphi)$ is $T_{\pi_Q(\varphi(\beta))}Q$, so that letting $\beta = 0$ in (6.3.2) implies that $\varphi(0) = 0$. Arguing similarly for φ^{-1} instead of φ, we conclude that φ restricted to the zero section S of T^*S is a diffeomorphism onto the zero section Q of T^*Q. Define $f : Q \to S$ by $f = \varphi^{-1}|Q$. We will show below that φ is fiber-preserving or, equivalently, that $f \circ \pi_Q = \pi_S \circ \varphi^{-1}$. For this we need the following:

Lemma 6.3.3 *Define the flow F_t^Q on T^*Q by $F_t^Q(\alpha) = e^t\alpha$ and let V_Q be the vector field it generates. Then*

$$\langle \Theta_Q, V_Q \rangle = 0, \quad \pounds_{V_Q}\Theta_Q = \Theta_Q, \quad \textit{and} \quad \mathbf{i}_{V_Q}\Omega_Q = -\Theta_Q. \tag{6.3.3}$$

Proof Since F_t^Q is fiber-preserving, V_Q will be tangent to the fibers and hence $T\pi_Q \circ V_Q = 0$. This implies by (6.2.1) that $\langle \Theta_Q, V_Q \rangle = 0$. To prove the second formula, note that $\pi_Q \circ F_t^Q = \pi_Q$. Let $\alpha \in T_q^*Q, v \in T_\alpha(T^*Q)$, and Θ_α denote Θ_Q evaluated at α. We have

$$\begin{aligned}
((F_t^Q)^*\Theta)_\alpha \cdot v &= \Theta_{F_t^Q(\alpha)} \cdot TF_t^Q(v) \\
&= \left\langle F_t^Q(\alpha), (T\pi_Q \circ TF_t^Q)(v) \right\rangle \\
&= \left\langle e^t\alpha, T(\pi_Q \circ F_t^Q)(v) \right\rangle \\
&= e^t \langle \alpha, T\pi_Q(v) \rangle = e^t\Theta_\alpha \cdot v,
\end{aligned}$$

that is,

$$(F_t^Q)^*\Theta_Q = e^t\Theta_Q.$$

Taking the derivative relative to t at $t = 0$ yields the second formula. Finally, the first two formulae imply

$$\mathbf{i}_{V_Q}\Omega_Q = -\mathbf{i}_{V_Q}\mathbf{d}\Theta_Q = -\pounds_{V_Q}\Theta_Q + \mathbf{d}\mathbf{i}_{V_Q}\Theta_Q = -\Theta_Q. \quad \blacktriangledown$$

Continuing the proof of the proposition, note that by (6.3.3) we have

$$\begin{aligned}
\mathbf{i}_{\varphi^*V_Q}\Omega_S &= \mathbf{i}_{\varphi^*V_Q}\varphi^*\Omega_Q = \varphi^*(\mathbf{i}_{V_Q}\Omega_Q) \\
&= -\varphi^*\Theta_Q = -\Theta_S = \mathbf{i}_{V_S}\Omega_S,
\end{aligned}$$

so that weak nondegeneracy of Ω_S implies $\varphi^*V_Q = V_S$. Thus φ commutes with the flows F_t^Q and F_t^S, that is, for any $\beta \in T^*S$ we have $\varphi(e^t\beta) = e^t\varphi(\beta)$. Letting $t \to -\infty$ in this equality implies $(\varphi \circ \pi_S)(\beta) = (\pi_Q \circ \varphi)(\beta)$ since $e^t\beta \to \pi_S(\beta)$ and $e^t\varphi(\beta) \to (\pi_Q \circ \varphi)(\beta)$ for $t \to -\infty$. Thus $\pi_Q \circ \varphi = \varphi \circ \pi_S$, or $f \circ \pi_Q = \pi_S \circ \varphi^{-1}$.

Finally, we show that $T^*f = \varphi$. For $\beta \in T^*S$, $v \in T_\beta(T^*S)$, (6.3.2) gives

$$\begin{aligned}
\langle T^*f(\beta), T(\pi_Q \circ \varphi)(v)\rangle &= \langle \beta, T(f \circ \pi_Q \circ \varphi)(v)\rangle \\
&= \langle \beta, T\pi_S v)\rangle = (\Theta_S)_\beta \cdot v \\
&= (\varphi^*\Theta_Q)_\beta \cdot v = (\Theta_Q)_{\varphi(\beta)} \cdot T_\beta\varphi(v) \\
&= \langle \varphi(\beta), T_\beta(\pi_Q \circ \varphi)(v)\rangle ,
\end{aligned}$$

which shows that $T^*f = \varphi$ since the range of $T_\beta(\pi_Q \circ \varphi)$ is the whole tangent space at $(\pi_Q \circ \varphi)(\beta)$ to Q. ∎

In finite dimensions, the first part of this proposition can be seen in coordinates as follows. Write $(s^1, \ldots, s^n) = f(q^1, \ldots, q^n)$ and define

$$p_j = \frac{\partial s^i}{\partial q^j} r_i, \tag{6.3.4}$$

where $(q^1, \ldots, q^n, p_1, \ldots, p_n)$ are cotangent bundle coodinates on T^*Q and $(s^1, \ldots, s^n, r_1, \ldots, r_n)$ on T^*S. Since f is a diffeomorphism, it determines the q^i in terms of the s^j, say $q^i = q^i(s^1, \ldots, s^n)$, so both q^i and p_j are functions of $(s^1, \ldots, s^n, r_1, \ldots, r_n)$. The map T^*f is given by

$$(s^1, \ldots, s^n, r_1, \ldots, r_n) \mapsto (q^1, \ldots, q^n, p_1, \ldots, p_n). \tag{6.3.5}$$

To see that (6.3.5) preserves the canonical one-form, use the chain rule and (6.3.4):

$$r_i\, ds^i = r_i \frac{\partial s^i}{\partial q^k}\, dq^k = p_k\, dq^k. \tag{6.3.6}$$

Note that if f and g are diffeomorphisms of Q, then

$$T^*(f \circ g) = T^*g \circ T^*f, \tag{6.3.7}$$

that is, the cotangent lift switches the order of compositions; in fact, *it is useful to think of T^*f as the* ***adjoint*** *of Tf.*

Exercises

6.3-1 ***The Lorentz group*** *$\pounds$ is the group of invertible linear transformations of $\mathbb{R}^4$ to itself that preserve the quadratic form $x^2 + y^2 + z^2 - c^2t^2$, where c is a constant, the speed of light. Describe all elements of this group. Let Λ_0 denote one of these transformations. Map $\pounds$ to itself by $\Lambda \mapsto \Lambda_0\Lambda$. Calculate the cotangent lift of this map.*

6.3-2 *We have shown that a transformation of T^*Q is the cotangent lift of a diffeomorphism of configuration space if and only if it preserves the canonical one-form. Find this result in Whittaker's book.*

6.4 Lifts of Actions

A ***left action*** of a group G on M associates to each group element $g \in G$ a diffeomorphism Φ_g of M, such that $\Phi_{gh} = \Phi_g \circ \Phi_h$. Thus, the collection of Φ_g's is a *group of transformations of* M. If we replace the condition $\Phi_{gh} = \Phi_g \circ \Phi_h$ by $\Psi_{gh} = \Psi_h \circ \Psi_g$ we speak of a ***right action***. We often write $\Phi_g(m) = g \cdot m$ and $\Psi_g(m) = m \cdot g$ for $m \in M$.

Definition 6.4.1 *Let Φ be an action of a group G on a manifold Q. The* ***right lift*** *Φ^* of the action Φ to the symplectic manifold T^*Q is the right action defined by the rule*

$$\Phi_g^*(\alpha) = (T^*_{g^{-1}\cdot q}\Phi_g)(\alpha), \tag{6.4.1}$$

where $g \in G, \alpha \in T_q^*Q$, and $T^*\Phi_g$ is the cotangent lift of the diffeomorphism $\Phi_g : Q \to Q$.

By (6.3.7), we see that

$$\Phi^*_{gh} = T^*\Phi_{gh} = T^*(\Phi_g \circ \Phi_h) = T^*\Phi_h \circ T^*\Phi_g = \Phi^*_h \circ \Phi^*_g \tag{6.4.2}$$

so Φ^* is a right action. To get a ***left action***, denoted Φ_* and called the ***left lift*** of Φ, one sets

$$(\Phi_*)_g = T^*_{g\cdot q}(\Phi_{g^{-1}}). \tag{6.4.3}$$

In either case these lifted actions are actions by canonical transformations because of Proposition **6.3.2**.

Examples

(a) To describe N particles in $\mathbb{R}^3$, choose the configuration space $Q = \mathbb{R}^{3N}$. We write $(\mathbf{q}_j)$ for an N-tuple of vectors labeled by $j = 1, \ldots, N$. Similarly, elements of the momentum phase space $P = T^*\mathbb{R}^{3N} \cong \mathbb{R}^{6N} \cong \mathbb{R}^{3N} \times \mathbb{R}^{3N}$ are denoted $(\mathbf{q}_j, \mathbf{p}^j)$. Let the additive group $G = \mathbb{R}^3$ of translations act on Q according to

$$\Phi_{\mathbf{x}}(\mathbf{q}_j) = \mathbf{q}_j + \mathbf{x}, \quad \text{where} \quad \mathbf{x} \in \mathbb{R}^3. \tag{6.4.4}$$

Each of the N position vectors $\mathbf{q}_j$ is translated by the same vector $\mathbf{x}$.

Lifting the diffeomorphism $\Phi_{\mathbf{x}} : Q \to Q$, we obtain an action Φ^* of G on P. We assert that

$$\Phi^*_{\mathbf{x}}(\mathbf{q}_j, \mathbf{p}^j) = (\mathbf{q}_j - \mathbf{x}, \mathbf{p}^j). \tag{6.4.5}$$

To verify (6.4.5), observe that $T\Phi_{\mathbf{x}} : TQ \to TQ$ is given by

$$(\mathbf{q}_i, \dot{\mathbf{q}}_j) \mapsto (\mathbf{q}_i + \mathbf{x}, \dot{\mathbf{q}}_j) \tag{6.4.6}$$

so its dual is $(\mathbf{q}_i, \mathbf{p}^j) \mapsto (\mathbf{q}_i - \mathbf{x}, \mathbf{p}^j)$. ♦

(b) Consider the action of $GL(n, \mathbb{R})$, the group of $n \times n$ invertible matrices, or more properly, the group of invertible linear transformations of $\mathbb{R}^n$ to itself, on $\mathbb{R}^n$ given by

$$\Phi_A(\mathbf{q}) = A\mathbf{q}. \tag{6.4.7}$$

The group of induced canonical transformations of $T^*\mathbb{R}^n$ to itself is given by

$$\Phi_A^*(\mathbf{q}, \mathbf{p}) = (A^{-1}\mathbf{q}, A^T\mathbf{p}), \tag{6.4.8}$$

which is readily verified. Notice that this reduces to the same transformation of $\mathbf{q}$ and $\mathbf{p}$ when A is orthogonal. ♦

Exercise

6.4-1 *Let the multiplicative group* $\mathbb{R}\setminus\{0\}$ *act on* $\mathbb{R}^n$ *by* $\Phi_\lambda(\mathbf{q}) = \lambda\mathbf{q}$. *Calculate the cotangent lift of this action.*

6.5 Generating Functions

Consider a symplectic diffeomorphism $\varphi : T^*Q_1 \to T^*Q_2$ described by functions

$$p_i = p_i(q^j, s^j), \quad r_i = r_i(q^j, s^j), \tag{6.5.1}$$

where (q^i, p_i) and (s^j, r_j) are cotangent coordinates on T^*Q_1 and on T^*Q_2, respectively. In other words, assume that we have a map

$$\Gamma : Q_1 \times Q_2 \to T^*Q_1 \times T^*Q_2 \tag{6.5.2}$$

whose image is the graph of φ. Let Θ_1 be the canonical one-form on T^*Q_1 and Θ_2 be that on T^*Q_2. By definition,

$$\mathbf{d}(\Theta_1 - \varphi^*\Theta_2) = 0. \tag{6.5.3}$$

This implies, in view of (6.5.1), that

$$p_i\, dq^i - r_i\, ds^i \tag{6.5.4}$$

is closed. Restated, $\Gamma^*(\Theta_1 - \Theta_2)$ is closed. This holds if (and implies locally by the Poincaré lemma)

$$\Gamma^*(\Theta_1 - \Theta_2) = \mathbf{d}S \tag{6.5.5}$$

for a function $S(q, s)$. In coordinates, (6.5.5) reads

$$p_i\, dq^i - r_i\, ds^i = \frac{\partial S}{\partial q^i}\, dq^i + \frac{\partial S}{\partial s^i}\, ds^i \tag{6.5.6}$$

which is equivalent to

$$p_i = \frac{\partial S}{\partial q^i}, \quad r_i = -\frac{\partial S}{\partial s^i}. \tag{6.5.7}$$

One calls S a ***generating function*** for the canonical transformation. Of course, presupposed relations other than (6.5.1) lead to different conclusions than (6.5.7). Point transformations are generated in this sense; if $S(q^i, r_j) = s^j(q)r_j$, then

$$s^i = \frac{\partial S}{\partial r_i} \quad \text{and} \quad p_i = \frac{\partial S}{\partial q^i}. \tag{6.5.8}$$

(Here one writes $p_i \, dq^i + s^i \, dr_i = \mathbf{d}S$.)

In general, consider a diffeomorphism $\varphi : P_1 \to P_2$ of one symplectic manifold (P_1, Ω_1) to another (P_2, Ω_2) and denote the graph of φ, by $\Gamma(\varphi) \subset P_1 \times P_2$. Let $i_\varphi : \Gamma(\varphi) \to P_1 \times P_2$ be the inclusion and let $\Omega = \pi_1^*\Omega_1 - \pi_2^*\Omega_2$, where $\pi_i : P_1 \times P_2 \to P_i$ is the projection. One verifies that φ *is symplectic if and only if* $i_\varphi^*\Omega = 0$. (Indeed, since $\pi_1 \circ i_\varphi$ is the projection restricted to $\Gamma(\varphi)$ and $\pi_2 \circ i_\varphi = \varphi \circ \pi_1$ on $\Gamma(\varphi)$, it follows that

$$i_\varphi^*\Omega = (\pi_1 | \Gamma(\varphi))^*(\Omega_1 - \varphi^*\Omega_2),$$

and hence $i_\varphi^*\Omega = 0$ iff φ is symplectic because $(\pi_1|\Gamma(\varphi))^*$ is injective.) In this case, one says $\Gamma(\varphi)$ is an ***isotropic*** submanifold of $P_1 \times P_2$ (equipped with the symplectic form Ω); in fact, since $\Gamma(\varphi)$ has half the dimension of $P_1 \times P_2$, it is *maximally* isotropic, or a ***Lagrangian manifold***.

Now suppose one *chooses* a form Θ such that $\Omega = -\mathbf{d}\Theta$. Then $i_\varphi^*\Omega = -\mathbf{d}i_\varphi^*\Theta = 0$, so *locally* on $\Gamma(\varphi)$ there is a function $S : \Gamma(\varphi) \to \mathbb{R}$ such that

$$i_\varphi^*\Theta = \mathbf{d}S. \tag{6.5.9}$$

This defines the ***generating function*** of the canonical transformation φ. Since $\Gamma(\varphi)$ is diffeomorphic to P_1 and also to P_2 we can regard S as a function on P_1 or P_2. If $P_1 = T^*Q_1$ and $P_2 = T^*Q_2$, we can equally well regard (at least locally) S as defined on $Q_1 \times Q_2$. In this way, the general construction of generating functions reduces to the case in equations (6.5.7) and (6.5.8) above. By making other choices of Q, the reader can construct other generating functions and reproduce formulas in, for instance, Goldstein [1980] or Whittaker [1927]. The approach here is based on Sniatycki and Tulczyjew [1971].

Generating functions play an important role in Hamilton-Jacobi theory, in the quantum-classical mechanical relationship (where S plays the role of the quantum mechanical phase), and in numerical integration schemes for Hamiltonian systems. We shall see a few of these aspects later on.

Exercises

6.5-1 *Show that*

$$S(q^i, s^j, t) = \frac{1}{2t}\|\mathbf{q} - \mathbf{s}\|^2$$

generates a canonical transformation that is the identity at $t = 0$.

6.5-2 *(A first-order symplectic integrator) Given* H*, let* $S(q^i, r_j, t) = r_k q^k - tH(q^i, r_j)$. *Show that* S *generates a canonical transformation which is a first-order approximation to the flow of* X_H *for small* t.

6.6 Fiber Translations and Magnetic Terms

We saw above that cotangent lifts provide a basic construction of canonical transformations. Fiber translations provide a second.

Proposition 6.6.1 (Momentum Shifting Lemma) *Let* A *be a one-form on* Q *and let* $t_A : T^*Q \to T^*Q$ *be defined by* $\alpha_q \mapsto \alpha_q + A(q)$*, where* $\alpha_q \in T_q^*Q$. *Let* Θ *be the canonical one-form on* T^*Q. *Then*

$$t_A^*\Theta = \Theta + \pi_Q^* A, \tag{6.6.1}$$

where $\pi_Q : T^*Q \to Q$ *is the projection. Hence*

$$t_A^*\Omega = \Omega - \pi_Q^* \mathbf{d}A, \tag{6.6.2}$$

where $\Omega = -\mathbf{d}\Theta$ *is the canonical symplectic form. Thus,* t_A *is a canonical transformation if and only if* $\mathbf{d}A = 0$.

Proof We prove this using a finite-dimensional coordinate computation. The reader is asked to supply the coordinate-free and infinite-dimensional proofs as an exercise. In coordinates, t_A is the map

$$t_A(q^i, p_j) = (q^i, p_j + A_j). \tag{6.6.3}$$

Thus,

$$t_A^*\Theta = t_A^*(p_i \mathbf{d}q^i) = (p_i + A_i)\mathbf{d}q^i = p_i \mathbf{d}q^i + A_i \mathbf{d}q^i, \tag{6.6.4}$$

which is the coordinate expression for $\Theta + \pi_Q^* A$. The remaining assertions follow directly from this. ■

In particular, fiber translation by the differential of a function $A = \mathbf{d}f$ is a canonical transformation; in fact, f induces, in the sense of the preceding section, a generating function (see Exercise **6.6-2**). The two basic classes of canonical transformations, lifts, and fiber translations, play an important part in mechanics.

A symplectic form on T^*Q, different from the canonical one, is obtained in the following way. Let B be a closed two-form on Q. Then $\Omega - \pi_Q^* B$ is a closed two-form on T^*Q, where Ω is the canonical two-form. To see that $\Omega - \pi_Q^* B$ is (weakly) nondegenerate, use the fact that in a local chart this form is given at the point (w, α) by

$$((u, \beta), (v, \gamma)) \mapsto \langle \gamma, u \rangle - \langle \beta, v \rangle - B(w)(u, v). \tag{6.6.5}$$

Proposition 6.6.2

i *Let Ω be the canonical two-form on T^*Q and let $\pi_Q : T^*Q \to Q$ be the projection. If B is a closed two-form on Q, then*

$$\Omega_B = \Omega - \pi_Q^* B \tag{6.6.6}$$

is a (weak) symplectic form on T^*Q.

ii *Let B and B' be closed two-forms on Q and assume that $B - B' = \mathbf{d}A$. Then the mapping t_A (fiber translation by A) is a symplectic diffeomorphism of $(T^*Q, \Omega - \pi_Q^* B)$ with $(T^*Q, \Omega - \pi_Q^* B')$.*

Proof Part **i** follows from the momentum shifting lemma. For **ii**, use formula (6.6.2) to get

$$t_A^* \Omega = \Omega - \pi_Q^* \mathbf{d}A = \Omega - \pi_Q^* B + \pi_Q^* B', \tag{6.6.7}$$

so that

$$t_A^*(\Omega - \pi_Q^* B') = \Omega - \pi_Q^* B$$

since $\pi_Q \circ t_A = \pi_Q$. ■

In Volume II, we shall see symplectic forms of the type Ω_B arise in the reduction process. In the following examples, we explain why the extra term $\pi_Q^* B$ is often called a ***magnetic term***. Terms of this sort arise in a variety of situations; in Yang-Mills theory, see, for example, Nill [1983] and Gozzi and Thacker [1987].

Exercises

6.6-1 *Provide the intrinsic proof of Proposition* **6.6.1**.

6.6-2 *If $A = \mathbf{d}f$, use a coordinate calculation to check that $S(q^i, r_i) = r_i q^i - f(q^i)$ is a generating function for t_A.*

6.7 A Particle in a Magnetic Field

Let B be a closed two-form on $\mathbb{R}^3$ and let $\mathbf{B} = B_x\mathbf{i} + B_y\mathbf{j} + B_z\mathbf{k}$ be the associated divergence-free vector field, that is, $\mathbf{i}_\mathbf{B}(dx \wedge dy \wedge dz) = B$, so that

$$B = B_x \, dy \wedge dz - B_y \, dx \wedge dz + B_z \, dx \wedge dy.$$

Thinking of $\mathbf{B}$ as a magnetic field, the equations of motion for a particle with charge e and mass m are given by the ***Lorentz force law***:

$$m\frac{d\mathbf{v}}{dt} = \frac{e}{c}\mathbf{v} \times \mathbf{B}, \tag{6.7.1}$$

where $\mathbf{v} = (\dot{x}, \dot{y}, \dot{z})$. On $\mathbb{R}^3 \times \mathbb{R}^3$, that is, $(\mathbf{x}, \mathbf{v})$-space, consider the symplectic form

$$\Omega_B = m(dx \wedge d\dot{x} + dy \wedge d\dot{y} + dz \wedge d\dot{z}) - \frac{e}{c}B \tag{6.7.2}$$

that is, (6.6.6). As Hamiltonian, take the kinetic energy:

$$H = \frac{m}{2}(\dot{x}^2 + \dot{y}^2 + \dot{z}^2) \tag{6.7.3}$$

writing $X_H(u, v, w) = (u, v, w, \dot{u}, \dot{v}, \dot{w})$, the condition

$$\mathbf{d}H = \mathbf{i}_{X_H}\Omega_B \tag{6.7.4}$$

is

$$\begin{aligned} m(\dot{x} \, d\dot{x} + \dot{y} \, d\dot{y} + \dot{z} \, d\dot{z}) \quad = \quad & m(u \, d\dot{x} - \dot{u} \, dx + v \, d\dot{y} - \dot{v} \, dy \\ & + w \, d\dot{z} - \dot{w} \, dz) - \frac{e}{c}[B_x v \, dz - B_x w \, dy \\ & - B_y u \, dz + B_y w \, dx + B_z u \, dy - B_z v \, dx], \end{aligned}$$

which is equivalent to $u = \dot{x}, v = \dot{y}, w = \dot{z}, m\dot{u} = e(B_z v - B_y w)/c, m\dot{v} = e(B_x w - B_z u)/c$, and $m\dot{w} = e(B_y u - B_x v)/c$, that is, to

$$\left.\begin{aligned} m\ddot{x} &= \frac{e}{c}(B_z\dot{y} - B_y\dot{z}), \\ m\ddot{y} &= \frac{e}{c}(B_x\dot{z} - B_z\dot{x}), \\ m\ddot{z} &= \frac{e}{c}(B_y\dot{x} - B_x\dot{y}), \end{aligned}\right\} \tag{6.7.5}$$

which is the same as (6.7.1). Thus *the equations of motion for a particle in a magnetic field are Hamiltonian, with energy equal to the kinetic energy and with the symplectic form* Ω_B.

If $B = \mathbf{d}A$; that is, $\mathbf{B} = \nabla \times \mathbf{A}$ where $\mathbf{A}^\flat = A$, then the map $t_A : (\mathbf{x}, \mathbf{v}) \mapsto (\mathbf{x}, \mathbf{p})$ where $\mathbf{p} = m\mathbf{v} + e\mathbf{A}/c$ pulls back the canonical form to Ω_B by the momentum shifting lemma. Thus, equations (6.7.1) are also Hamiltonian relative to the canonical bracket on $(\mathbf{x}, \mathbf{p})$-space with the Hamiltonian

$$H_A = \frac{1}{2m} \|\mathbf{p} - \frac{e}{c}\mathbf{A}\|^2. \tag{6.7.6}$$

Remarks

1. Not every magnetic field can be written as $\mathbf{B} = \nabla \times \mathbf{A}$ on Euclidean space. For example, the ***field of a magnetic monopole of strength*** $g \neq 0$, namely

$$\mathbf{B}(\mathbf{r}) = g \frac{\mathbf{r}}{\|\mathbf{r}\|^3}, \tag{6.7.7}$$

cannot be written this way since the flux of $\mathbf{B}$ through the unit sphere is $4\pi g$, yet Stokes' theorem applied to the two-sphere would give zero; see Exercise **4.4-1**. Thus, one might think that the Hamiltonian formulation involving only $\mathbf{B}$ (that is, using Ω_B and H) is preferable. However, there is a way to recover the magnetic potential $\mathbf{A}$ by regarding it as a connection on a *nontrivial bundle* over $\mathbb{R}^3 \setminus \{0\}$. (This bundle over the sphere S^2 is the ***Hopf fibration*** $S^3 \to S^2$.) This same construction can be carried out using reduction and we shall do so later. For a readable account of some aspects of this situation, see Yang [1985].

2. When one studies the motion of a particle in a Yang-Mills field, one finds a beautiful generalization of this construction and related ideas using the theory of principal bundles; see Sternberg [1977], Weinstein [1978], and Montgomery [1984].

3. In Chapter 8 we shall study centrifugal and Coriolis forces and discover some structures analogous to those here. ♦

Exercises

6.7-1 *Show that particles in constant magnetic fields move in helixes.*

6.7-2 *Verify "by hand" that* $\frac{1}{2}m\|\mathbf{v}\|^2$ *is conserved for a particle moving in a magnetic field.*

6.8 Linearization of Hamiltonian Systems

One process of linearizing a system is by doubling its dimension using the tangent operation. A second is that of linearizing along a given solution. For example, to linearize a Hamiltonian system on a symplectic manifold at a fixed point, one usually wants the linearized Hamiltonian to be the second variation of the original Hamiltonian at the fixed point. The tangent linearization does not give this; in canonical coordinates q^i, p_i, the tangent linearized symplectic structure is

$$dq^i \wedge d(\delta p_i) + d(\delta q^i) \wedge dp_i \tag{6.8.1}$$

in the variables $(q^i, p_i, \delta q^i, \delta p_i)$. However, one really wants to use

$$dq^i \wedge dp_i + d(\delta q^i) \wedge d(\delta p_i) \tag{6.8.2}$$

which restricts to $d(\delta q^i) \wedge d(\delta p_i)$ at a fixed point, while (6.8.1) restricts to zero.

One can use symplectic connections to compare tangent spaces at different points along the unperturbed curve and thus make the linearization process meaningful. A useful class of intrinsic symplectic connections on cotangent bundles of Lie groups is constructed in Marsden, Ratiu, and Raugel [1991]. For systems with a symmetry group G, they use a G-invariant connection and this gives, via reduction, a linearization theory for Lie-Poisson systems, for example. For instance, ideal fluid flow is linearized in this fashion. One also gets a generalization of the linearization procedure at a fixed point noted in Holm, Marsden, Ratiu, and Weinstein [1985] and Abarbanel, Holm, Marsden, and Ratiu [1986].

We begin by reviewing the special case of a Hamiltonian system in $\mathbb{R}^{2n}$. Let $H : \mathbb{R}^{2n} \to \mathbb{R}$ be a Hamiltonian function, which in canonical coordinates (q^i, p_j) gives rise to Hamilton's equations

$$\dot{q}^i = \frac{\partial H}{\partial p_i}, \qquad \dot{p}_i = -\frac{\partial H}{\partial q^i}. \tag{6.8.3}$$

Linearizing along a solution curve $(q^i(t), p_i(t))$ and calling the new variables $(\delta q^i, \delta p_i)$ we get the equations

$$\begin{aligned} (\delta q^i)^{\cdot} &= \frac{\partial^2 H}{\partial q^j \, \partial p_i} \delta q^j + \frac{\partial^2 H}{\partial p_j \, \partial p_i} \delta p_j, \\ (\delta p_i)^{\cdot} &= -\frac{\partial^2 H}{\partial q^j \, \partial q^i} \delta q^j - \frac{\partial^2 H}{\partial q^i \, \partial p_j} \delta p_j. \end{aligned} \tag{6.8.4}$$

The matrix of the canonical symplectic form $d(\delta q^i) \wedge d(\delta p_i)$ is

$$\mathbb{J} = \begin{bmatrix} 0 & \mathbb{I} \\ -\mathbb{I} & 0 \end{bmatrix}.$$

Recall (see §2.7) that a linear operator with matrix

$$T = \begin{bmatrix} A & B \\ C & D \end{bmatrix}$$

is infinitesimally symplectic, that is, $T^t\mathbb{J} + \mathbb{J}T = 0$, or equivalently, T is ω-skew, if and only if B and C are symmetric matrices and $D = -A^t$. The linear system (6.8.4) has a matrix clearly satisfying these conditions and, therefore, it defines a Hamiltonian system in the $(\delta q^i, \delta p_i)$-variables, whose Hamiltonian function is verified to be the second variation:

$$\frac{1}{2}\omega(T(\delta q^i, \delta p_i), (\delta q^i, \delta p_i)) = \frac{1}{2}\delta^2 H(q^i(t), p_i(t))(\delta q^i, \delta p_i)^2. \qquad (6.8.5)$$

The same argument and formulae hold for infinite-dimensional weak symplectic vector spaces $E \times E'$, where E' and E are (weakly) paired. One of the goals of Marsden, Ratiu, and Raugel [1991] is to generalize this to arbitrary symplectic manifolds. Formula (6.8.5) cannot be correct, in general, since the second variation of a function does not make intrinsic sense, except at critical points. Additional structure is needed to correct the second variation by the addition of terms making the resulting formula invariant. One of the motivations for working in this general context is to deal with Hamiltonian systems in Lie-Poisson spaces, which, as we shall see later, is equivalent to G-invariant Hamiltonian systems on T^*G, where G is a Lie group. At critical points of $H + C$, where C is a Casimir on $\mathfrak{g}^*$ ($\mathfrak{g}^*$ is the dual of the Lie algebra $\mathfrak{g}$ of G), such a linearization has been carried out in Holm, Marsden, Ratiu, and Weinstein [1985] and Abarbanel, Holm, Marsden, and Ratiu [1986]; as expected, the Hamiltionian function of the linearized equations is the second variation of $H+C$, but the Poisson structure instead of being Lie-Poisson is a "constant coefficient" Poisson bracket.

There are a number of several interesting infinite-dimensional systems whose phase spaces are of the form $U \times E'$ where U is open in a Banach space E weakly paired with E'. In all of these cases the linearized equations are infinite-dimensional versions of (6.8.4) and the Hamiltonian function is given by the second variation of the original Hamiltonian along a given integral curve. As we have mentioned, one of the purposes of Marsden, Ratiu, and Raugel [1991] is to generalize this to the nontrivial case. The latter include systems like the rigid body and fluids, charged fluids, Maxwell-Vlasov equations, etc. However, the case with a trivial connection still includes a surprisingly large number of interesting systems. Here are some examples:

1. *The Sine-Gordon equation* $u_{tt} - u_{xx} = \sin u$ has phase space $E \times E'$ where E consists of maps $u : \mathbb{R} \to \mathbb{R}$ (one can also use maps $u : \mathbb{R} \to S^1$, but use of the universal covering space $\mathbb{R}$ of S^1 gives a *linear* space) and E' consists of maps $\dot{u} : \mathbb{R} \to \mathbb{R}$; $E \times E'$ has the canonical symplectic structure. The Hamiltonian has the form kinetic plus potential energy (see Chernoff and Marsden [1974] for details).

2. *The Yang-Mills equations* have phase space $T^*\mathcal{A}$ where $\mathcal{A}$ is the space of connections on a given principal bundle, which is an affine space, so again we can put the trivial symplectic connection on $T^*\mathcal{A}$. The

Yang-Mills equations are Hamiltonian on $T^*\mathcal{A}$ relative to the canonical symplectic structure, so again (6.8.4) is applicable and the Hamiltonian is the second variation of H. See, for example, Arms, Marsden, and Moncrief [1982] for the explicit formula. One of the interesting complications in this example is the presence of a gauge symmetry; the statements above are valid in any gauge. Interestingly, the symplectic form is always canonical, but the Hamiltonian is linear in the so-called atlas fields, representing the gauge freedom (the coefficients of the atlas fields are the momentum map for the gauge group).

3. *General relativity* (in dynamical form) has phase space $T^*\mathrm{Riem}(M)$ where $\mathrm{Riem}(M)$ is the space of Riemannian metrics on a *fixed* hypersurface M. Again the dynamical equations are Hamiltonian on $T^*\mathrm{Riem}(M)$ relative to the *canonical* symplectic structure (for any choice of gauge). Thus, again we can put the trivial symplectic connection on $T^*\mathrm{Riem}(M)$ and formulas (6.8.4) and (6.8.5) (in their obvious infinite-dimensional generalization) apply. These linearized equations are studied in some detail, for the purpose of getting results on the *space* of nonlinear solutions, in Fischer, Marsden, and Moncrief [1980] and Arms, Marsden, and Moncrief [1982]. ♦

An interesting question here is to *couple* these systems to ones with nontrivial phase space. For instance, charged fluids, general relativistic fluids or elasticity, the Maxwell-Vlasov equations, etc., are such systems. All of these will produce nontrivial linearizations by these methods.

Let us begin by studying the symplectic properties of the tangent bundle of a symplectic manifold (P,Ω). Then TP becomes an exact symplectic manifold if the map $\Omega^\flat : TP \to T^*P$ is used to pull back the canonical one-form on T^*P. This one-form, denoted Θ_T, has the expression

$$(\Theta_T)_v \cdot w = \Omega_z(v, T\tau_P(w)), \tag{6.8.6}$$

where $v \in T_zP$, $w \in T_v(TP)$, $\tau_P : TP \to P$ is the projection, and $\langle\,,\rangle$ denotes the pairing between $T^*(TP)$ and $T(TP)$. If $f : P \to P$ is a diffeomorphism one verifies that $Tf : TP \to TP$ is symplectic iff f is symplectic.

If $P = T^*Q$ and (q^i, p_i) denote standard coordinates in a chart of T^*Q, inducing the coordinates (q^i, p_i, v^i, w_i) on $T(T^*Q)$, then Θ_T has the local expression

$$\Theta_T = v_i\, dp_i - w_i\, dq^i. \tag{6.8.7}$$

Let $\pi : T^*(TQ) \to T^*Q$, $\iota : T(T^*Q) \to T^*(TQ)$, and $\kappa : T^*(TQ) \to T(T^*Q)$ be given in charts by

$$\begin{aligned} \pi(u,e,\alpha,\beta) &= (u,\alpha), && (6.8.8)\\ \iota(u,\alpha,e,\beta) &= (u,e,\alpha,\beta), && (6.8.9)\\ \kappa(u,e,\alpha,\beta) &= (u,\alpha,e,\beta), && (6.8.10) \end{aligned}$$

where $u, e \in E, \alpha, \beta \in E^*$, and E is the model space of Q. These formulae actually define the maps π, ι, and κ globally and the following diagram commutes:

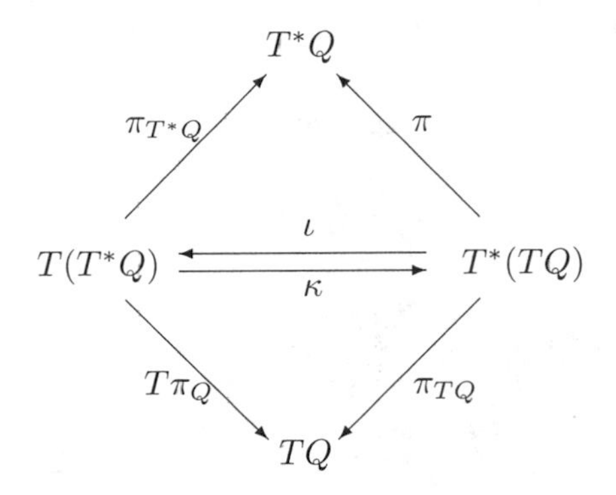

Let Θ_ι be the pull back of the canonical one-form on $T^*(TQ)$ by ι to $T(T^*Q)$. Then

$$\Theta_\iota = v^i \, dq^i + w_i \, dp_i. \tag{6.8.11}$$

Note that $\mathbf{d}\Theta_T = dq^i \wedge dw_i - dp_i \wedge dv^i$ is different from $\mathbf{d}\Theta_\iota = -dq^i \wedge dv^i - dp_i \wedge dw^i$ and thus $T(T^*Q)$ *is endowed with two different symplectic structures*.

We remark in passing that a vector field X is locally Hamiltonian if and only if $X(P)$ is a Lagrangian submanifold of $(TP, \mathbf{d}\Theta_T)$ (see Abraham and Marsden [1978], §5.3, and Sanchez de Alvarez [1986, 1989]).

Let φ_t be the flow of a Hamiltonian vector field X_H on a symplectic manifold P and let $\psi_t = T\varphi_t$ be the tangent flow and Y be its generating vector field. Let $s_P : T(TP) \to T(TP)$ be the ***canonical involution*** given locally by $s_P(u, v, \dot{u}, \dot{v}) = (u, \dot{u}, v, \dot{v})$. One verifies that $Y = s_P \circ TX_H$ is Hamiltonian with respect to the symplectic form $\mathbf{d}\Theta_T$ on TP with the energy $\mathcal{H}(v) = -\Omega(X_H(p), v), v \in T_pP$, which is given in coordinates by the formula

$$\mathcal{H}(q^i, p_i, v^i, w_i) = v^i \frac{\partial H}{\partial q^i} + w_j \frac{\partial H}{\partial p_j}. \tag{6.8.12}$$

The Hamiltonian system $X_{\mathcal{H}}$ on TP is called the ***linearized Hamiltonian system*** or ***first variation equation of*** X_H. We remark that Y is *not* Hamiltonian relative to $\mathbf{d}\Theta_\iota$.

If Q is a pseudo-Riemannian manifold and $P = TQ$ with the symplectic form induced by the metric, the linearized Hamiltonian $\mathcal{H}$ of the Hamiltonian given by the kinetic energy of the metric on Q gives rise to the Hamiltonian vector field $X_{\mathcal{H}}$, which coincides with the first variation equation for geodesics, which is an important construction in geometry (see, for instance, Milnor [1965]).

Next let H_ϵ be a family of Hamiltonian functions on P depending smoothly on a parameter $\epsilon \in \mathbb{R}$. Let H_0 denote the value of H_ϵ at $\epsilon = 0$ and

$$H_1 = \left.\frac{dH_\epsilon}{d\epsilon}\right|_{\epsilon=0}.$$

Let φ_t^ϵ be the flow of the Hamiltonian vector field with Hamiltonian H_ϵ and let

$$\left.\frac{d}{d\epsilon}\varphi_t^\epsilon(p)\right|_{\epsilon=0} = \varphi_t'(p) \in T_{\varphi_t(p)}P.$$

Since $\varphi_0^\epsilon(p) = p$, we have $\varphi_0'(p) = 0$. Thus φ_t' *is an integral curve of the Hamiltonian vector field* $X_{\mathcal{H}_1}$ *on* $(TP, \mathbf{d}\Theta_T)$, where

$$\mathcal{H}_1 = \langle \mathbf{d}H_0, \cdot\rangle + \tau_P^* H_1, \tag{6.8.13}$$

with $\tau_P : TP \to P$ *the canonical tangent bundle projection,* $\langle\, , \rangle$ *the pairing between* T^*P *and* TP*, and* $\langle \mathbf{d}H_0, \cdot\rangle : TP \to \mathbb{R}$ *is given by*

$$\langle \mathbf{d}H_0, \cdot\rangle (v_p) := \langle \mathbf{d}H_0(p), v_p\rangle \tag{6.8.14}$$

for $v_p \in T_pP$. In local coordinates (q^i, p_i, v^i, w_i) on TP,

$$\begin{aligned}
\mathcal{H}_1(q^i, p_i, v^i, w_i) &= v^i \frac{\partial H_0}{\partial q^i} + w_i \frac{\partial H_0}{\partial p_i} + H_1(q^i, p_i) \\
&= \mathcal{H}_0(q^i, p_i, v^i, w_i) + H_1(q^i, p_i),
\end{aligned} \tag{6.8.15}$$

where $\mathcal{H}_0$ is given in terms of H_0 by (6.8.12). Hamilton's equations for $\mathcal{H}_1$ on TP relative to the symplectic form $\mathbf{d}\Theta_T$ are

$$\left.\begin{aligned}
\frac{dq^i}{dt} &= \frac{\partial H_0}{\partial p_i}, \quad \frac{\partial p_i}{dt} = -\frac{\partial H_0}{\partial q^i}, \\
\frac{dv^i}{dt} &= \left(v^j \frac{\partial}{\partial q^j} + w_j \frac{\partial}{\partial p_j}\right)\frac{\partial H_0}{\partial p_i} + \frac{\partial H_1}{\partial p_i}, \\
\frac{dw^i}{dt} &= -\left(v^j \frac{\partial}{\partial q^j} + w_j \frac{\partial}{\partial p_j}\right)\frac{\partial H_0}{\partial q^i} - \frac{\partial H_1}{\partial q^i}.
\end{aligned}\right\} \tag{6.8.16}$$

One calls this the ***first variation equation relative to a parameter***. If we set $H_1 = 0$ we recover the first variation equation for $X_{\mathcal{H}_0}$ discussed earlier.

Further details on the linearization of Hamiltonian systems and the use of symplectic connections to accomplish this may be found in Marsden, Ratiu, and Raugel [1991].

7

Lagrangian Mechanics

Our approach so far has emphasized the Hamiltonian point of view. However, in many instances, this formalism can be derived from the Euler-Lagrange equations. This situation, computational convenience, and the fact that the Lagrangian is very useful in covariant relativistic theories, can be used as arguments for the importance of the Lagrangian formulation.

7.1 The Principle of Critical Action

Much of mechanics can be based on variational principles. Indeed, it is the variational formulation that is the most covariant, being useful for relativistic systems as well. We shall see the utility of the Lagrangian approach in the study of rotating frames and moving systems in §8.8.

Consider a ***configuration manifold*** Q and the velocity phase space TQ. We consider a function $L : TQ \to \mathbb{R}$ called the ***Lagrangian***. Speaking informally, Hamilton's ***principle of critical action*** states that

$$\delta \int L\left(q^i, \frac{dq^i}{dt}\right) dt = 0, \tag{7.1.1}$$

where we take variations amongst paths $q^i(t)$ in Q with fixed endpoints. (We will study this process a little more carefully in §8.1.) Taking the variation in (7.1.1), we get

$$\int \left[\frac{\partial L}{\partial q^i} \delta q^i + \frac{\partial L}{\partial \dot{q}^i} \frac{d}{dt} \delta q^i \right] dt \tag{7.1.2}$$

for the left-hand side. Integrating the second term by parts and using the boundary conditions $\delta q^i = 0$ at the endpoints of the time interval in question, we get

$$\int \left[\frac{\partial L}{\partial q^i} - \frac{d}{dt} \left(\frac{\partial L}{\partial \dot{q}^i} \right) \right] \delta q^i \, dt = 0. \tag{7.1.3}$$

If this is to hold for all such variations $\delta q^i(t)$, then

$$\frac{\partial L}{\partial q^i} - \frac{d}{dt} \frac{\partial L}{\partial \dot{q}^i} = 0, \tag{7.1.4}$$

which are the ***Euler-Lagrange equations***.

We set $p_i = \partial L / \partial \dot{q}^i$, assume that the transformation $(q^i, \dot{q}^j) \mapsto (q^i, p_j)$ is invertible and we define the ***Hamiltonian*** by

$$H(q^i, p_j) = p_i \dot{q}^i - L(q^i, \dot{q}^i). \tag{7.1.5}$$

Note that

$$\dot{q}^i = \frac{\partial H}{\partial p_i},$$

since

$$\frac{\partial H}{\partial p_i} = \dot{q}^i + p_j \frac{\partial \dot{q}^j}{\partial p_i} - \frac{\partial L}{\partial \dot{q}^j} \frac{\partial \dot{q}^j}{\partial p_i} = \dot{q}^i$$

from (7.1.5) and the chain rule. Likewise,

$$\dot{p}_i = -\frac{\partial H}{\partial q^i}$$

from (7.1.4) and

$$\frac{\partial H}{\partial q^j} = p_i \frac{\partial \dot{q}^i}{\partial q^j} - \frac{\partial L}{\partial q^j} - \frac{\partial L}{\partial \dot{q}^i} \frac{\partial \dot{q}^i}{\partial q^j} = -\frac{\partial L}{\partial q^j}.$$

In other words, the *Euler-Lagrange equations are equivalent to Hamilton's equations*.

Thus, it is reasonable to explore the geometry of the Euler-Lagrange equations using the canonical form on T^*Q pulled back to TQ using $p_i = \partial L / \partial \dot{q}^i$. We do this in the next sections.

Exercises

7.1-1 *Verify that the Euler-Lagrange and Hamilton equations are equivalent, even if L is time-dependent.*

7.1-2 *Show that the conservation of energy equation results if, in Hamilton's principle, variations corresponding to reparametrizations of the given curve $q(t)$ are chosen.*

7.2 The Legendre Transform

Given a Lagrangian $L : TQ \to \mathbb{R}$, define a map $\mathbb{F}L : TQ \to T^*Q$, called the ***fiber derivative***, by

$$\mathbb{F}L(v) \cdot w = \left.\frac{d}{dt}\right|_{t=0} L(v + tw), \tag{7.2.1}$$

where $v, w \in T_qQ$. Thus $\mathbb{F}L(v) \cdot w$ is the derivative of L at v along the fiber T_qQ in the direction w. Note that $\mathbb{F}L$ is fiber-preserving. In a local chart $U \times E$ for TQ, where U is open in the model space E for Q, the fiber derivative is given by

$$\mathbb{F}L(u, e) = (u, \mathbf{D}_2 L(u, e)), \tag{7.2.2}$$

where $\mathbf{D}_2 L$ denotes the partial derivative of L with respect to its second argument. For finite-dimensional manifolds, with (q^i) denoting coordinates on Q and $(q^i, \dot{q}^i)$ the induced coordinates on TQ, the fiber derivative has the expression

$$\mathbb{F}L(q^i, \dot{q}^i) = \left(q^i, \frac{\partial L}{\partial \dot{q}^i}\right), \tag{7.2.3}$$

that is, $\mathbb{F}L$ is given by

$$p_i = \frac{\partial L}{\partial \dot{q}^i}. \tag{7.2.4}$$

The associated energy function is $E(v) = \mathbb{F}L(v) \cdot v - L(v)$.

In many examples it is the relationship (7.2.4) that gives physical meaning to the momentum variables. We call $\mathbb{F}L$ the ***Legendre transform***.

Let Ω denote the canonical symplectic form on T^*Q. Using $\mathbb{F}L$, we obtain a one-form Θ_L and a closed two-form Ω_L on TQ by setting

$$\Theta_L = (\mathbb{F}L)^*\Theta \quad \text{and} \quad \Omega_L = (\mathbb{F}L)^*\Omega. \tag{7.2.5}$$

We call Θ_L the ***Lagrangian one-form*** and Ω_L the ***Lagrangian two-form***. Since $\mathbf{d}$ commutes with pull back, we get $\Omega_L = -\mathbf{d}\Theta_L$. Using the local expressions for Θ and Ω, a straightforward pull-back computation yields the following local formula for Θ_L and Ω_L: if E is the model space for Q, U is the range in E of a chart on Q, and $U \times E$ is the corresponding range of the induced chart on TQ, then for $(u, e) \in U \times E$ and tangent vectors $(e_1, e_2), (f_1, f_2)$ in $E \times E$, we have

$$\begin{aligned} T_{(u,e)}\mathbb{F}L \cdot (e_1, e_2) &= (u, \mathbf{D}_2 L(u,e), e_1, \mathbf{D}_1(\mathbf{D}_2 L(u,e)) \cdot e_1 \\ &\quad + \mathbf{D}_2(\mathbf{D}_2 L(u,e)) \cdot e_2), \end{aligned} \tag{7.2.6}$$

so that

$$\Theta_L(u,e) \cdot (e_1, e_2) = \mathbf{D}_2 L(u,e) \cdot e_1 \tag{7.2.7}$$

and

$$\begin{aligned} &\Omega_L(u,e) \cdot ((e_1,e_2),(f_1,f_2)) \\ &\quad = \mathbf{D}_1(\mathbf{D}_2 L(u,e) \cdot e_1) \cdot f_1 - \mathbf{D}_1(\mathbf{D}_2 L(u,e) \cdot f_1) \cdot e_1 \\ &\qquad + \mathbf{D}_2\mathbf{D}_2 L(u,e) \cdot e_1 \cdot f_2 - \mathbf{D}_2\mathbf{D}_2 L(u,e) \cdot f_1 \cdot e_2, \end{aligned} \tag{7.2.8}$$

where $\mathbf{D}_1$ and $\mathbf{D}_2$ denote the first and second (Fréchet) partial derivatives. In finite dimensions, formulae (7.2.6) and (7.2.7) yield

$$\Theta_L = \frac{\partial L}{\partial \dot{q}^i} dq^i \tag{7.2.9}$$

and

$$\Omega_L = \frac{\partial^2 L}{\partial \dot{q}^i \, \partial q^j} dq^i \wedge dq^j + \frac{\partial^2 L}{\partial \dot{q}^i \, \partial \dot{q}^j} dq^i \wedge d\dot{q}^j \quad \text{(sum on all } i, j\text{)}, \tag{7.2.10}$$

that is, as a $2n \times 2n$ skew-symmetric matrix,

$$\Omega_L = \begin{bmatrix} A & \left[\dfrac{\partial^2 L}{\partial \dot{q}^i \partial \dot{q}^j}\right] \\ \left[-\dfrac{\partial^2 L}{\partial \dot{q}^i \partial \dot{q}^j}\right] & 0 \end{bmatrix}, \tag{7.2.11}$$

where A is the skew-symmetrization of $\partial^2 L / \partial \dot{q}^i \, \partial q^j$. From these expressions, it follows that Ω_L *is (weakly) nondegenerate if and only if* $\mathbf{D}_2\mathbf{D}_2 L(u,e)$ *is (weakly) nondegenerate*. In this case, we say L is a ***regular*** or ***nondegenerate*** Lagrangian. The implicit function theorem shows that the fiber derivative is locally invertible if and only if L is regular.

Exercise

7.2-1 *Let*

$$L(q^1, q^2, q^3, \dot{q}^1, \dot{q}^2, \dot{q}^3) = \frac{m}{2}((\dot{q}^1)^2 + (\dot{q}^2)^2 + (\dot{q}^3)^2) + q^1\dot{q}^1 + q^2\dot{q}^2 + q^3\dot{q}^3.$$

Calculate Θ_L, Ω_L *and the corresponding Hamiltonian.*

7.3 Lagrange's Equations

Given a Lagrangian L, the ***action*** of L is the map $A : TQ \to \mathbb{R}$ that is defined by $A(v) = \mathbb{F}L(v) \cdot v$, and the ***energy*** of L is $E = A - L$. In charts,

$$
\begin{aligned}
A(u,e) &= \mathbf{D}_2 L(u,e) \cdot e, &(7.3.1)\\
E(u,e) &= \mathbf{D}_2 L(u,e) \cdot e - L(u,e), &(7.3.2)
\end{aligned}
$$

and in finite dimensions, (7.3.1) and (7.3.2) read

$$
\begin{aligned}
A(q^i, \dot{q}^i) &= \dot{q}^i \frac{\partial L}{\partial \dot{q}^i} = p_i \dot{q}^i, &(7.3.3)\\
E(q^i, \dot{q}^i) &= \dot{q}^i \frac{\partial L}{\partial \dot{q}^i} - L(q^i, \dot{q}^i) = p_i \dot{q}^i - L(q^i, \dot{q}^i). &(7.3.4)
\end{aligned}
$$

If L is a Lagrangian such that $\mathbb{F}L : TQ \to T^*Q$ is a diffeomorphism, we say L is a ***hyperregular*** Lagrangian. In this case, set $H = E \circ (\mathbb{F}L)^{-1}$. Then X_H and X_E are $\mathbb{F}L$-related since $\mathbb{F}L$ is, by construction, symplectic. Thus, hyperregular Lagrangians on TQ induce Hamiltonian systems on T^*Q. Conversely, one can show that hyperregular Hamiltonians on T^*Q come from Lagrangians on TQ (see §**7.4** for definitions and details).

More generally, a vector field Z on TQ is called a ***Lagrangian vector field*** or a ***Lagrangian system*** for L, if the ***Lagrangian condition***

$$
\Omega_L(v)(Z(v), w) = \mathbf{d}E(v) \cdot w \qquad (7.3.5)
$$

holds for all $v \in T_qQ$ and $w \in T_v(TQ)$. If L is ***regular***, that is, Ω_L is a (weak) symplectic form, there would exist at most one such Z, which would be the Hamiltonian vector field of E with respect to the (weak) symplectic form Ω_L. In this case we know that E is conserved on the flow of Z. In fact the same result holds, even if L is degenerate.

Proposition 7.3.1 *Let Z be a Lagrangian vector field for L and let $v(t) \in TQ$ be an integral curve of Z. Then $E(v(t))$ is constant in t.*

Proof By the chain rule,

$$
\begin{aligned}
\frac{d}{dt} E(v(t)) &= \mathbf{d}E(v(t)) \cdot \dot{v}(t) = \mathbf{d}E(v(t)) \cdot Z(v(t))\\
&= \Omega_L(v(t))(Z(v(t))), Z(v(t)) = 0 &(7.3.6)
\end{aligned}
$$

by skew-symmetry of Ω_L. ∎

Note For our needs, we usually assume Ω_L is nondegenerate, but the degenerate case comes up in the Dirac theory of constraints (see Dirac [1950, 1964], Kunzle [1969], Hansen, Regge, and Teitelboim [1976], Gotay, Nester, and Hinds [1979], and Gotay, Isenberg, Marsden, and Montgomery [1994], and references therein, and §8.5). ♦

The vector field Z often has a special property, not shared by the flow of a general Hamiltonian system, namely, Z is a second-order equation.

Definition 7.3.2 *A vector field V on TQ is called a* ***second-order equation*** *provided $T\tau_Q \circ V =$ identity, where $\tau_Q : TQ \to Q$ is the canonical projection. If $c(t)$ is an integral curve of V, $(\tau_Q \circ c)(t)$ is called the* ***base integral curve*** *of $c(t)$.*

It is easy to see that the condition for V being second-order is equivalent to the following: for any chart $U \times E$ on TQ, we can write $V(u,e) = ((u,e),(e,V_2(u,e)))$, for some map $V_2 : U \times E \to E$. Thus, the dynamics is determined by $\dot{u} = e, \dot{e} = V_2(u,e)$; that is, $\ddot{u} = V_2(u,\dot{u})$, a second-order equation in the standard sense. This local computation also shows that the base integral curve uniquely determines an integral curve of V through a given initial condition in TQ.

Theorem 7.3.3 *Let Z be a Lagrangian system for L and suppose Z is a second-order equation. Then in a chart $U \times E$, an integral curve $(u(t), v(t)) \in U \times E$ of Z satisfies* ***Lagrange's equations****; that is:*

$$\begin{aligned} \frac{du(t)}{dt} &= v(t), \\ \frac{d}{dt}\mathbf{D}_2 L(u(t),v(t)) \cdot w &= \mathbf{D}_1 L(u(t),v(t)) \cdot w \end{aligned} \tag{7.3.7}$$

for all $w \in E$. In finite dimensions, Lagrange's equations take the form

$$\begin{aligned} \frac{dq^i}{dt} &= \dot{q}^i, \\ \frac{d}{dt}\left(\frac{\partial L}{\partial \dot{q}^i}\right) &= \frac{\partial L}{\partial q^i}, \quad i = 1, \ldots, n. \end{aligned} \tag{7.3.8}$$

If L is regular, i.e., Ω_L is (weakly) nondegenerate, then Z is automatically second order and if it is strongly nondegenerate, then

$$\frac{d^2u}{dt^2} = \frac{dv}{dt} = [\mathbf{D}_2\mathbf{D}_2 L(u,v)]^{-1}(\mathbf{D}_1 L(u,v) - \mathbf{D}_1\mathbf{D}_2 L(u,v) \cdot v), \tag{7.3.9}$$

or in finite dimensions

$$\ddot{q}^j = G^{ij}\left(\frac{\partial L}{\partial q^i} - \frac{\partial^2 L}{\partial q^i \partial \dot{q}^j}\dot{q}^j\right), \quad i, j = 1, \ldots, n, \tag{7.3.10}$$

where $[G^{ij}]$ is the inverse of the matrix $(\partial^2 L/\partial q^i \partial \dot{q}^j)$. Thus $u(t)$ and $q^i(t)$ are base integral curves of the Lagrangian vector field Z if and only if they satisfy Lagrange's equations.

Proof From the definition of the energy E we have the local expression

$$\begin{aligned}\mathbf{D}E(u,e)\cdot(e_1,e_2) &= \mathbf{D}_1(\mathbf{D}_2L(u,e)\cdot e)\cdot e_1+\mathbf{D}_2(\mathbf{D}_2L(u,e)\cdot e)\cdot e_2\\ &\quad -\mathbf{D}_1L(u,e)\cdot e_1 \qquad (7.3.11)\end{aligned}$$

(the term $\mathbf{D}_2L(u,e)\cdot e_2$ has cancelled). Locally, we may write

$$Z(u,e)=(u,e,Y_1(u,e),Y_2(u,e)).$$

Using formula (7.2.8) for Ω_L the condition (7.3.5) on Z may be written

$$\begin{aligned}&\mathbf{D}_1(\mathbf{D}_2L(u,e)\cdot Y_1(u,e))\cdot e_1-\mathbf{D}_1(\mathbf{D}_2L(u,e)\cdot e_1)\cdot Y_1(u,e)\\ &\qquad+\mathbf{D}_2\mathbf{D}_2L(u,e)\cdot Y_1(u,e)\cdot e_2-\mathbf{D}_2\mathbf{D}_2L(u,e)\cdot e_1\cdot Y_2(u,e)\\ &\quad= \mathbf{D}_1(\mathbf{D}_2L(u,e)\cdot e)\cdot e_1-\mathbf{D}_1L(u,e)\cdot e_1\\ &\qquad+\mathbf{D}_2\mathbf{D}_2L(u,e)\cdot e\cdot e_2. \qquad (7.3.12)\end{aligned}$$

Thus if Ω_L is a weak symplectic form, then $\mathbf{D}_2\mathbf{D}_2L(u,e)$ is weakly nondegenerate, so setting $e_1=0$ we get $Y_1(u,e)=e$; that is, Z is a second-order equation. In any case, if we assume that Z is second order, condition (7.3.12) becomes

$$\mathbf{D}_1L(u,e)\cdot e_1=\mathbf{D}_1(\mathbf{D}_2L(u,e)\cdot e_1)\cdot e+\mathbf{D}_2\mathbf{D}_2L(u,e)\cdot e_1\cdot Y_2(u,e) \quad (7.3.13)$$

for all $e_1\in E$. If $(u(t),v(t))$ is an integral curve of Z and using dots to denote time differentiation, then $\dot u=v$ and $\ddot u=Y_2(u,v)$, so (7.3.13) becomes

$$\begin{aligned}\mathbf{D}_1L(u,\dot u)\cdot e_1 &= \mathbf{D}_1(\mathbf{D}_2L(u,\dot u)\cdot e_1)\cdot\dot u+\mathbf{D}_2\mathbf{D}_2L(u,\dot u)\cdot e_1\cdot\ddot u\\ &= \frac{d}{dt}\mathbf{D}_2L(u,\dot u)\cdot e_1 \qquad (7.3.14)\end{aligned}$$

by the chain rule.

The last statement follows by using the chain rule on the left-hand side of Lagrange's equation and using nondegeneracy of L to solve for $\dot v$, that is, $\ddot q^j$. ■

Exercise

7.3-1 *Give an explicit example of a degenerate Lagrangian L that has a second-order Lagrangian system Z.*

7.4 Hyperregular Lagrangians and Hamiltonians

Above we said that a smooth Lagrangian $L:TQ\to\mathbb{R}$ is ***hyperregular*** if $\mathbb{F}L:TQ\to T^*Q$ is a diffeomorphism. From (7.2.8) or (7.2.11) it follows that the symmetric bilinear form $\mathbf{D}_2\mathbf{D}_2L(u,e)$ is strongly nondegenerate. As before, let $\pi_Q : T^*Q \to Q$ and $\tau_Q : TQ \to Q$ denote the canonical projections.

Proposition 7.4.1 *Let L be a hyperregular Lagrangian on TQ and let $H = E \circ (\mathbb{F}L)^{-1} \in \mathcal{F}(T^*Q)$, where E is the energy of L. Then the Lagrangian vector field Z on TQ and the Hamiltonian vector field X_H on T^*Q are $\mathbb{F}L$-related, i.e.,*

$$(\mathbb{F}L)^* X_H = Z.$$

Furthermore, if $c(t)$ is an integral curve of Z and $d(t)$ an integral curve of X_H with $\mathbb{F}L(c(0)) = d(0)$, then $\mathbb{F}L(c(t)) = d(t)$ and $(\tau_Q \circ c)(t) = (\pi_Q \circ d)(t)$. The curve $(\tau_Q \circ c)(t)$ is called the ***base integral curve*** *of $c(t)$ and similarly $(\pi_Q \circ d)(t)$ is the* ***base integral curve*** *of $d(t)$.*

Proof For $v \in TQ$ and $w \in T_v(TQ)$, we have

$$\begin{aligned}
&\Omega(\mathbb{F}L(v))(T_v\mathbb{F}L(Z(v)), T_v\mathbb{F}L(w)) \\
&\quad = ((\mathbb{F}L)^*\Omega)(v)(Z(v), w) \\
&\quad = \Omega_L(v)(Z(v), w) \\
&\quad = \mathbf{d}E(v) \cdot w \\
&\quad = \mathbf{d}(H \circ \mathbb{F}L)(v) \cdot w \\
&\quad = \mathbf{d}H(\mathbb{F}L(v)) \cdot T_v\mathbb{F}L(w) \\
&\quad = \Omega(\mathbb{F}L(v))(X_H(\mathbb{F}L(v)), T_v\mathbb{F}L(w)),
\end{aligned}$$

so that by weak nondegeneracy of Ω and the fact that $T_v\mathbb{F}L$ is an isomorphism, it follows that $T_v\mathbb{F}L(Z(v)) = X_H(\mathbb{F}L(v))$. Thus $T\mathbb{F}L \circ Z = X_H \circ \mathbb{F}L$, that is, $Z = (\mathbb{F}L)^* X_H$.

If φ_t denotes the flow of Z and ψ_t the flow of X_H, the relation $Z = (\mathbb{F}L)^* X_H$ is equivalent to $\mathbb{F}L \circ \varphi_t = \psi_t \circ \mathbb{F}L$. Thus, if $c(t) = \varphi_t(v)$, then $\mathbb{F}L(c(t)) = \psi_t(\mathbb{F}L(v))$ is an integral curve of X_H which at $t = 0$ passes through $\mathbb{F}L(v) = \mathbb{F}L(c(0))$, whence $\psi_t(\mathbb{F}L(v)) = d(t)$ by uniqueness of integral curves of smooth vector fields. Finally, since $\tau_Q = \pi_Q \circ \mathbb{F}L$, we get $(\tau_Q \circ c)(t) = (\pi_Q \circ \mathbb{F}L \circ c)(t) = (\pi_Q \circ d)(t)$. ■

The action A of L is related to the Lagrangian vector field Z of L by

$$A(v) = \langle \Theta_L(v), Z(v) \rangle, \quad v \in TQ. \tag{7.4.1}$$

We prove this formula under the assumption that Z is a second-order equation, even if L is not regular. In fact,

$$\begin{aligned}
\langle \Theta_L(v), Z(v) \rangle &= \langle ((\mathbb{F}L)^*\Theta)(v), Z(v) \rangle \\
&= \langle \Theta(\mathbb{F}L(v)),\ T_v\mathbb{F}L(Z(v)) \rangle \\
&= \langle \mathbb{F}L(v), T\pi_Q \cdot T_v\mathbb{F}L(Z(v)) \rangle \\
&= \langle \mathbb{F}L(v), T_v(\pi_Q \circ \mathbb{F}L)(Z(v)) \rangle \\
&= \langle \mathbb{F}L(v), T_v\tau_Q(Z(v)) \rangle = \langle \mathbb{F}L(v), v \rangle = A(v),
\end{aligned}$$

by definition of a second-order equation and the definition of the action. If L is hyperregular and $H = E \circ (\mathbb{F}L)^{-1}$, then

$$A \circ (\mathbb{F}L)^{-1} = \langle \Theta, X_H \rangle . \tag{7.4.2}$$

Indeed, by (7.4.1), the properties of push-forward, and the previous proposition, we have

$$A \circ (\mathbb{F}L)^{-1} = (\mathbb{F}L)_* A = (\mathbb{F}L)_*(\langle \Theta_L, Z \rangle) = \langle (\mathbb{F}L)_* \Theta_L, (\mathbb{F}L)_* Z \rangle = \langle \Theta, X_H \rangle .$$

If $H : T^*Q \to \mathbb{R}$ is a smooth Hamiltonian, the function $G : T^*Q \to \mathbb{R}$ given by $G = \langle \Theta, X_H \rangle$ is called the ***action*** of H. Thus (7.4.2) says that the push-forward of the action A of L equals the action G of $H = E \circ (\mathbb{F}L)^{-1}$. The Hamiltonian H is called ***hyperregular***, if $\mathbb{F}H : T^*Q \to TQ$ given by

$$\mathbb{F}H(\alpha) \cdot \beta = \left. \frac{d}{dt} \right|_{t=0} H(\alpha + t\beta), \quad \text{where} \quad \alpha, \beta \in T_q^*Q, \tag{7.4.3}$$

is a diffeomorphism; here we must assume that either the model space E of Q is reflexive so that $T_q^{**}Q = T_qQ$ for all $q \in Q$ or, what is more reasonable, that $\mathbb{F}H(\alpha)$ lies in $T_qQ \subset T_q^{**}Q$. As in the case of Lagrangians, hyperregularity of H implies the strong nondegeneracy of $\mathbf{D}_2\mathbf{D}_2H(u, \alpha)$ and the curve $t \mapsto \alpha + t\beta$ appearing in (7.4.3) can be replaced by an arbitrary smooth curve $\alpha(t)$ in T_q^*Q such that $\alpha(0) = \alpha$ and $\alpha'(0) = \beta$.

Proposition 7.4.2

i *Let $H \in \mathcal{F}(T^*Q)$ be a hyperregular Hamiltonian and define $E = H \circ (\mathbb{F}H)^{-1}, A = G \circ (\mathbb{F}H)^{-1}$ and $L = A - E \in \mathcal{F}(TQ)$. Then L is a hyperregular Larangian and $\mathbb{F}L = \mathbb{F}H^{-1}$. Furthermore, A is the action of L and E the energy of L.*

ii *Let $L \in \mathcal{F}(TQ)$ be a hyperregular Lagrangian and define $H = E \circ (\mathbb{F}L)^{-1}$. Then H is a hyperregular Hamiltonian and $\mathbb{F}H = (\mathbb{F}L)^{-1}$.*

Proof

i Locally $G(u, \alpha) = \langle \alpha, \mathbf{D}_2H(u, \alpha) \rangle$, so that

$$A(u, \mathbf{D}_2H(u, \alpha)) = (A \circ \mathbb{F}H)(u, \alpha) = G(u, \alpha) = \langle \alpha, \mathbf{D}_2H(u, \alpha) \rangle ,$$

whence

$$(L \circ \mathbb{F}H)(u, \alpha) = L(u, \mathbf{D}_2H(u, \alpha)) = \langle \alpha, \mathbf{D}_2H(u, \alpha) \rangle - H(u, \alpha).$$

Let $e = \mathbf{D}_2(\mathbf{D}_2H(u, \alpha)) \cdot \beta$ and let $e(t) = \mathbf{D}_2H(u, \alpha + t\beta)$ be a curve which at $t = 0$ passes through $e(0) = \mathbf{D}_2H(u, \alpha)$ and whose derivative at $t = 0$ equals $e'(0) = \mathbf{D}_2(\mathbf{D}_2 H(u,\alpha)) \cdot \beta = e$. Therefore

$$\begin{aligned}
&\langle (\mathbb{F}L \circ \mathbb{F}H)(u,\alpha), e\rangle \\
&\quad = \langle \mathbb{F}L(u, \mathbf{D}_2 H(u,\alpha)), e\rangle \\
&\quad = \left.\frac{d}{dt}\right|_{t=0} L(u, e(t)) \\
&\quad = \left.\frac{d}{dt}\right|_{t=0} L(u, \mathbf{D}_2 H(u, \alpha + t\beta)) \\
&\quad = \left.\frac{d}{dt}\right|_{t=0} \langle \alpha + t\beta, \mathbf{D}_2 H(u, \alpha + t\beta)\rangle - \left.\frac{d}{dt}\right|_{t=0} H(u, \alpha + t\beta) \\
&\quad = \langle \alpha, \mathbf{D}_2(\mathbf{D}_2 H(u,\alpha)) \cdot \beta\rangle = \langle \alpha, e\rangle .
\end{aligned}$$

Since $\mathbf{D}_2\mathbf{D}_2 H(u,\alpha)$ is strongly nondegenerate this implies that $e \in E$ is arbitrary and hence $\mathbb{F}L \circ \mathbb{F}H =$ identity. Since $\mathbb{F}H$ is a diffeomorphism, this says that $\mathbb{F}L = (\mathbb{F}H)^{-1}$ and hence that L is hyperregular.

To see that A is the action of L note that since $\mathbb{F}H^{-1} = \mathbb{F}L$ we have by definition of G

$$A = G \circ (\mathbb{F}H)^{-1} = \langle \Theta, X_H\rangle \circ \mathbb{F}L,$$

which by (7.4.2) implies that A is the action of L. Therefore $E = A - L$ is the energy of L.

ii Locally, since we define $H = E \circ (\mathbb{F}L)^{-1}$ we have

$$\begin{aligned}
(H \circ \mathbb{F}L)(u,e) &= H(u, \mathbf{D}_2 L(u,e)) \\
&= A(u,e) - L(u,e) = \mathbf{D}_2 L(u,e) \cdot e - L(u,e)
\end{aligned}$$

and proceed as before. Let $\alpha = \mathbf{D}_2(\mathbf{D}_2 L(u,e)) \cdot f$, where $f \in E$, and $\alpha(t) = \mathbf{D}_2 L(u, e + tf)$; then $\alpha(0) = \mathbf{D}_2 L(u,e)$, and $\alpha'(0) = \alpha$, so that

$$\begin{aligned}
&\langle \alpha, (\mathbb{F}H \circ \mathbb{F}L)(u,e)\rangle \\
&\quad = \langle \alpha, \mathbb{F}H(u, \mathbf{D}_2 L(u,e))\rangle \\
&\quad = \left.\frac{d}{dt}\right|_{t=0} H(u, \alpha(t)) \\
&\quad = \left.\frac{d}{dt}\right|_{t=0} H(u, \mathbf{D}_2 L(u, e + tf)) \\
&\quad = \left.\frac{d}{dt}\right|_{t=0} \langle \mathbf{D}_2 L(u, e + tf), e + tf\rangle - \left.\frac{d}{dt}\right|_{t=0} L(u, e + tf) \\
&\quad = \langle \mathbf{D}_2(\mathbf{D}_2 L(u,e)) \cdot f, e\rangle = \langle \alpha, e\rangle ,
\end{aligned}$$

which shows by strong nondegeneracy of $\mathbf{D}_2\mathbf{D}_2 L$ that $\mathbb{F}H \circ \mathbb{F}L =$ identity. Since $\mathbb{F}L$ is a diffeomorphism it follows that $\mathbb{F}H = (\mathbb{F}L)^{-1}$ and H is hyperregular. ∎

The main result is summarized in the following.

Theorem 7.4.3 *Hyperregular Lagrangians $L \in \mathcal{F}(TQ)$ and hyperregular Hamiltonians $H \in F(T^*Q)$ correspond in a bijective manner by the prior constructions. The following diagram commutes:*

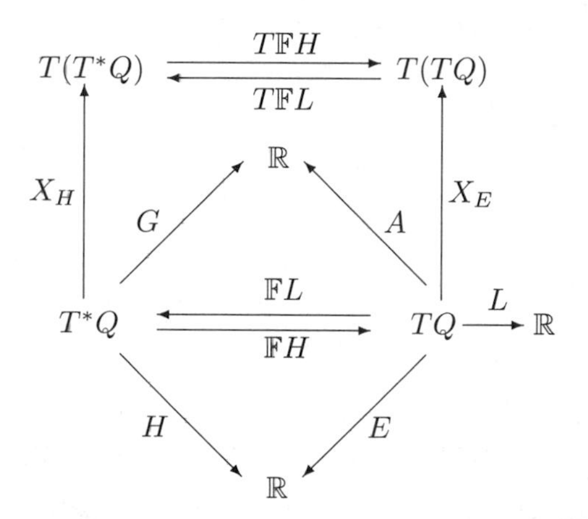

Proof Let L be a hyperregular Lagrangian and let H be the associated Hamiltonian which is hyperregular, that is,

$$H = E \circ (\mathbb{F}L)^{-1} = (A - L) \circ (\mathbb{F}L)^{-1} = G - L \circ \mathbb{F}H$$

by Propositions **7.4.1** and **7.4.2**. From H we construct a Lagrangian L' by

$$L' = G \circ (\mathbb{F}H)^{-1} - H \circ (\mathbb{F}H)^{-1} = G \circ (\mathbb{F}H)^{-1} - (G - L \circ \mathbb{F}H) \circ (\mathbb{F}H)^{-1} = L.$$

Conversely, if H is a given hyperregular Hamiltonian, then the associated Lagrangian L is hyperregular and is given by

$$L = G \circ (\mathbb{F}H)^{-1} - H \circ (\mathbb{F}H)^{-1} = A - H \circ \mathbb{F}L.$$

Thus the corresponding hyperregular Hamiltonian induced by L is

$$H' = E \circ (\mathbb{F}L)^{-1} = (A - L) \circ (\mathbb{F}L)^{-1} = A \circ (\mathbb{F}L)^{-1} - (A - H \circ \mathbb{F}L) \circ (\mathbb{F}L)^{-1} = H.$$

The commutativity of the two diagrams is now a direct consequence of the above and Propositions **7.4.1** and **7.4.2**. ■

Exercise

7.4-1 *Write down the Lagrangian and the equations of motion for a spherical pendulum with S^2 as configuration space. Convert the equations to Hamiltonian form using the Legendre transformation. Find the conservation law corresponding to angular momentum about the axis of gravity by "bare hands" methods.*

7.5 Geodesics

As an illustration of the Lagrangian formalism, we consider geodesics. Let Q be a weak pseudo-Riemannian manifold whose metric evaluated at $q \in Q$ is denoted interchangeably by $\langle \cdot , \cdot \rangle$ or $g(q)$ or g_q. Consider on TQ the Lagrangian given by the kinetic energy of the metric, that is,

$$L(v) = \frac{1}{2} \langle v, v \rangle_q , \tag{7.5.1}$$

or in finite dimensions

$$L(v) = \frac{1}{2} g_{ij} v^i v^j . \tag{7.5.2}$$

The fiber derivative of L is given for $v, w \in T_qQ$ by

$$\mathbb{F}L(v) \cdot w = \langle v, w \rangle \tag{7.5.3}$$

or in finite dimensions by

$$\mathbb{F}L(v) \cdot w = g_{ij} v^i w^j \quad \text{i.e.,} \quad p_i = g_{ij} \dot{q}^j . \tag{7.5.4}$$

From here we see that in any chart U in Q, $\mathbf{D}_2\mathbf{D}_2 L(u, e) \cdot (e_1, e_2) = \langle e_1, e_2 \rangle$, where $\langle \, , \rangle$ denotes the inner product on E induced by the chart. Thus L is automatically weakly nondegenerate. Note that the action is given by $A = 2L$, so $E = L$.

The Lagrangian vector field Z in this case is denoted by $S : TQ \to T^2Q$ and is called the ***Christoffel map*** or ***geodesic spray*** of the metric $\langle \, , \rangle_q$. Thus S is a second-order equation and hence has a local expression of the form

$$S(q, v) = ((q, v), (v, \gamma(q, v))) \tag{7.5.5}$$

for any chart on Q. To determine the map $\gamma : U \times E \to E$ from Lagrange's equations, note that

$$\mathbf{D}_1 L(q, v) \cdot w = \frac{1}{2} \mathbf{D}_q \langle v, v \rangle_q \cdot w \quad \text{and} \quad \mathbf{D}_2 L(q, v) \cdot w = \langle v, w \rangle_q \tag{7.5.6}$$

so that Lagrange's equations (7.3.7) are

$$\dot{q} = v, \tag{7.5.7}$$

$$\frac{d}{dt}(\langle v, w \rangle_q) = \frac{1}{2} \mathbf{D}_q \langle v, v \rangle_q \cdot w. \tag{7.5.8}$$

Expanding the left-hand side of (7.5.8) yields

$$\mathbf{D}_q \langle v, w \rangle_q \cdot \dot{q} + \langle \dot{v}, w \rangle_q . \tag{7.5.9}$$

Taking into account $\dot{q} = v$, we get

$$\langle \ddot{q}, w \rangle_q = \frac{1}{2} \mathbf{D}_q \langle v, v \rangle_q \cdot w - \mathbf{D}_q \langle v, w \rangle_q \cdot v. \tag{7.5.10}$$

Hence $\gamma : U \times E \to E$ is defined by the equality

$$\langle \gamma(q,v), w \rangle_q = \frac{1}{2}\mathbf{D}_q \langle v, v \rangle_q \cdot w - \mathbf{D}_q \langle v, w \rangle_q \cdot v; \tag{7.5.11}$$

note that $\gamma(q,v)$ is a quadratic form in v. If Q is finite dimensional, we define the ***Christoffel symbols*** Γ^i_{jk} by putting

$$\gamma^i(q,v) = -\Gamma^i_{jk}(q)v^j v^k \tag{7.5.12}$$

and demanding $\Gamma^i_{jk} = \Gamma^i_{kj}$. With this notation, relation (7.5.11) is equivalent to

$$-g_{il}\Gamma^i_{jk}v^j v^k w^l = \frac{1}{2}\frac{\partial g_{jk}}{\partial q^l} v^j v^k w^l - \frac{\partial g_{jl}}{\partial q^k} v^j w^l v^k. \tag{7.5.13}$$

Taking into account the symmetry of Γ^i_{jk}, this gives

$$\Gamma^h_{jk} = \frac{1}{2}g^{hl}\left(\frac{\partial g_{jl}}{\partial q^k} + \frac{\partial g_{kl}}{\partial q^j} - \frac{\partial g_{jk}}{\partial q^l}\right). \tag{7.5.14}$$

In infinite dimensions, since the metric $\langle \, , \rangle$ is only weakly nondegenerate (7.5.11) guarantees the uniqueness of γ but not its existence. It exists whenever the Lagrangian vector field S exists.

The integral curves of S projected to Q are called ***geodesics*** of the metric g. By (7.5.5) they have the local expression

$$\ddot{q} = \gamma(q, \dot{q}), \tag{7.5.15}$$

which, in finite dimensions, reads

$$\ddot{q}^i + \Gamma^i_{jk}\dot{q}^j \dot{q}^k = 0, \quad i,j,k = 1, \ldots, n. \tag{7.5.16}$$

Note that the definition of γ makes sense both in the finite- and infinite-dimensional case. The Christoffel symbols Γ^i_{jk} are defined only for finite-dimensional manifolds. Working with g provides a way to deal with geodesics of weak Riemannian (and pseudo-Riemannian) metrics on infinite-dimensional manifolds. Taking the Lagrangian approach as basic, as opposite to the pseudo-Riemannian, we have seen where the Γ^i_{jk} live as geometric objects: in $T(TQ)$ since they encode the principal part of the Lagrangian vector field Z. If one writes down the transformation properties of Z on $T(TQ)$ in natural charts, the transformation rule for the Γ^i_{jk} will result:

$$\overline{\Gamma}^k_{ij} = \frac{\partial q^p}{\partial \overline{q}^i}\frac{\partial q^m}{\partial \overline{q}^j}\Gamma^r_{pm}\frac{\partial \overline{q}^k}{\partial q^r} + \frac{\partial \overline{q}^k}{\partial q^l}\frac{\partial^2 q^l}{\partial \overline{q}^i \, \partial \overline{q}^j}, \tag{7.5.17}$$

where $(q^1, \ldots, q^n), (\overline{q}^1, \ldots, \overline{q}^n)$ are two different coordinate systems on an open set of Q. We leave this exercise to the reader.

The Lagrangian approach also yields invariant manifolds for the geodesic flow. For example, let $\Sigma_e = \{v \in TQ \mid \|v\| = e\}, e \in \mathbb{R}$, be the ***pseudo-sphere bundle*** of radius $\sqrt{e}$ in TQ. *Then Σ_e is a smooth submanifold of TQ invariant under the geodesic flow.* Indeed, if we show that Σ_e is a smooth submanifold, its invariance under the geodesic flow, that is, under the flow of Z, follows by conservation of energy. To show that Σ_e is a smooth submanifold we prove that e is a regular value of the function $2L$. This is done locally by (7.5.6)

$$\begin{aligned} \mathbf{D}L(u,v)\cdot(w_1,w_2) &= \mathbf{D}_1L(u,v)\cdot w_1 + \mathbf{D}_2L(u,v)\cdot w_2 \\ &= \frac{1}{2}\mathbf{D}_u\left\langle v,v\right\rangle_u \cdot w_1 + \left\langle v,w_2\right\rangle_u \\ &= \left\langle v,w_2\right\rangle_u, \end{aligned} \tag{7.5.18}$$

since $\langle v,v\rangle = 2e = \text{constant}$. By weak nondegeneracy of the pseudo-metric $\langle\,,\rangle$, this shows that $\mathbf{D}L(u,v) : E\times E \to \mathbb{R}$ is a surjective linear map, that is, e is a regular value of L.

Let us reconcile the present approach to geodesics via Lagrangian systems to a common approach in differential geometry. Define the ***covariant derivative*** $\nabla : \mathfrak{X}(Q)\times\mathfrak{X}(Q)\to\mathfrak{X}(Q); (X,Y)\mapsto \nabla_X Y$ locally by

$$(\nabla_X Y)(u) = -\gamma(u)(X(u),Y(u)) + \mathbf{D}Y(u)\cdot X(u), \tag{7.5.19}$$

where X, Y are the local representatives of X and Y and $\gamma(u) : E\times E\to E$ denotes the symmetric bilinear form defined by the polarization of the quadratic form $\gamma(u,v)$ in v. If $E = \mathbb{R}^n$ then $\gamma(u,v)$ is an $\mathbb{R}^n$-valued quadratic form on $\mathbb{R}^n$ determined from the Γ^i_{jk} by

$$\nabla_X Y = X^j Y^k \Gamma^i_{jk}\frac{\partial}{\partial q^i} + X^j\frac{\partial Y^k}{\partial q^j}\frac{\partial}{\partial q^k}. \tag{7.5.20}$$

It is straightforward to check that this definition is chart independent and that ∇ satisfies the following conditions:

i ∇ is $\mathbb{R}$-bilinear;

ii for $f : Q\to\mathbb{R}$,

$$\nabla_{fX}Y = f\nabla_X Y \quad \text{and} \quad \nabla_X fY = f\nabla_X Y + X[f]Y;$$

and

iii for vector fields X and Y,

$$\begin{aligned} (\nabla_X Y - \nabla_Y X)(u) &= \mathbf{D}Y(u)\cdot X(u) - \mathbf{D}X(u)\cdot Y(u) \\ &= [X,Y](u). \end{aligned} \tag{7.5.21}$$

If $c(t)$ is a curve in Q and $X \in \mathfrak{X}(Q)$, the ***covariant derivative of*** X ***along*** c is defined by

$$\frac{DX}{dt} = \nabla_u X, \tag{7.5.22}$$

where u is a vector field coinciding with $\dot{c}(t)$ at $c(t)$. This is possible since, by (7.5.19) or (7.5.20), $\nabla_X Y$ depends only on the point values of X. Explicitly, in a local chart, we have

$$\frac{DX}{dt}(c(t)) = -\gamma_{c(t)}(u(c(t)), X(c(t))) + \frac{d}{dt}X(c(t)), \tag{7.5.23}$$

which shows that DX/dt depends only on $\dot{c}(t)$ and not on how $\dot{c}(t)$ is extended to a vector field. In finite dimensions,

$$\left(\frac{DX}{dt}\right)^i = \Gamma^i_{jk}(c(t))\dot{c}^j(t)X^k(c(t)) + \frac{d}{dt}X^i(c(t)). \tag{7.5.24}$$

The vector field X is called ***autoparallel*** or ***parallel transported*** along c if $DX/dt = 0$. Thus $\dot{c}$ is autoparallel along c if and only if $\ddot{c}(t) - \gamma(t)(\dot{c}(t), \dot{c}(t)) = 0$, that is, $c(t)$ is a geodesic. In finite dimensions, this reads $\ddot{c}^i + \Gamma^i_{jk}\dot{c}^j\dot{c}^k = 0$.

Exercises

7.5-1 *Consider the Lagrangian*

$$L_\epsilon(x, y, z, \dot{x}, \dot{y}, \dot{z}) = \frac{1}{2}(\dot{x}^2 + \dot{y}^2 + \dot{z}^2) - \frac{1}{2\epsilon}[1 - (x^2 + y^2 + z^2)]^2$$

for a particle in $\mathbb{R}^3$. *Let* $\gamma_\epsilon(t)$ *be the curve in* $\mathbb{R}^3$ *obtained by solving the Euler-Lagrange equations for* L_ϵ *with the initial conditions* $\mathbf{x}_0, \mathbf{v}_0 = \dot{\gamma}_\epsilon(0)$. *Show that*

$$\lim_{\epsilon \to 0} \gamma_\epsilon(t)$$

is a great circle on the two-sphere S^2, *provided that* $\mathbf{x}_0$ *has length one and that* $\mathbf{x}_0 \cdot \mathbf{v}_0 = 0$.

7.5-2 *Write out the geodesic equations in terms of* q^i *and* p_i *and check directly that Hamilton's equations are satisfied.*

7.6 The Kaluza-Klein Approach to Charged Particles

In §6.7 we studied the motion of a charged particle in a magnetic field as a Hamiltonian system. This description is, in fact, the reduction of a larger and, in some sense, simpler system called the ***Kaluza-Klein system***. After learning reduction theory (see Volume II or Marsden [1992]); the reader can verify this, but here all the constructions are done directly.

Physically, we are motivated as follows: since charge is a basic conserved quantity, we would like to introduce a new cyclic variable whose conjugate momentum is the charge. This process is applicable to other situations as well; for example, in fluid dynamics one can profitably introduce a variable conjugate to the conserved mass density or entropy; see Marsden, Ratiu, and Weinstein [1984a,b]. For a charged particle, the resultant system is in fact geodesic motion! Recall from §6.7 that if $\mathbf{B} = \nabla \times \mathbf{A}$ is a given magnetic field on $\mathbb{R}^3$, then with respect to canonical variables $(\mathbf{q}, \mathbf{p})$, the Hamiltonian is

$$H(\mathbf{q}, \mathbf{p}) = \frac{1}{2m}\|\mathbf{p} - \frac{e}{c}\mathbf{A}\|^2. \tag{7.6.1}$$

First observe that we can obtain (7.6.1) via the Legendre transform if we choose

$$L(\mathbf{q}, \dot{\mathbf{q}}) = \frac{1}{2}m\|\dot{\mathbf{q}}\|^2 + \frac{e}{c}\mathbf{A} \cdot \dot{\mathbf{q}}, \tag{7.6.2}$$

for then

$$\mathbf{p} = \frac{\partial L}{\partial \dot{\mathbf{q}}} = m\dot{\mathbf{q}} + \frac{e}{c}\mathbf{A} \tag{7.6.3}$$

and

$$\begin{aligned} H(\mathbf{q}, \mathbf{p}) &= \mathbf{p} \cdot \dot{\mathbf{q}} - L(\mathbf{q}, \dot{\mathbf{q}}) \\ &= (m\dot{\mathbf{q}} + \frac{e}{c}\mathbf{A}) \cdot \dot{\mathbf{q}} - \frac{1}{2}m\|\dot{\mathbf{q}}\|^2 - \frac{e}{c}\mathbf{A} \cdot \dot{\mathbf{q}} \\ &= \frac{1}{2}m\|\dot{\mathbf{q}}\|^2 = \frac{1}{2m}\|\mathbf{p} - \frac{e}{c}\mathbf{A}\|^2. \end{aligned} \tag{7.6.4}$$

Thus, the Euler-Lagrange equations for (7.6.2) reproduce the equations for a particle in a magnetic field. (If an electric field $E = -\nabla\varphi$ is present as well, one simply subtracts $e\varphi$ from L, treating $e\varphi$ as a potential energy, as in the next section.) Let the configuration space be

$$Q_K = \mathbb{R}^3 \times S^1 \tag{7.6.5}$$

with variables $(\mathbf{q}, \theta)$, define $A = \mathbf{A}^\flat$, a one-form on $\mathbb{R}^3$, and consider the one-form

$$\omega = A + \mathbf{d}\theta \tag{7.6.6}$$

on Q_K called the ***connection one-form***. Let the ***Kaluza-Klein Lagrangian*** be defined by

$$\begin{aligned} L_K(\mathbf{q}, \dot{\mathbf{q}}, \theta, \dot{\theta}) &= \frac{1}{2}m\|\dot{\mathbf{q}}\|^2 + \frac{1}{2}\left\|\left\langle \omega, (\mathbf{q}, \dot{\mathbf{q}}, \theta, \dot{\theta}\right\rangle\right\|^2 \\ &= \frac{1}{2}m\|\dot{\mathbf{q}}\|^2 + \frac{1}{2}(\mathbf{A} \cdot \dot{\mathbf{q}} + \dot{\theta})^2. \end{aligned} \tag{7.6.7}$$

The corresponding momenta are

$$\mathbf{p} = m\dot{\mathbf{q}} + (\mathbf{A} \cdot \dot{\mathbf{q}} + \dot{\theta})\mathbf{A} \tag{7.6.8}$$

and

$$p = \mathbf{A} \cdot \dot{\mathbf{q}} + \dot{\theta}. \tag{7.6.9}$$

Since (7.6.7) is quadratic and positive-definite in $\dot{\mathbf{q}}$ and $\dot{\theta}$, *the Euler-Lagrange equations are the geodesic equations on* $\mathbb{R}^3 \times S^1$ *for the metric for which* L_K *is the kinetic energy*. Since p is constant in time as can be seen from the Euler-Lagrange equation for $(\theta, \dot{\theta})$, we can define the ***charge*** e by setting

$$p = e/c; \tag{7.6.10}$$

then (7.6.8) coincides with (7.6.3). The corresponding Hamiltonian on T^*Q_K endowed with the canonical symplectic form is

$$H_K(\mathbf{q}, \mathbf{p}, \theta, p) = \frac{1}{2m}\|\mathbf{p} - p\mathbf{A}\|^2 + \frac{1}{2}p^2. \tag{7.6.11}$$

With (7.6.10), (7.6.11) differs from (7.6.1) only by the constant $p^2/2$.

These constructions generalize to the case of a particle in a Yang-Mills field where ω becomes the ***connection*** of a Yang-Mills field and its ***curvature*** measures the field strength which, for an electromagnetic field, reproduces the relation $\mathbf{B} = \nabla \times \mathbf{A}$. Also, the possibility of putting the interaction in the Hamiltonian, or via a momentum shift, into the symplectic structure, also generalizes. We refer to Wong [1970], Sternberg [1977], Weinstein [1978], and Montgomery [1984] for details and further references. Finally, we remark that the relativistic context is the most natural to introduce the full electromagnetic field. In that setting the construction we have given for the magnetic field will include both electric and magnetic effects. Consult Misner, Thorne, and Wheeler [1973] for additional information.

Exercise

7.6-1 *The bob on a spherical pendulum has a charge* e*, mass* m*, and moves under the influence of a constant gravitational field with acceleration* g*, and a magnetic field* $\mathbf{B}$*. Write down the Lagrangian, the Euler-Lagrange equations, and the variational principle for this system. Transform the system to Hamiltonian form. Find a conserved quantity if the field* $\mathbf{B}$ *is symmetric about the axis of gravity.*

7.7 Motion in a Potential Field

We now generalize geodesic motion to include potentials $V : Q \to \mathbb{R}$. Recall that the ***gradient*** of V is the vector field grad $V = \nabla V$ defined by the equality

$$\langle \operatorname{grad} V(q), v \rangle_q = \mathbf{d}V(q) \cdot v, \tag{7.7.1}$$

for all $v \in T_qQ$. In finite dimensions, this becomes

$$(\operatorname{grad} V)^i = g^{ij} \frac{\partial V}{\partial q^j}. \tag{7.7.2}$$

Define the (weakly nondegenerate) Lagrangian $L(v) = \frac{1}{2} \langle v, v \rangle_q - V(q)$. A computation similar to the one in §**7.5** shows that Lagrange's equations are

$$\ddot{q} = \gamma(q, \dot{q}) - \operatorname{grad} V(q), \tag{7.7.3}$$

or in finite dimensions

$$\ddot{q}^i + \Gamma^i_{jk} \dot{q}^j \dot{q}^k + g^{il} \frac{\partial V}{\partial q^l} = 0. \tag{7.7.4}$$

The action of L is given by

$$A(v) = \langle v, v \rangle_q, \tag{7.7.5}$$

so that the energy is

$$E(v) = A(v) - L(v) = \frac{1}{2} \langle v, v \rangle_q + V(q). \tag{7.7.6}$$

The equations (7.7.3) written as

$$\dot{q} = v, \ \dot{v} = \gamma(q, v) - \operatorname{grad} V(q) \tag{7.7.7}$$

are thus Hamiltonian with Hamiltonian function E with respect to the symplectic form Ω_L.

To write equations (7.7.7) in an invariant form, we use the following terminology:

Definition 7.7.1 *Let $v, w \in T_qQ$. The* ***vertical lift*** *of w with respect to v is defined by*

$$\operatorname{ver}(w, v) = \left. \frac{d}{dt} \right|_{t=0} (v + tw) \in T_v(TQ).$$

The ***horizontal part*** *of a vector $U \in T_v(TQ)$ is $T_v\tau_Q(U) \in T_qQ$. A vector field is called* ***vertical*** *if its horizontal part is zero.*

In charts, if $v = (u, e)$, $w = (u, f)$, and $U = ((u, e), (e_1, e_2))$, the definition says that

$$\operatorname{ver}(w, v) = ((u, e), (0, f)) \quad \text{and} \quad T_v\tau_Q(U) = (u, e_1).$$

So, U is vertical iff $e_1 = 0$. Thus, *any vertical vector $U \in T_v(TQ)$ is the vertial lift of some vector w (which in a natural local chart is (u, e_2)) with respect to v.*

If S denotes the geodesic spray of the metric $\langle\, , \rangle$ on TQ, equations (7.7.7) say that the Lagrangian vector field Z defined by $L(v) = \frac{1}{2}\langle v, v\rangle_q - V(q)$ where $v \in T_qQ$, is given by

$$Z = S - \operatorname{ver}(\nabla V), \tag{7.7.8}$$

that is,

$$Z(v) = S(v) - \operatorname{ver}((\nabla V)(q), v). \tag{7.7.9}$$

Remark In general, there is *no* canonical way to take the *vertical part* of a vector $U \in T_v(TQ)$ without extra structure. Having such a structure is what one means by a ***connection***. In case Q is pseudo-Riemannian, such a projection can be constructed in the following manner. Suppose, in natural charts, that $U = ((u, e), (e_1, e_2))$. Define

$$U_{\text{ver}} = ((u, e),\ (0, \gamma(u)(e_1, e_2) + e_2))$$

where $\gamma(u)$ is the bilinear symmetric form associated to the quadratic form $\gamma(u, e)$ in e. ♦

We conclude with some miscellaneous remarks connecting motion in a potential field with geodesic motion. We confine ourselves to the finite-dimensional case for simplicity.

Definition 7.7.2 *Let $g = \langle\, , \rangle$ be a pseudo-Riemannian metric on Q and let $V : Q \to \mathbb{R}$ be bounded above. If $e > V(q)$ for all $q \in Q$ define the **Jacobi metric** g_e by $g_e = (e - V)g$, i.e., $g_e(v, w) = (e - V(q))\langle v, w\rangle$ for all $v, w \in T_qQ$.*

Theorem 7.7.3 *Let Q be finite dimensional. The base integral curves of the Lagrangian $L(v) = \frac{1}{2}\langle v, v\rangle - V(q)$ with energy e are the same as geodesics of the Jacobi metric with energy 1, up to a reparametrization.*

The proof is based on the following of separate interest.

Proposition 7.7.4 *Let (P, Ω) be a (finite-dimensional) symplectic manifold, $H, K \in \mathcal{F}(P)$, and assume that $\Sigma = H^{-1}(h) = K^{-1}(k)$ for $h, k \in \mathbb{R}$ regular values of H and K, respectively. Then the integral curves of X_H and X_K on the invariant submanifold Σ of both X_H and X_K coincide up to a reparametrization.*

Proof From $\Omega(X_H(z), v) = \mathbf{d}H(z) \cdot v$, we see that

$$X_H(z) \in (\ker \mathbf{d}H(z))^{\Omega} = (T_z\Sigma)^{\Omega},$$

the symplectic orthogonal complement of $T_z\Sigma$. Since Ω is nondegenerate, it defines an isomorphism between $T_zP/(T_z\Sigma)^\Omega$ and $(T_z\Sigma)^*$. Since these spaces are finite dimensional and $T_z\Sigma$ has codimension one, $(T_z\Sigma)^\Omega$ has dimension one. Thus, the nonzero vectors $X_H(z)$ and $X_K(z)$ are multiples of each other at every point $z \in \Sigma$, that is, there is a smooth nowhere vanishing function $\lambda : \Sigma \to \mathbb{R}$ such that $X_H(z) = \lambda(z)X_K(z)$ for all $z \in \Sigma$. Let $c(t)$ be the integral curve of X_K with initial condition $c(0) = z_0 \in \Sigma$. The function $\varphi \mapsto \int_0^\varphi dt/(\lambda \circ c)(t)$ is a smooth monotone function and therefore has an inverse $t \mapsto \varphi(t)$. If $d(t) = (c \circ \varphi)(t)$, then $d(0) = z_0$ and

$$\begin{aligned} d'(t) &= \varphi'(t)c'(\varphi(t)) = \frac{1}{t'(\varphi)}X_K(c(\varphi(t))) = (\lambda \circ c)(\varphi)X_K(d(t)) \\ &= \lambda(d(t))X_K(d(t)) = X_H(d(t)) \end{aligned}$$

that is, the integral curve of X_H through z_0 is obtained by reparametrizing the integral curve of X_K through z_0. ■

Proof of Theorem 7.7.3 Let H be the Hamiltonian for L, namely

$$H(q,p) = \frac{1}{2}\|p\|^2 + V(q)$$

and H_e be that for the Jacobi metric:

$$H_e(q,p) = \frac{1}{2}(e - V(q))^{-1}\|p\|^2.$$

The factor $(e - V(q))^{-1}$ occurs because the inverse metric is used for the momenta. Clearly $H = e$ defines the same set as $H_e = 1$, so the result follows from Proposition **7.7.4** if we show that e is a regular value of H and 1 is a regular value of H_e. Note that if $(q,p) \in H^{-1}(e)$, then $p \neq 0$ since $e > V(q)$ for all $q \in Q$. Therefore $\mathbb{F}H(q,p) \neq 0$ for any $(q,p) \in H^{-1}(e)$ and hence $\mathbf{d}H(q,p) \neq 0$, that is, e is a regular value of H. Since $\mathbb{F}H_e(q,\dot{p}) = \frac{1}{2}(e - V(q))^{-1}\mathbb{F}H(q,p)$, this also shows that $\mathbb{F}H_e(q,p) \neq 0$ for all $(q,p) \in H^{-1}(e) = H_e^{-1}(1)$ and thus 1 is a regular value of H_e. ■

7.8 The Lagrange-d'Alembert Principle

In this section we study a generalization of Lagrange's equations for mechanical systems with exterior forces. A special class of such forces is dissipative forces, which will be studied at the end of this section.

Let $L : TQ \to \mathbb{R}$ be a Lagrangian function, let Z be the Lagrangian vector field associated to L, assumed to be a second-order equation, and denote by $\tau_Q : TQ \to Q$ the canonical projection. Recall that a vector field Y on TQ is *vertical* if $T\tau_Q \circ Y = 0$. Such a vector field Y defines a one-form on TQ by

$$\Delta^Y = -\mathbf{i}_Y\Omega_L.$$

Proposition 7.8.1 *If Y is vertical, then Δ^Y is a horizontal one-form, i.e., $\Delta^Y(U) = 0$ for any vertical vector field U on TQ. Conversely, given a horizontal one-form Δ on TQ, and assuming that L is regular, the vector field Y on TQ, defined by $\Delta = -\mathbf{i}_Y \Omega_L$, is vertical.*

Proof This follows from a straightforward calculation in local coordinates. We use the fact that a vector field $Y(u,e) = (Y_1(u,e), Y_2(u,e))$ is vertical if and only if the first component Y_1 is zero and the local formula for Ω_L derived earlier:

$$\begin{aligned}
&\Omega_L(u,e)(Y_1,Y_2),(U_1,U_2)) \\
&\quad = \mathbf{D}_1(\mathbf{D}_2 L(u,e)\cdot Y_1)\cdot U_1 - \mathbf{D}_1(\mathbf{D}_2 L(u,e)\cdot U_1)\cdot Y_1 \\
&\qquad + \mathbf{D}_2\mathbf{D}_2 L(u,e)\cdot Y_1 \cdot U_2 - \mathbf{D}_2\mathbf{D}_2 L(u,e)\cdot U_1\cdot Y_2. \qquad (7.8.1)
\end{aligned}$$

This shows that $(\mathbf{i}_Y\Omega_L)(U) = 0$ for all vertical U is equivalent to

$$\mathbf{D}_2\mathbf{D}_2 L(u,e)(U_2, Y_1) = 0.$$

If Y is vertical, this is clearly true. Conversely if L is regular, and the last displayed equation is true, then $Y_1 = 0$, so Y is vertical. ■

Proposition 7.8.2 *Any fiber-preserving map $F : TQ \to T^*Q$ over the identity induces a horizontal one-form $\tilde{F}$ on TQ by*

$$\tilde{F}(v)\cdot V_v = \langle F(v), T_v\tau(V_v)\rangle, \qquad (7.8.2)$$

where $v \in TQ$ and $V_v \in T_v(TQ)$. Conversely, formula (7.8.2) defines, for any horizontal one-form $\tilde{F}$ a fiber-preserving map F over the identity. Any such F is called a ***force field*** *and thus, in the regular case, any vertical vector field Y is induced by a force field.*

Proof Given F, formula (7.8.2) clearly defines a smooth one-form $\tilde{F}$ on TQ. If V_v is vertical, then the right-hand side of formula (7.8.2) vanishes, and so $\tilde{F}$ is a horizontal one-form. Conversely, given a horizontal one-form $\tilde{F}$ on TQ, and given $v, w \in T_qQ$, let $V_v \in T_v(TQ)$ be such that $T_v\tau(V_v) = w$. Then define F by formula (7.8.2); that is, $\langle F(v), w\rangle = \tilde{F}(v)\cdot V_v$. Since $\tilde{F}$ is horizontal, we see that F is well defined, and its expression in charts shows that it is smooth. ■

Treating Δ^Y as the exterior force one-form acting on a mechanical system with a Lagrangian L, we now will write the governing equations of motion. The basic principle is of course the Lagrange-d'Alembert principle. First, we recall the definition from Vershik and Faddeev [1981] and Wang and Krishnaprasad [1992].

Definition 7.8.3 *The* ***Lagrangian force*** *associated with a Lagrangian L and a given second-order vector field (the ultimate equations of motion) X is the horizontal one-form on TQ defined by*

$$\Phi_L(X) = \mathbf{i}_X\Omega_L - \mathbf{d}E. \qquad (7.8.3)$$

Given a horizontal one-form ω (referred to as the ***exterior force one-form****), the* ***local Lagrange d'Alembert principle*** *associated with the second-order vector field X on TQ states that*

$$\Phi_L(X) + \omega = 0. \tag{7.8.4}$$

It is easy to check that $\Phi_L(X)$ is indeed horizontal if X is second order. Conversely, if L is regular and if $\Phi_L(X)$ is horizontal, then X is second order.

One can also formulate an equivalent principle in terms of variational principles.

Definition 7.8.4 *Given a Lagrangian L and a force field F, as defined in Proposition* **7.8.2**, *the* ***integral Lagrange d'Alembert principle*** *for a curve $q(t)$ in Q is*

$$\delta \int_a^b L(q(t), \dot{q}(t))\, dt + \int_a^b F(q(t), \dot{q}(t)) \cdot \delta q \, dt = 0, \tag{7.8.5}$$

where the variation is given by the usual expression

$$\begin{aligned} \delta \int_a^b L(q(t), \dot{q}(t))\, dt &= \int_a^b \left(\frac{\partial L}{\partial q^i} \delta q^i + \frac{\partial L}{\partial \dot{q}^i} \frac{d}{dt} \delta q^i \right) dt. \\ &= \int_a^b \left(\frac{\partial L}{\partial q^i} - \frac{d}{dt} \frac{\partial L}{\partial \dot{q}^i} \right) \delta q^i \, dt. \end{aligned} \tag{7.8.6}$$

for a given variation δq (vanishing at the endpoints).

The two forms of the Lagrange-d'Alembert principle are in fact equivalent. This will follow from the fact that both give the Euler-Lagrange equations with forcing in local coordinates (provided that Z is second order). We shall see this in the following development.

Proposition 7.8.5 *Let the exterior force one-form ω be associated to a vertical vector field Y, that is, let $\omega = \Delta^Y = -\mathbf{i}_Y \Omega_L$. Then $X = Z + Y$ satisfies the local Lagrange-d'Alembert principle. Conversely, if, in addition, L is regular, the only second-order vector field X satisfying the local Lagrange-d'Alembert principle is $X = Z + Y$.*

Proof For the first part, the equality $\Phi_L(X) + \omega = 0$ is a simple verification. For the converse, we already know that X is a solution, and uniqueness is guaranteed by regularity. ■

To develop the differential equations associated to $X = Z + Y$, we take $\omega = -\mathbf{i}_Y \Omega_L$ and note that, in a coordinate chart, $Y(q, v) = (0, Y_2(q, v))$ since Y is vertical, that is, $Y_1 = 0$. From the local formula for Ω_L, we get

$$\omega(q, v) \cdot (u, w) = \mathbf{D}_2 \mathbf{D}_2 L(q, v) \cdot Y_2(q, v) \cdot u. \tag{7.8.7}$$

Letting $X(q,v) = (v, X_2(q,v))$, one finds that

$$
\begin{aligned}
&\Phi_L(X)(q,v)\cdot(u,w)\\
&= \left(-\mathbf{D}_1(\mathbf{D}_2L(q,v)\cdot)\cdot v - \mathbf{D}_2\mathbf{D}_2L(q,v)\cdot X_2(q,v) + \mathbf{D}_1L(q,v)\right)\cdot u.
\end{aligned} \tag{7.8.8}
$$

Thus, the local Lagrange-d'Alembert principle becomes

$$
\begin{aligned}
&\left(-\mathbf{D}_1(\mathbf{D}_2L(q,v)\cdot)\cdot v - \mathbf{D}_2\mathbf{D}_2L(q,v)\cdot X_2(q,v) + \mathbf{D}_1L(q,v)\right.\\
&\qquad \left.+\mathbf{D}_2\mathbf{D}_2L(q,v)\cdot Y_2(q,v)\right) = 0.
\end{aligned} \tag{7.8.9}
$$

Setting $v = dq/dt$ and $X_2(q,v) = dv/dt$, the preceding relation and the chain rule gives

$$
\frac{d}{dt}(\mathbf{D}_2L(q,v) - \mathbf{D}_1L(q,v)) = \mathbf{D}_2\mathbf{D}_2L(q,v)\cdot Y_2(q,v), \tag{7.8.10}
$$

which, in finite dimensions, reads

$$
\frac{d}{dt}\left(\frac{\partial L}{\partial \dot{q}^i}\right) - \frac{\partial L}{\partial q^i} = \frac{\partial^2 L}{\partial \dot{q}^i\,\partial \dot{q}^j}Y^j(q^k,\dot{q}^k). \tag{7.8.11}
$$

The force one-form Δ^Y is therefore given by

$$
\Delta^Y(q^k,\dot{q}^k) = \frac{\partial^2 L}{\partial \dot{q}^i\,\partial \dot{q}^j}Y^j(q^k,\dot{q}^k)\,dq^i \tag{7.8.12}
$$

and the corresponding force field is

$$
F^Y = \left(q^i, \frac{\partial^2 L}{\partial \dot{q}^i\,\partial \dot{q}^j}Y^j(q^k,\dot{q}^k)\right). \tag{7.8.13}
$$

Thus, the condition for an integral curve takes the form of the standard Euler-Lagrange equations with forces:

$$
\frac{d}{dt}\left(\frac{\partial L}{\partial \dot{q}^i}\right) - \frac{\partial L}{\partial q^i} = F^Y_i(q^k\dot{q}^k). \tag{7.8.14}
$$

Since the integral Lagrange-d'Alembert principle gives the same equations, it follows that the two principles are equivalent. From now on, we will refer to either one as simply the ***Lagrange-d'Alembert principle.***

We summarize the results obtained so far in the following:

Theorem 7.8.6 *Given a regular Lagrangian and a force field $F : TQ \to T^*Q$, for a curve $q(t)$ in Q, the following are equivalent:*

(a) *$q(t)$ satisfies the local Lagrange-d'Alembert principle;*

(b) *$q(t)$ satisfies the integral Lagrange-d'Alembert principle; and*

(c) *$q(t)$ is the base integral curve of the second-order equation $Z + Y$, where Y is the vertical vector field on Q inducing the force field F by (7.8.15), and Z is the Lagrangian vector field on L.*

Let E denote the energy defined by L, that is, $E = A - L$, where $A(v) = \langle \mathbb{F}L(v), v\rangle$ is the action of L.

Definition 7.8.7 *A vertical vector field Y on TQ is called* ***weakly dissipative*** *if $\langle \mathbf{d}E, Y\rangle \leq 0$ at all points of TQ. If the inequality is strict off the zero section of TQ, Y is called* ***dissipative****. A* ***dissipative Lagrangian system*** *on TQ is a vector field $Z+Y$, for Z a Lagrangian vector field and Y a dissipative vector field.*

Corollary 7.8.8 *A vertical vector field Y on TQ is dissipative if and only if the force field F^Y that it induces satisfies $\left\langle F^Y(v), v\right\rangle < 0$ for all nonzero $v \in TQ$ (≤ 0 for the weakly dissipative case).*

Proof Let Y be a vertical vector field. By Proposition **7.8.1**, Y induces a horizontal one-form $\Delta^Y = -\mathbf{i}_Y\Omega_L$ on TQ, and by Proposition **7.8.2** Δ^Y in turn induces a force field F^Y given by

$$\left\langle F^Y(v), w\right\rangle = \Delta^Y(v)\cdot V_v = -\Omega_L(v)(Y(v), V_v), \tag{7.8.15}$$

where $T\tau_Q(V_v) = w$ and $V_v \in T_v(TQ)$. If Z denotes the Lagrangian system defined by L, we get

$$\begin{aligned}(\mathbf{d}E\cdot Y)(v) &= (\mathbf{i}_Z\Omega_L)(Y)(v) = \Omega_L(Z,Y)(v)\\ &= -\Omega_L(v)(Y(v), Z(v))\\ &= \left\langle F^Y(v), T_v\tau(Z(v))\right\rangle\\ &= \left\langle F^Y(v), v\right\rangle,\end{aligned}$$

since Z is a second-order equation. Thus, $\mathbf{d}E\cdot Y < 0$ if and only if $\left\langle F^Y(v), v\right\rangle < 0$ for all $v\in TQ$. ■

Definition 7.8.9 *Given a dissipative vector field Y on TQ, let $F^Y : TQ \to T^*Q$ be the induced force field. If there is a function $R : TQ \to \mathbb{R}$ such that F^Y is the fiber derivative of $-R$, then R is called a* ***Rayleigh dissipation function****.*

Note that in this case, $\mathbf{D}_2R(q,v)\cdot v > 0$ for the dissipativity of Y. Thus, if R is linear in the fiber variable, the Rayleigh dissipation function takes on the classical form $\langle \mathcal{R}(q)v, v\rangle$, where $\mathcal{R}(q) : TQ \to T^*Q$ is a bundle map over the identity that defines a symmetric positive-definite form on each fiber of TQ.

Finally, if the force field is given by a Rayleigh dissipation function R, then the Euler-Lagrange equations with forcing become

$$\frac{d}{dt}\left(\frac{\partial L}{\partial \dot{q}^i}\right) - \frac{\partial L}{\partial q^i} = -\frac{\partial R}{\partial \dot{q}^i}. \tag{7.8.16}$$

Combining Corollary **7.8.8** with the fact that the differential of E along Z is zero, we find that under the flow of the Euler-Lagrange equations with forcing of Rayleigh dissipation type

$$\frac{d}{dt}E(q, v) = F(v) \cdot v = -\mathbb{F}R(q, v) \cdot v < 0. \tag{7.8.17}$$

Exercises

7.8-1 *What is the power or rate of work equation (see* §2.1*) for a system with forces on a Riemannian manifold?*

7.8-2 *Write the equations for a ball in a rotating hoop, including friction, in the language of this section. (See* §2.10*). Compute the Rayleigh dissipation function.*

7.9 The Hamilton-Jacobi Equation

In §6.5 we studied generating functions of canonical transformations. Here we link them with the flow of a Hamiltonian system via the Hamilton-Jacobi equation .

We begin with the gauge invariance of Lagrangian systems. Starting with Hamilton's principle

$$\delta \int_a^b L(q^i(t), \dot{q}^i(t), t)\, dt = 0, \tag{7.9.1}$$

we see that if a total time derivative of a function is added to L, the condition (7.9.1) is unchanged, as long as the function itself is held fixed at $t = a$ and $t = b$. If $S(q, q_0, t - t_0)$ is a function of $q, q_0 \in Q$ with q_0, t_0 appearing parametrically, and we substitute a curve $q(t)$ for q, then we can change L by

$$L \mapsto \overline{L} := L - \frac{dS}{dt} = L - \frac{\partial S}{\partial q^i}\dot{q}^i - \frac{\partial S}{\partial t} \tag{7.9.2}$$

without affecting (7.9.1). This is consistent with the fact that the Euler-Lagrange equations for dS/dt are identically satisfied, so the Euler-Lagrange equations for L and $L - dS/dt$ are the same. (One says that dS/dt is a ***null Lagrangian***.)

The momentum for L is $p_i = \partial L/\partial \dot{q}^i$ as usual, while that for $\overline{L}$ is

$$\overline{p}_i := \frac{\partial \overline{L}}{\partial \dot{q}^i} = p_i - \frac{\partial S}{\partial q^i}. \tag{7.9.3}$$

The Hamiltonian for $\overline{L}$ is $\overline{H} = \overline{p}_i \dot{q}^i - \overline{L}$, which gives, using (7.9.2) and (7.9.3),

$$\overline{H} = H + \frac{\partial S}{\partial t}. \tag{7.9.4}$$

The new Hamiltonian will be especially simple if we ask that $\overline{H} = 0$ and $\overline{p} = 0$. Note that the condition $\overline{p} = 0$ means $p_i = \partial S/\partial q^i$, which is one of the equations defining a generating function. Then (7.9.4) becomes

$$H\left(q^1, \ldots, q^n, \frac{\partial S}{\partial q^1}, \ldots, \frac{\partial S}{\partial q^n}\right) + \frac{\partial S}{\partial t} = 0, \tag{7.9.5}$$

which is called the ***Hamilton-Jacobi equation***.

Let us now consider the *time-dependent* canonical transformation ψ generated by S. For notational convenience, we write the transformation as

$$\psi : (q^i, p_i, t) \mapsto (\overline{q}^i, \overline{p}_i, t),$$

where

$$\overline{p}_i = -\frac{\partial S}{\partial \overline{q}^i} \quad \text{and} \quad p_i = \frac{\partial S}{\partial q^i}. \tag{7.9.6}$$

For each fixed t, the map $(q^i, p_i) \mapsto (\overline{q}^i, \overline{p}_i)$ has S as a generating function and is therefore a canonical transformation, if we assume that (7.9.6) defines an invertible transformation. Write (7.9.6) as

$$\begin{aligned} p_i\, dq^i &= \overline{p}_i\, d\overline{q}^i + \frac{\partial S}{\partial q^i} dq^i + \frac{\partial S}{\partial \overline{q}^i} d\overline{q}^i \\ &= \overline{p}_i\, d\overline{q}^i - \frac{\partial S}{\partial t} dt + \mathbf{d}S, \end{aligned} \tag{7.9.7}$$

where $\mathbf{d}S$ is the full differential of S:

$$\mathbf{d}S = \frac{\partial S}{\partial q^i} dq^i + \frac{\partial S}{\partial \overline{q}^i} d\overline{q}^i + \frac{\partial S}{\partial t} dt.$$

From (7.9.7), we get the following basic relationship on $(T^*Q) \times \mathbb{R}$:

$$p_i\, dq^i - H\, dt = \overline{p}_i\, d\overline{q}^i - \overline{H}\, dt + \mathbf{d}S \tag{7.9.8}$$

where $\overline{H} = H + \partial S/\partial t$. It follows that the transformation ψ satisfies

$$dq^i \wedge dp_i + dH \wedge dt = d\overline{q}^i \wedge d\overline{p}_i + d\overline{H} \wedge dt. \tag{7.9.9}$$

When one transforms Hamilton's equations under a *time-dependent* transformation, it is important to do this transformation in the *extended* phase space $P \times \mathbb{R}$ with variables (q^i, p_j, t). This is discussed intrinsically in Abraham and Marsden [1978] but here we proceed by the above coordinate calculations for simplicity.

From Exercise **6.2-3** we know that the vector field $\overline{X}_H = (X_H, 1)$ on $T^*Q \times \mathbb{R}$ is uniquely determined by the equation $\mathbf{i}_{\overline{X}_H}\omega_H = 0$, where $\omega_H = dq^i \wedge dp_i + dH \wedge dt$. The transformation ψ sends the two-form ω_H to $\omega_{\overline{H}}$ and thus $\overline{X}_H$ will be sent to a vector field Y in the variables $(\overline{q}^i, \overline{p}_i, t)$ satisfying $\mathbf{i}_Y \omega_{\overline{H}} = 0$. However, again by Exercise **6.2-3**, this vector field is uniquely determined by this equation and $\overline{X}_{\overline{H}} = (X_{\overline{H}}, 1)$ satisfies this equation. Therefore $Y = \overline{X}_H$ and thus ψ transforms the equations

$$\dot{q}^i = \frac{\partial H}{\partial p_i}, \quad \dot{p}_i = -\frac{\partial H}{\partial q^i} \tag{7.9.10}$$

into

$$\dot{\overline{q}}^i = \frac{\partial \overline{H}}{\partial \overline{p}_i}, \quad \dot{\overline{p}}_i = -\frac{\partial \overline{H}}{\partial \overline{q}^i}. \tag{7.9.11}$$

Again, if $\overline{H} = 0$, then $\overline{q}^i$ and $\overline{p}_i$ are constants of integration, so that (7.9.6) provides the integration of the problem. We now summarize these calculations.

Theorem 7.9.1 (Hamilton-Jacobi) *If $S(q^i, \overline{q}^i, t)$ is a solution of the Hamilton-Jacobi equation, if (7.9.6) defines an invertible transformation ψ, and if $\overline{q}^i$ and $\overline{p}_i$ are constants in time, then (q^i, p_i) defined by (7.9.6) solve Hamilton's equations. Conversely, if (7.9.6) defines an invertible transformation $\psi : (q^i, p_i, t) \mapsto (\overline{q}^i, \overline{p}_i, t)$ such that $(\overline{q}^i, \overline{p}_i)$ are independent of time and if (q^i, p_i) satisfy Hamilton's equations for some Hamiltonian H, then the generating function S in (7.9.6) satisfies the Hamilton-Jacobi equation (7.9.5) for H, provided we adjust the function H by a suitable constant.*

The converse is proved in the following way. Since (q^i, p_i) satisfy Hamilton's equations for H, by Exercise **6.2-3** we know $\mathbf{i}_{\overline{X}_H}\omega_H = 0$. As before, the transformation (7.9.6) sends $\overline{X}_H = (X_H, 1)$ to $(\overline{X}_{\overline{H}}, 1)$, $\overline{H} = H + \partial S/\partial t$ and ω_H to $\omega_{\overline{H}}$. Therefore we must have $\mathbf{i}_{\overline{X}_{\overline{H}}}\omega_{\overline{H}} = 0$ which is equivalent to Hamilton's equations for $(\overline{q}^i, \overline{p}_i)$ with Hamiltonian $\overline{H}$. However, we assume that $(\overline{q}^i, \overline{p}_i)$ are time-independent and thus

$$\frac{\partial \overline{H}}{\partial \overline{p}_i} = 0, \quad \frac{\partial \overline{H}}{\partial \overline{q}^i} = 0,$$

that is, $\overline{H}$ is a function of t alone and thus $X_{\overline{H}} = 0$. Now apply again $\mathbf{i}_{\overline{X}_{\overline{H}}}\omega_{\overline{H}} = 0$ for $\overline{H}_{\overline{H}} = (0, 1)$ to conclude that $d\overline{H} = 0$, that is, $\overline{H}$ is a constant. The relationship $H - \overline{H} + \partial S/\partial t = 0$ computed at q^i and $p_i = \partial S/\partial q^i$ (given by (7.9.6) is the Hamilton-Jacobi equation.

Remarks

1. In general, the function S develops *singularities* or *caustics* as time increases, so it must be used with care. This process is, however, fundamental in geometric optics. Moreover, one has to be careful with the sense in which S generates the identity at $t = 0$ as it might have singular behavior in t.

2. Here is another link between the Lagrangian and Hamiltonian view of the Hamilton-Jacobi theory: If we define S for short time by the ***action integral***

$$S(q, q_0, t, t_0) = \int_\gamma L\, dt,$$

where γ is the extremal between (q_0, t_0) and (q, t); one verifies that $\mathbf{d}S = p\, dq - H\, dt$, and so S satisfies the Hamilton-Jacobi equation. See Arnold [1989], §4.6, and Abraham and Marsden [1978], §5.2, for more information.

3. If H is time-independent and W satisfies the time-independent Hamilton-Jacobi equation

$$H\left(q, \frac{\partial W}{\partial q}\right) = E,$$

then $S(q, q_0, t, t_0) = W(q, q_0) - E(t - t_0)$ satisfies the time-dependent one.

4. The Hamilton-Jacobi equation is fundamental in the study of the quantum-classical relationship; see the following section.

5. The action function S is a key tool used in the proof of the ***Liouville-Arnold theorem*** which gives the existence of action angle coordinates for systems with integrals in involution; see Arnold [1989] and Abraham and Marsden [1978], for details.

6. The Hamilton-Jacobi equation plays an important role in numerical integrators that preserve the symplectic structure (see deVogelaére [1956], Channell [1983], Feng [1986], Channell and Scovel [1990], Ge and Marsden [1988], and Marsden [1992]).

Now, we describe in geometric terms the relation between solutions of the (first order partial differential) equation of Hamilton-Jacobi (7.9.5) and solutions of Hamilton's equations (7.9.10); that is, the geometrical content of Theorem **7.9.1**. First we notice that in the equation (7.9.5) the unknown function S appears only through its derivatives. Thus, we won't lose anything if instead of considering S we consider its differential $\mathbf{d}S$. For each $x \in \tilde{M} = M \times \mathbb{R}$ with coordinates $x = (x^1 = q^1, \ldots, x^n = q^n, x^0 = t)$,

$\mathbf{d}S(x)$ is an element of the cotangent bundle $T^*\tilde{M}$. As x varies in $\tilde{M}$ the set $\{\mathbf{d}S(x) \mid x \in \tilde{M}\}$ gives a submanifold of $T^*\tilde{M}$ which in terms of coordinates is given by $\xi_j = \partial S/\partial x^j$. It is natural to call this submanifold *the graph of* $\mathbf{d}S$ and denote it by graph $\mathbf{d}S \subset T^*\tilde{M}$. The restriction of the symplectic form on $T^*\tilde{M}$ to graph $\mathbf{d}S$ is zero since

$$\sum_{j=0}^{n} dx^j \wedge d\xi_j = \sum_{j=0}^{n} dx^j \wedge d\frac{\partial S}{\partial x_j} = \sum_{j=0}^{n} dx^j \wedge dx^k \frac{\partial^2 S}{\partial x^j \partial x^k} = 0.$$

Moreover, the dimension of the submanifold graph $\mathbf{d}S$ is half of the dimension of the symplectic manifold $T^*\tilde{M}$. Such a submanifold is called ***Lagrangian***, as we already mentioned in connection with generating functions (§6.5). What is important here is that the projection from graph $\mathbf{d}S$ to $\tilde{M}$ is a diffeomorphism, and even more, the converse holds: if $\Lambda \subset T^*\tilde{M}$ is a Lagrangian submanifold of $T^*\tilde{M}$ such that the projection on $\tilde{M}$ is a diffeomorphism around $\lambda \in \Lambda$, then in the neighborhood of λ, $\Lambda =$ graph $\mathbf{d}\varphi$ for some function φ. To show this, notice that because the projection is a diffeomorphism, Λ is given (around λ) as a submanifold of the form $(x^j, \rho_j(x))$. The condition for Λ to be Lagrangian requires that

$$\sum_{j=0}^{n} dx^j \wedge d\xi_j = 0 \text{ on } \Lambda,$$

that is,

$$\sum_{j=0}^{n} dx^j \wedge d\rho_j(x) = 0, \quad \text{i.e.} \quad \frac{\partial \rho_j}{\partial x^k} - \frac{\partial \rho_k}{\partial x^j} = 0;$$

thus, there is a φ such that $\rho_j = \partial\varphi/\partial x^j$, which is the same as $\Lambda =$ graph $\mathbf{d}\varphi$. The conclusion of these remarks is that Lagrangian submanifolds of $T^*\tilde{M}$ are natural generalizations of graphs of differentials of functions on $\tilde{M}$. They make sense even if the projection to $\tilde{M}$ is not a diffeomorphism. From this point of view, *a solution of Hamilton-Jacobi equation is a Lagrangian submanifold of* $T^*\tilde{M}$ *which is contained in the surface* $\tilde{H} \subset T^*\tilde{M}$ *defined by* $\tilde{H} = \xi_0 + H(p, q, t) = 0$. This point of view is more convenient since now we include solutions which are singular from the usual point of view. This is not the only benefit: We also get more insight in the content of the Theorem **7.9.1**. The tangent space to $\tilde{H}$ has dimension 1 less than the dimension of the symplectic manifold $T^*\tilde{M}$ and it is given by the set of vectors X such that $(d\xi_0 + dH)(X) = 0$. If a vector Y is in the symplectic orthogonal of $T_z(\tilde{H})$, that is,

$$\sum_{j=0}^{n} (dx^j \wedge d\xi_j)(X, Y) = 0$$

for all $X \in T_z(\tilde{H})$, then Y is a multiple of the nonvanishing vector $X_{\tilde{H}}$ given by

$$X_{\tilde{H}} = \frac{\partial}{\partial t} - \frac{\partial H}{\partial t}\frac{\partial}{\partial \xi_0} + X_H.$$

Moreover, the integral curves of $X_{\tilde{H}}$ are exactly the solutions of the Hamilton equations. Now, the key observation which links the Hamilton equations and Hamilton-Jacobi equation is that *the vector field $X_{\tilde{H}}$ which is obviously tangent to $\tilde{H}$ is, moreover, tangent to any Lagrangian submanifold contained in $\tilde{H}$* (the reason for this is a very simple algebraic fact given in Exercise **7.9-3**). This is the same as saying that a solution of Hamilton's equations for H is either disjoint from a Lagrangian submanifold contained in $\tilde{H}$ or completely contained in it. But this actually gives a way to construct a solution of Hamilton-Jacobi equation starting from an initial condition at $t = t_0$. Namely, give a Lagrangian submanifold Λ_0 in T^*M and embed it in $T^*\tilde{M}$ at $t = 0$ using

$$(q, p) \mapsto (q, p, \xi_0 = -H(q, p, t_0), t = t_0).$$

It will be an isotropic submanifold of $T^*\tilde{M}$. This corresponds (in the usual setting) to giving the intitial condition $S(q, p, t_0)$. Then take all integral curves of $X_{\tilde{H}}$ starting from Λ_0. The collection of these curves spans a manifold Λ with one dimension more than Λ_0. It is obtained by flowing Λ_0 along $X_{\tilde{H}}$; that is, $\Lambda = \cup_t \Lambda_t$ where $\Lambda_t = \Phi_t(\Lambda_0)$ and Φ_t is the flow of $X_{\tilde{H}}$. Since $X_{\tilde{H}}$ is tangent to $\tilde{H}$ and $\Lambda_0 \subset \tilde{H}$, we get $\Lambda_t \subset \tilde{H}$ and hence $\Lambda \subset \tilde{H}$. Since the flow along $X_{\tilde{H}}$ is a canonical map, it leaves the symplectic form of $T^*\tilde{M}$ invariant and therefore takes an isotropic submanifold into an isotropic one; in particular Λ_t is an isotropic submanifold of $T^*\tilde{M}$. The tangent space of Λ at some $\lambda \in \Lambda_t$ is a direct sum of the tangent space of Λ_t and the subspace generated by $X_{\tilde{H}}$; since the first subspace is contained in $T_\lambda \tilde{H}$ and the second is symplectically orthogonal on $T_\lambda \tilde{H}$, we see that Λ is also an isotropic submanifold of $T^*\tilde{M}$. But it has the dimension half as $T^*\tilde{M}$ and therefore Λ is a Lagrangian submanifold contained in $\tilde{H}$, that is, it is a solution of Hamilton-Jacobi equation with initial condition Λ_0 at $t = t_0$.

Using the above point of view it is easy to understand the singularities of a solution of Hamilton-Jacobi equation. They correspond to those points of the Lagrangian manifold solution where the projection to $\tilde{M}$ is not a local diffeomorphism. These singularities might be present in the initial condition (that is, Λ_0 might not locally project diffeomorphically to M) or they might appear at later times by folding the submanifolds Λ_t as t varies. The projection of such a singular point to $\tilde{M}$ is called a *caustic point of the solution.* Caustic points are of fundamental importance in geometric optics and the semiclassical approximation of quantum mechanics. We refer to Abraham and Marsden [1978] §5.3 and Guillemin and Sternberg [1984] for further information.

Exercises

7.9-1 *Solve the Hamilton-Jacobi equation for the harmonic oscillator. Check directly the validity of the Hamilton-Jacobi theorem (connecting the solution of the Hamilton-Jacobi equation and the flow of the Hamiltonian vector field) for this case.*

7.9-2 *Verify by* ***direct calculation*** *the following. Let* $S(q, q_0)$ *and*

$$H(q,p) = \frac{p^2}{2m} + V(q)$$

be given, where $q, p \in \mathbb{R}$. *Show that*

$$\frac{1}{2m}(S_q)^2 + V = E$$

and $\dot{q} = p/m$ *if and only if* $(q, S_q(q, q_0))$ *satisfies Hamilton's equation with energy* E. *What goes wrong?*

7.9-3 *Let* (V, Ω) *be a symplectic vector space and* $W \subset V$ *be a linear subspace. Recall from §2.4 that* $W^\Omega = \{v \in V \mid \Omega(v, w) = 0$ *for all* $w \in W\}$ *denotes the symplectic orthogonal of* W. *A subspace* $L \subset V$ *is called Lagrangian if* $L = L^\Omega$. *Show that if* $\dim W = \dim V - 1$ *and* $L \subset W$ *is a Lagrangian subspace, then* $W^\Omega \subset L$.

7.10 The Classical Limit and the Maslov Index

Here we give a brief supplementary introduction through the simplest examples, of the quantum-classical relationship and the Maslov index. For further information and generalizations, the reader may consult Guillemin and Sternberg [1977], [1984] and Woodhouse [1980]. We also refer to Littlejohn [1988] for an interpretation of the Maslov index in terms of Berry's phase.

We begin with the ***one-dimensional Schrödinger equation***. Let $V : \mathbb{R} \to \mathbb{R}$ be a given potential, let $\psi : \mathbb{R} \to \mathbb{C}$ be a wave function, and let $E, \hbar, m$ be constants (energy, Planck's constant, and mass, respectively). Consider the ***stationary Schrödinger equation***:

$$L\psi = E\psi, \tag{7.10.1}$$

where

$$L\psi = -\frac{\hbar^2}{2m}\psi'' + V\psi \tag{7.10.2}$$

and the time-independent Hamilton-Jacobi equation for the function $S : \mathbb{R} \to \mathbb{R}$:

$$\frac{1}{2m}(S')^2 + V = E. \tag{7.10.3}$$

Recall from Exercise **7.8-2** that the Hamilton-Jacobi equation is related to Hamilton's equations

$$\dot{q} = \frac{\partial H}{\partial p}, \quad \dot{p} = -\frac{\partial H}{\partial q}, \tag{7.10.4}$$

where $H(q,p) = p^2/2m + V(q)$, as follows: *the equation* (7.10.3) *holds for* S *and* $\dot{q} = p/m$ *iff* $\dot{p} = -\partial H/\partial q$.

Two related central questions are:

1. How does one pass from classical objects to quantum objects? Here, "objects" can refer to the equations themselves, to solutions, or to properties of the equations or solutions.

2. In what sense are solutions of the Hamilton-Jacobi equation a limit of solutions of the Schrödinger equation as $\hbar \to 0$?

Progress with these questions was made with the basic work of Weyl, Birkhoff, Van Hove, and, amongst many others, Keller, Maslov, Souriau, and Kostant (see the preceding references for the literature citations). Van Hove showed that there is no general quantization having all the properties one would want. (Van Hove's theorem in $\mathbb{R}^n$ is proved in Abraham and Marsden [1978], §5.4.) Van Hove also found some positive results that were rediscovered and extended by Souriau and Kostant in a procedure now called prequantization. In studying question 2 using the WKB method, Keller and Maslov discovered the topological meaning of the corrected Bohr-Sommerfeld quantization rules. The invariant they discovered is called the ***Keller-Maslov-Arnold-Hörmander*** index. (Arnold's article [1967] was instrumental in explaining Maslov's ideas to mathematicians.) Our one-dimensional example will contain many of the features of the general case, in terms of understandable concepts without a lot of preliminaries.

If S is a solution of (7.10.3), we try to solve (7.10.1) with

$$\psi = \exp(iS/\hbar). \tag{7.10.5}$$

Substitution of (7.10.5) in (7.10.2) gives

$$E\psi = L\psi + \frac{i\hbar}{2m}\psi S'' \tag{7.10.6}$$

by using (7.10.3). Equation (7.10.6) differs from (7.10.1) by a term of order $\hbar$. Next, try

$$\psi = a\exp(iS/\hbar) \tag{7.10.7}$$

for $a : \mathbb{R} \to \mathbb{R}$. Then, substituting into (7.10.2) and using the Hamilton-Jacobi equation, we get

$$L\psi = E\psi - \hbar\frac{i}{2m}(S''a + 2S'a')\frac{\psi}{a} - \frac{\hbar^2}{2m}\frac{a''}{a}\psi$$

or

$$E\psi = L\psi + \hbar\frac{i}{2m}(S''a + 2S'a')\frac{\psi}{a} + \frac{\hbar^2}{2m}\frac{a''}{a}\psi. \tag{7.10.8}$$

We can arrange for this equation to differ from (7.10.1) by a term of order $\hbar^2$ by requiring a to satisfy the ***transport equation***

$$2a'S' + aS'' = 0, \tag{7.10.9}$$

whose solution is $a = (\text{constant})/|S'|^{1/2}$. Thus, (7.10.8) becomes

$$E\psi = L\psi + \frac{\hbar^2}{2m}\frac{a''}{a}\psi \tag{7.10.10}$$

which differs from (7.10.1) by a term of order $\hbar^2$. The idea is now to continue this process by writing

$$\psi = \left(\sum_{k=0}^{N} a_k (i\hbar)^k\right)\exp(iS/\hbar) \tag{7.10.11}$$

for some functions $a_k : \mathbb{R} \to \mathbb{R}$ and requiring it to satisfy (7.10.1) up to an error term of order $\hbar^{N+2}$. This procedure is usually called the ***WKB method*** (after G. Wentzel, H. A. Kramers, and L. Brillouin, although it probably goes back to Liouville, Green, and Lord Rayleigh). The computation below shows that this is indeed possible. Substituting (7.10.11) into (7.10.2) and using, as before, the Hamilton-Jacobi equation (7.10.3) yields

$$\begin{aligned} L\psi &= E\psi - \frac{\exp(iS/\hbar)}{2m}\Bigg[(S''a_0 + 2S'a_0')i\hbar \\ &\quad + i\hbar\sum_{k=1}^{N}(S''a_k + 2S'a_k' - a_{k-1}'')(i\hbar)^k + i^N a_N''\hbar^{N+2}\Bigg]. \end{aligned} \tag{7.10.12}$$

Requiring the ***transport equations***

$$S''a_k + 2S'a_k' - a_{k-1}'' = 0, \quad k = 0, 1, \ldots, N, \quad a_{-1} \equiv 0, \tag{7.10.13}$$

to hold, which can be solved recursively, we see that (7.10.12) reduces to

$$E\psi = L\psi + \frac{i^N \exp(iS/\hbar)}{2m} a_N''\hbar^{N+2}. \tag{7.10.14}$$

Thus, we have "solved" (7.10.1) up to an error of order $\hbar^{N+2}$. Therefore, if we let $N \to \infty$ we have found an asymptotic solution

$$\psi \sim \left(\sum_{k=0}^{\infty} a_k \hbar^k\right)\exp(iS/\hbar) \tag{7.10.15}$$

of (7.10.1). *The key observation in this procedure is that once S is determined, the coefficients a_k are obtained recursively as solutions of linear ordinary differential equations.* The solutions are *a fortiori* only local since S given by (7.10.3) is only local, as we shall see below.

Suppose the energy surface for the classical system has the form shown in Figure 7.10.1.

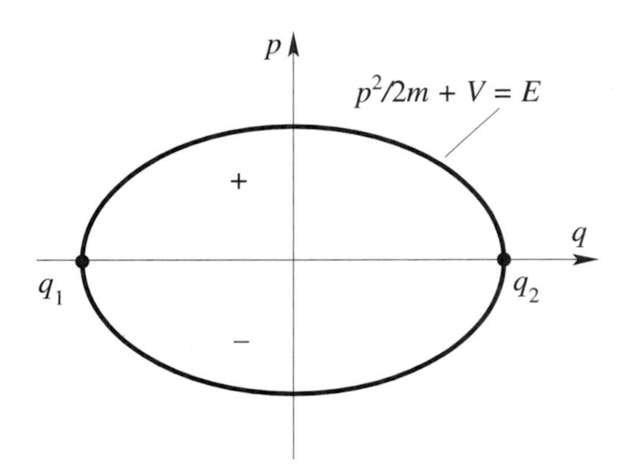

FIGURE 7.10.1.

There correspond two solutions of (7.10.3):

$$S = \pm \int p(q)\, dq + C_\pm, \tag{7.10.16}$$

where $p(q) = \sqrt{2m(E - V(q))}$, and $C_\pm$ are constants. Thus if ψ is given by (7.10.11), or asymptotically by (7.10.15), then the first transport equation (7.10.9) for $k = 0$ yields

$$a_{0\pm} = \frac{d_\pm}{[2m(E - V(q))]^{1/4}} \tag{7.10.17}$$

for some constants $d_\pm$. This expression diverges at q_1 and q_2 and becomes imaginary outside the interval $[q_1, q_2]$.

The subtlety of questions 1 and 2 centers on the multiple valuedness of S and the presence of the turning points at q_1 and q_2. To get around these difficulties there have been several approaches.

1. Use analytic continuation methods to avoid the turning points. This approach was developed by Zwaan.

2. Approximate the potential by a linear one near each turning point. Schrödinger's equation then yields an Airy function which is asymptotically matched by Bessel functions (Langer and Jeffreys).

3. Use a modified WKB method near the turning point and an asymptotic expansion (Maslov). We shall describe this method shortly.

There are other approaches too. For instance, Miller and Good [1953] effectively used area-preserving maps to deform Figure 7.10.1 into that for a harmonic oscillator. The same idea was used by Maslov [1965] for higher superpositions of such expressions.

To study the behavior near q_1 and q_2, we replace $\psi = a \exp(iS/\hbar)$ by a superposition of such expressions, that is, by

$$\psi(q) = \int_{-\infty}^{\infty} a(q,p) \exp(i\varphi(q,p)/\hbar)\, dp \tag{7.10.18}$$

where $\varphi : \mathbb{R}^2 \to \mathbb{R}$ is positive homogeneous of degree one in p, that is, $\varphi(q, rp) = r\varphi(q,p)$ for all $r > 0$. This integral is called an ***oscillatory function***; the theory of such integrals parallels that of Fourier integrals. Let us take $\varphi(q,p) = qp - T(p)$ for some real-valued function T defined in a neighborhood of the origin whose second derivative never vanishes, that is,

$$\psi(q) = \int_{-\infty}^{\infty} a(q,p) \exp\left[i\frac{pq - T(p)}{\hbar}\right] dp, \tag{7.10.19}$$

and try to solve (7.10.1). A direct computation shows that

$$\begin{aligned} L\psi - E\psi \quad = \quad & \int_{-\infty}^{\infty} \left[a\left(\frac{p^2}{2m} + V(q) - E\right) \right. \\ & \left. - \left(\frac{i\hbar p}{m}\frac{\partial a}{\partial q} + \frac{\hbar^2}{2m}\frac{\partial^2 a}{\partial q^2}\right)\right] \exp\left[i\frac{pq - T(p)}{\hbar}\right] dp. \end{aligned} \tag{7.10.20}$$

To evaluate the right-hand side of (7.10.20) asymptotically in $\hbar$ we need the following:

Theorem 7.10.1 (Stationary Phase Formula) *Let $a, \varphi : \mathbb{R} \to \mathbb{R}$ be C^∞ functions, φ having finitely many nondegenerate critical points. Then*

$$\begin{aligned} & \int_{-\infty}^{\infty} a(x) \exp(i\varphi(x)/\hbar)\, dx \\ & = \sqrt{2\pi\hbar} \sum_{\varphi'(y)=0} \exp\left(\frac{i\pi}{4}\operatorname{sgn}\varphi''(y)\right) \frac{a(y)\exp(i\varphi(y)/\hbar)}{|\varphi''(y)|^{\frac{1}{2}}} + O(\hbar^{\frac{3}{2}}) \end{aligned} \tag{7.10.21}$$

where the sum is taken over all critical points y of φ. (Recall that a critical point y of φ is nondegenerate iff $\varphi''(y) \neq 0$.)

Proof (After Guillemin and Sternberg [1977]) Let $\{\chi_n\}$ be a C^∞ partition of unity on the real line, that is, each χ_n is $C^\infty, 0 \leq \chi_n \leq 1$, $\operatorname{supp} \chi_n =$ closure of $\{x \in \mathbb{R} \mid \chi_n(x) \neq 0\}$ is compact, each $x \in \mathbb{R}$ has a neighborhood intersecting only finitely many of $\operatorname{supp} \chi_n$, and $\sum_n \chi_n(x) = 1$ for each $x \in \mathbb{R}$. Since there are only finitely many critical points of φ, we can arrange the supports of χ_n such that each $\operatorname{supp} \chi_n$ contains at most one critical point of φ. Writing

$$\int_{-\infty}^{\infty} a(x) \exp\left[\frac{i\varphi(x)}{\hbar}\right] dx = \sum_n \int_{-\infty}^{\infty} \chi_n(x) a(x) \exp\left[\frac{i\varphi(x)}{\hbar}\right] dx,$$

we see that each integral on the right-hand side is a definite integral on $\operatorname{supp} \chi_n$ and that there are only a finite number of integrals that have overlapping domains of integration. Some of these integrals have domains which contain critical points of f, others do not.

We begin by studying those integrals that do not have a critical point of φ in their domain. Thus, we can assume that supp a is compact and that $\varphi' \neq 0$ on supp a. Integrating by parts,

$$\begin{aligned} \int_{-\infty}^{\infty} a(x) \exp\left[\frac{i\varphi(x)}{\hbar}\right] dx &= \int_{-\infty}^{\infty} a(x) \frac{\hbar}{i\varphi'(x)} \frac{d}{dx}\left(\exp\left[\frac{i\varphi(x)}{\hbar}\right]\right) dx \\ &= i\hbar \int_{-\infty}^{\infty} \frac{d}{dx}\left(\frac{a(x)}{\varphi'(x)}\right) \exp\left[\frac{i\varphi(x)}{\hbar}\right] dx, \end{aligned}$$

which is an integral of the same type since $\frac{d}{dx}[a(x)/\varphi'(x)]$ is again C^∞ with compact support inside supp a. Thus the procedure can be repeated any number of times yielding

$$\int_{-\infty}^{\infty} a(x) \exp\left[\frac{i\varphi(x)}{\hbar}\right] dx = O(\hbar^N)$$

for any $N \in \mathbb{N}$. Thus, to prove (7.10.21), it suffices to establish it if supp a is compact and contains exactly one critical point x_0 of φ. This will be carried out in several steps.

Step 1 *(Morse Lemma) There is a change of variables $x \mapsto z$ such that $\varphi(x(z)) = \varphi(x_0) + \frac{1}{2}(\operatorname{sgn} \varphi''(x_0))(z - z_0)^2$, where $x(z_0) = x_0$.*

To show this, we can clearly assume that $x_0 = 0, \varphi(x_0) = 0, \varphi'(x_0) = 0, \varphi''(x_0) \neq 0$. Write first

$$\varphi(x) = \int_0^1 \frac{d}{dt}\varphi(tx)\, dt = x \int_0^1 \varphi'(tx)\, dt = x\alpha(x),$$

where $\alpha(x) = \int_0^1 \varphi'(tx)\, dt$ is again a C^∞ function. Since $\varphi'(x) = \alpha(x) + x\alpha'(x)$ and $\varphi'(0) = \alpha(0) = 0$, the same argument shows that $\alpha(x) =$

$x\beta(x)$ for some C^∞ function $\beta(x)$. Therefore $\varphi(x) = x^2\beta(x)$ and $\beta(x) = \int_0^1 \alpha'(tx)\,dt$ whence $\beta(0) = \alpha'(0) = \varphi''(0)/2$. Define $z(x) = \sqrt{2}|\beta(x)|^{\frac{1}{2}}x$ which is C^∞ in a neighborhood of 0, since $\beta(0) \neq 0$, and satisfies $z'(0) = \sqrt{2}|\beta(0)|^{\frac{1}{2}} \neq 0$. Therefore $x \mapsto z$ is a diffeomorphism in a neighborhood of 0 and in this neighborhood, suitably shrunk if necessary, $\beta(x)$ does not change sign. Thus

$$\varphi(x) = x^2\beta(x) = (\operatorname{sgn}\beta(x))x^2|\beta(x)| = \frac{1}{2}(\operatorname{sgn}\beta(0))z^2 = \frac{1}{2}(\operatorname{sgn}\varphi''(0))z^2.$$

Step 2 *Performing the change of variables $x \mapsto z$ given in* ***Step 1*** *we get*

$$\begin{aligned}\int_{-\infty}^{\infty} & a(x)\exp(i\varphi(x)/\hbar)\,dx \\ &= \frac{a(x_0)\exp(i\varphi(x_0)/\hbar)}{|\varphi''(x_0)|^{\frac{1}{2}}}\int_{-\infty}^{\infty}\exp\left[\frac{\pm i(z-z_0)^2}{2\hbar}\right]dz \\ &\quad + \exp\left[\frac{i\varphi(x_0)}{\hbar}\right]\int_{-\infty}^{\infty}(z-z_0)\gamma(z)\exp\left[\frac{\pm i(z-z_0)^2}{2\hbar}\right]dz,\end{aligned}$$

where + or − is taken in accordance with $\operatorname{sgn}\varphi''(x_0)$ *and $\gamma(z)$ is C^∞ with $(z-z_0)\gamma(z)$ bounded together with all its derivatives. (The bound for each derivative may be different.)*

Indeed,

$$\int_{-\infty}^{\infty} a(x)\exp\left[\frac{i\varphi(x_0)}{\hbar}\right]dx = \int_{-\infty}^{\infty} a(x(z))\exp\left[\frac{i\varphi(x_0)}{\hbar} \pm \frac{i(z-z_0)^2}{2\hbar}\right]\left|\frac{dx}{dz}\right|dz$$

and note that

$$\frac{dx}{dz}(z_0) = 1/\sqrt{2}|\beta(x_0)|^{\frac{1}{2}} = |\varphi''(x_0)|^{-\frac{1}{2}}$$

so that proceeding as in ***Step 1*** we can write

$$a(x(z))\left|\frac{dx}{dz}\right| - a(x_0)\left|\frac{dx}{dz}(z_0)\right| = (z-z_0)\gamma(z)$$

for some C^∞ function $\gamma(z)$ (z_0 denotes the point given by $x(z_0) = x_0$), that is,

$$a(x(z))\left|\frac{dx}{dz}\right| = a(x_0)\frac{1}{|\varphi''(x_0)|^{\frac{1}{2}}} + (z-z_0)\gamma(z).$$

Since

$$(z-z_0)\gamma(z) = (z-z_0)\int_0^1 \frac{d}{dt}\left(a(x(z_t)\left|\frac{dx}{dz}(z_t)\right|\right)dt,$$

where $z_t = tz + (1-t)z_0$, we see that on its domain of definition $\gamma(z)$ is smooth and has itself and all its derivatives bounded because $a(x)$ has compact support.

To show that each integral in ***Step 2*** is well defined, we prove:

Step 3 *Let $h(z)$ be a C^2 function of a real variable such that the three functions $|h(z)|, |h'(z)|, |h''(z)|$ are all bounded by $M > 0$. If $\lambda \in \mathbb{C}$, the integral*

$$\int_{-\infty}^{\infty} e^{-\lambda z^2/2} h(z)\, dz$$

is uniformly convergent for $\operatorname{Re}\lambda \geq 0, |\lambda| \geq 1$, *bounded by a constant depending on M only, holomorphic for* $\operatorname{Re}\lambda > 0$, *and continuous for* $\operatorname{Re}\lambda \geq 0$.

It suffices to prove this for $\int_0^\infty e^{-\lambda z^2/2} h(z)\, dz$ for then, by changing variables $z \mapsto -z$, the same result holds for the integral from $-\infty$ to 0 and hence for the sum. Let $0 < A < B$. Then

$$\begin{aligned}
&\int_A^B e^{-\lambda z^2/2} h(z)\, dz \\
&= -\lambda^{-1} \int_A^B \frac{1}{z} (e^{-\lambda z^2/2})' h(z)\, dz \\
&= -(\lambda z)^{-1} e^{-\lambda z^2/2} h(z) \Big|_A^B + \lambda^{-1} \int_A^B e^{-\lambda z^2/2} \left(\frac{h(z)}{z}\right)' dz \\
&= -(\lambda z)^{-1} e^{-\lambda z^2/2} h(z) \Big|_A^B - \lambda^{-2} \int_A^B (e^{-\lambda z^2/2})' \frac{1}{z} \left(\frac{h(z)}{z}\right)' dz \\
&= -(\lambda z)^{-1} e^{-\lambda z^2/2} \left(h(z) + \lambda^{-1} \left(\frac{h(z)}{z}\right)' \right) \Bigg|_A^B \\
&\quad + \lambda^{-2} \int_A^B e^{-\lambda z^2/2} \left[\frac{1}{z} \left(\frac{h(z)}{z}\right)' \right]' dz.
\end{aligned}$$

The first term tends to zero as $A \to \infty$ by boundedness of $|h|, |h'|$ if $\operatorname{Re}\lambda \geq 0, |\lambda| \geq 1$. The integral in the second term can also be bounded in absolute value for the same range of λ by a constant depending only on M since $|h''|$ is bounded. In particular, the integral $\int_0^\infty e^{-\lambda z^2/2} h(z)\, dz$ is uniformly convergent.

Arguing in the same manner for the λ-derivative, we conclude that the integral is holomorphic for $\operatorname{Re}\lambda > 0$. Similarly one shows continuity for $\operatorname{Re}\lambda \geq 0$.

Step 4

$$\int_{-\infty}^{\infty} \exp\left[\frac{\pm i(z - z_0)^2}{2\hbar}\right] dz = \sqrt{2\pi\hbar}\, e^{\pm\pi i/4}.$$

From the previous step it follows that this integral exists, by taking $\lambda = \mp i$ and $h(z) = 1$. Moreover, the classical formula $\int_{-\infty}^{\infty} e^{-u^2/2}\, du = \sqrt{2\pi}$

implies that for real positive λ we have

$$\int_{-\infty}^{\infty} e^{-\lambda u^2/2}\, du = \sqrt{2\pi/\lambda}.$$

By analytically continuing both sides for $\operatorname{Re}\lambda > 0$, the same formula holds for complex λ in the right half-plane. Now let $\lambda \to \mp i/\hbar$ to obtain

$$\begin{aligned}\int_{-\infty}^{\infty} \exp\left[\frac{\pm i(z-z_0)^2}{2\hbar}\right] dz &= \sqrt{2\pi}\sqrt{\hbar}\sqrt{\exp\left(\pm\frac{\pi}{2}i\right)} \\ &= \sqrt{2\pi\hbar}\exp\left(\pm\frac{\pi i}{4}\right).\end{aligned}$$

Step 5 *The second integral in* ***Step 2*** *is* $O(\hbar^{3/2})$.

Indeed, the integral exists by ***Step 3*** and

$$\begin{aligned}&\int_{-\infty}^{\infty} (z-z_0)\gamma(z)\exp\left[\frac{\pm i(z-z_0)^2}{2\hbar}\right] dz \\ &= \int_{-\infty}^{\infty} z\gamma(z+z_0)\exp\left(\frac{\pm iz^2}{2\hbar}\right) dz \\ &= \pm\frac{\hbar}{i}\int_{-\infty}^{\infty}\left[\exp\left(\frac{\pm iz^2}{2\hbar}\right)\right]' \gamma(z+z_0)\, dz \\ &= \pm\, i\hbar\int_{-\infty}^{\infty} \gamma'(z+z_0)\exp\left(\frac{\pm iz^2}{2\hbar}\right) dz.\end{aligned}$$

The boundary terms vanish if γ vanishes sufficiently fast at ∞. This integral has exactly the form of the original integral and therefore can be written as a sum of two integrals, the first of order $O(\hbar^{1/2})$ by ***Step 4*** and the second $\hbar$ times again an integral of the same type. Thus this integral is of order $\hbar\hbar^{1/2} = \hbar^{3/2}$.

From ***Steps 2, 4,*** and ***5*** we conclude that if there is a single critical point x_0 of φ in supp a we get

$$\begin{aligned}&\int_{-\infty}^{\infty} a(x)\exp\left[\frac{i\varphi(x)}{\hbar}\right] dx \\ &= \sqrt{2\pi\hbar}\exp\left(\pm i\frac{\pi}{4}\right)\frac{a(x_0)\exp(i\varphi(x_0)/\hbar)}{|\varphi''(x_0)|^{1/2}} + O(\hbar^{3/2}). \quad \blacksquare\end{aligned}$$

The previous proof shows that the same formula holds if all functions depend smoothly on additional parameters. In particular, we shall use the following expression in analyzing the right-hand side of (7.10.19):

$$\begin{aligned}
&\int_{-\infty}^{\infty} c(q,p)\exp\left[i\frac{f(q,p)}{\hbar}\right]dp \\
&\quad = \sqrt{2\pi\hbar}\sum_{f_p=0}\exp\left(\frac{i\pi}{4}\operatorname{sgn} f_{pp}\right)\frac{c(q,p)\exp(if/\hbar)}{\mid f_{pp}\mid^{1/2}} + O(\hbar^{3/2}),
\end{aligned} \tag{7.10.22}$$

where the sum is over all p such that $f_p = \partial f/\partial p$ vanishes; these critical points are assumed to be finite in number and nondegenerate, that is, $f_{pp} = \partial^2 f/\partial p^2 \neq 0$.

Applying (7.10.22) to (7.10.20) gives

$$\begin{aligned}
&L\psi - E\psi \\
&\quad = \sqrt{2\pi\hbar}\sum_{q=T'(p)}\frac{\exp[-i\pi \operatorname{sgn} T''(p)/4]}{|T''(p)|^{1/2}}a(q,p)\left(\frac{p^2}{2m}+V(q)-E\right) \\
&\quad\quad \times \exp\left[i\frac{pq-T(p)}{\hbar}\right] + O(\hbar^{3/2}),
\end{aligned} \tag{7.10.23}$$

provided the number of critical points in p of the q-dependent function $f(q,p) = qp - T(p)$ is finite and all these p-critical points are nondegenerate. By assumption, T'' never vanishes for p near zero and thus $T'(p)$ is either strictly increasing or strictly decreasing. Thus, for a fixed q, there is exactly one p such that $q = T'(p)$, that is, (7.10.23) reads

$$\begin{aligned}
L\psi - E\psi &= \sqrt{2\pi\hbar}\frac{\exp[-i\pi \operatorname{sgn} T''(p)/4]}{|T''(p)|^{1/2}}a(T'(p),p) \\
&\quad \times \left(\frac{p^2}{2m}+V(T'(p))-E\right)\exp\left[i\frac{pT'(p)-T(p)}{\hbar}\right] + O(\hbar^{3/2}).
\end{aligned}$$

Now we require that $L\psi - E\psi = O(\hbar^{3/2})$, which forces the first term to vanish (since $a(q,p)$ is not the zero function), that is,

$$\frac{p^2}{2m} + V(T'(p)) = E. \tag{7.10.24}$$

Thus the graph of $q = T'(p)$ (as a function of p) is contained in the energy surface. Equation (7.10.24) is the Hamilton-Jacobi equation in the variable p, which is approximated near the turning points q_1 and q_2.

Applying (7.10.22) to formula (7.10.19) gives

$$\begin{aligned}
\psi(q) &= \sqrt{2\pi\hbar}\sum_{q=T'(p)}\frac{1}{|T''(p)|^{1/2}}\exp[-i\pi \operatorname{sgn} T''(p)/4] \\
&\quad \times \exp\left[i\frac{pq-T(p)}{\hbar}\right]a(q,p) + O(\hbar^{3/2})
\end{aligned}$$

$$\begin{aligned} &= \sqrt{2\pi\hbar}\frac{1}{|T''(p)|^{1/2}}\exp[-i\pi\operatorname{sgn}T''(p)/4] \\ &\quad \times \exp\left[i\frac{pT'(p)-T(p)}{\hbar}\right]a(T'(p),p)+O(\hbar^{3/2}) \\ &= O(\hbar^{1/2}). \end{aligned} \tag{7.10.25}$$

We now seek to represent ψ near q_1 and q_2 using functions T_1 and T_2 given by (7.10.24) and seek to represent ψ on the $\pm$ portions in the form (7.10.19). We are, in effect, using a superposition of two WKB approximations.

Notice that if $q - T'(p) = 0$, as above, then

$$\frac{d}{dq}(pq - T(p)) = p + \frac{dp}{dq}q - T'(p)\frac{dp}{dq} = p, \tag{7.10.26}$$

so both S and $pq - T(p)$ are given by integrating p with respect to q; that is, they are both *actions*.

Since $q = T'(p)$ along the energy curve in Figure 7.10.1, we see that $T''(p) > 0$ on the $+$ side and $T''(p) < 0$ on the $-$ side of the p-axis. Thus the term

$$e^{-i\pi\operatorname{sgn}T''(p)/4} \tag{7.10.27}$$

in (7.10.25) jumps, or suffers a phase shift, as p crosses the q-axis. In Figure 7.10.2 we show the different regions and functions being considered.

So now we have obtained ψ on four different regions: The upper and lower part of the energy surface and the parts around the two turning points $(q_1, 0), (q_2, 0)$; See Figure 7.10.2. The structure of this function is that of a product of an amplitude times an exponential plus higher-order terms. We shall require that they all match on the overlaps at first order. Since there are constants of integration in these formulae (as in (7.10.17), for example), matching at points A, B, and C determines all the constants. Thus, the consistency condition is the match of these solutions at the point D. This will happen only if the phases in (7.10.25) match.

The phase changes in the exponentials $\exp(iS/\hbar)$ and $\exp[i(pq - T(p))/\hbar]$ are given by

$$\frac{1}{\hbar}\oint p\,dq, \tag{7.10.28}$$

since both S and $pq - T(p)$ are given by integrating p and the line integral is over the energy curve. On the other hand, the phase change due to the term (7.10.27) is

$$-2 \times \left[\frac{\pi}{4} - \left(-\frac{\pi}{4}\right)\right] = -\pi, \tag{7.10.29}$$

so the consistency condition is

$$\frac{1}{\hbar}\oint p\,dq - \pi = 2\pi n, \quad \text{i.e.,} \quad \oint \frac{p\,dq}{2\pi\hbar} = n + \frac{1}{2}. \tag{7.10.30}$$

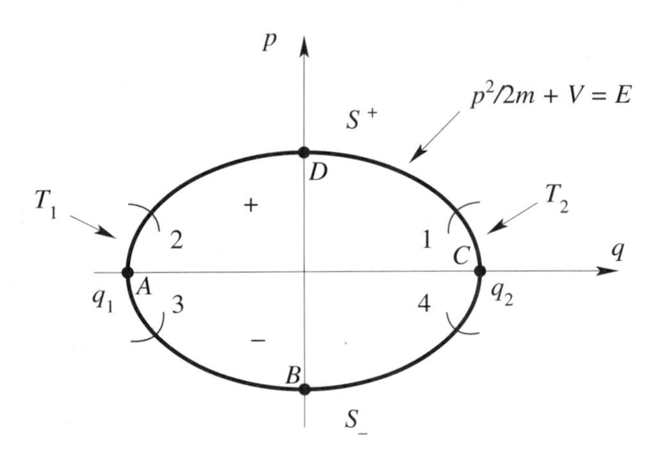

FIGURE 7.10.2.

The $\frac{1}{2}$ is the correction to the Bohr-Sommerfeld rules which one sees, for example, in the harmonic oscillator solution. Equation (7.10.30) is the ***quantization condition***. Its generalization to arbitrary manifolds reads

$$\frac{1}{2\pi\hbar}\oint_\gamma p_i\, dq^i - \frac{1}{4} I_\gamma = \text{integer}, \tag{7.10.31}$$

where I_γ is the ***Keller-Maslov-Arnold-Hörmander index*** of a closed curve γ. This topological invariant is thus arrived at via the WKB method. To properly understand it in higher dimensions requires a lengthy excursion into the theory of Lagrangian submanifolds. However, our simplified example shows that starting with a study of the asymptotic limit $\hbar \to 0$, one is led to quantization conditions; that is, questions 1 and 2, formulated at the beginning of this section are intimately related.

The overall aims of quantization and geometric asymptotics become clearer if one has in mind some of the classical-quantum correspondences. To this end, we present the table below (see Slawianowski [1971]). The basic classical object is a symplectic manifold (T^*Q, Ω) and the quantum object is the intrinsic Hilbert space $\mathcal{H} = L^2(Q)$ of half densities on Q. The dictionary sets up a correspondence between operations on each.

Classical Mechanics	*Quantum Mechanics*
immersed Lagrangian manifold $\Lambda \to (T^*Q, \Omega)$	element of $L^2(Q)$ or $\mathcal{D}'(Q)$
$\Lambda =$ graph of $\mathbf{d}S$	$\psi = \exp(iS/\hbar)$
multiplication by (-1) on fibers	complex conjugation
T^*Q	Hilbert space
$(T^*Q, -\Omega)$	dual space
Cartesian product	tensor product
disjoint union	direct product
Lagrangian manifold $\Omega \subset (T^*Q, \Omega_Q) \times (T^*R, -\Omega_R)$	(unbounded) operator from $L^2(R)$ to $L^2(Q)$
composition of canonical relations	composition of operators
graphs of canonical relations	unitary operators
Hamilton-Jacobi equation	Schrödinger equation
coisotropic submanifold $C \subset T^*Q$	involutive system of linear differential equations
reduced space C/C^Ω	solution space
reduction of Lagrangian submanifolds	projection onto solution space
symplectic action (Hamiltonian G-space)	unitary representation
coadjoint orbits (homogeneous Hamiltonian G-spaces)	irreducible representations
reduction of phase space by a symmetry group	multiplicities of irreducibles
momentum mapping	associated representation of the group algebra
polarization	complete set of observables
special symplectic structure	representation of a complete set of observables
change of special symplectic structure (Tulczyjew [1977])	Fourier integral operator

Some of the unexplained notions used in this table will be defined later in the text.

8
Variational Principles, Constraints, and Rotating Systems

This chapter deals with two related topics: constrained Lagrangian (and Hamiltonian) systems and rotating systems. Constrained systems are illustrated by a particle constrained to move on a sphere. Such constraints are called "holonomic." (In this volume we shall not discuss "nonholonomic" constraints such as rolling constraints.) For rotating systems, one needs to distinguish systems that are viewed from rotating coordinate systems (passively rotating systems) and systems which themselves are rotated (actively rotating systems—such as a Foucault pendulum rotating with the Earth). We begin with a more detailed look at variational principles and then we turn to a version of the Lagrange multiplier theorem that will be useful.

8.1 A Return to Variational Principles

In this section we make some general remarks about variational principles. Technicalities involving infinite-dimensional manifolds prevent us from presenting that point of view. For these, we refer to, for example, Smale [1964], Palais [1968], and Klingenberg [1978]. For the classical geometric theory without the infinite-dimensional framework, the reader may consult, for example, Bolza [1973], Whittaker [1927], Gelfand and Fomin [1963], or Hermann [1968].

Definition 8.1.1 *Let Q be a manifold and let $L : TQ \to \mathbb{R}$ be a regular Lagrangian. Fix two points q_1 and q_2 in Q and an interval $[a, b]$ and let*

$$\Omega(q_1, q_2, [a, b]) = \{c : [a, b] \to Q \mid c \text{ is a } C^2 \text{ curve, } c(a) = q_1, c(b) = q_2\} \tag{8.1.1}$$

called the ***path space*** *from q_1 to q_2. Define the map*

$$I : \Omega(q_1, q_2, [a, b]) \to \mathbb{R} \quad \text{by} \quad I(c) = \int_a^b L(c(t)), \dot{c}(t))\, dt.$$

What we shall *not* prove is that $\Omega(q_1, q_2, [a, b])$ is a C^∞ infinite-dimensional manifold. This is a special case of a general result in the topic of manifolds of mappings, wherein spaces of C^r (or H^s) maps from one manifold to another are shown to be smooth infinite-dimensional manifolds.

Proposition 8.1.2 *The tangent space to the manifold $\Omega(q_1, q_2, [a, b])$ at a point, i.e., a curve $c \in \Omega(q_1, q_2, [a, b])$, is given as follows:*

$$T_c\Omega(q_1, q_2, [a, b]) = \{v : [a, b] \to TQ \mid v \text{ is a } C^2 \text{ map, } \tau_Q \circ v = c \text{ and } v(a) = 0, v(b) = 0\}, \tag{8.1.2}$$

where $\tau_Q : TQ \to Q$ denotes the canonical projection.

Proof The tangent space to a manifold consists of tangents to curves in the manifold. The tangent vector to a curve $c_\lambda \in \Omega(q_1, q_2, [a, b])$ with $c_0 = c$ is

$$v = \left. \frac{d}{d\lambda} c_\lambda \right|_{\lambda=0}. \tag{8.1.3}$$

However $c_\lambda(t)$, for each fixed t, is a curve through $c_0(t) = c(t)$. Hence $\frac{d}{d\lambda} c_\lambda(t)\big|_{\lambda=0}$ is a tangent vector to Q based at $c(t)$. Hence $v(t) \in T_{c(t)}Q$, that is $\tau_Q \circ v = c$. The restrictions $c_\lambda(a) = q_1$ and $c_\lambda(b) = q_2$ lead to $v(a) = 0$ and $v(b) = 0$, but otherwise v is an arbitrary C^2 function. ■

One refers to v as an ***infinitesimal variation*** of the curve c subject to fixed endpoints. Classically, the notation $v = \delta c$ is used. Now we can state and sketch the proof of a main result in the calculus of variations.

Theorem 8.1.3 (Variational Principle of Hamilton) *Let L be a Lagrangian on TQ. A curve $c_0 : [a, b] \to Q$ joining $q_1 = c_0(a)$ to $q_2 = c_0(b)$ satisfies the Euler-Lagrange equations*

$$\frac{d}{dt}\left(\frac{\partial L}{\partial \dot{q}^i}\right) = \frac{\partial L}{\partial q^i}, \tag{8.1.4}$$

if and only if c_0 is a critical point of the function $I : \Omega(q_1, q_2, [a, b]) \to \mathbb{R}$, that is, $\mathbf{d}I(c_0) = 0$. If L is regular, either condition is equivalent to c_0 being a base integral curve of X_E.

As in §7.1, the condition $\mathbf{d}I(c_0) = 0$ is denoted

$$\delta \int_a^b L(c_0(t), \dot{c}_0(t))\, dt = 0; \tag{8.1.5}$$

that is, the integral is stationary when it is differentiated with c as the independent variable.

Proof We work out $\mathbf{d}I(c) \cdot v$ just as in §7.1. Write v as the tangent to the curve c_λ in $\Omega(q_1, q_2, [a, b])$ as in (8.1.3). By the chain rule,

$$\mathbf{d}I(c) \cdot v = \left.\frac{d}{d\lambda} I(c_\lambda)\right|_{\lambda=0} = \left.\frac{d}{d\lambda} \int_a^b L(c_\lambda(t), \dot{c}_\lambda(t))\, dt\right|_{\lambda=0}. \tag{8.1.6}$$

Differentiating (8.1.6) under the integral sign, and using local coordinates, we get

$$\mathbf{d}I(c) \cdot v = \int_a^b \left(\frac{\partial L}{\partial q^i} v^i + \frac{\partial L}{\partial \dot{q}^i} \dot{v}^i \right) dt. \tag{8.1.7}$$

Since v vanishes at both ends, the second term in (8.1.7) can be integrated by parts to give

$$\mathbf{d}I(c) \cdot v = \int_a^b \left(\frac{\partial L}{\partial q^i} - \frac{d}{dt} \frac{\partial L}{\partial \dot{q}^i} \right) v^i\, dt. \tag{8.1.8}$$

Now $\mathbf{d}I(c) = 0$ means $\mathbf{d}I(c) \cdot v = 0$ for all $v \in T_c\Omega(q_1, q_2, [a, b])$. This holds if and only if

$$\frac{\partial L}{\partial q^i} - \frac{d}{dt} \left(\frac{\partial L}{\partial \dot{q}^i} \right) = 0, \tag{8.1.9}$$

since the integrand is continuous and v is arbitrary, except for $v = 0$ at the ends.

The last statement was proved in Theorem **7.3.3**. ■

Next we discuss variational principles with the constraint of constant energy imposed. To compensate for this constraint, we let the interval $[a, b]$ be variable.

Definition 8.1.4 *Let L be a regular Lagrangian and let Σ_e be a regular energy surface for the energy E of L, that is, e is a regular value of E and $\Sigma_e = E^{-1}(e)$. Let $q_1, q_2 \in Q$ and let $[a, b]$ be a given interval. Set*

$$\begin{aligned}\Omega(q_1, q_2, [a, b], e) = \{(\tau, c) \mid\ & \tau : [a, b] \to \mathbb{R} \text{ is } C^2, \dot{\tau} > 0, \\ & c : [\tau(a), \tau(b)] \to Q \text{ is a } C^2 \text{ curve, } c(\tau(a)) = q_1, \\ & c(\tau(b)) = q_2 \text{ and } E(c(\tau(t)), \dot{c}(\tau(t))) = e \text{ for all } t \in [a, b]\}.\end{aligned} \tag{8.1.10}$$

Arguing as in Proposition **8.1.2**, differentiation of curves $(\tau(\lambda), c(\lambda))$ in $\Omega(q_1, q_2, [a,b], e)$ shows that the tangent space to $\Omega(q_1, q_2, [a,b], e)$ at (τ, c) consists of pairs of C^2 maps $\alpha : [a,b] \to \mathbb{R}$ and $v : [\tau(a), \tau(b)] \to TQ$ such that

$$v(t) \in T_{c(t)}Q, \dot{c}(\tau(a))\alpha(a)+v(\tau(a)) = 0, \dot{c}(\tau(b))\alpha(b)+v(\tau(b)) = 0, \quad (8.1.11)$$

and

$$\mathbf{d}E[c(\tau(t)), \dot{c}(\tau(t))] \cdot [\dot{c}(\tau(t))\alpha(t) + v(\tau(t)), \ddot{c}(\tau(t))\dot{\alpha}(t) + \dot{v}(\tau(t))] = 0. \quad (8.1.12)$$

Theorem 8.1.5 (Principle of Critical Action) *Let $c_0(t)$ be a solution of Lagrange's equations and let $q_1 = c_0(a)$ and $q_2 = c_0(b)$. Let e be the energy of $c_0(t)$ and assume it is a regular value of E. Define*

$$\mathcal{A} : \Omega(q_1, q_2, [a,b], e) \to \mathbb{R} \quad \text{by} \quad \mathcal{A}(\tau, c) = \int_{\tau(a)}^{\tau(b)} A(c(t), \dot{c}(t))\, dt, \quad (8.1.13)$$

where A is the action of L. Then

$$\mathbf{d}\mathcal{A}(Id, c_0) = 0, \quad \text{where } Id \text{ is the identity map.} \quad (8.1.14)$$

Conversely, if (Id, c_0) is a critical point of $\mathcal{A}$ and c_0 has energy e, a regular value of E, then c_0 is a solution of Lagrange's equations.

In coordinates, (8.1.14) reads

$$\mathcal{A}(\tau, c) = \int_{\tau(a)}^{\tau(b)} \frac{\partial L}{\partial \dot{q}^i} \dot{q}^i\, dt = \int_{\tau(a)}^{\tau(b)} p_i\, dq^i, \quad (8.1.15)$$

the integral of the canonical one-form along the curve $\gamma = (c, \dot{c})$.

Proof If the curve e has energy e,

$$\mathcal{A}(\tau, c) = \int_{\tau(a)}^{\tau(b)} [L(q^i, \dot{q}^i) + e]\, dt.$$

Differentiating $\mathcal{A}$ with respect to τ and c by the method of Theorem **8.1.3** gives

$$\begin{aligned} &\mathbf{d}\mathcal{A}(Id, c_0) \cdot (\alpha, v) \\ &\quad = \alpha(b)(L(c_0(b), \dot{c}_0(b)) + e) - \alpha(a)(L(c_0(a), \dot{c}_0(a)) + e) \\ &\qquad + \int_a^b \left(\frac{\partial L}{\partial q^i}(c_0(t), \dot{c}_0(t)) v^i(t) + \frac{\partial L}{\partial \dot{q}^i}(c_0(t), \dot{c}_0(t)) \dot{v}^i(t) \right) dt. \end{aligned} \quad (8.1.16)$$

Integrating by parts gives

$$\begin{aligned}
&\mathbf{d}\mathcal{A}(Id, c_0) \cdot (\alpha, v) \\
&= \left[\alpha(t)(L(c_0(t), \dot{c}_0(t)) + e) + \frac{\partial L}{\partial \dot{q}^i}(c_0(t), \dot{c}_0(t)) v^i(t) \right]_a^b \\
&\quad + \int_a^b \left(\frac{\partial L}{\partial q^i}(c_0(t), \dot{c}_0(t)) - \frac{d}{dt}\frac{\partial L}{\partial \dot{q}^i}(c_0(t), \dot{c}_0(t)) \right) v^i(t)\, dt.
\end{aligned} \tag{8.1.17}$$

Using the boundary conditions $v = -\dot{c}\alpha$, noted in the description of the tangent space $T_{(Id,c_0)}\Omega(q_1, q_2, [a,b], e)$, and the energy constraint $(\partial L/\partial \dot{q}^i)\dot{c}^i - L = e$, the boundary terms cancel, leaving

$$\mathbf{d}\mathcal{A}(Id, c_0) \cdot (\alpha, v) = \int_a^b \left(\frac{\partial L}{\partial q^i} - \frac{d}{dt}\frac{\partial L}{\partial \dot{q}^i} \right) v^i\, dt. \tag{8.1.18}$$

However, we can choose v arbitrarily; notice that the presence of α in the linearized energy constraint means that no restrictions are placed on the variations v^i on the open set where $\dot{c} \neq 0$. The result therefore follows. ∎

If $L = K - V$, where K is the kinetic energy of a Riemannian metric, then Theorem **8.1.5** states that a curve c_0 is a solution of Lagrange's equations if and only if

$$\delta_e \int_a^b 2K(c_0, \dot{c}_0)\, dt = 0, \tag{8.1.19}$$

where δ_e indicates a variation holding the energy and endpoints but not the parametrization fixed; that is, symbolic notation for the precise statement in Theorem **8.1.5**. Since $K \geq 0$, this is the same as

$$\delta_e \int_a^b \sqrt{2K(c_0, \dot{c}_0)}\, dt = 0, \tag{8.1.20}$$

that is, arc length is extremized (subject to constant energy). This is ***Jacobi's form of the principle of least action*** and represents a key to linking mechanics and geometric optics, which was one of Hamilton's original motivations. In particular, geodesics are characterized as extremals of arc length. Using the Jacobi metric (see §7.3) one gets yet another variational principle.

The above variational principles for Lagrangian systems carry over to some extent to Hamiltonian systems.

Theorem 8.1.6 (Hamilton's Variational Principle in Phase Space) *Consider a Hamiltonian H on a given cotangent bundle T^*Q. A curve $(q^i(t), p_i(t))$ in T^*Q satisfies Hamilton's equations iff*

$$\delta \int_a^b [p_i \dot{q}^i - H(q^i, p_i)]\, dt = 0 \tag{8.1.21}$$

for variations over curves $(q^i(t), p_i(t))$ *in phase space, where* $\dot{q}^i = dq^i/dt$ *and where* q^i *are fixed at the endpoints.*

Proof Computing as in (8.1.7), we find that

$$\delta \int_a^b \left[p_i \dot{q}^i - H(q^i, p_i)\right] dt = \int_a^b \left[(\delta p_i)\dot{q}^i + p_i(\delta \dot{q}^i) - \frac{\partial H}{\partial q^i}\delta q^i - \frac{\partial H}{\partial p_i}\delta p_i\right] dt. \tag{8.1.22}$$

Since $q^i(t)$ are fixed at the two ends, we have $p_i \delta q^i = 0$ at the two ends, and hence the second term of (8.1.22) can be integrated by parts to give

$$\int_a^b \left[\dot{q}^i(\delta p_i) - \dot{p}_i(\delta q^i) - \frac{\partial H}{\partial q^i}\delta q^i - \frac{\partial H}{\partial p_i}\delta p_i\right] dt, \tag{8.1.23}$$

which vanishes for all $\delta p_i, \delta q^i$ exactly when Hamilton's equations hold. ■

Hamilton's principle (8.1.21) on an exact symplectic manifold $(P, \Omega = -\mathbf{d}\Theta)$ reads

$$\delta \int_a^b (\Theta - H dt) = 0, \tag{8.1.24}$$

again with suitable boundary conditions. Likewise, if we impose the constraint $H =$ constant, the principle of least action reads

$$\delta \int_{\tau(a)}^{\tau(b)} \Theta = 0. \tag{8.1.25}$$

In Cendra and Marsden [1987], Cendra, Ibort, and Marsden [1987], and Marsden and Scheurle [1993a,b], it is shown how to form variational principles on certain symplectic and Poisson manifolds even when Ω is not exact, but does arise by a reduction process. We shall come to this point in Chapter 14 and in Volume II.

The one form $\rho := \Theta - H dt$ in (8.1.24), regarded as a one-form on $P \times \mathbb{R}$ is often called a ***contact form*** and plays an important role in time-dependent and relativistic mechanics. Let $\sigma = -\mathbf{d}\rho = \Omega + dH \wedge dt$ and observe that the vector field X_H is characterized by the statement that its suspension $Z_H = (X_H, 1)$, a vector field on $P \times \mathbb{R}$, lies in the kernel of σ:

$$\mathbf{i}_{Z_H}\sigma = 0.$$

Some History of the Euler-Lagrange Equations

In the following paragraphs we make a few historical remarks concerning the Euler-Lagrange equations.[1] Naturally, much of the story focuses on

[1] Many of these interesting historical points were conveyed by Hans Duistermaat to whom we are very grateful. The reader can also profitably consult the standard texts such as those of Whittaker [1927], Wintner [1941], and Lanczos [1949].

Lagrange. Section V of Lagrange's *Mécanique Analytique* [1788] contains the equations of motion in Euler-Lagrange form (8.1.4). Lagrange writes $Z = T - V$ for what we would call the Lagrangian today. In the previous section Lagrange came to these equations by asking for a coordinate invariant expression for mass times acceleration. His conclusion is that it is given (in abbreviated notation) by $(d/dt)(\partial T/\partial v) - \partial T/\partial q$, which transforms under arbitrary substitutions of position variables as a one-form. Lagrange does *not* recognize the equations of motion as being equivalent to the variational principle

$$\delta \int L \, dt = 0$$

—this was observed only a few decades later by Hamilton. The peculiar fact about this is that Lagrange *did* know the general form of the differential equations for variational problems and he actually had commented on Euler's proof of this—his early work on this in 1759 was admired very much by Euler. He immediately applied it to give a proof of the Maupertuis principle of least action, as a consequence of Newton's equations of motion. This principle, apparently having its roots in the early work of Leibnitz, is a less natural principle in the sense that the curves are only varied over those which have a constant energy. It is also Hamilton's principle that applies in the *time-dependent* case, when H is *not* conserved and which also generalizes to allow for certain external forces as well.

This discussion in the *Mécanique Analytique precedes* the equations of motion in general coordinates, and so is written in the case that the kinetic energy is of the form $\sum_i m_i v_i^2$, where the m_i are positive constants. Wintner [1941] is also amazed by the fact that the more complicated Maupertuis principle precedes Hamilton's principle. One possible explanation is that Lagrange did not consider L as an interesting physical quantity—for him it was only a convenient function for writing down the equations of motion in a coordinate-invariant fashion. The time span between his work on variational calculus and the *Mécanique Analytique* (1788, 1808) could also be part of the explanation—he may not have been thinking of the variational calculus when he addressed the question of a coordinate invariant formulation of the equations of motion.

Section V starts by discussing the evident fact that the position and velocity at time t depend on the initial position and velocity, which can be chosen freely. We might write this as (suppressing the coordinate indices for simplicity): $q = q(t, q_0, v_0), v = v(t, q_0, v_0)$, and in modern terminology we would talk about the flow in $x = (q, v)$-space. One problem in reading Lagrange is that he does not explicitly write the variables on which his quantities depend. In any case, he then makes an infinitesimal variation in the initial condition and looks at the corresponding variations of position and velocity at time t. In our notation: $\delta x = (\partial x/\partial x_0)(t, x_0)\delta x_0$. We would say that he considers the tangent mapping of the flow on the tangent bundle of $X = TQ$. Now comes the first interesting result. He makes two such variations, one denoted by δx and the other by Δx, and he writes down a bilinear form $\omega(\delta x, \Delta x)$, in which we recognize ω as the pull back of the canonical symplectic form on the cotangent bundle of Q, by means of the fiber derivative $\mathbb{F}L$. What he then shows is that this symplectic product is constant as a function of t. This is nothing other than the *invariance of the symplectic form* ω *under the flow in* TQ.

It is striking that Lagrange obtains the invariance of the symplectic form in TQ and not in T^*Q. In fact, Lagrange does *not* look at the equations of motion in the cotangent bundle via the transformation $\mathbb{F}L$; again it is Hamilton who observes that these take the canonical Hamiltonian form. This is retrospectively puzzling since, later on in Section V, Lagrange states very explicitly that it is useful to pass to the (q, p)-coordinates by means of the coordinate transformation $\mathbb{F}L$ and one even sees written down a system of ordinary differential equations *in Hamiltonian form*, but with the total energy function H replaced by some other mysterious function $-\Omega$. Lagrange does use the letter H for the constant value of energy, apparently in honor of Huygens. He also knew about the conservation of momentum as a result of translational symmetry.

The part where he does this deals with the case in which he perturbs the system by perturbing the potential from $V(q)$ to $V(q) - \Omega(q)$, leaving the kinetic energy unchanged. To this perturbation problem, he applies his famous method of variation of constants, which is presented here in a truly nonlinear framework! In our notation, he keeps $t \mapsto x(t, x_0)$ as the solution of the unperturbed system, and then looks at the differential equations for $x_0(t)$ that make $t \mapsto x(t, x_0(t))$ a solution of the perturbed system. The result is that, if V is the vector field of the unperturbed system and $V + W$ is the vector field of the perturbed system, then $dx_0/dt = ((e^{tV})^* W)(x_0)$. In words, $x_0(t)$ is the solution of the time-dependent system, the vector field of which is obtained by pulling back W by means of the flow of V after time t. In the case that Lagrange considers, the dq/dt-component of the perturbation is equal to zero, and the dp/dt-component is equal $\partial\Omega/\partial q$. Thus, it is obviously in a Hamiltonian form; here one does not use anything about Legendre-transformations (which Lagrange does not seem to know). But Lagrange knows already that the flow of the unperturbed system preserves the symplectic form, and he shows that the pull back of his W under such a transformation is a vector field in Hamiltonian form. Actually, this is a time-dependent vector field, defined by the function $G(t, q_0, p_0) = -\Omega(q(t, q_0, p_0))$. A potential point of confusion is that Lagrange denotes this by $-\Omega$, and writes down expressions like $d\Omega/dp$, and one might first think these are zero because Ω was assumed to depend only on q. Lagrange presumably means that $dq_0/dt = \partial G/\partial p_0, dp_0/dt = -\partial G/\partial q_0$.

Most classical textbooks on mechanics, for example, Routh [1877, 1884], correctly point out that Lagrange has the invariance of the symplectic form in (q, v) coordinates (rather than in the canonical (q, p) coordinates). Less attention is usually paid to the variation of constants equation in

Hamiltonian form, but it must have been generally known that Lagrange derived these—see, for example, Weinstein [1981]. In fact, we should point out that the whole question of linearizing the Euler-Lagrange and Hamilton equations and retaining the mechanical structure is remarkably subtle (see Marsden, Ratiu, and Raugel [1991], for example).

Lagrange continues by introducing the *Poisson brackets* for arbitrary functions, arguing that these are useful in writing the time derivative of arbitrary functions of arbitrary variables, along solutions of systems in Hamiltonian form. He also continues by saying that if Ω is small, then $x_0(t)$ in zero-order approximation is a constant and he obtains the next order approximation by an integration over t; here Lagrange introduces the first steps of the so-called *method of averaging*. When Lagrange discovered (in 1808) the invariance of the symplectic form, the variations-of-constants equations in Hamiltonian form, and the Poisson brackets, he was already 73 years old. It is quite probable that Lagrange gave some of these bracket ideas to Poisson at this time. In any case, it is clear that Lagrange had a surprisingly large part of the symplectic picture of classical mechanics.

Exercises

8.1-1 *In Hamilton's principle, show that the boundary conditions of fixed $q(a)$ and $q(b)$ can be changed to $p(b) \cdot \delta q(b) = p(a) \cdot \delta q(a)$. What is the corresponding statement for Hamilton's principle in phase space?*

8.1-2 *Show that the equations for a particle in a magnetic field B and a potential V can be written as*

$$\delta \int (K - V)\, dt = \frac{e}{c} \int \delta q \cdot (v \times B).$$

8.2 The Lagrange Multiplier Theorem

We recall the Lagrange multiplier theorem for purposes of studying constrained dynamics in the next section. We state the result with a sketch of the proof, referring to Abraham, Marsden, and Ratiu [1988] for details. As in the previous section, we shall not be absolutely precise about the technicalities (such as how to interpret dual spaces). The reader versed in applied analysis can supply these without difficulty.

First consider the case of functions defined on linear spaces. Let V and Λ be Banach spaces and let $\varphi : V \to \Lambda$ be a smooth map. Suppose 0 is a regular value of φ so that $C := \varphi^{-1}(0)$ is a submanifold. Let $h : V \to \mathbb{R}$ be a smooth function and define

$$\overline{h} : V \times \Lambda^* \to \mathbb{R} \quad \text{by} \quad \overline{h}(x, \lambda) = h(x) - \langle \lambda, \varphi(x) \rangle . \tag{8.2.1}$$

Theorem 8.2.1 (Lagrange Multiplier Theorem for Linear Spaces) *The following are equivalent conditions on $x_0 \in C$:*

i *x_0 is a critical point of $h|C$; and*

ii *there is a $\lambda_0 \in \Lambda^*$ such that (x_0, λ_0) is a critical point of $\overline{h}$.*

Sketch of Proof Since $\mathbf{D}\overline{h}(x_0, \lambda_0) \cdot (x, \lambda) = \mathbf{D}h(x_0) \cdot x - \langle \lambda_0, \mathbf{D}\varphi(x_0) \cdot x \rangle - \langle \lambda, \varphi(x_0) \rangle$ and $\varphi(x_0) = 0$, the condition $\mathbf{D}\overline{h}(x_0, \lambda_0) \cdot (x, \lambda) = 0$ is equivalent to

$$\mathbf{D}h(x_0) \cdot x = \langle \lambda_0, \mathbf{D}\varphi(x_0) \cdot x \rangle \tag{8.2.2}$$

for all $x \in V$ and $\lambda \in \Lambda^*$. The tangent space to C at x_0 is $\ker \mathbf{D}\varphi(x_0)$, so (8.2.2) implies $h|C$ has a critical point at x_0. Conversely, if $h|C$ has a critical point at x_0, then $\mathbf{D}h(x_0) \cdot x = 0$ for all x satisfying $\mathbf{D}\varphi(x_0) \cdot x = 0$. By the implicit function theorem, there is a smooth coordinate change that straightens out C; that is, it allows us to assume that $V = W \oplus \Lambda, x_0 = 0, C$ is (in a neighborhood of 0) equal to W, and φ (in a neighborhood of the origin) is the projection to Λ. With these simplifications, condition **i** means the first partial derivative of h vanishes. We choose λ_0 to be $\mathbf{D}_2 h(x_0)$ regarded as an element of Λ^*; then(8.2.2) clearly holds. ■

The Lagrange multiplier theorem is a convenient test for constrained critical points, as we know from calculus. It also leads to a convenient test for constrained maxima and minima. For instance, to test for a minimum, let $\alpha > 0$ be a constant, let (x_0, λ_0) be a critical point of $\overline{h}$, and consider

$$h_\alpha(x, \lambda) = h(x) - \langle \lambda, \varphi(x) \rangle + \alpha \|\lambda - \lambda_0\|^2, \tag{8.2.3}$$

which also has a critical point at (x_0, λ_0). Clearly, if h_α has a minimum at (x_0, λ_0), then $h|C$ has a minimum at x_0. Conversely, if $h|C$ has a nondegenerate minimum at x_0, then for sufficiently large $\alpha > 0$, h_α has a minimum at (x_0, λ_0). This observation is convenient since one can use the unconstrained second derivative test on h_α, which leads to the theory of ***bordered Hessians***. (For an elementary discussion, see Marsden and Tromba [1988], p. 377ff.)

A second remark concerns the generalization of the Lagrange multiplier theorem to the case where V is a manifold but h is still real-valued. Such a context is as follows. Let M be a manifold and let $N \subset M$ be a submanifold. Suppose $\pi : E \to M$ is a vector bundle over M and φ is a section of E that is transverse to fibers. Assume $N = \varphi^{-1}(0)$.

Theorem 8.2.2 (Lagrange Multiplier Theorem for Manifolds) *The following are equivalent for $x_0 \in N$ and $h : M \to \mathbb{R}$ smooth:*

i *x_0 is a critical point of $h|N$; and*

ii *there is a section λ_0 of the dual bundle E^* such that $\lambda_0(x_0)$ is a critical point of $\overline{h} : E^* \to \mathbb{R}$ defined by*

$$\overline{h}(\lambda_x) = h(x) - \langle \lambda_x, \varphi(x) \rangle . \tag{8.2.4}$$

In (8.2.4), λ_x denotes an arbitrary element of E^*_x. We leave it to the reader to adapt the proof of the previous theorem to this situation.

Exercise

8.2-1 *Write out the second derivative of h_α at (x_0, λ_0) and relate your answer to the bordered Hessian.*

8.3 Holonomic Constraints

Many mechanical systems are obtained from higher-dimensional ones by adding constraints. Rigidity in rigid body mechanics and incompressibility in fluid mechanics are two such examples, while constraining a free particle to move on a sphere is another.

Typically, constraints are of two types. Holonomic contraints are those imposed on the configuration of a system, such as those mentioned in the preceding paragraph. Others, such as *rolling constraints* involve the conditions on the velocities and are termed nonholonomic. (We refer to Bloch, Krishnaprasad, Marsden, and Murray [1994] for more information and literature in the nonholonomic case.)

A ***holonomic constraint*** can be defined for our purposes as the specification of a submanifold $N \subset Q$ of a given configuration manifold Q. (More generally a holonomic constraint is an integrable subbundle of TQ.) Since we have the natural inclusion $TN \subset TQ$, a given Lagrangian $L : TQ \to \mathbb{R}$ can be restricted to TN to give a Lagrangian L_N. We now have two Lagrangian systems, namely those associated to L and to L_N, assuming both are regular. We now relate the associated variational principles and the Hamiltonian vector fields.

Suppose that $N = \varphi^{-1}(0)$ for a section $\varphi : Q \to E^*$, the dual of a vector bundle E over Q. The variational principle for L_N can be phrased as

$$\delta \int L_N(q, \dot{q}) \, dt = 0, \tag{8.3.1}$$

where the variation is over curves with fixed endpoints and subject to the constraint $\varphi(q(t)) = 0$. By the Lagrange multiplier theorem, (8.3.1) is equivalent to

$$\delta \int [L(q(t), \dot{q}(t)) - \langle \lambda(q(t), t), \varphi(q(t)) \rangle] \, dt = 0 \tag{8.3.2}$$

for some function $\lambda(q, t)$ taking values in the bundle E and where the variation is over curves q in Q and curves λ in E. In coordinates, (8.3.2) reads

$$\delta \int [L(q^i, \dot{q}^i) - \lambda^a(q^i, t)\varphi_a(q^i)] \, dt = 0. \tag{8.3.3}$$

The corresponding Euler-Lagrange equations in the variables q^i, λ^a are

$$\frac{d}{dt}\frac{\partial L}{\partial \dot{q}^i} = \frac{\partial L}{\partial q^i} - \lambda^a \frac{\partial \varphi_a}{\partial q^i} \tag{8.3.4}$$

and

$$\varphi_a = 0, \tag{8.3.5}$$

which are viewed as equations in the unknowns $q^i(t)$ and $\lambda^a(q^i, t)$; if E is a trivial bundle we can take λ to be a function only of t.

The combination $\mathcal{L} = L - \lambda^a \varphi_a$ is related to the Routhian construction for a Lagrangian with cyclic variables—we shall study this in Volume II; see also Marsden and Scheurle [1993a,b].

Notice that $\mathcal{L} = L - \lambda^a \varphi_a$ as a Lagrangian in q and λ is degenerate in λ; that is, the time derivative of λ does not appear, so its conjugate momentum π_a is constrained to be zero. We think of $\mathcal{L}$ as defined on TE. Formally, the corresponding Hamiltonian on T^*E is

$$\mathcal{H}(q, p, \lambda, \pi) = H(q, p) + \lambda^a \varphi_a, \tag{8.3.6}$$

where H is the Hamiltonian corresponding to L.

One has to be a little careful in interpreting Hamilton's equations because $\mathcal{L}$ is degenerate; the general theory appropriate for this situation is the *Dirac theory of constraints*, which we mention in §8.5; see Dirac [1950], [1964], Sudarshan and Mukunda [1974], Hansen, Regge, and Teitelboim [1976], and Gotay, Nester, and Hinds [1978]. However, in the present context this theory is quite simple and proceeds as follows. One calls $C \subset T^*E$ defined by $\pi_a = 0$, the ***primary constraint set***; it is the image of the Legendre transform provided the original L was regular. The canonical form Ω is pulled back to C to give a presymplectic form (a closed but possibly degenerate two-form) Ω_C and one seeks $X_{\mathcal{H}}$ such that

$$\mathbf{i}_{X_{\mathcal{H}}} \Omega_C = \mathbf{d}\mathcal{H}. \tag{8.3.7}$$

In this case, the degeneracy of Ω_C gives no equation for λ; that is, the evolution of λ is indeterminate. The other Hamiltonian equations are equivalent to (8.3.4) and (8.3.5), so in this sense the Lagrangian and Hamiltonian pictures are still equivalent.

8.4 Constrained Motion in a Potential Field

We saw in the preceding section how to write the equations for a constrained system in terms of variables on the containing space. We continue this line of investigation here by specializing to the case of motion in a potential field. In fact, we shall determine by geometric methods, the extra terms that need to be added to the Euler-Lagrange equations to ensure that the constraints are maintained.

Let Q be a (weak) Riemannian manifold and let $N \subset Q$ be a submanifold. Let

$$\mathbb{P} : (TQ)|N \to TN \tag{8.4.1}$$

be the orthogonal projection of TQ to TN defined pointwise on N.

Consider a Lagrangian $L : TQ \to \mathbb{R}$ of the form $L = K - V \circ \tau_Q$; that is, kinetic minus potential energy. The Riemannian metric associated to the kinetic energy is denoted by $\langle\langle\, , \rangle\rangle$. The restriction $L_N = L|TN$ is also of the form kinetic minus potential, using the metric induced on N and the potential $V_N = V|N$. We know from §7.7 that if E_N is the energy of L_N, then

$$X_{E_N} = S_N - \operatorname{ver}(\nabla V_N), \tag{8.4.2}$$

where S_N is the spray of the metric on N and ver( ) denotes vertical lift. Recall that integral curves of (8.4.2) are solutions of the Euler-Lagrange equations. Let S be the geodesic spray on Q.

Now ∇V_N and ∇V are related in a very simple way: for $q \in N$,

$$\nabla V_N(q) = \mathbb{P} \cdot [\nabla V(q)].$$

Thus, the main complication is in the geodesic spray.

Proposition 8.4.1 *$S_N = T\mathbb{P} \circ S$ at points of TN.*

Proof For the purpose of this proof we can ignore the potential and let $L = K$. Let $R = TQ|N$, so that $\mathbb{P} : R \to TN$ and therefore $T\mathbb{P} : TR \to T(TN)$, $S : R \to T(TQ)$, and $T\tau_Q \circ S =$ identity since S is second order. But $TR = \{w \in T(TQ) \mid T\tau_Q(w) \in TN\}$, so $S(TN) \subset TR$ and hence $T\mathbb{P} \circ S$ makes sense at points of TN.

If $v \in TQ$ and $w \in T_v(TQ)$, then $\Theta_L(v) \cdot w = \langle\langle v, T_v\tau_Q(w)\rangle\rangle$. Letting $i : R \to TQ$ be the inclusion, we claim that

$$\mathbb{P}^*\Theta_{L|TN} = i^*\Theta_L. \tag{8.4.3}$$

Indeed, for $v \in R$ and $w \in T_vR$, the definition of pull backs gives

$$\mathbb{P}^*\Theta_{L|TN}(v) \cdot w = \langle\langle \mathbb{P}v, (T\tau_Q \circ T\mathbb{P})(w)\rangle\rangle = \langle\langle \mathbb{P}v, T(\tau_Q \circ \mathbb{P})(w)\rangle\rangle. \tag{8.4.4}$$

Since on $R, \tau_Q \circ \mathbb{P} = \tau_Q, \mathbb{P}^* = \mathbb{P}$, and $w \in T_vR$, (8.4.4) becomes

$$\begin{aligned}\mathbb{P}^*\Theta_{L|TN}(v) \cdot w &= \langle\langle \mathbb{P}v, T\tau_Q(w)\rangle\rangle = \langle\langle v, \mathbb{P}T\tau_Q(w)\rangle\rangle = \langle\langle v, T\tau_Q(w)\rangle\rangle \\ &= \Theta_L(v) \cdot w = (i^*\Theta_L)(v) \cdot w.\end{aligned}$$

Taking the exterior derivative of (8.4.3) gives

$$\mathbb{P}^*\Omega_{L|TN} = i^*\Omega_L. \tag{8.4.5}$$

In particular, for $v \in TN, w \in T_vR$, and $z \in T_v(TN)$, the definition of pull back and (8.4.5) gives

$$\begin{aligned}\Omega_L(v)(w,z) &= (i^*\Omega_L)(v)(w,z) = (\mathbb{P}^*\Omega_{L|TN})(v)(w,z)\\ &= \Omega_{L|TN}(\mathbb{P}v)(T\mathbb{P}(w), T\mathbb{P}(z))\\ &= \Omega_{L|TN}(v)(T\mathbb{P}(w), z).\end{aligned} \tag{8.4.6}$$

But

$$\mathbf{d}E(v)\cdot z = \Omega_L(v)(S(v),z) = \Omega_{L|TN}(v)(S_N(v),z)$$

since S and S_N are Hamiltonian vector fields for E and $E|TN$, respectively. From (8.4.6),

$$\Omega_{L|TN}(v)(T\mathbb{P}(S(v)),z) = \Omega_L(v)(S(v),z) = \Omega_{L|TN}(v)(S_N(v),z),$$

so by weak nondegeneracy of $\Omega_{L|TN}$ we get the desired relation $S_N = T\mathbb{P}\circ S$. ■

Corollary 8.4.2 *For $v \in T_qN$:*

i *$(S - S_N)(v)$ is the vertical lift of a vector $Z(v) \in T_qQ$ with respect to v;*

ii *$Z(v) \perp T_qN$; and*

iii *$Z(v) = -\nabla_v v + \mathbb{P}(\nabla_v v)$ is minus the normal component of $\nabla_v v$, where in $\nabla_v v, v$ is extended to a vector field on Q tangent to N.*

Proof

i Since $T\tau_Q(S(v)) = v = T\tau_Q(S_N(v))$ we have $T\tau_Q(S - S_N)(v) = 0$, that is, $(S - S_N)(v)$ is vertical. The statement now follows from the comments following Definition **7.7.1**.

ii For $u \in T_qQ$, we have $T\mathbb{P}\cdot \mathrm{ver}(u,v) = \mathrm{ver}(\mathbb{P}u, v)$ since

$$\mathrm{ver}(\mathbb{P}u,v) = \left.\frac{d}{dt}(v + t\mathbb{P}u)\right|_{t=0} = \left.\frac{d}{dt}\mathbb{P}(v+tu)\right|_{t=0} = T\mathbb{P}\cdot\mathrm{ver}(u,v). \tag{8.4.7}$$

By part **i**, $S(v) - S_N(v) = \mathrm{ver}(Z(v), v)$ for some $Z(v) \in T_qQ$, so that using the previous theorem and $\mathbb{P}\circ\mathbb{P} = \mathbb{P}$, we get

$$\begin{aligned}\mathrm{ver}(\mathbb{P}Z(v),v) &= T\mathbb{P}\cdot\mathrm{ver}(Z(v),v)\\ &= T\mathbb{P}(S(v) - S_N(v))\\ &= T\mathbb{P}(S(v) - T\mathbb{P}\circ S(v)) = 0.\end{aligned}$$

Therefore, $\mathbb{P}Z(v) = 0$, that is, $Z(v) \perp T_qN$.

iii Let $v(t)$ be a curve of tangents to $N; v(t) = \dot{c}(t)$, where $c(t) \in N$. Then in a chart,

$$S(c(t), v(t)) = \big(c(t), v(t), v(t), \gamma_{c(t)}(v(t), v(t))\big)$$

by (7.5.5). Extending $v(t)$ to a vector field v on Q tangent to N we get, in a standard chart,

$$\nabla_v v = -\gamma_c(v, v) + \mathbf{D}v(c) \cdot v = -\gamma_c(v, v) + \frac{dv}{dt}$$

by (7.5.19), so on TN,

$$S(v) = \frac{dv}{dt} - \operatorname{ver}(\nabla_v v, v).$$

Since $dv/dt \in TN$, (8.4.7) and the previous proposition give

$$S_N(v) = T\mathbb{P}\frac{dv}{dt} - \operatorname{ver}(\mathbb{P}(\nabla_v v), v) = \frac{dv}{dt} - \operatorname{ver}(\mathbb{P}(\nabla_v v), v).$$

Thus by part **i**,

$$\operatorname{ver}(Z(v), v) = S(v) - S_N(v) = \operatorname{ver}(-\nabla_v v + \mathbb{P}\nabla_v v, v). \quad \blacksquare$$

The vector field Z is called the ***force of constraint***. It equals the negative of the quadratic part of the second fundamental form of N in Q if the codimension of Q is one. We interpret $Z(v)$ as the constraining force needed to keep particles in N. Notice that N is totally geodesic (that is, geodesics in N are geodesics in Q) iff $Z = 0$.

Exercises

8.4-1 *Compute the force of constraint Z for $S^2 \subset \mathbb{R}^3$.*

8.4-2 *Assume L is a regular Lagrangian on TQ and $N \subset Q$. Let $i : TN \to TQ$ be the embedding obtained from $N \subset Q$ and let Ω_L be the Lagrange two-form on TQ. Show that $i^*\Omega_L$ is the Lagrange two-form $\Omega_{L|TN}$ on TN. Assuming L is hyperregular, show that the Legendre transform defines a symplectic embedding $T^*N \subset T^*Q$.*

8.4-3 *In $\mathbb{R}^3$, let*

$$H(\mathbf{q}, \mathbf{p}) = \frac{1}{2m}\left[\|p\|^2 - (p \cdot q)^2\right] + mgq^3.$$

Show that Hamilton's equations in $\mathbb{R}^3$ automatically preserve T^*S^2 and give the pendulum equations when restricted to this invariant (symplectic) submanifold.

8.5 Dirac Constraints

If (P, Ω) is a symplectic manifold, a submanifold $S \subset P$ is called a ***symplectic submanifold*** when $\omega := i^*\Omega$ is a symplectic form on S, $i : S \to P$ being the inclusion. Thus, S inherits a Poisson bracket structure; its relationship to the bracket structure on P is given by a formula of Dirac [1950] that will be derived in this section. Dirac's work was motivated by the study of constrained systems, especially relativistic ones, where one thinks of S as a constraint subspace of phase space (see the preceding section and Gotay, Isenberg, Marsden, and Montgomery [1994] for more information). Let us work in the finite-dimensional case; the reader is invited to study the intrinsic infinite-dimensional version using Remark **(a)** below. Let $\dim P = 2n$ and $\dim S = 2k$. In a neighborhood of a point z_0 of S, choose coordinates $z^1, \ldots, z^{2n}$ on P such that S is given by $z^{2k+1} = 0, \ldots, z^{2n} = 0$ and such that $z^1, \ldots, z^{2n}$ provide canonical coordinates for Ω on the tangent space to P at the point z_0 and $z^1, \ldots, z^{2k}$ provide canonical coordinates for ω on the tangent space to S at z_0.

Consider the matrix whose entries are

$$C^{ij}(z) = \{z^i, z^j\}, \quad i, j = 2k+1, \ldots, 2n.$$

By our choice of coordinates, $[C^{ij}(z)]$ is nonsingular at z_0 (since the coordinates are canonical at the point z_0) and hence also in a neighborhood of z_0. Let its inverse be denoted $[C_{ij}(z)]$. Let F be a smooth function on P and $F|S$ its restriction to S. We are interested in relating $X_{F|S}$ and X_F as well as the brackets $\{F, G\}|S$ and $\{F|S, G|S\}$.

Proposition 8.5.1 (Dirac's Bracket Formula) *In a coordinate neighborhood as described above, and for $z \in S$, we have*

$$X_{F|S}(z) = X_F(z) - \sum_{i,j=2k+1}^{2n} \{F, z^i\} C_{ij}(z) X_{z^j}(z) \tag{8.5.1}$$

and

$$\{F|S, G|S\}(z) = \{F, G\}(z) - \sum_{i,j=2k+1}^{2n} \{F, z^i\} C_{ij}(z) \{z^j, G\}. \tag{8.5.2}$$

Proof To verify (8.5.1), we show that the right-hand side satisfies the condition required for $X_{F|S}(z)$, namely that it be a vector field on S and that

$$\omega_z(X_{F|S}(z), v) = \mathbf{d}(F|S)_z \cdot v \tag{8.5.3}$$

for $v \in T_zS$. Since S is symplectic, $T_zS \cap (T_zS)^\Omega = \{0\}$, where $(T_zS)^\Omega$ denotes the Ω-orthogonal complement. Since $\dim(T_zS) + \dim(T_zS)^\Omega = 2n$, we get

$$T_zP = T_zS \oplus (T_zS)^\Omega. \tag{8.5.4}$$

If $\pi_z : T_zP \to T_zS$ is the associated projection operator one can verify that

$$X_{F|S}(z) = \pi_z \cdot X_F(z), \tag{8.5.5}$$

so, in fact, (8.5.1) is a formula for π_z in coordinates; equivalently,

$$(\mathrm{Id} - \pi_z)X_F(z) = \sum_{i,j=2k+1}^{2n} \{F, z^i\}C_{ij}(z)X_{z^j}(z) \tag{8.5.6}$$

gives the projection to $(T_zS)^\Omega$. To verify (8.5.6), we need to check that the right-hand side

i is an element of $(T_zS)^\Omega$;

ii equals $X_F(z)$ if $X_F(z) \in (T_zS)^\Omega$; and

iii equals 0 if $X_F(z) \in T_zS$.

To prove **i**, observe that $X_K(z) \in (T_zS)^\Omega$ means $\Omega(X_K(z), v) = 0$ for all $v \in T_zS$; that is, $\mathbf{d}K_z \cdot v = 0$ for all $v \in T_zS$. But for $K = z^j, j = 2k+1, \ldots, 2n, K \equiv 0$ on S, and hence $\mathbf{d}K_z \cdot v = 0$. Thus $X_{z^j}(z) \in (T_zS)^\Omega$, so **i** holds. For **ii**, if $X_F(z) \in (T_zS)^\Omega$, then $\mathbf{d}F_z \cdot v = 0$ for all $v \in T_zS$ and, in particular, for $v = \partial/\partial z^i, i = 1, \ldots, 2k$. Therefore, for $z \in S$, we can write

$$\mathbf{d}F_z = \sum_{j=2k+1}^{2n} a_j \, dz^j \tag{8.5.7}$$

and

$$X_F(z) = \sum_{j=2k+1}^{2n} a_j X_{z^j}(z). \tag{8.5.8}$$

The a_j are determined by pairing (8.5.8) with $dz^i, i = 2k+1, \ldots, 2n$, to give

$$-\left\langle dz^i, X_F(z)\right\rangle = \{F, z^i\} = \sum_{j=2k+1}^{2n} a_j\{z^j, z^i\} = \sum_{j=2k+1}^{2n} a_j C^{ji},$$

or

$$a_j = \sum_{i=2k+1}^{2n} \{F, z^i\}C_{ij}, \tag{8.5.9}$$

which proves **ii**. Finally, for **iii**, $X_F(z) \in T_zS = ((T_zS)^\Omega)^\Omega$ means $X_F(z)$ is Ω orthogonal to each $X_{z^j}, j = 2k+1, \ldots, 2n$. Thus $\{F, z^j\} = 0$, so the right-hand side of (8.5.6) vanishes.

Formula (8.5.6) is therefore proved, and so, equivalently (8.5.1) holds. Formula (8.5.2) follows by writing $\{F|S, G|S\} = \omega(X_{F|S}, X_{G|S})$ and substituting (8.5.1). In doing this, the last two terms cancel. ■

In (8.5.2) notice that $\{F|S, G|S\}(z)$ is intrinsic to $F|S, G|S$, and S. The bracket does not depend on how $F|S$ and $G|S$ are extended off S to functions F, G on P. This is not true for just $\{F, G\}(z)$, which *does* depend on the extensions, but the extra term in (8.5.2) cancels this dependence.

Remarks

(a) A coordinate-free way to write (8.5.2) is as follows. Write $S = \psi^{-1}(m_0)$, where $\psi : P \to M$ is a submersion on S. For $z \in S$, and $m = \psi(z)$, let

$$C_m : T_m^* M \times T_m^* M \to \mathbb{R} \tag{8.5.10}$$

be given by

$$C_m(\mathbf{d}F_m, \mathbf{d}G_m) = \{F \circ \psi, G \circ \psi\}(z) \tag{8.5.11}$$

for $F, G \in \mathcal{F}(M)$. Assume C_m is invertible, with "inverse" $C_m^{-1} : T_m M \times T_m M \to \mathbb{R}$. Then

$$\{F|S, G|S\}(z) = \{F, G\}(z) - C_m^{-1}(T_z\psi \cdot X_F(z), T_z\psi \cdot X_G(z)). \tag{8.5.12}$$

(b) There is another way to derive and write Dirac's formula using complex structures. Suppose $\langle\!\langle\, , \rangle\!\rangle_z$ is an inner product on T_zP and $\mathbb{J}_z : T_zP \to T_zP$ is an orthogonal transformation satisfying $\mathbb{J}_z^2 = -$ Identity and as in §5.3,

$$\Omega_z(u, v) = \langle\!\langle \mathbb{J}_z u, v \rangle\!\rangle \tag{8.5.13}$$

for all $u, v \in T_zP$. With the inclusion $i : S \to P$ as before, we get corresponding structures induced on S; let

$$\omega = i^*\Omega. \tag{8.5.14}$$

If ω is nondegenerate, then (8.5.14) and the induced metric defines an associated complex structure $\mathbb{K}$ on S. At a point $z \in S$, suppose one has arranged to choose $\mathbb{J}_z$ to map T_zS to itself, and that $\mathbb{K}_z$ is the restriction of $\mathbb{J}_z$ to T_zS. At z, we then get

$$(T_zS)^\perp = (T_zS)^\Omega$$

and thus symplectic projection coincides with orthogonal projection. From (8.5.5), and using coordinates as described earlier, but for which the $X_{z^j}(z)$ are also orthogonal, we get

$$\begin{aligned} X_{F|S}(z) &= X_F(z) - \sum_{j=2k+1}^{2n} \langle X_F(z), X_{z^j}(z)\rangle\, X_{z^j}(z) \\ &= X_F(z) + \sum_{j=2k+1}^{2n} \Omega(X_F(z), \mathbb{J}^{-1} X_{z^j}(z)) X_{z^j}. \end{aligned} \tag{8.5.15}$$

This is equivalent to (8.5.1) and so also gives (8.5.2); to see this, one shows that

$$\mathbb{J}^{-1} X_{z^j}(z) = - \sum_{i=2k+1}^{2n} X_{z^i}(z) C_{ij}(z). \tag{8.5.16}$$

Indeed, the symplectic pairing of each side with X_{z^p} gives δ^p_j.

(c) For a relationship between Poisson reduction and Dirac's formula, see Marsden and Ratiu [1986]. ♦

Examples

(a) ***Holonomic Constraints*** To do holonomic constraints by the Dirac formula, proceed as follows. Let $N \subset Q$ as in §8.4, so that $TN \subset TQ$; with $i : N \to Q$ the inclusion, one finds $(Ti)^* \Theta_L = \Theta_{L_N}$ by considering the following commutative diagram:

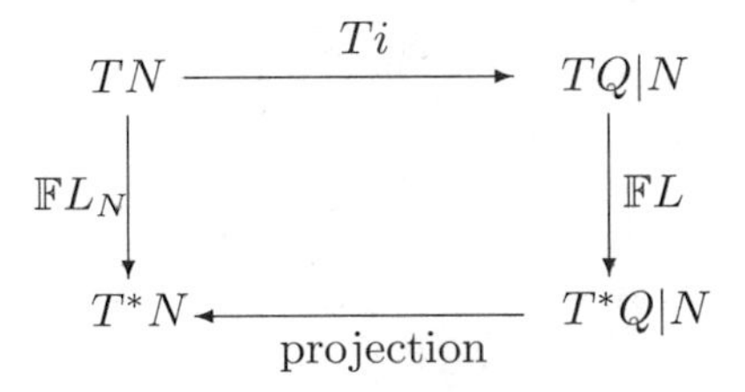

This realizes TN as a symplectic submanifold of TQ and so Dirac's formula can be applied, reproducing (8.4.2). See Exercise **8.5-1**.

(b) ***KdV Equation*** [2] Suppose one starts with a Lagrangian of the form

$$L(v_q) = \langle \alpha(q), v \rangle - h(q), \tag{8.5.17}$$

where α is a one-form on Q and h is a function on Q. In coordinates, (8.5.17) reads

$$L(q^i, \dot{q}^i) = \alpha_i(q) \dot{q}^i - h(q^i). \tag{8.5.18}$$

The corresponding momenta are

$$p_i = \frac{\partial L}{\partial \dot{q}^i} = \alpha_i; \quad \text{i.e.,} \quad p = \alpha(q), \tag{8.5.19}$$

while the Euler-Lagrange equations are

$$\frac{d}{dt}(\alpha_i(q^j)) = \frac{\partial L}{\partial q^i} = \frac{\partial \alpha_j}{\partial q^i} \dot{q}^j - \frac{\partial h}{\partial q^i},$$

---

[2]We thank P. Morrison and M. Gotay for the following comment on how to view the KdV equation using constraints; see Gotay [1988].

that is,

$$\frac{\partial \alpha_i}{\partial q^j}\dot{q}^j - \frac{\partial \alpha_j}{\partial q^i}\dot{q}^j = -\frac{\partial h}{\partial q^i}. \tag{8.5.20}$$

In other words, with $v^i = \dot{q}^i$,

$$\mathbf{i}_v \mathbf{d}\alpha = -\mathbf{d}h. \tag{8.5.21}$$

If $\mathbf{d}\alpha$ is nondegenerate on Q then (8.5.21) defines Hamilton's equations for a vector field v on Q with Hamiltonian h and symplectic form $\Omega_\alpha = -\mathbf{d}\alpha$.

This collapse, or reduction, from TQ to Q is another instance of the Dirac theory and how it deals with degenerate Lagrangians in attempting to form the corresponding Hamiltonian system. Here the primary constraint manifold is the graph of α. Note that if we form the Hamiltonian on the primaries,

$$H = p_i\dot{q}^i - L = \alpha_i\dot{q}^i - \alpha_i\dot{q}^i + h(q) = h(q), \tag{8.5.22}$$

that is, $H = h$, as expected from (8.5.21).

To put the KdV equation $u_t + 6uu_x + u_{xxx} = 0$ in this context, let $u = \psi_x$; that is, ψ is an indefinite integral for u. Observe that the KdV equation is the Euler-Lagrange equation for

$$L(\psi, \psi_t) = \int \left[\frac{1}{2}\psi_t\psi_x + \psi_x^3 - \frac{1}{2}(\psi_{xx})^2\right] dx, \tag{8.5.23}$$

that is, $\delta \int L\,dt = 0$ gives $\psi_{xt} + 6\psi_x\psi_{xx} + \psi_{xxxx} = 0$ which is the KdV equation for u. Here α is given by

$$\langle \alpha(\psi), \varphi\rangle = \frac{1}{2}\int \psi_x\varphi\,dx \tag{8.5.24}$$

and so by formula 6 in the table in §4.4,

$$-\mathbf{d}\alpha(\psi)(\psi_1, \psi_2) = \frac{1}{2}\int (\psi_1\psi_{2x} - \psi_2\psi_{1x})\,dx \tag{8.5.25}$$

which equals the KdV symplectic structure (3.2.9). Moreover, (8.5.22) gives the Hamiltonian

$$H = \int \left[\frac{1}{2}(\psi_{xx})^2 - \psi_x^3\right] dx = \int \left[\frac{1}{2}(u_x)^2 - u^3\right] dx \tag{8.5.26}$$

also coinciding with Example **(c)** of §3.2. ◆

Exercises

8.5-1 *Derive formula* (8.4.2) *from* (8.5.1).

8.5-2 *Work out Dirac's formula for* $T^*S^2 \subset T^*\mathbb{R}^3$ *(note that the embedding makes use of the metric). Reconcile your analysis with what you found in Exercise* **8.4-2**.

8.6 Centrifugal and Coriolis Forces

Let V be a three-dimensional oriented inner product space that we regard as "inertial space." Let ψ_t be a curve in $SO(V)$, the group of orientation-preserving orthogonal linear transformations of V to V, and let X_t be the (possibly time-dependent) vector field generating ψ_t; that is,

$$X_t(\psi_t(\mathbf{v})) = \frac{d}{dt}\psi_t(\mathbf{v}), \tag{8.6.1}$$

or, equivalently,

$$X_t(\mathbf{v}) = (\dot{\psi}_t \circ \psi_t^{-1})(\mathbf{v}). \tag{8.6.2}$$

Since $X_t(\mathbf{v})$ is in $so(3)$, we can write

$$X_t(\mathbf{v}) = \omega(t) \times \mathbf{v} \tag{8.6.3}$$

that is, $\omega(t)$ is the instantaneous axis of rotation.

Let $\{\mathbf{e}_1, \mathbf{e}_2, \mathbf{e}_3\}$ be a fixed (inertial) orthonormal frame in V and let $\{\boldsymbol{\xi}_i = \psi_t(\mathbf{e}_i) \mid i = 1,2,3\}$ be the corresponding ***rotating frame***. Given a point $\mathbf{v} \in V$, let $\mathbf{q} = (q^1, q^2, q^3)$ denote the vector in $\mathbb{R}^3$ defined by $\mathbf{v} = q^i\mathbf{e}_i$ and let $\mathbf{q}_R \in \mathbb{R}^3$ be the corresponding coordinate vector representing the components of the same vector $\mathbf{v}$ in the rotating frame, so $\mathbf{v} = q_R^i\boldsymbol{\xi}_i$. Let $A_t = A(t)$ be the *matrix* of ψ_t relative to the basis $\mathbf{e}_i$, that is, $\boldsymbol{\xi}_i = A_i^j\mathbf{e}_j$; then

$$\mathbf{q} = A_t\mathbf{q}_R; \quad \text{i.e.,} \quad q^j = A_i^j q_R^i, \tag{8.6.4}$$

and (8.6.2) in matrix notation becomes

$$\hat{\omega} = \dot{A}_t A_t^{-1}, \tag{8.6.5}$$

where $\hat{\omega}$ is the linear transformation satisfying $\hat{\omega}(\mathbf{v}) = \omega \times \mathbf{v}$. Assume that the point $\mathbf{v}(t)$ moves in V according to Newton's second law with a potential energy $U(\mathbf{v})$. Using $U(\mathbf{q})$ for the corresponding function induced on $\mathbb{R}^3$, Newton's law reads

$$m\ddot{\mathbf{q}} = -\nabla U(\mathbf{q}), \tag{8.6.6}$$

which are the Euler-Lagrange equations for

$$L(\mathbf{q}, \dot{\mathbf{q}}) = \frac{m}{2} \langle \dot{\mathbf{q}}, \dot{\mathbf{q}} \rangle - U(\mathbf{q}) \tag{8.6.7}$$

or Hamilton's equations for

$$H(\mathbf{q}, \dot{\mathbf{q}}) = \frac{m}{2} \langle \dot{\mathbf{q}}, \dot{\mathbf{q}} \rangle + U(\mathbf{q}). \tag{8.6.8}$$

To find the equation satisfied by $\mathbf{q}_R$, differentiate (8.6.4) with respect to time

$$\dot{\mathbf{q}} = \dot{A}_t \mathbf{q}_R + A_t \dot{\mathbf{q}}_R = \dot{A}_t A_t^{-1} \mathbf{q} + A_t \dot{\mathbf{q}}_R, \tag{8.6.9}$$

that is,

$$\dot{\mathbf{q}} = \omega(t) \times \mathbf{q} + A_t \dot{\mathbf{q}}_R, \tag{8.6.10}$$

where, by abuse of notation, ω is also used for the representation of ω in the inertial frame $\mathbf{e}_i$. Differentiating (8.6.10),

$$\begin{aligned} \ddot{\mathbf{q}} &= \dot{\omega} \times \mathbf{q} + \omega \times \dot{\mathbf{q}} + \dot{A}_t \dot{\mathbf{q}}_R + A_t \ddot{\mathbf{q}}_R \\ &= \dot{\omega} \times \mathbf{q} + \omega \times (\omega \times \mathbf{q} + A_t \dot{\mathbf{q}}_R) + \dot{A}_t A_t^{-1} A_t \dot{\mathbf{q}}_R + A_t \ddot{\mathbf{q}}_R, \end{aligned}$$

that is,

$$\ddot{\mathbf{q}} = \dot{\omega} \times \mathbf{q} + \omega \times (\omega \times \mathbf{q}) + 2(\omega \times A_t \dot{\mathbf{q}}_R) + A_t \ddot{\mathbf{q}}_R. \tag{8.6.11}$$

The ***angular velocity*** in the rotating frame is (see (8.6.4)):

$$\omega_R = A_t^{-1} \omega, \quad \text{i.e.,} \quad \omega = A_t \omega_R. \tag{8.6.12}$$

Differentiating (8.6.12) with respect to time gives

$$\dot{\omega} = \dot{A}_t \omega_R + A_t \dot{\omega}_R = \dot{A}_t A_t^{-1} \omega + A_t \dot{\omega}_R = A_t \dot{\omega}_R, \tag{8.6.13}$$

since $\dot{A}_t A_t^{-1} \omega = \omega \times \omega = 0$. Multiplying (8.6.11) by A_t^{-1} gives

$$A_t^{-1} \ddot{\mathbf{q}} = \dot{\omega}_R \times \mathbf{q}_R + \omega_R \times (\omega_R \times \mathbf{q}_R) + 2(\omega_R \times \dot{\mathbf{q}}_R) + \ddot{\mathbf{q}}_R. \tag{8.6.14}$$

Since $m\ddot{\mathbf{q}} = -\nabla U(\mathbf{q})$, we have

$$m A_t^{-1} \ddot{\mathbf{q}} = -\nabla U_R(\mathbf{q}_R), \tag{8.6.15}$$

where the ***rotated potential*** U_R is the *time-dependent* potential defined by

$$U_R(\mathbf{q}_R, t) = U(A_t \mathbf{q}_R) = U(\mathbf{q}), \tag{8.6.16}$$

so that $\nabla U(\mathbf{q}) = A_t \nabla U_R(\mathbf{q}_R)$. Therefore, (8.6.15) becomes

$$m\ddot{\mathbf{q}}_R + 2(\omega_R \times m\dot{\mathbf{q}}_R) + m\omega_R \times (\omega_R \times \mathbf{q}_R) + m\dot{\omega}_R \times \mathbf{q}_R = -\nabla U_R(\mathbf{q}_R, t),$$

that is,

$$m\ddot{\mathbf{q}}_R = -\nabla U_R(\mathbf{q}_R, t) - m\omega_R \times (\omega_R \times \mathbf{q}_R) - 2m(\omega_R \times \dot{\mathbf{q}}_R) - m\dot{\omega}_R \times \mathbf{q}_R, \tag{8.6.17}$$

which expresses the equations of motion entirely in terms of rotated quantities.

There are three types of "fictitious forces" that appear in (8.6.17) if we try to identify (8.6.17) with $m\mathbf{a} = \mathbf{F}$:

i	***centrifugal force***	$m\omega_R \times (\mathbf{q}_R \times \omega_R)$;
ii	***Coriolis force***	$2m\dot{\mathbf{q}}_R \times \omega_R$; and
iii	***Euler force***	$m\mathbf{q}_R \times \dot{\omega}_R$.

Note that the Coriolis force $2m\omega_R \times \dot{\mathbf{q}}_R$ is orthogonal to ω_R and $m\dot{\mathbf{q}}_R$ while the centrifugal force

$$m\omega_R \times (\omega_R \times \mathbf{q}_R) = m[(\omega_R \cdot \mathbf{q}_R)\omega_R - \|\omega_R\|^2 \mathbf{q}_R]$$

is in the plane of ω_R and $\mathbf{q}_R$. Also note that the Euler force is due to the nonuniformity of the rotation rate.

It is of interest to ask the sense in which (8.6.17) is Lagrangian and Hamiltonian. To answer this, it is useful to begin with the Lagrangian approach, which, we will see, is more covariant. Substitute (8.6.10) into (8.6.7) to express the Lagrangian in terms of rotated quantities:

$$\begin{aligned} L &= \frac{m}{2}\langle \omega \times \mathbf{q} + A_t\dot{\mathbf{q}}_R, \omega \times \mathbf{q} + A_t\dot{\mathbf{q}}_R \rangle - U(\mathbf{q}) \\ &= \frac{m}{2}\langle \omega_R \times \mathbf{q}_R + \dot{\mathbf{q}}_R, \omega_R \times \mathbf{q}_R + \dot{\mathbf{q}}_R \rangle - U_R(\mathbf{q}_R, t), \end{aligned} \tag{8.6.18}$$

which defines a new (time-dependent!) Lagrangian $L_R(\mathbf{q}_R, \dot{\mathbf{q}}_R, t)$. Remarkably, (8.6.17) are precisely the Euler-Lagrange equations for L_R; that is, (8.6.17) are equivalent to

$$\frac{d}{dt}\frac{\partial L_R}{\partial \dot{\mathbf{q}}_R^i} = \frac{\partial L_R}{\partial \mathbf{q}_R^i},$$

as is readily verified. If one thinks about performing a time-dependent transformation in the variational principle, then in fact, one sees that this is reasonable.

To find the sense in which (8.6.17) is Hamiltonian, perform a Legendre transformation on L_R. The conjugate momentum is

$$\mathbf{p}_R = \frac{\partial L_R}{\partial \dot{\mathbf{q}}_R} = m(\omega_R \times \mathbf{q}_R + \dot{\mathbf{q}}_R) \tag{8.6.19}$$

and so the Hamiltonian has the expression

$$H_R(\mathbf{q}_R, \mathbf{p}_R) = \langle \mathbf{p}_R, \dot{\mathbf{q}}_R \rangle - L_R$$

$$
\begin{aligned}
&= \frac{1}{m}\left\langle \mathbf{p}_R, \mathbf{p}_R - m\omega_R \times \mathbf{q}_R \right\rangle - \frac{1}{2m}\left\langle \mathbf{p}_R, \mathbf{p}_R \right\rangle + U_R(\mathbf{q}_R, t) \\
&= \frac{1}{2m}\left\langle \mathbf{p}_R, \mathbf{p}_R \right\rangle + U_R(\mathbf{q}_R, t) - \left\langle \mathbf{p}_R, \omega_R \times \mathbf{q}_R \right\rangle . \qquad (8.6.20)
\end{aligned}
$$

Thus, (8.6.17) are equivalent to Hamilton's canonical equations with Hamiltonian (8.6.20) and with the canonical symplectic form. In general, H_R is time-dependent. Alternatively, if we perform the momentum shift

$$
\mathfrak{p}_R = \mathbf{p}_R - m\omega_R \times \mathbf{q}_R = m\dot{\mathbf{q}}_R, \qquad (8.6.21)
$$

then we get

$$
\begin{aligned}
\tilde{H}_R(\mathbf{q}_R, \mathfrak{p}_R) : &= H_R(\mathbf{q}_R, \mathbf{p}_R) \\
&= \frac{1}{2m}\left\langle \mathfrak{p}_R, \mathfrak{p}_R \right\rangle + U_R(\mathbf{q}_R) - \frac{m}{2}\|\omega_R \times \mathbf{q}_R\|^2, \qquad (8.6.22)
\end{aligned}
$$

which is in the usual form of kinetic plus potential energy, but now the potential is *amended* by the centrifugal potential $m\|\omega_R \times \mathbf{q}_R\|^2/2$ and the canonical symplectic structure

$$
\Omega_{\text{can}} = d\mathbf{q}_R^i \wedge d(\mathbf{p}_R)_i
$$

gets transformed, by the momentum shifting lemma, or directly, to

$$
d\mathbf{q}_R^i \wedge d\mathbf{p}_{Ri} = d\mathbf{q}_R^i \wedge d\mathfrak{p}_{Ri} + \epsilon_{ijk}\omega_R^i d\mathbf{q}_R^i \wedge d\mathbf{q}_R^j,
$$

where ϵ_{ijk} is the alternating tensor. Note that

$$
\tilde{\Omega}_R = \tilde{\Omega}_{\text{can}} + *\omega_R, \qquad (8.6.23)
$$

where $*\omega_R$ means the two-form associated to the vector ω_R and that (8.6.23) has the same form as the corresponding expression for a particle in a magnetic field (§6.7).

In general, the momentum shift (8.6.21) is time-dependent, so care is needed in interpreting the sense in which the equations for $\mathfrak{p}_R$ and $\mathbf{q}_R$ are Hamiltonian. In fact, the equations should be computed as follows. Let X_H be a Hamiltonian vector field on P and let $\zeta_t : P \to P$ be a *time-dependent* map with generator Y_t:

$$
\frac{d}{dt}\zeta_t(z) = Y_t(\zeta_t(z)). \qquad (8.6.24)
$$

Assume that ζ_t is symplectic for each t. If $\dot{z}(t) = X_H(z(t))$ and we let $w(t) = \zeta_t(z(t))$, then w satisfies

$$
\dot{w} = T\zeta_t \cdot X_H(z(t)) + Y_t(\zeta_t(z(t)), \qquad (8.6.25)
$$

that is,

$$
\dot{w} = X_K(w) + Y_t(w) \qquad (8.6.26)
$$

where $K = H \circ \zeta_t^{-1}$. The extra term Y_t in (8.6.26) is, in the example under consideration, the Euler force.

So far we have been considering a fixed system as seen from different rotating observers. Analogously, one can consider systems that themselves are subjected to a superimposed rotation, an example being the Foucault pendulum. It is clear that the physical behavior in the two cases can be different—in fact, the Foucault pendulum and the example in the next section show that one can get a real physical effect from rotating a system—obviously rotating observers can cause nontrivial changes in the *description* of a system, but cannot make any *physical* difference. Nevertheless, the strategy for the analysis of rotating systems is analogous to the above. The easiest approach, as we have seen, is to transform the Lagrangian. The reader may wish to reread §2.10 for an easy and specific instance of this.

8.7 The Geometric Phase for a Particle in a Hoop

This discussion follows Berry [1985] with some modifications (due to Anandan and Marsden, Montgomery, and Ratiu [1990]) necessary for a geometric interpretation of the results. Consider Figure 8.7.1, which shows a planar hoop (not necessarily circular) in which a bead slides without friction. As the bead is sliding, the hoop is rotated in its plane through an angle $\theta(t)$ with angular velocity $\omega(t) = \dot{\theta}(t)\mathbf{k}$. Let s denote the arc length along the hoop, measured from a reference point on the hoop and let $\mathbf{q}(s)$ be the vector from the origin to the corresponding point on the hoop; thus the shape of the hoop is determined by this function $\mathbf{q}(s)$. The unit tangent vector is $\mathbf{q}'(s)$ and the position of the reference point $\mathbf{q}(s(t))$ relative to an inertial frame in space is $R_{\theta(t)}\mathbf{q}(s(t))$, where R_θ is the rotation in the plane of the hoop through an angle θ. Note that $\dot{R}_\theta R_\theta^{-1}\mathbf{q} = \omega \times \mathbf{q}$ and $R_\theta \omega = \omega$.

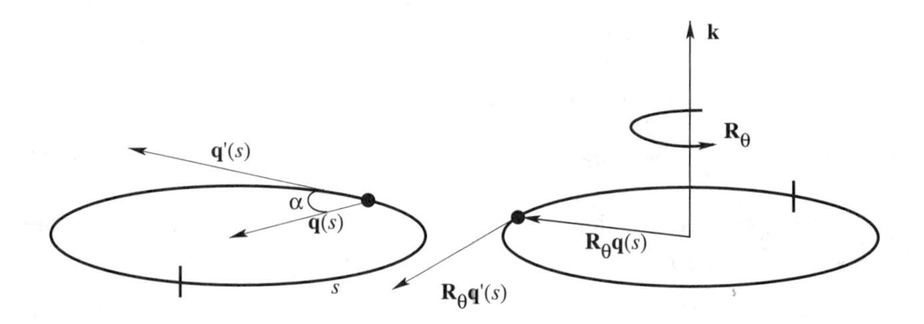

FIGURE 8.7.1. A particle sliding in a rotating hoop.

The configuration space is a fixed closed curve (the hoop) in the plane with length L. The Lagrangian $L(s, \dot{s}, t)$ is simply the kinetic energy of the particle. Since we have

$$\frac{d}{dt} R_{\theta(t)} \mathbf{q}(s(t)) = R_{\theta(t)} \mathbf{q}'(s(t)) \dot{s}(t) + R_{\theta(t)} [\omega(t) \times \mathbf{q}(s(t))],$$

the Lagrangian is

$$L(s, \dot{s}, t) = \frac{1}{2} m \|\mathbf{q}'(s) \dot{s} + \omega \times \mathbf{q}\|^2. \tag{8.7.1}$$

Note that the momentum conjugate to s is $p = \partial L / \partial \dot{s}$; that is,

$$p = m\mathbf{q}' \cdot [\mathbf{q}' \dot{s} + \omega \times \mathbf{q}] = mv, \tag{8.7.2}$$

where v is the component of the velocity *with respect to the inertial frame* tangent to the curve. The Euler-Lagrange equations

$$\frac{d}{dt} \frac{\partial L}{\partial \dot{s}} = \frac{\partial L}{\partial s}$$

become

$$\frac{d}{dt} [\mathbf{q}' \cdot (\mathbf{q}' \dot{s} + \omega \times \mathbf{q})] = (\mathbf{q}' \dot{s} + \omega \times \mathbf{q}) \cdot (\mathbf{q}'' \dot{s} + \omega \times \mathbf{q}').$$

Using $\|\mathbf{q}'\|^2 = 1$, its consquence $\mathbf{q}' \cdot \mathbf{q}'' = 0$, and simplifying, we get

$$\ddot{s} + \mathbf{q}' \cdot (\dot{\omega} \times \mathbf{q}) - (\omega \times \mathbf{q}) \cdot (\omega \times \mathbf{q}') = 0. \tag{8.7.3}$$

The second and third terms in (8.7.3) are the Euler and centrifugal forces, respectively. Since $\omega = \dot{\theta} \mathbf{k}$, we can rewrite (8.7.3) as

$$\ddot{s} = \dot{\theta}^2 \mathbf{q} \cdot \mathbf{q}' - \ddot{\theta} q \sin \alpha, \tag{8.7.4}$$

where α is as in Figure 8.7.1 and $q = \|\mathbf{q}\|$. From (8.7.4) and Taylor's formula with remainder, we get

$$\begin{aligned} s(t) &= s_0 + \dot{s}_0 t + \int_0^t (t - \tau) \{ \dot{\theta}(\tau)^2 \mathbf{q}(s(\tau)) \cdot \mathbf{q}'(s(\tau)) \\ &\qquad - \ddot{\theta}(\tau) q(s(\tau)) \sin \alpha(s(\tau)) \} \, d\tau. \end{aligned} \tag{8.7.5}$$

The angular velocity $\dot{\theta}$ and acceleration $\ddot{\theta}$ are assumed small with respect to the particle's velocity, so by the averaging theorem (see, for example, Hale [1963]), the s-dependent quantities in (8.7.5) can be replaced by their averages round the hoop

$$\begin{aligned} s(t) &\approx s_0 + \dot{s}_0 t \\ &+ \int_0^t (t - \tau) \left\{ \dot{\theta}(\tau)^2 \frac{1}{L} \int_0^L \mathbf{q} \cdot \mathbf{q}' \, ds - \ddot{\theta}(\tau) \frac{1}{L} \int_0^L q(s) \sin \alpha(s) \, ds \right\} d\tau. \end{aligned} \tag{8.7.6}$$

Aside The essence of averaging in this case can be seen as follows. Suppose $g(t)$ is a rapidly varying function whose oscillations are bounded in magnitude by a constant C and $f(t)$ is slowly varying on an interval $[a, b]$. Over one period of g, say $[\alpha, \beta]$, we have

$$\int_\alpha^\beta f(t)g(t)\,dt \approx \overline{g}\int_\alpha^\beta f(t)\,dt, \tag{8.7.7}$$

where

$$\overline{g} = \frac{1}{\beta-\alpha}\int_\alpha^\beta g(t)\,dt$$

is the average of g. The error in (8.7.7) is $\int_\alpha^\beta f(t)(g(t)-\overline{g})\,dt$, whose absolute value is bounded as follows. Let M be the maximum value of f on $[\alpha, \beta]$ and m be the minimum. Then

$$\begin{aligned}\left|\int_\alpha^\beta f(t)[g(t)-\overline{g}]dt\right| &= \left|\int_\alpha^\beta (f(t)-m)[g(t)-\overline{g}]dt\right| \\ &\leq (\beta-\alpha)(M-m)C \\ &\leq (\beta-\alpha)^2 DC,\end{aligned}$$

where D is the maximum of $|f'(t)|$ for $\alpha \leq t \leq \beta$. Now these errors over each period are added up over $[a, b]$. Since the error estimate has the *square* of $\beta - \alpha$ as a factor, one still gets something small as the period of g tends to 0. In (8.7.5) we change variables from t to s, do the averaging, and then change back. ♦

The first inner integral in (8.7.6) over s vanishes (since the integrand is $\frac{d}{ds}\|\mathbf{q}(s)\|^2$) and the second is $2A$ where A is the area enclosed by the hoop. Integrating by parts,

$$\int_0^T (T-\tau)\ddot{\theta}(\tau)\,d\tau = -T\dot{\theta}(0) + \int_0^T \dot{\theta}(\tau)\,d\tau = -T\dot{\theta}(0) + 2\pi \tag{8.7.8}$$

assuming the hoop makes one complete revolution in time T. Substituting (8.7.8) in (8.7.6) gives

$$s(T) \approx s_0 + \dot{s}_0 T + \frac{2A}{L}\dot{\theta}_0 T - \frac{4\pi A}{L}, \tag{8.7.9}$$

where $\dot{\theta}_0 = \dot{\theta}(0)$. The initial velocity of the bead *relative to the hoop* is $\dot{s}_0$, while its component along the curve *relative to the inertial frame* is (see (8.7.2)),

$$v_0 = \mathbf{q}'(0)\cdot[\mathbf{q}'(0)\dot{s}_0 + \omega_0 \times \mathbf{q}(0)] = \dot{s}_0 + \omega_0 q(s_0)\sin\alpha(s_0). \tag{8.7.10}$$

Now we replace $\dot{s}_0$ in (8.7.9) by its expression in terms of v_0 from (8.7.10) and average over all initial conditions to get

$$\langle s(T) - s_0 - v_0 T \rangle = -\frac{4\pi A}{L}, \tag{8.7.11}$$

which means that *on average*, the shift in position is by $4\pi A/L$ between the rotated and nonrotated hoop. Note that if $\dot{\theta}_0 = 0$ (the situation assumed by Berry [1985]), then averaging over initial conditions is not necessary.

This extra length $4\pi A/L$ is sometimes called the geometric phase or the ***Berry-Hannay phase***. This example is related to a number of interesting effects, both classically and quantum mechanically, such as the Foucault pendulum and the Aharanov-Bohm effect. The effect is known as *holonomy* and can be viewed as an instance of ***reconstruction*** in the context of symmetry and reduction to be developed in Volume II. For further information and additional references, see Aharanov and Anandan[1987], Montgomery [1988], [1990], and Marsden, Montgomery, and Ratiu [1989, 1990]. For related ideas in soliton dynamics, see Alber and Marsden [1992].

Exercises

8.7-1 *Consider the dynamics of a ball in a slowly rotating planar hoop, as in the text. However, this time, consider rotating the hoop about an axis that is not perpendicular to the plane of the hoop, but makes an angle θ with the normal. Compute the geometric phase for this problem.*

8.7-2 *Study the geometric phase for a particle in a general spatial hoop that is moved through a closed curve in $SO(3)$.*

8.7-3 *Consider the dynamics of a ball in a slowly rotating planar hoop, as in the text. However, this time, consider a charged particle with charge e and a fixed magnetic field $\mathbf{B} = \nabla \times \mathbf{A}$ in the vicinity of the hoop. Compute the geometric phase for this problem.*

8.8 The General Theory of Moving Systems

The particle in the rotating hoop is an example of a rotated or, more generally, a *moving system*. Other examples are a pendulum on a merry-go-round and a fluid on a rotating sphere (like the Earth's ocean and atmosphere). As we have emphasized, systems of this type are not to be confused with rotating observers! This section gives a general context for such systems. Our general purpose is to show how to derive Lagrangians systematically and the resulting equations of motion for moving systems, like the bead in the hoop of the last section. This will also set up the reader who wants to pursue the question of how moving systems fit in the context of phases (Marsden, Montgomery, and Ratiu [1990]).

Consider a Riemannian manifold $\mathcal{S}$, a submanifold Q, and a space M of embeddings of Q into $\mathcal{S}$. Let $m_t \in M$ be a given curve. If a particle in Q is following a curve $q(t)$, and if Q moves by superposing the motion m_t, then the path of the particle in $\mathcal{S}$ is given by $m_t(q(t))$. Thus its velocity in $\mathcal{S}$ is given by

$$T_{q(t)}m_t \cdot \dot{q}(t) + \mathcal{Z}_t(m_t(q(t))), \tag{8.8.1}$$

where $\mathcal{Z}_t(m_t(q)) = \frac{d}{dt}m_t(q)$. Consider a Lagrangian on TQ of the form

$$L_{m_t}(q,v) = \frac{1}{2}\|T_{q(t)}m_t \cdot v + \mathcal{Z}_t(m_t(q))^T\|^2 - V(q) - U(m_t(q)), \tag{8.8.2}$$

where V is a given potential on Q, U is a given potential on $\mathcal{S}$, and ^{T} denotes the projection onto the tangent space to the (moving) image of Q. Taking the Legendre transform of (8.8.2), we get a Hamiltonian with momentum

$$p = (T_{q(t)}m_t \cdot v + \mathcal{Z}_t(m_t(q))^T)^\flat, \tag{8.8.3}$$

where $^\flat$ denotes the index lowering operation determined by the metric on $\mathcal{S}$. Physically, if $\mathcal{S}$ is $\mathbb{R}^3$, then p is the inertial momentum (see the hoop example in the preceding section). This extra term $\mathcal{Z}_t(m_t(q))^T$ is associated with a connection called the ***Cartan connection*** on the bundle $Q \times M \to M$, with horizontal lift defined to be $\mathcal{Z}(m) \mapsto (Tm^{-1} \cdot \mathcal{Z}(m)^T, \mathcal{Z}(m))$. (See for example, Marsden and Hughes [1983] for an account of some aspects of Cartan's contributions.) The Hamiltonian picks up a cross term and takes the form

$$H_{\mathcal{Z}_t}(q,p) = \frac{1}{2}\|p\|^2 - \mathcal{P}(\mathcal{Z}_t) + \frac{1}{2}\|\mathcal{Z}_t^T\|^2 + V(q) + U(m_t(q)), \tag{8.8.4}$$

where the cross term is $\mathcal{P}(\mathcal{Z}_t)(q,p) = \langle p, \mathcal{Z}_t^T(q)\rangle$. The Hamiltonian vector field of this cross term $X_{\mathcal{P}(\mathcal{Z}_t)}$ represents the noninertial forces and also has the natural interpretation as a horizontal lift of the vector field $\mathcal{Z}_t$ relative to a certain connection on the bundle $T^*Q \times M \to M$, naturally derived from the Cartan connection.

Let G be a Lie group which acts on T^*Q in a Hamiltonian fashion and leaves H_0 (defined by setting $\mathcal{Z} = 0$ in (8.8.4)) invariant. In our examples G is either $\mathbb{R}$ acting on T^*Q by the flow of H_0 (the hoop), or a subgroup of the isometry group of Q which leaves V and U invariant, and acts on T^*Q by cotangent lift (this is appropriate for the Foucault pendulum). In any case, we assume G has an invariant measure relative to which we can average.

Assuming the "averaging principle" (see Arnold [1989], for example) we replace $H_{\mathcal{Z}_t}$ by its G-average,

$$\langle H_{\mathcal{Z}_t}\rangle(q,p) = \frac{1}{2}\|p\|^2 - \langle\mathcal{P}(\mathcal{Z}_t)\rangle + \frac{1}{2}\langle\|\mathcal{Z}_t^T\|^2\rangle + V(q) + \langle U(m_t(q))\rangle\,. \tag{8.8.5}$$

In (8.8.5) we shall assume the term $\frac{1}{2}\left\langle \|\mathcal{Z}_t^T\|^2 \right\rangle$ is small and discard it. Thus, define

$$\begin{aligned}\mathcal{H}(q,p,t) &= \frac{1}{2}\|p\|^2 - \langle \mathcal{P}(\mathcal{Z}_t)\rangle + V(q) + \langle U(m_t(q))\rangle \\ &= \mathcal{H}_0(q,p) - \langle \mathcal{P}(\mathcal{Z}_t)\rangle + \langle U(m_t(q))\rangle, \end{aligned} \tag{8.8.6}$$

where $\mathcal{H}_0 = \frac{1}{2}\|p\|^2 + V(q)$. Consider the dynamics on $T^*Q \times M$ given by the vector field

$$(X_{\mathcal{H}}, \mathcal{Z}_t) = (X_{\mathcal{H}_0} - X_{\langle \mathcal{P}(\mathcal{Z}_t)\rangle} + X_{\langle U \circ m_t\rangle}, \mathcal{Z}_t). \tag{8.8.7}$$

The vector field, consisting of the extra terms in this representation due to the superposed motion of the system, namely

$$\operatorname{hor}(\mathcal{Z}_t) = (-X_{\langle \mathcal{P}(\mathcal{Z}_t)\rangle}, \mathcal{Z}_t), \tag{8.8.8}$$

has a natural interpretation as the horizontal lift of $\mathcal{Z}_t$ relative to a connection on $T^*Q \times M$, which is obtained by averaging the Cartan connection and is called the ***Cartan-Hannay-Berry connection***. The holonomy of this connection is the ***Hannay-Berry phase*** of a slowly moving constrained system. For details of this approach, see Marsden, Montgomery, and Ratiu [1990].

Exercise

8.8-1 *Set up the equations for the Foucault pendulum using the ideas in this section.*

9

An Introduction to Lie Groups

To prepare for the next chapters, we present some basic facts about Lie groups. Alternative expositions and additional details can be obtained from Abraham and Marsden [1978], Olver [1986], and Sattinger and Weaver [1986]. In particular, in this book we shall require only elementary facts about the general theory and a knowledge of a few of the more basic groups, such as the rotation and Euclidean groups.

Recall how some of the basic groups arise in mechanics:

1. ***Linear and Angular Momentum***
   These arise as conserved quantities associated with the groups of translations and rotations in space.

2. ***Rigid Body***
   Consider a free rigid body rotating about a fixed point. "Free" means that there are no external forces, and "rigid" means that the distance between any two points of the body is unchanged during the motion. Consider a point X of the body at time $t = 0$, and denote its position at time t by $f(X, t)$. Rigidity of the body and the assumption of a smooth motion imply that $f(X, t) = \mathbf{A}(t) \cdot X$ where $\mathbf{A}(t)$ is a proper rotation, that is, $\mathbf{A}(t) \in SO(3)$, the proper rotation group of $\mathbb{R}^3$, the 3×3 orthogonal matrices with determinant 1. The set $SO(3)$ will be shown to be a three-dimensional Lie group and, since it describes any possible position of the body, it serves as the *configuration space*. The group $SO(3)$ also plays a dual role of a *symmetry group* since the same physical motion is described if we rotate our coordinate axes.

Used as a symmetry group, $SO(3)$ leads to conservation of angular momentum.

3. ***Heavy Top***
   Consider a rigid body moving with a fixed point but under the influence of gravity. This problem still has a configuration space $SO(3)$, but the symmetry group is only the circle group S^1, consisting of rotations about the direction of gravity. One says that gravity has ***broken*** the symmetry from $SO(3)$ to S^1. This time, "eliminating" the S^1 symmetry leads one the Euclidean group $SE(3)$ of rigid motion of $\mathbb{R}^3$. This is a manifestation of the general theory of semidirect products (see the Introduction and Marsden, Ratiu, and Weinstein [1984a,b]).

4. ***Incompressible Fluid***
   Let Ω be a region in $\mathbb{R}^3$ that is filled with a moving incompressible fluid, and is free of external forces. Denote by $\eta(X, t)$ the trajectory of a fluid particle which at time $t = 0$ is at $X \in \Omega$. For fixed t the map η_t defined by $\eta_t(X) = \eta(X, t)$ is a diffeomorphism of Ω. In fact, since the fluid is incompressible we have $\eta_t \in \mathrm{Diff}_{\mathrm{vol}}(\Omega)$, the group of volume-preserving diffeomorphisms of Ω. Thus, the configuration space for the problem is the infinite-dimensional Lie group $\mathrm{Diff}_{\mathrm{vol}}(\Omega)$. Using $\mathrm{Diff}_{\mathrm{vol}}(\Omega)$ as a symmetry group leads to Kelvin's circulation theorem as a conservation law. See Marsden and Weinstein [1983].

5. ***Compressible Fluids***
   In this case the configuration space is the whole diffeomorphism group $\mathrm{Diff}(\Omega)$. The symmetry group consists of density-preserving diffeomorphisms $\mathrm{Diff}_\rho(\Omega)$. The density plays a role similar to that of gravity in the heavy top and again leads to semidirect products, as does the next example.

6. ***Magnetohydrodynamics (MHD)***
   This example is that of a compressible fluid consisting of charged particles with the dominant electromagnetic force being the magnetic field produced by the particles themselves (possibly together with an external field). The configuration space remains $\mathrm{Diff}(\Omega)$ but the fluid motion is coupled with the magnetic field (regarded as a two-form on Ω).

7. ***Maxwell-Vlasov Equation***
   Let $f(\mathbf{x}, \mathbf{v}, t)$ denote the density function of a collisionless plasma. The function f evolves in time by means of a time-dependent canonical transformation on $\mathbb{R}^6$, that is, $(\mathbf{x}, \mathbf{v})$-space. In other words, the evolution of f can be described by $f_t = \eta_t^* f_0$ where f_0 is the initial value of f, f_t its value at time t, and η_t is a canonical transformation.

Thus the underlying configuration space in this case is $\mathrm{Diff}_{\mathrm{can}}(\mathbb{R}^6)$, the group of canonical transformations.

8. ***Maxwell's Equations***
Maxwell's equations for electrodynamics are invariant under gauge transformations that transform the magnetic (or 4) potential by $\mathbf{A} \mapsto \mathbf{A} + \nabla\varphi$. This gauge group is an infinite-dimensional Lie group. The conserved quantity associated with the gauge symmetry in this case is the charge.

9.1 Basic Definitions and Properties

Definition 9.1.1 *A* ***Lie group*** *is a (Banach) manifold G that has a group structure consistent with its manifold structure in the sense that group multiplication*

$$\mu : G \times G \to G; \quad (g, h) \mapsto gh$$

is a C^∞ map.

The maps $L_g : G \to G$; $h \mapsto gh$, and $R_h : G \to G$; $g \mapsto gh$ are called the ***left and right translation maps***. Note that

$$L_{g_1} \circ L_{g_2} = L_{g_1 g_2} \quad \text{and} \quad R_{h_1} \circ R_{h_2} = R_{h_2 h_1}.$$

If $e \in G$ denotes the identity element, then $L_e = \mathrm{Id} = R_e$ and so $(L_g)^{-1} = L_{g^{-1}}$ and $(R_h)^{-1} = R_{h^{-1}}$. Thus L_g and R_h are diffeomorphisms for each g and h. Notice that $L_g \circ R_h = R_h \circ L_g$, that is, left and right translation commute. By the chain rule, $T_{gh} L_{g^{-1}} \circ T_h L_g = T_h(L_{g^{-1}} \circ L_g) = \mathrm{Id}$. Thus, $T_h L_g$ is invertible. Likewise, $T_g R_h$ is an isomorphism.

We now show that the ***inversion map*** $I : G \to G$; $g \mapsto g^{-1}$ is C^∞. Indeed, consider solving

$$\mu(g, h) = e$$

for h as a function of g. The partial derivative with respect to h is just $T_h L_g$, which is an isomorphism. Thus, the solution g^{-1} is a smooth function of g by the implicit function theorem.

Lie groups can be finite- or infinite-dimensional. For a first reading of this section, the reader may wish to assume G is finite dimensional. We caution that some interesting infinite-dimensional groups (such as groups of diffeomorphisms) are *not* Banach-Lie groups.

Examples

(a) Any Banach space V is an abelian Lie group with group operations $\mu : V \times V \to V$, $\mu(x, y) = x + y$, and $I : V \to V$, $I(x) = -x$. The identity is just the zero vector. We call such a Lie group a ***vector group***. ♦

(b) The group of linear isomorphisms of $\mathbb{R}^n$ to $\mathbb{R}^n$ is a Lie group of dimension n^2, called the ***general linear group*** and denoted $GL(n, \mathbb{R})$. It is a smooth manifold, since it is an open subset of the vector space $L(\mathbb{R}^n, \mathbb{R}^n)$ of all linear maps of $\mathbb{R}^n$ to $\mathbb{R}^n$. Indeed, $GL(n, \mathbb{R})$ is the inverse image of $\mathbb{R}\backslash\{0\}$ under the continuous map $A \mapsto \det A$ of $L(\mathbb{R}^n, \mathbb{R}^n)$ to $\mathbb{R}$. For $A, B \in GL(n, \mathbb{R})$, the group operation is composition

$$\mu : GL(n, \mathbb{R}) \times GL(n, \mathbb{R}) \to GL(n, \mathbb{R})$$

given by

$$(A, B) \mapsto A \circ B,$$

and the inversion map is

$$I : GL(n, \mathbb{R}) \to GL(n, \mathbb{R}),$$

defined by

$$I(A) = A^{-1}.$$

Group multiplication is the restriction of the continuous bilinear map $(A, B) \in L(\mathbb{R}^n, \mathbb{R}^n) \times L(\mathbb{R}^n, \mathbb{R}^n) \mapsto A \circ B \in L(\mathbb{R}^n, \mathbb{R}^n)$. Thus μ is C^∞ and so $GL(n, \mathbb{R})$ is a Lie group.

The group identity element e is the identity map on $\mathbb{R}^n$. If we choose a basis in $\mathbb{R}^n$, we can represent each $A \in GL(n, \mathbb{R})$ by an invertible $(n \times n)$-matrix. The group operation is then matrix multiplication $\mu(A, B) = AB$ and $I(A) = A^{-1}$ is matrix inversion. The identity element e is the $n \times n$ identity matrix. The group operations are obviously smooth since the formulas for the product and inverse of matrices are smooth in the matrix components. ♦

(c) In the same way, one sees that for a Banach space $V, GL(V, V)$, the group of invertible elements of $L(V, V)$ is a Banach Lie group. For the proof that this is open in $L(V, V)$, see Abraham, Marsden, and Ratiu [1988]. Further examples are given in the next section. ♦

Given any local chart on G, one can construct an entire atlas on the Lie group G by use of left (or right) translations. Suppose, for example, that (U, φ) is a chart about $e \in G$, and that $\varphi : U \to V$. Define a chart (U_g, φ_g) about $g \in G$ by letting

$$U_g = L_g(U) = \{L_g h \mid h \in U\}$$

and defining

$$\varphi_g = \varphi \circ L_{g^{-1}} : U_g \to V, \; h \mapsto \varphi(g^{-1}h).$$

The set of charts $\{(U_g, \varphi_g)\}$ forms an atlas provided one can show that the transition maps

$$\varphi_{g_1} \circ \varphi_{g_2}^{-1} = \varphi \circ L_{g_1^{-1} g_2} \circ \varphi^{-1} : \varphi_2(U_{g_1} \cap U_{g_2}) \to \varphi_1(U_{g_1} \cap U_{g_2})$$

are differentiable. But this follows from the smoothness of group multiplication and inversion.

A vector field X on G is called ***left invariant*** if for every $g \in G$, $L_g^* X = X$, that is, if

$$(T_h L_g) X(h) = X(gh)$$

for every $h \in G$. We have the commutative diagram in Figure 9.1.1 and illustrate the geometry in Figure 9.1.2.

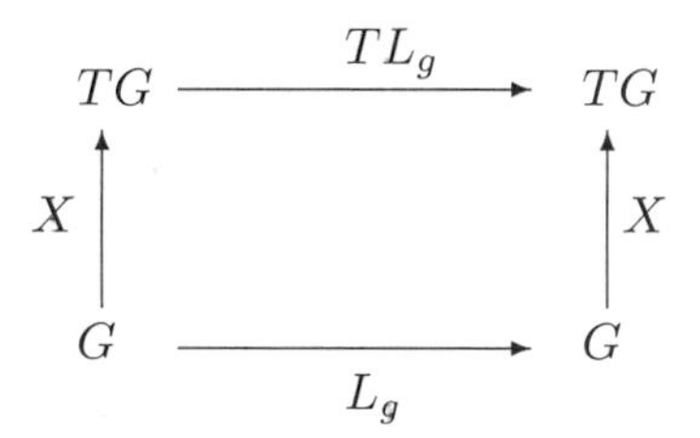

FIGURE 9.1.1. The commutative diagram for a left invariant vector field.

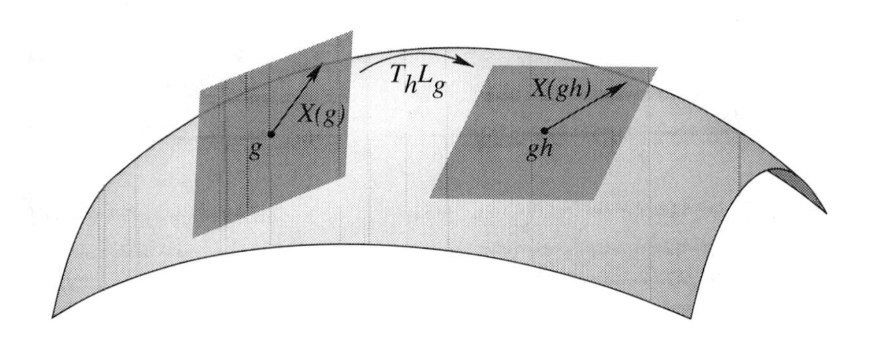

FIGURE 9.1.2. A left invariant vector field.

Let $\mathfrak{X}_L(G)$ denote the set of left invariant vector fields on G. If $X, Y \in \mathfrak{X}_L(G)$ and $g \in G$, then $L_g^*[X,Y] = [L_g^* X, L_g^* Y] = [X,Y]$, so $[X,Y] \in \mathfrak{X}_L(G)$. Therefore $\mathfrak{X}_L(G)$ is a Lie subalgebra of $\mathfrak{X}(G)$, the set of all vector fields on G.

For each $\xi \in T_e G$, we define a vector field X_ξ on G by letting

$$X_\xi(g) = T_e L_g(\xi).$$

Then

$$\begin{aligned} X_\xi(gh) &= T_e L_{gh}(\xi) = T_e(L_g \circ L_h)(\xi) \\ &= T_h L_g(T_e L_h(\xi)) = T_h L_g(X_\xi(h)), \end{aligned}$$

which shows that X_ξ is left invariant. The linear maps

$$\zeta_1 : \mathfrak{X}_L(G) \to T_eG,\ X \mapsto X(e)$$

and

$$\zeta_2 : T_eG \to \mathfrak{X}_L(G),\ \xi \mapsto X_\xi$$

satisfy $\zeta_1 \circ \zeta_2 = \mathrm{id}_{T_eG}$ and $\zeta_2 \circ \zeta_1 = \mathrm{id}_{\mathfrak{X}_L(G)}$. Therefore, *$\mathfrak{X}_L(G)$ and T_eG are isomorphic as vector spaces.*

Define the ***Lie bracket*** in T_eG by

$$[\xi, \eta] := [X_\xi, X_\eta](e),$$

where $\xi, \eta \in T_eG$ and where $[X_\xi, X_\eta]$ is the Jacobi-Lie bracket of vector fields. This makes T_eG into a Lie algebra. We say that this defines a bracket in T_eG via ***left-extension***. Note that by construction,

$$[X_\xi, X_\eta] = X_{[\xi,\eta]},$$

for all $\xi, \eta \in T_eG$.

Definition 9.1.2 *The vector space T_eG with this Lie algebra structure is called the **Lie algebra of** G and is denoted by $\mathfrak{g}$.*

Defining the set $\mathfrak{X}_R(G)$ of ***right invariant*** vector fields on G in the analogous way, we get a vector space isomorphism $\xi \mapsto Y_\xi$, where $Y_\xi(g) = (T_eR_g)(\xi)$, between $T_eG = \mathfrak{g}$ and $\mathfrak{X}_R(G)$. In this way, each $\xi \in \mathfrak{g}$ defines an element $Y_\xi \in \mathfrak{X}_R(G)$, and also an element $X_\xi \in \mathfrak{X}_L(G)$. We will prove that a relation between X_ξ and Y_ξ is given by

$$I_* X_\xi = -Y_\xi \tag{9.1.1}$$

where $I : G \to G$ is the inversion map: $I(g) = g^{-1}$. Since I is a diffeomorphism, (9.1.1) shows that $I_* : \mathfrak{X}_L(G) \to \mathfrak{X}_R(G)$ is a vector space isomorphism. To prove (9.1.1) notice first that for $u \in T_gG$ and $v \in T_hG$, the derivative of the multiplication map has the expression

$$T_{(g,h)}\mu(u, v) = T_hL_g(v) + T_gR_h(u). \tag{9.1.2}$$

In addition, differentiating the map $g \mapsto \mu(g, I(g)) = e$ gives

$$T_{(g,g^{-1})}\mu(u, T_gI(u)) = 0$$

for all $u \in T_gG$. This and (9.1.2) yields

$$T_gI(u) = -(T_eR_{g^{-1}} \circ T_gL_{g^{-1}})(u) \tag{9.1.3}$$

for all $u \in T_gG$. Consequently, if $\xi \in \mathfrak{g}$, and $g \in G$, we have

$$\begin{aligned}
(I_*X_\xi)(g) &= (TI \circ X_\xi \circ I^{-1})(g) = T_{g^{-1}}I(X_\xi(g^{-1})) \\
&= -(T_eR_g \circ T_{g^{-1}}L_g)(X_\xi(g^{-1})) && \text{(by (9.1.3))} \\
&= -T_eR_g(\xi) = -Y_\xi(g) && \text{(since } X_\xi(g^{-1}) = T_eL_{g^{-1}}(\xi))
\end{aligned}$$

and (9.1.1) is proved. Hence for $\xi, \eta \in \mathfrak{g}$,

$$\begin{aligned} -Y_{[\xi,\eta]} &= I_* X_{[\xi,\eta]} = I_*[X_\xi, X_\eta] = [I_* X_\xi, I_* X_\eta] \\ &= [-Y_\xi, -Y_\eta] = [Y_\xi, Y_\eta], \end{aligned}$$

so that

$$-[Y_\xi, Y_\eta](e) = Y_{[\xi,\eta]}(e) = [\xi, \eta] = [X_\xi, X_\eta](e).$$

Therefore, the Lie algebra bracket $[\,,]^R$ in $\mathfrak{g}$ defined by ***right extension*** of elements in $\mathfrak{g}$:

$$[\xi, \eta]^R := [Y_\xi, Y_\eta](e)$$

is the *negative* of the one defined by left extension, that is,

$$[\xi, \eta]^R := -[\xi, \eta].$$

Examples

(a) For a vector group V, $T_eV \cong V$; it is easy to see that the left invariant vector field defined by $u \in T_eV$ is the constant vector field: $X_u(v) = u$ for all $v \in V$. Therefore, the Lie algebra of a vector group V is V itself, with the trivial bracket $[v, w] = 0$ for all $v, w \in V$. We say that the Lie algebra is ***abelian*** in this case. ♦

(b) The Lie algebra of $GL(n, \mathbb{R})$ is $L(\mathbb{R}^n, \mathbb{R}^n)$, the vector space of all linear transformations of $\mathbb{R}^n$, with the commutator bracket

$$[A, B] = AB - BA.$$

To see this, we recall that $GL(n, \mathbb{R})$ is open in $L(\mathbb{R}^n, \mathbb{R}^n)$ and so the Lie algebra as a vector space is $L(\mathbb{R}^n, \mathbb{R}^n)$. To compute the bracket, note that for any $\xi \in L(\mathbb{R}^n, \mathbb{R}^n)$, $X_\xi : GL(n, \mathbb{R}) \to L(\mathbb{R}^n, \mathbb{R}^n)$ given by $A \mapsto A\xi$, is a left invariant vector field on $GL(n, \mathbb{R})$, because for every $B \in GL(n, \mathbb{R})$, the map $L_B : GL(n, \mathbb{R}) \to GL(n, \mathbb{R})$ defined by $L_B(A) = BA$ is a linear mapping and hence $X_\xi(L_B A) = BA\xi = T_A L_B X_\xi(A)$. Therefore, by the local formula $[X, Y](x) = \mathbf{D}Y(x) \cdot X(x) - \mathbf{D}X(x) \cdot Y(x)$, we get

$$[\xi, \eta] = [X_\xi, X_\eta](I) = \mathbf{D}X_\eta(I) \cdot X_\xi(I) - \mathbf{D}X_\xi(I) \cdot X_\eta(I).$$

But $X_\eta(A) = A\eta$ is linear in A, so $\mathbf{D}X_\eta(I) \cdot B = B\eta$. Hence $\mathbf{D}X_\eta(I) \cdot X_\xi(I) = \xi\eta$, and similarly $\mathbf{D}X_\xi(I) \cdot X_\eta(I) = \eta\xi$. Thus $L(\mathbb{R}^n, \mathbb{R}^n)$ has the bracket

$$[\xi, \eta] = \xi\eta - \eta\xi. \quad ♦ \tag{9.1.4}$$

(c) We can also establish (9.1.4) by a coordinate calculation. Choosing a basis on $\mathbb{R}^n$, each $A \in GL(n, \mathbb{R})$ is specified by its components A^i_j such that $(Av)^i = A^i_j v^j$ (sum on j). Thus, a vector field X on $GL(n, \mathbb{R})$ has the form $X(A) = \sum_{i,j} C^i_j(A)(\partial/\partial A^i_j)$. It is checked to be left invariant provided there is a matrix (ξ^i_j) such that for all A,

$$X(A) = \sum_{i,j} A^i_k \xi^k_j \frac{\partial}{\partial A^i_j}.$$

If $Y(A) = \sum_{i,j} A^i_k \eta^k_j (\partial/\partial A^i_j)$ is another left invariant vector field, we have

$$\begin{aligned}
(XY)[f] &= \sum A^i_k \xi^k_j \frac{\partial}{\partial A^i_j} \left[\sum A^l_m \eta^m_p \frac{\partial f}{\partial A^l_p} \right] \\
&= \sum A^i_k \xi^k_j \delta^l_i \delta^j_m \eta^m_p \frac{\partial f}{\partial A^l_p} + \text{(second derivatives)} \\
&= \sum A^i_k \xi^k_j \eta^j_m \frac{\partial f}{\partial A^i_j} + \text{(second derivatives)},
\end{aligned}$$

where we used $\partial A^s_m / \partial A^k_j = \delta^k_s \delta^j_m$. Therefore, the bracket is the left invariant vector field $[X, Y]$ given by

$$[X, Y][f] = (XY - YX)[f] = \sum A^i_k (\xi^k_j \eta^j_m - \eta^k_j \xi^j_m) \frac{\partial f}{\partial A^i_m}.$$

This shows that the vector field bracket is the usual commutator bracket of $(n \times n)$-matrices, as before. ◆

If X_ξ is the left invariant vector field corresponding to $\xi \in \mathfrak{g}$, there is a unique integral curve $\gamma_\xi : \mathbb{R} \to G$ of X_ξ starting at e; $\gamma_\xi(0) = e$ and $\gamma'_\xi(t) = X_\xi(\gamma_\xi(t))$. We claim that

$$\gamma_\xi(s + t) = \gamma_\xi(s)\gamma_\xi(t),$$

which means that $\gamma_\xi(t)$ is a ***one-parameter subgroup***. Indeed, as functions of t, both sides equal $\gamma_\xi(s)$ at $t = 0$ and both satisfy the differential equation $\sigma'(t) = X_\xi(\sigma(t))$ by left invariance of X_ξ, so they are equal. Left invariance or $\gamma_\xi(t + s) = \gamma_\xi(t)\gamma_\xi(s)$ also shows that $\gamma_\xi(t)$ is defined for all $t \in \mathbb{R}$.

Definition 9.1.3 *The* ***exponential map*** $\exp : \mathfrak{g} \to G$ *is defined by*

$$\exp(\xi) = \gamma_\xi(1).$$

We claim that

$$\exp(s\xi) = \gamma_\xi(s).$$

Indeed, for fixed $s \in \mathbb{R}$, the curve $t \mapsto \gamma_\xi(ts)$ which at $t = 0$ passes through e, satisfies the differential equation

$$\frac{d}{dt}\gamma_\xi(ts) = sX_\xi(\gamma_\xi(ts)) = X_{s\xi}(\gamma_\xi(ts)).$$

Since $\gamma_{s\xi}(t)$ satisfies the same differential equation and passes through e at $t = 0$, it follows that $\gamma_{s\xi}(t) = \gamma_\xi(ts)$. Putting $t = 1$ yields $\exp(s\xi) = \gamma_\xi(s)$.

Hence the exponential mapping maps the line $s\xi$ in $\mathfrak{g}$ onto the one-parameter subgroup $\gamma_\xi(s)$ of G, which is tangent to ξ at e. It follows from left invariance that the flow F_t^ξ of X_ξ satisfies $F_t^\xi(g) = gF_t^\xi(e) = g\gamma_\xi(t)$, so

$$F_t^\xi(g) = g\exp(t\xi) = R_{\exp t\xi}\, g.$$

Let $\gamma(t)$ be a one-parameter subgroup of G, so $\gamma(0) = e$ in particular. We claim that $\gamma = \gamma_\xi$, where $\xi = \gamma'(0)$. Indeed, taking the derivative at $s = 0$ in the relation $\gamma(t + s) = \gamma(t)\gamma(s)$ gives

$$\frac{d\gamma(t)}{dt} = \left.\frac{d}{ds}\right|_{s=0} L_{\gamma(t)}\gamma(s) = T_e L_{\gamma(t)}\gamma'(0) = X_\xi(\gamma(t)),$$

so that $\gamma = \gamma_\xi$ since both equal e at $t = 0$. In other words, *all one-parameter subgroups of G are of the form $\exp t\xi$ for some $\xi \in \mathfrak{g}$*. Since everything proved above for X_ξ can be repeated for Y_ξ, it follows that *the exponential map is the same for the left and right Lie algebras of a Lie group.*

From smoothness of the group operations and smoothness of the solutions of differential equations with respect to initial conditions, it follows that exp is a C^∞ map. Differentiating the identity $\exp(s\xi) = \gamma_\xi(s)$ in s at $s = 0$ shows that $T_0 \exp = \mathrm{id}_{\mathfrak{g}}$. Therefore, by the inverse function theorem, exp is a local diffeomorphism from a neighborhood of zero in $\mathfrak{g}$ onto a neighborhood of e in G. In other words, the exponential map defines a local chart for G at e; in finite dimensions, the coordinates associated to this chart are called the ***canonical coordinates*** of G. By left translation, this chart provides an atlas of all G. (For typical infinite-dimensional groups like diffeomorphism groups, exp is *not* locally onto. It is not true that the exponential map is a local diffeomorphism at any $\xi \neq 0$, even for finite-dimensional Lie groups.)

Computing Brackets

Here is a *computationally useful formula for the bracket.* One follows these three steps:

1. Calculate the ***inner automorphisms***

$$I_g : G \to G, \quad \text{where } I_g(h) = ghg^{-1}.$$

2. Differentiate $I_g(h)$ with respect to h at $h = e$ to produce the ***adjoint operators***

$$\mathrm{Ad}_g : \mathfrak{g} \to \mathfrak{g}; \quad \mathrm{Ad}_g \cdot \eta = T_e I_g \cdot \eta.$$

Note that (see Figure 9.1.3);

$$\mathrm{Ad}_g \, \eta = T_{g^{-1}} L_g \cdot T_e R_{g^{-1}} \cdot \eta.$$

3. Differentiate $\mathrm{Ad}_g \, \eta$ with respect to g at e in the direction ξ to get $[\xi, \eta]$, that is,

$$T_e \varphi^\eta \cdot \xi = [\xi, \eta], \tag{9.1.5}$$

where $\varphi^\eta(g) = \mathrm{Ad}_g \eta$.

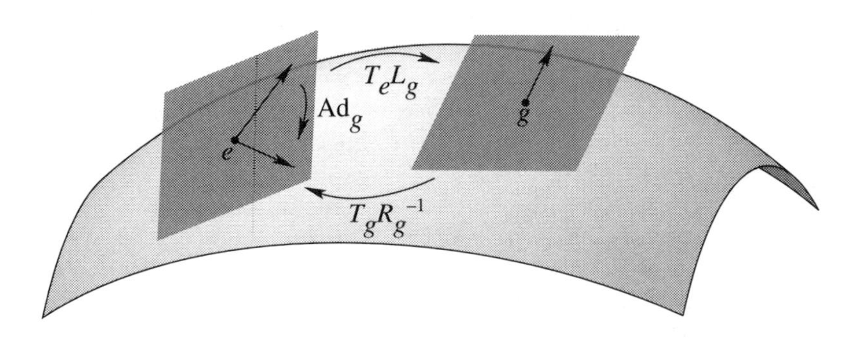

FIGURE 9.1.3. The adjoint mapping is the linearization of conjugation.

Proposition 9.1.4 *Formula* (9.1.5) *is valid.*

Proof Denote by $\varphi_t(g) = g \exp t\xi = R_{\exp t\xi} \, g$, the flow of X_ξ. Then

$$\begin{aligned} [\xi, \eta] &= [X_\xi, X_\eta](e) = \left. \frac{d}{dt} T_{\varphi_t(e)} \varphi_t^{-1} \cdot X_\eta(\varphi_t(e)) \right|_{t=0} \\ &= \left. \frac{d}{dt} T_{\exp t\xi} \, R_{\exp(-t\xi)} \, X_\eta(\exp t\xi) \right|_{t=0} \end{aligned}$$

$$= \left. \frac{d}{dt} T_{\exp t\xi} R_{\exp(-t\xi)} T_e L_{\exp t\xi} \, \eta \right|_{t=0}$$

$$= \left. \frac{d}{dt} T_e(L_{\exp t\xi} \circ R_{\exp(-t\xi)})\eta \right|_{t=0} = \left. \frac{d}{dt} \mathrm{Ad}_{\exp t\xi} \, \eta \right|_{t=0},$$

which is (9.1.5). ■

Another way of expressing (9.1.5) is

$$[\xi, \eta] = \left. \frac{d}{dt} \frac{d}{ds} g(t)h(s)g(t)^{-1} \right|_{s=0, t=0}, \tag{9.1.6}$$

where $g(t)$ and $h(s)$ are curves in G with $g(0) = e, h(0) = e$ and where $g'(0) = \xi$ and $h'(0) = \eta$.

Examples

(a) Let $G = V$ be a vector group. Then $\mathfrak{g} = V$ and $\exp : V \to V$ is the identity mapping. ◆

(b) Let $G = GL(n, \mathbb{R})$; so $\mathfrak{g} = L(\mathbb{R}^n, \mathbb{R})$. For every $A \in L(\mathbb{R}^n, \mathbb{R}^n)$, the mapping $\gamma_A : \mathbb{R} \to GL(n, \mathbb{R})$ defined by

$$t \mapsto \sum_{i=0}^{\infty} \frac{t^i}{i!} A^i$$

is a one-parameter subgroup, because $\gamma_A(0) = I$ and

$$\gamma_A'(t) = \sum_{i=0}^{\infty} \frac{t^{i-1}}{(i-1)!} A^i = \gamma_A(t) A.$$

Therefore, the exponential mapping is given by

$$\exp : L(\mathbb{R}^n, \mathbb{R}^n) \to GL(n, \mathbb{R}^n), \quad A \mapsto \gamma_A(1) = \sum_{i=0}^{\infty} \frac{A^i}{i!}.$$

As is customary, we will write

$$e^A = \sum_{i=0}^{\infty} \frac{A^i}{i!}.$$

We sometimes write $\exp_G : \mathfrak{g} \to G$ when there is more than one group involved. ◆

(c) Formula (9.1.4) also follows from (9.1.5). Here, $I_A B = ABA^{-1}$ and so

$$\mathrm{Ad}_A \cdot \eta = A\eta A^{-1}.$$

Differentiating this with respect to A at $A =$ Identity in the direction ξ gives

$$[\xi, \eta] = \xi\eta - \eta\xi. \quad \blacklozenge$$

Proposition 9.1.5 *Let G and H be Lie groups with Lie algebras $\mathfrak{g}$ and $\mathfrak{h}$. Let $f : G \to H$ be a smooth homomorphism of Lie groups, that is, $f(gh) = f(g)f(h)$ for all $g, h \in G$. Then $T_e f : \mathfrak{g} \to \mathfrak{h}$ is a Lie algebra homomorphism, that is, $(T_e f)[\xi, \eta] = [T_e f(\xi), T_e f(\eta)]$ for all $\xi, \eta \in \mathfrak{g}$. In addition,*

$$f \circ \exp_G = \exp_H \circ T_e f.$$

Proof Since f is a group homomorphism, $f \circ L_g = L_{f(g)} \circ f$. Thus, $Tf \circ TL_g = TL_{f(g)} \circ Tf$ from which it follows that

$$X_{T_e f(\xi)}(f(g)) = T_g f(X_\xi(g)),$$

i.e., that X_ξ and $X_{T_e f(\xi)}$ are f-***related***. It follows that the vector fields $[X_\xi, X_\eta]$ and $[X_{T_e f(\xi)}, X_{T_e f(\eta)}]$ are also f-related for all $\xi, \eta \in \mathfrak{g}$ (see Abraham, Marsden, and Ratiu [1986], §4.2). Hence

$$\begin{aligned} T_e f([\xi, \eta]) &= (Tf \circ [X_\xi, X_\eta])(e) && (\text{where } e = e_G) \\ &= [X_{T_e f(\xi)}, X_{T_e f(\eta)}](\bar{e}) && (\text{where } \bar{e} = e_H = f(e)) \\ &= [T_e f(\xi), T_e f(\eta)]. \end{aligned}$$

Thus $T_e f$ is a Lie algebra homomorphism.

Fixing $\xi \in \mathfrak{g}$, note that $\alpha : t \mapsto f(\exp_G(t\xi))$ and $\beta : t \mapsto \exp_H(tT_e f(\xi))$ are one-parameter subgroups of H. Moreover, $\alpha'(0) = T_e f(\xi) = \beta'(0)$, and so $\alpha = \beta$. In particular, $f(\exp_G(\xi)) = \exp_H(T_e f(\xi))$ for all $\xi \in \mathfrak{g}$. ■

Corollary 9.1.6 *Assume that $f_1, f_2 : G \to H$ are homomorphisms of connected Lie groups and that $T_e f_1 = T_e f_2$. Then $f_1 = f_2$.*

This follows from Proposition **9.1.5** since a connected Lie group G is generated by a neighborhood of the identity element. This latter fact may be proved following these steps:

1. Show that any open subgroup of a Lie group is closed (since its complement is a union of sets homeomorphic to it).

2. Show that a subgroup of a Lie group is open if and only if it contains a neighborhood of the identity element.

3. Conclude that a Lie group is connected if and only if it is generated by arbitrarily small neighborhoods of the identity element.

From Proposition **9.1.5** and the fact that the inner automorphisms are group homomorphisms, we get

Corollary 9.1.7

i $\exp(\mathrm{Ad}_g \xi) = g(\exp \xi) g^{-1}$, *for every* $\xi \in \mathfrak{g}$ *and* $g \in G$*; and*

ii $\mathrm{Ad}_g[\xi, \eta] = [\mathrm{Ad}_g \xi, \mathrm{Ad}_g \eta]$.

Definition 9.1.8 *A* ***Lie subgroup*** H *of a Lie group* G *is a subgroup of* G *which is also an injectively immersed submanifold of* G*. If* H *is a submanifold of* G*, then* H *is called a* **regular** *Lie subgroup.*

For example, the one-parameter subgroups of the torus T^2 that wind densely on the torus are Lie subgroups that are *not* regular.

One can prove the following result (see Abraham and Marsden [1978], p. 259, and Varadarajan [1974], p. 99, Theorem 2.12.6):

Proposition 9.1.9 *If* H *is a closed subgroup of a Lie group* G*, then* H *is a regular Lie subgroup.*

The Lie algebras $\mathfrak{g}$ and $\mathfrak{h}$ of G and a Lie subgroup H, respectively, are related in the following way:

Proposition 9.1.10 *Let* H *be a Lie subgroup of* G*. Then* $\mathfrak{h}$ *is a Lie subalgebra of* $\mathfrak{g}$*. Moreover,* $\mathfrak{h} = \{\xi \in \mathfrak{g} \mid \exp t\xi \in H \ for\ all\ t \in \mathbb{R}\}$.

Proof The first statement is a consequence of Proposition **9.1.5**, which also shows that $\exp t\xi \in H$ for all $\xi \in \mathfrak{h}$ and $t \in \mathbb{R}$. Conversely, if $\exp t\xi \in H$ for all $t \in \mathbb{R}$, we have, $(d/dt) \exp t\xi|_{t=0} \in \mathfrak{h}$ since H is a Lie subgroup; but this equals ξ by definition of the exponential map. ∎

A Lie subalgebra $\mathfrak{h}$ of $\mathfrak{g}$ determines a connected Lie group H injectively immersed in G, a result of Sophus Lie (see Abraham and Marsden [1978], Ex. 4.1E, p. 273, and Varadarajan [1974], p. 58, Theorem 2.5.2). Another important result (known as the third fundamental theorem of Lie) is that for any finite-dimensional Lie algebra there is a connected, simply connected Lie group whose Lie algebra coincides with the given one (Varadarajan [1974], p. 230, Theorem 3.15.1). (Simply connected means that every closed curve based at a point can be continuously contracted to that point, that is, it is homotopic to that point.)

We remind the reader that the Lie algebras appropriate to fluid dynamics and plasma physics are infinite dimensional. Nevertheless, there is still, with the appropriate technical conditions, a correspondence between Lie groups and Lie algebras, analogous to the preceding theorems. The reader should be warned, however, that these theorems as well as Proposition **9.1.9** do not *naively* generalize to the infinite-dimensional situation and to prove them for special cases, specialized analytical theorems may be required.

We conclude with some miscellaneous remarks that we leave as exercises for the reader.

Remarks

1. Proposition **9.1.5** applied to the determinant map gives the identity $\det(\exp A) = \exp(\operatorname{trace} A)$ for $A \in GL(n, \mathbb{R})$.

2. Let G_1 and G_2 be Lie groups with Lie algebras $\mathfrak{g}_1$ and $\mathfrak{g}_2$. Then $G_1 \times G_2$ is a Lie group with Lie algebra $\mathfrak{g}_1 \times \mathfrak{g}_2$, and the exponential map is given by
$$\exp : \mathfrak{g}_1 \times \mathfrak{g}_2 \to G_1 \times G_2; \quad (\xi_2, \xi_2) \mapsto (\exp_1(\xi_1), \exp_2(\xi_2)).$$

3. An ***abelian Lie group*** G satisfies $gh = hg$ for all $g, h \in G$. Its Lie algebra is abelian, that is $[\xi, \eta] = 0$, for all $\xi, \eta \in \mathfrak{g}$. For instance, the unit circle in the complex plane $S^1 = \{z \in \mathbb{C} \mid |z| = 1\}$ is an abelian Lie group under multiplication. The tangent space $T_e S^1$ is the imaginary axis and we identify $\mathbb{R}$ with $T_e S^1$ by $t \mapsto 2\pi i t$. With this identification, the exponential map $\exp : \mathbb{R} \to S^1$ is given by $\exp(t) = e^{2\pi i t}$. Note that $\exp^{-1}(1) = \mathbb{Z}$. ♦

We close this section with the proof of the ***Maurer-Cartan structure equations*** on a Lie group G. Define λ, $\rho \in \Omega^1(G; \mathfrak{g})$, the space of $\mathfrak{g}$-valued one-forms on G, by
$$\lambda(u_g) = T_g L_{g^{-1}}(u_g), \quad \rho(u_g) = T_g R_{g^{-1}}(u_g).$$

Thus, λ and ρ are Lie algebra valued one-forms on G that are defined by left and right translation to the identity respectively. Define the two-form $[\lambda, \lambda]$ by
$$[\lambda, \lambda](u, v) = [\lambda(u), \lambda(v)],$$
and similarly for $[\rho, \rho]$.

Theorem 9.1.11 (Maurer-Cartan Structure Equations)
$$\mathbf{d}\lambda + [\lambda, \lambda] = 0, \quad \mathbf{d}\rho - [\rho, \rho] = 0.$$

Proof We use identity 6 from the table in §4.4. Let $X, Y \in \mathfrak{X}(G)$ and let, for fixed $g \in G$, $\xi = T_g L_{g^{-1}}(X(g)), \eta = T_g L_{g^{-1}}(Y(g))$. Thus,
$$(\mathbf{d}\lambda)(X_\xi, X_\eta) = X_\xi[\lambda(X_\eta)] - X_\eta[\lambda(X_\xi)] - \lambda([X_\xi, X_\eta]).$$

Since $\lambda(X_\eta)(h) = T_h L_{h^{-1}}(X_\eta(h)) = \eta$ is constant, the first term vanishes. Similarly, the second term vanishes too. The third term equals $\lambda([X_\xi, X_\eta]) = \lambda(X_{[\xi,\eta]}) = [\xi, \eta]$ and hence $(\mathbf{d}\lambda)(X_\xi, X_\eta) = -[\xi, \eta]$. Therefore,
$$\begin{aligned} &(\mathbf{d}\lambda + [\lambda, \lambda])\,(X_\xi, X_\eta) \\ &\quad = -[\xi, \eta] + [\lambda, \lambda](X_\xi, X_\eta) \\ &\quad = -[\xi, \eta] + [\lambda(X_\xi), \lambda(X_\eta)] \\ &\quad = -[\xi, \eta] + [\xi, \eta] = 0. \end{aligned}$$

This proves that

$$(\mathbf{d}\lambda + [\lambda, \lambda])\,(X, Y)(g) = 0.$$

Since $g \in G$ was arbitrary as well as X and Y, it follows that $\mathbf{d}\lambda + [\lambda, \lambda] = 0$.

The second relation is proved in the same way but working with the right invariant vector fields Y_ξ, Y_η. The sign in front of the second term changes since $[Y_\xi, Y_\eta] = Y_{-[\xi,\eta]}$. ■

Remark If α is a $(0, k)$-tensor with values in the Banach space E_1, and β is a $(0, l)$-tensor with values in the banach space E_2, if $B : E_1 \times E_2 \to E_3$ is a bilinear map, then replacing in (4.2.1) the multiplication on $\mathbb{R}$ by B, the same formula defines an E_3-valued $(0, k + l)$-tensor on M. Therefore, using definitions (4.2.2)–(4.2.4) if $\alpha \in \Omega^k(M, E_1)$ and $\beta \in \Omega^k(M, E_2)$, then $[(k + l)!/k!\,l!]\mathbf{A}(\alpha \otimes \beta) \in \Omega^{k+l}(M, E_3)$. We shall call this expression the ***wedge product associated to*** B and denote it either by $\alpha \wedge_B \beta$ or $B^\wedge(\alpha, \beta)$.

In particular, if $E_1 = E_2 = E_3 = \mathfrak{g}$ and $B = [\ ,\ ]$ is the Lie algebra bracket, then for $\alpha, \beta \in \Omega^1(M; \mathfrak{g})$ we have

$$\begin{aligned} [\alpha, \beta]^\wedge(u, v) &= [\alpha(u), \beta(v)] - [\alpha(v), \beta(u)] \\ &= -[\beta, \alpha]^\wedge(u, v) \end{aligned}$$

for any vectors u, v tangent to M. Thus, one can write the structure equations as

$$\mathbf{d}\lambda + \frac{1}{2}[\lambda, \lambda]^\wedge = 0, \quad \mathbf{d}\rho - \frac{1}{2}[\rho, \rho]^\wedge = 0.$$

Exercises

9.1-1 *Verify* $\mathrm{Ad}_g[\xi, \eta] = [\mathrm{Ad}_g\, \xi, \mathrm{Ad}_g\, \eta]$ *directly for* $GL(n)$.

9.1-2 *Let G be a Lie group with group operations $\mu : G \times G \to G$ and $I : G \to G$. The tangent bundle TG is also a Lie group, called the* ***tangent group*** *of G with group operations $T\mu : TG \times TG \mapsto TG$, $TI : TG \to TG$.*

9.1-3 *(Defining a Lie group by a chart at the identity.) Let G be a group and suppose that $\varphi : U \to V$ is a one-to-one map from a subset U of G containing the identity element to an open subset V in a Banach space (or Banach manifold). The following conditions are necessary and sufficient for φ to be a chart in a Hausdorff Banach Lie group structure on G:*

(a) *The set $W = \{(x, y) \in V \times V \mid \varphi^{-1}(y) \in U\}$ is open in $V \times V$ and the map $(x, y) \in W \mapsto \varphi(\varphi(x)\varphi^{-1}(y)) \in V$ is smooth.*

(b) *For every $g \in G$, the set $V_g = \varphi(gUg^{-1} \cap U)$ is open in V and the map $x \in V_g \mapsto \varphi(g\varphi^{-1}(x)g^{-1}) \in V$ is smooth.*

9.2 Some Classical Lie Groups

We have already discussed the classical matrix Lie group $GL(n, \mathbb{R})$. In this section we will show that a number of classical matrix groups are Lie subgroups of $GL(n, \mathbb{R})$.

The Special Linear Group $SL(n, \mathbb{R})$

Let $\det : L(\mathbb{R}^n, \mathbb{R}^n) \to \mathbb{R}$ be the determinant map and observe that $GL(n, \mathbb{R}) = \{A \in L(\mathbb{R}^n, \mathbb{R}^n) \mid \det A \neq 0\}$, so $GL(n, \mathbb{R})$ is open in $L(\mathbb{R}^n, \mathbb{R}^n)$. Notice that $\mathbb{R}\backslash\{0\}$ is a group under multiplication and that

$$\det : GL(n, \mathbb{R}) \to \mathbb{R}\backslash\{0\}$$

is a Lie group homomorphism because

$$\det(AB) = (\det A)(\det B).$$

Lemma 9.2.1 *The map* $\det : GL(n, \mathbb{R}) \to \mathbb{R}\backslash\{0\}$ *is* C^∞ *and its derivative is given by* $\mathbf{D}\det_A \cdot B = (\det A)\,\mathrm{trace}(A^{-1}B)$.

Proof The smoothness of det is clear from its formula in terms of matrix elements. Using the identity

$$\det(A + \lambda B) = (\det A)\det(I + \lambda A^{-1}B),$$

it suffices to prove

$$\left.\frac{d}{d\lambda}\det(I + \lambda C)\right|_{\lambda=0} = \mathrm{tr}\, C.$$

This follows from the identity

$$\det(I + \lambda C) = 1 + \lambda\,\mathrm{tr}\, C + \cdots + \lambda^n \det C. \quad \blacksquare$$

Define the ***special linear group*** $SL(n, \mathbb{R})$ by

$$SL(n, \mathbb{R}) = \{A \in GL(n, \mathbb{R}) \mid \det A = 1\} = \det^{-1}(1). \tag{9.2.1}$$

From Proposition **9.1.9** it follows that $SL(n, \mathbb{R})$ is a closed Lie subgroup of $GL(n, \mathbb{R})$. However, this method invokes a rather subtle result to prove something that is actually straightforward. In fact, it follows from Lemma **9.2.1** that $\det : GL(n, \mathbb{R}) \to \mathbb{R}$ is a submersion, so $SL(n, \mathbb{R}) = \det^{-1}(1)$ is a *smooth* closed submanifold and hence a closed subgroup.

The tangent space to $SL(n, \mathbb{R})$ at $A \in SL(n, \mathbb{R})$ therefore consists of all matrices B such that $\mathrm{tr}(A^{-1}B) = 0$. In particular, the tangent space at the identity consists of the matrices with trace zero. We have seen that the Lie algebra of $GL(n, \mathbb{R})$ is $L(\mathbb{R}^n, \mathbb{R}^n)$ with the Lie bracket given by

$[A, B] = AB - BA$. It follows that the *Lie algebra* $\mathfrak{sl}(n, \mathbb{R})$ *of* $SL(n, \mathbb{R})$ *consists of the set of* $n \times n$ *matrices having trace zero, with the bracket*

$$[A, B] = AB - BA.$$

Since $\operatorname{tr}(B) = 0$ imposes one condition on B, it follows that

$$\dim[\mathfrak{sl}(n, \mathbb{R})] = n^2 - 1.$$

We leave it to the reader to check that $SL(n, \mathbb{R})$ is a noncompact, connected Lie group, although $GL(n, \mathbb{R})$ is not connected. The latter has two connected components, one defined by $\det > 0$ and the other by $\det < 0$. Summarizing:

Proposition 9.2.2 *The group* $SL(n, \mathbb{R})$ *is a noncompact connected* $(n^2 - 1)$*-dimensional noncompact Lie group whose Lie algebra consists of the* $(n \times n)$ *matrices with trace zero (or linear maps of* $\mathbb{R}^n$ *to* $\mathbb{R}^n$ *with trace zero) with the bracket*

$$[A, B] = AB - BA.$$

The Orthogonal Group $O(n)$

On $\mathbb{R}^n$ we use the standard inner product

$$\langle \mathbf{x}, \mathbf{y} \rangle = \sum_{i=1}^{n} x^i y^i,$$

where $\mathbf{x} = (x^1, \ldots, x^n) \in \mathbb{R}^n$ and $\mathbf{y} = (y^1, \ldots, y^n) \in \mathbb{R}^n$. A linear map $A \in L(\mathbb{R}^n, \mathbb{R}^n)$ is called ***orthogonal*** if

$$\langle A\mathbf{x}, A\mathbf{y} \rangle = \langle \mathbf{x}, \mathbf{y} \rangle, \tag{9.2.2}$$

for all $\mathbf{x}, \mathbf{y} \in \mathbb{R}$. In terms of the norm $\|\mathbf{x}\| = \langle \mathbf{x}, \mathbf{x} \rangle^{1/2}$, one sees from the polarization identity that A is orthogonal iff

$$\|A\mathbf{x}\| = \|\mathbf{x}\|,$$

for all $\mathbf{x} \in \mathbb{R}^n$, or in terms of the transpose A^T, which is defined by $\langle A\mathbf{x}, \mathbf{y} \rangle = \langle \mathbf{x}, A^T\mathbf{y} \rangle$, we see that A is orthogonal iff

$$AA^T = I.$$

Let $O(n)$ denote the orthogonal elements of $L(\mathbb{R}^n, \mathbb{R}^n)$. For $A \in O(n)$, we see that $1 = \det(AA^T) = (\det A)(\det A^T) = (\det A)^2$; hence $\det A = \pm 1$ and so $A \in GL(n, \mathbb{R})$. Furthermore, if $A, B \in O(n)$ then $\langle AB\mathbf{x}, AB\mathbf{y} \rangle = \langle B\mathbf{x}, B\mathbf{y} \rangle = \langle \mathbf{x}, \mathbf{y} \rangle$ and so $AB \in O(n)$. Letting $\mathbf{x}' = A^{-1}\mathbf{x}$ and $\mathbf{y}' = A^{-1}\mathbf{y}$, we see that $\langle \mathbf{x}, \mathbf{y} \rangle = \langle A\mathbf{x}', A\mathbf{y}' \rangle = \langle \mathbf{x}', \mathbf{y}' \rangle$, that is, $\langle \mathbf{x}, \mathbf{y} \rangle = \langle A^{-1}\mathbf{x}, A^{-1}\mathbf{y} \rangle$; hence $A^{-1} \in O(n)$.

Let $S(n)$ denote the vector space of symmetric linear maps of $\mathbb{R}^n$ to itself, and let $\psi : GL(n, \mathbb{R}) \to S(n)$ be defined by $\psi(A) = AA^T$. We claim ψ is a submersion. Indeed, its derivative is

$$\mathbf{D}\psi(A) \cdot B = AB^T + BA^T$$

which is onto (to hit C, take $B = CA/2$). Thus, $\psi^{-1}(I) = O(n)$ is a closed Lie subgroup of $GL(n, \mathbb{R})$, called the ***orthogonal group***. Since $O(n)$ is closed and bounded in $L(\mathbb{R}^n, \mathbb{R}^n)$, it is compact. We shall see in §**9.3** that $O(n)$ is not connected, but has two connected components, one where $\det = +1$ and the other where $\det = -1$.

The Lie algebra $\mathfrak{o}(n)$ of $O(n)$ is $\ker \mathbf{D}\psi(I)$, namely, the skew-symmetric linear maps with the usual bracket $[A, B] = AB - BA$. The space of skew-symmetric $n \times n$ matrices have dimension equal to the number of entries above the diagonal, namely, $n(n-1)/2$. Thus, $\dim[O(n)] = n(n-1)/2$.

The ***special orthogonal group*** is defined as

$$SO(n) = O(n) \cap SL(n, \mathbb{R}), \quad \text{i.e.,} \quad SO(n) = \{A \in O(n) \mid \det A = +1\}. \tag{9.2.3}$$

Since $SO(n)$ is the kernel of $\det : O(n) \to \{-1, 1\}$, that is, $SO(n) = \det^{-1}(1)$, it is an open and closed Lie subgroup of $O(n)$, hence is compact. We also note that $SO(n)$ is the connected component of $O(n)$ containing the identity I, and so has the same Lie algebra as $O(n)$. We summarize:

Proposition 9.2.3 *The group $O(n)$ is a compact Lie group of dimension $n(n-1)/2$. Its Lie algebra $\mathfrak{o}(n)$ is the space of skew-symmetric $n \times n$ matrices with bracket $[A, B] = AB - BA$. The connected component of the identity in $O(n)$ is the compact Lie group $SO(n)$ which has the same Lie algebra $\mathfrak{so}(n) = \mathfrak{o}(n)$.*

Rotations in the Plane $SO(2)$

May be identified with the unit circle $S^1 = \{\mathbf{x} \in \mathbb{R}^2 \mid \|\mathbf{x}\| = 1\}$. We parametrize S^1 by the angle θ, $0 \leq \theta < 2\pi$. For each $\theta \in [0, 2\pi]$, let

$$A_\theta = \begin{bmatrix} \cos\theta & -\sin\theta \\ \sin\theta & \cos\theta \end{bmatrix},$$

using the standard basis of $\mathbb{R}^2$. Then $A_\theta \in SO(2)$ and represents a counterclockwise rotation through the angle θ. Conversely, if

$$A = \begin{bmatrix} a_1 & a_2 \\ a_3 & a_4 \end{bmatrix}$$

is orthogonal, the relations $a_1^2 + a_2^2 = 1$, $a_4^2 + a_4^2 = 1$, $a_1 a_3 + a_2 a_4 = 0$, and $\det A = a_1 a_4 - a_2 a_3 = 1$ show that $A = A_\theta$ for some θ. Thus $SO(2)$ can be identified with S^1; that is, with rotations in the plane.

Rotations in Space $SO(3)$

The Lie algebra $\mathfrak{so}(3)$ of $SO(3)$ may be identified with $\mathbb{R}^3$ as follows. We define the vector space isomorphism $\hat{}: \mathbb{R} \to \mathfrak{so}(3)$ called the ***hat map***, by

$$\mathbf{v} = (v_1, v_2, v_3) \mapsto \hat{\mathbf{v}} = \begin{bmatrix} 0 & -v_3 & v_2 \\ v_3 & 0 & -v_1 \\ -v_2 & v_1 & 0 \end{bmatrix}. \tag{9.2.4}$$

Note that

$$\hat{\mathbf{v}} \cdot \mathbf{w} = \mathbf{v} \times \mathbf{w}$$

and therefore that

$$\begin{aligned} (\hat{\mathbf{u}}\hat{\mathbf{v}} - \hat{\mathbf{v}}\hat{\mathbf{u}})\,\mathbf{w} &= \hat{\mathbf{u}}(\mathbf{v} \times \mathbf{w}) - \hat{\mathbf{v}}(\mathbf{u} \times \mathbf{w}) \\ &= \mathbf{u} \times (\mathbf{v} \times \mathbf{w}) - \mathbf{v} \times (\mathbf{u} \times \mathbf{w}) \\ &= (\mathbf{u} \times \mathbf{v}) \times \mathbf{w} = (\mathbf{u} \times \mathbf{v})\hat{} \cdot \mathbf{w}. \end{aligned}$$

Thus if we put the cross product on $\mathbb{R}^3$, $\hat{}$ becomes a Lie algebra isomorphism and so *we can identify* $\mathfrak{so}(3)$ *with* $\mathbb{R}^3$ *with the cross product as Lie bracket.*

We also note that the standard dot product may be written

$$\mathbf{v} \cdot \mathbf{w} = \frac{1}{2} \operatorname{trace}\left(\hat{\mathbf{v}}^T \hat{\mathbf{w}}\right) = -\frac{1}{2} \operatorname{trace}\left(\hat{\mathbf{v}}\hat{\mathbf{w}}\right).$$

Theorem 9.2.4 (Euler's Theorem) *Every element $A \in SO(3)$ is a rotation through an angle θ about an axis $\mathbf{w}$.*

To prove this, we use the following lemma:

Lemma 9.2.5 *Every $A \in SO(3)$ has an eigenvalue equal to* 1.

Proof The eigenvalues of A are given by roots of the third degree polynomial $\det(A - \lambda I) = 0$. Roots occur in conjugate pairs, so at least one is real. If λ is a real root and x is a nonzero real eigenvector, $A\mathbf{x} = \lambda\mathbf{x}$, so $\|A\mathbf{x}\|^2 = \|\mathbf{x}\|^2$ and $\|A\mathbf{x}\|^2 = |\lambda|\,\|\mathbf{x}\|^2$ imply $\lambda = \pm 1$. If all three roots are real, they are $(1,1,1)$ or $(1,-1,-1)$ since $\det A = 1$. If there is one real and two complex conjugate roots, they are $(1, \omega, \bar{\omega})$ since $\det A = 1$. In any case one real root must be $+1$. ■

Proof of Theorem 9.2.4 By Lemma **9.2.5**, the matrix A has an eigenvector $\mathbf{w}$ with eigenvalue 1, say $A\mathbf{w} = \mathbf{w}$. The line spanned by $\mathbf{w}$ is also invariant under A. Let P be the plane perpendicular to $\mathbf{w}$; that is,

$$P = \{\mathbf{y} \mid \langle \mathbf{w}, \mathbf{y} \rangle = 0\}.$$

Since A is orthogonal, $A(P) = P$. Let $\mathbf{e}_1, \mathbf{e}_2$ be an orthogonal basis in P. Then relative to $(\mathbf{w}, \mathbf{e}_1, \mathbf{e}_2)$, A has the matrix

$$A = \begin{bmatrix} 1 & 0 & 0 \\ 0 & a_1 & a_2 \\ 0 & a_3 & a_4 \end{bmatrix}.$$

Since

$$\begin{bmatrix} a_1 & a_2 \\ a_3 & a_4 \end{bmatrix}$$

lies in $SO(2)$, A is a rotation about the axis $\mathbf{w}$ by some angle. ■

Corollary 9.2.6 *Any $A \in SO(3)$ can be written in some orthonormal basis as the matrix*

$$A = \begin{bmatrix} 1 & 0 & 0 \\ 0 & \cos\theta & -\sin\theta \\ 0 & \sin\theta & \cos\theta \end{bmatrix}.$$

The infinitesimal version of Euler's theorem is the following:

Proposition 9.2.7 *Identifying the Lie algebra $\mathfrak{so}(3)$ of $SO(3)$ with the Lie algebra $\mathbb{R}^3$, $\exp(t\mathbf{w})$ is a rotation about $\mathbf{w}$ by the angle $t\|\mathbf{w}\|$, where $\mathbf{w} \in \mathbb{R}^3$.*

Proof To simplify the computation, we pick an orthonormal basis $(\mathbf{e}_1, \mathbf{e}_2, \mathbf{e}_3)$ of $\mathbb{R}^3$, with $\mathbf{e}_1 = \mathbf{w}/\|\mathbf{w}\|$. Relative to this basis, $\hat{\mathbf{w}}$ has the matrix

$$\hat{\mathbf{w}} = \|\mathbf{w}\| \begin{bmatrix} 0 & 0 & 0 \\ 0 & 0 & -1 \\ 0 & 1 & 0 \end{bmatrix}.$$

Let

$$c(t) = \begin{bmatrix} 1 & 0 & 0 \\ 0 & \cos t\|\mathbf{w}\| & -\sin t\|\mathbf{w}\| \\ 0 & \sin t\|\mathbf{w}\| & \cos t\|\mathbf{w}\| \end{bmatrix}.$$

Then

$$\begin{aligned} c'(t) &= \begin{bmatrix} 0 & 0 & 0 \\ 0 & -\|\mathbf{w}\| \sin t\|\mathbf{w}\| & -\|\mathbf{w}\| \cos t\|\mathbf{w}\| \\ 0 & -\|\mathbf{w}\| \cos t\|\mathbf{w}\| & -\|\mathbf{w}\| \sin t\|\mathbf{w}\| \end{bmatrix} \\ &= c(t)\hat{\mathbf{w}} = T_I L_{c(t)}(\hat{\mathbf{w}}) = X_{\hat{\mathbf{w}}}(c(t)), \end{aligned}$$

where $X_{\hat{\mathbf{w}}}$ is the left invariant vector field corresponding to $\hat{\mathbf{w}}$. Therefore $c(t)$ is an integral curve of $X_{\hat{\mathbf{w}}}$; but $\exp(t\hat{\mathbf{w}})$ is also an integral curve of $X_{\hat{\mathbf{w}}}$. Since both agree at $t = 0$, $\exp(t\hat{\mathbf{w}}) = c(t)$, for all $t \in \mathbb{R}$. But the matrix definition of $c(t)$ expresses it as a rotation by an angle $t\|\mathbf{w}\|$ about the axis $\mathbf{w}$. ■

Despite Euler's theorem, it might be good to recall now that $SO(3)$ *cannot* be written as $S^2 \times S^1$; see Exercise **1.2-4**.

Amplifying on Proposition **9.2.7**, we give the following explicit formula for $\exp \xi$, where $\xi \in \mathfrak{so}(3)$, due to Rodrigues [1840]:

$$\exp[\hat{\mathbf{v}}] = I + \frac{\sin \|\mathbf{v}\|}{\|\mathbf{v}\|}\hat{\mathbf{v}} + \frac{1}{2}\left[\frac{\sin\left(\frac{\|\mathbf{v}\|}{2}\right)}{\frac{\|\mathbf{v}\|}{2}}\right]^2 \hat{\mathbf{v}}^2. \tag{9.2.5}$$

(See also Helgason [1978], Exercise 1, p. 249 and see Altmann [1986] for some interesting history of this formula.)

Proof of Rodrigues' Formula By (9.2.4),

$$\hat{\mathbf{v}}^2\mathbf{w} = \mathbf{v} \times (\mathbf{v} \times \mathbf{w}) = \langle \mathbf{v}, \mathbf{w}\rangle\, \mathbf{v} - \|\mathbf{v}\|^2\mathbf{w}. \tag{9.2.6}$$

Consequently, we have the recurrence relations

$$\hat{\mathbf{v}}^3 = -\|\mathbf{v}\|^2\hat{\mathbf{v}}, \quad \hat{\mathbf{v}}^4 = -\|\mathbf{v}\|^2\hat{\mathbf{v}}^2, \quad \hat{\mathbf{v}}^5 = \|\mathbf{v}\|^4\hat{\mathbf{v}}, \quad \hat{\mathbf{v}}^6 = \|\mathbf{v}\|^4\hat{\mathbf{v}}^2, \ldots.$$

Splitting the exponential series in odd and even powers,

$$\begin{aligned}
\exp[\hat{\mathbf{v}}] &= I + \left[I - \frac{\|\mathbf{v}\|^2}{3!} + \frac{\|\mathbf{v}\|^4}{5!} - \cdots + (-1)^{n+1}\frac{\|\mathbf{v}\|^{2n}}{2n+1!} + \cdots\right]\hat{\mathbf{v}} \\
&\quad + \left[\frac{1}{2!} - \frac{\|\mathbf{v}\|^2}{4!} + \frac{\|\mathbf{v}\|^4}{6!} + \cdots + (-1)^{2n-1}\frac{\|\mathbf{v}\|^{2n-2}}{(2n)!} + \cdots\right]\hat{\mathbf{v}}^2 \\
&= I + \frac{\sin\|\mathbf{v}\|}{\|\mathbf{v}\|}\hat{\mathbf{v}} + \frac{1-\cos\|\mathbf{v}\|}{\|\mathbf{v}\|^2}\hat{\mathbf{v}}^2,
\end{aligned} \tag{9.2.7}$$

and so the result follows from identity $2\sin^2(\|\mathbf{v}\|/2) = 1 - \cos\|\mathbf{v}\|$. ■

The following alternative expression, equivalent to (9.2.5), is often useful. Set $\mathbf{n} = \mathbf{v}/\|\mathbf{v}\|$ so that $\|\mathbf{n}\| = 1$. From (9.2.6) and (9.2.7) we obtain

$$\exp[\hat{\mathbf{v}}] = I + (\sin\|\mathbf{v}\|)\hat{\mathbf{n}} + (1 - \cos\|\mathbf{v}\|)[\mathbf{n} \otimes \mathbf{n} - I]. \tag{9.2.8}$$

Therefore, we obtain a rotation about the unit vector $\mathbf{n} = \mathbf{v}/\|\mathbf{v}\|$ of magnitude $\|\mathbf{v}\|$.

The results (9.2.5) and (9.2.8) are useful in computational solid mechanics, along with their quaternionic counterparts. We shall return to this point below in connection with $SU(2)$; see Whittaker [1927] and Simo and Fox [1989] for more information.

The Symplectic Group $Sp(2n, \mathbb{R})$

Let

$$\mathbb{J} = \begin{bmatrix} 0 & I \\ -I & 0 \end{bmatrix},$$

where I is the $n \times n$ identity matrix. Recall that $A \in L(\mathbb{R}^{2n}, \mathbb{R}^{2n})$ is ***symplectic*** if $A^T \mathbb{J} A = \mathbb{J}$. Let $Sp(2n, \mathbb{R})$ be the set of symplectic matrices. For A a symplectic matrix, $\mathbb{J} = A^T \mathbb{J} A$ gives

$$1 = \det \mathbb{J} = (\det A^T) \cdot (\det \mathbb{J}) \cdot (\det A) = (\det A)^2.$$

Hence $\det A = \pm 1$, and so $A \in GL(2n, \mathbb{R})$. Furthermore, if $A, B \in Sp(2n, \mathbb{R})$, then $(AB)^T \mathbb{J}(AB) = B^T A^T \mathbb{J} AB = \mathbb{J}$, hence $AB \in Sp(2n, \mathbb{R})$, and if $A^T \mathbb{J} A = \mathbb{J}$, then $\mathbb{J} A = (A^T)^{-1} \mathbb{J} = (A^{-1})^T \mathbb{J}$, so $\mathbb{J} = (A^{-1})^T \mathbb{J} A^{-1}$, or $A^{-1} \in Sp(2n, \mathbb{R})$. Thus $Sp(2n, \mathbb{R})$ is a group. If

$$A = \begin{bmatrix} a & b \\ c & d \end{bmatrix} \in GL(2n, \mathbb{R}),$$

then $A \in Sp(2n, \mathbb{R})$ iff $a^T c$ and $b^T d$ are symmetric and $a^T d - c^T b = I$.

A similar submersion argument to the one we used for SL and SO shows that $Sp(2n, \mathbb{R})$ is a Lie subgroup of $GL(2n, \mathbb{R})$, called the ***symplectic group***. One can show that $Sp(2n, \mathbb{R})$ is not compact by considering cotangent lifts of translations, for example. The Lie algebra of $Sp(2n, \mathbb{R})$ is clearly

$$\mathfrak{sp}(2n, \mathbb{R}) = \left\{ A \in L(\mathbb{R}^{2n}, \mathbb{R}^{2n}) \mid A^T \mathbb{J} + \mathbb{J} A = 0 \right\}.$$

If

$$A = \begin{bmatrix} a & b \\ c & d \end{bmatrix} \in \mathfrak{sl}(2n, \mathbb{R}),$$

then $A \in \mathfrak{sp}(2n, \mathbb{R})$ iff $d = -a^T$, $c = c^T$, and $b = b^T$. The dimension of $\mathfrak{sp}(2n, \mathbb{R})$ can be readily calculated to be $2n^2 + n$.

Proposition 9.2.8 *$Sp(2n, \mathbb{R})$ is a noncompact, nonconnected Lie group of dimension $2n^2 + n$. Its Lie algebra $\mathfrak{sp}(2n, \mathbb{R})$ consists of the $2n \times 2n$ matrices A satisfying $A^T \mathbb{J} + \mathbb{J} A = 0$, where*

$$\mathbb{J} = \begin{bmatrix} 0 & I \\ -I & 0 \end{bmatrix}$$

with I the $n \times n$ identity matrix.

Recall that the symplectic group is related to classical mechanics as follows. Consider a particle of mass m moving in a potential $V(\mathbf{q})$, where $\mathbf{q} = (q^1, q^2, q^3) \in \mathbb{R}^3$. Newton's second law states that the particle moves along a curve $\mathbf{q}(t)$ in $\mathbb{R}^3$ in such a way that $m\ddot{\mathbf{q}} = -\operatorname{grad} V(\mathbf{q})$. Introduce the momentum $p_i = m\dot{q}^i$, $i = 1, 2, 3$, and the energy $H(\mathbf{q}, \mathbf{p}) = (1/2m) \sum_{i=1}^3 p_i^2 + V(\mathbf{q})$. Compute $\partial H / \partial q^i = \partial V / \partial q^i = -m\ddot{\mathbf{q}}^i = -\dot{p}_i$ and $\partial H / \partial p_i = (1/m) p_i = \dot{q}^i$; hence *Newton's law $\mathbf{F} = m\mathbf{a}$ is equivalent to Hamilton's equations*

$$\dot{q}^i = \frac{\partial H}{\partial p_i}, \quad \dot{p}_i = -\frac{\partial H}{\partial q^i}, \quad i = 1, 2, 3.$$

Writing $z = (\mathbf{q}, \mathbf{p})$,

$$\mathbb{J} \cdot \operatorname{grad} H(z) = \begin{bmatrix} 0 & I \\ -I & 0 \end{bmatrix} \begin{bmatrix} \dfrac{\partial H}{\partial \mathbf{q}} \\ \dfrac{\partial H}{\partial \mathbf{p}} \end{bmatrix} = (\dot{\mathbf{q}}, \dot{\mathbf{p}}) = \dot{z},$$

so Hamilton's equations read $\dot{z} = \mathbb{J} \cdot \operatorname{grad} H(z)$. Now let $f : \mathbb{R}^3 \times \mathbb{R}^3 \to \mathbb{R}^3 \times \mathbb{R}^3$ and write $w = f(z)$. If $z(t)$ satisfies Hamilton's equations $\dot{z} = \mathbb{J} \cdot \operatorname{grad} H(z)$, then $w(t) = f(z(t))$ satisfies $\dot{w} = A^T \dot{z}$, where $A^T = [\partial z^i / \partial w^j]$ is the Jacobian matrix of f. By the chain rule,

$$\dot{w} = A^T \mathbb{J} \operatorname{grad}_z H(z) = A^T \mathbb{J} A \operatorname{grad}_w H(z(w)).$$

Thus the equations for $w(t)$ have the form of Hamilton's equations with energy $K(w) = H(z(w))$ if and only if $A^T \mathbb{J} A = \mathbb{J}$; that is, iff A is symplectic. A nonlinear transformation f is ***canonical*** iff its Jacobian is symplectic.

As a special case, consider a linear map $A \in Sp(2n, \mathbb{R})$ and let $w = Az$. Suppose H is quadratic, that is, of the form $H(z) = \langle z, Bz \rangle /2$ where B is a symmetric $(2n \times 2n)$ matrix. Then

$$\operatorname{grad} H(z) \cdot \delta z = \frac{1}{2} \langle \delta z, Bz \rangle + \langle z, B\delta z \rangle = \frac{1}{2} (\langle \delta z, Bz \rangle + \langle Bz, \delta z \rangle) = \langle \delta z, Bz \rangle ,$$

so $\operatorname{grad} H(z) = Bz$ and thus the equations of motion become the linear equations $\dot{z} = \mathbb{J} Bz$. Now

$$\dot{w} = A\dot{z} = A\mathbb{J}Bz = \mathbb{J}(A^T)^{-1}Bz = \mathbb{J}(A^T)^{-1}BA^{-1}Az = \mathbb{J}B'w,$$

where $B' = (A^T)^{-1}BA^{-1}$ is symmetric. For the new Hamiltonian we get

$$H'(w) = \langle w, (A^T)^{-1}BA^{-1}w \rangle = \langle A^{-1}w, BA^{-1}w \rangle = H(A^{-1}w) = H(z).$$

Thus, $Sp(2n, \mathbb{R})$ *is the linear invariance group of classical mechanics.*

Complex Groups

Many important Lie groups involve *complex* matrices. It is proved, as in the real case, that $GL(n, \mathbb{C}) = \{n \times n \text{ invertible complex matrices}\}$ is an open set in $L(\mathbb{C}^n, \mathbb{C}^n) = \{n \times n \text{ complex matrices}\}$. Clearly $GL(n, \mathbb{C})$ is a group under matrix multiplication. Therefore $GL(n, \mathbb{C})$ is a Lie group, and has a Lie algebra $\mathfrak{gl}(n, \mathbb{C}) = \{n \times n \text{ complex matrices}\} = L(\mathbb{C}^n, \mathbb{C}^n)$. Hence $GL(n, \mathbb{C})$ has complex dimension n^2, that is, real dimension $2n^2$. The group $GL(n, \mathbb{C})$ is connected, while $GL(n, \mathbb{R})$ is not.

The ***complex special linear group*** $SL(n, \mathbb{C}) = \{A \in GL(n, \mathbb{C}) \mid \det A = 1\}$ is a Lie subgroup of $GL(n, \mathbb{C})$ of (real) dimension $2(n^2 - 1)$. Its Lie algebra is $\mathfrak{sl}(n, \mathbb{C}) = \{A \in \mathfrak{gl}(n, \mathbb{C}) \mid \operatorname{tr} A = 0\}$.

The ***unitary group*** $U(n)$ will now be defined. Recall that $\mathbb{C}^n$ has the Hermitian inner product:

$$\langle \mathbf{x}, \mathbf{y} \rangle = \sum_{i=0}^{n} x^i \bar{y}^i,$$

where $\mathbf{x} = (x^1, \ldots, x^n) \in \mathbb{C}^n$, and $\mathbf{y} = (y^1, \ldots, y^n) \in \mathbb{C}^n$, and $\bar{y}^i$ denotes the complex conjugate. Let

$$U(n) = \{A \in GL(n, \mathbb{C}) \mid \langle A\mathbf{x}, A\mathbf{y} \rangle = \langle \mathbf{x}, \mathbf{y} \rangle\}.$$

The orthogonality condition $\langle A\mathbf{x}, A\mathbf{y} \rangle = \langle \mathbf{x}, \mathbf{y} \rangle$ is equivalent to $AA^\dagger = I$, where $A^\dagger = \bar{A}^T$, that is, $\langle A\mathbf{x}, \mathbf{y} \rangle = \langle \mathbf{x}, A^\dagger \mathbf{y} \rangle$. From $|\det A| = 1$, we see that det maps $U(n)$ into the unit circle $S^1 = \{z \in \mathbb{C} \mid |z| = 1\}$. As is to be expected by now, $U(n)$ is a closed Lie subgroup of $GL(n, \mathbb{C})$ with Lie algebra

$$\mathfrak{u}(n) = \{A \in L(\mathbb{C}^n, \mathbb{C}^n) \mid \langle A\mathbf{x}, \mathbf{y} \rangle = - \langle \mathbf{x}, A\mathbf{y} \rangle\};$$

$U(n)$ is compact and connected, and has (real) dimension n^2. In the special case $n = 1$, a complex linear map $\varphi : \mathbb{C} \to \mathbb{C}$ is multiplication by some complex number z, and φ is an isometry if and only if $|z| = 1$. In this way the group $U(1)$ is identified with the unit circle S^1.

The ***special unitary group***

$$SU(n) = \{A \in U(n) \mid \det A = 1\}$$

is a closed Lie subgroup of $U(n)$ with Lie algebra

$$\mathfrak{su}(n) = \{A \in L(\mathbb{C}^n, \mathbb{C}^n) \mid \langle A\mathbf{x}, \mathbf{y} \rangle = - \langle \mathbf{x}, A\mathbf{y} \rangle \text{ and } \operatorname{tr} A = 0\}.$$

$SU(n)$ is compact and connected, and has (real) dimension $n^2 - 1$.

In the special case $n = 2$, $\dim SU(2) = 3$. Also, $SU(2)$ is diffeomorphic to the three-sphere $S^3 = \{x \in \mathbb{R}^4 \mid \|\mathbf{x}\| = 1\}$, with the diffeomorphism given by

$$x = (x^1, x^2, x^3, x^4) \in S^3 \subset \mathbb{R}^4 \mapsto \begin{bmatrix} x^1 + ix^2 & x^3 + ix^4 \\ -x^3 + ix^4 & x^1 - ix^2 \end{bmatrix} \in SU(2).$$

Therefore $SU(2)$ is simply connected. The group $SU(2)$ is used in the construction of the (nonabelian) gauge group for the Yang-Mills equations in elementary particle physics.

Under the identification $\mathbb{C}^n = \mathbb{R}^n \oplus i\mathbb{R}^n$, we can consider the complex matrix groups $GL(n, \mathbb{C})$, $U(n)$, and $SU(n)$ as Lie subgroups of the real matrix group $GL(2n, \mathbb{R})$. *The symplectic group is related to the unitary group $U(n)$ by*

$$Sp(2n, \mathbb{R}) \cap O(2n, \mathbb{R}) = U(n, \mathbb{C}).$$

More on the Group $SU(2)$

Next we outline the relationship between $SU(2)$ and $SO(3)$. We begin by noting that $SO(3)$ is diffeomorphic to $\mathbb{RP}^3$. To see this, map the unit ball D in $\mathbb{R}^3$ to $SO(3)$ by sending (x, y, z) to the rotation about (x, y, z) through angle $\pi\sqrt{x^2 + y^2 + z^2}$ (and $(0, 0, 0)$ to the identity). Then D with antipodal points on the boundary identified is diffeomorphic to $SO(3)$ by mapping D to the upper hemisphere of S^3 by

$$(x, y, z) \mapsto \left(x, y, z, \sqrt{1 - x^2 - y^2 - x^2}\right).$$

We see that D with antipodal points on the boundary identified is diffeomorphic to the upper hemisphere in S^3 with antipodal points on the equator identified; this latter manifold is $\mathbb{RP}^3$. This construction thus induces a diffeomorphism of $\mathbb{RP}^3$ with $SO(3)$.

Let $\sigma_1, \sigma_2, \sigma_3$ be the ***Pauli spin matrices***, defined by

$$\sigma_1 = \begin{bmatrix} 0 & 1 \\ 1 & 0 \end{bmatrix}, \quad \sigma_2 = \begin{bmatrix} 0 & -i \\ i & 0 \end{bmatrix}, \quad \text{and} \quad \sigma_3 = \begin{bmatrix} 1 & 0 \\ 0 & -1 \end{bmatrix},$$

and let $\sigma = (\sigma_1, \sigma_2, \sigma_3)$. Then one checks that $[\sigma_1, \sigma_2] = 2i\sigma_3$ (+cyclic permutations) from which one finds that the map

$$x \mapsto \tilde{x} = \frac{1}{2i} x \cdot \sigma = \frac{1}{2}\begin{pmatrix} -ix^3 & -ix^1 - x^2 \\ -ix^1 + x^2 & ix^3 \end{pmatrix},$$

where $\mathbf{x} \cdot \sigma = x^1\sigma_1 + x^2\sigma_2 + x^3\sigma_3$ is a Lie algebra isomorphism between $\mathbb{R}^3$ and the (2×2) skew-Hermitian traceless matrices (the Lie algebra of $SU(2)$); that is, $[\tilde{x}, \tilde{y}] = (x \times y)\tilde{}$. Note that

$$-\det(\mathbf{x} \cdot \sigma) = \|\mathbf{x}\|^2, \quad \text{and} \quad \operatorname{trace}(\tilde{x}\tilde{y}) = -\frac{1}{2} x \cdot y.$$

Define $\pi : SU(2) \to GL(3, \mathbb{R})$ by

$$(\pi(A)(\mathbf{x})) \cdot \sigma = A(\mathbf{x} \cdot \sigma)A^\dagger = A(\mathbf{x} \cdot \sigma)A^{-1}.$$

Since $\det(A(\mathbf{x} \cdot \sigma)A^{-1}) = \det(\mathbf{x} \cdot \sigma)$, it follows that

$$\pi(SU(2)) \subset O(3).$$

But $\pi(SU(2))$ is connected, being the continuous image of a connected space, and so

$$\pi(SU(2)) \subset SO(3).$$

From the definition, one sees that $\pi(A) = \pi(B)$ iff $A = \pm B$. In fact, π is onto and is a local diffeomorphism. To show it is a local diffeomorphism, use the inverse function theorem. To show it is onto, observe that $T_e\pi : \alpha \mapsto \tilde{\alpha}$, where $\tilde{\alpha}(\mathbf{x} \cdot \sigma) = (\mathbf{x} \cdot \sigma)\alpha^\dagger + \alpha(\mathbf{x} \cdot \sigma)$, and use the fact that an open subgroup is also closed (its complement is a union of open cosets), and connectivity of $SO(3)$ to obtain $\pi(SU(2)) = SO(3)$. Therefore,

$$\pi : SU(2) \to SO(3)$$

is a 2 to 1 surjective submersion. Summarizing, we have the commutative diagram in Figure 9.2.1. Regarding S^3 as the unit sphere in $\mathbb{C}^2$ and letting S^1 act on $\mathbb{C}^2$ by rotating each factor, taking the quotient space gives a map $h : S^3 \to \mathbb{CP}^1$ called the ***Hopf fibration***.

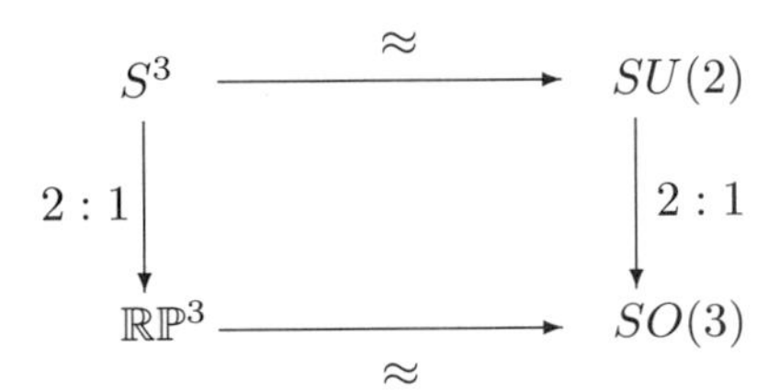

FIGURE 9.2.1. The link between $SU(2)$ and $SO(3)$.

This relation between $SU(2)$ and $SO(3)$ determined by the map π is related to the quaternionic representation of rotations, and is usually referred to as the ***Euler-Rodriguez parametrization***. This is important because it, unlike Euler angles, gives a singularity free representation that is of crucial importance in computational mechanics. We outline a few key points. We consider elements $(q^0, q^1, q^2, q^3) = (q^0, \mathbf{q}) \in \mathbb{R} \times \mathbb{R}^3$ with unit length; that is, $(q^0)^2 + \|\mathbf{q}\|^2 = 1$, defining $S^3 \subset \mathbb{R}^4$. (As above, $S^3 \cong SU(2)$.) The four-tuple (q^i) is a quaternion with ***scalar part*** q^0 and ***vector part*** $\mathbf{q}$. One usually writes

$$(q^0, \mathbf{q}) = q^0 + q^1\mathbf{i} + q^2\mathbf{j} + q^3\mathbf{k},$$

where $\mathbf{i}^2 = \mathbf{j}^2 = \mathbf{k}^2 = -1$ and $\mathbf{ij} = \mathbf{k}$ (and cyclic permutations thereof) defining the multiplicative structure. Let ω be given along with a unit vector $\mathbf{n}$. Then let

$$q^0 = \cos(\omega/2) \quad \text{and} \quad \mathbf{q} = \sin(\omega/2)\mathbf{n}. \tag{9.2.9}$$

Then Rodrigues' formula (9.2.5) reads

$$\exp(\omega\mathbf{n}) = [(q^0)^2 - \|\mathbf{q}\|]I + 2q^0\hat{\mathbf{q}} + 2\mathbf{q} \otimes \mathbf{q}, \tag{9.2.10}$$

where $\omega\mathbf{n} \in \mathbb{R}^3$ is thought of as an infinitesimal rotation. This expression then produces a rotation associated to each unit quaternion $(q^0, \mathbf{q})$. In addition, using this parametrization, Rodrigues [1840] found a beautiful way of expressing the product of two rotations $\exp(\omega_1 \boldsymbol{\eta}_1) \cdot \exp(\omega_2 \boldsymbol{\eta}_2)$ in terms of the given data. In fact, this was an early exploration of the spin group! We refer to Whittaker [1927], §7, Altmann [1986], Enos [1993], Simo and Lewis [1994] and references therein for further information.

9.3 Actions of Lie Groups

In this section we develop some basic facts about actions of Lie groups on manifolds. One of our main applications later will be the description of Hamiltonian systems with symmetry groups.

Definition 9.3.1 *Let M be a manifold and let G be a Lie group. A **(left) action** of a Lie group G on M is a smooth mapping $\Phi : G \times M \to M$ such that:*

i $\Phi(e, x) = x$ *for all* $x \in M$*; and*

ii $\Phi(g, \Phi(h, x)) = \Phi(gh, x)$ *for all* $g, h \in G$ *and* $x \in M$.

A ***right action*** is a map $\Psi : M \times G \to M$ that satisfies $\Psi(x, e) = x$ and $\Psi(\Psi(x, g), h) = \Psi(x, gh)$. We sometimes use the notation $g \cdot x = \Phi(g, x)$ for left actions, and $x \cdot g = \Psi(x, g)$ for right actions. In the infinite-dimensional case there are important situations where care with the smoothness is needed. For the formal development we assume we are in the Banach-Lie group context.

For every $g \in G$ let $\Phi_g : M \to M$ be given by $x \mapsto \Phi(g, x)$. Then **i** becomes $\Phi_e = \mathrm{id}_M$ while **ii** becomes $\Phi_{gh} = \Phi_g \circ \Phi_h$. Definition **9.3.1** can now be rephrased by saying that the map $g \mapsto \Phi_g$ is a homomorphism of G into $\mathrm{Diff}(M)$, the group of diffeomorphisms of M. In the special but important case where M is a Banach space V and each $\Phi_g : V \to V$ is a continuous linear transformation, the action Φ of G on V is called a ***representation*** of G on V.

Examples

(a) $SO(3)$ acts on $\mathbb{R}^3$ by $(A, x) \mapsto Ax$. This action leaves the two-sphere S^2 invariant, so the same formula defines an action of $SO(3)$ on S^2. ♦

(b) $GL(n, \mathbb{R})$ acts on $\mathbb{R}^n$ by $(A, x) \mapsto Ax$. ♦

(c) Let X be a complete vector field on M, that is, one for which the flow F_t of X is defined for all $t \in \mathbb{R}$. Then $F_t : M \to M$ defines an action of $\mathbb{R}$ on M. ♦

If Φ is an action of G on M and $x \in M$, the ***orbit*** of x is defined by

$$\mathrm{Orb}(x) = \{\Phi_g(x) \mid g \in G\} \subset M.$$

In finite dimensions one can show that $\mathrm{Orb}(x)$ is an immersed submanifold of M (Abraham and Marsden [1978, p. 265]). For $x \in M$, the ***isotropy*** (or ***stabilizer*** or ***symmetry***) group of Φ at x is given by

$$G_x := \{g \in G \mid \Phi_g(x) = x\} \subset G.$$

Since the map $\Phi_x : G \to M$ defined by $\Phi_x(g) = \Phi(g, x)$ is continuous, $G_x = \Phi_x^{-1}(x)$ is a closed subgroup and hence a Lie subgroup of G. The manifold structure of $\mathrm{Orb}(x)$ is defined by requiring the bijective map $[g] \in G/G_x \mapsto g \cdot x \in \mathrm{Orb}(x)$ to be a diffeomorphism. That G/G_x is a smooth manifold follows from Proposition **9.3.2**, which is discussed below.

An action is said to be:

1. ***transitive*** if there is only one orbit or, equivalently, if for every $x, y \in M$ there is a $g \in G$ such that $g \cdot x = y$;

2. ***effective*** (or ***faithful***) if $\Phi_g = \mathrm{id}_M$ implies $g = e$; that is, $g \mapsto \Phi_g$ is one-to-one; and

3. ***free*** if it has no fixed points, that is, $\Phi_g(x) = x$ implies $g = e$ or, equivalently, if for each $x \in M$, $g \mapsto \Phi_g(x)$ is one-to-one. Note that an action is free iff $G_x = \{e\}$ for all $x \in M$, and that every free action is faithful.

Examples

(a) Left translation $L_g : G \to G$; $h \mapsto gh$, defines a transitive and free action of G on itself. Note that right multiplication $R_g : G \to G$, $h \mapsto hg$, does not define a left action because $R_{gh} = R_h \circ R_g$, so that $g \mapsto R_g$ is an antihomomorphism. However, $g \mapsto R_g$ does define a right action, while $g \mapsto R_{g^{-1}}$ defines a left action of G on itself. ♦

(b) G acts on G by conjugation, $g \mapsto I_g = R_{g^{-1}} \circ L_g$. The map $I_g : G \to G$ given by $h \mapsto ghg^{-1}$ is the ***inner automorphism*** associated with g. Orbits of this action are called ***conjugacy classes*** or, in the case of matrix groups, ***similarity classes***. ♦

(c) Differentiating conjugation at e, we get the ***adjoint representation*** of G on $\mathfrak{g}$:

$$\mathrm{Ad}_g := T_e I_g : T_e G = \mathfrak{g} \to T_e G = \mathfrak{g}.$$

Explicitly, the adjoint action of G on $\mathfrak{g}$ is given by

$$\mathrm{Ad} : G \times \mathfrak{g} \to \mathfrak{g}, \quad \mathrm{Ad}_g(\xi) = T_e(R_{g^{-1}} \circ L_g)\xi.$$

For example, for $SO(3)$ we have $I_A(B) = ABA^{-1}$, so differentiating with respect to B at B = identity gives $\mathrm{Ad}_A \hat{\mathbf{v}} = A\hat{\mathbf{v}}A^{-1}$. However, $(\mathrm{Ad}_A \hat{\mathbf{v}})(\mathbf{w}) = A\hat{\mathbf{v}}(A^{-1}\mathbf{w}) = A(\mathbf{v} \times A^{-1}\mathbf{w}) = A\mathbf{v} \times \mathbf{w}$, so $(\mathrm{Ad}_A \hat{\mathbf{v}}) = (A\mathbf{v})\hat{}$. Identifying $\mathfrak{so}(3) \cong \mathbb{R}^3$, we get $\mathrm{Ad}_A \mathbf{v} = A\mathbf{v}$. ♦

(d) The ***coadjoint action*** of G on $\mathfrak{g}^*$, the dual of the Lie algebra $\mathfrak{g}$ of G, is defined as follows. Let $\mathrm{Ad}_g^* : \mathfrak{g}^* \to \mathfrak{g}^*$ be the dual of Ad_g, defined by

$$\left\langle \mathrm{Ad}_g^* \alpha, \xi \right\rangle = \langle \alpha, \mathrm{Ad}_g \xi \rangle$$

for $\alpha \in \mathfrak{g}^*$, and $\xi \in \mathfrak{g}$. Then the map

$$\Phi^* : G \times \mathfrak{g}^* \to \mathfrak{g}^* \quad \text{given by} \quad (g, \alpha) \mapsto \mathrm{Ad}_{g^{-1}}^* \alpha$$

is the coadjoint action of G on $\mathfrak{g}^*$. The corresponding ***coadjoint representation*** of G on $\mathfrak{g}^*$ is denoted

$$\mathrm{Ad}^* : G \to GL(\mathfrak{g}^*, \mathfrak{g}^*), \quad \mathrm{Ad}_{g^{-1}}^* = \left(T_e(R_g \circ L_{g^{-1}})\right)^* .$$

We will avoid the introduction of yet another * by writing $(\mathrm{Ad}_{g^{-1}})^*$ or simply $\mathrm{Ad}_{g^{-1}}^*$, where * denotes the usual linear-algebraic dual, rather than $\mathrm{Ad}^*(g)$, in which * is simply part of the name of the function Ad^*. Any representation of G on a vector space V similarly induces a ***contragredient representation*** of G on V^*. ♦

An action of Φ of G on a manifold M defines an equivalence relation on M by the relation of belonging to the same orbit; explicitly, for $x, y \in M$, we write $x \sim y$ if there exists a $g \in G$ such that $g \cdot x = y$, that is if $y \in \mathrm{Orb}(x)$ (and hence $x \in \mathrm{Orb}(y)$). We let M/G be the set of these equivalence classes, that is, the set of orbits, sometimes called the ***orbit space***. Let $\pi : M \to M/G : x \mapsto \mathrm{Orb}(x)$, and give M/G the quotient topology by defining $U \subset M/G$ to be open if and only if $\pi^{-1}(U)$ is open in M. To guarantee that the orbit space M/G has a smooth manifold structure, further conditions on the action are required.

An action $\Phi : G \times M \to M$ is called ***proper*** if the mapping $\tilde{\Phi} : G \times M \to M \times M$, defined by $\tilde{\Phi}(g, x) = (x, \Phi(g, x))$, is proper. In finite dimensions this means that if $K \subset M \times M$ is compact, then $\tilde{\Phi}^{-1}(K)$ is compact. In general, this means that if $\{x_n\}$ is a convergent sequence in M and $\Phi_{g_n} x_n$ converges in M, then $\{g_n\}$ has a convergent subsequence in G. For instance, if G is compact, this condition is automatically satisfied. Orbits of proper Lie group actions are closed and hence embedded submanifolds. The next proposition gives a useful sufficient condition for M/G to be a smooth manifold.

Proposition 9.3.2 *If $\Phi : G \times M \to M$ is a proper and free action, then M/G is a smooth manifold and $\pi : M \to M/G$ is a smooth submersion.*

For the proof, we refer to Abraham and Marsden [1978], Proposition 4.2.23. (In infinite dimensions one uses these ideas but additional technicalities often arise; see Ebin [1970] and Isenberg and Marsden [1982].) The idea of the chart construction for M/G is based on the following observation. If $x \in M$, then there is an isomorphism φ_x of $T_{\pi(x)}(M/G)$ with the quotient space $T_xM/T_x\operatorname{Orb}(x)$. Moreover, if $y = \Phi_g(x)$, then $T_x\Phi_g$ induces an isomorphism $\psi_{x,y} : T_xM/T_x\operatorname{Orb}(x) \to T_yM/T_y\operatorname{Orb}(y)$ satisfying $\varphi_y \circ \psi_{x,y} = \varphi_x$.

Examples

(a) $G = \mathbb{R}$ acts on $M = \mathbb{R}$ by translations; explicitly, $\Phi : G \times M \to M$, $\Phi(s,x) = x + s$. Then for $x \in \mathbb{R}$, $\operatorname{Orb}(x) = \mathbb{R}$. Hence M/G is a single point and the action is transitive, proper, and free. ♦

(b) $G = SO(3), M = \mathbb{R}^3$ $(\cong \mathfrak{so}(3)^*)$. Consider the action for $\mathbf{x} \in \mathbb{R}^3$ and $A \in SO(3)$ given by $\Phi_A\mathbf{x} = A\mathbf{x}$ for $\mathbf{x} \in \mathbb{R}^3$ and $A \in SO(3)$. Then $\operatorname{Orb}(x) = \{\mathbf{y} \in \mathbb{R}^3 \mid \|\mathbf{y}\| = \|\mathbf{x}\|\}$ = a sphere of radius $\|\mathbf{x}\|$. Hence $M/G \cong \mathbb{R}^+$. The set $\mathbb{R}^+ = \{r \in \mathbb{R} \mid r \geq 0\}$ is not a manifold because it includes the endpoint $r = 0$. Indeed, the action is not free, since it has the fixed point $\mathbf{0} \in \mathbb{R}^3$. ♦

(c) Let G be abelian. Then $\operatorname{Ad}_g = \operatorname{id}_{\mathfrak{g}}$, $\operatorname{Ad}^*_{g^{-1}} = \operatorname{id}_{\mathfrak{g}^*}$ and the adjoint and coadjoint orbits of $\xi \in \mathfrak{g}$ and $\alpha \in \mathfrak{g}^*$, respectively, are the one-point sets $\{\xi\}$ and $\{\alpha\}$. ♦

We will see later that coadjoint orbits can be natural phase spaces for some mechanical systems like the rigid body; in particular, they are always even dimensional.

Next we turn to the infinitesimal description of an action, which will be a crucial concept for mechanics.

Definition 9.3.3 *Suppose $\Phi : G \times M \to M$ is an action. For $\xi \in \mathfrak{g}$, the map $\Phi^\xi : \mathbb{R} \times M \to M$, defined by $\Phi^\xi(t,x) = \Phi(\exp t\xi, x)$, is an $\mathbb{R}$-action on M. In other words, $\Phi_{\exp t\xi} : M \to M$ is a flow on M. The corresponding vector field on M, given by*

$$\xi_M(x) := \left.\frac{d}{dt}\right|_{t=0} \Phi_{\exp t\xi}(x),$$

*is called the **infinitesimal generator** of the action corresponding to ξ.*

Proposition 9.3.4 *The tangent space at x to an orbit $\operatorname{Orb}(x_0)$ is*

$$T_x\operatorname{Orb}(x_0) = \{\xi_M(x) \mid \xi \in \mathfrak{g}\},$$

where $\operatorname{Orb}(x_0)$ is endowed with the manifold structure making $G/G_{x_0} \to \operatorname{Orb}(x_0)$ into a diffeomorphism.

The idea is as follows: Let $\sigma_\xi(t)$ be a curve in G tangent to ξ at $t = 0$. Then the map $\Phi^\xi_x(t) = \Phi_{\sigma_\xi(t)}(x)$ is a smooth curve in $\operatorname{Orb}(x_0)$ with $\Phi^\xi_x(0) = x$. Hence

$$\left.\frac{d}{dt}\right|_{t=0} \Phi^\xi_x(t) = \left.\frac{d}{dt}\right|_{t=0} \Phi_{\sigma_\xi(t)}(x) = \xi_M(x)$$

is a tangent vector at x to $\operatorname{Orb}(x_0)$. Furthermore, each tangent vector is obtained in this way since tangent vectors are equivalence classes of such curves.

The Lie algebra of the isotropy group G_x, $x \in M$, called the ***isotropy*** (or ***stabilizer***, or ***symmetry***) ***algebra at*** x equals, by Proposition **9.1.10**, $\mathfrak{g}_x = \{\xi \in \mathfrak{g} \mid \xi_M(x) = 0\}$.

Examples

(a) The infinitesimal generators for the adjoint action are computed as follows. Let $\operatorname{Ad} : G \times \mathfrak{g} \to \mathfrak{g}$, $\operatorname{Ad}_g(\eta) = T_e(R_{g^{-1}} \circ L_g)(\eta)$. For $\xi \in \mathfrak{g}$, we compute the corresponding infinitesimal generator $\xi_{\mathfrak{g}}$. By definition, $\xi_{\mathfrak{g}}(\eta) = (d/dt)|_{t=0} \operatorname{Ad}_{\exp t\xi}(\eta)$. By (9.1.5), this equals $[\xi, \eta]$. Thus, for the adjoint action,

$$\xi_{\mathfrak{g}} = \operatorname{ad}_\xi; \quad \text{i.e.,} \quad \xi_{\mathfrak{g}}(\eta) = [\xi, \eta]. \quad \blacklozenge \tag{9.3.1}$$

(b) We illustrate **(a)** for the group $SO(3)$ as follows. Let $A(t) = \exp(tC)$, where $C \in \mathfrak{so}(3)$; then $A(0) = I$ and $A'(0) = C$. Thus with $B \in \mathfrak{so}(3)$,

$$\begin{aligned} \left.\frac{d}{dt}\right|_{t=0} (\operatorname{Ad}_{\exp tC} B) &= \left.\frac{d}{dt}\right|_{t=0} (\exp(tC))B(\exp(tC))^{-1}) \\ &= \left.\frac{d}{dt}\right|_{t=0} (A(t)BA(t)^{-1}) \\ &= A'(0)BA^{-1}(0) + A(0)BA^{-1\prime}(0). \end{aligned}$$

Differentiating $A(t)A^{-1}(t) = I$, we find

$$\frac{d}{dt}(A^{-1}(t)) = -A^{-1}(t)A'(t)A^{-1}(t),$$

so that $A^{-1\prime}(0) = -A'(0) = -C$. Then the preceding equation becomes

$$\left.\frac{d}{dt}\right|_{t=0} (\operatorname{Ad}_{\exp tC} B) = CB - BC = [C, B],$$

as expected. $\blacklozenge$

(c) Let $\operatorname{Ad}^* : G \times \mathfrak{g}^* \to \mathfrak{g}^*$ be the coadjoint action $(g, \alpha) \mapsto \operatorname{Ad}^*_{g^{-1}} \alpha$. If $\xi \in \mathfrak{g}$, we compute for $\alpha \in \mathfrak{g}^*$ and $\eta \in \mathfrak{g}$

$$\begin{aligned} \langle \xi_{\mathfrak{g}^*}(\alpha), \eta \rangle &= \left\langle \left.\frac{d}{dt}\right|_{t=0} \operatorname{Ad}^*_{\exp(-t\xi)}(\alpha), \eta \right\rangle \\ &= \left.\frac{d}{dt}\right|_{t=0} \left\langle \operatorname{Ad}^*_{\exp(-t\xi)}(\alpha), \eta \right\rangle \\ &= \left.\frac{d}{dt}\right|_{t=0} \langle \alpha, \operatorname{Ad}_{\exp(-t\xi)} \eta \rangle = \left\langle \alpha, \left.\frac{d}{dt}\right|_{t=0} \operatorname{Ad}_{\exp(-t\xi)} \eta \right\rangle \\ &= \langle \alpha, -[\xi, \eta] \rangle = -\langle \alpha, \operatorname{ad}_\xi(\eta) \rangle = -\langle \operatorname{ad}^*_\xi(\alpha), \eta \rangle. \end{aligned}$$

Hence

$$\xi_{\mathfrak{g}^*} = -\mathrm{ad}^*_\xi, \quad \text{or} \quad \xi_{\mathfrak{g}^*}(\alpha) = -\langle \alpha, [\xi, \cdot] \rangle . \quad \blacklozenge \tag{9.3.2}$$

(d) Identifying $\mathfrak{so}(3) \cong (\mathbb{R}^3, \times)$ and $\mathfrak{so}(3)^* \cong \mathbb{R}^{3^*}$, using the pairing given by the standard Euclidean inner product, (9.3.2) reads

$$\xi_{\mathfrak{so}(3)^*}(l) = -l \cdot (\xi \times \cdot),$$

for $l \in \mathfrak{so}(3)^*$ and $\xi \in \mathfrak{so}(3)$. For $\eta \in \mathfrak{so}(3)$, we have

$$\left\langle \xi_{\mathfrak{so}(3)^*}(l), \eta \right\rangle = -l \cdot (\xi \times \eta) = -(l \times \xi) \cdot \eta = -\langle l \times \xi, \eta \rangle,$$

so that

$$\xi_{\mathbb{R}^3}(l) = -l \times \xi = \xi \times l.$$

As expected, $\xi_{\mathbb{R}^3}(l) \in T_l \operatorname{Orb}(l)$ is tangent to $\operatorname{Orb}(l)$ (see Figure 9.3.1). Allowing ξ to vary in $\mathfrak{so}(3) \cong \mathbb{R}^3$, one obtains all of $T_l \operatorname{Orb}(l)$, consistent with Proposition **9.3.4**. $\blacklozenge$

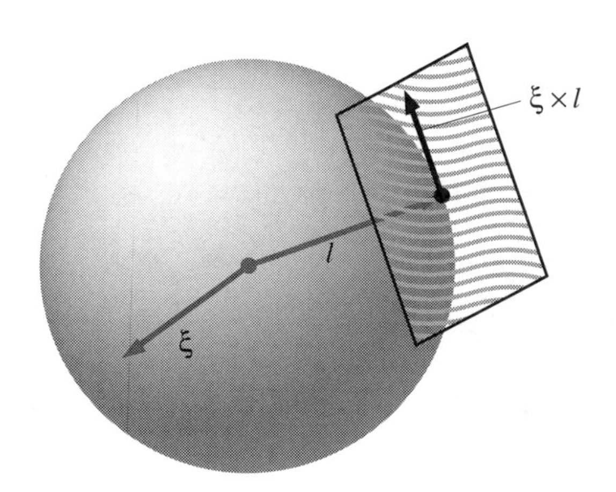

FIGURE 9.3.1. $\xi_{\mathbb{R}^3}(l)$ is tangent to $\operatorname{Orb}(l)$.

Definition 9.3.5 *Let M and N be manifolds and let G be a Lie group which acts on M by $\Phi_g : M \to M$, and on N by $\Psi_g : N \to N$. A smooth map $f : M \to N$ is called* ***equivariant*** *with respect to these actions if, for all $g \in G$,*

$$f \circ \Phi_g = \Psi_g \circ f, \tag{9.3.3}$$

that is, if the diagram in Figure 9.3.2 *commutes.*

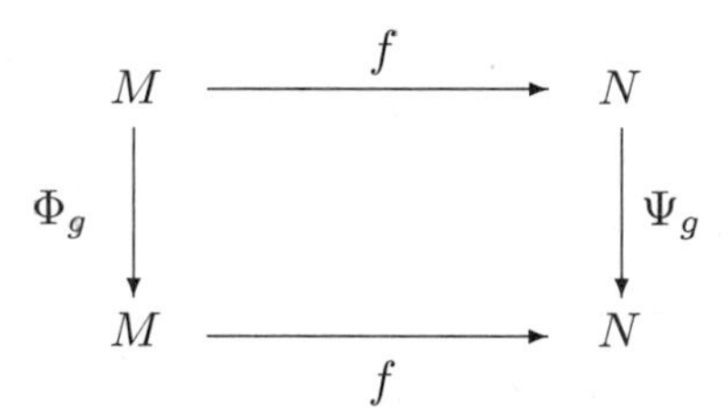

FIGURE 9.3.2. Commutative diagram for equivariance.

Setting $g = \exp(t\xi)$ and differentiating (9.3.3) with respect to t at $t = 0$ gives $Tf \circ \xi_M = \xi_N \circ f$. In other words, ξ_M and ξ_N are f-related. In particular, *if f is an equivariant diffeomorphism, then $f^*\xi_N = \xi_M$.*

Also note that if M/G and N/G are both smooth manifolds with the canonical projections smooth submersions, an equivariant map $f : M \to N$ induces a smooth map $f_G : M/G \to N/G$.

Proposition 9.3.6 *Let the Lie group G act on the left on the manifold M. Then the infinitesimal generator map $\xi \mapsto \xi_M$ of the Lie algebra $\mathfrak{g}$ of G into the Lie algebra $\mathfrak{X}(M)$ of vector fields of M is a Lie algebra antihomomorphism; i.e., $(a\xi + b\eta)_M = a\xi_M + b\eta_M$ and*

$$[\xi_M, \eta_M] = -[\xi, \eta]_M$$

for all $\xi, \eta \in \mathfrak{g}$, and $a, b \in \mathbb{R}$.

To prove this, we use the following lemma:

Lemma 9.3.7

i *Let $c(t)$ be a curve in G, $c(0) = e$, $c'(0) = \xi \in \mathfrak{g}$. Then*

$$\xi_M(x) = \left.\frac{d}{dt}\right|_{t=0} \Phi_{c(t)}(x).$$

ii *For every $g \in G$,*

$$(\mathrm{Ad}_g\, \xi)_M = \Phi^*_{g^{-1}} \xi_M.$$

Proof

i Let $\Phi_x : G \to M$ be the map $\Phi_x(g) = \Phi(g, x)$. Then since Φ_x is smooth, the definition of the infinitesimal generator says that $T_e\Phi_x(\xi) = \xi_M(x)$. Thus **i** follows by the chain rule.

ii We have

$$\begin{aligned}(\mathrm{Ad}_g \xi)_M(x) &= \left.\frac{d}{dt}\right|_{t=0} \Phi(\exp(t\,\mathrm{Ad}_g \xi), x) \\ &= \left.\frac{d}{dt}\right|_{t=0} \Phi(g(\exp t\xi)g^{-1}, x) \quad \text{(by Corollary \textbf{9.1.7})} \\ &= \left.\frac{d}{dt}\right|_{t=0} \Phi_g(\exp t\xi, \Phi_{g^{-1}}(x)) \\ &= T_{\Phi_g^{-1}(x)}\Phi_g\left(\xi_M\left(\Phi_{g^{-1}}(x)\right)\right) \\ &= \left(\Phi^*_{g^{-1}}\xi_M\right)(x). \quad \blacksquare\end{aligned}$$

Proof of Proposition 9.3.6 Linearity follows since $\xi_M(x) = T_e\Phi_x(\xi)$. To prove the second relation, put $g = \exp t\eta$ in **ii** of the lemma to get

$$(\mathrm{Ad}_{\exp t\eta}\,\xi)_M = \Phi^*_{\exp(-t\eta)}\xi_M.$$

But $\Phi_{\exp(-t\eta)}$ is the flow of $-\eta_M$, so differentiating at $t = 0$ the right-hand side gives $[\xi_M, \eta_M]$. The derivative of the left-hand side at $t = 0$ equals $[\eta, \xi]_M$ by the preceding Example **(a)**. ■

In view of this theorem one defines a left ***Lie algebra action*** of a manifold M as a Lie algebra antihomomorphism $\xi \in \mathfrak{g} \mapsto \xi_M \in \mathfrak{X}(M)$, such that the mapping $(\xi, x) \in \mathfrak{g} \times M \mapsto \xi_M(x) \in TM$ is smooth.

Next we discuss the connectivity of some classical groups. First we state two facts about homogeneous spaces:

1. If H is a closed normal subgroup of the Lie group G (that is, if $h \in H$ and $g \in G$, then $ghg^{-1} \in H$), then the quotient G/H is a Lie group and the natural projection $\pi : G \to G/H$ is a smooth group homomorphism. (This follows from Proposition **9.3.2**; see also Varadarajan [1974] Theorem 2.9.6, p. 80.) Moreover, if H and G/H are connected then G is connected.

2. Let G, M be finite-dimensional and second countable and let $\Phi : G \times M \to M$ be a transitive action of G on M and for $x \in M$, let G_x be the isotropy subgroup of x. Then the map $gG_x \mapsto \Phi_g(x)$ is a diffeomorphism of G/G_x onto M. (This follows from Proposition **9.3.2**; see also Varadarajan [1974], Theorem 2.9.4, p. 77.)

The action $\Phi : GL(n, \mathbb{R}) \times \mathbb{R}^n \to \mathbb{R}^n$, $\Phi(A, x) = Ax$, restricted to $O(n) \times S^{n-1}$ induces a transitive action. The isotropy subgroup of $O(n)$ at $e_n \in S^{n-1}$ is $O(n-1)$. Clearly $O(n-1)$ is a closed subgroup of $O(n)$ by embedding any $A \in O(n-1)$ as

$$\tilde{A} = \begin{bmatrix} A & 0 \\ 0 & 1 \end{bmatrix} \in O(n),$$

and the elements of $O(n-1)$ leave e_n fixed. On the other hand, if $A \in O(n)$ and $A(e_n) = e_n$, then $A \in O(n-1)$. It follows from 2 that the map

$$O(n)/O(n-1) \to S^{n-1} : A \cdot O(n-1) \mapsto A(e_n)$$

is a diffeomorphism. By a similar argument, there is a diffeomorphism

$$S^{n-1} \cong SO(n)/SO(n-1).$$

The natural action of $GL(n, \mathbb{C})$ on $\mathbb{C}^n$ similarly induces a diffeomorphism of $S^{2n-1} \subset \mathbb{R}^{2n}$ with the homogeneous space $U(n)/U(n-1)$. Moreover, we get $S^{2n-1} \cong SU(n)/SU(n-1)$. In particular, since $SU(1)$ consists only of the 1×1 identity matrix, S^3 is diffeomorphic with $SU(2)$, a fact already proved at the end of §9.2.

Proposition 9.3.8 *Each of the Lie groups $SO(n)$, $SU(n)$, and $U(n)$ is connected for $n \geq 1$, and $O(n)$ has two components.*

Proof $SO(1)$ and $SU(1)$ are connected since both consist only of the 1×1 identity matrix and $U(1)$ is connected since $U(1) = \{z \in C \mid |z| = 1\} = S^1$. That $SO(n)$, $SU(n)$, and $U(n)$ are connected for all n now follows from fact 1 above, using induction on n and the representation of the spheres as homogeneous spaces. Since every matrix A in $O(n)$ has determinant ± 1, the orthogonal group can be written as the union of two nonempty disjoint connected open subsets as follows: $O(n) = SO(n) \cup A \cdot SO(n)$ where $A = \operatorname{diag}(-1, 1, 1, \ldots, 1)$. Thus $O(n)$ has two components. ■

Proposition 9.3.9 *$GL(n, \mathbb{R})$ has two components.*

Proof Consider the following two disjoint homeomorphic open subsets of $GL(n, \mathbb{R})$:

$$GL(n, \mathbb{R})^+ = \{A \in GL(n, \mathbb{R}) \mid \det A > 0\}$$

and

$$GL(n, \mathbb{R})^- = \{B \in GL(n, \mathbb{R}) \mid \det B < 0\}.$$

It suffices to prove that (the subgroup) $GL(n, \mathbb{R})^+$ is connected. To do this we show that each element of $GL(n, \mathbb{R})^+$ can be joined to the identity matrix I by a continuous curve. Recall that each $A \in GL(n, \mathbb{R})$ has a polar decomposition $A = PR$, where P is a positive-definite symmetric matrix and $R \in O(n)$. If $A \in GL(n, \mathbb{R})^+$, then R must have positive determinant, that is, $R \in SO(n)$. Let $P_t = tI + (1-t)P$, $t \in [0, 1]$. Then P_t is positive-definite for each t, so the path $t \mapsto P_t R$ is a continuous curve in $GL(n, \mathbb{R})^+$ joining A to R. Since $SO(n)$ is connected, and therefore pathwise connected, R can be joined to I by a continuous curve. Thus $GL(n, \mathbb{R})^+$ is pathwise connected. ■

Remarks

1. Let $\Phi : G \times G \to G$ be a left translation, $\Phi(g,h) = L_g h$. For $\xi \in \mathfrak{g}$, let Y_ξ be the corresponding *right* invariant vector field on G. Then $\xi_G(g) = Y_\xi(g) = T_e R_g(\xi)$, and similarly, the infinitesimal generator of a *right translation* is the *left invariant vector field* $g \mapsto T_e L_g(\xi)$.

2. Let $x(t)$ be a smooth curve in the Lie group G and consider the curve $\xi(t) = \operatorname{Ad}_{x(t)} \xi$ for $\xi \in \mathfrak{g}$. Then

$$\frac{d\xi(t)}{dt} = \left[T_{x(t)} R_{x(t)^{-1}} \frac{dx(t)}{dt}, \xi(t) \right].$$

Thus

$$T_g(\operatorname{Ad}_{\cdot}\, \eta)(v_g) = \left[T_g R_{g^{-1}}(v_g), \operatorname{Ad}_g \eta \right],$$

for $\eta \in \mathfrak{g}$, $g \in G$, and $v_g \in T_g G$. To see this, use $(d/dt)\operatorname{Ad}_{\exp t\xi}(\eta) = [\xi, \eta]$ and take the derivative of the relation $\operatorname{Ad}_{hg} \eta = \operatorname{Ad}_h(\operatorname{Ad}_g \eta)$ at $h = e$.

3. If $x(t)$ is a smooth curve in G and $\mu \in \mathfrak{g}^*$, the derivative of the curve $\mu(t) = \operatorname{Ad}^*_{x(t)} \mu$ in $\mathfrak{g}^*$ has the expression

$$\frac{d\mu(t)}{dt} = \operatorname{ad}\left(T_x L_{x(t)^{-1}} \frac{dx(t)}{dt} \right)^* \mu(t).$$

We conclude that

$$T_g(\operatorname{Ad}^*_{\cdot} \mu)(v_g) = \operatorname{Ad}(T_g L_{g^{-1}}(v_g))^* \operatorname{Ad}^*_g \mu = \operatorname{Ad}^*_g \operatorname{ad}(T_g L_{g^{-1}}(v_g))^* \mu,$$

for $\mu \in \mathfrak{g}^*$, $g \in G$, and $v_g \in T_g G$. [One uses the identity $\operatorname{Ad}_g \circ \operatorname{ad}\xi = \operatorname{ad}(\operatorname{Ad}_g \xi) \circ \operatorname{Ad}_g$.]

4. The formulae in 2 and 3 change in the following manner, if g is replaced by g^{-1}:

$$\begin{aligned}
\frac{d}{dt}(\operatorname{Ad}_{x(t)^{-1}} \xi) &= -\left[T_{x(t)^{-1}} L_{x(t)^{-1}} \frac{dx(t)}{dt}, \operatorname{Ad}_{x(t)^{-1}} \xi \right] \\
T_g(\operatorname{Ad}^{-1}_{\cdot} \xi)(v_g) &= -\left[T_g L_{g^{-1}}(v_g), \operatorname{Ad}_{g^{-1}} \xi \right] \\
\frac{d}{dt}(\operatorname{Ad}^*_{x(t)^{-1}} \mu) &= -\operatorname{ad}\left[T_{x(t)} R_{x(t)^{-1}} \frac{dx(t)}{dt} \right]^* \operatorname{Ad}^*_{x(t)^{-1}} \mu \\
T_g(\operatorname{Ad}^{-1}_{\cdot} \mu) &= -\operatorname{ad}\left(T_g R_{g^{-1}}(v_g) \right)^* \operatorname{Ad}^*_{g^{-1}} \mu \\
&= \operatorname{ad}\left(T_g L_{g^{-1}}(v_g) \right)^* \mu.
\end{aligned}$$

[One can deduce these using the fact that if $I : G \to G$ denotes the inversion mapping $g \mapsto g^{-1}$, then $T_g I(v_g) = -\left(T_e L_{g^{-1}} \circ T_g R_{g^{-1}} \right)(v_g)$.] ♦

Examples

(a) Let E be a finite-dimensional vector space with a bilinear form $\langle\, ,\rangle$. Let G be the group of *isometries* of E, that is, F is an isomorphism of E onto E and $\langle Fe, Fe'\rangle = \langle e, e'\rangle$ for all e, and $e' \in E$. Then G is a subgroup and a closed submanifold of $GL(E)$. The Lie algebra of G is

$$\{K \in L(E) \mid \langle Ke, e'\rangle + \langle e, Ke'\rangle = 0 \quad \text{for all} \quad e, e' \in E\}.$$

(b) If $\langle\, ,\rangle$ denotes the Minkowski metric on $\mathbb{R}^4$, that is,

$$\langle x, y\rangle = \sum_{i=1}^{3} x^i y^i - x^4 y^4,$$

then the group of linear isometries is called the ***Lorentz group*** L. The dimension of L is six and L has four connected components. If

$$S = \begin{bmatrix} I_3 & 0 \\ 0 & -1 \end{bmatrix} \in GL(4, \mathbb{R}),$$

then

$$L = \{A \in GL(4, \mathbb{R}) \mid A^T S A = S\}$$

and so the Lie algebra of L is

$$\mathfrak{l} = \{B \in L(\mathbb{R}^4, \mathbb{R}^4) \mid SA + A^T S = 0\}.$$

The identity component of L is $\{A \in L \mid \det A > 0 \text{ and } A_{44} > 0\} = L_{\uparrow}^{+}$; L and $L_{\uparrow}^{+}$ are not compact.

(c) Consider the (closed) subgroup G of $GL(5, \mathbb{R})$ that consists of matrices with the following block structure:

$$\{R, v, a, \tau\} := \begin{bmatrix} R & v & a \\ 0 & 1 & \tau \\ 0 & 0 & 1 \end{bmatrix},$$

where $R \in SO(3)$, $v, a \in \mathbb{R}^3$, and $\tau \in \mathbb{R}$. This group is called the ***Galilean group***. Its Lie algebra is a subalgebra of $L(\mathbb{R}^5, \mathbb{R}^5)$ given by the set of matrices of the form

$$\{\omega, u, \alpha, \theta\} := \begin{bmatrix} \hat{\omega} & u & \alpha \\ 0 & 0 & \theta \\ 0 & 0 & 0 \end{bmatrix},$$

where $\omega, u, \alpha \in \mathbb{R}^3$, and $\theta \in \mathbb{R}$. Obviously the Galilean group acts naturally on $\mathbb{R}^5$; moreover it acts naturally on $\mathbb{R}^4$, embedded as the following G-invariant subset of $\mathbb{R}^5$:

$$\begin{bmatrix} x \\ t \end{bmatrix} \mapsto \begin{bmatrix} x \\ t \\ 1 \end{bmatrix},$$

where $x \in \mathbb{R}^3$ and $t \in \mathbb{R}$. Concretely, the action of $\{R, v, a, \tau\}$ on (x, t) is given by

$$(x, t) \mapsto (Rx + tv + a, t + \tau).$$

Thus, the Galilean group gives a change of frame of reference (unaffecting the "absolute time" variable) by rotations (R), space translations (a), time translations (τ), and going to a moving frame, or boosts (v).

Coadjoint Isotropy Subalgebras Are Generically Abelian (Optional)

The aim of this supplement is to prove a theorem of Duflo and Vergne [1969] showing that, generically, the isotropy algebras for the coadjoint action are abelian. A very simple example is $G = SO(3)$. Here $\mathfrak{g}^* \cong \mathbb{R}^3$ and $G_\mu = S^1$ for $\mu \in \mathfrak{g}^*$ and $\mu \neq 0$, and $G_0 = SO(3)$. Thus, G_μ is abelian on the open dense set $\mathfrak{g}^* \backslash \{0\}$.

To prepare for the proof, we shall develop some tools.

If V is a finite-dimensional vector space, a subset $A \subset V$ is called ***algebraic*** if it is the common zero set of a finite number of polynomial functions on V. It is easy to see that if V_i is the zero set of a finite collection of polynomials C_i, for $i = 1, 2$, then $A_1 \cup A_2$ is the zero set of the collection $C_1 C_2$ formed by all products of an element in C_1 with an element in C_2. The whole space V is the zero set of the constant polynomial equal to 1. Finally, if A_α is the algebraic set given as the common zeros of some finite collection of polynomials C_α, where α ranges over some index set, then $\bigcap_\alpha A_\alpha$ is the zero set of the collection $\bigcup_\alpha C_\alpha$. This zero set can also be given as the common zeros of a *finite* collection of polynomials since the zero set of any collection of polynomials coincides with the zero set of the ideal in the polynomial ring generated by this collection and any ideal in the polynomial ring over $\mathbb{R}$ is finitely generated (we accept this from algebra). Thus, the collection of algebraic sets in V satisfies the axioms of the collection of closed sets of a topology which is called the ***Zariski topology*** of V.

Thus, the open sets of this topology are the complements of the algebraic sets. For example, the algebraic sets of $\mathbb{R}$ are just the finite sets, since every polynomial in $\mathbb{R}[X]$ has finitely many real roots (or none at all). Granting that we have a topology (the hard part), let us show that *any Zariski open set in V is open and dense in the usual topology.* Openness is clear, since algebraic sets are necessarily closed in the usual topology as inverse images of 0 by a continuous map. To show that a Zariski open set U is also dense, suppose the contrary, namely, that if $x \in V \backslash U$, then there is a neighborhood $U_1 \times U_2$ of x in the usual topology such that $(U_1 \times U_2) \cap U = \varnothing$ and $U_1 \subset \mathbb{R}, U_2 \subset V_2$ are open, where $V = \mathbb{R} \times V_2$, the splitting being achieved by the choice of a basis. Since $x \in V \backslash U$, there is a finite collection of polynomials $p_1, \ldots, p_N \in \mathbb{R}[X_1, \ldots, X_n], n = \dim V$, that vanishes identically on $U_1 \times U_2$. If $x = (x_1, \ldots, x_n) \in V$, then the polynomials $q_i(X_1) = p_i(X_1, x_2, \ldots, x_n) \in \mathbb{R}[X_1]$ all vanish identically on the open set $U_1 \subset \mathbb{R}$, which is impossible since each q_i has at most a finite number of roots. Therefore $(U_1 \times U_2) \cap U = \varnothing$ is absurd and hence U must be dense in V.

Theorem 9.3.10 (Duflo and Vergne [1969]) *Let $\mathfrak{g}$ be a finite-dimensional Lie algebra with dual $\mathfrak{g}^*$ and let $r = \min\{\dim \mathfrak{g}_\mu \mid \mu \in \mathfrak{g}^*\}$. The set $\{\mu \in \mathfrak{g}^* \mid \dim \mathfrak{g}_\mu = r\}$ is Zariski open and thus open and dense in the usual topology of $\mathfrak{g}^*$. If $\dim \mathfrak{g}_\mu = r$, then $\mathfrak{g}_\mu$ is abelian.*

Proof (Due to J. Carmona, as presented in Rais [1972]) Define $\varphi_\mu : g \in G \mapsto \mathrm{Ad}^*_{g^{-1}} \mu \in \mathfrak{g}^*$. This is a smooth map whose range is the coadjoint orbit $\mathcal{O}_\mu$ through μ and whose tangent map at the identity is $T_e\varphi_\mu(\xi) = -\mathrm{ad}^*_\xi \mu$. Note that $\ker T_e\varphi_\mu = \mathfrak{g}_\mu$ and $\mathrm{range}\, T_e\varphi_\mu = T_\mu\mathcal{O}_\mu$. Thus, if $n = \dim \mathfrak{g}$, we have $\mathrm{rank}\, T_e\varphi_\mu = n - \dim \mathfrak{g}_\mu \leq n - r$ since $\dim \mathfrak{g}_\mu \geq r$ for all $\mu \in \mathfrak{g}^*$. Therefore $U = \{\mu \in \mathfrak{g}^* \mid \dim \mathfrak{g}_\mu = r\} = \{\mu \in \mathfrak{g}^* \mid \mathrm{rank}(T_e\varphi_\mu) = n - r\}$ and $n - r$ is the maximal possible rank of all the linear maps $T_e\varphi_\mu : \mathfrak{g} \to \mathfrak{g}^*, \mu \in \mathfrak{g}^*$. Now choose a basis in $\mathfrak{g}$ and induce the natural bases on $\mathfrak{g}^*$ and $L(\mathfrak{g}, \mathfrak{g}^*)$. Let $S_i = \{\mu \in \mathfrak{g}^* \mid \mathrm{rank}\, T_e\varphi_\mu = n - r - i\}, 1 \leq i \leq n - r$. Then S_i is the zero set of the polynomials in μ obtained by taking all determinants of the $(n - r - i + 1)$-minors of the matrix representation of $T_e\varphi_\mu$ in these bases. Thus S_i is an algebraic set. Since $\bigcup_{i=1}^{n-r} S_i$ is the complement of U, if follows that U is a Zariski open set in $\mathfrak{g}^*$, and hence open and dense in the usual topology of $\mathfrak{g}^*$.

Now let $\mu \in \mathfrak{g}^*$ be such that $\dim \mathfrak{g}_\mu = r$ and let V be a complement to $\mathfrak{g}_\mu$ in $\mathfrak{g}$, that is, $\mathfrak{g} = V \oplus \mathfrak{g}_\mu$. Then $T_e\varphi_\mu | V$ is injective. Fix $\nu \in \mathfrak{g}^*$ and define

$$S = \{t \in \mathbb{R} \mid T_e\varphi_{\mu+t\nu} | V \text{ is injective.}\}$$

Note that $0 \in S$ and that S is open in $\mathbb{R}$ because the set of injective linear maps is open in $L(\mathfrak{g}, \mathfrak{g}^*)$ and $\mu \mapsto T_e\varphi_\mu$ is continuous. Thus S contains an open neighborhood of 0 in $\mathbb{R}$. Since the rank of a linear map can only increase by slight perturbations, we have $\mathrm{rank}\, T_e\varphi_{\mu+t\nu}|V \geq \mathrm{rank}\, T_e\varphi_\mu|V = n - r$, for $|t|$ small, and by maximality of $n - r$, this forces $\mathrm{rank}\, T_e\varphi_{\mu+t\nu} = n - r$ for t in a neighborhood of 0 contained in S. Thus, for $|t|$ small, $T_e\varphi_{\mu+t\nu}|V : V \to T_{\mu+t\nu}\mathcal{O}_{\mu+t\nu}$ is an isomorphism. Hence, if $\xi \in \mathfrak{g}_\mu$, $\mathrm{ad}^*_\xi(\mu + t\nu) \in T_{\mu+t\nu}\mathcal{O}_{\mu+t\nu}$ is the image of a unique $\xi(t) \in V$ under $T_e\varphi_{\mu+t\nu}|V$, that is, $\xi(t) = (T_e\varphi_{\mu+t\nu}|V)^{-1}(\mathrm{ad}^*_\xi(\mu+t\nu))$. This formula shows that for $|t|$ small, $t \mapsto \xi(t)$ is a smooth curve in V and $\xi(0) = 0$. However, since $\mathrm{ad}^*_\xi(\mu+t\nu) = -T_e\varphi_{\mu+t\nu}(\xi)$, the definition of $\xi(t)$ is equivalent to $T_e\varphi_{\mu+t\nu}(\xi(t) + \xi) = 0$, that is, $\xi(t) + \xi \in \mathfrak{g}_{\mu+t\nu}$. Similarly, given $\eta \in \mathfrak{g}_\mu$, there exists a unique $\eta(t) \in V$ such that $\eta(t) + \eta \in \mathfrak{g}_{\mu+t\nu}, \eta(0) = \eta$, and $t \mapsto \eta(t)$ is smooth for small $|t|$. Therefore $t \mapsto \langle \mu + t\nu, [\xi(t) + \xi, \eta(t) + \eta]\rangle$ is identically zero for small $|t|$. In particular, its derivative at $t = 0$ is also zero. But this derivative equals $\langle \nu, [\xi, \eta] \rangle + \langle \mu, [\xi'(0), \eta] \rangle + \langle \mu, [\xi, \eta'(0)] \rangle = \langle \nu, [\xi, \eta] \rangle - \langle \mathrm{ad}^*_\eta \mu, \xi'(0) \rangle + \langle \mathrm{ad}^*_\xi \mu, \eta'(0) \rangle = \langle \nu, [\xi, \eta] \rangle$, since $\xi, \eta \in \mathfrak{g}_\mu$. Thus $\langle \nu, [\xi, \eta] \rangle = 0$ for any $\nu \in \mathfrak{g}^*$, that is, $[\xi, \eta] = 0$. Since $\xi, \eta \in \mathfrak{g}_\mu$ are arbitrary, it follows that $\mathfrak{g}_\mu$ is abelian. ■

Remarks on Infinite Dimensional Groups

We can use a slight reinterpretation of the formulae in this section to calculate the Lie algebra structure of some infinite-dimensional groups. Here we will treat this topic only formally, that is, we assume that the spaces involved are manifolds and do not specify the function space topologies. For the formal calculations, these structures are not needed, but the reader should be aware that there is a mathematical gap here. (See Ebin and Marsden [1970] and Adams, Ratiu, and Schmid [1986a,b] for more information.)

Given a manifold M, let $\mathrm{Diff}(M)$ denote the group of all diffeomorphisms of M. The group operation is composition. The Lie algebra of $\mathrm{Diff}(M)$, as a vector space, consists of vector fields on M; indeed the flow of a vector field is a curve in $\mathrm{Diff}(M)$ and its tangent vector at $t = 0$ is the given vector field.

To determine the Lie algebra bracket we consider the action of an arbitrary Lie group G on M. Such an action of G on M may be regarded as a homomorphism $\Phi : G \to \mathrm{Diff}(M)$. By Proposition **9.1.5**, its derivative at the identity $T_e\Phi$ should be a Lie algebra homomorphism. From the definition of infinitesimal generator, we see that

$$T_e\Phi \cdot \xi = \xi_M.$$

Thus, **9.1.5** suggests that

$$[\xi_M, \eta_M]_{\text{Lie bracket}} = [\xi, \eta]_M.$$

However, by Proposition **9.3.6**,

$$[\xi, \eta]_M = -[\xi_M, \eta_M].$$

Thus,

$$[\xi_M, \eta_M]_{\text{Lie bracket}} = -[\xi_M, \eta_M].$$

This suggests that the Lie algebra bracket on $\mathfrak{X}(M)$ is minus the Jacobi-Lie bracket.

Another way to arrive at the same conclusion is to use the method of computing brackets in the table in §9.1. To do this, we first compute, according to step 1, the inner automorphism to be

$$I_\eta(\varphi) = \eta \circ \varphi \circ \eta^{-1}.$$

By step 2, we differentiate with respect to φ to compute the Ad map. Letting

$$X = \left.\frac{d}{dt}\right|_{t=0} \varphi_t,$$

where φ_t is a curve in Diff (M) with $\varphi_0 =$ Identity, we have

$$\begin{aligned}
\mathrm{Ad}_\eta(X) &= (T_e I_\eta)(X) = T_e I_\eta \left[\left.\frac{d}{dt}\right|_{t=0} \varphi_t \right] = \left.\frac{d}{dt}\right|_{t=0} I_\eta(\varphi_t) \\
&= \left.\frac{d}{dt}\right|_{t=0} (\eta \circ \varphi_t \circ \eta^{-1}) = T\eta \circ X \circ \eta^{-1} = \eta_* X.
\end{aligned}$$

Hence $\mathrm{Ad}_\eta(X) = \eta_* X$. Thus, *the adjoint action of* Diff(M) *on its Lie algebra is just the push-forward operation*. Finally, as in step 3, we compute the bracket by differentiating $\mathrm{Ad}_\eta(X)$ with respect to η. But by the Lie derivative characterization of brackets and the fact that push forward is the inverse of pull back, we arrive at the same conclusion. In summary, either method suggests that:

> *The Lie algebra bracket on* Diff (M) *is minus the Jacobi-Lie bracket of vector fields.*

One can also say that the Jacobi-Lie bracket gives the *right* (as opposed to *left*) Lie algebra structure on Diff (M).

If one restricts to the group of volume-preserving (or symplectic) diffeomorphisms, then the Lie bracket is again minus the Jacobi-Lie bracket on the space of divergence-free vector fields.

Here are three examples of actions of Diff (M). Firstly, Diff (M) acts on M by evaluation

$$\Phi : \mathrm{Diff}\,(M) \times M \to M$$

given by

$$(\varphi, x) \mapsto \varphi(x).$$

Secondly, the calculations we did for Ad_η show that the adjoint action of Diff (M) on its Lie algebra is given by push forward. Thirdly, if we identify the dual space $\mathfrak{X}(M)^*$ with one-form densities by means of integration, then the change of variables formula shows that the *coadjoint action is given by push forward of one-form densities.*

Another basic example of an infinite-dimensional group is the unitary group $U(\mathcal{H})$ of a complex Hilbert space $\mathcal{H}$. If G is a Lie group and $\rho : G \to U(\mathcal{H})$ is a group homomorphism, we call ρ a *unitary representation*. In other words, ρ is an action of G on $\mathcal{H}$ by unitary maps.

As with the diffeomorphism group, questions of smoothness regarding $U(\mathcal{H})$ need to be dealt with carefully and in this book we shall only give a brief indication of what is involved. The reason the care is needed is, for one thing, because one ultimately is dealing with PDE's rather than ODE's and the hypotheses made must be such that PDE's are not excluded. For example, for a unitary representation one assumes that for each $\psi, \varphi \in \mathcal{H}$, the map

$$g \mapsto \langle \psi, \rho(g)\varphi \rangle$$

of G to $\mathbb{C}$ is continuous. In particular, for $G = \mathbb{R}$ one has the notion of a continuous one-parameter group $U(t)$ so that $U(0) =$ identity and

$$U(t+s) = U(t) \circ U(s).$$

Stone's theorem says that in an appropriate sense we can write

$$U(t) = e^{tA}$$

where A is an (unbounded) skew-adjoint operator defined on a dense domain $D(A) \subset \mathcal{H}$. See, for example, Abraham, Marsden and Ratiu [1988, §**7.4B**] for the proof. Conversely each skew-adjoint operator defines a one parameter subgroup. Thus, Stone's theorem gives precise meaning to the statement: the Lie algebra $\mathfrak{u}(\mathcal{H})$ of $U(\mathcal{H})$ consists of the skew adjoint operators. The Lie bracket is the commutator, as long as one is careful with domains.

If ρ is a unitary representation of a finite dimensional Lie group G on $\mathcal{H}$, then $\rho(\exp(t\xi))$ is a one-parameter subgroup of $U(\mathcal{H})$, so Stone's theorem guarantees that there is a map $\xi \mapsto A(\xi)$ associating a self-adjoint operator $A(\xi)$ to each $\xi \in \mathfrak{g}$. Formally we have

$$[A(\xi), A(\eta)] = [\xi, \eta].$$

Results like this are aided by a theorem of Nelson [1959] guaranteeing a dense subspace $D_G \subset \mathcal{H}$ such that $A(\xi)$ is well-defined on D_G, $A(\xi)$ maps D_G to D_G, and for $\psi \in D_G$, $[\exp tA(\xi)]$ is C^∞ in t with derivative at $t = 0$ given by $A(\xi)\psi$. This space is called an ***essential G-smooth part of $\mathcal{H}$*** and on D_G the above commutator relation and the linearity $A(\alpha\xi + \beta\eta) = \alpha A(\xi) + \beta A(\eta)$ become *literally* true. Moreover, we loose little by using D_G since $A(\xi)$ is uniquely determined by what it is on D_G.

We identify $U(1)$ with the unit circle in $\mathbb{C}$ and each such complex number determines an element of $U(\mathcal{H})$ by multiplication. Thus, we regard $U(1) \subset U(\mathcal{H})$. As such, it is a normal subgroup (in fact, elements of $U(1)$ commute with elements of $U(\mathcal{H})$), so the quotient is a group called the ***projective unitary group of $\mathcal{H}$***. We write it as

$$U(\mathbb{P}\mathcal{H}) = U(\mathcal{H})/U(1).$$

We write elements of $U(\mathbb{P}\mathcal{H})$ as $[U]$ regarded as an equivalence class of $U \in U(\mathcal{H})$. The group $U(\mathbb{P}\mathcal{H})$ acts on projective Hilbert space $\mathbb{P}\mathcal{H} = \mathcal{H}/\mathbb{C}$ by

$$[U][\varphi] = [U\varphi]$$

as in §5.3.

One parameter subgroups of $U(\mathbb{P}\mathcal{H})$ are of the form $[U(t)]$ for a one parameter subgroup $U(t)$ of $U(\mathcal{H})$. This is a particularly simple case of the general problem considered by Bargmann and Wigner of lifting projective representations, a topic we return to later. In any case, this means we can identify the Lie algebra as

$$\mathfrak{u}(\mathbb{P}\mathcal{H}) = \mathfrak{u}(\mathcal{H})/i\mathbb{R},$$

where we identify the two skew adjoint operators A and $A + \lambda i$, for λ real.

A ***projective representation*** of a group G is a homomorphism $\tau : G \to U(\mathbb{P}\mathcal{H})$; we require continuity of $|\langle \psi, \tau(g)\varphi \rangle|$, which is well defined for $[\psi], [\varphi] \in \mathbb{P}\mathcal{H}$. There is an analogue of Nelson's theorem that guarantees an ***essential G-smooth part*** $\mathbb{P}D_G$ of $\mathbb{P}\mathcal{H}$ with properties like those of D_G.

Exercises

9.3-1 *Let a Lie group G act linearly on a vector space V. Define a group structure on $G \times V$ by $(g_1, v_1) \cdot (g_2, v_2) = (g_1g_2, g_1v_2 + v_1)$. Show that this makes $G \times V$ into a Lie group—it is called the* ***semidirect product*** *and is denoted $G \circledS V$. Determine its Lie algebra $\mathfrak{g} \circledS V$.*

9.3-2 (a) *Show that the Euclidian group $E(3)$ can be written as $O(3) \circledS \mathbb{R}^3$ in the sense of the preceding exercise.*

(b) *Show that $E(3)$ is isomorphic to the group of (4×4)-matrices of the form $\begin{bmatrix} A & \mathbf{b} \\ 0 & 1 \end{bmatrix}$ where $A \in O(3)$ and $\mathbf{b} \in \mathbb{R}^3$.*

9.3-3 *If G is a Lie group, show that TG is isomorphic (as a Lie group) with $G \circledS \mathfrak{g}$ (see Exercise* **9.1-2***).*

9.3-4 *Compute the* Ad *action of* Diff(M) *on its Lie algebra.*

10
Poisson Manifolds

The dual $\mathfrak{g}^*$ of any Lie algebra $\mathfrak{g}$ carries a Poisson bracket given by

$$\{F, G\}(\mu) = \left\langle \mu, \left[\frac{\delta F}{\delta \mu}, \frac{\delta G}{\delta \mu}\right]\right\rangle$$

for $\mu \in \mathfrak{g}^*$, a formula found by Lie [1890], §75. The explication of this ***Lie-Poisson bracket*** is one of the goals of this chapter and of Chapter 13. The bracket is not the bracket associated with any symplectic structure on $\mathfrak{g}^*$, so we need to study the more general concept of a *Poisson manifold*. However, the Lie-Poisson bracket *is* associated with a symplectic structure on coadjoint orbits. This fact and the reduction approach are given in Chapters 13 and 14. We saw in the introduction that the Lie-Poisson bracket plays an important role in the Hamiltonian description of many physical systems. Chapter 15 shows how this works in detail for the rigid body.

10.1 The Definition of Poisson Manifolds

This section generalizes the notion of a symplectic manifold by keeping just enough of the properties of Poisson brackets to describe Hamiltonian systems. The history of Poisson manifolds is complicated by the fact that the notion was rediscovered many times under different names; they occur in the works of Lie [1890], Dirac [1930], [1964], Pauli [1953], Martin [1959], Jost [1964], Arens [1970], Hermann [1973], Sudarshan and Mukunda [1974], Vinogradov and Krasilshchik [1975], and Lichnerowicz [1975b]. The name

Poisson manifold was coined by Lichnerowicz. Further historical comments are given in §10.3

Definition 10.1.1 *A* ***Poisson bracket*** *(or a* ***Poisson structure****) on a manifold P is a bilinear operation $\{\,,\}$ on $\mathcal{F}(P) = C^\infty(P)$ such that:*

i *$(\mathcal{F}(P), \{\,,\})$ is a Lie algebra; and*

ii *$\{\,,\}$ is a derivation in each factor, that is,*

$$\{FG, H\} = \{F, H\} G + F \{G, H\}$$

for all F, G, and $H \in \mathcal{F}(P)$.

A manifold P endowed with a Poisson bracket on $\mathcal{F}(P)$ is called a ***Poisson manifold****.*

A Poisson manifold is denoted by $(P, \{\,,\})$ or simply by P if there is no danger of confusion. Note that any manifold has the ***trivial Poisson structure*** which is defined by setting $\{F, G\} = 0$ for all $F, G \in \mathcal{F}(P)$. Occasionally we consider two different Poisson brackets $\{\,,\}_1$ and $\{\,,\}_2$ on the same manifold; the two distinct Poisson manifolds are then denoted by $(P, \{\,,\}_1)$ and $(P, \{\,,\}_2)$. The notation $\{\,,\}_P$ for the bracket on P is also used when confusion might arise.

Exercise

10.1-1 *If P_1 and P_2 are Poisson manifolds, show how to make $P_1 \times P_2$ into a Poisson manifold.*

10.2 Examples

(a) Symplectic Bracket

Any symplectic manifold is a Poisson manifold. The Poisson bracket is defined by the symplectic form as was shown in §**5.5**. Condition **ii** of the definition is satisfied as a consequence of the derivation property of vector fields:

$$\{FG, H\} = X_H[FG] = FX_H[G] + GX_H[F] = F\{G, H\} + G\{F, H\}. \quad \blacklozenge$$

(b) Lie-Poisson Bracket

If $\mathfrak{g}$ is a Lie algebra, then its dual $\mathfrak{g}^*$ is a Poisson manifold with respect to each of the ***Lie-Poisson brackets*** $\{\,,\}_+$ and $\{\,,\}_-$ defined by

$$\{F, G\}_\pm(\mu) = \pm \left\langle \mu, \left[\frac{\delta F}{\delta \mu}, \frac{\delta G}{\delta \mu} \right] \right\rangle \tag{10.2.1}$$

for $\mu \in \mathfrak{g}^*$ and $F, G \in \mathcal{F}(\mathfrak{g}^*)$. The properties of a Poisson bracket can be easily verified. Bilinearity and skew-symmetry are obvious. The derivation property of the bracket follows from the Leibniz rule for functional derivatives

$$\frac{\delta(FG)}{\delta\mu} = F(\mu)\frac{\delta G}{\delta\mu} + \frac{\delta F}{\delta\mu}G(\mu).$$

The Jacobi identity for the Lie-Poisson bracket follows from the Jacobi identity for the Lie algebra bracket and the formula

$$\pm\frac{\delta}{\delta\mu}\{F, G\}_\pm = \left[\frac{\delta F}{\delta\mu}, \frac{\delta G}{\delta\mu}\right] - \mathbf{D}^2F(\mu)\left(\operatorname{ad}^*_{\delta G/\delta\mu}\mu, \cdot\right) + \mathbf{D}^2G\left(\operatorname{ad}^*_{\delta F/\delta\mu}\mu, \cdot\right), \tag{10.2.2}$$

where for each $\xi \in \mathfrak{g}$, $\operatorname{ad}_\xi : \mathfrak{g} \to \mathfrak{g}$ denotes the map $\operatorname{ad}_\xi(\eta) = [\xi, \eta]$ and $\operatorname{ad}^*_\xi : \mathfrak{g}^* \to \mathfrak{g}^*$ is its dual. We give a different proof that (10.2.1) is a Poisson bracket in Chapter 13. ♦

(c) Rigid Body Bracket
Specializing Example **(b)** to the Lie algebra of the rotation group, $\mathfrak{so}(3) \cong \mathbb{R}^3$, and identifying $\mathbb{R}^3$ and $(\mathbb{R}^3)^*$ via the standard inner product, we get the following Poisson structure on $\mathbb{R}^3$:

$$\{F, G\}_-(\mathbf{\Pi}) = -\mathbf{\Pi} \cdot (\nabla F \times \nabla G), \tag{10.2.3}$$

where $\mathbf{\Pi} \in \mathbb{R}^3$ and ∇F, the gradient of F, is evaluated at $\mathbf{\Pi}$. The Poisson bracket properties can be verified by direct computation in this case, see Exercise **1.2-1**. We call (10.2.3) the ***rigid body bracket***. ♦

(d) Ideal Fluid Bracket
Specialize the Lie-Poisson bracket to the Lie algebra $\mathfrak{X}_{\mathrm{div}}(\Omega)$ of divergence-free vector fields defined in a region Ω of $\mathbb{R}^3$ and tangent to $\partial\Omega$, with the Lie bracket being the Jacobi-Lie bracket. Identify $\mathfrak{X}^*_{\mathrm{div}}(\Omega)$ with $\mathfrak{X}_{\mathrm{div}}(\Omega)$ using the L^2 pairing

$$\langle \mathbf{v}, \mathbf{w} \rangle = \int_\Omega \mathbf{v} \cdot \mathbf{w}\, d^3x, \tag{10.2.4}$$

where $\mathbf{v} \cdot \mathbf{w}$ is the ordinary dot product in $\mathbb{R}^3$. Thus, the plus Lie-Poisson bracket is

$$\{F, G\}(\mathbf{v}) = -\int_\Omega \mathbf{v} \cdot \left[\frac{\delta F}{\delta \mathbf{v}}, \frac{\delta G}{\delta \mathbf{v}}\right] d^3x, \tag{10.2.5}$$

where the functional derivative $\delta F/\delta \mathbf{v}$ is the element of $\mathfrak{X}_{\mathrm{div}}(\Omega)$ defined by

$$\lim_{\varepsilon \to 0} \frac{1}{\varepsilon} [F(\mathbf{v} + \varepsilon \delta \mathbf{v}) - F(\mathbf{v})] = \int_\Omega \frac{\delta F}{\delta \mathbf{v}} \cdot \delta \mathbf{v} \, d^3x. \quad \blacklozenge \tag{10.2.6}$$

(e) Poisson-Vlasov Bracket

Let $(P, \{\,,\}_P)$ be a Poisson manifold and let $\mathcal{F}(P)$ be the Lie algebra of functions under the Poisson bracket. Identify $\mathcal{F}(P)^*$ with densities f on P. Then the Lie-Poisson bracket has the expression

$$\{F, G\}(f) = \int_P f \left\{ \frac{\delta F}{\delta f}, \frac{\delta G}{\delta f} \right\}_P. \tag{10.2.7}$$

We shall explain how this bracket is derived in Chapter 13. $\blacklozenge$

(f) Linearized Lie-Poisson Bracket

Fix $\nu \in \mathfrak{g}^*$ and define for any $F, G \in \mathcal{F}(\mathfrak{g}^*)$ the bracket

$$\{F, G\}^\nu_\pm(\mu) = \pm \left\langle \nu, \left[\frac{\delta F}{\delta \mu}, \frac{\delta G}{\delta \mu} \right] \right\rangle. \tag{10.2.8}$$

The properties of a Poisson bracket are verified as in the case of the Lie-Poisson bracket, the only difference being that (10.2.2) is replaced by

$$\pm \frac{\delta}{\delta \mu} \{F, G\}^\nu_\pm = -\mathbf{D}^2 F(\nu) \left(\mathrm{ad}^*_{\delta G/\delta \mu} \mu, \cdot \right) + \mathbf{D}^2 G(\nu) \left(\mathrm{ad}^*_{\delta F/\delta \mu} \mu, \cdot \right) \tag{10.2.9}$$

This bracket is useful in the description of the linearized Lie-Poisson equations. $\blacklozenge$

(g) KdV Bracket

Let $S = [S^{ij}]$ be a symmetric matrix. On $\mathcal{F}(\mathbb{R}^n, \mathbb{R}^n)$, set

$$\{F, G\}(u) = \int_{-\infty}^{\infty} \sum_{i,j=1}^{n} S^{ij} \left[\frac{\delta F}{\delta u^i} \frac{d}{dx} \left(\frac{\delta G}{\delta u^j} \right) - \frac{d}{dx} \left(\frac{\delta G}{\delta u^j} \right) \frac{\delta F}{\delta u^i} \right] dx \tag{10.2.10}$$

for functions F, G satisfying $\delta F/\delta u$, and $\delta G/\delta u \to 0$ as $x \to \pm\infty$. This is a Poisson structure that is useful for the KdV equation and for gas dynamics (see Benjamin [1984]); it is actually a particular case of Example **(f)**, the Lie algebra being the pseudo-differential operators on the line of order ≤ -1 and $\nu = S d/dx$. If S is invertible and $S^{-1} = [S_{ij}]$, then (10.2.10) is the Poisson bracket associated with the weak symplectic form

$$\Omega(u, v) = \frac{1}{2} \int_{-\infty}^{\infty} \sum_{i,j=l}^{n} S_{ij} \left[\left(\int_{-\infty}^{y} u^i(x)\, dx \right) v^j(y) \right.$$

$$- \left(\int_{-\infty}^{y} v^j(x)\, dx \right) u^i(y) \bigg] \, dy. \tag{10.2.11}$$

This is easily seen by noting that $X_H(u)$ is given by

$$X_H^i(u) = S^{ij} \frac{d}{dx} \frac{\delta H}{\delta u^j}. \quad \blacklozenge \tag{10.2.12}$$

(h) Toda Lattice Bracket
Let $P = \{(\mathbf{a}, \mathbf{b}) \in \mathbb{R}^{2n} \mid a^i > 0,\ i = 1, \ldots, n\}$ and consider the bracket

$$\{F, G\}(\mathbf{a}, \mathbf{b}) = \left[\left(\frac{\partial F}{\partial \mathbf{a}} \right)^T, \left(\frac{\partial F}{\partial \mathbf{b}} \right)^T \right] \mathbf{W} \begin{bmatrix} \dfrac{\partial G}{\partial \mathbf{a}} \\ \dfrac{\partial G}{\partial \mathbf{a}} \end{bmatrix}, \tag{10.2.13}$$

where $(\partial F/\partial \mathbf{a})^T$ is the row vector $(\partial F/\partial a^1, \ldots, \partial F/\partial a^n)$, etc., and

$$\mathbf{W} = \begin{bmatrix} 0 & \mathbf{A} \\ -\mathbf{A} & 0 \end{bmatrix}, \quad \text{where} \quad \mathbf{A} = \begin{bmatrix} a^1 & & 0 \\ & \ddots & \\ 0 & & a^n \end{bmatrix}. \tag{10.2.14}$$

In terms of the coordinate functions a_i, b_j, the bracket (10.2.13) is given by

$$\left. \begin{aligned} \{a^i, a^j\} &= 0, \\ \{b^i, b^j\} &= 0, \\ \{a^i, b^j\} &= 0 \quad \text{if } i \neq j, \\ \{a^i, b^j\} &= a^i \quad \text{if } i = j. \end{aligned} \right\} \tag{10.2.15}$$

This Poisson bracket is determined by the symplectic form

$$\Omega = -\sum_{i=1}^{n} \frac{1}{a^i} da^i \wedge db^i \tag{10.2.16}$$

as easy verification shows. The mapping $(\mathbf{a}, \mathbf{b}) \mapsto (\log \mathbf{a}^{-1}, \mathbf{b})$ is a symplectic diffeomorphism of P with $\mathbb{R}^{2n}$ endowed with the canonical symplectic structure. This symplectic structure is known as the *first Poisson structure of the non-periodic Toda lattice*. We shall not study this example in any detail in this book, but we point out that its bracket is the restriction of a Lie-Poisson bracket to a certain coadjoint orbit of the group of lower triangular matrices; we refer the interested reader to §14.5, Kostant [1979], and Symes [1980, 1982a,b] for further information. $\blacklozenge$

Exercises

10.2-1 *Verify directly that the Lie-Poisson bracket satisfies Jacobi's identity.*

10.2-2 (A Quadratic Bracket) *Let $A = [A^{ij}]$ be a skew-symmetric matrix. On $\mathbb{R}^n$, define $B^{ij} = A^{ij}x^i x^j$ (no sum). Show that the following defines a Poisson structure:*

$$\{F, G\} = \sum_{i,j=1}^{n} B^{ij} \frac{\partial F}{\partial x^i} \frac{\partial G}{\partial x^j}. \tag{10.2.17}$$

10.2-3 (A Cubic Bracket) *For $\mathbf{x} = (x^1, x^2, x^3) \in \mathbb{R}^3$, put*

$$\left.\begin{array}{rcl} \{x^1, x^2\} &=& \|\mathbf{x}\|^2 x^3, \\ \{x^2, x^3\} &=& \|\mathbf{x}\|^2 x^1, \\ \{x^3, x^1\} &=& \|\mathbf{x}\|^2 x^2. \end{array}\right\} \tag{10.2.18}$$

Let $B^{ij} = \{x^i, x^j\}$, for $i < j$ and $i, j = 1, 2, 3$, set $B^{ji} = -B^{ij}$, and define

$$\{F, G\} = \sum_{i,j=1}^{n} B^{ij} \frac{\partial F}{\partial x^i} \frac{\partial G}{\partial x^j}. \tag{10.2.19}$$

Check that this makes $\mathbb{R}^3$ into a Poisson manifold.

10.2-4 *Let $\Phi : \mathfrak{g}^* \to \mathfrak{g}^*$ be a smooth function and define for $F, G : \mathfrak{g}^* \to \mathbb{R}$,*

$$\{F, H\}_\Phi (\mu) = \left\langle \Phi(\mu), \left[\frac{\delta F}{\delta \mu}, \frac{\delta H}{\delta \mu}\right]\right\rangle.$$

(a) *Show that this rule defines a Poisson bracket on $\mathfrak{g}^*$ if and only if Φ satisfies the following identity.*

$$\begin{aligned} &\langle \mathbf{D}\Phi(\mu) \cdot \mathrm{ad}^*_\zeta(\mu), [\eta, \xi]\rangle \\ &\quad + \langle \mathbf{D}\Phi(\mu) \cdot \mathrm{ad}^*_\eta \Phi(\mu), [\xi, \zeta]\rangle \\ &\quad + \langle \mathbf{D}\Phi(\mu) \cdot \mathrm{ad}^*_\xi \Phi(\mu), [\zeta, \eta]\rangle = 0 \end{aligned}$$

for all $\xi, \eta, \zeta \in \mathfrak{g}$, and all $\mu \in \mathfrak{g}^*$.

(b) *Show that this relation holds if $\Phi(\mu) = \mu$ and $\Phi(\mu) = \nu$, a fixed element of $\mathfrak{g}^*$, thereby obtaining the Lie-Poisson structure* (10.2.1) *and the linearized Lie-Poisson structure* (10.2.8) *on $\mathfrak{g}^*$. Show that it also holds if $\Phi(\mu) = a\mu + \nu$ for some $a \in \mathbb{R}$.*

(c) *Assume* $\mathfrak{g}$ *has a weakly nondegenerate invariant bilinear form* $K : \mathfrak{g} \times \mathfrak{g} \to \mathbb{R}$ *and identity* $\mathfrak{g}^*$ *with* $\mathfrak{g}$ *by* K*. If* $\Psi : \mathfrak{g} \to \mathfrak{g}$ *is smooth, show that*

$$\{F, H\}_\Psi (\xi) = K(\Psi(\xi), [\nabla F(\xi), \nabla H(\xi)])$$

is a Poisson bracket if and only if

$$\begin{aligned} &K(\mathbf{D}\Psi(\lambda) \cdot [\Psi(\lambda), \zeta], [\eta, \xi]) \\ &\quad + K(\mathbf{D}\Psi(\lambda) \cdot [\Psi(\lambda), \eta], [\xi, \zeta]) \\ &\quad + K(\mathbf{D}\Psi(\lambda) \cdot [\Psi(\lambda), \xi], [\zeta, \eta]) = 0 \end{aligned}$$

for all $\lambda, \xi, \eta, \zeta \in \mathfrak{g}$*. Here,* $\nabla F(\xi), \nabla H(\xi) \in \mathfrak{g}$ *are the gradients of* F *and* H *at* $\xi \in \mathfrak{g}$ *relative to* K*.*

Conclude as in **(b)** *that this relation holds if* $\Psi(\lambda) = a\lambda + \chi$ *for* $a \in \mathbb{R}$ *and* $\chi \in \mathfrak{g}$*.*

(d) *In the hypothesis of* **(c)**, *let* $\Psi(\lambda) = \nabla\psi(\lambda)$ *for some smooth* $\psi : \mathfrak{g} \to \mathbb{R}$*. Show that* $\{\ ,\ \}_\Psi$ *is a Poisson bracket if and only if*

$$\begin{aligned} &\mathbf{D}^2\psi(\lambda)([\nabla\psi(\lambda), \zeta], [\eta, \xi]) - \mathbf{D}^2\psi(\lambda)(\nabla\psi(\lambda), [\zeta, [\eta, \xi]]) \\ &\quad + \mathbf{D}^2\psi(\lambda)([\nabla\psi(\lambda), \eta], [\xi, \zeta]) - \mathbf{D}^2\psi(\lambda)(\nabla\psi(\lambda), [\eta, [\xi, \zeta]]) \\ &\quad + \mathbf{D}^2\psi(\lambda)([\nabla\psi(\lambda), \xi], [\zeta, \eta]) - \mathbf{D}^2\psi(\lambda)(\nabla\psi(\lambda), [\xi, [\zeta, \eta]]) = 0 \end{aligned}$$

for all $\lambda, \xi, \eta, \zeta \in \mathfrak{g}$*. In particular, if* $\mathbf{D}^2\psi(\lambda)$ *is an invariant bilinear form for all* λ*, this condition holds. However, if* $\mathfrak{g} = \mathfrak{so}(3)$ *and* ψ *is arbitrary, then this condition also holds (see Exercise* **1.3-2.***)*

10.3 Hamiltonian Vector Fields and Casimir Functions

Proposition 10.3.1 *Let* P *be a Poisson manifold. If* $H \in \mathcal{F}(P)$*, then there is a unique vector field* X_H *on* P *such that*

$$X_H[G] = \{G, H\} \tag{10.3.1}$$

for all $G \in \mathcal{F}(P)$*. We call* X_H *the* ***Hamiltonian vector field*** *of* H*.*

Proof This is a consequence of the fact that any derivation on $\mathcal{F}(P)$ is represented by a vector field. Fixing H, the map $G \mapsto \{G, H\}$ is a derivation, and so it uniquely determines X_H satisfying (10.3.1). (In infinite dimensions some technical conditions are needed for this proof, which are deliberately ignored here; see Abraham, Marsden, and Ratiu [1988], §4.2.) ■

Notice that (10.3.1) agrees with our definition of Poisson brackets in the symplectic case, so if the Poisson manifold P is symplectic, X_H defined here agrees with the definition in §5.5.

Proposition 10.3.2 *The map $H \mapsto X_H$ of $\mathcal{F}(P)$ to $\mathfrak{X}(P)$ is a Lie algebra antihomomorphism; i.e.,*

$$[X_H, X_K] = -X_{\{H,K\}}.$$

Proof Using Jacobi's identity, we find that

$$\begin{aligned} [X_H, X_K][F] &= X_H[X_K[F]] - X_K[X_H[F]] \\ &= \{\{F, K\}, H\} - \{\{F, H\}, K\} \\ &= -\{F, \{H, K\}\} \\ &= -X_{\{H,K\}}[F]. \quad \blacksquare \end{aligned}$$

Proposition 10.3.3 *Let $H \in \mathcal{F}(P)$ and φ_t be the flow of X_H. Then:*

i
$$\frac{d}{dt}(F \circ \varphi_t) = \{F, H\} \circ \varphi_t = \{F \circ \varphi_t, H\},$$

or for short

$$\dot{F} = \{F, H\}.$$

ii $H \circ \varphi_t = H$ *(**conservation of energy**).*

Proof

i For any $z \in P$ we have

$$\begin{aligned} \frac{d}{dt} F(\varphi_t(z)) &= \mathbf{d}F(\varphi_t(z)) \cdot X_H(\varphi_t(z)) = \{F, H\}(\varphi_t(z)) \\ &= \mathbf{d}F(\varphi_t(z)) \cdot T_z\varphi_t(X_H(z)) = \mathbf{d}(F \circ \varphi_t)(z) \cdot X_H(z) \\ &= \{F \circ \varphi_t, H\}(z), \end{aligned}$$

since $X_H(\varphi_t(z)) = T_z\varphi_t(X_H(z))$.

ii For the proof of **ii**, let $H = F$ in **i**. $\blacksquare$

Corollary 10.3.4 *Let $G, H \in \mathcal{F}(P)$. Then G is constant along the integral curves of X_H if and only if $\{G, H\} = 0$, if and only if H is constant along the integral curves of X_G.*

Among the elements of $\mathcal{F}(P)$ are functions C such that $\{C, F\} = 0$ for all $F \in \mathcal{F}(P)$, that is, C is constant along the flow of all Hamiltonian vector fields or, equivalently, $X_C = 0$, that is, C generates trivial dynamics. Such functions are called ***Casimir functions*** of the Poisson structure. This terminology is used in, for example, Sudarshan and Mukunda [1974]. H.B.G. Casimir is a prominent physicist who wrote his thesis (Casimir [1931]) was on the quantum mechanics of the rigid body, under the direction of Paul Ehrenfest. Recall that it was Ehrenfest who, in *his* thesis, worked on the variational structure of ideal flow in Lagrangian or material representation.

Some History of Poisson Structures[1]

Following from the work of Lagrange and Poisson discussed at the end of §8.1, the general concept of the Poisson manifold should be credited to Sophus Lie in his treatise on transformation groups about 1880 in the chapter on "function groups." Specifically, on page 237, Lie defines what today is called a Poisson structure. The title of Chapter 19 is *The Coadjoint Group*, which is explicitly identified on page 334. Chapter 17, pages 294-298, defines a linear Poisson structure on the dual of a Lie algebra, today called the Lie-Poisson structure, and "Lie's Third Theorem" is proved for the set of regular elements. On page 349, together with a remark on page 367, it is shown that the Lie-Poisson structure naturally induces a symplectic structure on each coadjoint orbit. As we shall point out in §11.2, Lie also had many of the ideas of momentum maps. For many years this work appears to have been forgotten.

Because of the above history, Marsden and Weinstein [1983] coined the phrase "Lie-Poisson bracket" for this object, and this terminology is now in common use. However, it is not clear that Lie understood the fact that the Lie-Poisson bracket is obtained by a simple reduction process, namely, that it is induced from the canonical cotangent Poisson bracket on T^*G by passing to $\mathfrak{g}^*$ regarded as the quotient T^*G/G, as will be explained in Chapter 13. The link between the closedness of the symplectic form and the Jacobi identity is a little harder to trace explicitly; some comments in this direction are given in Souriau [1970], who gives credit to Maxwell.

Lie's work starts by taking functions $F_1, \ldots, F_r$ on a symplectic manifold M, with the property that there exist functions G_{ij} of r variables, such that

$$\{F_i, F_j\} = G_{ij}(F_1, \ldots, F_r).$$

In Lie's time, all functions in sight are implicitly assumed to be analytic. The collection of all functions ϕ of $F_1, \ldots, F_r$ is the "function group"; it is provided with the bracket

$$[\phi, \psi] = \sum_{ij} G_{ij} \phi_i \psi_j, \tag{10.3.2}$$

where

$$\phi_i = \frac{\partial \phi}{\partial F_i} \qquad \text{and} \qquad \psi_j = \frac{\partial \psi}{\partial F_j}.$$

Considering $F = (F_1, \ldots, F_r)$ as a map from M to an r-dimensional space P, and ϕ and ψ as functions on P, one may formulate this as $[\phi, \psi]$

[1]We thank Hans Duistermaat and Alan Weinstein for their comments on this section; see Weinstein [1983a]. Note that Lie uses the word "group" for both "group" and "algebra." For example, a "function group" should be translated as "function algebra."

is a Poisson structure on P, with the property that

$$F^*[\phi, \psi] = \{F^*\phi, F^*\psi\}.$$

Lie writes down the equations for the G_{ij} that follow from the antisymmetry and the Jacobi identity for the bracket $\{\,,\}$ on M. He continues with the question: Suppose we have given a system of functions G_{ij} in r variables that satisfy these equations, is it induced as above from a function group of functions of $2n$ variables? He shows that under suitable rank conditions the answer is yes. As we shall see below, this result is the precursor to many of the fundamental results about the geometry of Poisson manifolds.

It is obvious that if G_{ij} is a system that satisfies the equations that Lie writes down, then (10.3.2) is a Poisson structure in r-dimensional space. Vice versa, for any Poisson structure $[\phi, \psi]$, the functions

$$G_{ij} = [F_i, F_j]$$

satisfy Lie's equations.

Lie continues with more remarks on local normal forms of function groups, that is, of Poisson structures), under suitable rank conditions, which are not always stated as explicitly as one would like. These amount to: a Poisson structure of constant rank is the same as a foliation with symplectic leaves. It is this characterization that Lie uses to get the symplectic form on the coadjoint orbits. On the other hand, Lie does not apply the symplectic form on the coadjoint orbits to representation theory—representation theory of Lie groups started only later with Schur on $GL(n)$, Elie Cartan on representations of semisimple Lie algebras, and in the 1930s by Weyl on compact Lie groups. The coadjoint orbit symplectic structure was connected with representation theory in the work of Kirillov and Kostant. On the other hand, Lie *did* apply the Poisson structure on the dual of the Lie algebra to prove that every abstract Lie algebra can be realized as a Lie algebra of Hamiltonian vector fields, or as a Lie subalgebra of the Poisson algebra of functions on some symplectic manifold. This is "Lie's third fundamental theorem" in the form as given by Lie.

Of course, in geometry, people like Engel, Study and, in particular, Elie Cartan studied Lie's work intensely and propagated it very actively. However, through the tainted glasses of retrospection, Lie's work on Poisson structures did not appear to receive as much attention in mechanics as it deserved; for example, even though Cartan himself did very important work in mechanics (such as, Cartan [1923, 1928a,b]), he did not seem to realize that the Lie-Poisson bracket was central to the Hamiltonian description of some of the rotating fluid systems he was studying. However, others, such as Hamel [1904, 1949], did study Lie intensively and used it to make substantial contributions and extensions (such as to the study of nonholonomic systems, including rolling constraints), but many other active schools seem to have missed it. Even more surprising in this context is the contribution of Poincaré [1901b, 1910] to the Lagrangian side of the story, a tale that we shall come to in Chapter 13.

Exercise

10.3-1 *Verify the relation* $[X_H, X_K] = -X_{\{H,K\}}$ *directly for the rigid body bracket.*

10.4 Examples

(a) Symplectic Case
On a symplectic manifold P, any Casimir function is constant on connected components of P. This holds since in the symplectic case, $X_C = 0$ implies $\mathbf{d}C = 0$ and hence C is locally constant. ♦

(b) Rigid Body Casimirs
In Example **(c)** of §10.2, let $C(\mathbf{\Pi}) = \|\mathbf{\Pi}\|^2/2$. Then $\nabla C(\mathbf{\Pi}) = \mathbf{\Pi}$ and by the properties of the triple product, we have for any $F \in \mathcal{F}(\mathbb{R}^3)$,

$$\begin{aligned} \{C, F\}(\mathbf{\Pi}) &= -\mathbf{\Pi} \cdot (\nabla C \times \nabla F) = -\mathbf{\Pi} \cdot (\mathbf{\Pi} \times \nabla F) \\ &= -\nabla F \cdot (\mathbf{\Pi} \times \mathbf{\Pi}) = 0. \end{aligned}$$

This shows that $C(\mathbf{\Pi}) = \|\mathbf{\Pi}\|^2/2$ is a Casimir function. A similar argument shows that

$$C_\Phi(\mathbf{\Pi}) = \Phi(\|\mathbf{\Pi}\|^2) \tag{10.4.1}$$

is a Casimir function, where Φ is an arbitrary (differentiable) function of one variable; this is proved by noting that

$$\nabla C_\Phi(\mathbf{\Pi}) = \Phi'(\|\mathbf{\Pi}\|^2)\mathbf{\Pi}. \quad ♦$$

(c) Helicity
In Example **(d)** of §10.2, the ***helicity***

$$C(\mathbf{v}) = \int_\Omega \mathbf{v} \cdot (\nabla \times \mathbf{v})\, d^3x \tag{10.4.2}$$

can be checked to be a Casimir function if $\partial\Omega = \varnothing$. ♦

(d) Poisson-Vlasov Casimirs
In Example **(e)** of §10.2, given a differentiable function $\Phi : \mathbb{R} \to \mathbb{R}$, the map $C : \mathcal{F}(P) \to \mathbb{R}$ defined by

$$C(f) = \int \Phi(f(q,p))\, dq\, dp \tag{10.4.3}$$

is a Casimir function. Here we choose P to be symplectic, have written $dq\,dp = dz$ for the Liouville measure, and have used it to identify functions and densities. ♦

Exercises

10.4-1 *Verify that* (10.4.3) *defines a Casimir function.*

10.4-2 *Let P be a Poisson manifold and let $M \subset P$ be a connected submanifold with the property that for each $v \in T_xM$ there is a Hamiltonian vector field X_H on P such that $v = X_H(x)$; that is, T_xM is spanned by Hamiltonian vector fields. Prove that any Casimir function is constant on M.*

10.5 Properties of Hamiltonian Flows

Proposition 10.5.1 *If φ_t is the flow of X_H, then*

$$\varphi_t^* \{F, G\} = \{\varphi_t^* F, \varphi_t^* G\};$$

in other words,

$$\{F, G\} \circ \varphi_t = \{F \circ \varphi_t, G \circ \varphi_t\}.$$

Thus, the flows of Hamiltonian vector fields preserve the Poisson structure.

Proof This is actually true even for time-dependent Hamiltonian systems (as we will see later), but here we will prove it only in the time-independent case. Let $F, K \in \mathcal{F}(P)$ and let φ_t be the flow of X_H. Let

$$u = \{F \circ \varphi_t, K \circ \varphi_t\} - \{F, K\} \circ \varphi_t.$$

Because of the bilinearity of the Poisson bracket,

$$\frac{du}{dt} = \left\{\frac{d}{dt} F \circ \varphi_t, K \circ \varphi_t\right\} + \left\{F \circ \varphi_t, \frac{d}{dt} K \circ \varphi_t\right\} - \frac{d}{dt}\{F, K\} \circ \varphi_t.$$

Using **10.3.3**, this becomes

$$\frac{du}{dt} = \{\{F \circ \varphi_t, H\}, K \circ \varphi_t\} + \{F \circ \varphi_t, \{K \circ \varphi_t, H\}\} - \{\{F, K\} \circ \varphi_t, H\},$$

which, by Jacobi's identity, gives

$$\frac{du}{dt} = \{u, H\} = X_H[u].$$

By Exercise **4.3-2**, the solution is $u_t = u_0 \circ \varphi_t$. Since $u_0 = 0$, we get $u = 0$, which is the result. ■

As in the symplectic case, with which this is of course consistent, this argument shows how Jacobi's identity plays a crucial role.

A smooth mapping $f : P_1 \to P_2$ between the two Poisson manifolds $(P_1, \{\,,\}_1)$ and $(P_2, \{\,,\}_2)$ is called ***canonical*** or ***Poisson*** if

$$f^* \{F, G\}_2 = \{f^*F, f^*G\}_1$$

for all $F, G \in \mathcal{F}(P_2)$. Proposition **10.5.1** shows that flows of Hamiltonian vector fields are canonical maps. We saw already in Chapter 5 that if P_1 and P_2 are symplectic manifolds, a map $f : P_1 \to P_2$ is canonical if and only if it is symplectic. The next proposition shows that Poisson maps push Hamiltonian flows to Hamiltonian flows.

Proposition 10.5.2 *Let $f : P_1 \to P_2$ be a Poisson map and let $H \in \mathcal{F}(P_2)$. If φ_t is the flow of X_H and ψ_t is the flow of $X_{H \circ f}$, then $\varphi_t \circ f = f \circ \psi_t$ and $Tf \circ X_{H\circ f} = X_H \circ f$. Conversely, if f is a map from P_1 to P_2 and for any $H \in \mathcal{F}(P_2)$, the Hamiltonian vector fields $X_{H\circ f} = \mathfrak{X}(P_1)$ and $X_H \in \mathfrak{X}(P_2)$ are f-related, i.e., $Tf \circ X_{H\circ f} = X_H \circ f$, then f is canonical.*

Proof For any $G \in \mathcal{F}(P_2)$ and $z \in P_1$, Proposition **10.3.3i** and the definition of Poisson maps yield

$$\begin{aligned} \frac{d}{dt} G((f \circ \psi_t)(z)) &= \frac{d}{dt}(G \circ f)(\psi_t(z)) \\ &= \{G \circ f, H \circ f\}(\psi_t(z)) = \{G, H\}(f \circ \psi_t)(z), \end{aligned}$$

that is, $(f \circ \psi_t)(z)$ is an integral curve of X_H on P_2 through the point $f(z)$. Since $(\varphi_t \circ f)(z)$ is another such curve, uniqueness of integral curves implies that $(f \circ \psi_t)(z) = (\varphi_t \circ f)(z)$. The relation $Tf \circ X_{H\circ f} = X_H \circ f$ follows from $f \circ \psi_t = \varphi_t \circ f$ by taking the time derivative.

Conversely, assume that for any $H \in \mathcal{F}(P_2)$ we have $Tf \circ X_{H\circ f} = X_H \circ f$. Therefore, by the chain rule,

$$\begin{aligned} X_{H\circ f}[F \circ f](z) &= \mathbf{d}F(f(z)) \cdot T_z f(X_{H\circ f}(z)) \\ &= \mathbf{d}F(f(z)) \cdot X_H(f(z)) = X_H[F](f(z)), \end{aligned}$$

that is, $X_{H\circ f}[f^*F] = f^*(X_H[F])$. Thus for $G \in \mathcal{F}(P_2)$,

$$\{G, H\} \circ f = f^*(X_H[G]) = X_{H\circ f}[f^*G] = \{G \circ f, H \circ f\}$$

and so f is canonical. ■

Exercises

10.5-1 *Verify directly that a rotation $R : \mathbb{R}^3 \to \mathbb{R}^3$ is a Poisson map for the rigid body bracket.*

10.5-2 *If P_1 and P_2 are Poisson manifolds, show that the projection $\pi_1 : P_1 \times P_2 \to P_1$ is a Poisson map. Is the corresponding statement true for symplectic maps?*

10.6 The Poisson Tensor

By the derivation property of the Poisson bracket, the value of the bracket $\{F, G\}$ at $z \in P$ (and thus $X_F(z)$ as well), depends on F only through $\mathbf{d}F(z)$ (see Abraham, Marsden, and Ratiu [1988], Theorem 4.2.16 for this type of argument). Thus, there is a contravariant antisymmetric two-tensor $B \in \Lambda^2(T^*P)$ such that $B(z)(\alpha_z, \beta_z) = \{F, G\}(z)$, where $\mathbf{d}F(z) = \alpha_z$ and $\mathbf{d}G(z) = \beta_z \in T_z^*P$. This tensor B is called a ***cosymplectic*** or ***Poisson structure***. In local coordinates $(z^1, \ldots, z^n)$, B is determined by its matrix elements $\{z^I, z^J\} = B^{IJ}(z)$ and the bracket becomes

$$\{F, G\} = B^{IJ}(z)\frac{\partial F}{\partial z^I}\frac{\partial G}{\partial z^J}. \tag{10.6.1}$$

Let $B^\sharp : T^*P \to TP$ be the vector bundle map associated to B, that is,

$$B(z)(\alpha_z, \beta_z) = \left\langle \alpha_z, B^\sharp(\beta_z)\right\rangle.$$

Consistent with our conventions $\dot{F} = \{F, H\}$, the Hamiltonian vector field is given by $X_H(z) = B_z^\sharp \cdot \mathbf{d}H(z)$. Indeed, $\dot{F}(z) = \mathbf{d}F(z) \cdot X_H(z)$ and

$$\{F, H\}(z) = B(z)(\mathbf{d}F(z), \mathbf{d}H(z)) = \mathbf{d}F(z) \cdot B^\sharp(z) \cdot \mathbf{d}H(z).$$

Comparing these expressions gives the stated result.

A convenient way to specify a bracket in finite dimensions is by giving the coordinate relations $\{z^I, z^J\} = B^{IJ}(z)$. The Jacobi identity is then implied by the special cases

$$\{\{z^I, z^J\}, z^K\} + \{\{z^K, z^I\}, z^J\} + \{\{z^J, z^K\}, z^I\} = 0,$$

which are equivalent to the differential equations

$$B^{LI}\frac{\partial B^{JK}}{\partial z^L} + B^{LJ}\frac{\partial B^{KI}}{\partial z^L} + B^{LK}\frac{\partial B^{IJ}}{\partial z^L} = 0 \tag{10.6.2}$$

(the terms are cyclic in I, J, K). Writing $X_H[F] = \{F, H\}$ in coordinates gives

$$X_H^I \frac{\partial F}{\partial z^I} = B^{JK}\frac{\partial F}{\partial z^J}\frac{\partial H}{\partial z^K}$$

and so

$$X_H^I = B^{IJ}\frac{\partial H}{\partial z^J}. \tag{10.6.3}$$

This expression tells us that B^{IJ} should be thought of as the negative inverse of the symplectic matrix, which is literally correct in the nondegenerate case. Indeed, if we write out

$$\Omega(X_H, v) = \mathbf{d}H \cdot v$$

in coordinates, we get

$$\Omega_{IJ} X_H^I v^J = \frac{\partial H}{\partial z^J} v^J, \quad \text{i.e.,} \quad \Omega_{IJ} X_H^I = \frac{\partial H}{\partial z^J}.$$

If $[\Omega^{IJ}]$ denotes the inverse of $[\Omega_{IJ}]$, we get

$$X_H^I = \Omega^{JI} \frac{\partial H}{\partial z^J}, \tag{10.6.4}$$

so comparing (10.6.3) and (10.6.4) we see that

$$B^{IJ} = -\Omega^{IJ}.$$

Recalling that the matrix of $\Omega^\sharp$ is the inverse of that of $\Omega^\flat$ and that the matrix of $\Omega^\flat$ is the *negative* of that of Ω, we see that $B^\sharp = \Omega^\sharp$.

Let us prove this abstractly. The basic link between the Poisson tensor B and the symplectic form Ω is that they give the same Poisson bracket:

$$\{F, H\} = B(\mathbf{d}F, \mathbf{d}H) = \Omega(X_F, X_H),$$

that is,

$$\left\langle \mathbf{d}F, B^\sharp \mathbf{d}H \right\rangle = \langle \mathbf{d}F, X_H \rangle .$$

But

$$\Omega(X_H, v) = \mathbf{d}H \cdot v,$$

and so

$$\left\langle \Omega^\flat X_H, v \right\rangle = \langle \mathbf{d}H, v \rangle ,$$

whence,

$$X_H = \Omega^\sharp \mathbf{d}H$$

since $\Omega^\sharp = (\Omega^\flat)^{-1}$. Thus $B^\sharp \mathbf{d}H = \Omega^\sharp \mathbf{d}H$ for all H, and thus,

$$B^\sharp = \Omega^\sharp.$$

Proposition 10.6.1 *For any function $H \in \mathcal{F}(P)$, we have $\mathcal{L}_{X_H} B = 0$.*

Proof By definition, we have

$$B(\mathbf{d}F, \mathbf{d}G) = \{F, G\} = X_G[F]$$

for any locally defined functions F and G on P. Therefore,

$$\mathcal{L}_{X_H}(B(\mathbf{d}F, \mathbf{d}G)) = \mathcal{L}_{X_H} \{F, G\} = \{\{F, G\}, H\} .$$

However, since the Lie derivative is a derivation,

$$\begin{aligned}
&\mathcal{L}_{X_H}(B(\mathbf{d}F, \mathbf{d}G)) \\
&\quad = (\mathcal{L}_{X_H}B)(\mathbf{d}F, \mathbf{d}G) + B(\mathcal{L}_{X_H}\mathbf{d}F, \mathbf{d}G) + B(\mathbf{d}F, \mathcal{L}_{X_H}\mathbf{d}G) \\
&\quad = (\mathcal{L}_{X_H}B)(\mathbf{d}F, \mathbf{d}G) + B(\mathbf{d}\{F, H\}, \mathbf{d}G) + B(\mathbf{d}F, \mathbf{d}\{G, H\}) \\
&\quad = (\mathcal{L}_{X_H}B)(\mathbf{d}F, \mathbf{d}G) + \{\{F, H\}, G\} + \{F, \{G, H\}\} \\
&\quad = (\mathcal{L}_{X_H}B)(\mathbf{d}F, \mathbf{d}G) + \{\{F, G\}, H\},
\end{aligned}$$

by the Jacobi identity. It follows that $(\mathcal{L}_{X_H}B)(\mathbf{d}F, \mathbf{d}G) = 0$ for any locally defined functions $F, G \in \mathcal{F}(U)$. Since any element of T_z^*P can be written as $\mathbf{d}F(z)$ for some $F \in \mathcal{F}(U)$, U open in P, it follows that $\mathcal{L}_{X_H}B = 0$. ■

Suppose that the Poisson tensor B is strongly nondegenerate, that is, it it defines an isomorphism $B^\sharp : \mathbf{d}F(z) \mapsto X_F(z)$ of T_z^*P with T_zP for all $z \in P$. Then P is symplectic and the symplectic form Ω is defined by the formula $\Omega(X_F, X_G) = \{F, G\}$ for any locally defined Hamiltonian vector fields X_F and X_G. One gets $\mathbf{d}\Omega = 0$ from Jacobi's identity – see Exercise **5.5-1**. This is the ***Pauli-Jost Theorem***, due to Pauli [1953] and Jost [1964].

Suppose that the Poisson tensor B is weakly nondegenerate, that is, $B^\sharp$ defines an injective vector bundle map of T^*P to TP. In this case, $\{F, G\} = 0$ for all $G \in \mathcal{F}(U)$ and for all open sets U in P, implies that $\mathbf{d}F = 0$, that is, that F is constant on the connected components of P. It is not true that the condition above implies that P is symplectic as the following counterexample shows. Let $P = \mathbb{R}^2$ with Poisson bracket

$$\{F, G\}(x, y) = y\left(\frac{\partial F}{\partial x}\frac{\partial G}{\partial y} - \frac{\partial F}{\partial y}\frac{\partial G}{\partial x}\right).$$

If $\{F, G\} = 0$ for all G, then F must be constant on both the upper and lower half-planes and hence by continuity it must be constant on $\mathbb{R}^2$. However, $\mathbb{R}^2$ with this Poisson structure is clearly not symplectic.

The subset $B^\sharp(T^*P)$ of TP is called the ***characteristic field*** of the Poisson structure; it need not be a subbundle of TP, in general. Note that skew-symmetry of the tensor B is equivalent to $(B^\sharp)^* = -B^\sharp$, where $(B^\sharp)^* : T^*P \to TP$ is the dual of $B^\sharp$. If P is finite dimensional, the ***rank*** of the Poisson structure at a point $z \in P$ is defined to be the rank of $B^\sharp(z) : T_z^*P \to T_zP$; in local coordinates, it is the rank of the matrix $[B^{IJ}(z)]$. Since the flows of Hamiltonian vector fields preserve the Poisson structure, the rank is constant along such a flow. A Poisson structure for which the rank is everywhere equal to the dimension of the manifold is nondegenerate and hence symplectic.

An injectively immersed submanifold $i : S \to P$ is called a ***Poisson immersion*** if any Hamiltonian vector field defined on an open subset of P containing $i(S)$ is in the range of T_zi at all points $i(z)$ for $z \in S$. This is equivalent to the following assertion:

Proposition 10.6.2 *An immersion $i : S \to P$ is Poisson iff it satisfies the following condition. If $F, G : V \subset S \to \mathbb{R}$, where V is open in S, and if $\overline{F}, \overline{G} : U \to \mathbb{R}$ are extensions of $F \circ i^{-1}, G \circ i^{-1} : i(V) \to \mathbb{R}$ to an open neighborhood U of $i(V)$ in P, then $\{\overline{F}, \overline{G}\}|i(V)$ is well defined and independent of the extensions.*

Proof If $i : S \to P$ is an injectively immersed Poisson manifold, then

$$\begin{aligned}\{\overline{F}, \overline{G}\}(i(z)) &= \mathbf{d}\overline{F}(i(z)) \cdot X_{\overline{G}}(i(z)) = \mathbf{d}\overline{F}(i(z)) \cdot T_z i(v) \\ &= \mathbf{d}(\overline{F} \circ i)(z) \cdot v = \mathbf{d}F(z) \cdot v,\end{aligned}$$

where $v \in T_z S$ is the unique vector satisfying $X_{\overline{G}}(i(z)) = T_z i(v)$. Thus $\{\overline{F}, \overline{G}\}(i(z))$ is independent on the extension $\overline{F}$ of $F \circ i^{-1}$. By skew-symmetry of the bracket, it is also independent on the extension $\overline{G}$ of $G \circ i^{-1}$. Then one can define a Poisson structure on S by setting $\{F, G\} = \{\overline{F}, \overline{G}\}|i(V)$ for any open subset V of S. In this way $i : S \to P$ becomes a Poisson map and the computation above shows that $X_{\overline{G}}(i(z)) = T_z i(X_G)$. Conversely, assume that the condition on the bracket stated above holds and let $H : U \to P$ be a Hamiltonian defined on an open subset U of P intersecting $i(S)$. Then, by what was already shown, S is a Poisson manifold and $i : S \to P$ is a Poisson map. We claim that if $z \in S$ is such that $i(z) \in U$, we have $X_H(i(z)) = T_z i(X_{H \circ i}(z))$, and thus $X_H(i(z)) \in$ range $T_z i$, thereby showing that S is a Poisson manifold. To see this, let $K : U \to R$ be an arbitrary function. We have

$$\begin{aligned}\mathbf{d}K(i(z)) \cdot X_H(i(z)) &= \{K, H\}(i(z)) = \{K \circ i, H \circ i\}(z) \\ &= \mathbf{d}(K \circ i)(z) \cdot X_{H \circ i}(z) \\ &= \mathbf{d}K(i(z)) \cdot T_z i(X_{H \circ i}(z)).\end{aligned}$$

Since K is arbitrary, we conclude that $X_H(i(z)) = T_z i(X_{H \circ i}(z))$. ■

If $S \subset P$ is a submanifold of P and the inclusion i is Poisson, we say that S is a ***Poisson submanifold*** of P. Note that the only immersed Poisson submanifolds of a symplectic manifold are those whose range in P is open since for any (weak) symplectic manifold P, we have

$$T_z P = \{X_H(z) \mid H \in \mathcal{F}(U), \quad U \text{ open in } P\}.$$

In particular, the only Poisson submanifolds of a symplectic manifold P are its open sets.

Definition 10.6.3 *Let P be a Poisson manifold. We say that $z_1, z_2 \in P$ are* ***on the same symplectic leaf*** *of P if there is a piecewise smooth curve in P joining z_1 and z_2, each segment of which is a trajectory of a locally defined Hamiltonian vector field. This is clearly an equivalence relation and an equivalence class is called a* ***symplectic leaf***. *The symplectic leaf containing the point z is denoted Σ_z.*

Theorem 10.6.4 (Symplectic Stratification Theorem) *Let P be a finite dimensional Poisson manifold. Then P is the disjoint union of its symplectic leaves. Each symplectic leaf in P is an injectively immersed Poisson submanifold and the induced Poisson structure on the leaf is symplectic. The dimension of the leaf through a point z equals the rank of the Poisson structure at that point.*

The Poisson bracket on P can be alternatively described as follows.

> *To evaluate the Poisson bracket of F and G at $z \in P$, restrict F and G to the symplectic leaf Σ through z, take their bracket on Σ, (in the sense of brackets on a symplectic manifold), and evaluate at z.*

Also note that since the Casimir functions have differentials that annihilate the characteristic field, they are constant on symplectic leaves.

To get a feeling for the geometric content of the symplectic stratification theorem, let us first prove it under the assumption that the characteristic field is a smooth vector subbundle of TP which is the case considered originally by Lie [1890]. In finite dimensions, this is guaranteed if the rank of the Poisson structure is constant. Jacobi's identity shows that the characteristic field is involutive and thus by the Frobenius theorem, it is integrable. Therefore P is foliated by injectively immersed submanifolds whose tangent space at any point coincides with the subspace of all Hamiltonian vector fields evaluated at z. Thus each such leaf Σ is an immersed Poisson submanifold of P. Define the two-form Ω on Σ by

$$\Omega(z)(X_F(z), X_G(z)) = \{F, G\}(z)$$

for any functions F, G defined on a neighborhood of z in P. Note that Ω is closed by the Jacobi identity (Exercise **5.5-1**). Also, if $\{F, G\}(z) = -\mathbf{d}G(z) \cdot X_F(z) = 0$, for all locally defined G, then $X_F(z) = 0$ by the Hahn-Banach theorem, thus showing that Ω is weakly nondegenerate and thereby proving the theorem for the constant rank case.

The general case, proved by Kirillov [1976a], is more subtle since for differentiable distributions which are not subbundles, integrability and involutivity are not equivalent. To prove this case, we proceed in a series of technical propositions.[2]

Proposition 10.6.5 *Let P be a finite dimensional Poisson manifold with $B_z^\sharp : T_z^*P \to T_zP$ the Poisson tensor. Take $z \in P$ and functions $f_1, \ldots, f_k$ defined on P such that $\{B_z^\sharp df_j\}_{1 \le j \le k}$ is a basis of the range of $B_z^\sharp$. Let $\Phi_{j,t}$ be the local flow defined in a neighborhood of z generated by the Hamiltonian vector field $X_{f_j} = B^\sharp df_j$. Let*

$$\Psi^z_{f_1,\ldots,f_k}(t_1, \ldots, t_k) = (\Phi_{1,t_1} \circ \cdots \circ \Phi_{k,t_k})(z)$$

---

[2]This proof was kindly supplied by O. Popp

for small enough $t_1, \ldots, t_k$. Then:

(i) *There is an open neighborhood U_δ of $0 \in \mathbb{R}^k$ such that:*

$$\Psi^z_{f_1,\ldots,f_k} : U_\delta \to P$$

is an embedding.

(ii) *The ranges of $(T\Psi^z_{f_1,\ldots,f_k})(t)$ and $B_{\Psi^z_{f_1,\ldots,f_k}(t)}$ are equal for $t \in U_\delta$.*

(iii) $\Psi^z_{f_1,\ldots,f_k}(U_\delta) \subset \Sigma_z$.

(iv) *If $\Psi^y_{g_1,\ldots,g_k} : U_\eta \to P$ is another map constructed as above and $y \in \Psi^z_{f_1,\ldots,f_k}(U_\delta)$ then there is $U_\epsilon \subset U_\eta$ (an open subset) such that $\Psi^y_{g_1,\ldots,g_k}$ is a diffeomorphism from U_ϵ to an open subset in $\Psi^z_{f_1,\ldots,f_k}(U_\delta)$.*

Proof (i) The smoothness of $\Psi^z_{f_1,\ldots,f_k}$ follows from the smoothness of $\Phi_{j,t}$ in both the flow parameter and manifold variables. Then

$$T_0\Psi^z_{f_1,\ldots,f_k}(\partial/\partial x^j) = X_{f_j}(z) = B^\sharp_z df_j$$

which shows that $T_0\Psi^z_{f_1,\ldots,f_k}$ is injective. It follows that $\Psi^z_{f_1,\ldots,f_k}$ is an embedding on a sufficiently small neighborhood of 0, say U_δ. Notice also that the ranges of $T_0\Psi^z_{f_1,\ldots,f_k}$ and of $B^\sharp_z$ coincide.

(ii) From **10.5.2** we recall that for any invertible Poisson map Φ on P, we have $T\Phi \cdot X_f = X_{f\circ\Phi^{-1}} \circ \Phi$ and from **10.5.1** we know that the Hamiltonian flows are Poisson maps. Therefore, if $t = (t_1, \ldots, t_k)$,

$$\begin{aligned} &\frac{\partial}{\partial t_j} T_t\Psi^z_{f_1,\ldots,f_k} \\ &\quad= (T\Phi_{1,t_1} \circ \ldots \circ T\Phi_{j-1,t_{j-1}} \circ X_{f_j} \circ \Phi_{j+1,t_{j+1}} \circ \ldots \circ \Phi_{k,t_k})(z) \\ &\quad= (X_{h_j} \circ \Psi^z_{f_1,\ldots,f_k})(t), \end{aligned}$$

where

$$h_j = f_j \circ \left(\Phi_{1,t_1} \circ \ldots \circ \Phi_{j-1,t_{j-1})}\right)^{-1}.$$

This shows that

$$\operatorname{range} T_t\Psi^x_{f_1,\ldots,f_k} \subset \operatorname{range} B^\sharp_{\Psi^x_{f_1,\ldots,f_k}(t)}$$

if $t \in U_\delta$. Since $B^\sharp$ is invariant under Hamiltonian flows, it follows that

$$\dim\operatorname{range} B^\sharp_{\Psi^z_{f_1,\ldots,f_k}(t)} = \dim\operatorname{range} B^\sharp_z.$$

This last equality, the previous inclusion, and the last remark in the proof of (i) above conclude (ii).

(iii) This is obvious since $\Psi^z_{f_1,\ldots,f_k}$ is built from piecewise Hamiltonian curves starting from z.

(iv) Note that $X_g(z) \in \operatorname{range} B^\sharp_z$ for any $z \in P$ and any smooth function g. Using (ii), we see that X_g is tangent to the image of $\Psi^z_{f_1,\ldots,f_k}$. Therefore the integral curves of X_g remain tangent to $\Psi^z_{f_1,\ldots,f_k}(U_\delta)$ if they start from that set. To get $\Psi^y_{g_1,\ldots,g_k}$ we just have to find Hamiltonian curves which start from y. Therefore we can restrict ourselves to the submanifold $\Psi^z_{f_1,\ldots,f_k}(U_\delta)$ when computing the flows along the Hamiltonian vector fields X_{g_j}; therefore we can consider that the image of $\Psi^y_{g_1,\ldots,g_k}$ is in $\Psi^z_{f_1,\ldots,f_k}(U_\delta)$; now the derivative at $0 \in \mathbb{R}^k$ of $\Psi^y_{g_1,\ldots,g_k}$ is an isomorphism to the tangent space of $\Psi^z_{f_1,\ldots,f_k}(U_\delta)$ at y (that is, $\operatorname{range} B^\sharp_y$), using (ii) above. Thus, the existence of the neighborhood U_ϵ follows from the inverse function theorem. ■

Proposition 10.6.6 *Let P be a Poisson manifold and B its Poisson tensor. Then for each symplectic leaf $\Sigma \subset P$, the family of charts satisfying* (i) *in the previous proposition, namely,*

$$\{\Psi^z_{f_1,\ldots,f_k} \mid z \in \Sigma,\ \{B^\sharp_z\, df_j\}_{1 \le j \le k} \quad \text{a basis for } \operatorname{range} B^\sharp_z\ \},$$

gives Σ a structure of differentiable manifold such that the inclusion is an immersion. Then $T_z\Sigma = \operatorname{range} B^\sharp_z$ (so $\dim \Sigma = \operatorname{rank} B^\sharp_z$) for all $z \in \Sigma$. Moreover, Σ has a unique symplectic structure such that the inclusion is a Poisson map.

Proof Let $w \in \Psi^z_{f_1,\ldots,f_k}(U_\delta) \cap \Psi^y_{g_1,\ldots,g_k}(U_\epsilon)$ and consider $\Psi^w_{h_1,\ldots,h_k} : U_\gamma \to P$. Using (iv) in the proposition above, we can choose U_γ small enough so that

$$\Psi^w_{h_1,\ldots,h_k}(U_\gamma) \subset \Psi^z_{f_1,\ldots,f_k}(U_\delta) \cap \Psi^y_{g_1,\ldots,g_k}(U_\epsilon)$$

is a diffeomorphic embedding in both $\Psi^z_{f_1,\ldots,f_k}(U_\delta)$ and $\Psi^y_{g_1,\ldots,g_k}(U_\epsilon)$. This shows that the transition maps for the given charts are diffeomorphisms and so define the structure of a differentiable manifold on Σ. The fact that the inclusion is an immersion follows from (i) of the above proposition. We get the tangent space of Σ using (i), (ii) of the previous proposition; then the equality of dimensions follows.

It follows from the definition of an immersed Poisson submanifold that Σ is such a submanifold of P. Thus, if $i : \Sigma \to P$ is the inclusion,

$$\{f \circ i, g \circ i\}_\Sigma = \{f, g\} \circ i.$$

Hence if $\{f \circ i, g \circ i\}_\Sigma(z) = 0$ for all functions g then $\{f, g\}(z) = 0$ for all g, that is, $(X_g f)(z) = 0$ for all g. This implies that $df | T_z\Sigma = 0$ since the vectors $X_g(z)$ span $T_z\Sigma$. Therefore, $i^*\mathbf{d}f = \mathbf{d}(f \circ i) = 0$, which shows that the Poisson tensor on Σ is nondegenerate and thus Σ is a symplectic manifold. This proves the proposition and also completes the proof of the symplectic stratification theorem. ■

There is another proof of the above theorem (using the same idea as for the Darboux coordinates) in Weinstein [1983] (see Libermann and Marle [1987] also.) The proof given above is along the Frobenius integrability idea. Actually it can be used to produce a proof of the generalized Frobenius theorem.

Theorem 10.6.7 (Singular Frobenius Theorem) *Let D be a distribution of subspaces of the tangent bundle of a finite dimensional manifold M i.e., $D_x \subset T_xM$ as x varies in M. Suppose it is smooth in the sense that for each x there are smooth vector fields X_i in D such that $X_i(x)$ give a basis of D_x. Then D is integrable i.e., for each $x \in M$ there is an immersed submanifold $\Sigma_x \subset M$ with $T_x\Sigma_x = D_x$ if and only if the distribution D is invariant under the (local) flows along vector fields with values in D.*

Proof The only if part follows easily. For the if part we remark that the proof of the theorem above can be reproduced here replacing range $B_z^\sharp$ by D_x and the Hamiltonian vector fields with vector fields in D. The crucial property needed to prove (ii) in the above proposition (i.e. Hamiltonian fields remain Hamiltonian under Hamiltonian flows) is replaced by the invariance of D given in the hypothesis. ■

Remarks

(i) The conclusion of the above theorem is the same as the Frobenius integrability theorem but it is not supposed that the dimension of D_x is constant.

(ii) Analogous to the symplectic leaves of a Poisson manifold, we can define the *maximal integral manifolds* of the integrable distribution D using curves along vector fields in D instead of Hamiltonian vector fields. They are also immersed submanifolds in M.

(iii) The condition that (local) flows of the vector fields with values in D leave D invariant implies the involution property of D, that is, $[X, Y]$ is a vector field with values in D if both X and Y are vector fields with values in D (use (4.3.7)). But the involution property alone is not enough to guarantee that D is integrable (if the dimension of D is not constant).

(iv) This generalization of the Frobenius integrability theorem is due to Hermann [1964], Stefan [1974], Sussman [1973]; see also Libermann and Marle [1987].

Examples

(a) Let $P = \mathbb{R}^3$ with the rigid body bracket. Then the symplectic leaves are spheres centered at the origin. The single point at the origin is the singular leaf in the sense that the Poisson structure has rank zero there. As we shall see later, it is true more generally that the symplectic leaves in $\mathfrak{g}^*$ with the Lie-Poisson bracket are the coadjoint orbits. ♦

(b) Symplectic leaves need not be submanifolds and *one cannot conclude that if all the Casimir functions are constants then the Poisson structure is nondegenerate.* For example, consider $\mathbb{T}^3$ with a codimension 1 foliation with dense leaves, such as obtained by taking the leaves to be the product of $\mathbb{T}^1$ with a leaf of the irrational flow on $\mathbb{T}^2$. Put the usual area element on these leaves and define a Poisson structure on $\mathbb{T}^3$ by declaring these to be the symplectic leaves. Any Casimir function is constant, yet the Poisson structure is degenerate. ♦

Related to the stratification theorem is an analogue of Darboux' theorem. To state it, first recall from Exercise **10.5-2** that we define the product Poisson structure on $P_1 \times P_2$ where P_1, P_2 are Poisson manifolds by the requirements that the projections $\pi_1 : P_1 \times P_2 \to P$ and $\pi_2 : P_1 \times P_2 \to P_2$ are Poisson mappings, and $\pi_1^*(\mathcal{F}(P_1))$ and $\pi_2^*(\mathcal{F}(P_2))$ are commuting subalgebras of $\mathcal{F}(P_1 \times P_2)$. In terms of coordinates, if bracket relations $\{z^I, z^J\} = B^{IJ}(z)$ and $\{w^I, w^J\} = C^{IJ}(w)$ are given on P_1 and P_2, respectively, then these define a bracket on functions of z^I and w^J when augmented by the relations $\{z^I, w^J\} = 0$.

Theorem 10.6.8 (Lie-Weinstein) *Let z_0 be a point in a Poisson manifold P. There is a neighborhood U of z_0 in P and an isomorphism $\varphi = \varphi_S \times \varphi_N : U \to S \times N$, where S is symplectic, N is Poisson, and the rank of N at $\varphi_N(z_0)$ is zero. The factors S and N are unique up to local isomorphism. Moreover, if the rank of the Poisson manifold is constant near z_0, there are coordinates $(q^1, \ldots, q^k, p_1, \ldots, p_k, y^1, \ldots, y^l)$ near x_0 satisfying the canonical bracket relations $\{q^i, q^j\} = \{p_i, p_j\} = \{q^i, y^j\} = \{p_i, y^j\} = 0$, $\{q^i, p_j\} = \delta^i_j$.*

In this theorem, the manifold S can be taken to be the symplectic leaf of P through z_0 and N is, locally, any submanifold of P, transverse to S, and such that $S \cap N = \{z_0\}$. In many cases the transverse structure on N is of Lie-Poisson type. For the proof of this theorem and related results, see Weinstein [1983b]; the second part of the theorem is due to Lie [1890]. For the main examples in this book, we shall not require a detailed local analysis of their Poisson structure, so we shall forego a more detailed study of the local structure of Poisson manifolds.

We saw at the beginning of this section that the matrix $[B^{IJ}]$ of the Poisson tensor B converts the differential

$$\mathbf{d}H = \frac{\partial H}{\partial z^I} dz^I$$

of a function to the corresponding Hamiltonian vector field; this is consistent with our treatment in the Introduction and Overview. Another basic concept, that of a Poisson map, is also worthwhile working out in coordinates.

Let $f : P_1 \to P_2$ be a Poisson map, so $\{F \circ f, G \circ f\}_1 = \{F, G\}_2 \circ f$. In coordinates z^I on P_1 and w^K on P_2, and writing $w^K = w^K(z^I)$ for the map f, this reads

$$\frac{\partial}{\partial z^I}(F \circ f)\frac{\partial}{\partial z^J}(G \circ f)B_1^{IJ}(z) = \frac{\partial F}{\partial w^K}\frac{\partial G}{\partial w^L}B_2^{KL}(w).$$

By the chain rule, this is equivalent to

$$\frac{\partial F}{\partial w^K}\frac{\partial w^K}{\partial z^I}\frac{\partial G}{\partial w^L}\frac{\partial w^L}{\partial z^J}B_1^{IJ}(z) = \frac{\partial F}{\partial w^K}\frac{\partial G}{\partial w^L}B_2^{KL}(w).$$

Since F and G are arbitrary, *f is Poisson iff*

$$B_1^{IJ}(z)\frac{\partial w^K}{\partial z^I}\frac{\partial w^L}{\partial z^J} = B_2^{KL}(w).$$

Intrinsically, regarding $B_1(z)$ as a map $B_1(z) : T_z^*P_1 \times T_z^*P_1 \to \mathbb{R}$, this reads

$$B_1(z)(T_z^*f \cdot \alpha_w, T_z^*f \cdot \beta_w) = B_2(w)(\alpha_w, \beta_w), \tag{10.6.5}$$

where $\alpha_w, \beta_w \in T_w^*P_2$ and $f(z) = w$. In other words, *f is Poisson iff*

$$f^*B_2 = B_1. \tag{10.6.6}$$

Exercises

10.6-1 *If $H \in \mathcal{F}(P)$, where P is a Poisson manifold, show that the flow φ_t of X_H preserves the symplectic leaves of P.*

10.6-2 *Let $(P, \{\ ,\ \})$ be a Poisson manifold with Poisson tensor $B \in \Omega_2(P)$. Let $B^\sharp : T^*P \to TP, B^\sharp(\mathbf{d}H) = X_H$, be the induced bundle map. We shall denote by the same symbol $B^\sharp : \Omega^1(P) \to \mathfrak{X}(P)$ the induced map on the sections. The definitions introduced in §10.3 and §10.6 read*

$$B(\mathbf{d}F, \mathbf{d}H) = \left\langle \mathbf{d}F, B^\sharp(\mathbf{d}H)\right\rangle = \{F, H\}.$$

Define $\alpha^\sharp := B^\sharp(\alpha)$. In analogy with exercise **5.5-4**, *define for any $\alpha, \beta \in \Omega^1(P)$,*

$$\{\alpha, \beta\} = -\mathcal{L}_{\alpha^\sharp}\beta + \mathcal{L}_{\beta^\sharp}\alpha - \mathbf{d}(B(\alpha, \beta)).$$

(a) *Show that if the Poisson bracket on P is induced by a symplectic form Ω, that is, if $B^\sharp = \Omega^\sharp$, then*

$$B(\alpha, \beta) = \Omega(\alpha^\sharp, \beta^\sharp).$$

(b) *Show that, for any* $F, G \in \mathcal{F}(P)$*, we have*

$$\{F\alpha, G\beta\} = FG\{\alpha, \beta\} - \alpha^\sharp[G]\beta + \beta^\sharp[F]\alpha.$$

(c) *Show that, for any* $F, G \in \mathcal{F}(P)$ *we have*

$$\mathbf{d}\{F, G\} = \{\mathbf{d}F, \mathbf{d}G\}.$$

(d) *Show that, if* $\alpha, \beta \in \Omega^1(P)$ *are closed, then,* $\{\alpha, \beta\} = \mathbf{d}(B(\alpha, \beta))$.

(e) *Use* $\mathcal{L}_{X_H} B = 0$ *to show that* $\{\alpha, \beta\}^\sharp = -[\alpha^\sharp, \beta^\sharp]$.

(f) *Show that* $(\Omega^1(P), \{\ ,\ \})$ *is a Lie algebra; that is, prove Jacobi's identity.*

10.6-3 *Let* P *be a manifold and* X, Y *be two linearly independent commuting vector fields. Show that*

$$\{F, K\} = X[F]Y[K] - Y[F]X[K]$$

defines a Poisson bracket on P*. Show that*

$$X_H = Y[H]X - X[H]Y.$$

Show that the symplectic leaves are two-dimensional and have a tangent space spanned by X *and* Y*. Show how to get Example* **(b)** *preceding* **10.6.8** *from this construction.*

10.7 Quotients of Poisson Manifolds

Here we shall give the simplest version of a general construction of Poisson manifolds based on symmetry. This construction represents the first steps in a general procedure called ***reduction***.

Suppose that G is a Lie group that acts on a Poisson manifold and that each map $\Phi_g : P \to P$ is a Poisson map. Let us also suppose that the action is free and proper, so that the quotient space P/G is a smooth manifold and the projection $\pi : P \to P/G$ is a submersion (see the discussion of this point in §9.3).

Theorem 10.7.1 *Under these hypotheses, there is a unique Poisson structure on* P/G *such that* π *is a Poisson map.*

Proof Let us first assume P/G is Poisson and shows uniqueness. The condition that π be Poisson is that for two functions $f, k : P/G \to \mathbb{R}$,

$$\{f, k\} \circ \pi = \{f \circ \pi, k \circ \pi\}, \tag{10.7.1}$$

where the brackets are on P/G and P, respectively. The function $\overline{f} = f \circ \pi$ is the unique G-invariant function that projects to f. In other words, if $[z] \in P/G$ is an equivalence class, whereby $g_1 \cdot z$ and $g_2 \cdot z$ are equivalent, we let $\overline{f}(g \cdot z) = f([z])$ for all $g \in G$. Obviously, this defined $\overline{f}$ unambiguously, so that $\overline{f} = f \circ \pi$. We can also characterize this as saying $\overline{f}$ assigns the value $f([z])$ to the whole orbit $G \cdot z$. We can write (10.7.1) as $\{f, k\} \circ \pi = \{\overline{f}, \overline{g}\}$. Since π is onto, this determines $\{f, k\}$ uniquely.

We can also use (10.7.1) to *define* $\{f, k\}$. First, note that

$$\begin{aligned} \{\overline{f}, \overline{g}\}(gz) &= \{\overline{f}, \overline{g}\} \circ \Phi_g(z) \\ &= \{\overline{f} \circ \Phi_g, \overline{g} \circ \Phi_g\}(z) \\ &= \{\overline{f}, \overline{g}\}(z), \end{aligned}$$

since Φ_g is Poisson and since $\overline{f}$ and $\overline{g}$ are constant on orbits. Thus, $\{\overline{f}, \overline{g}\}$ is constant on orbits too, and so it defines $\{f, g\}$ uniquely.

It remains to show that $\{f, g\}$ so defined satisfies the properties of a Poisson structure. However, these all follow from their counterparts on P. For example, if we write Jacobi's identity on P, namely

$$0 = \{\{\overline{f}, \overline{k}\}, \overline{l}\} + \{\{\overline{l}, \overline{f}\}, \overline{k}\} + \{\{\overline{k}, \overline{l}\}, \overline{f}\},$$

it gives, by construction,

$$\begin{aligned} 0 &= \{\{f, k\} \circ \pi, l \circ \pi\} + \{\{l, f\} \circ \pi, k \circ \pi\} + \{\{k, l\} \circ \pi, f \circ \pi\} \\ &= \{\{f, k\}, l\} \circ \pi + \{\{l, f\}, k\} \circ \pi + \{\{k, l\}, f\} \circ \pi \end{aligned}$$

and thus Jacobi's identity holds on P/G. ■

This construction is just one of many that produce new Poisson and symplectic manifolds from old ones. We shall encounter more constructions of this type in Volume II; we also refer to Marsden and Ratiu [1986] for a generalization of the construction here.

If H is a G-invariant Hamiltonian on P, it defines a corresponding function h on P/G such that $H = h \circ \pi$. Since π is a Poisson map, it transforms X_H on P to X_h on P/G; that is, $T\pi \circ X_H = X_h \circ \pi$, or X_H and X_h are π-related. We say that the Hamiltonian system X_H on P ***reduces*** to that on P/G.

As we shall see in the next chapter, G-invariance of H may be associated with a conserved quantity $J : P \to \mathbb{R}$. If it is also G-invariant, the corresponding function j on P/G is conserved for X_h since

$$\{h, j\} \circ \pi = \{H, J\} = 0$$

and so $\{h, j\} = 0$.

Example Consider the differential equations on $\mathbb{C}^2$ given by

$$\left.\begin{aligned}\dot{z}_1 &= -i\omega_1 z_1 + i\epsilon p\bar{z}_2 + iz_1(s_{11}|z_1|^2 + s_{12}|z_2|^2),\\ \dot{z}_2 &= -i\omega_2 z_2 + i\epsilon q\bar{z}_1 - iz_2(s_{21}|z_1|^2 + s_{22}|z_2|^2).\end{aligned}\right\} \tag{10.7.2}$$

Use the standard Hamiltonian structure obtained by taking the real and imaginary parts of z_i as conjugate variables. For example, we write $z_1 = q_1 + ip_1$ and require $\dot{q}_1 = \partial H/\partial p_1$ and $\dot{p}_1 = -\partial H/\partial q_1$. Recall from Chapter 5 that a useful trick in this regard, that enables one to work in complex notation, is to write Hamilton's equations as $\dot{z}_k = -2i\partial H/\partial \bar{z}_k$. Using this, one readily finds that (see Exercise **5.4-3**): *The system* (10.7.2) *is Hamiltonian if and only if* $s_{12} = -s_{21}$ *and* $p = q$. In this case we can choose

$$H(z_1, z_2) = \frac{1}{2}(\omega_2|z_2|^2 + \omega_1|z_1|^2) - \epsilon p \operatorname{Re}(z_1 z_2) - \frac{s_{11}}{4}|z_1|^4 - \frac{s_{12}}{2}|z_1 z_2|^2 + \frac{s_{22}}{4}|z_2|^4. \tag{10.7.3}$$

Note that for (10.7.2) with $\epsilon = 0$ there are two separate S^1 actions acting on z_1 and z_2 independently; corresponding conserved quantities are $|z_1|^2$ and $|z_2|^2$. However, for $\epsilon \neq 0$, the symmetry action is

$$(z_1, z_2) \mapsto (e^{i\theta} z_1, e^{-i\theta} z_2) \tag{10.7.4}$$

with the conserved quantity (Exercise **5.5-3**)

$$J(z_1, z_2) = \frac{1}{2}(|z_1|^2 - |z_2|^2). \tag{10.7.5}$$

Let $\phi = \frac{\pi}{2} - \theta_1 - \theta_2$, where $z_1 = r_1 \exp(i\theta_1)$, $z_2 = r_2 \exp(i\theta_2)$. We know that the Hamiltonian structure for (10.7.2) on $\mathbb{C}^2$ described above induces one on $\mathbb{C}^2/S^1$ (exclude points where r_1 or r_2 vanishes), and that the two integrals (energy and the conserved quantity) descend to the quotient space, as does the Poisson bracket. The quotient space $\mathbb{C}^2/S^1$ is parametrized by (r_1, r_2, ϕ) and H and J can be dropped to the quotient. Concretely, the process of dropping to the quotient is very simple: if $F(z_1, z_2) = F(r_1, \theta_1, r_2, \theta_2)$ is S^1 invariant, then it can be written (uniquely) as a function f of (r_1, r_2, φ).

One can also drop the Poisson bracket to the quotient by Theorem **10.7.1**. Consequently, the equations in (r_1, r_2, ϕ) can be cast in Hamiltonian form $\dot{f} = \{f, h\}$ for the induced Poisson bracket. This bracket is obtained by using the chain rule to relate the complex variables and the polar coordinates. One finds that

$$\{f, k\}(r_1, r_2, \phi) = -\frac{1}{r_1}\left(\frac{\partial f}{\partial r_1}\frac{\partial k}{\partial \phi} - \frac{\partial f}{\partial \phi}\frac{\partial k}{\partial r_1}\right) - \frac{1}{r_2}\left(\frac{\partial f}{\partial r_2}\frac{\partial k}{\partial \phi} - \frac{\partial f}{\partial \phi}\frac{\partial k}{\partial r_2}\right). \tag{10.7.6}$$

The Poisson bracket (10.7.6) is, of course, nothing but the original *canonical* Poisson bracket on the space of q and p variables, but written in the new polar coordinate variables, and as such, is an example of a *noncanonical* bracket. Notice that by construction, Jacobi's identity is automatic for this reduced bracket. (See Knobloch, Mahalov, and Marsden [1994] for further examples of this type.) ♦

As we shall see in Chapter 13, a key example of the Poisson reduction given in **10.7.1** is when $P = T^*G$ and G acts on itself by left translations. Then $P/G \cong \mathfrak{g}^*$ and the reduced Poisson bracket is none other than the Lie-Poisson bracket!

Exercises

10.7-1 *Let $\mathbb{R}^3$ be equipped with the rigid body bracket and let $G = S^1$ act on $P = \mathbb{R}^3 \setminus$ (z-axis) by rotation about the z-axis. Compute the induced bracket on P/G.*

10.7-2 *Compute explicitly the reduced Hamiltonian h in the example in the text and verify directly that the equations for $\dot{r}_1, \dot{r}_2, \dot{\varphi}$ are Hamiltonian on $\mathbb{C}^2$ with Hamiltonian h. Also check that the function j induced by J is a constant of the motion.*

10.8 The Schouten Bracket

The goal of this subsection is to express the Jacobi identity for a Poisson structure in geometric terms analogous to $\mathbf{d}\Omega$ for symplectic structures. This will be done in terms of a bracket defined on contravariant antisymmetric tensors generalizing the Lie bracket of vector fields (see, for example, Schouten [1940], Nijenhuis [1953], Lichnerowicz [1978], Olver [1984, 1986], Koszul [1985], Libermann and Marle [1987], Bhaskara and Viswanath [1988], Kosman-Schwarzbach and Magri [1990], Vaisman [1994], and references therein).

A ***contravariant antisymmetric*** q***-tensor*** on a finite-dimensional vector space V is a q-linear map $A : V^* \times V^* \times \cdots \times V^*$ (q times) $\to \mathbb{R}$ that is antisymmetric in each pair of arguments. The space of these tensors will be denoted by $\Lambda_q(V)$. Thus each element $\Lambda_q(V)$ is a finite linear combination of terms of the form $v_1 \wedge \cdots \wedge v_q$, called a q***-vector***, for $v_1, \ldots, v_q \in V$. If V is an infinite-dimensional Banach space, we define $\Lambda_q(V)$ to be the span of all elements of the form $v_1 \wedge \cdots \wedge v_q$ with $v_1, \ldots, v_q \in V$, where the exterior product is defined in the usual manner relative to a weakly nondegenerate pairing $\langle\,,\rangle : V^* \times V \to \mathbb{R}$. Thus $\Lambda_0(V) = \mathbb{R}$ and $\Lambda_1(V) = V$. If P is a smooth manifold, let $\Lambda_q(P) = \bigcup_{z \in P} \Lambda_q(T_zP)$, a smooth vector bundle with fiber over $z \in P$ equal to $\Lambda_q(T_zP)$. Let $\Omega_q(P)$ denote the smooth sections of $\Lambda_q(P)$, that is, the elements of $\Omega_q(P)$ are smooth contravariant antisymmetric q-tensor fields on P. Let $\Omega_*(P)$ be the direct sum of the spaces $\Omega_q(P)$, where $\Omega_0(P) = \mathcal{F}(P)$. Note that $\Omega_q(P) = 0$ for $q > \dim(P)$ and that $\Omega_1(P) = \mathfrak{X}(P)$. If $X_1, \ldots, X_q \in \mathfrak{X}(P)$, $X_1 \wedge \cdots \wedge X_q$ is called a ***q-vector field***.

On the manifold P, consider a $(q+p)$-form α and a contravariant antisymmetric q-tensor A. The ***interior product*** $\mathbf{i}_A\alpha$ of A with α is defined as follows. If $q = 0$, so $A \in \mathbb{R}$, let $\mathbf{i}_A\alpha = A\alpha$. If $q \geq 1$ and if $A = v_1 \wedge \cdots \wedge v_q$, where $v_i \in T_zP$, $i = 1, \ldots, q$, define $\mathbf{i}_A\alpha \in \Omega^p(P)$ by

$$(\mathbf{i}_A\alpha)(v_{q+1}, \ldots, v_{q+p}) = \alpha(v_1, \ldots, v_{q+p}) \tag{10.8.1}$$

for arbitrary $v_{q+1}, \ldots, v_{q+p} \in T_zP$. One checks that the definition does not depend on the representation of A as a q-vector, so $\mathbf{i}_A\alpha$ is well defined on $\Lambda_q(P)$ by linear extension. In local coordinates, for finite-dimensional P,

$$(\mathbf{i}_A\alpha)_{i_{q+1}\ldots i_{q+p}} = A^{i_1\ldots i_q}\alpha_{i_1\ldots i_{q+p}}, \tag{10.8.2}$$

where all components are nonstrict. If P is finite dimensional and $p = 0$, (10.8.1) defines an isomorphism of $\Omega_q(P)$ with $\Omega^q(P)$. If P is a Banach manifold, (10.8.1) defines a weakly nondegenerate pairing of $\Omega_q(P)$ with $\Omega^q(P)$. If $A \in \Omega_q(P)$, q is called the ***degree*** of A and is denoted by $\deg A$. One checks that

$$\mathbf{i}_{A\wedge B}\alpha = \mathbf{i}_B\mathbf{i}_A\alpha. \tag{10.8.3}$$

The Lie derivative $\pounds_X$ is a derivation relative to $\wedge$, that is,

$$\pounds_X(A \wedge B) = (\pounds_X A) \wedge B + A \wedge (\pounds_X B)$$

for any $A, B \in \Omega_*(P)$.

Theorem 10.8.1 (Schouten Bracket Theorem) *There is a unique bilinear operation* $[,] : \Omega_*(P) \times \Omega_*(P) \to \Omega_*(P)$ *natural with respect to restriction to open sets, called the* **Schouten bracket**, *that satisfies the following properties:*

i *it is a* ***biderivation of degree*** -1, *i.e., it is bilinear,*

$$\deg[A, B] = \deg A + \deg B - 1, \tag{10.8.4}$$

and for $A, B, C \in \Omega_*(P)$,

$$[A, B \wedge C] = [A, B] \wedge C + (-1)^{(\deg A+1)\deg B} B \wedge [A, C]; \tag{10.8.5}$$

ii *it is determined on* $\mathcal{F}(P)$ *and* $\mathfrak{X}(P)$ *by*

(a) $[F, G] = 0$, *for all* $F, G \in \mathcal{F}(P)$;

(b) $[X, F] = X[F]$, *for all* $F \in \mathcal{F}(P)$, $X \in \mathfrak{X}(P)$;

(c) $[X, Y]$ *for all* $X, Y \in \mathfrak{X}(P)$ *is the usual Jacobi-Lie bracket of vector fields; and*

iii $[A, B] = (-1)^{\deg A \deg B}[B, A]$.

In addition, the Schouten bracket satisfies the ***graded Jacobi identity***

$$(-1)^{\deg A \deg C}[[A, B], C] + (-1)^{\deg B \deg A}[[B, C], A] + (-1)^{\deg C \deg B}[[C, A], B] = 0. \tag{10.8.6}$$

Proof The proof proceeds in standard fashion and is similar to that characterizing the exterior or Lie derivative by its properties, see (Abraham, Marsden, and Ratiu [1988]): on functions and vector fields it is given by **ii**; then **i** and linear extension determine it on any skew-symmetric contravariant tensor in the second variable and a function and vector field in the first; **iii** tells how to switch such variables and finally **i** again defines it on any pair of skew-symmetric contravariant tensors. The operation so defined satisfies **i**, **ii**, and **iii** by construction. Uniqueness is a consequence of the fact that the skew-symmetric contravariant tensors are generated as an exterior algebra locally by functions and vector fields and **ii**. The graded Jacobi identity is verified on an arbitrary triple of q-, p-, and r-vectors using **i**, **ii**, and **iii** and then invoking trilinearity of the identity. ■

The graded Jacobi identity and bilinearity say that $\Omega_*(P)$ together with the Schouten bracket form a ***super Lie algebra***.

The following formulae are useful in computing with the Schouten bracket. If $X \in \mathfrak{X}(P)$ and $A \in \Omega_p(P)$, induction on the degree of A and the use of property **i** show that

$$[X, A] = \mathcal{L}_X A. \tag{10.8.7}$$

An immediate consequence of this formula and the graded Jacobi identity is the *derivation property of the Lie derivative relative to the Schouten bracket*, that is,

$$\mathcal{L}_X[A, B] = [\mathcal{L}_X A, B] + [A, \mathcal{L}_X B], \tag{10.8.8}$$

for $A \in \Omega_p(P), B \in \Omega_q(P)$, and $X \in \mathfrak{X}(P)$. Using induction on the number of vector fields, (10.8.7), and the properties in Theorem **10.8.1**, one can prove that

$$[X_1 \wedge \cdots \wedge X_r, A] = \sum_{i=1}^{r}(-1)^{i+1} X_1 \wedge \cdots \wedge \check{X}_i \wedge \cdots \wedge X_r \wedge (\mathcal{L}_{X_i} A), \tag{10.8.9}$$

where $X_1, \ldots, X_r \in \mathfrak{X}(P)$ and $\check{X}_i$ means that X_i has been omitted. The last formula plus linear extension can be taken as the definition of the Schouten bracket and one can deduce Theorem **10.8.1** from it; see Vaisman [1994] for this approach. If $A = Y_1 \wedge \cdots \wedge Y_s$ for $Y_1, \cdots, Y_s \in \mathfrak{X}(P)$, the formula above plus the derivation property of the Lie derivative give

$$\begin{aligned}[X_1 \wedge \cdots \wedge X_r, Y_1 \wedge \cdots \wedge Y_s] \\ = \ (-1)^{r+1} \sum_{i=1}^{r} \sum_{j=1}^{s} (-1)^{i+j} [X_i, Y_j] \wedge X_1 \wedge \cdots \wedge \check{X}_i \wedge \cdots \\ \wedge X_r \wedge Y_1 \wedge \cdots \wedge \check{Y}_j \wedge \cdots \wedge Y_s. \qquad (10.8.10)\end{aligned}$$

Finally, if $A \in \Omega_p(P), B \in \Omega_q(P)$, and $\alpha \in \Omega^{p+q-1}(P)$, the formula

$$\mathbf{i}_{[A,B]}\alpha = (-1)^{q(p+1)}\mathbf{i}_A \mathbf{d}\, \mathbf{i}_B \alpha + (-1)^p \mathbf{i}_B \mathbf{d}\, \mathbf{i}_A \alpha - \mathbf{i}_B \mathbf{i}_A \mathbf{d}\alpha \qquad (10.8.11)$$

(which is a direct consequence of (10.8.10) and Cartan's formula for $\mathbf{d}\alpha$) can be taken as the definition of $[A, B] \in \Omega_{p+q-1}(P)$; this is the approach taken originally in Nijenhuis [1955].

In local coordinates, denoting $\partial/\partial z^i = \partial_i$, the formulae (10.8.9) and (10.8.10) imply that

1. $\left[f, \partial_{i_1} \wedge \ldots \wedge \partial_{i_p}\right] = \sum_{k=1}^{p} (-1)^{k-1} \left(\partial_{i_k} f\right) \partial_{i_1} \wedge \cdots \wedge \check{\partial}_{i_k} \wedge \cdots \wedge \partial_{i_p}$ for any functon f, where ˇ over a symbol means that it is deleted, and
2. $\left[\partial_{i_1} \wedge \cdots \wedge \partial_{i_p}, \partial_{j_1} \wedge \cdots \wedge \partial_{j_q}\right] = 0.$

Therefore, if $A = A^{i_1 \ldots i_p} \partial_{i_1} \wedge \cdots \wedge \partial_{i_p}$ and $B = B^{j_1 \ldots j_q} \partial_{j_1} \wedge \cdots \wedge \partial_{j_q}$, we get

$$\begin{aligned}[A, B] \ = \ & A^{\ell i_1 \ldots i_{\ell-1} i_{\ell+1} \ldots i_p} \partial_\ell B^{j_1 \ldots j_q} \partial_{i_1} \wedge \cdots \wedge \partial_{i_{\ell-1}} \wedge \partial_{i_{\ell+1}} \\ & \wedge \partial_{j_1} \wedge \cdots \wedge \partial_{j_q} + (-1)^p B^{\ell j_1 \ldots j_{\ell-1} j_{\ell+1} \ldots j_q} \partial_\ell A^{i_1 \ldots i_p} \partial_{i_1} \\ & \wedge \cdots \wedge \partial_{i_p} \wedge \partial_{j_1} \wedge \cdots \wedge \partial_{j_{\ell-1}} \wedge \partial_{j_{\ell+1}} \wedge \cdots \wedge \partial_{j_q} \qquad (10.8.12)\end{aligned}$$

or, more succinctly,

$$\begin{aligned}[A, B]^{k_2 \ldots k_{p+q}} \ = \ & \varepsilon^{k_2 \ldots k_{p+q}}_{i_2 \ldots i_p j_1 \ldots j_q} A^{\ell i_2 \ldots i_p} \frac{\partial}{\partial x^\ell} B^{j_1 \ldots j_q} \\ & + (-1)^p \varepsilon^{k_2 \ldots k_{p+q}}_{i_1 \ldots i_p j_2 \ldots j_q} B^{\ell j_2 \ldots j_p} \frac{\partial}{\partial x^\ell} A^{i_1 \ldots i_q} \qquad (10.8.13)\end{aligned}$$

where all components are nonstrict. Here

$$\varepsilon^{i_1 \ldots i_{p+q}}_{j_1 \ldots j_{p+q}}$$

is the ***Kronecker symbol***: it is zero if $(i_1, \ldots, i_{p+q}) \neq (j_1, \ldots, j_{p+q})$, and is 1 (resp., -1) if $j_1, \ldots, j_{p+q}$ is an even (resp., odd) permutation of $i_1, \ldots, i_{p+q}$.

From §10.6 the Poisson tensor $B \in \Omega_2(P)$ defined by a Poisson bracket $\{\,,\}$ on P satisfies $B(\mathbf{d}F, \mathbf{d}G) = \{F, G\}$ for any $F, G \in \mathcal{F}(P)$. By (10.8.2), this can be written

$$\{F, G\} = \mathbf{i}_B(\mathbf{d}F \wedge \mathbf{d}G), \qquad (10.8.14)$$

or in local coordinates,

$$\{F, G\} = B^{IJ} \frac{\partial F}{\partial z^I} \frac{\partial G}{\partial z^J}.$$

Writing B locally as a sum of terms of the form $X \wedge Y$ for some $X, Y \in \mathfrak{X}(P)$ and taking $Z \in \mathfrak{X}(P)$ arbitrary, by (10.8.1), we have for $F, G, H \in \mathcal{F}(P)$,

$$\begin{aligned}
&\mathbf{i}_B(\mathbf{d}F \wedge \mathbf{d}G \wedge \mathbf{d}H)(Z) \\
&\quad = (\mathbf{d}F \wedge \mathbf{d}G \wedge \mathbf{d}H)(X, Y, Z) \\
&\quad = \det \begin{bmatrix} \mathbf{d}F(X) & \mathbf{d}F(Y) & \mathbf{d}F(Z) \\ \mathbf{d}G(X) & \mathbf{d}G(Y) & \mathbf{d}G(Z) \\ \mathbf{d}H(X) & \mathbf{d}H(Y) & \mathbf{d}H(Z) \end{bmatrix} \\
&\quad = \det \begin{bmatrix} \mathbf{d}F(X) & \mathbf{d}F(Y) \\ \mathbf{d}G(X) & \mathbf{d}G(Y) \end{bmatrix} \mathbf{d}H(z) + \det \begin{bmatrix} \mathbf{d}H(X) & \mathbf{d}H(Y) \\ \mathbf{d}F(X) & \mathbf{d}F(Y) \end{bmatrix} \mathbf{d}G(Z) \\
&\qquad + \det \begin{bmatrix} \mathbf{d}G(X) & \mathbf{d}G(Y) \\ \mathbf{d}H(X) & \mathbf{d}H(Y) \end{bmatrix} \mathbf{d}F(Z) \\
&\quad = \mathbf{i}_B(\mathbf{d}F \wedge \mathbf{d}G)\mathbf{d}H(Z) + \mathbf{i}_B(\mathbf{d}H \wedge \mathbf{d}F)\mathbf{d}G(Z) + \mathbf{i}_B(\mathbf{d}G \wedge \mathbf{d}H)\mathbf{d}F(Z),
\end{aligned}$$

that is,

$$\begin{aligned}
&\mathbf{i}_B(\mathbf{d}F \wedge \mathbf{d}G \wedge \mathbf{d}H) \\
&\quad = \mathbf{i}_B(\mathbf{d}F \wedge \mathbf{d}G)\mathbf{d}H + \mathbf{i}_B(\mathbf{d}H \wedge \mathbf{d}F)\mathbf{d}G + \mathbf{i}_B(\mathbf{d}G \wedge \mathbf{d}H)\mathbf{d}F \qquad (10.8.15)
\end{aligned}$$

Therefore, (10.8.14) and (10.8.15) imply

$$\begin{aligned}
&\{\{F, G\}, H\} + \{\{H, F\}, G\} + \{\{G, H\}, F\} \\
&\quad = \mathbf{i}_B(\mathbf{d}\{F, G\} \wedge \mathbf{d}H) + \mathbf{i}_B(\mathbf{d}\{H, F\} \wedge \mathbf{d}G) + \mathbf{i}_B(\mathbf{d}\{G, H\} \wedge \mathbf{d}F) \\
&\quad = \mathbf{i}_B\mathbf{d}(\mathbf{i}_B(\mathbf{d}F \wedge \mathbf{d}G)\mathbf{d}H + \mathbf{i}_B(\mathbf{d}H \wedge \mathbf{d}F)\mathbf{d}G + \mathbf{i}_B(\mathbf{d}G \wedge \mathbf{d}H)\mathbf{d}F) \\
&\quad = \mathbf{i}_B\mathbf{d}\,\mathbf{i}_B(\mathbf{d}F \wedge \mathbf{d}G \wedge \mathbf{d}H) \\
&\quad = \frac{1}{2}\mathbf{i}_{[B,B]}(\mathbf{d}F \wedge \mathbf{d}G \wedge \mathbf{d}H),
\end{aligned}$$

the last equality being a consequence of (10.8.11). The identity we just obtained,

$$\{\{F, G\}, H\} + \{\{H, F\}, G\} + \{\{G, H\}, F\} = \frac{1}{2}\mathbf{i}_{[B,B]}(\mathbf{d}F \wedge \mathbf{d}G \wedge \mathbf{d}H) \qquad (10.8.16)$$

shows that Jacobi's identity for $\{\ ,\ \}$ is equivalent to $[B, B] = 0$. Thus *a Poisson structure is uniquely defined by a contravariant antisymmetric two-tensor whose Schouten bracket with itself vanishes.* The local formula (10.8.13) becomes

$$[B, B]^{IJK} = \sum_{L=1}^{n} \left(B^{LK} \frac{\partial B^{IJ}}{\partial z^L} + B^{LI} \frac{\partial B^{JK}}{\partial z^L} + B^{LJ} \frac{\partial B^{KI}}{\partial z^L} \right)$$

which coincides with our earlier expression (10.6.2).

Here is another interesting identity that gives the Lie derivative of the Poisson tensor along a Hamiltonian vector field:

$$\pounds_{X_H} B = \mathbf{i}_{[B,B]} \mathbf{d}H. \tag{10.8.17}$$

Indeed, in coordinates,

$$(\pounds_X B)^{IJ} = X^K \frac{\partial B^{IJ}}{\partial z^K} - B^{IK} \frac{\partial X^J}{\partial z^K} - B^{KJ} \frac{\partial X^I}{\partial z^K}$$

so if $X^I = B^{IJ}(\partial H / \partial z^J)$, this becomes

$$\begin{aligned}
(\pounds_{X_H} B)^{IJ} &= B^{KL} \frac{\partial B^{IJ}}{\partial z^K} \frac{\partial H}{\partial z^L} - B^{IK} \frac{\partial}{\partial z^K} \left(B^{JL} \frac{\partial H}{\partial z^L} \right) \\
&\quad + B^{JK} \frac{\partial}{\partial z^K} \left(B^{IL} \frac{\partial H}{\partial z^L} \right) \\
&= \left(B^{KL} \frac{\partial B^{IJ}}{\partial z^K} - B^{IK} \frac{\partial B^{JL}}{\partial z^K} - B^{KJ} \frac{\partial B^{IL}}{\partial z^K} \right) \frac{\partial H}{\partial z^L} \\
&= [B,B]^{LIJ} \frac{\partial H}{\partial z^L} = \left(\mathbf{i}_{[B,B]} \mathbf{d}H \right)^{IJ},
\end{aligned}$$

so (10.8.17) follows. This shows how Jacobi's identity $[B, B] = 0$ *is directly used to show that the flow φ_t of a Hamiltonian vector field is Poisson.* The above derivation shows that *the flow of a time-dependent Hamiltonian vector field consists of Poisson maps*; indeed, even in this case,

$$\frac{d}{dt} (\varphi_t^* B) = \varphi_t^* (\pounds_{X_H} B) = \varphi_t^* \left(\mathbf{i}_{[B,B]} \mathbf{d}H \right) = 0$$

is valid.

Exercises

10.8-1 *Prove the following formulae by the method indicated in the text.*

(a) *If $A \in \Omega_q(P)$ and $X \in \mathfrak{X}(P)$, then $[X, A] = \mathcal{L}_X A$.*

(b) *If $A \in \Omega_q(P)$ and $X_1, \ldots, X_r \in \mathfrak{X}(P)$, then*

$$[X_1 \wedge \cdots \wedge X_r, A] = \sum_{i=1}^{r} (-1)^{i+1} X_1 \wedge \cdots \wedge \check{X}_i \wedge \cdots \wedge X_r \wedge (\mathcal{L}_{X_i} A).$$

(c) *If $X_1, \ldots, X_r, Y_1, \ldots, Y_s \in \mathfrak{X}(P)$, then*

$$\begin{aligned}
&[X_1 \wedge \cdots \wedge X_r, Y_1 \wedge \cdots \wedge Y_s] \\
&= (-1)^{r+1} \sum_{i=1}^{r} \sum_{j=1}^{s} (-1)^{i+j} [X_i, Y_i] \wedge \\
&\qquad \wedge X_1 \wedge \cdots \wedge \check{X}_i \wedge \cdots \wedge X_r \wedge Y_1 \wedge \cdots \wedge \check{Y}_j \wedge \cdots \wedge Y_s.
\end{aligned}$$

(d) *If* $A \in \Omega_p(P), B \in \Omega_q(P)$, *and* $\alpha \in \Omega^{p+q-1}(P)$, *then*

$$\mathbf{i}_{[A,B]}\alpha = (-1)^{q(p+1)}\mathbf{i}_A\mathbf{d}\,\mathbf{i}_B\alpha + (-1)^p\mathbf{i}_B\mathbf{d}\,\mathbf{i}_A\alpha - \mathbf{i}_B\mathbf{i}_A\mathbf{d}\alpha.$$

10.8-2 *Let* M *be a finite-dimensional manifold. A skew-symmetric contravariant tensor field* $A(x) : T_x^*M \times \cdots \times T_x^*M \to \mathbb{R}$ *(*k *copies of* T_xM*) on* M *is called a* k*-vector field on* M*. Let* $x_0 \in M$ *be such that* $A(x_0) = 0$.

(a) *If* $X \in \mathfrak{X}(M)$, *show that* $(\mathcal{L}_X A)(x_0)$ *depends only on* $X(x_0)$, *thereby defining a map* $\mathbf{d}_{x_0}A : T_{x_0}M \to T_{x_0}M \wedge \cdots \wedge T_{x_0}M$ *(*k *times), called the* ***intrinsic derivative*** *of* A *at* x_0.

(b) *If* $\alpha_1, \ldots, \alpha_k \in T_x^*M, v_1, \ldots, v_k \in T_xM$, *show that*

$$\langle \alpha_1 \wedge \cdots \wedge \alpha_k, v_1 \wedge \cdots \wedge v_k \rangle := \det\left[\langle \alpha_i, v_j \rangle\right]$$

defines a nondegenerate pairing between $T_x^*M \wedge \cdots \wedge T_x^*M$ *and* $T_xM \wedge \cdots \wedge T_xM$. *Conclude that these two spaces are dual to each other, that the space* $\Omega^k(M)$ *of* k*-forms is dual to the space of* k*-contravariant skew-symmetric tensor fields* $\Omega_k(M)$, *and that the bases*

$$\left\{ \mathbf{d}x^{i_1} \wedge \cdots \wedge \mathbf{d}x^{i_k} \;\middle|\; i_1 < \cdots < i_k \right\}$$

and

$$\left\{ \frac{\partial}{\partial x^{i_1}} \wedge \cdots \wedge \frac{\partial}{\partial x^{i_k}} \;\middle|\; i_1 < \cdots < i_k \right\}$$

are dual to each other.

(c) *Show that the dual map*

$$(\mathbf{d}_{x_0}A)^* : T_{x_0}^*M \wedge \cdots \wedge T_{x_0}^*M \to T_{x_0}^*M, A(x_0) = 0,$$

is given by

$$(\mathbf{d}_{x_0}A)^*(\alpha_1 \wedge \cdots \wedge \alpha_k) = \mathbf{d}(A(\tilde{\alpha}_1, \ldots, \tilde{\alpha}_k))(x_0),$$

where $\tilde{\alpha}_1, \ldots, \tilde{\alpha}_k \in \Omega^1(M)$ *are arbitrary one-forms whose values at* x_0 *are* $\alpha_1, \ldots, \alpha_k$.

10.8-3 (Weinstein [1983]) *Let* $(P, \{\,,\})$ *be a finite-dimensional Poisson manifold with Poisson tensor* $B \in \Omega_2(P)$. *Let* $z_0 \in P$ *be such that* $B(z_0) = 0$. *For* $\alpha, \beta \in T_{z_0}^*P$, *define*

$$[\alpha, \beta]_B = (\mathbf{d}_{z_0}B)^*(\alpha \wedge \beta) = \mathbf{d}(B(\tilde{\alpha}, \tilde{\beta}))(z_0)$$

where $\mathbf{d}_{z_0}B$ *is the intrinsic derivative of* B *and* $\tilde{\alpha}, \tilde{\beta} \in \Omega^1(P)$ *are such that* $\tilde{\alpha}(z_0) = \alpha, \tilde{\beta}(z_0) = \beta$. *(See Exercise* **10.8-2**.*) Show that* $(\alpha, \beta) \mapsto [\alpha, \beta]_B$ *defines a bilinear skew-symmetric map* $T_{z_0}^*P \times T_{z_0}^*P \to T_{z_0}^*P$. *Show* *that the Jacobi identity for the Poisson bracket implies that* $[\,,\,]_B$ *is a Lie bracket on* $T^*_{z_0}P$. *Since* $(T^*_{z_0}P, [\,,\,]_B)$ *is a Lie algebra, its dual* $T_{z_0}P$ *naturally carries the induced Lie-Poisson structure, called the linearization of the given Poisson bracket at* z_0. *Show that the linearization in local coordinates has the expression*

$$\{F, G\}(v) = \frac{\partial B^{ij}(z_0)}{\partial z^k} \frac{\partial F}{\partial v^i} \frac{\partial G}{\partial v^j} v^k,$$

for $F, G : T_{z_0}P \to \mathbb{R}$ *and* $v \in T_{z_0}P$.

10.8-4 (Magri-Weinstein) *On the finite-dimensional manifold* P, *assume one has a symplectic form* Ω *and a Poisson structure* B. *Denote by* $K = B^\sharp \circ \Omega^\flat : TP \to TP$. *Show that* $(\Omega^\flat)^{-1} + B^\sharp : T^*P \to TP$ *defines a new Poisson structure on* P *if and only if* $\Omega^\flat \circ K^n$ *induces a presymplectic form on* P *for all* $n \in \mathbb{N}$.

10.9 Generalities on Lie-Poisson Structures

Proposition 10.9.1 *Let* G *be a Lie group. The equations of motion for the Hamiltonian* H *with respect to the* $\pm$ *Lie-Poisson brackets on* $\mathfrak{g}^*$ *are*

$$\frac{d\mu}{dt} = \mp \operatorname{ad}^*_{\delta H/\delta\mu} \mu. \tag{10.9.1}$$

Proof Let $F \in \mathcal{F}(\mathfrak{g}^*)$ be an arbitrary function. By the chain rule,

$$\frac{dF}{dt} = \mathbf{D}F(\mu) \cdot \dot{\mu} = \left\langle \frac{\delta F}{\delta \mu}, \dot{\mu} \right\rangle, \tag{10.9.2}$$

while

$$\begin{aligned} \{F, H\}_\pm (\mu) &= \pm \left\langle \mu, \left[\frac{\delta F}{\delta \mu}, \frac{\delta H}{\delta \mu} \right] \right\rangle = \pm \left\langle \mu, -\operatorname{ad}_{\delta H/\delta\mu} \frac{\delta F}{\delta \mu} \right\rangle \\ &= \mp \left\langle \operatorname{ad}^*_{\delta H/\delta\mu} \mu, \frac{\delta F}{\delta \mu} \right\rangle. \end{aligned} \tag{10.9.3}$$

Nondegeneracy of the pairing and arbitrariness of F imply the result. ■

Caution In infinite dimensions, $\mathfrak{g}^*$ does not necessarily mean the literal functional analytic dual of $\mathfrak{g}$, but rather a space in (nondegenerate) duality with $\mathfrak{g}$. In this case, care must be taken with the definition of $\delta F/\delta\mu$. ♦

Formula (10.9.1) says that on $\mathfrak{g}^*_\pm$, *the Hamiltonian vector field of* $H : \mathfrak{g}^* \to \mathbb{R}$ *is given by*

$$X_H(\mu) = \mp \operatorname{ad}^*_{\delta H/\delta\mu} \mu. \tag{10.9.4}$$

For example, for $G = SO(3)$, formula (10.2.3) for the Lie-Poisson bracket gives

$$X_H(\Pi) = \Pi \times \nabla H. \tag{10.9.5}$$

Lagrange devoted a good deal of attention in Volume 2 of *Mécanique Analytique* [1788] to the study of rotational motion of mechanical systems. In fact, in equation A on page 212 he gives the reduced Lie-Poisson equations for $SO(3)$ for a rather general Lagrangian. This equation is essentially the same as (10.9.5). His derivation was just how we would do it today—by reduction from material to spatial representation. Formula (10.9.5) actually hides a subtle point in that it identifies $\mathfrak{g}$ and $\mathfrak{g}^*$. Indeed, the way Lagrange wrote the equations, they are much more like their counterpart on $\mathfrak{g}$, which are called the *Euler-Poincaré equations*. We will come to these in Chapter 13, where additional historical information may be found.

In finite dimensions, if ξ_a, $a = 1, 2, \ldots, l$, is a basis for $\mathfrak{g}$, the structure constants C_{ab}^d are defined by

$$[\xi_a, \xi_b] = C_{ab}^d \xi_d \tag{10.9.6}$$

(a sum on "d" is understood). Thus, the Lie-Poisson bracket becomes

$$\{F, K\}_\pm (\mu) = \pm \mu_d \frac{\delta F}{\delta \mu_a} \frac{\delta K}{\delta \mu_b} C_{ab}^d, \tag{10.9.7}$$

where $\mu = \mu_a \xi^a, \{\xi^a\}$ is the basis of $\mathfrak{g}^*$ dual to $\{\xi_a\}$, and summation on repeated indices is understood. Taking F and K to be components of μ, (10.9.7) becomes

$$\{\mu_a, \mu_b\}_\pm = \pm C_{ab}^d \mu_d. \tag{10.9.8}$$

The equations of motion for a Hamiltonian H likewise become

$$\dot{\mu}_a = \mp \mu_d C_{ab}^d \frac{\delta H}{\delta \mu_b}. \tag{10.9.9}$$

In the Lie-Poisson reduction theorem in Chapter 13 we will show that the maps from T^*G to $\mathfrak{g}^*_-$ (resp., $\mathfrak{g}^*_+$) given by $\alpha_g \mapsto T_e^* L_g \cdot \alpha_g$ (resp., $\alpha_g \mapsto T_e^* R_g \cdot \alpha_g$) are Poisson maps. We will show in the next chapter that this is a general property of momentum maps. Here is another class of Poisson maps that will also turn out to be momentum maps.

Proposition 10.9.2 *Let G and H be Lie groups and let $\mathfrak{g}$ and $\mathfrak{h}$ be their Lie algebras. Let $\alpha : \mathfrak{g} \to \mathfrak{h}$ be a linear map. The map α is a homomorphism of Lie algebras if and only if its dual $\alpha^* : \mathfrak{h}^*_\pm \to \mathfrak{g}^*_\pm$ is a (linear) Poisson map.*

Proof Let $F, K \in \mathcal{F}(\mathfrak{g}^*)$. To compute $\delta(F \circ \alpha^*)/\delta\mu$, we let $\nu = \alpha^*(\mu)$ and use the definition of the functional derivative and the chain rule to get

$$\begin{aligned} \left\langle \frac{\delta}{\delta\mu}(F \circ \alpha^*), \delta\mu \right\rangle &= \mathbf{D}(F \circ \alpha^*)(\mu) \cdot \delta\mu = \mathbf{D}F(\alpha^*(\mu)) \cdot \alpha^*(\delta\mu) \\ &= \left\langle \frac{\delta F}{\delta\nu}, \alpha^*(\delta\mu) \right\rangle = \left\langle \alpha \cdot \frac{\delta F}{\delta\nu}, \delta\mu \right\rangle. \end{aligned} \tag{10.9.10}$$

Thus,

$$\frac{\delta}{\delta\mu}(F \circ \alpha^*) = \alpha \cdot \frac{\delta F}{\delta\nu}. \tag{10.9.11}$$

Hence,

$$\begin{aligned} \{F \circ \alpha^*, K \circ \alpha^*\}_+(\mu) &= \left\langle \mu, \left[\frac{\delta}{\delta\mu}(F \circ \alpha^*), \frac{\delta}{\delta\mu}(K \circ \alpha^*)\right]\right\rangle \\ &= \left\langle \mu, \left[\alpha \cdot \frac{\delta F}{\delta\nu}, \alpha \cdot \frac{\delta K}{\delta\nu}\right]\right\rangle. \end{aligned} \tag{10.9.12}$$

The expression (10.9.12) equals

$$\left\langle \mu, \alpha \cdot \left[\frac{\delta F}{\delta\mu}, \frac{\delta G}{\delta\mu}\right]\right\rangle \tag{10.9.13}$$

for all F and K if and only if α is a Lie algebra homomorphism. ■

This theorem applies to the case $\alpha = T_e\sigma$ for $\sigma : G \to H$ a Lie group homomorphism, by studying the reduction diagram in Figure 10.9.1 (and being cautious that σ need not be a diffeomorphism.)

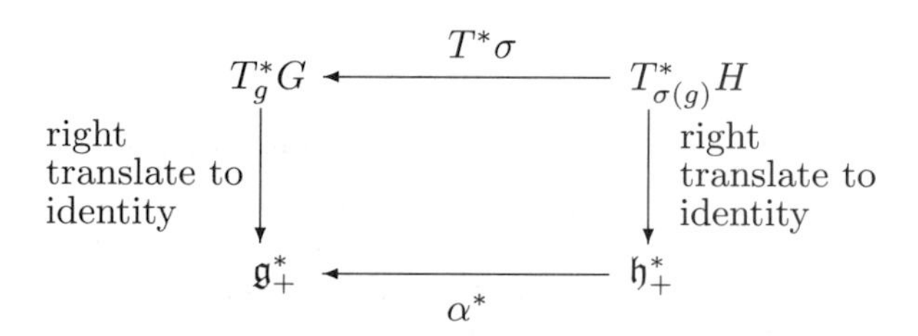

FIGURE 10.9.1. Lie group homomorphisms induce Poisson maps.

Examples

(a) Plasma to Fluid Poisson Map for the Momentum Variables Let G be the group of diffeomorphisms of a manifold Q and let H be the group of canonical transformations of $P = T^*Q$. We assume that $H^1(Q) = 0$ which implies that all locally Hamiltonian vector fields on T^*Q are globally Hamiltonian. Thus the Lie algebra $\mathfrak{h}$ consists of functions on T^*Q modulo constants. Its dual is identified with itself via the L^2-inner product relative to the Liouville measure $dq\,dp$ on T^*Q. Let $\sigma : G \to H$ be the map $\eta \mapsto T^*\eta^{-1}$, which is a group homomorphism and let $\alpha = T_e\sigma : \mathfrak{g} \to \mathfrak{h}$. We claim that $\alpha^* : \mathcal{F}(T^*Q)/\mathbb{R} \to \mathfrak{g}^*$ is given by

$$\alpha^*(F) = \int pf(q,p)\,dp, \tag{10.9.14}$$

where we regard $\mathfrak{g}^*$ as the space of one-form densities on Q and the integral denotes fiber integration for each fixed $q \in Q$. Indeed, α is the map taking vector fields X on Q to their lifts $X_{\mathcal{P}(X)}$ on T^*Q. Thus as a map of $\mathfrak{X}(Q)$ to $\mathcal{F}(T^*Q)/\mathbb{R}$, α is given by $X \mapsto \mathcal{P}(X)$. Its dual is given by

$$\begin{aligned} \langle \alpha^*(f), X\rangle &= \langle f, \alpha(X)\rangle = \int_P f\mathcal{P}(X)\,dq\,dp \\ &= \int_P f(q,p)p \cdot X(q)\,dq\,dp \end{aligned} \tag{10.9.15}$$

so $\alpha^*(F)$ is given by (10.9.14) as claimed. ♦

(b) Plasma to Fluid Map for the Density Variable
Let $G = \mathcal{F}(Q)$, regarded as an abelian group and let the map $\sigma : G \to \mathrm{Diff}_{\mathrm{can}}(T^*Q)$ be given by $\sigma(\varphi) =$ fiber translation by $\mathbf{d}\varphi$. A computation similar to that above gives the Poisson map

$$\alpha^*(f)(q) = \int f(q,p)\,dp \tag{10.9.16}$$

from $\mathcal{F}(T^*Q)$ to $\mathrm{Den}(Q) = \mathcal{F}(Q)^*$. The integral in (10.9.16) denotes the fiber integration of $f(q,p)$ for fixed $q \in Q$. ♦

Next we characterize Lie-Poisson brackets as the linear ones. Let V^* and V be Banach spaces and let $\langle\, , \rangle : V^* \times V \to \mathbb{R}$ be a weakly nondegenerate pairing of V^* with V. Think of elements of V as linear functionals on V^*. A Poisson bracket on V^* is called ***linear*** if the bracket of any two linear functionals on V is again linear. This condition is equivalent to the associated Poisson tensor $B(\mu) : V \to V^*$ being *linear* in $\mu \in V^*$.

Proposition 10.9.3 *Let $\langle\, , \rangle : V^* \times V \to \mathbb{R}$ be a (weakly) nondegenerate pairing of the Banach spaces V^* and V, and let V^* have a linear Poisson bracket. Assume that the bracket of any two linear functionals on V is in the range of $B(\mu)$ for all $\mu \in V^*$ (this condition is automatically satisfied if V is finite dimensional). Then V is a Lie algebra and the Poisson bracket on V^* is the corresponding Lie-Poisson bracket.*

Proof If $x \in V$, we denote by x' the functional $x'(\mu) = \langle \mu, x \rangle$ on V^*. By hypothesis, the Poisson bracket $\{x', y'\}$ is a linear functional on V^*. By assumption this bracket is represented by an element which we denote $[x, y]'$ in V, that is, we can write $\{x', y'\} = [x, y]'$. (The element $[x, y]$ is unique since $\langle \, , \rangle$ is weakly nondegenerate.) It is straightforward to check that the operation $[\, ,]$ on V so defined is a Lie algebra bracket. Thus V is a Lie algebra, and one then checks that the given Poisson bracket is the Lie-Poisson bracket for this algebra. ■

Exercise

10.9-1 *Let $\sigma : SO(3) \to GL(3)$ be the inclusion map. Identify $\mathfrak{so}(3)^* = \mathbb{R}^3$ with the rigid body bracket and identify $\mathfrak{gl}(3)^*$ with $\mathfrak{gl}(3)$ using $\langle A, B \rangle =$* trace (AB^t). *Compute the induced map $\alpha^* : \mathfrak{gl}(3) \to \mathbb{R}^3$ and verify directly that it is Poisson.*

11 Momentum Maps

In this chapter we show how to obtain conserved quantities for Lagrangian and Hamiltonian systems with symmetries. This is done using the concept of a momentum mapping, which is a geometric generalization of the classical linear and angular momentum. This concept is more than a mathematical reformulation of a concept that simply describes the well-known Noether theorem. Rather, it is a rich concept that is ubiquitous in the modern developments of geometric mechanics. It has led to surprising insights into many areas of mechanics and geometry.

11.1 Canonical Actions and Their Infinitesimal Generators

Let P be a Poisson manifold, let G be a Lie group, and let $\Phi : G \times P \to P$ be a smooth left action of G on P by canonical transformations. If we denote the action by $g \cdot z = \Phi_g(z)$, so that $\Phi_g : P \to P$, then the action being ***canonical*** means

$$\Phi_g^* \{F_1, F_2\} = \left\{\Phi_g^* F_1, \Phi_g^* F_2\right\} \tag{11.1.1}$$

for any $F_1, F_2 \in \mathcal{F}(P)$ and any $g \in G$. If P is a symplectic manifold with symplectic form Ω, then the action is canonical if and only if it is symplectic, that is, $\Phi_g^* \Omega = \Omega$ for all $g \in G$.

Recall from Chapter 9 on Lie groups, that the ***infinitesimal generator*** of the action corresponding to a Lie algebra element $\xi \in \mathfrak{g}$ is the vector field ξ_P on P obtained by differentiating the action with respect to g at the identity in the direction ξ. By the chain rule,

$$\xi_P(z) = \frac{d}{dt}\left[\exp(t\xi)\cdot z\right]\Big|_{t=0}. \tag{11.1.2}$$

We will need two general identities, both of which were proved in Chapter **9**. First, the flow of the vector field ξ_P is

$$\varphi_t = \Phi_{\exp t\xi}. \tag{11.1.3}$$

Second, we have

$$\Phi^*_{g^{-1}}\xi_P = (\mathrm{Ad}_g\,\xi)_P \tag{11.1.4}$$

and its differentiated companion

$$[\xi_P, \eta_P] = -\,[\xi, \eta]_P\,. \tag{11.1.5}$$

To illustrate these identities, consider the action of $SO(3)$ on $\mathbb{R}^3$. As was explained in Chapter 9, the Lie algebra $\mathfrak{so}(3)$ of $SO(3)$ is identified with $\mathbb{R}^3$ and the Lie bracket with the cross product. For the action of $SO(3)$ on $\mathbb{R}^3$ given by rotations, the infinitesimal generator of $\omega \in \mathbb{R}^3$ is

$$\omega_{\mathbb{R}^3}(\mathbf{x}) = \omega \times \mathbf{x} = \hat{\omega}(\mathbf{x}). \tag{11.1.6}$$

Then (11.1.4) becomes the identity

$$(\mathbf{A}\omega \times \mathbf{x}) = \mathbf{A}(\omega \times \mathbf{A}^{-1}\mathbf{x}) \tag{11.1.7}$$

for $A \in SO(3)$, while (11.1.5) becomes the Jacobi identity for the vector product.

Returning to the general case, differentiate (11.1.1) with respect to g in the direction ξ, giving

$$\xi_P[\{F_1, F_2\}] = \{\xi_P[F_1], F_2\} + \{F_1, \xi_P[F_2]\}\,. \tag{11.1.8}$$

In the symplectic case, differentiating $\Phi_g^*\Omega = \Omega$ gives

$$\pounds_{\xi_P}\Omega = 0, \tag{11.1.9}$$

that is, ξ_P is ***locally Hamiltonian***. For Poisson manifolds, a vector field satisfying (11.1.8) is called an ***infinitesimal Poisson automorphism***. Such a vector field need not be locally Hamiltonian (that is, locally of the form X_H). For example, consider the Poisson structure

$$\{F, H\} = x\left(\frac{\partial F}{\partial x}\frac{\partial H}{\partial y} - \frac{\partial H}{\partial x}\frac{\partial F}{\partial y}\right) \tag{11.1.10}$$

on $\mathbb{R}^2$ and $X = \partial/\partial y$ in a neighborhood of a point of the y-axis.

We are interested in the case in which ξ_P is globally Hamiltonian, a condition stronger than (11.1.8). Thus, *assume that there is a global Hamiltonian* $J(\xi) \in \mathcal{F}(P)$ for ξ_P, that is,

$$X_{J(\xi)} = \xi_P. \tag{11.1.11}$$

Does this equation determine $J(\xi)$? Obviously not, for if $J_1(\xi)$ and $J_2(\xi)$ both satisfy (11.1.11), then

$$X_{J_1(\xi)-J_2(\xi)} = 0; \quad \text{i.e.,} \quad J_1(\xi) - J_2(\xi) \in \mathcal{C}(P)$$

the space of Casimir functions on P. If P is symplectic and connected, then $J(\xi)$ is determined by (11.1.11) up to a constant.

Exercises

11.1-1 *Verify* (11.1.4) *and* (11.1.5) *for the action of* $GL(n)$ *on itself by conjugation.*

11.1-2 *Let* S^1 *act on* S^2 *by rotations about the* z*-axis. Compute* $J(\xi)$.

11.2 Momentum Maps

Since the right-hand side of (11.1.11) is linear in ξ, by using a basis we can modify any given $J(\xi)$ so it too is linear in ξ, and still retain condition (11.1.11). In this process, we can replace the left *Lie group* action by a canonical left *Lie algebra* action $\xi \mapsto \xi_P$, where canonical, in the Poisson manifold context, means that (11.1.8) is satisfied and, in the symplectic manifold context, that (11.1.9) is satisfied. (Recall that for a left Lie algebra action, the map $\xi \in \mathfrak{g} \mapsto \xi_P \in \mathfrak{X}(P)$ is a Lie algebra antihomomorphism.) Thus we make the following definition:

Definition 11.2.1 *Let a Lie algebra* $\mathfrak{g}$ *act canonically (on the left) on the Poisson manifold* P. *Suppose there is a linear map* $J : \mathfrak{g} \to \mathcal{F}(P)$ *such that*

$$X_{J(\xi)} = \xi_P \tag{11.2.1}$$

for all $\xi \in \mathfrak{g}$. *The map* $\mathbf{J} : P \to \mathfrak{g}^*$ *defined by*

$$\langle \mathbf{J}(z), \xi \rangle = J(\xi)(z) \tag{11.2.2}$$

for all $\xi \in \mathfrak{g}$ *and* $z \in P$ *is called a* ***momentum mapping*** *of the action.*

The preceding definition can be elucidated with the following remarks. Consider the angular momentum function for a particle in Euclidean three-space, $\mathbf{J}(z) = \mathbf{q} \times \mathbf{p}$, where $z = (\mathbf{q}, \mathbf{p})$. Let $\xi \in \mathbb{R}^3$ and consider the component of $\mathbf{J}$ around the axis ξ, namely, $\langle \mathbf{J}(z), \xi \rangle = \xi \cdot (\mathbf{q} \times \mathbf{p})$. One checks that Hamilton's equations determined by this function of $\mathbf{q}$ and $\mathbf{p}$ describe infinitesimal rotations about the axis ξ. The defining condition (11.2.1) is a generalization of this elementary statement about angular momentum.

Recalling that $X_H[F] = \{F, H\}$, we see that (11.2.1) can be phrased in terms of the Poisson bracket as follows: *for any function F on P and any $\xi \in \mathfrak{g}$,*

$$\{F, J(\xi)\} = \xi_P[F]. \tag{11.2.3}$$

Equation (11.2.2) defines an isomorphism between the space of smooth maps $\mathbf{J}$ from P to $\mathfrak{g}^*$ and the space of linear maps J from $\mathfrak{g}$ to $\mathcal{F}(P)$. We think of the collection of functions $J(\xi)$ as ξ varies in $\mathfrak{g}$ as the components of $\mathbf{J}$. Denote by

$$\mathcal{H}(P) = \{X_F \in \mathfrak{X}(P) \mid F \in \mathcal{F}(P)\} \tag{11.2.4}$$

the Lie algebra of Hamiltonian vector fields on P and by

$$\mathcal{P}(P) = \{X \in \mathfrak{X}(P) \mid X[\{F_1, F_2\}] = \{X[F_1], F_2\} + \{F_1, X[F_2]\}\}, \tag{11.2.5}$$

the Lie algebra of infinitesimal Poisson automorphisms of P. By (11.1.8), for any $\xi \in \mathfrak{g}$ we have $\xi_P \in \mathcal{P}(P)$. Therefore, giving a momentum map $\mathbf{J}$ is equivalent to specifying a linear map $J : \mathfrak{g} \to \mathcal{F}(P)$ making the diagram in Figure 11.2.1 commute.

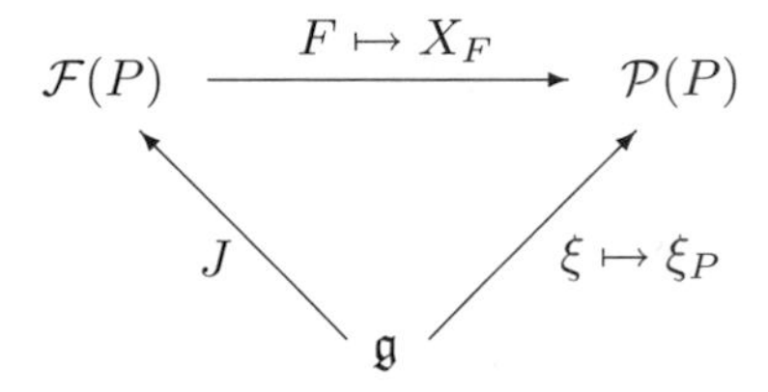

FIGURE 11.2.1. The momentum map diagram.

Since both $\xi \mapsto \xi_P$ and $F \mapsto X_F$ are Lie algebra antihomomorphisms, for $\xi, \eta \in \mathfrak{g}$ we get

$$\begin{aligned} X_{J([\xi,\eta])} &= [\xi, \eta]_P = -[\xi_P, \eta_P] \\ &= -\left[X_{J(\xi)}, X_{J(\eta)}\right] = X_{\{J(\xi), J(\eta)\}} \end{aligned} \tag{11.2.6}$$

and so we have the basic identity

$$X_{J([\xi,\eta])} = X_{\{J(\xi), J(\eta)\}}. \tag{11.2.7}$$

The preceding development *defines* momentum maps, but does not tell us how to *compute* them in examples. We shall concentrate on that aspect in Chapter **12**.

Building on the above commutative diagram, §11.3 discusses an alternative approach to the definition of the momentum map but it will not be used subsequently in the main text. Rather, we shall give the formulae that will be most important for later applications; the interested reader is referred to Souriau [1970], Weinstein [1977], Abraham and Marsden [1978], Guillemin and Sternberg [1984], and Libermann and Marle [1987] for more information.

Some History of the Momentum Map

The momentum map can be found in the second volume of Lie [1890], where it appears in the context of homogeneous canonical transformations, in which case its expression is given as the contraction of the canonical one-form with the infinitesimal generator of the action. On page 300 it is shown that the momentum map is canonical and on page 329 that it is equivariant with respect to some linear action whose generators are identified on page 331. On page 338 it is proved that if the momentum map has constant rank (a hypothesis that seems to be implicit in all of Lie's work in this area), its image is Ad^*-invariant, and on page 343, actions are classified by Ad^*-invariant submanifolds.

We now present the modern history of the momentum map based on information and references provided to us by B. Kostant and J.-M. Souriau. We would like to thank them for all their help.

In Kostant's 1965 Phillips lectures at Haverford (the notes of which were written by Dale Husemoller), and in the 1965 U.S.-Japan Seminar on Differential Geometry, Kostant [1966] introduced the momentum map to generalize a theorem of Wang and thereby classified all homogeneous symplectic manifolds; this is called today "Kostant's coadjoint orbit covering theorem." These lectures also contained the key points of geometric quantization. Meanwhile, Souriau [1965] introduced the momentum map in his Marseille lecture notes and put it in print in Souriau [1966]. The momentum map finally got its formal definition and its name, based on its physical interpretation, in Souriau [1967a] and its properties of equivariance were studied in Souriau [1967b], where the coadjoint orbit theorem is also formulated. In 1968, the momentum map appeared as a key tool in Kostant [1968] and from then on became a standard notion. Souriau [1969] discussed it at length in his book and Kostant [1970] (page 187, Theorem 5.4.1) dealt with it in his quantization lectures. Kostant and Souriau realized its importance for linear representations, a fact apparently not foreseen by Lie (Weinstein [1983a]). Independently, work on the momentum map and the coadjoint orbit covering theorem was done by A. Kirillov. This is described in Kirillov [1976]. This book was first published in 1972 and states that his work on the classification theorem was done about five years earlier (page 301). The modern formulation of the momentum map was developed in the context of classical mechanics in the work of Smale [1970], who applied it extensively in his topological program for the planar n-body problem.

Exercises

11.2-1 *Verify that Hamilton's equations determined by the function*

$$\langle J(z), \xi \rangle = \xi \cdot (\mathbf{q} \times \mathbf{p})$$

give the infinitesimal generator of rotations about the ξ-axis.

11.2-2 *Verify that $J([\xi, \eta]) = \{J(\xi), J(\eta)\}$ for angular momentum.*

11.3 An Algebraic Definition of the Momentum Map

This section gives an optional approach to momentum maps and may be skipped on a first reading. The point of departure is the commutative diagram in Figure 11.2.1 plus the observation that the following sequence is exact (that is, the range of each map lies in the kernel of the following one):

$$0 \longrightarrow \mathcal{C}(P) \xrightarrow{\;i\;} \mathcal{F}(P) \xrightarrow{\;\mathcal{H}\;} \mathcal{P}(P) \xrightarrow{\;\pi\;} \mathcal{P}(P)/\mathcal{H}(P) \longrightarrow 0$$

Here, i is the inclusion, π the projection, $\mathcal{H}(F) = X_F$, and $\mathcal{H}(P)$ denotes the Lie algebra of globally Hamiltonian vector fields on P. Let us investigate conditions under which a left Lie algebra action, that is, an antihomomorphism $\rho : \mathfrak{g} \to \mathcal{P}(P)$ lifts through $\mathcal{H}$ to a linear map $J : \mathfrak{g} \to \mathcal{F}(P)$. As we have already seen, this is equivalent to $\mathbf{J}$ being a momentum map. (The requirement that J be a Lie algebra homomorphism will be discussed later.)

If $\mathcal{H} \circ J = \rho$, then $\pi \circ \rho = \pi \circ \mathcal{H} \circ J = 0$. Conversely, if $\pi \circ \rho = 0$, then $\rho(\mathfrak{g}) \subset \mathcal{H}(P)$, so there is a linear map $J : \mathfrak{g} \to \mathcal{F}(P)$ such that $\mathcal{H} \circ J = \rho$. Thus, the obstruction to the existence of J is $\pi \circ \rho = 0$. If P is symplectic, then $\mathcal{P}(P)$ coincides with the Lie algebra of locally Hamiltonian vector fields and thus $\mathcal{P}(P)/\mathcal{H}(P)$ is isomorphic to the first cohomology space $H^1(P)$ regarded as an abelian group. *Thus in the symplectic case, $\pi \circ \rho = 0$ if and only if the induced mapping $\rho' : \mathfrak{g}/[\mathfrak{g}, \mathfrak{g}] \to H^1(P)$ vanishes.* Here is a list of cases which guarantee that $\pi \circ \rho = 0$:

1. *P is symplectic and $\mathfrak{g}/[\mathfrak{g}, \mathfrak{g}] = 0$.* By the first Whitehead lemma,

this is the case whenever $\mathfrak{g}$ is semisimple (see Jacobson [1962] and Guillemin and Sternberg [1984]).

2. $\mathcal{P}(P)/\mathcal{H}(P) = 0$. If P is symplectic this is equivalent to the vanishing of the first cohomology group $H^1(P)$.

3. *P is exact symplectic, i.e., $\Omega = -\mathbf{d}\Theta$, and Θ is invariant under the $\mathfrak{g}$ action, that is,*
$$\pounds_{\xi_P}\Theta = 0. \tag{11.3.1}$$

This case occurs, for example, when $P = T^*Q$ and the action is a lift.

In case 3, there is an explicit formula for the momentum map. Since

$$0 = \pounds_{\xi_P}\Theta = \mathbf{d}\mathbf{i}_{\xi_P}\Theta + \mathbf{i}_{\xi_P}\mathbf{d}\Theta, \tag{11.3.2}$$

it follows that

$$\mathbf{d}(\mathbf{i}_{\xi_P}\Theta) = \mathbf{i}_{\xi_P}\Omega, \tag{11.3.3}$$

that is, the interior product of ξ_P with Θ satisfies (11.2.1) and hence the momentum map $\mathbf{J} : P \to \mathfrak{g}^*$ is given by

$$\langle \mathbf{J}(z), \xi \rangle = (\mathbf{i}_{\xi_P}\Theta)(z). \tag{11.3.4}$$

In coordinates, write $\Theta = p_i\, dq^i$ and define $A^j{}_a$ and B_{aj} by

$$\xi_P = \xi^a A^j{}_a \frac{\partial}{\partial q^j} + \xi^a B_{aj} \frac{\partial}{\partial p_j}. \tag{11.3.5}$$

Then (11.3.4) reads

$$J_a(q,p) = p_i A^i{}_a(q,p). \tag{11.3.6}$$

The following example shows that ρ' does not always vanish. Consider the phase space $P = S^1 \times S^1$, with the symplectic form $\Omega = d\theta_1 \wedge d\theta_2$, the Lie algebra $\mathfrak{g} = \mathbb{R}^2$, and the action

$$\rho(x_1, x_2) = x_1 \frac{\partial}{\partial \theta_1} + x_2 \frac{\partial}{\partial \theta_2}. \tag{11.3.7}$$

In this case $[\mathfrak{g}, \mathfrak{g}] = 0$ and $\rho' : \mathbb{R}^2 \to H^1(S^1 \times S^1)$ is an isomorphism, as can be easily checked.

11.4 Conservation of Momentum Maps

One reason that momentum maps are important in mechanics is because they are conserved quantities.

Theorem 11.4.1 (Hamiltonian Version of Noether's Theorem) *If the Lie algebra $\mathfrak{g}$ acts canonically on the Poisson manifold P, admits a momentum mapping $\mathbf{J} : P \to \mathfrak{g}^*$, and $H \in \mathcal{F}(P)$ is $\mathfrak{g}$-invariant, i.e., $\xi_P[H] = 0$ for all $\xi \in \mathfrak{g}$, then $\mathbf{J}$ is a constant of the motion for H, i.e.,*

$$\mathbf{J} \circ \varphi_t = \mathbf{J},$$

where φ_t is the flow of X_H. If the Lie algebra action comes from a canonical left Lie group action Φ, then the invariance hypothesis on H is implied by the invariance condition: $H \circ \Phi_g = H$ for all $g \in G$.

Proof The condition $\xi_P[H] = 0$ implies that the Poisson bracket of $J(\xi)$, the Hamiltonian function for ξ_P, and H vanishes: $\{J(\xi), H\} = 0$. This implies that for each Lie algebra element ξ, $J(\xi)$ is a conserved quantity along the flow of X_H. This means that the values of the corresponding $\mathfrak{g}^*$-valued momentum map $\mathbf{J}$ are conserved. The last assertion of the theorem follows by differentiating the condition $H \circ \Phi_g = H$ with respect to g at the identity e in the direction ξ to obtain $\xi_P[H] = 0$. ■

Exercise

11.4-1 *For the action of S^1 on $\mathbb{C}^2$ given by $e^{i\theta}(z_1, z_2) = (e^{i\theta}z_1, e^{-i\theta}z_2)$, show that the momentum map is $J = (|z_1|^2 - |z_2|^2)/2$. Show that the Hamiltonian given in the example in* §10.6 *is invariant under S^1, so that Theorem* **11.4.1** *applies.*

11.5 Examples

(a) The Hamiltonian

On a Poisson manifold P, consider the $\mathbb{R}$-action given by the flow of a complete Hamiltonian vector field X_H. The corresponding momentum map $\mathbf{J} : P \to \mathbb{R}$ (where we identify $\mathbb{R}^*$ with $\mathbb{R}$ via the usual dot product) equals H. ♦

(b) Linear Momentum

In §6.4 we discussed the N-particle system and constructed the cotangent lift of the $\mathbb{R}^3$-action on $\mathbb{R}^{3N}$ (translation on every factor) to be the action on $T^*\mathbb{R}^{3N} \cong \mathbb{R}^{6N}$ given by

$$\mathbf{x} \cdot (\mathbf{q}_i, \mathbf{p}^j) = (\mathbf{q}_j + \mathbf{x}, \mathbf{p}^j), \quad j = 1, \ldots, N. \tag{11.5.1}$$

We show that this action has a momentum map and compute it from the definition. In the next chapter, we shall recompute it more easily utilizing further developments of the theory. Let $\xi \in \mathfrak{g} = \mathbb{R}^3$; the infinitesimal generator ξ_P at a point $(\mathbf{q}_j, \mathbf{p}^j) \in \mathbb{R}^{6N} = P$ is given by differentiating (11.5.1) with respect to $\mathbf{x}$ in the direction ξ

$$\xi_P(\mathbf{q}_j, \mathbf{p}^j) = (\xi, \xi, \ldots, \xi, \mathbf{0}, \mathbf{0}, \ldots, \mathbf{0}). \tag{11.5.2}$$

On the other hand, by definition of the canonical symplectic structure Ω on P, any candidate $J(\xi)$ has a Hamiltonian vector field given by

$$X_{J(\xi)}(\mathbf{q}_j, \mathbf{p}^j) = \left(\frac{\partial J(\xi)}{\partial \mathbf{p}^j}, -\frac{\partial J(\xi)}{\partial \mathbf{q}_j} \right). \tag{11.5.3}$$

Then, $X_{J(\xi)} = \xi_P$ implies that

$$\frac{\partial J(\xi)}{\partial \mathbf{p}^j} = \xi \quad \text{and} \quad \frac{\partial J(\xi)}{\partial \mathbf{q}_j} = 0, \quad 1 \leq j \leq N. \tag{11.5.4}$$

Solving these equations and choosing constants such that J is linear, we get

$$J(\xi)(\mathbf{q}_j, \mathbf{p}^j) = \left(\sum_{j=1}^{N} \mathbf{p}^j \right) \cdot \xi, \quad \text{i.e.,} \quad \mathbf{J}(\mathbf{q}_j, \mathbf{p}^j) = \sum_{j=1}^{N} \mathbf{p}^j. \tag{11.5.5}$$

This expression is called the ***total linear momentum*** of the N-particle system. In this example, Noether's theorem can be deduced directly as follows. Denote by $J_\alpha, q_j^\alpha, p_\alpha^j$, the αth components of $\mathbf{J}$, $\mathbf{q}_j$ and $\mathbf{p}^j$, $\alpha = 1, 2, 3$. Given a Hamiltonian H, determining the evolution of the N particle system by Hamilton's equations, we get

$$\frac{dJ_\alpha}{dt} = \sum_{j=1}^{N} \frac{dp_\alpha^j}{dt} = -\sum_{j=1}^{N} \frac{\partial H}{\partial q_\alpha^j} = \left[\sum_{j=1}^{N} \frac{\partial}{\partial q_\alpha^j} \right] H. \tag{11.5.6}$$

The bracket on the right is an operator that evaluates the variation of the scalar function H under a spatial translation, that is, under the action of the translation group $\mathbb{R}^3$ on each of the N coordinate directions. Obviously J_α is conserved if H is translation-invariant, which is exactly the statement of Noether's theorem. ◆

(c) Angular Momentum
Let $SO(3)$ act on the configuration space $Q = \mathbb{R}^3$ by $\Phi(\mathbf{A}, \mathbf{q}) = \mathbf{A}\mathbf{q}$. We show that the lifted action to $P = T^*\mathbb{R}^3$ has a momentum map and compute it. First note that if $(\mathbf{q}, \mathbf{v}) \in T_\mathrm{q}\mathbb{R}^3$, then $T_\mathrm{q}\Phi_\mathrm{A}(\mathbf{q}, \mathbf{v}) = (\mathbf{A}\mathbf{q}, \mathbf{A}\mathbf{v})$. Let $\mathbf{A} \cdot (\mathbf{q}, \mathbf{p}) = T^*_{\mathrm{Aq}}\Phi_{\mathrm{A}^{-1}}(\mathbf{p})$ denote the lift of the $SO(3)$ action to P, and identify covectors with vectors using the Euclidean inner product. If $(\mathbf{q}, \mathbf{p}) \in T^*_\mathrm{q}\mathbb{R}^3$, then $(\mathbf{A}\mathbf{q}, \mathbf{v}) \in T_{\mathrm{Aq}}\mathbb{R}^3$, so

$$\begin{aligned} \langle \mathbf{A} \cdot (\mathbf{q}, \mathbf{p}), (\mathbf{A}\mathbf{q}, \mathbf{v}) \rangle &= \langle (\mathbf{q}, \mathbf{p}), \mathbf{A}^{-1} \cdot (\mathbf{A}\mathbf{q}, \mathbf{v}) \rangle = \langle \mathbf{p}, \mathbf{A}^{-1}\mathbf{v} \rangle \\ &= \langle \mathbf{A}\mathbf{p}, \mathbf{v} \rangle = \langle (\mathbf{A}\mathbf{q}, \mathbf{A}\mathbf{p}), (\mathbf{A}\mathbf{q}, \mathbf{v}) \rangle, \end{aligned}$$

that is,

$$\mathbf{A} \cdot (\mathbf{q}, \mathbf{p}) = (\mathbf{A}\mathbf{q}, \mathbf{A}\mathbf{p}) . \tag{11.5.7}$$

Differentiating with respect to $\mathbf{A}$, we find that the infinitesimal generator corresponding to $\xi = \hat{\omega} \in \mathfrak{so}(3)$ is

$$\hat{\omega}_P(\mathbf{q}, \mathbf{p}) = (\xi\mathbf{q}, \xi\mathbf{p}) = (\omega \times \mathbf{q}, \omega \times \mathbf{p}) . \tag{11.5.8}$$

As in the previous example, to find the momentum map, we solve

$$\frac{\partial J(\xi)}{\partial \mathbf{p}} = \xi\mathbf{q} \quad \text{and} \quad -\frac{\partial J(\xi)}{\partial \mathbf{q}} = \xi\mathbf{p}, \tag{11.5.9}$$

such that $J(\xi)$ is linear in ξ. A solution is given by

$$J(\xi) = (\xi\mathbf{q}) \cdot \mathbf{p} = (\omega \times \mathbf{q}) \cdot \mathbf{p} = (\mathbf{q} \times \mathbf{p}) \cdot \omega,$$

so that

$$\mathbf{J}(\mathbf{q}, \mathbf{p}) = \mathbf{q} \times \mathbf{p}. \tag{11.5.10}$$

Of course, (11.5.10) is the standard formula for the *angular momentum* of a particle.

In this case, Noether's theorem states that a Hamiltonian that is rotationally invariant has the three components of $\mathbf{J}$ as constants of the motion. This example can be generalized as follows. ♦

(d) Momentum for Matrix Groups
Let $G \subset GL(n, \mathbb{R})$ be a subgroup of the general linear group of $\mathbb{R}^n$. We let G act on $\mathbb{R}^n$ by matrix multiplication on the left, that is, $\Phi_{\mathrm{A}}(\mathbf{q}) = \mathbf{A}\mathbf{q}$. As in the previous example, the induced action on $P = T^*\mathbb{R}^n$ is given by

$$\mathbf{A} \cdot (\mathbf{q}, \mathbf{p}) = (\mathbf{A}\mathbf{q}, (\mathbf{A}^T)^{-1}\mathbf{p}) \tag{11.5.11}$$

and the infinitesimal generator corresponding to $\xi \in \mathfrak{g}$ by

$$\xi_P(\mathbf{q}, \mathbf{p}) = (\xi\mathbf{q}, -\xi^T\mathbf{p}). \tag{11.5.12}$$

To find the momentum map, we solve

$$\frac{\partial J(\xi)}{\partial \mathbf{p}} = \xi\mathbf{q} \quad \text{and} \quad \frac{\partial J(\xi)}{\partial \mathbf{q}} = \xi^T\mathbf{p}, \tag{11.5.13}$$

which we can do by choosing $J(\xi)(\mathbf{q}, \mathbf{p}) = (\xi\mathbf{q}) \cdot \mathbf{p}$, that is,

$$\langle \mathbf{J}(\mathbf{q}, \mathbf{p}), \xi \rangle = (\xi\mathbf{q}) \cdot \mathbf{p}. \tag{11.5.14}$$

If $n = 3$ and $G = SO(3)$, (11.5.14) is equivalent to (11.5.10). In coordinates, $(\xi\mathbf{q}) \cdot \mathbf{p} = \xi^i{}_j q^j p_i$, so $[\mathbf{J}(\mathbf{q}, \mathbf{p})]^i_j = q^i p_j$. If we identify $\mathfrak{g}$ and $\mathfrak{g}^*$ using $\langle A, B \rangle = \operatorname{tr}(AB^T)$, then $\mathbf{J}(\mathbf{q}, \mathbf{p})$ is the projection of the matrix $q^j p_i$ onto the subspace $\mathfrak{g}$. ♦

(e) Canonical Momentum on $\mathfrak{g}^*$

Let the Lie group G with Lie algebra $\mathfrak{g}$ act by the coadjoint action on $\mathfrak{g}^*$ endowed with the $\pm$ Lie-Poisson structure. Since $\mathrm{Ad}_{g^{-1}} : \mathfrak{g} \to \mathfrak{g}$ is a Lie algebra isomorphism, its dual $\mathrm{Ad}^*_{g^{-1}} : \mathfrak{g}^* \to \mathfrak{g}^*$ is a canonical map by Proposition **10.9.2**. Let us prove this fact directly. A computation shows that

$$\frac{\delta F}{\delta(\mathrm{Ad}^*_{g^{-1}}\,\mu)} = \mathrm{Ad}_g \frac{\delta\left(F \circ \mathrm{Ad}^*_{g^{-1}}\right)}{\delta\mu}, \tag{11.5.15}$$

whence

$$\begin{aligned}
&\{F, H\}_{\pm}\left(\mathrm{Ad}^*_{g^{-1}}\,\mu\right) \\
&\quad = \pm\left\langle \mathrm{Ad}^*_{g^{-1}}\,\mu, \left[\frac{\delta F}{\delta\left(\mathrm{Ad}^*_{g^{-1}}\,\mu\right)}, \frac{\delta H}{\delta\left(\mathrm{Ad}^*_{g^{-1}}\,\mu\right)}\right]\right\rangle \\
&\quad = \pm\left\langle \mathrm{Ad}^*_{g^{-1}}\,\mu, \left[\mathrm{Ad}_g\frac{\delta\left(F\circ \mathrm{Ad}^*_{g^{-1}}\right)}{\delta\mu}, \mathrm{Ad}_g\frac{\delta\left(H\circ \mathrm{Ad}^*_{g^{-1}}\right)}{\delta\mu}\right]\right\rangle \\
&\quad = \pm\left\langle \mu, \left[\frac{\delta\left(F\circ \mathrm{Ad}^*_{g^{-1}}\right)}{\delta\mu}, \frac{\delta\left(H\circ \mathrm{Ad}^*_{g^{-1}}\right)}{\delta\mu}\right]\right\rangle \\
&\quad = \left\{F\circ \mathrm{Ad}^*_{g^{-1}}, H\circ \mathrm{Ad}^*_{g^{-1}}\right\}_{\pm}(\mu),
\end{aligned}$$

that is, the coadjoint action of G on $\mathfrak{g}^*$ is canonical. From Proposition **10.9.1**, the Hamiltonian vector field for $H \in \mathcal{F}(\mathfrak{g}^*)$ is given by

$$X_H(\mu) = \mp \mathrm{ad}^*_{(\delta H/\delta\mu)}\,\mu. \tag{11.5.16}$$

Since the infinitesimal generator of the coadjoint action corresponding to $\xi \in \mathfrak{g}^*$ is given by $\xi_{\mathfrak{g}^*} = -\mathrm{ad}^*_\xi$, it follows that the momentum map of the coadjoint action, if it exisits, must satisfy

$$\mp\, \mathrm{ad}^*_{(\delta J(\xi)/\delta\mu)}\,\mu = -\mathrm{ad}^*_\xi\,\mu \tag{11.5.17}$$

for every $\mu \in \mathfrak{g}^*$, that is, $J(\xi)(\mu) = \pm\langle\xi, \mu\rangle$, which means that

$$\mathbf{J} = \pm \text{ identity on } \mathfrak{g}^*. \quad \blacklozenge$$

(f) The Dual of a Lie Algebra Homomorphism

The plasma to fluid map and averaging over a symmetry group in fluid flows are duals of Lie algebra homomorphisms and provide examples of interesting Poisson maps (see §1.7). *Let us now show that all such maps are momentum maps.*

Let H and G be Lie groups, let $A : H \to G$ be a Lie group homomorphism and suppose that $\alpha : \mathfrak{h} \to \mathfrak{g}$ is the induced Lie algebra homomorphism, so its dual $\alpha^* : \mathfrak{g}^* \to \mathfrak{h}^*$ is a Poisson map. We assert that α^* *is also a momentum map*. Let H act on $\mathfrak{g}^*_+$ by

$$h \cdot \mu = \mathrm{Ad}^*_{A(h)^{-1}} \mu$$

that is,

$$\langle h \cdot \mu, \xi \rangle = \left\langle \mu, \mathrm{Ad}_{A(h)^{-1}} \xi \right\rangle . \tag{11.5.18}$$

Differentiating (11.5.18) with respect to h at e in the direction $\eta \in \mathfrak{h}$ gives the infinitesimal generator

$$\langle \eta_{\mathfrak{g}^*}(\mu), \xi \rangle = - \left\langle \mu, \mathrm{ad}_{\alpha(\eta)} \xi \right\rangle = - \left\langle \mathrm{ad}^*_{\alpha(\eta)} \mu, \xi \right\rangle . \tag{11.5.19}$$

Setting $\mathbf{J}(\mu) = \alpha^*(\mu)$, that is,

$$J(\eta)(\mu) = \langle \mathbf{J}(\mu), \eta \rangle = \langle \alpha^*(\mu), \eta \rangle = \langle \mu, \alpha(\eta) \rangle , \tag{11.5.20}$$

we get

$$\frac{\delta J(\eta)}{\delta \mu} = \alpha(\eta)$$

and so on $\mathfrak{g}^*_+$,

$$X_{J(\eta)}(\mu) = -\mathrm{ad}^*_{\delta J(\eta)/\delta\mu} \mu = -\mathrm{ad}^*_{\alpha(\eta)} \mu = \eta_{\mathfrak{g}^*}(\mu), \tag{11.5.21}$$

so we have proved the assertion. ◆

(g) Momentum Maps for Unitary Representations on Projective Space

Here we discuss the momentum map for an action of a finite dimensional Lie group G on projective space that is induced by a unitary representation on the underlying Hilbert space. Recall from §**5.3** that the unitary group $U(\mathcal{H})$ acts on $\mathbb{P}\mathcal{H}$ by symplectomorphisms. Due to the difficulties in defining the Lie algebra of $U(\mathcal{H})$ (see the remarks at the end of §9.3) we cannot define the momentum map for the whole unitary group.

Let $\rho : G \to U(\mathcal{H})$ be a unitary representation of G. We can define the infinitesimal action of its Lie algebra $\mathfrak{g}$ on $\mathbb{P}\mathcal{D}_G$, the essenial G-smooth part of $\mathbb{P}\mathcal{H}$ by

$$\xi_{\mathbb{P}\mathcal{H}}([\psi]) = \frac{d}{dt}[(\exp(tA(\xi)))\psi]\bigg|_{t=0} = T_\psi \pi(A(\xi)\psi), \tag{11.5.22}$$

where the infinitesimal generator $A(\xi)$ was defined in §9.3 and where $[\psi] \in \mathbb{P}\mathcal{D}_G$, and where the projection is denoted $\pi : \mathcal{H} \setminus \{0\} \to \mathbb{P}\mathcal{H}$. Let $\varphi \in (\mathbb{C}\psi)^\perp$ and $\|\psi\| = 1$. Since $A(\xi)\psi - \langle A(\xi)\psi, \psi\rangle\psi \in (\mathbb{C}\psi)^\perp$, we have

$$\begin{aligned} (\mathbf{i}_{\xi_{\mathbb{P}\mathcal{H}}}\Omega)(T_\psi\pi(\varphi)) &= -2\hbar \mathrm{Im} \left\langle A(\xi)\psi - \langle A(\xi)\psi, \psi\rangle\psi, \varphi \right\rangle \\ &= -2\hbar \mathrm{Im} \left\langle A(\xi)\psi, \varphi \right\rangle. \end{aligned}$$

On the other hand, if $\mathbf{J} : \mathbb{P}\mathcal{D}_G \to \mathfrak{g}^*$ is defined by

$$\langle \mathbf{J}([\psi]), \xi \rangle = J(\xi)([\psi]) = -i\hbar \frac{\langle \psi, A(\xi)\psi \rangle}{\|\psi\|^2}, \tag{11.5.23}$$

then, for $\varphi \in (\mathbb{C}\psi)^\perp$ and $\|\psi\| = 1$, a short computation gives

$$\begin{aligned} \mathbf{d}(J(\xi))([\psi])(T_\psi \pi(\varphi)) &= \frac{d}{dt} J(\xi)([\psi + t\varphi])\bigg|_{t=0} \\ &= -2\hbar \mathrm{Im} \left\langle A(\xi)\psi, \varphi \right\rangle . \end{aligned}$$

This shows that the map $\mathbf{J}$ defined in (11.5.23) is the momentum map of the G-action on $\mathbb{P}\mathcal{H}$. We caution that this momentum map is defined only on a dense subset of the symplectic manifold. Recall that a similar thing happened when we discussed the angular momentum for quantum mechanics in §3.3.

Exercises

11.5-1 (Momentum Maps Induced by Subgroups) *Consider a Poisson action of a Lie group G on the Poisson manifold P with a momentum map $\mathbf{J}$ and let H be a Lie subgroup of G. Denote by $i : \mathfrak{h} \to \mathfrak{g}$ the inclusion between the corresponding Lie algebras and $i^* : \mathfrak{g}^* \to \mathfrak{h}^*$ the dual map. Check that the induced H-action on P has a momentum map given by $\mathbf{K} = i^* \circ \mathbf{J}$, i.e., $K = J|\mathfrak{h}$.*

11.5-2 (Euclidean Group in the Plane) *The special Euclidean group $SE(2)$ consists of all transformations of $\mathbb{R}^2$ of the form $\mathbf{Az} + \mathbf{a}$, where $\mathbf{z}, \mathbf{a} \in \mathbb{R}^2$, and*

$$\mathbf{A} \in SO(2) = \left\{ \text{matrices of the form} \begin{bmatrix} \cos\theta & -\sin\theta \\ \sin\theta & \cos\theta \end{bmatrix} \right\}. \tag{11.5.24}$$

This group is three dimensional, with the composition law

$$(\mathbf{A}, \mathbf{a}) \cdot (\mathbf{B}, \mathbf{b}) = (\mathbf{AB}, \mathbf{Ab} + \mathbf{a}), \tag{11.5.25}$$

identity element $(\mathbf{I}, \mathbf{0})$, and inverse $(\mathbf{A}, \mathbf{a})^{-1} = \left(\mathbf{A}^{-1}, -\mathbf{A}^{-1}\mathbf{a}\right)$. We let $SE(2)$ act on $\mathbb{R}^2$ by $(\mathbf{A}, \mathbf{a}) \cdot \mathbf{z} = \mathbf{Az} + \mathbf{a}$. Let $\mathbf{z} = (\mathbf{q}, \mathbf{p})$ denote coordinates on $\mathbb{R}^2$. Since $\det \mathbf{A} = 1$, we get $\Phi^*_{(\mathbf{A},\mathbf{a})}(dq \wedge dp) = dq \wedge dp$, that is, $SE(2)$ acts canonically on the symplectic manifold $\mathbb{R}^2$. Show that this action has a momentum map given by

$$\mathbf{J}(q, p) = \left(-\frac{1}{2}(q^2 + p^2), p, -q \right).$$

11.6 Equivariance of Momentum Maps

Return to the commutative diagram in §**11.2** and the relations (11.1.8). Since two of the maps in the diagram are Lie algebra antihomomorphisms, it is natural to ask whether J is a Lie algebra homomorphism. Equivalently, since $X_{J[\xi,\eta]} = X_{\{J(\xi),J(\eta)\}}$, it follows that

$$J([\xi,\eta]) - \{J(\xi), J(\eta)\} =: \Sigma(\xi,\eta)$$

is a Casimir function on P and hence is constant on every symplectic leaf of P. As a function on $\mathfrak{g}\times\mathfrak{g}$ with values in the vector space $\mathcal{C}(P)$ of Casimir functions on P, Σ is bilinear, antisymmetric, and satisfies

$$\Sigma(\xi,[\eta,\zeta]) + \Sigma(\eta,[\zeta,\xi]) + \Sigma(\zeta,[\xi,\eta]) = 0 \tag{11.6.1}$$

for all $\xi,\eta,\zeta \in \mathfrak{g}$. One says that Σ is a ***$\mathcal{C}(P)$-valued 2-cocycle*** of $\mathfrak{g}$; see Souriau [1970] and Guillemin and Sternberg [1984], p. 170, for more information. It is natural to ask when $\Sigma(\xi,\eta) = 0$ for all $\xi,\eta\in\mathfrak{g}$. In general, this does not happen and one is led to the study of this invariant. We shall derive an equivalent condition for $J:\mathfrak{g}\to\mathcal{F}(P)$ to be a Lie algebra homomorphism; that is, for $\Sigma = 0$, or, in other words, for the following ***commutation relations*** to hold:

$$J([\xi,\eta]) = \{J(\xi), J(\eta)\}. \tag{11.6.2}$$

Differentiating relation (11.2.2) with respect to z in the direction $v_z \in T_zP$, we get

$$\mathbf{d}(J(\xi))(z)\cdot v_z = \langle T_z\mathbf{J}\cdot v_z, \xi\rangle \tag{11.6.3}$$

for all $z\in P$, $v_z\in T_zP$, and $\xi\in\mathfrak{g}$. Thus, for $\xi,\eta\in\mathfrak{g}$,

$$\begin{aligned}\{J(\xi),J(\eta)\}(z) &= X_{J(\eta)}\left[J(\xi)\right](z) = \mathbf{d}(J(\xi))(z)\cdot X_{J(\eta)}(z) \\ &= \left\langle T_z\mathbf{J}\cdot X_{J(\eta)}(z),\xi\right\rangle = \left\langle T_z\mathbf{J}\cdot \eta_P(z),\xi\right\rangle. \end{aligned} \tag{11.6.4}$$

Note that

$$\begin{aligned} J([\xi,\eta])(z) &= \langle \mathbf{J}(z), [\xi,\eta]\rangle = -\left\langle \mathbf{J}(z), \operatorname{ad}_\eta \xi\right\rangle \\ &= -\left\langle \operatorname{ad}_\eta^* \mathbf{J}(z),\xi\right\rangle. \end{aligned} \tag{11.6.5}$$

Consequently, J is a Lie algebra homomorphism if and only if

$$T_z\mathbf{J}\cdot\eta_P(z) = -\operatorname{ad}_\eta^*\mathbf{J}(z) \tag{11.6.6}$$

for all $\eta\in\mathfrak{g}$, that is, (11.6.2) and (11.6.6) are equivalent. Momentum maps satisfying (11.6.2) (or (11.6.6)) are called ***infinitesimally equivariant momentum maps*** and canonical (left) Lie algebra actions admitting infinitesimally equivariant momentum maps are called ***Hamiltonian actions***. With this terminology, we have proved the following proposition:

Theorem 11.6.1 *A canonical left Lie algebra action is Hamiltonian if and only if there is a Lie algebra homomorphism $\psi : \mathfrak{g} \to \mathcal{F}(P)$ such that $X_{\psi(\xi)} = \xi_P$ for all $\xi \in \mathfrak{g}$. If ψ exists, an infinitesimally equivariant momentum map $\mathbf{J}$ is determined by $J = \psi$. Conversely, if $\mathbf{J}$ is infinitesimally equivariant, we can take $\psi = J$.*

Let us justify the terminology "infinitesimally equivariant momentum map." Suppose the canonical left Lie algebra action of $\mathfrak{g}$ on P arises from a canonical left Lie group action of G on P, where $\mathfrak{g}$ is the Lie algebra of G and we assume G and P are connected. We say that $\mathbf{J}$ is ***equivariant*** if the diagram in Figure 11.6.1 commutes, that is,

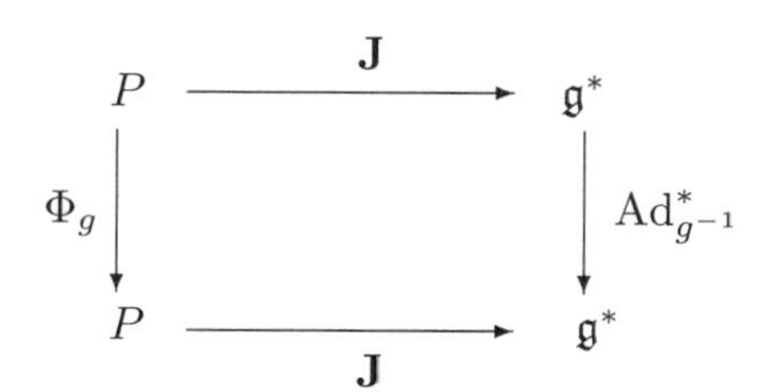

FIGURE 11.6.1. Equivariance of momentum maps.

$$\mathrm{Ad}^*_{g^{-1}} \circ \mathbf{J} = \mathbf{J} \circ \Phi_g \tag{11.6.7}$$

for all $g \in G$. Equivalently, equivariance can be reformulated as the identity

$$J(\mathrm{Ad}_g \xi)(g \cdot z) = J(\xi)(z) \tag{11.6.8}$$

for all $g \in G, \xi \in \mathfrak{g}$, and $z \in P$. A (left) canonical Lie group action is called ***globally Hamiltonian*** if it has an equivariant momentum map. Differentiating (11.6.7) with respect to g at $g = e$ in the direction $\eta \in \mathfrak{g}$ shows that *equivariance implies infinitesimal equivariance*. We shall see shortly that all the preceding examples (except the one in Exercise **11.5-2**) have equivariant momentum maps. Another case of interest occurs in Yang-Mills theory, where the 2-cocycle Σ is related to the ***anomaly*** (see Bao and Nair [1985] and references therein). The converse question, When does infinitesimal equivariance imply equivariance? is treated in §12.4.

Exercises

11.6-1 *Show that the map $J : S^2 \to \mathbb{R}$ given by $(x, y, z) \mapsto z$ is a momentum map.*

11.6-2 *Check directly that angular momentum is an equivariant momentum map, whereas the momentum map in Exercise* **11.5-2** *is* ***not*** *equivariant.*

11.6-3 *Prove that the momentum map determined by* (11.3.4) *is equivariant.*

12
Computation and Properties of Momentum Maps

The previous chapter gave the general theory of momentum maps. In this chapter, we develop techniques for computing them. One of the most important cases is when there is a group action on a cotangent bundle and this action is obtained from lifting an action on the base. These transformations are called *extended point transformations.*

12.1 Momentum Maps on Cotangent Bundles

Define the map $\mathcal{P} : \mathfrak{X}(Q) \to \mathcal{F}(T^*Q)$, by

$$\mathcal{P}(X)(\alpha_q) = \langle \alpha_q, X(q) \rangle ,$$

for $q \in Q$ and $\alpha_q \in T_q^*Q$, where $\langle\, , \rangle$ denotes the pairing between covectors $\alpha \in T_q^*Q$ and vectors. We call $\mathcal{P}(X)$ the ***momentum function of*** X.

Definition 12.1.1 *Given a manifold Q, let $\mathcal{L}(T^*Q)$ denote the space of smooth functions $F : T^*Q \to \mathbb{R}$ that are linear on fibers of T^*Q.*

Using coordinates and working in finite dimensions, we can write $F, H \in \mathcal{L}(T^*Q)$ as

$$F(q,p) = \sum_{i=1}^{n} X^i(q)p_i, \quad \text{and} \quad H(q,p) = \sum_{i=1}^{n} Y^i(q)p_i,$$

for functions X^i and Y^i. We claim that the standard Poisson bracket $\{F, H\}$ is again linear on the fibers. Indeed, using summations on repeated indices,

$$\{F, H\}(q, p) = \frac{\partial F}{\partial q^j}\frac{\partial H}{\partial p_j} - \frac{\partial H}{\partial q^j}\frac{\partial F}{\partial p_j} = \frac{\partial X^i}{\partial q^j} p_i Y^k \delta_k^j - \frac{\partial Y^i}{\partial q^j} p_i X^k \delta_k^j$$

and so

$$\{F, H\} = \left(\frac{\partial X^i}{\partial q^j} Y^j - \frac{\partial Y^i}{\partial q^j} X^j\right) p_i. \tag{12.1.1}$$

Hence $\mathcal{L}(T^*Q)$ is a Lie subalgebra of $\mathcal{F}(T^*Q)$. If Q is infinite dimensional, a similar proof, using canonical cotangent bundle charts, works.

Lemma 12.1.2 (Momentum Commutator Lemma) *The Lie algebras:*

i $(\mathfrak{X}(Q), [\,,])$ *of vector fields on* Q*; and*

ii *Hamiltonian vector fields* X_F *on* T^*Q *with* $F \in \mathcal{L}(T^*Q)$

are isomorphic. Moreover, each of these algebras is anti-isomorphic to $(\mathcal{L}(T^*Q), \{\,,\})$*. In particular, we have*

$$\{\mathcal{P}(X), \mathcal{P}(Y)\} = -\mathcal{P}([X, Y]). \tag{12.1.2}$$

Proof Since $\mathcal{P}(X) : T^*Q \to \mathbb{R}$ is linear on fibers, it follows that $\mathcal{P} : \mathfrak{X}(Q) \to \mathcal{L}(T^*Q)$. This map is linear and satisfies (12.1.2) since $[X, Y]^i = (\partial Y^i/\partial q^j)\, X^j - (\partial X^i/\partial q^j)\, Y^j$ implies that

$$-\mathcal{P}([X, Y]) = \left(\frac{\partial X^i}{\partial q^j} Y^j - \frac{\partial Y^i}{\partial q^j} X^j\right) p_i,$$

which coincides with $\{\mathcal{P}(X), \mathcal{P}(Y)\}$ by (12.1.1). (We leave it to the reader to write out the infinite-dimensional proof.) Furthermore, $\mathcal{P}(X) = 0$ implies that $X = 0$ by the Hahn-Banach theorem. Finally, for each $F \in \mathcal{L}(T^*Q)$, define $X(F) \in \mathfrak{X}(Q)$ by $\langle X(F)(q), \alpha_q \rangle = F(\alpha_q)$, where $\alpha_q \in T_q^*Q$. Then $\mathcal{P}(X(F)) = F$, so $\mathcal{P}$ is also surjective, thus proving that $(\mathfrak{X}(Q), [\,,])$ and $\mathcal{L}(T^*Q), \{\,,\})$ are anti-isomorphic Lie algebras.

The map $F \mapsto X_F$ is a Lie algebra antihomomorphism from the algebra $(\mathcal{L}(T^*Q), \{\,,\})$ to $(\{X_F \mid F \in \mathcal{L}(T^*Q)\}, [\,,])$ by (5.5.6). This map is surjective by definition. Moreover, if $X_F = 0$, then F is constant on T^*Q, hence equal to zero since it is linear on the fibers. ■

In quantum mechanics, the ***Dirac rule*** associates the differential operator

$$\mathrm{X} = \frac{\hbar}{i} X^j \frac{\partial}{\partial q^j} \tag{12.1.3}$$

with the momentum function $\mathcal{P}(X)$. (Dirac [1930], §21 and §22.) Thus if we define $P_{\mathrm{X}} = \mathcal{P}(X)$, (12.1.2) gives

$$i\hbar\{P_{\mathrm{X}}, P_{\mathrm{Y}}\} = i\hbar\{\mathcal{P}(X), \mathcal{P}(Y)\} = -i\hbar\mathcal{P}([X,Y]) = P_{[\mathrm{X},\mathrm{Y}]}. \tag{12.1.4}$$

One can augment (12.1.4) by including lifts of functions on Q. Given $f \in \mathcal{F}(Q)$, let $f^* = f \circ \pi_Q$ where $\pi_Q : T^*Q \to Q$ is the projection, so f^* is constant on fibers. One finds that

$$\{f^*, g^*\} = 0 \tag{12.1.5}$$

and

$$\{f^*, \mathcal{P}(X)\} = X[f]. \tag{12.1.6}$$

The Hamiltonian flow φ_t of X_{f^*} is fiber translation by $-t\,\mathbf{d}f$, that is, $(q,p) \mapsto (q, p - t\mathbf{d}f(q))$. The flow of $X_{\mathcal{P}(X)}$ is given by the following:

Proposition 12.1.3 *If $X \in \mathfrak{X}(Q)$ has flow φ_t, then the flow of $X_{\mathcal{P}(X)}$ on T^*Q is $T^*\varphi_{-t}$.*

Proof If $\pi_Q : T^*Q \to Q$ denotes the canonical projection, differentiating the relation

$$\pi_Q \circ T^*\varphi_{-t} = \varphi_t \circ \pi_Q \tag{12.1.7}$$

at $t = 0$ gives

$$T\pi_Q \circ Y = X \circ \pi_Q, \tag{12.1.8}$$

where

$$Y(\alpha_q) = \left.\frac{d}{dt} T^*\varphi_{-t}(\alpha_q)\right|_{t=0}, \tag{12.1.9}$$

so $T^*\varphi_{-t}$ is the flow of Y. Since $T^*\varphi_{-t}$ preserves the canonical one-form Θ on T^*Q, it follows that $\pounds_Y\Theta = 0$. Hence

$$\mathbf{i}_Y\Omega = -\mathbf{i}_Y\mathbf{d}\Theta = \mathbf{d}\mathbf{i}_Y\Theta. \tag{12.1.10}$$

By definition of the canonical one-form,

$$\begin{aligned} \mathbf{i}_Y\Theta(\alpha_q) &= \langle\Theta(\alpha_q), Y(\alpha_q)\rangle = \langle\alpha_q, T\pi_Q(Y(\alpha_q))\rangle \\ &= \langle\alpha_q, X(q)\rangle = \mathcal{P}(X)(\alpha_q), \end{aligned} \tag{12.1.11}$$

that is, $\mathbf{i}_Y\Omega = \mathbf{d}\mathcal{P}(X)$ so that $Y = X_{\mathcal{P}(X)}$. ■

Because of this proposition, the Hamiltonian vector field $X_{\mathcal{P}(X)}$ on T^*Q is called the ***cotangent lift*** of $X \in \mathfrak{X}(Q)$ to T^*Q. We also use the notation $X' := X_{\mathcal{P}(X)}$ for the cotangent lift of X. From $X_{\{F,H\}} = -[X_F, X_H]$ and (12.1.2), we get

$$\begin{aligned} [X', Y'] &= [X_{\mathcal{P}(X)}, X_{\mathcal{P}(Y)}] = -X_{\{\mathcal{P}(X),\, \mathcal{P}(Y)\}} \\ &= -X_{-\mathcal{P}[X,Y]} = [X,Y]'. \end{aligned} \tag{12.1.12}$$

For finite-dimensional Q, in local coordinates, we have

$$
\begin{aligned}
X' : &= X_{\mathcal{P}(X)} = \sum_{i=1}^{n} \left(\frac{\partial \mathcal{P}(X)}{\partial p_i} \frac{\partial}{\partial q^i} - \frac{\partial \mathcal{P}(X)}{\partial q^i} \frac{\partial}{\partial p_i} \right) \\
&= X^i \frac{\partial}{\partial q^i} - \frac{\partial X^i}{\partial q^j} p_i \frac{\partial}{\partial p_j}.
\end{aligned}
\tag{12.1.13}
$$

Theorem 12.1.4 (Momentum Maps for Lifted Actions) *Let the Lie algebra $\mathfrak{g}$ act on the left on the manifold Q, so that $\mathfrak{g}$ acts on $P = T^*Q$ on the left by the canonical action $\xi_P = \xi'_Q$, where ξ'_Q is the cotangent lift of ξ_Q to P and $\xi \in \mathfrak{g}$. This $\mathfrak{g}$-action on P is Hamiltonian with infinitesimally equivariant momentum map $\mathbf{J} : P \to \mathfrak{g}^*$ given by*

$$
\langle \mathbf{J}(\alpha_q), \xi \rangle = \langle \alpha_q, \xi_Q(q) \rangle = \mathcal{P}(\xi_Q)(\alpha_q). \tag{12.1.14}
$$

If $\mathfrak{g}$ is the Lie algebra of a Lie group G and G acts on Q and hence on T^*Q by cotangent lift, then $\mathbf{J}$ is equivariant.

In coordinates q^i, p_j on T^*Q and ξ^a on $\mathfrak{g}$, (12.1.14) reads

$$
J_a \xi^a = p_i \xi^i_Q = p_i A^i{}_a \xi^a,
$$

where $\xi^i_Q = \xi^a A^i{}_a$ are the components of ξ_Q; thus,

$$
J_a(q,p) = p_i A^i{}_a(q). \tag{12.1.15}
$$

Proof For $\xi, \eta \in \mathfrak{g}$, (12.1.12) gives

$$
[\xi, \eta]_P = [\xi, \eta]'_Q = -[\xi_Q, \eta_Q]' = -[\xi'_Q, \eta'_Q] = -[\xi_P, \eta_P]
$$

and hence $\xi \mapsto \xi_P$ is a left algebra action. This action is also canonical, for if $F, H \in \mathcal{F}(P)$,

$$
\begin{aligned}
\xi_P[\{F, H\}] &= X_{\mathcal{P}(\xi_Q)}[\{F, H\}] = \{X_{\mathcal{P}(\xi_Q)}[F], H\} + \{F, X_{\mathcal{P}(\xi_Q)}[H]\} \\
&= \{\xi_P[F], H\} + \{F, \xi_P[H]\}
\end{aligned}
$$

by the Jacobi identity for the Poisson bracket. If φ_t is the flow of ξ_Q, the flow of $\xi'_Q = X_{\mathcal{P}(\xi_Q)}$ is $T^*\varphi_{-t}$. Consequently, $\xi_P = X_{\mathcal{P}(\xi_Q)}$ and, thus, the $\mathfrak{g}$-action on P admits a momentum map given by $J(\xi) = P(\xi_Q)$. Since $\xi \in \mathfrak{g} \mapsto \mathcal{P}(\xi_Q) = J(\xi) \in \mathcal{F}(P)$ is a Lie algebra homomorphism by (11.1.5) and (12.1.12), it follows that $\mathbf{J}$ is an infinitesimally equivariant momentum map (Theorem **11.6.1**).

Equivariance under G is proved directly in the following way. For any $g \in G$, we have

$$
\begin{aligned}
\langle \mathbf{J}(g \cdot \alpha_q), \xi \rangle &= \langle g \cdot \alpha_q, \xi_Q(g \cdot \alpha_q) \rangle \\
&= \left\langle \alpha_q, (T_{g \cdot q} \Phi_g^{-1} \circ \xi_Q \circ \Phi_g)(q) \right\rangle
\end{aligned}
$$

$$\begin{aligned}
&= \left\langle \alpha_q, (\Phi_g^* \xi_Q)(q) \right\rangle \\
&= \left\langle \alpha_q, (\mathrm{Ad}_{g^{-1}} \xi)_Q(q) \right\rangle \quad \text{(by Lemma \textbf{9.3.7ii})} \\
&= \left\langle \mathbf{J}(\alpha_q), \mathrm{Ad}_{g^{-1}} \xi \right\rangle \\
&= \left\langle (\mathrm{Ad}_{g^{-1}}^* \mathbf{J})(\alpha_q), \xi \right\rangle . \quad \blacksquare
\end{aligned}$$

Remarks

1. Let $G = \mathrm{Diff}(Q)$ act on T^*Q by cotangent lift. Then the infinitesimal generator of $X \in \mathfrak{X}(Q) = \mathfrak{g}$ is $X_{\mathcal{P}(X)}$ by Proposition **12.1.3** so that the associated momentum map is $\mathbf{J} : T^*Q \to \mathfrak{X}(Q)^*$ which is defined through $J(X) = \mathcal{P}(X)$ by the above calculations.

2. ***Momentum Fiber Translations*** Let $G = \mathcal{F}(Q)$ act on T^*Q by fiber translations by $\mathbf{d}f$, that is,
$$f \cdot \alpha_q = \alpha_q + \mathbf{d}f(q). \tag{12.1.16}$$
Since the infinitesimal generator of $\xi \in \mathcal{F}(Q) = \mathfrak{g}$ is the vertical lift of $\mathbf{d}\xi(q)$ at α_q and this in turn equals the Hamiltonian vector field $-X_{\xi \circ \pi_Q}$, we see that the momentum map $\mathbf{J} : T^*Q \to \mathcal{F}(Q)^*$ is given by
$$J(\xi) = -\xi \circ \pi_Q. \tag{12.1.17}$$
This momentum map is equivariant since π_Q is constant on fiber translations.

3. The commutation relations
$$\left.\begin{aligned}
\{\mathcal{P}(X), \mathcal{P}(Y)\} &= -\mathcal{P}([X, Y]), \\
\{\mathcal{P}(X), \xi \circ \pi_Q\} &= -X[\xi] \circ \pi_Q, \\
\{\xi \circ \pi_Q, \eta \circ \pi_Q\} &= 0,
\end{aligned}\right\} \tag{12.1.18}$$
can be rephrased as saying that the pair $(\mathbf{J}(X), \mathbf{J}(f))$ fit together to form a momentum map for the semidirect product group
$$\mathrm{Diff}(Q) \circledS \mathcal{F}(Q).$$
This plays an important role in the general theory of semidirect products for which we refer the reader to Marsden, Weinstein, Ratiu, Schmid and Spencer [1983], and Marsden, Ratiu and Weinstein [1984a, b]. $\blacklozenge$

The terminology ***extended point transformations*** arises for the following reasons. Let $\Phi : G \times Q \to Q$ be a smooth action and consider its lift $\Phi : G \times T^*Q \to T^*Q$ to the cotangent bundle. The action Φ moves points in the configuration space Q, and Φ is its natural extension to phase space T^*Q; in coordinates, the action on configuration points $q^i \mapsto \overline{q}^i$ induces the following action on momenta:
$$p_i \mapsto \overline{p}_i = \frac{\partial \overline{q}^j}{\partial q^i} p_j. \tag{12.1.19}$$

Exercises

12.1-1 *What is the anaolgue of* (12.1.18) *for rotations and translations on* $\mathbb{R}^3$*?*

12.1-2 *Prove* (12.1.2) *in infinite dimensions.*

12.1-3 *Prove Theorem* **12.1.4** *as a consequence of formula* (11.3.4) *and Exercise* **11.6-3**.

12.2 Momentum Maps on Tangent Bundles

Proposition 12.2.1 *Let the Lie algebra* $\mathfrak{g}$ *act on the left on the manifold* Q *and assume that* $L : TQ \to \mathbb{R}$ *is a regular Lagrangian. Endow* TQ *with the symplectic form* $\Omega_L = (\mathbb{F}L)^*\Omega$*, where* $\Omega = -\mathbf{d}\Theta$ *is the canonical symplectic form on* T^*Q*. Then* $\mathfrak{g}$ *acts canonically on* $P = TQ$ *by* $\xi_P(p) = d/dt|_{t=0} T_p\varphi_t$*, where* φ_t *is the flow of* ξ_Q *and has the infinitesimally equivariant momentum map* $\mathbf{J} : TQ \to \mathfrak{g}^*$ *given by*

$$\langle \mathbf{J}(v_q), \xi \rangle = \langle \mathbb{F}L(v_q), \xi_Q(q) \rangle . \tag{12.2.1}$$

If $\mathfrak{g}$ *is the Lie algebra of a Lie group* G *and* G *acts on* Q *and hence on* TQ *by tangent lift, then* $\mathbf{J}$ *is equivariant.*

Proof Use (11.3.4), a direct calculation or, if L is hyperregular, the following argument. Since $\mathbb{F}L$ is a symplectic diffeomorphism, $\xi \mapsto \xi_P = (\mathbb{F}L)^*\xi_{T^*Q}$ is a canonical left Lie algebra action. Therefore, the composition of $\mathbb{F}L$ with the momentum map (12.1.14) is the momentum map of the $\mathfrak{g}$-action on TQ. ■

In coordinates $(q^i, \dot{q}^i)$ on TQ and (ξ^a) on $\mathfrak{g}$, (12.2.1) reads

$$J_a(q^i, \dot{q}^i) = \frac{\partial L}{\partial \dot{q}^i} A^i{}_a(q), \tag{12.2.2}$$

where $\xi^i_Q(q) = \xi^a A^i{}_a(q)$ are the components of ξ_Q.

Exercises

12.2-1 *Derive the conservation of* $\mathbf{J}$ *given by* $\langle \mathbf{J}(v_q), \xi \rangle = \langle \mathbb{F}L(v_q), \xi_Q(q) \rangle$ ***directly*** *from Hamilton's variational principle. (This is the way Noether originally derived conserved quantities).*

12.2-2 *If* L *is independent of one of the coordinates* q^i*, then it is clear that*

$p_i = \partial L/\partial \dot{q}^i$ *is a constant of the motion from the Euler-Lagrange equations. Derive this from Proposition* **12.2.1**.

12.3 Examples

(a) The Hamiltonian
A Hamiltonian $H : P \to \mathbb{R}$ on a Poisson manifold P having a complete vector field X_H is an equivariant momentum map for the $\mathbb{R}$-action given by the flow of X_H. ♦

(b) Linear Momentum
In the notations of Example **(b)** of §11.5 we recompute the linear momentum of the N-particle system. Since $\mathbb{R}^3$ acts on points $(\mathbf{q}_1, \ldots, \mathbf{q}_N)$ in $\mathbb{R}^{3N}$ by $\mathbf{x} \cdot (\mathbf{q}_j) = (\mathbf{q}_j + \mathbf{x})$, the infinitesimal generator is

$$\xi_{\mathbb{R}^N}(\mathbf{q}_j) = (\mathbf{q}_1, \ldots, \mathbf{q}_N, \xi, \ldots, \xi) \tag{12.3.1}$$

(this has the base point $(\mathbf{q}_1, \ldots, \mathbf{q}_N)$ and vector part $(\xi, \ldots, \xi)$ (N times)). Consequently, by (12.1.14), an equivariant momentum map $\mathbf{J} : T^*\mathbb{R}^{3N} \to \mathbb{R}^3$ is given by

$$J(\xi)(\mathbf{q}_j, \mathbf{p}^j) = \sum_{j=1}^{N} \mathbf{p}^j \cdot \xi, \quad \text{i.e.,} \quad \mathbf{J}(\mathbf{q}_j, \mathbf{p}^j) = \sum_{j=1}^{N} \mathbf{p}^j. \quad ♦ \tag{12.3.2}$$

(c) Angular Momentum
In the notation of Example **(c)** of §11.5, let $SO(3)$ act on $\mathbb{R}^3$ by matrix multiplication $\mathbf{A} \cdot \mathbf{q} = \mathbf{A}\mathbf{q}$. The infinitesimal generator is given by $\hat{\omega}_{\mathbb{R}^3}(\mathbf{q}) = \hat{\omega}\mathbf{q} = \omega \times \mathbf{q}$ where $\omega \in \mathbb{R}^3$. Consequently, by (12.1.14), an equivariant momentum map $\mathbf{J} : T^*\mathbb{R}^3 \to \mathfrak{so}(3)^* \cong \mathbb{R}^3$ is given by

$$\langle \mathbf{J}(\mathbf{q}, \mathbf{p}), \omega \rangle = \mathbf{p} \cdot \hat{\omega}\mathbf{q} = \omega \cdot (\mathbf{q} \times \mathbf{p}),$$

that is,

$$\mathbf{J}(\mathbf{q}, \mathbf{p}) = \mathbf{q} \times \mathbf{p}. \tag{12.3.3}$$

Equivariance in this case reduces to the relation $\mathbf{A}\mathbf{q} \times \mathbf{A}\mathbf{p} = \mathbf{A}(\mathbf{q} \times \mathbf{p})$ for any $A \in SO(3)$. If $A \in O(3) \setminus SO(3)$, such as a reflection, this relation is no longer satisfied; a minus sign appears on the right-hand side, a fact sometimes phrased by stating that *angular momentum is a pseudo-vector*. On the other hand, letting $O(3)$ act on $\mathbb{R}^3$ by matrix multiplication, $\mathbf{J}$ is given by the same formula and so is the momentum map of a lifted action and these are *always* equivariant. We have an apparent contradiction—What is wrong? The answer is that the adjoint action and the isomorphism $\hat{}: \mathbb{R}^3 \to \mathfrak{so}(3)$ are related for the component of $-$(Identity) in $O(3)$ by $\mathbf{A}\hat{x}\mathbf{A}^{-1} = -(\mathbf{A}x)\hat{}$.

Thus $\mathbf{J}(\mathbf{q}, \mathbf{p})$ is indeed equivariant as it stands. (One does not need a separate terminology like "pseudo-vector" to see what is going on.) ♦

(d) Momentum for Matrix Groups
In the notations of Example **(d)** of §11.5, let the Lie group $G \subset GL(n, \mathbb{R})$ act on $\mathbb{R}^n$ by $\mathbf{A} \cdot \mathbf{q} = \mathbf{Aq}$. The infinitesimal generator of this action is given by

$$\xi_{\mathbb{R}^N}(\mathbf{q}) = \xi \mathbf{q},$$

for $\xi \in \mathfrak{g}$, the Lie algebra of G, regarded as a subalgebra $\mathfrak{g} \subset \mathfrak{gl}(n, \mathbb{R})$. By (12.1.14), the lift of the G-action on $\mathbb{R}^3$ to $T^*\mathbb{R}^n$ has an equivariant momentum map $\mathbf{J} : T^*\mathbb{R}^n \to \mathfrak{g}^*$ given by

$$\langle \mathbf{J}(\mathbf{q}, \mathbf{p}), \xi \rangle = \mathbf{p} \cdot (\xi \mathbf{q}) \tag{12.3.4}$$

which coincides with (11.5.14). ♦

(e) Canonical Momentum on $\mathfrak{g}^*$
The coadjoint action of G on a coadjoint orbit in $\mathfrak{g}^*$ with the $\pm$ orbit symplectic structure has a momentum map which is $\pm$ the inclusion map, by Example **(e)** of §11.5. This momentum map is clearly equivariant, thus providing an example of a globally Hamiltonian action which is not an extended point transformation. ♦

(f) The Dual of a Lie Algebra Homomorphism
From Example **(f)** of §11.5 it follows that the dual of a Lie algebra homomorphism $\alpha : \mathfrak{h} \to \mathfrak{g}$ is an equivariant momentum map which does not arise from an action which is an extended point transformation. Recall that a *linear map $\alpha : \mathfrak{h} \to \mathfrak{g}$ is a Lie algebra homomorphism if and only if the dual map $\alpha^* : \mathfrak{g}^* \to \mathfrak{h}^*$ is Poisson.* ♦

(g) Momentum Maps Induced by Subgroups
If the Poisson Lie group action of G on P admits an equivariant momentum map $\mathbf{J}$ and if H is a Lie subgroup of G, then in the notation of Exercise **11.5-1**, $i^* \circ \mathbf{J} : P \to \mathfrak{h}$ is an equivariant momentum map of the induced H-action on P. ♦

(h) Products
Let P_1 and P_2 be Poisson manifolds and let $P_1 \times P_2$ be the product manifold endowed with the product Poisson structure, that is, if $F, G : P_1 \times P_2 \to \mathbb{R}$, then $\{F, G\}(z_1, z_2) = \{F_{z_2}, G_{z_2}\}_1(z_1) + \{F_{z_1}, G_{z_1}\}_2(z_2)$, where $\{\,,\}_i$ is the Poisson bracket on P_i, $F_{z_1} : P_2 \to \mathbb{R}$ is the function obtained by freezing $z_1 \in P_1$, and similarly for $F_{z_2} : P_1 \to \mathbb{R}$. Let the Lie algebra $\mathfrak{g}$ act canonically on P_1 and P_2 with (equivariant) momentum mappings $\mathbf{J}_1 : P_1 \to \mathfrak{g}^*$ and $\mathbf{J}_2 : P_2 \to \mathfrak{g}^*$. Then $\mathbf{J} = \mathbf{J}_1 + \mathbf{J}_2 : P_1 \times P_2 \to \mathfrak{g}^*$, $\mathbf{J}(z_1, z_2) = \mathbf{J}(z_1) + \mathbf{J}(z_2)$ is an (equivariant) momentum mapping of the canonical $\mathfrak{g}$-action on the product $P_1 \times$

P_2. There is an obvious generalization to the product of N Poisson manifolds. Note that example **(b)** is a special case of this, for $G = \mathbb{R}^3$ for all factors in the product manifold equal to $T^*\mathbb{R}^3$. ♦

(i) Cotangent Lift on T^*G

The momentum map for the cotangent lift of the *left* translation action of G on G is, by (12.1.14), equal to

$$\langle \mathbf{J}_L(\alpha_g), \xi \rangle = \langle \alpha_g, \xi_G(g) \rangle = \langle \alpha_g, T_e R_g(\xi) \rangle = \langle T_e^* R_g(\alpha_g), \xi \rangle ,$$

that is,

$$\mathbf{J}_L(\alpha_g) = T_e^* R_g(\alpha_g). \tag{12.3.5}$$

Similarly, the momentum map for the lift to T^*G of *right* translation of G on G equals

$$\mathbf{J}_R(\alpha_g) = T_e^* L_g(\alpha_g). \tag{12.3.6}$$

Notice that $\mathbf{J}_L$ is *right* invariant, whereas $\mathbf{J}_R$ is *left* invariant. Both are equivariant momentum maps ($\mathbf{J}_R$ with respect to Ad_g^*, which is a *right* action), so they are Poisson maps. The diagram in Figure 12.3.1 summarizes the situation.

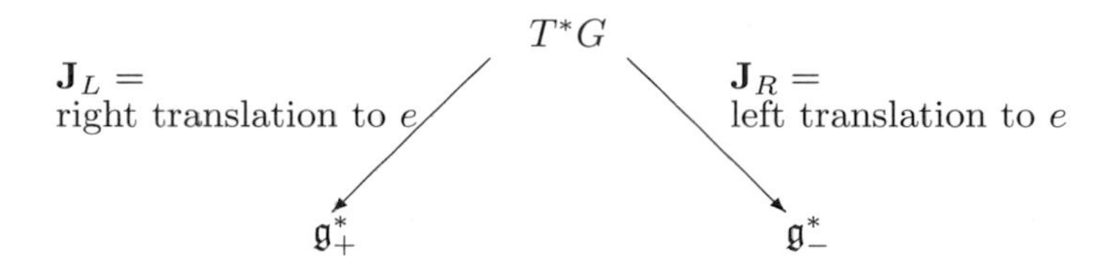

FIGURE 12.3.1. Momentum maps for left and right translations.

This diagram is an example of what is called a ***dual pair***; these illuminate the relation between the body and spatial description of rigid bodies and fluids; see Chapter 15 for more information. ♦

(j) Momentum Translation on Functions

Let $P = \mathcal{F}(T^*Q)^*$ with the Lie-Poisson bracket given in Example **(e)** of §**10.2**. Using the Liouville measure on T^*Q and assuming that elements of $\mathcal{F}(T^*Q)$ fall off rapidly enough at infinity, we identify $\mathcal{F}(T^*Q)^*$ with $\mathcal{F}(T^*Q)$ using the L^2-pairing. Let $G = \mathcal{F}(Q)$ act on P by

$$(\varphi \cdot f)(\alpha_q) = f(\alpha_q + \mathbf{d}\varphi(q)), \tag{12.3.7}$$

that is, in coordinates,

$$f(q^i, p_j) \mapsto f\left(q^i, p_j + \frac{\partial \varphi}{\partial q^i}\right).$$

The infinitesimal generator is

$$\xi_P(f)(\alpha_q) = \mathbb{F}f(\alpha_q) \cdot \mathbf{d}\xi(q), \tag{12.3.8}$$

where $\mathbb{F}f$ is the fiber derivative of f. In coordinates, (12.3.8) reads

$$\xi_P(f)(q^i, p_j) = \frac{\partial f}{\partial p_j} \cdot \frac{\partial \xi}{\partial q^j}.$$

As usual, we assume that all elements of $P = \mathcal{F}(T^*Q)$ fall off at infinity. Then, if $f, g, h \in \mathcal{F}(T^*Q)$ we have by Corollary **5.5.8**

$$\int_{T^*Q} f\{g, h\}\, dq\, dp = \int_{T^*Q} g\{h, f\}\, dq\, dp. \tag{12.3.9}$$

Next, note that if $F, H : P = \mathcal{F}(T^*Q) \to \mathbb{R}$, then we get by (12.3.9)

$$\begin{aligned} X_H[F](f) &= \{F, H\}(f) = \int_P f \left\{ \frac{\delta F}{\delta f}, \frac{\delta H}{\delta f} \right\} dq\, dp \\ &= \int_P \frac{\delta F}{\delta f} \left\{ \frac{\delta H}{\delta f}, f \right\} dq\, dp. \end{aligned}$$

On the other hand, by (12.3.8), we have

$$\xi_P[F](f) = \int_P \frac{\delta F}{\delta f} (\mathbb{F}f \cdot (\mathbf{d}\xi \circ \pi_Q))\, dq\, dp, \tag{12.3.10}$$

which suggests that the definition of $\mathbf{J}$ should be

$$\langle \mathbf{J}(f), \xi \rangle = \int_P f(\alpha_q)\xi(q)\, dq\, dp. \tag{12.3.11}$$

Indeed, by (12.3.11), we have $\delta J(\xi)/\delta f = \xi \circ \pi_Q$ so that

$$\left\{ \frac{\delta J(\xi)}{\delta f}, f \right\} = \{\xi \circ \pi_Q, f\} = \mathbb{F}f \cdot (\mathbf{d}\xi \circ \pi_Q)$$

and hence by (12.3.9)

$$\begin{aligned} X_{J(\xi)}[F](f) &= \int_P \frac{\delta F}{\delta f} \left\{ \frac{\delta J(\xi)}{\delta f}, f \right\} dq\, dp \\ &= \int_P \frac{\delta F}{\delta f} (\mathbb{F}f \cdot (\mathbf{d}\xi \circ \pi_Q))\, dq\, dp, \end{aligned}$$

which coincides with (12.3.10) thereby proving that $\mathbf{J}$ given by (12.3.11) is the momentum map. In other words, the fiber integral

$$\mathbf{J}(f) = \int_P f(q, p)\, dp \tag{12.3.12}$$

is the momentum map in this case. This momentum map is infinitesimally equivariant. Indeed, if $\xi, \eta \in \mathcal{F}(Q)$, we have for $f \in P$,

$$\begin{aligned} \{J(\xi), J(\eta)\}(f) &= \int_P f \left\{ \frac{\delta J(\xi)}{\delta f}, \frac{\delta J(\eta)}{\delta f} \right\} dq\, dp \\ &= \int_P f \{\xi \circ \pi_Q, \eta \circ \pi_Q\}\, dq\, dp \\ &= 0 = J([\xi, \eta])(f). \quad \blacklozenge \end{aligned}$$

(k) More Momentum Translations

Let $\mathrm{Diff}_{\mathrm{can}}(T^*Q)$ denote the group of symplectic diffeomorphisms of T^*Q and, as above, let $G = \mathcal{F}(Q)$ act on T^*Q by translation with $\mathbf{d}f$ along the fiber, that is, $f \cdot \alpha_q = \alpha_q + \mathbf{d}f(q)$. Since the action of the additive group $\mathcal{F}(Q)$ is Hamiltonian, $\mathcal{F}(Q)$ acts on $\mathrm{Diff}_{\mathrm{can}}(T^*Q)$ by composition on the right with translations, that is, the action is $(f, \varphi) \in \mathcal{F}(Q) \times \mathrm{Diff}_{\mathrm{can}}(T^*Q) \mapsto \varphi \circ \rho_f \in \mathrm{Diff}_{\mathrm{can}}(T^*Q)$, where $\rho_f(\alpha_q) = \alpha_q + \mathbf{d}f(q)$. The infinitesimal generator of this action is given by (see (12.1.17)):

$$\xi_{\mathrm{Diff}_{\mathrm{can}}(T^*Q)}(\varphi) = -T\varphi \circ X_{\xi \circ \pi_Q} \tag{12.3.13}$$

for $\xi \in \mathcal{F}(Q) = \mathfrak{g}$, so that the equivariant momentum map of the lifted action $\mathbf{J} : T^*(\mathrm{Diff}_{\mathrm{can}}(T^*Q)) \to \mathcal{F}(Q)^*$ given by (12.1.14) in this case

$$J(\xi)(\alpha_\varphi) = -\left\langle \alpha_\varphi, T\varphi \circ X_{\xi \circ \pi_Q} \right\rangle, \tag{12.3.14}$$

where the pairing on the right is between vector fields and one-form densities α_φ. $\blacklozenge$

(l) Maxwell's Equations

Let $\mathcal{A}$ be the space of vector potentials $\mathbf{A}$ on $\mathbb{R}^3$ and $P = T^*\mathcal{A}$, whose elements are denoted $(\mathbf{A}, -\mathbf{E})$ with $\mathbf{A}$ and $\mathbf{E}$ vector fields. Let $G = \mathcal{F}(\mathbb{R}^3)$ act on $\mathcal{A}$ by $\varphi \cdot \mathbf{A} = \mathbf{A} + \nabla\varphi$. Thus the infinitesimal generator is

$$\xi_{\mathcal{A}}(\mathbf{A}) = \nabla\xi.$$

Hence the momentum map is

$$\langle \mathbf{J}(\mathbf{A}, -\mathbf{E}), \xi \rangle = \int -\mathbf{E} \cdot \nabla\xi \, d^3x = \int (\mathrm{div}\, \mathbf{E})\xi \, d^3x \tag{12.3.15}$$

(assuming fast enough falloff to justify integration by parts). Thus,

$$\mathbf{J}(\mathbf{A}, -\mathbf{E}) = \mathrm{div}\, \mathbf{E} \tag{12.3.16}$$

is the equivariant momentum map. $\blacklozenge$

(m) Virtual Work
We usually think of covectors as momenta conjugate to configuration variables. However, covectors can also be thought of as forces. Indeed, if $\alpha_q \in T_q^*Q$ and $w_q \in T_qQ$, we think of

$$\langle \alpha_q, w_q \rangle = \text{force} \times \text{infinitesimal displacement}$$

as the ***virtual work***. We now give an example of a momentum map in this context.

Consider a region $\mathcal{B} \subset \mathbb{R}^3$ with boundary $\partial\mathcal{B}$. Let $\mathcal{C}$ be the space of maps $\varphi : \mathcal{B} \to \mathbb{R}^3$. Regard $T_\varphi^*\mathcal{C}$ as the space of ***loads***; that is, pairs of maps $\mathbf{b} : \mathcal{B} \to \mathbb{R}^3$, $\tau : \partial\mathcal{B} \to \mathbb{R}^3$ paired with a tangent vector $\mathbf{V} \in T_\varphi\mathcal{C}$ by

$$\langle (\mathbf{b}, \tau), \mathbf{V} \rangle = \iiint_{\mathcal{B}} \mathbf{b} \cdot \mathbf{V}\, d^3x + \iint_{\partial\mathcal{B}} \tau \cdot \mathbf{V}\, dA.$$

Let $\mathbf{A} \in GL(3, \mathbb{R})$ act on $\mathcal{C}$ by $\varphi \mapsto \mathbf{A} \circ \varphi$. The infinitesimal generator of this action is $\xi_{\mathcal{C}}(\varphi)(X) = \xi\varphi(X)$ for $\xi \in \mathfrak{gl}(3)$ and $X \in \mathcal{B}$. Pair $\mathfrak{gl}(3, \mathbb{R})$ with itself via $\langle \mathbf{A}, \mathbf{B} \rangle = \frac{1}{2}\mathrm{tr}\,(\mathbf{AB})$. The induced momentum map $\mathbf{J} : T^*\mathcal{C} \to \mathfrak{gl}(3, \mathbb{R})$ is given by

$$\mathbf{J}(\varphi, (\mathbf{b}, \tau)) = \iiint_{\mathcal{B}} \varphi \otimes \mathbf{b}\, d^3x + \iint_{\partial\mathcal{B}} \varphi \otimes \tau\, dA. \tag{12.3.17}$$

(This is the "astatic load," a concept from elasticity; see, for example, Marsden and Hughes [1983].) If we take $SO(3)$ rather than $GL(3, \mathbb{R})$, we get the angular momentum. ♦

(n) Momentum Maps for Unitary Representations on Projective Space
Here we show that the momentum map discussed in Example g of §11.5 is equivariant. Recall from the discussion at the end of §9.3 that associated to a unitary representation ρ of a Lie group G on a complex Hilbert space $\mathcal{H}$, there are skew adjoint operators $A(\xi)$ for each $\xi \in \mathfrak{g}$ depending linearly on ξ and such that $\rho(\exp(t\xi)) = \exp(tA(\xi))$. Thus, taking the t-derivative in the formula

$$\rho(g)\rho(\exp(t\xi))\rho(g^{-1}) = \exp(t\rho(g)A(\xi)\rho(g)^{-1}),$$

we get

$$A(\mathrm{Ad}_g\xi) = \rho(g)A(\xi)\rho(g)^{-1}. \tag{12.3.18}$$

Using the formula we derived in §11.5, namely

$$\langle \mathbf{J}([\psi]), \xi \rangle = J(\xi)([\psi]) = -i\hbar\frac{\langle \psi, A(\xi)\psi \rangle}{\|\psi\|^2}, \tag{12.3.19}$$

we get

$$J(\mathrm{Ad}_g\xi)([\psi]) = -i\hbar\frac{\langle\psi, \rho(g)A(\xi)\rho(g)^{-1}\psi\rangle}{\|\psi\|^2} =$$

$$= J(\xi)([\rho(g)^{-1}\psi]) = J(\xi)(g^{-1}\cdot[\psi]),$$

which shows that $\mathbf{J} : \mathbb{P}\mathcal{H} \to \mathfrak{g}^*$ is equivariant. ♦

Exercises

12.3-1 *Compute $\mathbf{J}_L$ and $\mathbf{J}_R$ for $G = SO(3)$.*

12.3-2 *Compute the momentum maps determined by spatial translations and rotations for Maxwell's equations.*

12.3-3 *Repeat Exercise* **12.3-2** *for elasticity (the context of Example* **(m)***).*

12.3-4 *Let P be a symplectic manifold and $\mathbf{J} : P \to \mathfrak{g}^*$ be an (equivariant) momentum map for the symplectic action of a group G on P. Let $\mathcal{F}$ be the space of (smooth) functions on P identified with its dual via integration and equipped with the Lie-Poisson bracket. Let $\mathcal{J} : \mathcal{F} \to \mathfrak{g}^*$ be defined by*

$$\mathcal{J}(f)(\xi) = \int f\,\langle J, \xi\rangle\,d\mu,$$

where μ is Liouville measure. Show that $\mathcal{J}$ is an (equivariant) momentum map.

12.4 Equivariance and Infinitesimal Equivariance

This optional section explores the equivariance of momentum maps a little deeper. We have just seen that equivariance implies infinitesimal equivariance. In this section, we prove, amongst other things, the converse if G and P are connected. For this purpose, introduce the map $\Gamma_\eta : G \times P \to \mathbb{R}$ defined by

$$\Gamma_\eta(g, z) = \langle\mathbf{J}(\Phi_g(z)), \eta\rangle - \left\langle\mathrm{Ad}^*_{g^{-1}}\mathbf{J}(z), \eta\right\rangle \quad \text{for } \eta \in \mathfrak{g}. \tag{12.4.1}$$

Since

$$\Gamma_{\eta,g}(z) := \Gamma_\eta(g, z) = \left(\Phi_g^* J(\eta)\right)(z) - J\left(\mathrm{Ad}_{g^{-1}}\eta\right)(z), \tag{12.4.2}$$

we get

$$\begin{aligned} X_{\Gamma_{\eta,g}} &= X_{\Phi_g^* J(\eta)} - X_{J\left(\mathrm{Ad}_{g^{-1}}\eta\right)} = \Phi_g^* X_{J(\eta)} - \left(\mathrm{Ad}_{g^{-1}}\eta\right)_P \\ &= \Phi_g^*\eta_P - \left(\mathrm{Ad}_{g^{-1}}\eta\right)_P = 0 \end{aligned} \tag{12.4.3}$$

by (11.1.4). Therefore $\Gamma_{\eta,g}$ is a Casimir function on P, and so is constant on every symplectic leaf of P. Since $\eta \mapsto \Gamma_\eta(g,z)$ is linear for every $g \in G$ and $z \in P$, we can define the map $\sigma : G \to L(\mathfrak{g}, \mathcal{C}(P))$, from G to the vector space of all linear maps of $\mathfrak{g}$ into the space of Casimir functions $\mathcal{C}(P)$ on P, by $\sigma(g) \cdot \eta = \Gamma_{\eta,g}$. The behavior of σ under group multiplication is the following. For $\xi \in \mathfrak{g}$, $z \in P$, and $g, h \in G$, we have

$$\begin{aligned}
(\sigma(gh)\cdot\xi)(z) &= \Gamma_\xi(gh,z) = \langle \mathbf{J}(\Phi_{gh}(z)), \xi\rangle - \left\langle \operatorname{Ad}^*_{(gh)^{-1}} \mathbf{J}(z), \xi \right\rangle \\
&= \langle \mathbf{J}(\Phi_g(\Phi_h(z))), \xi\rangle - \left\langle \operatorname{Ad}^*_{g^{-1}} \mathbf{J}((\Phi_h(z)), \xi \right\rangle \\
&\quad + \left\langle \mathbf{J}(\Phi_h(z)), \operatorname{Ad}_{g^{-1}}\xi \right\rangle - \left\langle \operatorname{Ad}^*_{h^{-1}} \mathbf{J}(z), \operatorname{Ad}_{g^{-1}}\xi \right\rangle \\
&= \Gamma_\xi(g, \Phi_h(z)) + \Gamma_{\operatorname{Ad}_{g^{-1}}\xi}(h,z) \\
&= (\sigma(g)\cdot\xi)(\Phi_h(z)) + \left(\sigma(h)\cdot \operatorname{Ad}_{g^{-1}}\xi\right)(z). \qquad (12.4.4)
\end{aligned}$$

Connected Lie group actions admitting momentum maps preserve symplectic leaves. This is because G is generated by a neighborhood of the identity in which each element has the form $\exp t\xi$; since $(t,z) \mapsto (\exp t\xi)\cdot z$ is a Hamiltonian flow, it follows that z and $\Phi_h(z)$ are on the same leaf. Thus, $(\sigma(g)\cdot\xi)(z) = (\sigma(g)\cdot\xi)(\Phi_h(z))$ because Casimir functions are constant on leaves. Therefore,

$$\sigma(gh) = \sigma(g) + \operatorname{Ad}^\dagger_{g^{-1}} \sigma(h), \qquad (12.4.5)$$

where $\operatorname{Ad}^\dagger_g$ denotes the action of G on $L(\mathfrak{g}, \mathcal{C}(P))$ induced via the adjoint action by

$$(\operatorname{Ad}^\dagger_g \lambda)(\xi) = \lambda(\operatorname{Ad}_g \xi) \qquad (12.4.6)$$

for $g \in G$, $\xi \in \mathfrak{g}$, and $\lambda \in L(\mathfrak{g}, \mathcal{C}(P))$. Mappings $\sigma : G \to L(\mathfrak{g}, \mathcal{C}(P))$, behaving under group multiplication as in (12.4.5), are called $L(\mathfrak{g}, \mathcal{C}(P))$-valued ***one-cocycles*** of the group G. A one-cocycle σ is called a ***one-coboundary*** if there is a $\lambda \in L(\mathfrak{g}, \mathcal{C}(P))$ such that

$$\sigma(g) = \lambda - \operatorname{Ad}^\dagger_{g^{-1}} \lambda \quad \text{for all } g \in G. \qquad (12.4.7)$$

The quotient space of one-cocycles modulo one-coboundaries is called the ***first*** $L(\mathfrak{g}, \mathcal{C}(P))$***-valued group cohomology of*** G and is denoted by $H^1(G, L(\mathfrak{g}, \mathcal{C}(P)))$; its elements are denoted by $[\sigma]$, for σ a one-cocycle.

At the Lie algebra level, bilinear skew-symmetric maps $\Sigma : \mathfrak{g}\times\mathfrak{g} \to \mathcal{C}(P)$ satisfying the Jacobi type identity (11.6.1) are called $\mathcal{C}(P)$***-valued two-cocycles of*** $\mathfrak{g}$. A cocycle Σ is called a ***coboundary*** if there is a $\lambda \in L(\mathfrak{g}, \mathcal{C}(P))$ such that

$$\Sigma(\xi,\eta) = \lambda([\xi,\eta]) \quad \text{for all } \xi, \eta \in \mathfrak{g}. \qquad (12.4.8)$$

The quotient space of two-cocycles by two-coboundaries is called the ***second cohomology of*** $\mathfrak{g}$ ***with values in*** $\mathcal{C}(P)$. It is denoted by $H^2(\mathfrak{g}, \mathcal{C}(P))$ and its elements by $[\Sigma]$. With these notations we have proved the first two parts of the following proposition:

Proposition 12.4.1 *Let the connected Lie group G act canonically on the Poisson manifold P and have a momentum map $\mathbf{J}$. For $g \in G$ and $\xi \in \mathfrak{g}$, define*

$$\Gamma_{\xi,g} : P \to \mathbb{R}, \quad \Gamma_{\xi,g}(z) = \langle \mathbf{J}(\Phi_g(z)), \xi \rangle - \left\langle \mathrm{Ad}^*_{g^{-1}} \mathbf{J}(z), \xi \right\rangle. \tag{12.4.9}$$

Then

i *$\Gamma_{\xi,g}$ is a Casimir on P for every $\xi \in \mathfrak{g}$ and $g \in G$.*

ii *Defining $\sigma : G \to L(\mathfrak{g}, \mathcal{C}(P))$ by $\sigma(g) \cdot \xi = \Gamma_{\xi,g}$, we have the identity*

$$\sigma(gh) = \sigma(g) + \mathrm{Ad}^\dagger_{g^{-1}} \sigma(h). \tag{12.4.10}$$

iii *Defining $\sigma_\eta : G \to \mathcal{C}(P)$ by $\sigma_\eta(g) := \sigma(g) \cdot \eta$ for $\eta \in \mathfrak{g}$, we have*

$$T_e\sigma_\eta(\xi) = \Sigma(\xi, \eta) := J([\xi, \eta]) - \{J(\xi), J(\eta)\}. \tag{12.4.11}$$

If $[\sigma] = 0$, then $[\Sigma] = 0$.

iv *If $\mathbf{J}_1$ and $\mathbf{J}_2$ are two momentum mappings of the same action with cocycles σ_1 and σ_2, then $[\sigma_1] = [\sigma_2]$.*

Proof Since $\sigma_\eta(g)(z) = J(\eta)(g \cdot z) - J(\mathrm{Ad}_{g^{-1}} \eta)(z)$, taking the derivative at $g = e$, we get

$$\begin{aligned} T_e\sigma_\eta(\xi)(z) &= \mathbf{d}J(\eta)(\xi_P(z)) + J([\xi, \eta])(z) \\ &= X_{J(\xi)}[J(\eta)](z) + J([\xi, \eta])(z) \\ &= -\{J(\xi), J(\eta)\}(z) + J([\xi, \eta])(z). \end{aligned} \tag{12.4.12}$$

This proves (12.4.11). The second statement in **iii** is a consequence of the definition. To prove **iv** we note that

$$\begin{aligned} \sigma_1(g)(z) - \sigma_2(g)(z) &= \mathbf{J}_1(g \cdot z) - \mathbf{J}_2(g \cdot z) \\ &\quad -\mathrm{Ad}^*_{g^{-1}}(\mathbf{J}_1(z) - \mathbf{J}_2(z)). \end{aligned} \tag{12.4.13}$$

However, $\mathbf{J}_1$ and $\mathbf{J}_2$ are momentum mappings of the same action and, therefore, $J_1(\xi)$ and $J_2(\xi)$ generate the same Hamiltonian vector field for all $\xi \in \mathfrak{g}$, so $J_1 - J_2$ is constant as an element of $L(\mathfrak{g}, \mathcal{C}(P))$. Calling this element λ, we have

$$\sigma_1(g) - \sigma_2(g) = \lambda - \mathrm{Ad}^\dagger_{g^{-1}} \lambda, \tag{12.4.14}$$

so $\sigma_1 - \sigma_2$ is a coboundary. ■

Remarks

1. Part **iv** of this proposition also holds for Lie algebra actions admitting momentum maps with all σ's replaced by Σ's; indeed, $\{J_1(\xi), J_1(\eta)\} = \{J_2(\xi), J_2(\eta)\}$ because $J_1(\xi) - J_2(\xi)$ and $J_1(\eta) - J_2(\eta)$ are Casimir functions.

2. If $[\Sigma] = 0$, *the momentum map* $\mathbf{J} : P \to \mathfrak{g}^*$ *of the canonical Lie algebra action of* $\mathfrak{g}$ *on* P *can be always chosen to be infinitesimally equivariant*, a result due to Souriau [1970] for the symplectic case. To see this, note first that momentum maps are determined only up to elements of $L(\mathfrak{g}, \mathcal{C}(P))$. Therefore, if $\lambda \in L(\mathfrak{g}, \mathcal{C}(P))$ denotes the element determined by the condition $[\Sigma] = 0$, then $J + \lambda$ is an infinitesimally equivariant momentum map.

3. *The cohomology class* $[\Sigma]$ *depends only on the Lie algebra action* $\rho : \mathfrak{g} \to \mathfrak{X}(P)$ *and not on the momentum map.* Indeed, because J is determined only up to the addition of a linear map $\lambda : \mathfrak{g} \to \mathcal{C}(P)$ and denoting

$$\Sigma_\lambda(\xi, \eta) := (J + \lambda)([\xi, \eta]) - \{(J + \lambda)(\xi), (J + \lambda)(\eta)\}, \qquad (12.4.15)$$

we obtain

$$\begin{aligned} \Sigma_\lambda(\xi, \eta) &= J([\xi, \eta]) + \lambda([\xi, \eta]) - \{J(\xi), J(\eta)\} \\ &= \Sigma(\xi, \eta) + \lambda([\xi, \eta]), \end{aligned} \qquad (12.4.16)$$

that is, $[\Sigma_\lambda] = [\Sigma]$. Letting $\rho' \in H^2(\mathfrak{g}, \mathcal{C}(P))$ denote this cohomology class, $\mathbf{J}$ is infinitesimally equivariant if and only if ρ' vanishes. There are some cases in which one can predict that ρ' is zero:

(a) *Assume* P *is symplectic and connected* (*so* $\mathcal{C}(P) = \mathbb{R}$)*and suppose that* $H^2(\mathfrak{g}, \mathbb{R}) = 0$. By the second Whitehead lemma (see Jacobson [1962] or Guillemin and Sternberg [1984]), this is the case whenever $\mathfrak{g}$ is semisimple; thus *semisimple, symplectic Lie algebra actions on symplectic manifolds are Hamiltonian.*

(b) *Suppose* P *is exact symplectic,* $-\mathbf{d}\theta = \Omega$, and

$$\pounds_{\xi_P}\theta = 0. \qquad (12.4.17)$$

The proof of equivariance in this case is the following. Assume first that the Lie algebra $\mathfrak{g}$ has an underlying Lie group G which leaves θ invariant. Since $\left(\mathrm{Ad}_{g^{-1}}\xi\right)_P = \Phi_g^*\xi_P$, we get from (11.3.4)

$$\begin{aligned} J(\xi)(g \cdot z) &= (\mathbf{i}_{\xi_P}\theta)(g \cdot z) = \left(\mathbf{i}_{(\mathrm{Ad}_{g^{-1}}\xi)_P}\theta\right)(z) \\ &= J\left(\mathrm{Ad}_{g^{-1}}\xi\right)(z). \end{aligned} \qquad (12.4.18)$$

The proof without the assumption of the existence of the group G is obtained by differentiating the above string of equalities with respect to g at $g = e$.

A simple example in which $\rho' \neq 0$ is provided by phase-space translations on $\mathbb{R}^2$ defined by $\mathfrak{g} = \mathbb{R}^2 = \{(a, b)\}$, $P = \mathbb{R}^2 = \{(q, p)\}$, and

$$(a, b)_P = a\frac{\partial}{\partial q} + b\frac{\partial}{\partial p}. \tag{12.4.19}$$

This action has a momentum map given by $\langle \mathbf{J}(q,p), (a,b)\rangle = ap - bq$ and

$$\begin{aligned} \Sigma\left((a_1, b_1), (a_2, b_2)\right) &= J\left(\left[(a_1, b_1), (a_2, b_2)\right]\right) \\ &\quad - \left\{J\left(a_1, b_1\right), J\left(a_2, b_2\right)\right\} \\ &= -\left\{a_1 p - b_1 q, a_2 p - b_2 q\right\} \\ &= b_1 a_2 - a_1 b_2. \end{aligned} \tag{12.4.20}$$

Since $[\mathfrak{g}, \mathfrak{g}] = \{0\}$, the only coboundary is zero, so $\rho' \neq 0$. This example is amplified in Example **(b)** of §12.6.

4. *If P is symplectic and connected and σ is a one-cocycle of the G-action on P, then:*

   (a) *$g \cdot \mu = \mathrm{Ad}^*_{g^{-1}} \mu + \sigma(g)$ is an action of G on $\mathfrak{g}^*$; and*

   (b) *$\mathbf{J}$ is equivariant with respect to this action.*

   Indeed, since P is symplectic and connected, $\mathcal{C}(P) = \mathbb{R}$, and thus $\sigma : G \to \mathfrak{g}^*$. By Proposition **12.4.1**,

$$\begin{aligned} (gh) \cdot \mu &= \mathrm{Ad}^*_{(gh)^{-1}} \mu + \sigma(gh) \\ &= \mathrm{Ad}^*_{g^{-1}} \mathrm{Ad}^*_{h^{-1}} \mu + \sigma(g) + \mathrm{Ad}^*_{g^{-1}} \sigma(h) \\ &= \mathrm{Ad}^*_{g^{-1}}(h \cdot \mu) + \sigma(g) = g \cdot (h \cdot \mu), \end{aligned} \tag{12.4.21}$$

   which proves (a); (b) is a consequence of the definition.

5. *If P is symplectic and connected, $\mathbf{J} : P \to \mathfrak{g}^*$ is a momentum map, and Σ is the associated real-valued Lie algebra two-cocycle, then the momentum map $\mathbf{J}$ can be explicitly adjusted to be infinitesimally equivariant by enlarging $\mathfrak{g}$ to the central extension defined by Σ.*

   Indeed, the ***central extension defined by*** Σ is the Lie algebra $\mathfrak{g}' := \mathfrak{g} \oplus \mathbb{R}$ with the bracket given by

$$[(\xi, a), (\eta, b)] = ([\xi, \eta], \Sigma(\xi, \eta)). \tag{12.4.22}$$

   Let $\mathfrak{g}'$ act on P by $\rho(\xi, a)(z) = \xi_P(z)$ and let $\mathbf{J}' : P \to (\mathfrak{g}')^* = \mathfrak{g}^* \oplus \mathbb{R}$ be the induced momentum map, that is, it satisfies

$$X_{J'(\xi,a)} = (\xi, a)_P = X_{J(\xi)}, \tag{12.4.23}$$

so that

$$J'(\xi, a) - J(\xi) = \ell(\xi, a), \tag{12.4.24}$$

where $\ell(\xi, a)$ is a constant on P and is linear in (ξ, a). Therefore

$$\begin{aligned} J' &([(\xi, a), (\eta, b)]) - \{J'(\xi, a), J'(\eta, a)\} \\ &= J'([\xi, \eta], \Sigma(\xi, \eta)) - \{J(\xi) + \ell(\xi, a), J(\eta) + \ell(\eta, b)\} \\ &= J([\xi, \eta]) + \ell([\xi, \eta], \Sigma(\xi, \eta)) - \{J(\xi), J(\eta)\} \\ &= \Sigma(\xi, \eta) + \ell([(\xi, a), (\eta, b)]) \\ &= (\lambda + \ell)([(\xi, a), (\eta, b)]), \end{aligned} \tag{12.4.25}$$

where $\lambda(\xi, a) = a$. Thus the real-valued two-cocycle of the $\mathfrak{g}'$ action is a coboundary and hence J' can be adjusted to become infinitesimally equivariant. Thus,

$$J'(\xi, a) = J(\xi) - a \tag{12.4.26}$$

is the desired infinitesimally equivariant momentum map of $\mathfrak{g}$ on P.

For example, the action of $\mathbb{R}^2$ on itself by translations has the nonequivariant momentum map $\langle \mathbf{J}(q,p), (\xi, \eta)\rangle = \xi p - \eta q$ with group one-cocycle $\sigma(x, y) \cdot (\xi, \eta) = \xi y - \eta x$; here we think of $\mathbb{R}^2$ endowed with the symplectic form $dq \wedge dp$. The corresponding infinitesimally equivariant momentum map of the central extension is given by (12.4.26), that is, by the expression $\langle \mathbf{J}(q,p), (\xi, \eta, a)\rangle = \xi p - \eta q - a$. For more examples, see §12.6.

Consider the situation for the corresponding action of the central extension G' of G on P if $G = E$, a topological vector space regarded as an abelian Lie group. Then $\mathfrak{g} = E$, $T\sigma_\eta = \sigma_\eta$ by linearity of σ_η, so that $\Sigma(\xi, \eta) = \sigma(\xi) \cdot \eta$, with ξ on the right-hand side thought of as an element of the Lie group, G. One defines the central extension G' of G by the circle group S^1 as the Lie group having an underlying manifold $E \times S^1$, and whose multiplication is given by (Souriau [1969])

$$\begin{aligned} &\left(q_1, e^{i\theta_1}\right) \cdot \left(q_2, e^{i\theta_2}\right) \\ &\quad = \left(q_1 + q_2, \exp\left\{i\left[\theta_1 + \theta_2 + \frac{1}{2}\Sigma(q_1, q_2)\right]\right\}\right), \end{aligned} \tag{12.4.27}$$

the identity element equal to $(0, 1)$, and the inverse given by $\left(q, e^{i\theta}\right)^{-1} = \left(-q, e^{-i\theta}\right)$. Then the Lie algebra of G' is $\mathfrak{g}' = E \oplus \mathbb{R}$ with the bracket given by (12.4.22) and thus the G'-action on P has an equivariant momentum map $\mathbf{J}$ given by (12.4.26). ♦

Let us return to the investigation of global equivariance. Assume J is a Lie algebra homomorphism. Since $\Gamma_{\eta, g}$ is a Casimir function on P for every $g \in G$ and $\eta \in \mathfrak{g}$, it follows that $\Gamma_\eta | G \times S$ is independent of $z \in S$, where S is a symplectic leaf. Denote this function that depends only on the leaf

S by $\Gamma_\eta^S : G \to \mathbb{R}$. Fixing $z \in S$, and taking the derivative of the map $g \mapsto \Gamma_\eta^S(g, z)$ at $g = e$ in the direction $\xi \in \mathfrak{g}$, gives

$$\langle -(\operatorname{ad} \xi)^* \mathbf{J}(z), \eta \rangle - \langle T_z \mathbf{J} \cdot \xi_P(z), \eta \rangle = 0, \tag{12.4.28}$$

that is, $T_e \Gamma_\eta^S = 0$ for all $\eta \in \mathfrak{g}$. By Proposition **12.4.1ii**, we have

$$\Gamma_\eta(gh) = \Gamma_\eta(g) + \Gamma_{\operatorname{Ad}_{g^{-1}}\eta}(h). \tag{12.4.29}$$

Taking the derivative of (12.4.29) with respect to g in the direction ξ at $h = e$ on the leaf S and using $T_e \Gamma_\eta^S = 0$, we get

$$T_g \Gamma_\eta^S (T_e L_g(\xi)) = T_e \Gamma^S_{\operatorname{Ad}_{g^{-1}}\eta}(\xi) = 0. \tag{12.4.30}$$

Thus, Γ_η is constant on $G \times S$ (recall that symplectic leaves are, by definition, connected). Since $\Gamma_\eta(e, z) = 0$, it follows that $\Gamma_\eta | G \times S = 0$ for any leaf S and hence $\Gamma_\eta = 0$ on $G \times P$. But $\Gamma_\eta = 0$ for every $\eta \in \mathfrak{g}$ is equivalent to equivariance. Together with Theorem **11.6.1** this proves the following:

Theorem 12.4.2 *Let the connected Lie group G act canonically on the left on the Poisson manifold P. The action of G is globally Hamiltonian if and only if there is a Lie algebra homomorphism $\psi : \mathfrak{g} \to \mathcal{F}(P)$ such that $X_{\psi(\xi)} = \xi_P$ for all $\xi \in \mathfrak{g}$ where ξ_P is the infinitesimal generator of the G-action. If $\mathbf{J}$ is the equivariant momentum map of the action, then we can take $\psi = J$.*

The converse question of the construction of a group action whose momentum map equals a given set of conserved quantities closed under bracketing is addressed in Fong and Meyer [1975]. See also Vinogradov and Krasilshchick [1975] and Conn [1984], [1985] for the related question of when the germs of Poisson vector fields are Hamiltonian.

12.5 Equivariant Momentum Maps Are Poisson

We next show that equivariant momentum maps are Poisson maps. This provides a fundamental method for finding canonical maps between Poisson manifolds. This result is partly contained in Lie's work [1890], is implicit in Guillemin and Sternberg [1980], and explicit in Holmes and Marsden [1983] and Guillemin and Sternberg [1984].

Theorem 12.5.1 (Canonical Momentum Maps) *If $\mathbf{J} : P \to \mathfrak{g}^*$ is an infinitesimally equivariant momentum map for a left Hamiltonian action of $\mathfrak{g}$ on a Poisson manifold P, then $\mathbf{J}$ is a Poisson map:*

$$\mathbf{J}^* \{F_1, F_2\}_+ = \{\mathbf{J}^* F_1, \mathbf{J}^* F_2\}, \quad \text{i.e.,} \quad \{F_1, F_2\}_+ \circ \mathbf{J} = \{F_1 \circ \mathbf{J}, F_2 \circ \mathbf{J}\} \tag{12.5.1}$$

for all $F_1, F_2 \in \mathcal{F}(\mathfrak{g}^*)$, where $\{ , \}_+$ denotes the "+" Lie-Poisson bracket.

Proof Infinitesimal equivariance means that $J([\xi,\eta]) = \{J(\xi), J(\eta)\}$. For $F_1, F_2 \in \mathcal{F}(\mathfrak{g}^*)$, let $z \in P$, $\xi = \delta F_1/\delta\mu$, and $\eta = \delta F_2/\delta\mu$ evaluated at the particular point $\mu = \mathbf{J}(z) \in \mathfrak{g}^*$. Then

$$\begin{aligned} \mathbf{J}^* \{F_1, F_2\}_+ (z) &= \left\langle \mu, \left[\frac{\delta F_1}{\delta\mu}, \frac{\delta F_2}{\delta\mu} \right] \right\rangle = \langle \mu, [\xi,\eta] \rangle \\ &= J([\xi,\eta])(z) = \{J(\xi), J(\eta)\}(z). \end{aligned}$$

But for any $z \in P$ and $v_z \in T_zP$,

$$\begin{aligned} \mathbf{d}(F_1 \circ \mathbf{J})(z) \cdot v_z &= \mathbf{d}F_1(\mu) \cdot T_z\mathbf{J}(v_z) \\ &= \left\langle \frac{\delta F_1}{\delta\mu}, T_z\mathbf{J}(v_z) \right\rangle = \mathbf{d}J(\xi)(z) \cdot v_z, \end{aligned}$$

that is, $(F_1 \circ \mathbf{J})(z)$ and $J(\xi)(z)$ have equal z-derivatives. Since the Poisson bracket on P depends only on the point values of the first derivatives, we conclude that

$$\{F_1 \circ \mathbf{J}, F_2 \circ \mathbf{J}\}(z) = \{J(\xi), J(\eta)\}(z). \quad \blacksquare$$

Theorem 12.5.2 (Collective Hamiltonian Theorem) *Let $\mathbf{J} : P \to \mathfrak{g}^*$ be a momentum map. Let $z \in P$ and $\mu = \mathbf{J}(z) \in \mathfrak{g}^*$. Then for any $F \in \mathcal{F}(\mathfrak{g}^*_+)$,*

$$X_{F\circ\mathbf{J}}(z) = X_{J(\delta F/\delta\mu)}(z) = \left(\frac{\delta F}{\delta\mu} \right)_P (z). \qquad (12.5.2)$$

Proof For any $H \in \mathcal{F}(P)$,

$$\begin{aligned} X_{F\circ\mathbf{J}}[H](z) &= -X_H[F \circ \mathbf{J}](z) = -\mathbf{d}(F \circ \mathbf{J})(z) \cdot X_H(z) \\ &= -\mathbf{d}F(\mu)(T_z\mathbf{J} \cdot X_H(z)) = -\left\langle T_z\mathbf{J}(X_H(z)), \frac{\delta F}{\delta\mu} \right\rangle \\ &= -\mathbf{d}J\left(\frac{\delta F}{\delta\mu} \right)(z) \cdot X_H(z) = -X_H \left[J\left(\frac{\delta F}{\delta\mu} \right) \right](z) \\ &= X_{J(\delta F/\delta\mu)}[H](z). \end{aligned}$$

This proves the first equality in (12.5.2) and the second results from the definition of the momentum map. $\blacksquare$

Functions on P of the form $F \circ \mathbf{J}$ are called ***collective***. Note that if F is the linear function determined by $\xi \in \mathfrak{g}$, (12.5.2) reduces to $X_{J(\xi)}(z) = \xi_P(z)$, the definition of the momentum map. To demonstrate the relation between these results, let us derive Theorem **12.5.1** from Theorem **12.5.2**. Let $\mu = \mathbf{J}(z)$, and $F, H \in \mathcal{F}(\mathfrak{g}^*_+)$. Then

$$\mathbf{J}^* \{F, H\}_+ (z) = \{F, H\}_+ (\mathbf{J}(z)) = \left\langle \mathbf{J}(z), \left[\frac{\delta F}{\delta\mu}, \frac{\delta H}{\delta\mu} \right] \right\rangle$$

$$\begin{aligned}
&= J\left(\left[\frac{\delta F}{\delta \mu}, \frac{\delta H}{\delta \mu}\right]\right)(z) = \left\{J\left(\frac{\delta F}{\delta \mu}\right), J\left(\frac{\delta H}{\delta \mu}\right)\right\}(z) \\
&\qquad \text{(by infinitesimal equivariance)} \\
&= X_{J(\delta H/\delta \mu)}\left[J\left(\frac{\delta F}{\delta \mu}\right)\right](z) = X_{H\circ \mathbf{J}}\left[J\left(\frac{\delta F}{\delta \mu}\right)\right](z) \\
&\qquad \text{(by the collective Hamiltonian theorem)} \\
&= -X_{J(\delta F/\delta \mu)}[H \circ \mathbf{J}](z) = -X_{F\circ \mathbf{J}}[H \circ \mathbf{J}](z) \\
&\qquad \text{(again by the collective Hamiltonian theorem)} \\
&= \{F \circ \mathbf{J}, H \circ \mathbf{J}\}(z). \quad \blacksquare
\end{aligned}$$

Remarks

1. Let $i : \mathfrak{g} \to \mathcal{F}(\mathfrak{g}^*)$ denote the natural embedding of $\mathfrak{g}$ in its bidual; that is, $i(\xi) \cdot \mu = \langle \mu, \xi \rangle$. Since $\delta i(\xi)/\delta \mu = \xi$, i is a Lie algebra homomorphism, that is,

$$i([\xi, \eta]) = \{i(\xi), i(\eta)\}_+ . \tag{12.5.3}$$

We claim that *a canonical left Lie algebra action of $\mathfrak{g}$ on a Poisson manifold P is Hamiltonian if and only if there is a Poisson algebra homomorphism $\chi : \mathcal{F}(\mathfrak{g}^*_+) \to \mathcal{F}(P)$ such that $X_{(\chi\circ i)(\xi)} = \xi_P$ for all $\xi \in \mathfrak{g}$.* Indeed, if the action is Hamiltonian, let $\chi = \mathbf{J}^*$ (pull back on functions) and the assertion follows from the definition of momentum maps. The converse relies on the following fact. Let M, N be finite dimensional manifolds and $\chi : \mathcal{F}(N) \to \mathcal{F}(M)$ be a ring homomorphism. Then there exists a unique smooth map $\varphi : M \to N$ such that $\chi = \varphi^*$. (A similar statement holds for infinite-dimensional manifolds in the presence of some additional technical conditions. See Abraham, Marsden, and Ratiu [1988], Supplement 4.2C.) Therefore, if a ring and Lie algebra homomorphism $\mathcal{F}(\mathfrak{g}^*_+) \to \mathcal{F}(P)$ is given, there is a unique map $\mathbf{J} : P \to \mathfrak{g}^*$ such that $\chi = \mathbf{J}^*$. But for $\xi, \mu \in \mathfrak{g}^*$ we have

$$\begin{aligned}
[(\chi \circ i)(\xi)](z) &= \mathbf{J}^*(i(\xi))(z) = i(\xi)(\mathbf{J}(z)) \\
&= \langle \mathbf{J}(z), \xi \rangle = J(\xi)(z),
\end{aligned} \tag{12.5.4}$$

that is, $\chi \circ i = J$ which is a Lie algebra homomorphism because χ is, by hypothesis. Since $X_{J(\xi)} = \xi_P$ again by hypothesis, it follows that $\mathbf{J}$ is an infinitesimally equivariant momentum map.

2. Here we have worked with left actions. If in all statements one changes left by right actions and "+" by "−" in the Lie-Poisson structures on $\mathfrak{g}^*$, the resulting statements are true.

Exercises

12.5-1 *Verify directly that angular momentum is a Poisson map.*

12.5-2 *What does the collective Hamiltonian theorem state for angular momentum? Is the result obvious?*

12.5-3 *If $z(t)$ is an integral curve of $F_{F \circ J}$, show that $\mu(t) = J(z(t))$ satisfies $\dot{\mu} = \operatorname{ad}^*_{\delta F/\delta \mu} \mu$.*

12.6 More Examples

(a) Phase Space Rotations

Let (P, Ω) be a linear symplectic space and let G be a subgroup of the linear symplectic group acting on P by matrix multiplication. The infinitesimal generator of $\xi \in \mathfrak{g}$ at $z \in P$ is

$$\xi_P(z) = \xi z, \tag{12.6.1}$$

where ξz is matrix multiplication. This vector field is Hamiltonian with Hamiltonian $\Omega(\xi z, z)/2$ by Proposition **2.7.1**. Thus a momentum map is

$$\langle \mathbf{J}(z), \xi \rangle = \frac{1}{2}\Omega(\xi z, z). \tag{12.6.2}$$

For $S \in G$, the adjoint action is

$$\operatorname{Ad}_S \xi = S\xi S^{-1}, \tag{12.6.3}$$

and hence

$$\begin{aligned} \left\langle \mathbf{J}(Sz), S\xi S^{-1} \right\rangle &= \frac{1}{2}\,\Omega(S\xi S^{-1}Sz, Sz) \\ &= \frac{1}{2}\,\Omega(S\xi z, Sz) = \frac{1}{2}\,\Omega(\xi z, z), \end{aligned} \tag{12.6.4}$$

so $\mathbf{J}$ is equivariant. Infinitesimal equivariance is a reformulation of (2.9.10). Notice that this momentum map is not of the cotangent lift type. ♦

(b) Phase Space Translations

Let (P, Ω) be a linear symplectic space and let G be a subgroup of the translation group of P, with $\mathfrak{g}$ identified with a linear subspace of P. Clearly

$$\xi_P(z) = \xi$$

in this case. The vector field is Hamiltonian with Hamiltonian given by the linear function

$$J(\xi)(z) = \Omega(\xi, z), \tag{12.6.5}$$

as is easily checked. This is therefore a momentum map for the action. This momentum map is not equivariant, however. The action of $\mathbb{R}^2$ on $\mathbb{R}^2$ by translation is a specific example; see the end of Remark 3 of §12.4. ♦

(c) Lifted Actions and Magnetic Terms
Another way nonequivariance of momentum maps come up is with lifted cotangent actions, but with symplectic forms which are the canonical ones modified by the addition of a magnetic term. For example, endow $P = T^*\mathbb{R}^2$ with the symplectic form

$$\Omega_B = dq^1 \wedge dp_1 + dq^2 \wedge dp_2 + B\, dq^1 \wedge dq^2$$

where B is a function of q^1 and q^2. Consider the action of $\mathbb{R}^2$ on $\mathbb{R}^2$ by translations and lift this to an action of $\mathbb{R}^2$ on P. Note that this action preserves Ω_B if and only if B is constant, which will be assumed from now on. By (12.6.5) the momentum map is

$$\langle \mathbf{J}(\mathbf{q}, \mathbf{p}), \xi \rangle = \mathbf{p} \cdot \xi + B(\xi^1 q^2 - \xi^2 q^1). \tag{12.6.6}$$

This momentum map is not equivariant; in fact, since $\mathbb{R}^2$ is abelian, its Lie algebra two-cocycle is given by

$$\Sigma(\xi, \eta) = -\{J(\xi), J(\eta)\} = -2B(\xi^1\eta^2 - \xi^2\eta^1).$$

Let us assume from now on that B is nonzero. Viewed in different coordinates, the form Ω_B can be made canonical and the action by $\mathbb{R}^2$ is still translation by a canonical transformations. To do this, one switches to ***guiding center coordinates*** $(\mathbf{R}, \mathbf{P})$ defined by $\mathbf{P} = \mathbf{p}$ and $\mathbf{R} = (q^1 - p_2/B,\ q^2 + p_1/B)$. The physical interpretation of these coordinates is the following: $\mathbf{P}$ is the momentum of the particle, while $\mathbf{R}$ is the center of the nearly circular orbit pursued by the particle with coordinates $(\mathbf{q}, \mathbf{p})$ when the magnetic field is strong (Littlejohn [1983, 1984]). In these coordinates, Ω_B takes the form

$$\Omega_B = B dR^1 \wedge dR^2 - \frac{1}{B} dP_1 \wedge dP_2$$

and the $\mathbb{R}^2$-action on $T^*\mathbb{R}^2$ becomes translation in the $\mathbf{R}$-variable. The momentum map (12.6.6) becomes

$$\langle \mathbf{J}(\mathbf{R}, \mathbf{P}), \xi \rangle = B(\xi^1 R^2 - \xi^2 R^1) \tag{12.6.7}$$

which is again a special case of (12.3.4).

The cohomology class $[\Sigma] \neq 0$, as the following argument shows. If Σ was exact, there would exist a linear functional $\lambda : \mathbb{R}^2 \to \mathbb{R}$ such that $\Sigma(\xi, \eta) = \lambda([\xi, \eta]) = 0$ for all ξ, η; this is clearly false. Thus, $\mathbf{J}$ *cannot* be adjusted to obtain an equivariant momentum map.

Following Remark 5 of §12.4, the nonequivariance of the momentum map can be removed by passing to a central extension of $\mathbb{R}^2$. Namely, let $G' = \mathbb{R}^2 \times S^1$ with multiplication given by

$$(\mathbf{a}, e^{i\theta})(\mathbf{b}, e^{i\varphi}) = \left(\mathbf{a} + \mathbf{b}, e^{i(\theta + \varphi + B(a^1 b^2 - a^2 b^1))}\right) \tag{12.6.8}$$

and letting G' act on $T^*\mathbb{R}^2$ as before by

$$(\mathbf{a}, e^{i\theta}) \cdot (\mathbf{q}, \mathbf{p}) = (\mathbf{q} + \mathbf{a}, \mathbf{p}).$$

Then the momentum map $\mathbf{J} : T^*\mathbb{R}^2 \to \mathfrak{g}'^* = \mathbb{R}^3$ is given by

$$\langle \mathbf{J}(\mathbf{q}, \mathbf{p}), (\xi, a) \rangle = \mathbf{p} \cdot \xi + B(\xi^1 q^2 - \xi^2 q^1) - a. \quad \blacklozenge \tag{12.6.9}$$

(d) Clairault's Theorem
Let M be a surface of revolution in $\mathbb{R}^3$ obtained by revolving a graph $r = f(z)$ about the z-axis, where f is a smooth positive function. Pull back the usual metric of $\mathbb{R}^3$ to M and note that it is invariant under rotations about the z-axis. Consider the geodesic flow on M. The momentum map associated with the S^1 symmetry is $\mathbf{J} : TM \to \mathbb{R}$ given by $\langle \mathbf{J}(\mathbf{q}, \mathbf{v}), \xi \rangle = \langle (\mathbf{q}, \mathbf{v}), \xi_M(\mathbf{q}) \rangle$, as usual. Here, ξ_M is the vector field on $\mathbb{R}^3$ associated with a rotation with angular velocity ξ about the z-axis, so $\xi_M(\mathbf{q}) = \xi \mathbf{k} \times \mathbf{q}$. Thus,

$$\langle \mathbf{J}(\mathbf{q}, \mathbf{v}), \xi \rangle = \xi r \|\mathbf{v}\| \cos \theta,$$

where r is the distance to the z-axis and θ is the angle between $\mathbf{v}$ and the horizontal plane. Thus, as $\|\mathbf{v}\|$ is conserved, by conservation of energy, $r \cos \theta$ is *conserved along any geodesic on a surface of revolution, a statement known as Clairault's Theorem.* $\blacklozenge$

(e) Mass of a nonrelativistic free quantum particle
Here we show by means of an example, the relation between (genuine) projective unitary representations and non equivariance of the momentum map for the action on the projective space. This complements the discussion in example **(n)** of §12.3 where we have shown that for unitary representations the momentum map is equivariant.

Let G be the Galilean group introduced in Remark 4 following **9.3.9**. Let $\mathcal{H} = L^2(\mathbb{R}^3)$ be the Hilbert space of square (Lebesgue) integrable complex functions on $\mathbb{R}^3$.

Fix a real number $m \neq 0$; for each $g = \{R, \boldsymbol{v}, \boldsymbol{a}, \tau\} \in G$, define the following unitary operator in $\mathcal{H}$:

$$(U_m(g)f)(\boldsymbol{p}) = \exp\left(i\left(\frac{\tau}{2m}|\boldsymbol{p}|^2 + (\boldsymbol{p} + m\boldsymbol{v}) \cdot \boldsymbol{a}\right)\right) f(R^{-1}(\boldsymbol{p} + m\boldsymbol{v})). \tag{12.6.10}$$

We can check by direct computation that:

$$U_m(g_1)U_m(g_2) = \exp(-im\sigma(g_1, g_2))U_m(g_1 g_2), \tag{12.6.11}$$

where (with $g_j = \{R_j, \boldsymbol{v}_j, \boldsymbol{a}_j, \tau_j\}$)

$$\sigma(g_1, g_2) = \frac{1}{2}|\boldsymbol{v}_1|^2 \tau_2 + (R_1 \boldsymbol{v}_2) \cdot (\boldsymbol{v}_1 \tau_2 + \boldsymbol{a}_1). \tag{12.6.12}$$

From (12.6.11) we see that the map $g \mapsto U_m(g)$ is not a group homomorphism, but when we project on $U(\mathbb{P}\mathcal{H})$ (see examples **(d)**-**(h)** after definition **9.3.5**) we get a group homomorphism, that is, a projective unitary representation of the Galilean group. Following the arguments in the examples quoted above and those in proposition **5.3.1**, we see that the Galilean group acts on $\mathbb{P}\mathcal{H}$ by symplectomorphisms. To get the momentum map, we notice that for any smooth $f \in \mathcal{H} = L^2(\mathbb{R}^3)$, the map $g \mapsto U_m(g)f \in \mathcal{H}$ is smooth, so it makes sense to define:

$$(a(\xi))f = \mathbf{d}(U_m(\cdot)f)(1)(\xi). \tag{12.6.13}$$

This shows that $a(\xi)$ is linear in ξ and defines a linear operator from $\mathcal{D} = C^\infty(\mathbb{R}^3)$ to $\mathcal{H} = L^2(\mathbb{R}^3)$. Because $U_m(g)$ is unitary and $U_m(1) = 1_{\mathcal{H}}$ it follows that $a(\xi)$ is formally skew adjoint on $\mathcal{D}$ for any $\xi \in \mathfrak{g}$. Explicitly, using the notations in remark 4 after **9.3.9**, these operators are:

$$(a(\boldsymbol{\omega})f)(\boldsymbol{p}) = -\boldsymbol{\omega} \cdot \left(\boldsymbol{p} \times \frac{\partial f}{\partial \boldsymbol{p}}\right), \quad (a(\boldsymbol{u})f)(\boldsymbol{p}) = m\boldsymbol{u} \cdot \frac{\partial f}{\partial \boldsymbol{p}},$$

$$(a(\boldsymbol{\alpha})f)(\boldsymbol{p}) = i(\boldsymbol{\alpha} \cdot \boldsymbol{p})f(\boldsymbol{p}), \quad (a(\theta)f)(\boldsymbol{p}) = i\theta\frac{|\boldsymbol{p}|^2}{2m}f(\boldsymbol{p}).$$

From these formulae we see that $\mathcal{D}$ is invariant under the group action and under $a(\xi)$ for each $\xi \in \mathfrak{g}$. From the theory of self adjoint operators one can show that $a(\xi)$ for each $\xi \in \mathfrak{g}$ is uniquely determined as a skew adjoint operator in $\mathcal{H}$. Therefore, $\mathbb{P}\mathcal{D}$ satisfies conditions (i)-(iii) of example **(f)** following **9.3.5** and is thus an essential G-smooth part of $\mathbb{P}\mathcal{H}$; on this set the momentum map can be defined. Following the computation in example **(g)** of §11.5 we can write the momentum map for the action of the Galilean group on the projective space induced by the projective unitary representation (12.6.11):

$$J(\xi)([f]) = -\frac{i}{2}\frac{\langle f, a(\xi)f\rangle}{\|f\|^2} \tag{12.6.14}$$

for $f \neq 0$.

In spite of the fact that (12.6.14) and (11.5.23) look practically the same, the corresponding momentum maps have different properties because the infinitesimal generators $a(\xi)$ have different properties: in (11.5.23) $a(\xi)$ is uniquely determined by ξ, but here $a(\xi)$ is given by the projective representation only up to a linear functional on $\mathfrak{g}$. More crucial, the relation (12.3.17) which holds when the representation is unitary, may not be true for projective representations. In our case this relation becomes:

$$a(\mathrm{Ad}_g \xi) = U_m(g) a(\xi) U_m(g)^{-1} - 2i\Gamma_\xi(g^{-1}) 1_{\mathcal{H}}, \tag{12.6.15}$$

for some function Γ depending on $\xi \in \mathfrak{g}$ and $g \in G$. To show this, notice that from (12.6.11) we get:

$$U_m(\exp(ta(\mathrm{Ad}_g \xi)) = U_m(g) U_m(\exp(t\xi)) U_m(g)^{-1} \exp(im\gamma(g, t\xi)),$$

with

$$\gamma(g, t\xi) = \sigma(g, \exp(t\xi) g^{-1}) + \sigma(\exp(t\xi), g^{-1}) - \sigma(g, g^{-1}).$$

Taking the derivative with respect to t at $t = 0$ and using (12.3.13) we get (12.3.15), where

$$\Gamma_\xi(g) = m \, \frac{1}{2} \frac{d}{dt} \gamma(g^{-1}, t\xi) \bigg|_{t=0}.$$

Using the notations in §9.3, we have for $\xi = \{\boldsymbol{\omega}, \boldsymbol{u}, \boldsymbol{\alpha}, \theta\}$ and $g = \{R, \boldsymbol{v}, \boldsymbol{a}, \tau\}$:

$$\Gamma_\xi(g) = -\frac{m}{2} \left(-\frac{1}{2} |\boldsymbol{v}|^2 \theta + (\boldsymbol{a} \times \boldsymbol{v}) \cdot \omega + (\tau \boldsymbol{v} - \boldsymbol{a}) \cdot \boldsymbol{u} + \boldsymbol{v} \cdot \boldsymbol{\alpha} \right).$$

The corresponding Lie algebra cocycle as defined in (12.4.11) is given by:

$$\Sigma(\xi, \xi') = \frac{m}{2} (\boldsymbol{u} \cdot \boldsymbol{\alpha}' - \boldsymbol{\alpha} \cdot \boldsymbol{u}'). \tag{12.6.16}$$

This cocycle on the Lie algebra is nontrivial, that is, its cohomology class is non zero (see exercise **12.6-3**). Therefore the mass of the particle measures the obstruction to equivariance for the momentum map (or for the projective representation to be a unitary representation) in $H^2(\mathfrak{g}, \mathbb{R})$. ♦

Exercises

12.6-1 *Consider an ellipsoid of revolution in $\mathbb{R}^3$ and a geodesic starting at the "equator" making an angle of α with the equator. Use Clairault's* *theorem to derive a bound on how high the geodesic climbs up the ellipse.*

12.6-2 *Consider the action of $SE(2)$ on $\mathbb{R}^2$ as in Exercise* **11.5-2**. *Since this action was not defined as a lift, Theorem* **12.1.4** *is not applicable. In fact, in Exercise* **11.6-2** *it was shown that this momentum map is not equivariant. Compute the group and Lie algebra cocycles defined by this momentum map. Find the Lie algebra central extension making the momentum map equivariant.*

12.6-3 *Using the Lie algebra representation in remark 4 after* **9.3.9** *for the Galilean group show that any 2-coboundary has the form:*

$$\lambda(\xi, \xi') = \boldsymbol{x} \cdot (\boldsymbol{\omega} \times \boldsymbol{\omega}') + \boldsymbol{y} \cdot (\boldsymbol{\omega} \times \boldsymbol{u}' - \boldsymbol{\omega}' \times \boldsymbol{u}) + \mathbf{z} \cdot (\boldsymbol{\omega} \times \boldsymbol{\alpha}' - \boldsymbol{\omega}' \times \boldsymbol{\alpha}) \,,$$

and conclude that the cocycle (12.3.16) *is not a coboundary. (Actually, it can be proven that $H^2(\mathfrak{g}, \mathbb{R}) \cong \mathbb{R}$, that is, it is 1-dimensional, but this requires more algebraic work (Guillemin and Sternberg* [1977, 1984])).

12.7 Poisson Automorphisms

Here are some miscellaneous facts about Poisson automorphisms, symplectic leaves, and momentum maps. For a Poisson manifold P, define the following Lie subalgebras of $\mathfrak{X}(P)$:

1. ***Infinitesimal Poisson Automorphisms***

   Let $\mathcal{P}(P)$ be the set of $X \in \mathfrak{X}(P)$ such that:

   $$X[\{F_1, F_2\}] = \{X[F_1], F_2\} + \{F_1, X[F_2]\}.$$

2. ***Infinitesimal Poisson Automorphisms Preserving Leaves***

   Let $\mathcal{PL}(P)$ be the set of $X \in \mathcal{P}(P)$ such that $X(z) \in T_zS$, where S is the symplectic leaf containing $z \in P$.

3. ***Locally Hamiltonian Vector Fields***

   Let $\mathcal{LH}(P)$ be the set of $X \in \mathfrak{X}(P)$ such that for each $z \in P$, there is an open neighborhood U of z and an $F \in \mathcal{F}(U)$ such that $X|U = X_F|U$.

4. ***Hamiltonian Vector Fields***

   Let $\mathcal{H}(P)$ be the set of Hamiltonian vector fields X_F for $F \in \mathcal{F}(P)$.

Then one has the following facts (references are given if the verification is not straightforward):

(a) $\mathcal{H}(P) \subset \mathcal{LH}(P) \subset \mathcal{PL}(P) \subset \mathcal{P}(P)$.

(b) If P is symplectic, then $\mathcal{LH}(P) = \mathcal{PL}(P) = \mathcal{P}(P)$ and if $H^1(P) = 0$, then $\mathcal{LH}(P) = \mathcal{H}(P)$.

(c) Let P be the trivial Poisson manifold, that is, $\{F, G\} = 0$ for all $F, G \in \mathcal{F}(P)$. Then $\mathcal{P}(P) \neq \mathcal{PL}(P)$.

(d) Let $P = \mathbb{R}^2$ with the bracket

$$\{F, G\}(x, y) = x\left(\frac{\partial F}{\partial x}\frac{\partial G}{\partial y} - \frac{\partial F}{\partial x}\frac{\partial F}{\partial y}\right).$$

This is, in fact, a Lie-Poisson bracket. The vector field

$$X(x, y) = \frac{\partial}{\partial y}$$

is an example of an element of $\mathcal{PL}(P)$ which is not in $\mathcal{LH}(P)$.

(e) $\mathcal{H}(P)$ is an ideal in any of the three Lie algebras including it.

(f) If P is symplectic, then $[\mathcal{LH}(P), \mathcal{LH}(P)] \subset \mathcal{H}(P)$. (The Hamiltonian for $[X, Y]$ is $\Omega(X, Y)$.) This is false for Poisson manifolds in general. If P is symplectic Calabi [1970] and Lichnerowicz [1973] showed that $[\mathcal{LH}(P), \mathcal{LH}(P)] = \mathcal{H}(P)$.

(g) If the Lie aglebra $\mathfrak{g}$ admits a momentum map on P, then $\mathfrak{g}_P \subset \mathcal{H}(P)$.

(h) Let G be a connected Lie group. If the action admits a momentum map, it preserves the leaves of P. The proof was given in §**12.4**.

12.8 Momentum Maps and Casimir Functions

In this section we return to Casimir functions studied in Chapter 10 and link them with momentum maps. We will do this in the context of the Poisson manifolds P/G studied in §10.7.

We start with a Poisson manifold P and a free and proper Poisson action of a Lie group G on P admitting an equivariant momentum mapping $\mathbf{J} : P \to \mathfrak{g}^*$. We want to link $\mathbf{J}$ with a Casimir function $C : P/G \to \mathbb{R}$.

Proposition 12.8.1 *Let $\Phi : \mathfrak{g}^* \to \mathbb{R}$ be a function that is invariant under the coadjoint action. Then:*

i *Φ is a Casimir function for the Lie-Poisson bracket;*

ii *$\Phi \circ \mathbf{J}$ is G-invariant on P and so defines a function $C : P/G \to \mathbb{R}$ such that $\Phi \circ \mathbf{J} = C \circ \pi$, as in Figure 12.8.1; and*

iii *the function C is a Casimir function on P/G.*

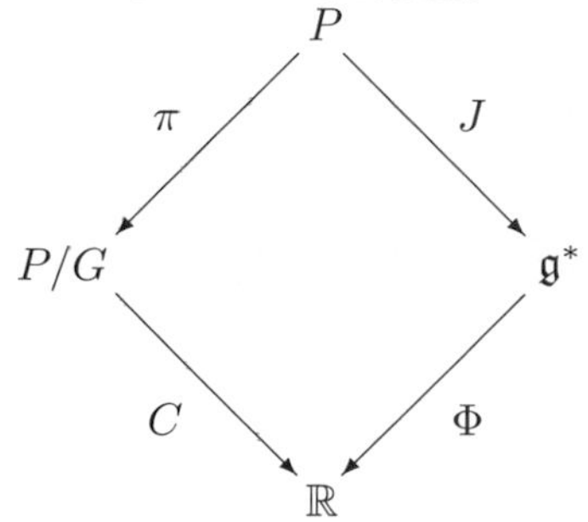

FIGURE 12.8.1. Casimir functions and momentum maps.

Proof To prove the first part, we write down the condition of Ad^*-invariance as

$$\Phi(\mathrm{Ad}^*_{g^{-1}} \mu) = \Phi(\mu). \tag{12.8.1}$$

Now $\langle \mathrm{Ad}^*_{g^{-1}} \mu, \xi \rangle = \langle \mu, \mathrm{Ad}_{g^{-1}} \xi \rangle$, so differentiating with respect to g at $g = e$, in the direction $\eta \in \mathfrak{g}$, we get

$$\left\langle (\mathbf{D}_g \mathrm{Ad}^*_{g^{-1}} \mu)\big|_{g=e} \cdot \eta, \xi \right\rangle = \langle \mu, [\xi, \eta] \rangle = -\langle \mathrm{ad}^*_\eta \mu, \xi \rangle. \tag{12.8.2}$$

Thus, from (12.8.1) and (12.8.2), we have

$$\mathbf{D}\Phi(\mu) \cdot \mathrm{ad}^*_\eta \mu = 0$$

for all $\eta \in \mathfrak{g}$. Thus, by definition of $\delta\Phi/\delta\mu$,

$$0 = \left\langle \mathrm{ad}^*_\eta \mu, \frac{\delta\Phi}{\delta\mu} \right\rangle = \left\langle \mu, \mathrm{ad}_\eta \frac{\delta\Phi}{\delta\mu} \right\rangle = -\langle \mathrm{ad}^*_{\delta\Phi/\delta\mu} \mu, \eta \rangle$$

for all $\eta \in \mathfrak{g}$. In other words,

$$\mathrm{ad}^*_{\delta\Phi/\delta\mu} \mu = 0$$

so by Proposition **10.9.1**, $X_\Phi = 0$ and thus Φ is a Casimir function.

To prove the second part, note that, by *equivariance* of $\mathbf{J}$ and *invariance* of Φ,

$$\Phi(\mathbf{J}(g \cdot z)) = \Phi(\mathrm{Ad}^*_{g^{-1}} \mathbf{J}(z)) = \Phi(\mathbf{J}(z)),$$

so $\Phi \circ \mathbf{J}$ is G-invariant.

Finally, for the third part, we use the collective Hamiltonian theorem **12.5.2** to get for $\mu = \mathbf{J}(z)$,

$$X_{\Phi \circ J}(z) = \left(\frac{\delta\Phi}{\delta\mu} \right)_P (z)$$

and so $T_z\pi \cdot X_{\Phi \circ J}(z) = 0$ since infinitesimal generators are tangent to orbits, so project to zero under π. But π is Poisson, so

$$0 = T_z\pi \cdot X_{\Phi \circ J}(z) = T_z\pi \cdot X_{C \circ \pi}(z) = X_C(\pi(z)).$$

Thus, C is a Casimir function on P/G. ■

Corollary 12.8.2 *If G is Abelian and $\Phi : \mathfrak{g}^* \to \mathbb{R}$ is any smooth function, then $\Phi \circ \mathbf{J} = C \circ \pi$ defines a Casimir function on P/G.*

This follows because for abelian groups, the Ad^*-action is trivial, so any function on $\mathfrak{g}^*$ is Ad^*-invariant.

Exercises

12.8-1 *Verify that $\Phi(\Pi) = \|\Pi\|^2$ is an invariant function on $\mathfrak{so}(3)^*$.*

12.8-2 *Use Corollary* **12.8.2** *to find the Casimir functions for the bracket* (10.7.6).

13

Euler-Poincaré and Lie-Poisson Reduction

Besides the Poisson structure on a symplectic manifold, the Lie-Poisson bracket on $\mathfrak{g}^*$, the dual of a Lie algebra, is perhaps the most fundamental example of a Poisson structure. We shall obtain it in the following manner. Given two smooth functions $F, H \in \mathcal{F}(\mathfrak{g}^*)$, we extend them to functions, F_L, H_L (respectively, F_R, H_R) on all T^*G by left (respectively, right) translations. The bracket $\{F_L, H_L\}$ (respectively, $\{F_R, H_R\}$) is taken in the canonical symplectic structure Ω on T^*G. The result is then restricted to $\mathfrak{g}^*$ regarded as the cotangent space at the identity; this defines $\{F, H\}$. We shall prove that one get the Lie-Poisson bracket this way. In §14.6 we show that the symplectic leaves of this bracket are the coadjoint orbits in $\mathfrak{g}^*$.

There is another side to the story too, where the basic objects that are reduced are not Poisson brackets, but rather are variational principles. This aspect of the story, which takes place on $\mathfrak{g}$ rather than on $\mathfrak{g}^*$, will be told as well.

13.1 The Lie-Poisson Reduction Theorem

We begin by studying the way the canonical Poisson bracket on T^*G is related to the Lie-Poisson bracket on $\mathfrak{g}^*$.

Theorem 13.1.1 (Lie-Poisson Reduction Theorem) *Identifying the set of functions on $\mathfrak{g}^*$ with the set of left (respectively, right) invariant* *functions on T^*G endows $\mathfrak{g}^*$ with Poisson structures given by*

$$\{F, H\}_\pm(\mu) = \pm\left\langle \mu, \left[\frac{\delta F}{\delta \mu}, \frac{\delta H}{\delta \mu}\right]\right\rangle. \tag{13.1.1}$$

The space $\mathfrak{g}^*$ with this Poisson structure is denoted $\mathfrak{g}^*_-$ (respectively, $\mathfrak{g}^*_+$). In contexts where the choice of left or right is clear, we shall drop the "$-$" or "$+$" from $\{F, H\}_-$ and $\{F, H\}_+$.

Following Marsden and Weinstein [1983], this bracket on $\mathfrak{g}^*$ is called the ***Lie-Poisson bracket*** after Lie [1890], p. 204. See Weinstein [1983a] and §13.8 below for more historical information. In fact, there are already some hints of this structure in Jacobi [1866], p. 7. It was rediscovered several times since Lie's work. For example, it appears explicitly in Berezin [1967]. It is closely related to results of Arnold, Kirillov, Kostant, and Souriau in the 1960s.

Before proving the theorem, we explain the terminology used in its statement. First, recall from Chapter 9 how the Lie algebra of a Lie group G is constructed. We define $\mathfrak{g} = T_eG$, the tangent space at the identity. For $\xi \in \mathfrak{g}$, we define a left invariant vector field $\xi_L = X_\xi$ on G by setting

$$\xi_L(g) = T_eL_g \cdot \xi \tag{13.1.2}$$

where $L_g : G \to G$ denotes left translation by $g \in G$ and is defined by $L_gh = gh$. Given $\xi, \eta \in \mathfrak{g}$, define

$$[\xi, \eta] = [\xi_L, \eta_L](e), \tag{13.1.3}$$

where the bracket on the right-hand side is the Jacobi-Lie bracket on vector fields. The bracket (13.1.3) makes $\mathfrak{g}$ into a Lie algebra, that is, $[\,,]$ is bilinear, antisymmetric, and satisfies Jacobi's identity. For example, if G is a subgroup of $GL(n)$, the group of invertible $n \times n$ matrices, we identify $\mathfrak{g} = T_eG$ with a vector space of matrices and then as we calculated in Chapter 9,

$$[\xi, \eta] = \xi\eta - \eta\xi, \tag{13.1.4}$$

the usual commutator of matrices.

A function $F_L : T^*G \to \mathbb{R}$ is called ***left invariant*** if, for all $g \in G$,

$$F_L \circ T^*L_g = F_L, \tag{13.1.5}$$

where T^*L_g denotes the cotangent lift of L_g, so T^*L_g is the pointwise adjoint of TL_g. Given $F : \mathfrak{g}^* \to \mathbb{R}$ and $\alpha_g \in T^*G$, set

$$F_L(\alpha_g) = F(T_e^*L_g \cdot \alpha_g) \tag{13.1.6}$$

which is the ***left invariant extension*** of F from $\mathfrak{g}^*$ to T^*G. One similarly defines the ***right invariant extension*** by

$$F_R(\alpha_g) = F(T_e^*R_g \cdot \alpha_g). \tag{13.1.7}$$

The main content of the Lie-Poisson reduction theorem is the pair of formulae

$$\{F, H\}_{-} = \{F_L, H_L\} |\mathfrak{g}^* \tag{13.1.8}$$

and

$$\{F, H\}_{+} = \{F_R, H_R\} |\mathfrak{g}^*, \tag{13.1.9}$$

where $\{\,,\}_{\pm}$ is the Lie-Poisson bracket on $\mathfrak{g}^*$ and $\{\,,\}$ is the canonical bracket on T^*G. Another way of saying this is that *the map* $\lambda : T^*G \to \mathfrak{g}^*_{-}$ *(respectively,* $\rho : T^*G \to \mathfrak{g}^*_{+}$ *on* T^*G*) given by*

$$\alpha_g \mapsto T_e^* L_g \cdot \alpha_g \text{ (respectively, } T_e^* R_g \cdot \alpha_g) \tag{13.1.10}$$

is a Poisson map.

Note that the correspondence between ξ and ξ_L identifies $\mathcal{F}(\mathfrak{g}^*)$ with the left invariant functions on T^*G, which is a subalgebra of $\mathcal{F}(T^*G)$ (since lifts are canonical), so (13.1.1) indeed defines a Poisson structure (although this fact may also be readily verified directly).

To prove the Lie-Poisson reduction theorem, first prove the following.

Lemma 13.1.2 *Let G act on itself by left translations. Then*

$$\xi_G(g) = T_e R_g \cdot \xi. \tag{13.1.11}$$

Proof By definition of infinitesimal generator,

$$\begin{aligned} \xi_G(g) &= \left.\frac{d}{dt} \Phi_{\exp(t\xi)}(g)\right|_{t=0} \\ &= \left.\frac{d}{dt} R_g(\exp(t\xi))\right|_{t=0} \\ &= T_e R_g \cdot \xi \end{aligned}$$

by the chain rule. ▼

Proof of the Theorem Let $J_L : T^*G \to \mathfrak{g}^*$ be the momentum map for the left action. From §12.1 we have

$$\begin{aligned} \langle J_L(\alpha_g), \xi \rangle &= \langle \alpha_g, \xi_Q(g) \rangle \\ &= \langle \alpha_g, T_e R_g \cdot \xi \rangle \\ &= \langle T_e^* R_g \cdot \alpha_g, \xi \rangle . \end{aligned}$$

Thus,

$$J_L(\alpha_g) = T_e^* R_g \cdot \alpha_g,$$

so $J_L = \rho$. Similarly $J_R = \lambda$. However, the momentum maps J_L and J_R are equivariant being the momentum maps for cotangent lifts, and so from §**12.5**, they are Poisson maps. The theorem now follows. ■

To gain further insight, we reprove the theorem in three special cases in the next three sections and in §**13.5** we prove it using momentum functions.

13.2 Proof of the Lie-Poisson Reduction Theorem for $GL(n)$

We now prove the Lie-Poisson reduction theorem for the special case of the matrix group $G = GL(n)$ of real invertible $n \times n$ matrices. Left translation by $\mathbf{U} \in G$ is given by matrix multiplication: $L_{\mathrm{U}}\mathbf{A} = \mathbf{UA}$. Identify the tangent space to G at $\mathbf{A}$ with the vector space of all $n \times n$ matrices, so for $\mathbf{B} \in T_{\mathrm{A}}G$, $T_A L_{\mathrm{U}} \cdot \mathbf{B} = \mathbf{UB}$ as well, since $L_{\mathrm{U}}\mathbf{A}$ is linear in $\mathbf{A}$. The cotangent space is identified with the tangent space via the pairing

$$\langle \pi, \mathbf{B} \rangle = \operatorname{trace}(\pi^T \mathbf{B}), \tag{13.2.1}$$

where π^T is the transpose of π. The cotangent lift of L_{U} is thus given by

$$\langle T^* L_{\mathrm{U}}\pi, \mathbf{B} \rangle = \langle \pi, T L_{\mathrm{U}} \cdot \mathbf{B} \rangle = \operatorname{trace}(\pi^T \mathbf{UB}), \quad \text{i.e.,} \quad T^* L_{\mathrm{U}}\pi = \mathbf{U}^T \pi. \tag{13.2.2}$$

Given functions $F, G : \mathfrak{g}^* \to \mathbb{R}$, let

$$F_L(\mathbf{A}, \pi) = F(\mathbf{A}^T \pi) \quad \text{and} \quad G_L(\mathbf{A}, \pi) = G(\mathbf{A}^T \pi) \tag{13.2.3}$$

be their left invariant extensions. By the chain rule, letting $\mu = \mathbf{A}^T \pi$, we get

$$\begin{aligned} \mathbf{D}_{\mathrm{A}} F_L(\mathbf{A}, \pi) \cdot \delta\mathbf{A} &= \mathbf{D}F(\mathbf{A}^T \pi) \cdot (\delta\mathbf{A})^T \pi \\ &= \left\langle \frac{\delta F}{\delta \mu}, (\delta\mathbf{A})^T \pi \right\rangle \\ &= \operatorname{trace}\left(\pi^T \delta\mathbf{A} \frac{\delta F}{\delta \mu} \right). \end{aligned} \tag{13.2.4}$$

The canonical bracket is therefore

$$\begin{aligned} \{F_L, G_L\} &= \left\langle \frac{\delta F_L}{\delta \mathbf{A}}, \frac{\delta G_L}{\delta \pi} \right\rangle - \left\langle \frac{\delta G_L}{\delta \mathbf{A}}, \frac{\delta F_L}{\delta \pi} \right\rangle \\ &= \mathbf{D}_{\mathrm{A}} F_L(\mathbf{A}, \pi) \cdot \frac{\delta G_L}{\delta \pi} - \mathbf{D}_{\mathrm{A}} G_L(\mathbf{A}, \pi) \cdot \frac{\delta F_L}{\delta \pi}. \end{aligned} \tag{13.2.5}$$

Since $\delta F_L / \delta \pi = \delta F / \delta \mu$ at the identity $\mathbf{A} = \mathrm{Id}$, where $\pi = \mu$, using (13.2.4), the Poisson bracket (13.2.5) becomes

$$\begin{aligned} \{F_L, G_L\}(\mu) &= \operatorname{trace}\left(\mu^T \frac{\delta G}{\delta \mu} \frac{\delta F}{\delta \mu} - \mu^T \frac{\delta F}{\delta \mu} \frac{\delta G}{\delta \mu} \right) \\ &= -\left\langle \mu, \frac{\delta F}{\delta \mu} \frac{\delta G}{\delta \mu} - \frac{\delta G}{\delta \mu} \frac{\delta F}{\delta \mu} \right\rangle \\ &= -\left\langle \mu, \left[\frac{\delta F}{\delta \mu}, \frac{\delta G}{\delta \mu} \right] \right\rangle, \end{aligned} \tag{13.2.6}$$

which is the $(-)$Lie-Poisson bracket. This derivation can be adapted for other matrix groups, including the rotation group $SO(3)$ as special cases. However, in the latter case, one has to be extremely careful to treat the orthogonality constraint properly.

13.3 Proof of the Lie-Poisson Reduction Theorem for $\mathrm{Diff}_{\mathrm{vol}}(M)$

Another special case is $G = \mathrm{Diff}_{\mathrm{vol}}(\Omega)$, the subgroup of the group of diffeomorphisms $\mathrm{Diff}(\Omega)$ of a region $\Omega \subset \mathbb{R}^3$, consisting of the volume-preserving diffeomorphisms. We shall treat $\mathrm{Diff}(\Omega)$ and $\mathrm{Diff}_{\mathrm{vol}}(\Omega)$ formally, although it is known how to handle the functional analysis issues involved (see Ebin and Marsden [1970] and Adams, Ratiu, and Schmid [1986a,b] and references therein). We shall prove (13.1.9) for this case.

For $\eta \in \mathrm{Diff}(\Omega)$, the tangent space at η is given by the set of maps $V : \Omega \to T\Omega$ satisfying $V(X) \in T_{\eta(X)}\Omega$, that is, vector fields over η. We think of V as a material velocity field. Thus, the tangent space at the identity is the space of vector fields on Ω (tangent to $\partial\Omega$). Given two such vector fields, their left Lie algebra bracket is related to the Jacobi-Lie bracket by (see Chapter **9**):

$$[V, W]_{LA} = -\,[V, W]_{JL}\,,$$

that is,

$$[V, W]_{LA} = (W \cdot \nabla)V - (V \cdot \nabla)W, \tag{13.3.1}$$

as one finds using the definitions. Let us compute the *right* Lie-Poisson bracket on $\mathfrak{g}^*$. Right translation by φ on G is given by

$$R_\varphi \eta = \eta \circ \varphi. \tag{13.3.2}$$

Differentiating (13.3.2) with respect to η gives

$$TR_\varphi \cdot V = V \circ \varphi. \tag{13.3.3}$$

Identify $T_\eta G$ with those V's such that the vector field on $\mathbb{R}^3$ given by $\mathbf{v} = V \circ \eta^{-1}$, is divergence-free and identify $T^*_\eta G$ with $T_\eta G$ via the pairing

$$\langle \pi, V \rangle = \int_\Omega \pi \cdot V\, dx\, dy\, dz, \tag{13.3.4}$$

where $\pi \cdot V$ is the dot product on $\mathbb{R}^3$. By the change of variables formula, and the fact that $\varphi \in G$ has unit Jacobian,

$$\begin{aligned}\langle T^* R_\varphi \cdot \pi, V \rangle &= \langle \pi, TR_\varphi \cdot V \rangle \\ &= \int_\Omega \pi \cdot (V \circ \varphi)\, dx\, dy\, dz = \int_\Omega (\pi \circ \varphi^{-1}) \cdot V\, dx\, dy\, dz,\end{aligned}$$

so

$$T^* R_\varphi \cdot \pi = \pi \circ \varphi^{-1}. \tag{13.3.5}$$

If $F : \mathfrak{g}^* \to \mathbb{R}$ is given, its right invariant extension is

$$F_R(\eta, \pi) = F(\pi \circ \eta^{-1}). \tag{13.3.6}$$

Let us denote elements of $\mathfrak{g}^*$ by $\mathbf{M}$, so we are investigating the relation between the canonical bracket of F_R and H_R and the Lie-Poisson bracket of F and H via the relation

$$\mathbf{M} \circ \eta = \pi.$$

From (13.3.6) and the chain rule, we get

$$\begin{aligned} \mathbf{D}_\eta F_R(\mathrm{Id}, \pi) \cdot \mathbf{v} &= -\mathbf{D}_{\mathrm{M}} F(\mathbf{M}) \cdot \mathbf{D}_\eta \pi(\mathrm{Id}) \cdot \mathbf{v} \\ &= -\int_\Omega ((\mathbf{v} \cdot \nabla)\mathbf{M}) \cdot \frac{\delta F}{\delta \mathbf{M}}\, dx\, dy\, dz, \qquad (13.3.7) \end{aligned}$$

where $\delta F / \delta \mathbf{M}$ is a divergence-free vector field parallel to the boundary. Since T^*G is not given as a product space, one has to worry about what it means to hold π constant in (13.3.7). We leave it to the ambitious reader to justify this formal calculation. Thus, the canonical bracket at the identity becomes

$$\begin{aligned} \{F_R, H_R\}(\mathrm{Id}, \pi) &= \int_\Omega \left(\frac{\delta F_R}{\delta \eta} \frac{\delta H_R}{\delta \pi} - \frac{\delta H_R}{\delta \eta} \frac{\delta F_R}{\delta \pi} \right) dx\, dy\, dz \\ &= \mathbf{D}_\eta F_R(\mathrm{Id}, \pi) \cdot \frac{\delta H_R}{\delta \pi} \\ &\quad - \mathbf{D}_\eta H_R(\mathrm{Id}, \pi) \cdot \frac{\delta F_R}{\delta \pi}. \qquad (13.3.8) \end{aligned}$$

At the identity, $\pi = \mathbf{M}$ and $\delta F_R / \delta \pi = \delta F / \delta \mathbf{M}$, so substituting this and (13.3.7) into (13.3.8), we get

$$\begin{aligned} &\{F_R, H_R\}(\mathrm{Id}, \mathbf{M}) \\ &= -\int_\Omega \left[\left(\frac{\delta H}{\delta \mathbf{M}} \cdot \nabla \right) \mathbf{M} \cdot \frac{\delta F}{\delta \mathbf{M}} - \left(\frac{\delta F}{\delta \mathbf{M}} \cdot \nabla \right) \mathbf{M} \cdot \frac{\delta H}{\delta \mathbf{M}} \right] dx\, dy\, dz. \end{aligned} \qquad (13.3.9)$$

Equation (13.3.9) may be integrated by parts to give

$$\begin{aligned} &\{F_R, H_R\}(\mathrm{Id}, \mathbf{M}) \\ &= \int \mathbf{M} \cdot \left[\left(\frac{\delta H}{\delta \mathbf{M}} \cdot \nabla \right) \frac{\delta F}{\delta \mathbf{M}} - \left(\frac{\delta F}{\delta \mathbf{M}} \cdot \nabla \right) \frac{\delta H}{\delta \mathbf{M}} \right] dx\, dy\, dz \\ &= \int \mathbf{M} \cdot \left[\frac{\delta F}{\delta \mathbf{M}}, \frac{\delta H}{\delta \mathbf{M}} \right]_{LA} dx\, dy\, dz \qquad (13.3.10) \end{aligned}$$

which is the "+" Lie-Poisson bracket. In doing this step note $\operatorname{div}(\delta H / \delta \mathbf{M}) = 0$ and since $\delta H / \delta \mathbf{M}$ and $\delta F / \delta \mathbf{M}$ are parallel to the boundary, no boundary term appears. When doing free boundary problems, these boundary terms are essential to retain (see Lewis, Marsden, Montgomery, and Ratiu [1986]).

For other diffeormorphism groups, it may be convenient to treat $\mathbf{M}$ as a one-form density rather than a vector field.

13.4 Proof of the Lie-Poisson Reduction Theorem for $\text{Diff}_{\text{can}}(P)$

This optional section discusses $G = \text{Diff}_{\text{can}}(P)$, the group of canonical transformations of a boundaryless symplectic (or Poisson) manifold P. The Lie algebra of $\text{Diff}_{\text{can}}(P)$ is the algebra of infinitesimal Poisson automorphisms, or Poisson derivations, that is, vector fields X on P for which

$$X[\{f,h\}] = \{X[f],h\} + \{f,X[h]\}$$

for any $f, h \in \mathcal{F}(P)$. To avoid complications, we work with the globally Hamiltonian vector fields by suitably restricting P or $\text{Diff}_{\text{can}}(P)$. Each Hamiltonian vector field can be identified with its generator (modulo additive constants being understood). From the formula

$$[X_k, X_h]_{LA} = -[X_k, X_h]_{JL} = X_{\{k,h\}} \tag{13.4.1}$$

we see that $\mathfrak{g}$ may be identified with $\mathcal{F}(P)$ with the Lie bracket given by the Poisson bracket. One could then identify $\mathfrak{g}^*$ with functions f on P via the pairing

$$\langle f, h\rangle = \int_P fh\, d\mu, \tag{13.4.2}$$

where $d\mu$ is the Liouville measure. If P is only a Poisson manifold, identify $\mathfrak{g}^*$ with the densities on P. As in the last section, $T_\eta G$ consists of vector fields of the form $X_k \circ \eta$.

To identify the dual space of $T_\eta G$, we need objects to pair with $T_\eta G$ in a nondegenerate way. Since $X_k \circ \eta = T\eta \circ X_{k\circ\eta}$, we cannot simply use the pairing (13.4.2) to identify $T_\eta^* G$ with $\mathcal{F}(P)$; such a procedure would not account for the extra factor $T\eta$. Instead, regard $\pi \in \mathfrak{g}^*$ as a one-form on P and pair it with $X_k \in \mathfrak{g}$ by

$$\langle \pi, X_k\rangle = \int_P \pi \cdot X_k\, d\mu. \tag{13.4.3}$$

This pairing is degenerate; for example, if $\pi = \mathbf{d}f$, then $\langle \pi, X_k\rangle = 0$ by Stokes' theorem. To simplify matters, let us work in coordinates, and write

$$\pi = \pi_i\, dq^i + \pi^i\, dp_i$$

so, integrating by parts,

$$\begin{aligned}\int_P \pi \cdot X_k\, d\mu &= \int_P \left(\pi_i \frac{\partial k}{\partial p_i} - \pi^i \frac{\partial k}{\partial q^i}\right) dq\, dp \\ &= \int_P \left(-\frac{\partial \pi_i}{\partial p_i} + \frac{\partial \pi^i}{\partial q^i}\right) k\, dq\, dp. \end{aligned} \tag{13.4.4}$$

Thus if we work modulo π's satisfying the divergence-like condition

$$\frac{\partial \pi^i}{\partial q^i} - \frac{\partial \pi_i}{\partial p_i} = 0, \tag{13.4.5}$$

then the pairing (13.4.3) is nondegenerate. Now let $f \in \mathcal{F}(P)$ be given and define π by requiring

$$\int_P \pi \cdot X_k \, d\mu = \int_P f k \, d\mu \tag{13.4.6}$$

for all $k \in \mathcal{F}(P)$. Thus, from (13.4.4), we need

$$\frac{\partial \pi^i}{\partial q^i} - \frac{\partial \pi_i}{\partial p_i} = f. \tag{13.4.7}$$

Note that if $\pi = (\partial h/\partial q^i)\, dq^i + (\partial h/\partial p_i)\, dp_i$, the left side of (13.4.7) is identically zero since $\partial^2 h/\partial q^i \, \partial p_i = \partial^2 h/\partial p_i \, \partial q^i$. If we take $\pi = (\partial \psi/\partial p_i)\, dq^i - (\partial \psi/\partial q^i)\, dp_i$, then ψ is determined by $-\Delta\psi = f$, so ψ is now uniquely determined modulo π's satisfying (13.4.5). In two-dimensional incompressible flow, which corresponds to the special case $\dim P = 2$, ψ is the stream function and f the vorticity.

Identify $T^*_\eta G$ with one-forms modulo exact one-forms over η ; that is, objects of the form $\pi_\eta = \pi \circ \eta$. Given $F \in \mathcal{F}(P)$, define F on $\mathfrak{g}^*$ by $F(\pi) = F(f)$, where π and f are related by (13.4.6) and extend it to be right invariant by

$$F_R(\eta, \pi_\eta) = F(\pi_\eta \circ \eta^{-1}).$$

As in §**13.3** and using vector analysis notation,

$$\begin{aligned} \mathbf{D}_\eta F_R(\mathrm{Id}, \pi) \cdot X_k &= -\mathbf{D}F(\pi) \cdot \mathbf{D}_\eta \pi_\eta(\mathrm{Id}) \cdot X_k \\ &= -\int \frac{\delta F}{\delta \pi} \cdot (X_k \cdot \nabla \pi) d\mu. \end{aligned} \tag{13.4.8}$$

Also, $\delta F_R/\delta\pi = \delta F/\delta\pi$ at $\eta = \mathrm{Id}$, as before. Thus the canonical bracket at $\eta = \mathrm{Id}$ is

$$\begin{aligned} \{F_R, H_R\}(\mathrm{Id}, \pi) &= \int_P \left(\frac{\delta F_R}{\delta \eta} \frac{\delta H_R}{\delta \pi} - \frac{\delta H_R}{\delta \eta} \frac{\delta F_R}{\delta \pi} \right) d\mu \\ &= -\int_P \left[\frac{\delta F}{\delta \pi} \cdot \left(\frac{\delta H}{\delta \pi} \cdot \nabla \pi \right) - \frac{\delta H}{\delta \pi} \cdot \left(\frac{\delta F}{\delta \pi} \cdot \nabla \pi \right) \right] d\mu, \end{aligned}$$

which may be integrated by parts to give

$$\begin{aligned} \{F_R, H_R\}(\mathrm{Id}, \pi) &= \int_P \pi \cdot \left\{ \left(\frac{\delta H}{\delta \pi} \cdot \nabla \right) \frac{\delta F}{\delta \pi} - \left(\frac{\delta F}{\delta \pi} \cdot \nabla \right) \frac{\delta F}{\delta \pi} \right) d\mu \\ &= \int_P \pi \cdot \left[\frac{\delta F}{\delta \pi}, \frac{\delta H}{\delta \pi} \right]_{LA} d\mu. \end{aligned} \tag{13.4.9}$$

To write this in terms of $\mathcal{F}(P)$ we use (13.4.6) to write

$$\left\langle \delta\pi, \frac{\delta F}{\delta\pi} \right\rangle = \int_P \delta\pi \cdot X_k \, d\mu$$

for some k to be determined. By the chain rule,

$$\begin{aligned} \left\langle \delta\pi, \frac{\delta F}{\delta\pi} \right\rangle &= \mathbf{D}_\pi F \cdot \delta\pi = \mathbf{D}_f F \cdot (\mathbf{D}_\pi f \cdot \delta\pi) \\ &= \int_P \frac{\delta F}{\delta f} (\mathbf{D}_\pi f \cdot \delta\pi) \, d\mu. \end{aligned} \tag{13.4.10}$$

Differentiating (13.4.6) implicitly relative, π we get

$$\int_P \delta\pi \cdot X_k \, d\mu = \int_P (\mathbf{D}_\pi f \cdot \delta\pi) \, k \, d\mu,$$

so by (13.4.10)

$$\left\langle \delta\pi, \frac{\delta F}{\delta\pi} \right\rangle = \int_P \delta\pi \cdot X_{\delta F/\delta f} \, d\mu, \quad \text{i.e.,} \quad \frac{\delta F}{\delta\pi} = X_{\delta F/\delta f}. \tag{13.4.11}$$

Thus (13.4.9) becomes, with the aid of (13.4.1) and (13.4.6),

$$\begin{aligned} \{F_R, H_R\}(\mathrm{Id}, \pi) &= \int_P \pi \cdot \left[X_{\delta F/\delta f}, X_{\delta H/\delta f}\right]_{LA} d\mu \\ &= \int_P \pi \cdot X_{\{\delta F/\delta f, \delta H/\delta f\}} \, d\mu \\ &= \int_P f \left\{ \frac{\delta F}{\delta f}, \frac{\delta H}{\delta f} \right\} d\mu \end{aligned} \tag{13.4.12}$$

which is the "+" Lie-Poisson bracket on $\mathfrak{g}^*$ identified with $\mathcal{F}(P)$.

Remarks

1. This derivation is related to one given by Kaufman and Dewar [1984].

2. The bracket (13.4.12) can be understood as a limit of the canonical bracket for a larger number of particles moving in P by taking f to be a sum of delta functions at the particle positions. This derivation is due to Bialynicki-Birula, Hubbard, and Turski [1984]; see also Kaufman [1982] and Marsden, Morrison, and Weinstein [1984]. ♦

13.5 The Lie-Poisson Reduction Theorem Using Momentum Functions

Now we turn to another general proof of the Lie-Poisson reduction theorem using properties of momentum functions. It will be useful to recall that the

Poisson bracket $\{F, H\}$ depends only on the linearization of F and H at each point, so in determining the canonical bracket on T^*G, we can assume the functions in question are linear on fibers.

Proof of the Lie-Poisson Reduction Theorem Recall that the space $\mathcal{F}_L(T^*G)$ of left invariant functions on T^*G is isomorphic (as a vector space) to $\mathcal{F}(\mathfrak{g}^*)$, the space of all functions on the dual $\mathfrak{g}^*$ of the Lie algebra $\mathfrak{g}$ of G. This isomorphism is given by $F \in \mathcal{F}(\mathfrak{g}^*) \leftrightarrow F_L \in \mathcal{F}_L(T^*G)$, where

$$F_L(\alpha_g) = F(T^*L_g \cdot \alpha_g). \tag{13.5.1}$$

Since $\mathcal{F}_L(T^*G)$ is closed under bracketing (T^*L_g is a symplectic map), $\mathcal{F}(\mathfrak{g}^*)$ gets endowed with a unique Poisson structure. As we remarked before, it is enough to consider the case in which F is replaced by its linearization at a particular point. This means it is enough to prove the Lie-Poisson reduction theorem for linear functions on $\mathfrak{g}^*$. If F is linear, we can write $F(\mu) = \langle \mu, \delta F/\delta\mu \rangle$, where $\delta F/\delta\mu$ is a constant on $\mathfrak{g}$, so that letting $\mu = T^*L_g \cdot \alpha_g$, we get

$$\begin{aligned} F_L(\alpha_g) &= F(T^*L_g \cdot \alpha_g) = \left\langle T^*L_g \cdot \alpha_g, \frac{\delta F}{\delta \mu} \right\rangle \\ &= \left\langle \alpha_g, TL_g \cdot \frac{\delta F}{\delta \mu} \right\rangle = \mathcal{P}\left(\left(\frac{\delta F}{\delta \mu} \right)_L \right)(\alpha_g), \end{aligned} \tag{13.5.2}$$

where $\xi_L(g) = T_eL_g(\xi)$ is the left invariant vector field on G whose value at e is $\xi \in \mathfrak{g}$. Thus, by (12.1.2), (13.5.2), and the definition of the Lie algebra bracket, we have

$$\begin{aligned} \{F_L, H_L\}(\mu) &= \left\{ \mathcal{P}\left(\left(\frac{\delta F}{\delta \mu} \right)_L \right), \mathcal{P}\left(\left(\frac{\delta H}{\delta \mu} \right)_L \right) \right\}(\mu) \\ &= -\mathcal{P}\left(\left[\left(\frac{\delta F}{\delta \mu} \right)_L, \left(\frac{\delta H}{\delta \mu} \right)_L \right] \right)(\mu) \\ &= -\mathcal{P}\left(\left[\frac{\delta F}{\delta \mu}, \frac{\delta H}{\delta \mu} \right]_L \right)(\mu) \\ &= -\left\langle \mu, \left[\frac{\delta F}{\delta \mu}, \frac{\delta H}{\delta \mu} \right] \right\rangle \end{aligned} \tag{13.5.3}$$

as required.

The formula with "+" follows by using right invariant extensions of linear functions since the Lie bracket of two right invariant vector fields equals minus the Lie algebra bracket of their generators.

Finally, we observe that since T^*G is a Poisson manifold, formulae (13.1.8) and (13.1.9) show that $\mathfrak{g}^*_-$ and $\mathfrak{g}^*_+$ inherit the same properties, so they are Poisson as well. ■

13.6 Reduction and Reconstruction of Dynamics

In the examples in subsequent sections, we will use the Lie-Poisson reduction theorem in the following way:

Theorem 13.6.1 (Lie-Poisson Reduction of Dynamics) *Let G be a Lie group and $H : T^*G \to \mathbb{R}$. Assume H is left (respectively, right) invariant. Then the function $H^- := H|\mathfrak{g}^*$ (respectively, $H^+ := H|\mathfrak{g}^*$) on $\mathfrak{g}^*$ satisfies*

$$H(\alpha_g) = H^-(\pi_L(\alpha_g)) \quad \textit{for all} \quad \alpha_g \in T_g^*G \tag{13.6.1}$$

where $\pi_L : T^*G \to \mathfrak{g}_-^*$ is given by $\pi_L(\alpha_g) = T^*L_g \cdot \alpha_g$ (respectively,

$$H(\alpha_g) = H^+(\pi_R(\alpha_g)) \quad \textit{for all} \quad \alpha_g \in T_g^*G, \tag{13.6.2}$$

where $\pi_R : T^*G \to \mathfrak{g}_+^*$ is given by $\pi_R(\alpha_g) = T^*R_g \cdot \alpha_g$).

The flow F_t of H on T^*G and the flow F_t^- (respectively, F_t^+) of H^- (respectively, H^+) on $\mathfrak{g}_-^*$ (respectively, $\mathfrak{g}_+^*$) are related by

$$\pi_L(F_t(\alpha_g)) = F_t^-(\pi_L(\alpha_g)), \tag{13.6.3}$$

$$\pi_R(F_t(\alpha_g)) = F_t^+(\pi_R(\alpha_g)). \tag{13.6.4}$$

In other words, a left invariant Hamiltonian on T^*G induces Lie-Poisson dynamics on $\mathfrak{g}_-^*$, while a right invariant one induces Lie-Poisson dynamics on $\mathfrak{g}_+^*$. The result is a direct consequence of the Lie-Poisson reduction theorem and the fact that a Poisson map relates Hamiltonian systems and their integral curves to Hamiltonian systems. As we shall see in Volume II, this is a special case of the reduction procedure.

As we have remarked, the maps λ and ρ induce Poisson isomorphisms between $(T^*G)/G$ and $\mathfrak{g}^*$ (with the $-$ and $+$ brackets, respectively) and this is a special instance of Poisson reduction. The following result is one useful way of formulating the general relation between T^*G and $\mathfrak{g}^*$. We treat the left invariant case for simplicity.

Theorem 13.6.2 *Let G be a Lie group and let $H : T^*G \to \mathbb{R}$ be a left invariant Hamiltonian. Let $h : \mathfrak{g}^* \to \mathbb{R}$ be the restriction of H to T_e^*G. For a curve $p(t) \in T_{g(t)}^*G$, let $\mu(t) = (T_{g(t)}^*L) \cdot p(t) = \lambda(p(t))$ be the induced curve in $\mathfrak{g}^*$. Assuming that $g(t)$ satisfies the differential equation*

$$\dot{g} = T_e L_g \frac{\delta h}{\delta \mu},$$

where $\mu = p(0)$, the following are equivalent:

i *$p(t)$ is an integral curve of X_H; i.e., Hamilton's equations on T^*G hold;*

ii *for any $F \in \mathcal{F}(T^*G)$, $\dot{F} = \{F, H\}$, where $\{\,,\}$ is the canonical bracket on T^*G;*

iii *$\mu(t)$ satisfies the* ***Lie-Poisson equations***

$$\frac{d\mu}{dt} = \mathrm{ad}^*_{\delta h/\delta\mu}\mu, \tag{13.6.5}$$

where $\mathrm{ad}_\xi : \mathfrak{g} \to \mathfrak{g}$ *is defined by* $\mathrm{ad}_\xi \eta = [\xi, \eta]$ *and* ad^*_ξ *is its dual, i.e.,*

$$\dot{\mu}_a = C^d_{ba} \frac{\delta h}{\delta \mu_b} \mu_d; \tag{13.6.6}$$

iv *for any $f \in \mathcal{F}(\mathfrak{g}^*)$, we have*

$$\dot{f} = \{f, h\}_-, \tag{13.6.7}$$

where $\{\,,\}_-$ is the minus Lie-Poisson bracket.

Proof First of all, the equivalence of **i** and **ii** is general for any cotangent bundle, as we know. The equivalence of **ii** and **iv** follows from the fact that λ is a Poisson map and $H = h \circ \lambda$. Finally, we establish the equivalence of **iii** and **iv**. Indeed, $\dot{f} = \{f, h\}_-$ means

$$\begin{aligned}
\left\langle \dot{\mu}, \frac{\delta f}{\delta \mu} \right\rangle &= -\left\langle \mu, \left[\frac{\delta f}{\delta \mu}, \frac{\delta h}{\delta \mu}\right] \right\rangle \\
&= \left\langle \mu, \mathrm{ad}_{\delta h/\delta\mu} \frac{\delta f}{\delta \mu} \right\rangle \\
&= \left\langle \mathrm{ad}^*_{\delta h/\delta\mu}\mu, \frac{\delta f}{\delta \mu} \right\rangle.
\end{aligned}$$

Since f is arbitrary, this is equivalent to **iii**. ■

It is useful to keep in mind that the Hamiltonian H on T^*G generally arises from a Lagrangian $L : TG \to \mathbb{R}$ via a Legendre transform $\mathbb{F}L$. In fact, many of the constructions and verifications are simpler using the Lagrangian. Assume that L is left invariant (respectively, right invariant); that is,

$$L(TL_g \cdot v) = L(v), \tag{13.6.8}$$

respectively,

$$L(TR_g \cdot v) = L(v) \tag{13.6.9}$$

for all $g \in G$ and $v \in T_hG$. Differentiating (13.6.8) and (13.6.9), we find

$$\mathbb{F}L(TL_g \cdot v) \cdot (TL_g \cdot w) = \mathbb{F}L(v) \cdot w, \tag{13.6.10}$$

respectively,

$$\mathbb{F}L(TR_g \cdot v) \cdot (TR_g \cdot w) = \mathbb{F}L(v) \cdot w \tag{13.6.11}$$

for all $v, w \in T_hG$ and $g \in G$. In other words,

$$T^*L_g \circ \mathbb{F}L \circ TL_g = \mathbb{F}L, \tag{13.6.12}$$

respectively,

$$T^*R_g \circ \mathbb{F}L \circ TR_g = \mathbb{F}L. \tag{13.6.13}$$

Note that the action of L is left (respectively, right) invariant

$$A(TL_g \cdot v) = A(v), \tag{13.6.14}$$

respectively,

$$A(TR_g \cdot v) = A(v) \tag{13.6.15}$$

since $A(TL_g \cdot v) = \mathbb{F}L(TL_g \cdot v) \cdot (TL_g \cdot v) = \mathbb{F}L(v) \cdot v = A(v)$ by (13.6.10). Thus the energy $E = A - L$ is left (respectively, right) invariant on TG. If L is hyperregular, so $\mathbb{F}L : TG \to T^*G$ is a diffeomorphism, then $H = E \circ (\mathbb{F}L)^{-1}$ is left (respectively, right) invariant on T^*G.

The Lagrangian formalism is also useful for the reconstruction process in which we reconstruct the dynamics on T^*G or TG from that on $\mathfrak{g}^*$:

$$T^*G \underset{\text{reconstruction}}{\overset{\text{Lie-Poisson reduction}}{\rightleftarrows}} \mathfrak{g}^*$$

Theorem 13.6.3 (Lie-Poisson Reconstruction Theorem) *Let $L : TG \to \mathbb{R}$ be a hyperregular Lagrangian which is left (respectively, right) invariant on TG. Let $H : T^*G \to \mathbb{R}$ be the associated Hamiltonian and $H^- : \mathfrak{g}^*_- \to \mathbb{R}$ (respectively, $H^+ : \mathfrak{g}^*_+ \to \mathbb{R}$) be the induced Hamiltonian on $\mathfrak{g}^*$. Let $\mu(t) \in \mathfrak{g}^*$ be an integral curve for H^- (respectively, H^+) with initial condition $\mu(0) = T_e^* L_{g_0} \cdot \alpha_{g_0}$ (respectively, $\mu(0) = T_e^* R_{g_0} \cdot \alpha_{g_0}$) and let $\xi(t) = \mathbb{F}L^{-1}\mu(t) \in \mathfrak{g}$. Let*

$$v_0 = TL_{g_0} \cdot \xi(0) \in T_{g_0}G.$$

Then the integral curve for the Langangian L with initial condition (g_0, v_0) is given by

$$V_L(t) = TL_{g(t)} \cdot \xi(t), \tag{13.6.16}$$

respectively,

$$V_R(t) = TR_g(t) \cdot \xi(t), \tag{13.6.17}$$

where $g(t)$ solves the equation

$$\frac{dg}{dt} = TL_{g(t)} \cdot \xi(t), \tag{13.6.18}$$

respectively,

$$\frac{dg}{dt} = TR_{g(t)} \cdot \xi(t). \tag{13.6.19}$$

The corresponding integral curve of X_H on T^*G with initial condition α_{g_0} and covering $\mu(t)$ is

$$\alpha(t) = \mathbb{F}L(V_L(t)) = T^*L_{(g_0 g(t))^{-1}}\mu(t), \tag{13.6.20}$$

respectively,

$$\alpha(t) = \mathbb{F}L(V_R(t)) = T^*R_{(g_0(g(t))^{-1}}\mu(t). \tag{13.6.21}$$

Proof According to Lie-Poisson reduction of dynamics, the integral curve of X_{H^-} on $\mathfrak{g}^*_-$ associated to an integral curve $\alpha(t) \in T^*_{g(t)}G$ of X_H is

$$\mu(t) = T^*L_{g(t)} \cdot \alpha(t). \tag{13.6.22}$$

Applying $\mathbb{F}L^{-1}$ to (13.6.22) gives

$$\xi(t) = TL_{g(t)^{-1}} \cdot V(t) \tag{13.6.23}$$

for the corresponding integral curve of the Lagrangian, that is, $V(t) = TL_{g(t)} \cdot \xi(t)$. Since X_E is a second-order equation, $dg/dt = V$, so we get the result. ∎

Thus, given $\xi(t)$, one solves (13.6.18) for $g(t)$ and then constructs $V(t)$ or $\alpha(t)$ from (13.6.16) and (13.6.20). As we shall see in the examples, this procedure has a natural physical interpretation. The previous theorem generalizes to general Lagrangian systems in the following way. In fact, Theorem **13.6.3** is a corollary of the next theorem.

Theorem 13.6.4 (Lagrangian Lie-Poisson Reconstruction) *Let $L : TG \to \mathbb{R}$ be a left invariant Lagrangian such that its Lagrangian vector field $Z \in \mathfrak{X}(TG)$ is a second-order equation and is left invariant. Let $Z_G \in \mathfrak{X}(\mathfrak{g})$ be the induced vector field on $(TG)/G \approx \mathfrak{g}$ and let $\xi(t)$ be an integral curve of Z_G. If $g(t) \in G$ is the solution of the nonautonomous ordinary differential equation $\dot{g}(t) = T_eL_{g(t)}\xi(t)$, $g(0) = e$, and $g \in G$, then $V(t) = T_eL_{gg(t)}\xi(t)$ is the integral curve of Z satisfying $V(0) = T_eL_g\xi(0)$ and $V(t)$ projects to $\xi(t)$, i.e., $TL_{\tau(V(t))^{-1}}V(t) = \xi(t)$, where $\tau : TG \to G$ is the tangent bundle projection.*

Proof Let $V(t)$ be the integral curve of Z satisfying $V(0) = T_eL_g\xi(0)$ for a given element $\xi(0) \in \mathfrak{g}$. Since $\xi(t)$ is the integral curve of Z_G whose flow is conjugated to the flow of Z by left translation, we have $TL_{\tau(V(t))^{-1}}V(t) = \xi(t)$. If $h(t) = \tau(V(t))$, since Z is a second-order equation, we have

$$V(t) = \dot{h}(t) = T_eL_{h(t)}\xi(t), \quad h(0) = \tau(V(0)) = g,$$

so that letting $g(t) = g^{-1}h(t)$ we get $g(0) = e$ and

$$\dot{g}(t) = TL_{g^{-1}}\dot{h}(t) = TL_{g^{-1}}TL_{h(t)}\xi(t) = TL_{g(t)}\xi(t).$$

This determines $g(t)$ uniquely from $\xi(t)$ and so

$$V(t) = T_e L_{h(t)} \xi(t) = T_e L_{gg(t)} \xi(t). \quad \blacksquare$$

These calculations suggest rather strongly that one should examine the Lagrangian (rather than the Hamiltonian) side of the story on an independent footing. We will do exactly that shortly.

Since Poisson brackets and Hamilton's equations naturally drop from T^*G to $\mathfrak{g}^*$, it is natural to ask if other structures do too, such as Hamilton-Jacobi theory. We investigate this question now. We shall just state the results of Ge and Marsden [1988], omitting the proofs.

Let H be a G invariant function on T^*G and let H_L be the corresponding left reduced Hamiltonian on $\mathfrak{g}^*$. (To be specific, we deal with left actions; of course, there are similar statements for right reduced Hamiltonians). If S is invariant, there is a unique function S_L such that $S(g, g_0) = S_L(g^{-1}g_0)$. (One gets a slightly different representation for S by writing $g_0^{-1}g$ in place of $g^{-1}g_0$.)

Proposition 13.6.5 *The* ***left*** *reduced Hamilton-Jacobi equation is the following equation for a function $S_L : G \to \mathbb{R}$:*

$$\frac{\partial S_L}{\partial t} + H_L(-TR_g^* \cdot \mathbf{d}S_L(g)) = 0, \tag{13.6.24}$$

which we call the ***Lie-Poisson Hamilton-Jacobi equation****. The Lie-Poisson flow of the Hamiltonian H_L is generated by the solution S_L of (13.6.24) in the sense that the flow is given by the Poisson transformation of $\mathfrak{g}^* : \Pi_0 \mapsto \Pi$ defined as follows. Define $g \in G$ by solving the equation*

$$\Pi_0 = -TL_g^* \cdot \mathbf{d}_g S_L \tag{13.6.25}$$

for $g \in G$ and then set

$$\Pi = \operatorname{Ad}_{g^{-1}}^* \Pi_0. \tag{13.6.26}$$

Here Ad denotes the adjoint action and so the action in (13.6.26) is the coadjoint action. Note that (13.6.26) and (13.6.25) give $\Pi = -TR_g^* \cdot \mathbf{d}S_L(g)$.

13.7 The Linearized Lie-Poisson Bracket

Here we show that the equations linearized about an equilibrium solution of a Lie-Poisson system (such as the ideal fluid equations) are Hamiltonian with respect to a "constant coefficient" Lie-Poisson bracket. The Hamiltonian for these linearized equations is $\frac{1}{2}\delta^2 \left(H + C\right)\big|_e$, the quadratic functional obtained by taking one-half of the second variation of the Hamiltonian plus conserved quantities and evaluating it at the equilibrium solution where the conserved quantity C (often a Casimir) is chosen so that the first variation $\delta(H+C)$ vanishes at the equilibrium. A consequence is that the linearized dynamics preserves $\frac{1}{2}\delta^2(H+C)|_e$. This is useful for studying stability of the linearized equations.

For a Lie algebra $\mathfrak{g}$, the Lie-Poisson bracket is defined on $\mathfrak{g}^*$, the dual of $\mathfrak{g}$ with respect to a weakly nondegenerate pairing $\langle\,,\rangle$ between $\mathfrak{g}^*$ and $\mathfrak{g}$ by the usual formula

$$\{F,G\}(\mu)=\left\langle \mu,\left[\frac{\delta F}{\delta\mu},\frac{\delta G}{\delta\mu}\right]\right\rangle, \tag{13.7.1}$$

where $\delta F/\delta\mu\in\mathfrak{g}$ is determined by

$$\mathbf{D}F(\mu)\cdot\delta\mu=\left\langle \delta\mu,\frac{\delta F}{\delta\mu}\right\rangle \tag{13.7.2}$$

when such an element $\delta F/\delta\mu$ exists, for any μ, $\delta\mu\in\mathfrak{g}^*$. The equations of motion are

$$\frac{d\mu}{dt}=-\operatorname{ad}\left(\frac{\delta H}{\delta\mu}\right)^*\mu, \tag{13.7.3}$$

where $H:\mathfrak{g}^*\to\mathbb{R}$ is the Hamiltonian, $\operatorname{ad}(\xi):\mathfrak{g}\to\mathfrak{g}$ is the adjoint action, $\operatorname{ad}(\xi)\cdot\eta=[\xi,\eta]$ for $\xi,\eta\in\mathfrak{g}$, and $\operatorname{ad}(\xi)^*:\mathfrak{g}^*\to\mathfrak{g}^*$ is its dual. Let $\mu_e\in\mathfrak{g}^*$ be an equilibrium solution of (13.7.3). The linearized equations of (13.7.3) at μ_e are obtained by expanding in a Taylor expansion with small parameter ε using $\mu=\mu_e+\varepsilon\delta\mu$, and taking $(d/d\varepsilon)|_{\varepsilon=0}$ of the resulting equations. This gives

$$\frac{\delta H}{\delta\mu}=\frac{\delta H}{\delta\mu_e}+\varepsilon\mathbf{D}\left(\frac{\delta H}{\delta\mu}\right)(\mu_e)\cdot\delta\mu+O(\varepsilon^2), \tag{13.7.4}$$

where $\langle\delta H/\delta\mu_e,\delta\mu\rangle:=\mathbf{D}H(\mu_e)\cdot\delta\mu$, and the derivative $\mathbf{D}(\delta H/\delta\mu)(\mu_e)\cdot\delta\mu$ is the linear functional

$$\nu\in\mathfrak{g}^*\mapsto\mathbf{D}^2H(\mu_e)\cdot(\delta\mu,\nu)\in\mathbb{R} \tag{13.7.5}$$

by using the definition (13.7.2). Since $\delta^2H:=\mathbf{D}^2H(\mu_e)\cdot(\delta\mu,\delta\mu)$, it follows that the functional (13.7.5) equals $\frac{1}{2}\delta(\delta^2H)/\delta(\delta\mu)$. Consequently, (13.7.4) becomes

$$\frac{\delta H}{\delta\mu}=\frac{\delta H}{\delta\mu_e}+\frac{1}{2}\varepsilon\frac{\delta(\delta^2H)}{\delta(\delta\mu)}+O(\varepsilon^2) \tag{13.7.6}$$

and the Lie-Poisson equations (13.7.3) yield

$$\begin{aligned}\frac{d\mu_e}{dt}+\varepsilon\frac{d(\delta\mu)}{dt} &= -\operatorname{ad}\left(\frac{\delta H}{\delta\mu_e}\right)^*\mu_e\\ &\quad-\frac{1}{2}\varepsilon\left[\operatorname{ad}\left(\frac{\delta(\delta^2H)}{\delta(\delta\mu)}\right)^*\mu_e-\operatorname{ad}\left(\frac{\delta H}{\delta\mu_e}\right)^*\delta\mu\right]+O(\varepsilon^2).\end{aligned}$$

Thus, the linearized equations are

$$\frac{d(\delta\mu)}{dt} = -\frac{1}{2}\operatorname{ad}\left(\frac{\delta(\delta^2 H)}{\delta(\delta\mu)}\right)^* \mu_e - \operatorname{ad}\left(\frac{\delta H}{\delta\mu_e}\right)^* \delta\mu. \tag{13.7.7}$$

If H is replaced by $H_C := H + C$, with the function C chosen to satisfy $\delta H_C/\delta\mu_e = 0$, we get $\operatorname{ad}(\delta H_C/\delta\mu_e)^*\delta\mu = 0$, and so

$$\frac{d(\delta\mu)}{dt} = -\frac{1}{2}\operatorname{ad}\left(\frac{\delta(\delta^2 H_C)}{\delta(\delta\mu)}\right)^* \mu_e. \tag{13.7.8}$$

Equation (13.7.8) is Hamiltonian with respect to the linearized Poisson bracket (see Example **(f)** of §10.2):

$$\{F, G\}(\mu) = \left\langle \mu_e, \left[\frac{\delta F}{\delta\mu}, \frac{\delta G}{\delta\mu}\right]\right\rangle. \tag{13.7.9}$$

Ratiu [1982] interprets this bracket in terms of a Lie-Poisson structure of a loop extension of $\mathfrak{g}$. The Poisson bracket (13.7.9) differs from the Lie-Poisson bracket (13.7.1) in that it is *constant* in μ. With respect to the Poisson bracket (13.7.9), Hamilton's equations given by $\delta^2 H_C$ are (13.7.8), as an easy verification shows. Note that the critical points of $\delta^2 H_C$ are stationary solutions of the linearized equation (13.7.8), that is, they are *neutral modes* for (13.7.8).

If $\delta^2 H_C$ is definite, then either $\delta^2 H_C$ or $-\delta^2 H_C$ is positive-definite and hence defines a norm on the space of perturbations $\delta\mu$ (which is $\mathfrak{g}^*$). Being twice the Hamiltonian function for (13.7.8), $\delta^2 H_C$ is conserved. So, any solution of (13.7.8) starting on an energy surface of $\delta^2 H_C$ (that is, on a sphere in this norm) stays on it and hence the zero solution of (13.7.8) is (Liapunov) stable. *Thus, formal stability,* (that is, $\delta^2 H_C$ definite) *implies linearized stability.* It should be noted, however, that the conditions for definiteness of $\delta^2 H_C$ are entirely different from the conditions for "normal mode stability," that is, that the operator acting on $\delta\mu$ given by (13.7.8) have a purely imaginary spectrum. In particular, having a purely imaginary spectrum for the linearized equation does *not* produce Liapunov stability of the linearized equations. The difference between $\delta^2 H_C$ and the operator in (13.7.8) can be made explicit, as follows. Assume that the pairing $\langle\,,\rangle$ identifies the dual $\mathfrak{g}^*$ with $\mathfrak{g}$ itself, that is, there is a weak Ad-invariant metric $\langle\langle\,,\rangle\rangle$ on $\mathfrak{g}$ and a linear operator $L : \mathfrak{g} \to \mathfrak{g}$ such that

$$\delta^2 H_C = \langle\langle \delta\mu, L\delta\mu \rangle\rangle\,; \tag{13.7.10}$$

L is symmetric with respect to the metric $\langle\langle\,,\rangle\rangle$, that is, $\langle\langle \xi, L\eta \rangle\rangle = \langle\langle L\xi, \eta \rangle\rangle$ for all $\xi, \eta \in \mathfrak{g}$. Then the linear operator in (13.7.8) becomes

$$\delta\mu \mapsto [L\delta\mu, \mu_e] \tag{13.7.11}$$

which, of course, differs from L, in general. However, note that the kernel of L is included in the kernel of the linear operator (13.7.11), that is, the zero eigenvalues of L give rise to "neutral modes" in the spectral analysis of (13.7.11). There is a remarkable coincidence of the zero-eigenvalue equations for these operators in fluid mechanics: for the Rayleigh equation describing plane-parallel shear flow in an inviscid homogeneous fluid, taking normal modes makes the zero-eigenvalue equations corresponding to L and to (13.7.11) coincide (see Abarbanel, Holm, Marsden, And Ratiu [1986]).

For additional applications of the stability method, see the Introduction and Holm, Marsden, Ratiu, and Weinstein [1985], Abarbanel and Holm [1987], Simo, Posbergh, and Marsden [1990, 1991], and Simo, Lewis, and Marsden [1991]. For a more general treatment of the linearization process, see Marsden, Ratiu, and Raugel [1991].

Some History of Lie-Poisson and Euler-Poincaré Equations

We continue with some comments on the history of Poisson structures that we began in §10.3. Recall that we pointed out how Lie, in his work up to 1890 on function groups, had many of the essential ideas of general Poisson manifolds and, in particular, had explicitly studied the Lie-Poisson bracket on duals of Lie algebras.

The theory developed so far in this chapter describes the adaptation of the concepts of Hamiltonian mechanics to the context of the duals of Lie algebras. This theory could easily have been given shortly after Lie's work, but evidently it was not observed for the rigid body or ideal fluids until the work of Pauli [1953], Martin [1959], Arnold [1966a], Ebin and Marsden [1970], Nambu [1973], and Sudarshan and Mukunda [1974], all of whom were apparently unaware of Lie's work on the Lie-Poisson bracket. It seems that even Elie Cartan was unaware of this aspect of Lie's work, which does seem surprising. Perhaps it is less surprising when one thinks for a moment about how many other things Cartan was involved in at the time. Nevertheless, one is struck by the amount of rediscovery and confusion in this subject. Evidently, this situation is not unique to mechanics.

Meanwhile, as Arnold [1988] and Chetaev [1989] pointed out, one can also write the equations directly on the Lie algebra, bypassing the Lie-Poisson equations on the dual. The resulting equations were first written down on a general Lie algebra by Poincaré [1901b]; we refer to these as the Euler-Poincaré equations. We shall develop them from a modern point of view in the next section. Poincaré [1910] goes on to study the effects of the deformation of the earth on its precession—he apparently recognizes the equations as Euler equations on a semidirect product Lie algebra. In general, the command that Poincaré had of the subject is most impressive, and is hard to match in his near contemporaries, except perhaps Riemann

[1860, 1861] and Routh [1877, 1884]. It is noteworthy that Poincaré [1901b] has no references, so it is rather hard to trace his train of thought or his sources; compare this style with that of Hamel [1904]! In particular, he gives no hint that he understood the work of Lie on the Lie-Poisson structure, but, of course, Poincaré understood the Lie group and the Lie algebra machine very well indeed.

Our derivation of the Euler-Poincaré equations in the next section is based on a reduction of variational principles, not on a reduction of the symplectic or Poisson structure, which is natural for the dual. We also show that the Lie-Poisson equations are related to the Euler-Poincaré equations by the "fiber derivative," in the same way as one gets from the ordinary Euler-Lagrange equations to the Hamilton equations. Even though this is relatively trivial, it does not appear to have been written down before. In the dynamics of ideal fluids, the resulting variational principle is related to what has been known as "Lin constraints" (see also Newcomb [1962] and Bretherton [1970].) This itself has an interesting history, going back to Ehrenfest, Boltzman, and Clebsch, but again, there was little if any contact with the heritage of Lie and Poincaré on the subject. One person who was well aware of the work of both Lie and Poincaré was Hamel.

How does Lagrange fit into this story? In *Mecanique Analytique*, Volume 2, equations A on page 212 are the Euler-Poincaré equations for the rotation group written out explicitly for a reasonably general Lagrangian. He eventually specializes them to the rigid body equations of course. We should remember that Lagrange also developed the key concept of the Lagrangian representation of fluid motion, but it is not clear that he understood that both systems are special instances of one theory. Lagrange spends a large number of pages on his derivation of the Euler-Poincaré equations for $SO(3)$, in fact, a good chunk of Volume 2. His derivation is not as clean as we would give today, but it seems to have the right spirit of a reduction method. That is, he tries to get the equations from the Euler-Lagrange equations on $TSO(3)$ by passing to the Lie algebra.

In view of the historical situation described above, one might argue that the term "Euler-Lagrange-Poincaré" equations is right for these equations. Since Poincaré noted the generalization to arbitrary Lie algebras, and applied it to interesting fluid problems, it is clear that his name belongs, but in light of other uses of the term "Euler-Lagrange," it seems that "Euler-Poincaré" is a reasonable choice.

Marsden and Scheurle [1993a,b] and Weinstein [1994] have studied a more general version of Lagrangian reduction whereby one drops the Euler-Lagrange equations from TQ to TQ/G. This is a nonabelian generalization of the classical Routh method, and leads to a very interesting coupling of the Euler-Lagrange and Euler-Poincaré equations that we shall briefly sketch in the next section. This problem was also studied by Hamel [1904] in connection with his work on nonholonomic systems (see Koiller [1992] and Bloch, Krishnaprasad, Marsden, and Murray [1994] for more information).

The current vitality of mechanics, including the investigation of fundamental questions, is quite remarkable, given its long history and development. This vitality comes about through rich interactions with both pure mathematics (from topology and geometry to group representation theory), and through new and exciting applications to areas like control theory. It is perhaps even more remarkable that absolutely fundamental points, such as a clear and unambiguous linking of Lie's work on the Lie-Poisson bracket on the dual of a Lie algebra and Poincaré's work on the Euler-Poincaré equations on the Lie algebra itself, with the most basic of examples in mechanics, such as the rigid body and the motion of ideal fluids, took nearly a century to complete. The attendant lessons to be learned about communication between pure mathematics and the other mathematical sciences are, hopefully, obvious.

Exercises

13.7-1 *Write out the linearized rigid body equations about an equilibrium explicitly.*

13.7-2 *Let $\mathfrak{g}$ be finite dimensional. Let $e_1, \ldots, e_n$ be a basis for $\mathfrak{g}$ and $e^1, \ldots, e^n$ a dual basis for $\mathfrak{g}^*$. Let $\mu = \mu_a e^a \in \mathfrak{g}^*$ and $H(\mu) = H(\mu_1, \ldots, \mu_n) : \mathfrak{g}^* \to \mathbb{R}$. Let $[\mu_a, \mu_b] = C^d_{ab}\mu_d$. Derive a co-ordinate expression for the linearized equations* (13.7.7)*:*

$$\frac{d(\delta\mu)}{dt} = -\frac{1}{2}\operatorname{ad}\left(\frac{\delta(\delta^2 H)}{\delta\mu}\right)^* \mu_e - \operatorname{ad}\left(\frac{\delta H}{\delta\mu_e}\right)^* \delta\mu.$$

13.8 The Euler-Poincaré Equations

To understand this section, it will be helpful to develop some more of the basics about rigid body dynamics from the Introduction (further details are given in Chaper 15). We regard an element $\mathbf{R} \in SO(3)$ giving the configuration of the body as a map of a reference configuration $\mathcal{B} \subset \mathbb{R}^3$ to the current configuration $\mathbf{R}(\mathcal{B})$; the map $\mathbf{R}$ takes a reference or label point $X \in \mathcal{B}$ to a current point $x = \mathbf{R}(X) \in \mathbf{R}(\mathcal{B})$. When the rigid body is in motion, the matrix $\mathbf{R}$ is time-dependent and the velocity of a point of the body is $\dot{x} = \dot{\mathbf{R}}X = \dot{\mathbf{R}}\mathbf{R}^{-1}x$. Since $\mathbf{R}$ is an orthogonal matrix, $\mathbf{R}^{-1}\dot{\mathbf{R}}$ and $\dot{\mathbf{R}}\mathbf{R}^{-1}$ are skew matrices, and so we can write

$$\dot{x} = \dot{\mathbf{R}}\mathbf{R}^{-1}x = \omega \times x, \tag{13.8.1}$$

which defines the ***spatial angular velocity vector*** ω. Thus, ω is essentially given by *right* translation of $\dot{\mathbf{R}}$ to the identity.

The corresponding body angular velocity is defined by

$$\Omega = \mathbf{R}^{-1}\omega, \tag{13.8.2}$$

so that Ω is the angular velocity relative to a body fixed frame. Notice that

$$\begin{aligned} \mathbf{R}^{-1}\dot{\mathbf{R}}X &= \mathbf{R}^{-1}\dot{\mathbf{R}}\mathbf{R}^{-1}x = \mathbf{R}^{-1}(\omega \times x) \\ &= \mathbf{R}^{-1}\omega \times \mathbf{R}^{-1}x = \Omega \times X \end{aligned} \tag{13.8.3}$$

so that Ω is given by *left* translations of $\dot{\mathbf{R}}$ to the identity. The kinetic energy is obtained by summing up $m\|\dot{x}\|^2/2$ over the body:

$$K = \frac{1}{2}\int_{\mathcal{B}} \rho(X)\|\dot{\mathbf{R}}X\|^2\, d^3X, \tag{13.8.4}$$

where ρ is a given mass density in the reference configuration. Since

$$\|\dot{\mathbf{R}}X\| = \|\omega \times x\| = \|\mathbf{R}^{-1}(\omega \times x)\| = \|\Omega \times X\|,$$

K is a quadratic function of Ω. Writing

$$K = \frac{1}{2}\Omega^T \mathbb{I}\Omega \tag{13.8.5}$$

defines the ***moment of inertia tensor*** $\mathbb{I}$, which, if the body does not degenerate to a line, is a positive-definite (3×3)-matrix, or better, a quadratic form. This quadratic form can be diagonalized, and this defines the principal axes and moments of inertia. In this basis, we write $\mathbb{I} = \operatorname{diag}(I_1, I_2, I_3)$. The function K is taken to be the Lagrangian of the system on $TSO(3)$ (and by means of the Legendre transformation we get the corresponding Hamiltonian description on $T^*SO(3)$). Notice directly from (13.8.4) that K is *left* (not right) invariant on $TSO(3)$. It follows that the corresponding Hamiltonian is also *left* invariant.

From the Lagrangian point of view, the relation between the motion in $\mathbf{R}$ space and that in body angular velocity (or Ω) space is as follows:

Theorem 13.8.1 *The curve $\mathbf{R}(t) \in SO(3)$ satisfies the Euler-Lagrange equations for*

$$L(\mathbf{R}, \dot{\mathbf{R}}) = \frac{1}{2}\int_{\mathcal{B}} \rho(X)\|\dot{\mathbf{R}}X\|^2\, d^3X, \tag{13.8.6}$$

if and only if $\Omega(t)$ defined by $\mathbf{R}^{-1}\dot{\mathbf{R}}v = \Omega \times v$ for all $v \in \mathbb{R}^3$ satisfies Euler's equations

$$\mathbb{I}\dot{\Omega} = \mathbb{I}\Omega \times \Omega. \tag{13.8.7}$$

One instructive way to prove this *indirectly* is to pass to the Hamiltonian formulation and use Lie-Poisson reduction, as outlined above. One way to do it *directly* is to use variational principles. By Hamilton's principle, $R(t)$ satisfies the Euler-Lagrange equations if and only if

$$\delta \int L\, dt = 0.$$

Let $l(\Omega) = \frac{1}{2}(\mathbb{I}\Omega)\cdot\Omega$, so that $l(\Omega) = L(\mathbf{R}, \dot{\mathbf{R}})$ if $\mathbf{R}$ and Ω are related as above. To see how we should transform Hamilton's principle, we differentiate the relation $\mathbf{R}^{-1}\dot{\mathbf{R}} = \hat{\Omega}$ with respect to $\mathbf{R}$ to get

$$- \mathbf{R}^{-1}(\delta\mathbf{R})\mathbf{R}^{-1}\dot{\mathbf{R}} + \mathbf{R}^{-1}(\delta\dot{\mathbf{R}}) = \widehat{\delta\Omega}. \tag{13.8.8}$$

Let the skew matrix $\hat{\Sigma}$ be defined by

$$\hat{\Sigma} = \mathbf{R}^{-1}\delta\mathbf{R} \tag{13.8.9}$$

and define the vector Σ by

$$\hat{\Sigma}v = \Sigma \times v. \tag{13.8.10}$$

Note that

$$\dot{\hat{\Sigma}} = -\mathbf{R}^{-1}\dot{\mathbf{R}}\mathbf{R}^{-1}\delta\mathbf{R} + \mathbf{R}^{-1}\delta\dot{\mathbf{R}},$$

so

$$\mathbf{R}^{-1}\delta\dot{\mathbf{R}} = \dot{\hat{\Sigma}} + \mathbf{R}^{-1}\dot{\mathbf{R}}\hat{\Sigma} \tag{13.8.11}$$

substituting (13.8.11) and (13.8.9) into (13.8.8) gives

$$-\hat{\Sigma}\hat{\Omega} + \dot{\hat{\Sigma}} + \hat{\Omega}\hat{\Sigma} = \widehat{\delta\Omega},$$

that is,

$$\widehat{\delta\Omega} = \dot{\hat{\Sigma}} + [\hat{\Omega}, \hat{\Sigma}]. \tag{13.8.12}$$

The identity $[\hat{\Omega}, \hat{\Sigma}] = (\Omega \times \Sigma)\hat{}$ holds by Jacobi's identity for the cross product, and so

$$\delta\Omega = \dot{\Sigma} + \Omega \times \Sigma. \tag{13.8.13}$$

These calculations prove the following:

Theorem 13.8.2 *Hamilton's variational principle*

$$\delta \int_a^b L\, dt = 0 \tag{13.8.14}$$

on $TSO(3)$ is equivalent to the ***reduced variational principle***

$$\delta \int_a^b l\, dt = 0 \tag{13.8.15}$$

on $\mathbb{R}^3$ where the variations $\delta\Omega$ are of the form (13.8.13) with $\Sigma(a) = \Sigma(b) = 0$.

To complete the proof of Theorem **13.8.1**, it suffices to work out the equations equivalent to the reduced variational principle (13.8.15). Since $l(\Omega) = \frac{1}{2}\langle \mathbb{I}\Omega, \Omega \rangle$, and $\mathbb{I}$ is symmetric, we get

$$\begin{aligned}
\delta \int_a^b l\, dt &= \int_a^b \langle \mathbb{I}\Omega, \delta\Omega \rangle\, dt \\
&= \int_a^b \langle \mathbb{I}\Omega, \dot{\Sigma} + \Omega \times \Sigma \rangle\, dt \\
&= \int_a^b \left[\left\langle -\frac{d}{dt}\mathbb{I}\Omega, \Sigma \right\rangle + \langle \mathbb{I}\Omega, \Omega \times \Sigma \rangle \right] \\
&= \int_a^b \left\langle -\frac{d}{dt}\mathbb{I}\Omega + \mathbb{I}\Omega \times \Omega, \Sigma \right\rangle dt,
\end{aligned}$$

where we have integrated by parts and used the boundary conditions $\Sigma(b) = \Sigma(a) = 0$. Since Σ is otherwise arbitrary, (13.8.15) is equivalent to

$$-\frac{d}{dt}(\mathbb{I}\Omega) + \mathbb{I}\Omega \times \Omega = 0,$$

which are Euler's equations. ■

We now generalize this procedure to an arbitrary Lie group and later will make the direct link with the Lie-Poisson equations.

Theorem 13.8.3 *Let G be a Lie group and let $L : TG \to \mathbb{R}$ be a left invariant Lagrangian. Let $l : \mathfrak{g} \to \mathbb{R}$ be its restriction to the identity. For a curve $g(t) \in G$, let $\xi(t) = g(t)^{-1} \cdot \dot{g}(t)$; i.e., $\xi(t) = T_{g(t)}L_{g(t)^{-1}}\dot{g}(t)$. Then the following are equivalent:*

i *$g(t)$ satisfies the Euler-Lagrange equations for L on G;*

ii *the variational principle*

$$\delta \int L(g(t), \dot{g}(t))\, dt = 0 \tag{13.8.16}$$

holds, for variations with fixed endpoints;

iii *the **Euler-Poincaré equations** hold:*

$$\frac{d}{dt}\frac{\delta l}{\delta \xi} = \operatorname{ad}^*_\xi \frac{\delta l}{\delta \xi}; \tag{13.8.17}$$

iv *the variational principle*

$$\delta \int l(\xi(t))\, dt = 0 \tag{13.8.18}$$

holds on $\mathfrak{g}$, using variations of the form

$$\delta\xi = \dot{\eta} + [\xi, \eta], \tag{13.8.19}$$

where η vanishes at the endpoints.

Proof First of all, the equivalence of **i** and **ii** holds on the tangent bundle of any configuration manifold Q, as we know from Chapter 8. To see that **ii** and **iv** are equivalent, one needs to compute the variations $\delta\xi$ induced on $\xi = g^{-1}\dot{g} = TL_{g^{-1}}\dot{g}$ by a variation of g. We will do this for matrix groups; see Bloch, Krishnaprasad, Marsden, and Ratiu [1994b] for the general case. To calculate this, we need to differentiate $g^{-1}\dot{g}$ in the direction of a variation δg. If $\delta g = dg/d\epsilon$ at $\epsilon = 0$, where g is extended to a curve g_ϵ, then,

$$\delta\xi = \frac{d}{d\epsilon} g^{-1} \frac{d}{dt} g,$$

while if $\eta = g^{-1}\delta g$, then

$$\dot{\eta} = \frac{d}{dt} g^{-1} \frac{d}{d\epsilon} g.$$

The difference $\delta\xi - \dot{\eta}$ is thus the commutator $[\xi, \eta]$.

To complete the proof, we show the equivalence of **iii** and **iv**. Indeed, using the definitions and integrating by parts,

$$\begin{aligned}
\delta \int l(\xi) dt &= \int \frac{\delta l}{\delta \xi} \delta\xi \, dt \\
&= \int \frac{\delta l}{\delta \xi} (\dot{\eta} + \mathrm{ad}_\xi \eta) \, dt \\
&= \int \left[-\frac{d}{dt} \left(\frac{\delta l}{\delta \xi} \right) + \mathrm{ad}_\xi^* \frac{\delta l}{\delta \xi} \right] \eta \, dt
\end{aligned}$$

so the result follows. ■

There is of course a right invariant version of this theorem in which $\xi = \dot{g}g^{-1}$ and when (13.8.17), (13.8.19) aquire minus signs.

In coordinates, (13.8.17), reads as follows

$$\frac{d}{dt} \frac{\partial l}{\partial \xi^a} = C^b_{da} \xi^d \frac{\partial l}{\partial x^b}. \tag{13.8.20}$$

Since the Euler-Lagrange and Hamilton equations on TQ and T^*Q are equivalent, it follows that the Lie-Poisson and Euler-Poincaré equations are also equivalent. To see this *directly*, we make the following Legendre transformation from $\mathfrak{g}$ to $\mathfrak{g}^*$:

$$\mu = \frac{\delta l}{\delta \xi}, \quad h(\mu) = \langle \mu, \xi \rangle - l(\xi).$$

Note that

$$\frac{\delta h}{\delta \mu} = \xi + \left\langle \mu, \frac{\delta \xi}{\delta \mu} \right\rangle - \left\langle \frac{\delta l}{\delta \xi}, \frac{\delta \xi}{\delta \mu} \right\rangle = \xi$$

and so it is now clear that the Lie-Poisson and Euler-Poincaré equations are equivalent.

We close this section by showing that the periodic KdV equation, (see example **(c)** in §3.2)

$$u_t + 6uu_x + u_{xxx} = 0$$

is an Euler-Poincaré equation on a certain Lie algebra called the ***Virasoro algebra*** $\mathfrak{v}$. These results were obtained in the Lie-Poisson context by Gelfand and Dorfman [1979], Kirillov [1981], Ovsienko and Khesin [1987], and Segal [1991]. See also Pressley and Segal [1986] and references therein.

We begin with the construction of the Virasoro algebra $\mathfrak{v}$. If one identifies elements of $\mathfrak{X}(S^1)$ with periodic functions of period 1 endowed with the Jacobi-Lie bracket

$$[u, v] = uv' - u'v,$$

the ***Gelfand-Fuchs cocycle*** is defined by the expression

$$\Sigma(u, v) = \gamma \int_0^1 u'(x)v''(x)dx,$$

where $\gamma \in \mathbb{R}$ is an arbitrary constant. The Lie algebra $\mathfrak{X}(S^1)$ of vector fields on the circle has a unique central extension by $\mathbb{R}$ determined by the Gelfand-Fuchs cocycle. Therefore (see (12.4.22) in remark 5 of §12.4) the Lie algebra bracket on

$$\mathfrak{v} := \{(u, a) \mid u \in \mathfrak{X}(S^1), \quad a \in \mathbb{R}\}$$

is given by

$$[(u, a), (v, b)] = \left(-uv' + u'v, \gamma \int_0^1 u'(x)v''(x)\, dx \right)$$

since the *left* Lie bracket on $\mathfrak{X}(S^1)$ is given by the negative of the Jacobi-Lie bracket for vector fields. Identify the dual of $\mathfrak{v}$ with $\mathfrak{v}$ by the L^2-inner product

$$\langle (u, a), (v, b) \rangle = ab + \int_0^1 u(x)v(x)\, dx.$$

We claim that the coadjoint action $\mathrm{ad}^*_{(u,a)}$ is given by

$$\mathrm{ad}^*_{(u,a)}(v, b) = (b\gamma u''' + 2u'v + uv', 0).$$

Indeed, if $(u, a), (v, b), (w, c) \in \mathfrak{v}$, we have

$$\begin{aligned}
&\left\langle \operatorname{ad}^*_{(u,a)}(v, b), (w, c)\right\rangle \\
&\quad= \langle (v, b), [(u, a), (w, c)]\rangle \\
&\quad= \left\langle (v, b), \left(-uw' + u'w, \gamma \int_0^1 u'(x)w''(x)\, dx\right)\right\rangle \\
&\quad= b\gamma \int_0^1 u'(x)w''(x)\, dx - \int_0^1 v(x)u(x)w'(x)\, dx + \int_0^1 v(x)u'(x)w(x)\, dx.
\end{aligned}$$

Integrating the first term twice and the second term once by parts and remembering that the boundary terms vanish by periodicity, this expresison becomes

$$\begin{aligned}
b\gamma \int_0^1 u'''(x)w(x)dx &+ \int_0^1 (v(x)u(x))'w(x)dx + \int_0^1 v(x)u'(x)w(x)dx \\
&= \int_0^1 (b\gamma u'''(x) + 2u'(x)v(x) + u(x)v'(x))w(x)dx \\
&= \langle (b\gamma u''' + 2u'v + uv', 0), (w, c)\rangle .
\end{aligned}$$

If $F : \mathfrak{v} \to \mathbb{R}$, its functional derivative relative to the L^2-pairing is given by

$$\frac{\delta F}{\delta(u, a)} = \left(\frac{\delta F}{\delta u}, \frac{\partial F}{\partial a}\right)$$

where $\delta F/\delta u$ is the usual L^2-functional derivative of F keeping $a \in \mathbb{R}$ fixed and $\partial F/\partial a$ is the standard partial derivative of F keeping u fixed. The Euler-Poincaré equations for right invariant systems given by $l : \mathfrak{v} \to \mathbb{R}$ becomes

$$\frac{d}{dt}\frac{\delta l}{\delta(u, a)} = -\operatorname{ad}^*_{(u,a)}\frac{\delta l}{\delta(u, a)}.$$

However,

$$\begin{aligned}
\operatorname{ad}^*_{(u,a)}\frac{\delta l}{\delta(u, a)} &= \operatorname{ad}^*_{(u,a)}\left(\frac{\delta l}{\delta u}, \frac{\partial l}{\partial a}\right) \\
&= \left(\gamma\frac{\partial l}{\partial a}u''' + 2u'\frac{\delta l}{\delta u} + u\left(\frac{\delta l}{\delta u}\right)', 0\right),
\end{aligned}$$

so that we get the system

$$\begin{aligned}
\frac{d}{dt}\frac{\partial l}{\partial a} &= 0 \\
\frac{d}{dt}\frac{\delta l}{\delta u} &= -\gamma\frac{\partial l}{\partial a}u''' - 2u'\frac{\delta l}{\delta u} - u\left(\frac{\delta l}{\delta u}\right)'.
\end{aligned}$$

If

$$l(u,a) = \frac{1}{2}\left(a^2 + \int_0^1 u^2(x)\,dx\right),$$

then $\partial l/\partial a = a, \delta l/\delta u = u$ and the above equations become

$$\begin{aligned} \frac{da}{dt} &= 0 \\ \frac{du}{dt} &= -\gamma a u''' - 3u'u. \end{aligned} \tag{13.8.21}$$

Since a is constant, we get

$$u_t + 3u_x u + \gamma a u''' = 0. \tag{13.8.22}$$

This equation is equivalent to the KdV equation upon rescaling time and choosing the constant a appropriately. Indeed, let $u(t,x) = v(\tau(t),x)$ for $\tau(t) = t/2$. Then $u_x = v_x$ and $u_t = -v_\tau/2$ so that (13.8.22) becomes

$$v_\tau + 6vv_x + 2\gamma a v_{xxx} = 0,$$

which becomes the KdV equation (see §3.2) if we choose $a = 1/2\gamma$.

The Lie-Poisson formulation goes the following way. The (+) Lie-Poisson bracket is given by

$$\begin{aligned} \{f,h\}(u,a) &= \left\langle (u,a), \left[\frac{\delta f}{\delta(u,a)}, \frac{\delta h}{\delta(u,a)}\right]\right\rangle \\ &= \int \left[u\left(\left(\frac{\delta f}{\delta u}\right)'\frac{\delta h}{\delta u} - \frac{\delta f}{\delta u}\left(\frac{\delta h}{\delta u}\right)'\right)\right. \\ &\qquad \left. + a\gamma\left(\frac{\delta f}{\delta u}\right)'\left(\frac{\delta h}{\delta u}\right)''\right] dx \end{aligned}$$

so that the Lie-Poisson equations $\dot{f} = \{f,h\}$ become

$$\begin{aligned} \frac{da}{dt} &= 0 \\ \frac{du}{dt} &= -u'\left(\frac{\delta h}{\delta u}\right) - 2u\left(\frac{\delta h}{\delta u}\right)' - a\gamma\left(\frac{\delta h}{\delta u}\right)'''. \end{aligned} \tag{13.8.23}$$

Taking

$$h(u,a) = \frac{1}{2}a^2 + \frac{1}{2}\int_0^1 u^2(x)\,dx,$$

we get $\partial h/\partial a = a, \delta h/\delta u = u$ and so (13.8.23) becomes (13.8.22) as was to be expected and could have directly obtained by a Legendre transform.

The conclusion is that the KdV equation is the expression in space coordinates of the geodesic equations on the Virasoro group V endowed with the right invariant metric whose value at the identity is the L^2-inner product. We shall not describe here the Virasoro group which is a central extension of the diffeomorphism group on S^1; we refer the reader to Pressley and Segal [1986].

Exercise

13.8-1 *Verify the coordinate form of the Euler-Poincaré equations.*

13.8-2 *Show that the Euler equations for a perfect fluid are Euler-Poincaré equations. Find the variational principle* (13.8.3) *in Newcomb* [1962] *and Bretherton* [1970].

13.9 The Reduced Euler-Lagrange Equations

As we have mentioned, the Lie-Poisson and Euler-Poincaré equations occur for many systems besides the rigid body equations. They include the equations of fluid and plasma dynamics, for example. For many other systems, such as a rotating molecule or a spacecraft with movable internal parts, one has a combination of equations of Euler-Poincaré type and Euler-Lagrange type. Indeed, on the Hamiltonian side, this process has undergone development for quite some time, and is discussed at length in Volume II. On the Lagrangian side, this process is also very interesting, and has been recently developed by, amongst others, Marsden and Scheurle [1993a,b]. The general problem is to drop Euler-Lagrange equations and variational principles from a general velocity phase-space TQ to the quotient TQ/G by a Lie group action of G on Q. If L is a G-invariant Lagrangian on TQ, it induces a reduced Lagrangian l on TQ/G. We give a brief *preview* of the general theory in this section. In fact, the material below can also act as motivation for the general theory of connections, also introduced in Volume II.

An important ingredient in this work is to introduce a connection A on the principal bundle $Q \to S = Q/G$, assuming that this quotient is nonsingular. For example, the mechanical connection (see Kummer [1981], Marsden [1992] and references therein), may be chosen for A. This connection allows one to split the variables into a horizontal and vertical part. Let x^α, also called "internal variables," be coordinates for shape-space Q/G, let η^a be coordinates for the Lie algebra $\mathfrak{g}$ relative to a chosen basis, let l be the Lagrangian regarded as a function of the variables $x^\alpha, \dot{x}^\alpha, \eta^a$, and let C^a_{db} be the structure constants of the Lie algebra $\mathfrak{g}$ of G.

If one writes the Euler-Lagrange equations on TQ in a local principal bundle trivialization, with coordinates x^α on the base and η^a in the fiber, then one gets the following system of ***Hamel equations***:

$$\frac{d}{dt}\frac{\partial l}{\partial \dot{x}^\alpha} - \frac{\partial l}{\partial x^\alpha} = 0, \tag{13.9.1}$$

$$\frac{d}{dt}\frac{\partial l}{\partial \eta^b} - \frac{\partial l}{\partial \eta^a} C^a_{db}\eta^d = 0. \tag{13.9.2}$$

However, this representation of the equations does not make global intrinsic sense (unless $Q \to S$ admits a global flat connection). The introduction of a connection overcomes this and one can intrinsically and globally split the original variational principle relative to horizontal and vertical variations. One gets from one-form to the other by means of the velocity shift given by replacing η by the vertical part relative to the connection

$$\xi^a = A^a_\alpha \dot{x}^\alpha + \eta^a.$$

Here, A^d_α are the local coordinates of the connection A. This change of coordinates is motivated from the mechanical point of view since the variables ξ have the interpretation of the locked angular velocity. The resulting ***reduced Euler-Lagrange equations*** have the following form:

$$\frac{d}{dt}\frac{\partial l}{\partial \dot{x}^\alpha} - \frac{\partial l}{\partial x^\alpha} = \frac{\partial l}{\partial \xi^a}\left(B^a_{\alpha\beta}\dot{x}^\beta + B^a_{\alpha d}\xi^d\right), \tag{13.9.3}$$

$$\frac{d}{dt}\frac{\partial l}{\partial \xi^b} = \frac{\partial l}{\partial \xi^a}(B^a_{b\alpha}\dot{x}^\alpha + C^a_{db}\xi^d). \tag{13.9.4}$$

In these equations, $B^a_{\alpha\beta}$ are the coordinates of the curvature B of A, $B^a_{d\alpha} = C^a_{bd}A^b_\alpha$ and $B^a_{b\alpha} = -B^a_{\alpha b}$.

It is interesting to note that the matrix

$$\begin{bmatrix} B^a_{\alpha\beta} & B^a_{\alpha d} \\ B^a_{d\alpha} & C^a_{bd} \end{bmatrix}$$

is itself the curvature of the connection regarded as residing on the bundle $TQ \to TQ/G$. Regarding the structure constants as a curvature tensor in the special case $Q = G$ may be regarded as a reformulation of the Mauer-Cartan equations (see Theorem **9.1.11**).

The variables ξ^a may be regarded as the rigid part of the variables on the original configuration space, while x^α are the internal variables. As in Simo, Lewis, and Marsden [1991], the division of variables into internal and rigid parts has deep implications for both stability theory and for bifurcation theory, again, continuing along lines developed originally by Riemann, Poincaré, and others. The main way this new insight is achieved is through a careful split of the variables, using the (mechanical) connection as one of the main ingredients. This split puts the second variation of the augmented Hamiltonian at a relative equilibrium as well as the symplectic form into "normal form." It is somewhat remarkable that they are *simultaneously* put into a simple form. This link helps considerably with an eigenvalue analysis of the linearized equations, and in Hamiltonian bifurcation theory; see, for example, Bloch, Krishnaprasad, Marsden, and Ratiu [1994a].

One of the key results in Hamiltonian reduction theory says that the reduction of a cotangent bundle T^*Q by a symmetry group G is a bundle over T^*S, where $S = Q/G$ is shape-space, and where the fiber is either $\mathfrak{g}^*$, the dual of the Lie algebra of G, or is a coadjoint orbit, depending on whether one is doing Poisson or symplectic reduction. We refer to Montgomery, Marsden, and Ratiu [1984] and Marsden [1992] and Volume II for details and references. The reduced Euler-Lagrange equations give the analogue of this structure on the tangent bundle.

Remarkably, equations (13.9.3) are very close in form to the equations for a mechanical system with classical nonholonomic velocity constraints (see Naimark and Fufaev [1972] and Koiller [1992].) The connection chosen in that case is the one-form that determines the constraints. This link is made precise in Bloch, Krishnaprasad, Marsden, and Murray [1994]. In addition, this structure appears in several control problems, especially the problem of stabilizing controls considered by Bloch, Krishnaprasad, Marsden, and Sanchez [1992].

For systems with a momentum map $\mathbf{J}$ constrained to a specific value μ, the key to the construction of a reduced Lagrangian system is the modification of the Lagrangian L to the Routhian R^μ, which is obtained from the Lagrangian by subtracting off the mechanical connection paired with the constraining value μ of the momentum map. On the other hand, a basic ingredient needed for the reduced Euler-Lagrange equations is a velocity shift in the Lagrangian, the shift being determined by the connection, so this velocity-shifted Lagrangian plays the role that the Routhian does in the constrained theory.

14
Coadjoint Orbits

In this chapter we prove, amongst other things, that *the coadjoint orbits of a Lie group are symplectic manifolds*. These symplectic manifolds are, in fact, the symplectic leaves for the Lie-Poisson bracket. This result was developed and used by Kirillov, Arnold, Kostant, and Souriau in the early to mid-1960s, although it had important roots going back to the work of Lie, Borel, and Weil. (See Kirillov [1962, 1976b], Arnold [1966a], Kostant [1970], and Souriau [1969].) Here we give a direct proof. In Volume II we shall see a more "natural" proof using reduction.

Recall from Chapter 9 that the ***adjoint representation*** of a Lie group G is defined by $\mathrm{Ad}_g = T_e I_g : \mathfrak{g} \to \mathfrak{g}$ where $I_g : G \to G$ is the inner automorphism $I_g(h) = ghg^{-1}$. The ***coadjoint action*** is given by $\mathrm{Ad}^*_{g^{-1}} : \mathfrak{g}^* \to \mathfrak{g}^*$, where $\mathrm{Ad}^*_{g^{-1}}$ is the dual of the linear map $\mathrm{Ad}_{g^{-1}}$, that is, it is defined by

$$\langle \mathrm{Ad}^*_{g^{-1}}(\mu), \xi \rangle = \langle \mu, \mathrm{Ad}_{g^{-1}}(\xi) \rangle,$$

where $\mu \in \mathfrak{g}^*, \xi \in \mathfrak{g}$, and $\langle \, , \rangle$ denotes the pairing between $\mathfrak{g}^*$ and $\mathfrak{g}$. The ***coadjoint orbit***, $\mathrm{Orb}(\mu)$, through $\mu \in \mathfrak{g}^*$ is the subset of $\mathfrak{g}^*$ defined by

$$\mathrm{Orb}(\mu) := \{\mathrm{Ad}^*_{g^{-1}}(\mu) \mid g \in G\} := G \cdot \mu.$$

Like the orbit of any group action, $\mathrm{Orb}(\mu)$ *is an immersed submanifold of* $\mathfrak{g}^*$ and *if G is compact,* $\mathrm{Orb}(\mu)$ *is a closed embedded submanifold.*

14.1 Examples of Coadjoint Orbits

(a) Rotation Group As we saw in §9.3, the adjoint action for $SO(3)$ is

$$\mathrm{Ad}_{\mathrm{A}}(\mathbf{v}) = \mathbf{A}\mathbf{v}, \quad \text{where} \quad \mathbf{A} \in SO(3) \quad \text{and} \quad \mathbf{v} \in \mathbb{R}^3 \cong \mathfrak{so}(3).$$

Identify $\mathfrak{so}(3)^*$ with $\mathbb{R}^3$ by the usual dot product, that is, if $\mathbf{\Pi}, \mathbf{v} \in \mathbb{R}^3$, we have $\langle \mathbf{\Pi}, \hat{\mathbf{v}} \rangle = \mathbf{\Pi} \cdot \mathbf{v}$. Thus, for $\mathbf{\Pi} \in \mathfrak{so}(3)^*$ and $\mathbf{A} \in SO(3)$,

$$\begin{aligned} \langle \mathrm{Ad}^*_{\mathrm{A}^{-1}}(\mathbf{\Pi}), \hat{\mathbf{v}} \rangle &= \langle \mathbf{\Pi}, \mathrm{Ad}_{\mathrm{A}^{-1}}(\hat{\mathbf{v}}) \rangle = \langle \mathbf{\Pi}, (\mathbf{A}^{-1}\hat{\mathbf{v}}) \rangle = \mathbf{\Pi} \cdot \mathbf{A}^{-1}\mathbf{v} \\ &= (\mathbf{A}^{-1})^T \mathbf{\Pi} \cdot \mathbf{v} = \mathbf{A}\mathbf{\Pi} \cdot \mathbf{v} \end{aligned} \tag{14.1.1}$$

since $\mathbf{A}$ is orthogonal. Hence, with $\mathfrak{so}(3)^*$ identified with $\mathbb{R}^3$, $\mathrm{Ad}^*_{\mathrm{A}^{-1}} = \mathbf{A}$, and so

$$\mathrm{Orb}(\mathbf{\Pi}) = \{ \mathrm{Ad}^*_{\mathrm{A}^{-1}}(\mathbf{\Pi}) \mid \mathbf{A} \in SO(3) \} = \{ \mathbf{A}\mathbf{\Pi} \mid \mathbf{A} \in SO(3) \}, \tag{14.1.2}$$

which is the sphere in $\mathbb{R}^3$ of radius $\|\mathbf{\Pi}\|$.

(b) Affine Group on $\mathbb{R}$ Consider the Lie group of transformations of $\mathbb{R}$ of the form $T(x) = ax + b$ where $a \neq 0$. Identify G with the set of pairs $(a, b) \in \mathbb{R}^2$ with $a \neq 0$. Since

$$(T_1 \circ T_2)(x) = a_1(a_2 x + b_2) + b_1 = a_1 a_2 x + a_1 b_2 + b_1$$

and

$$T^{-1}(x) = \frac{1}{a}(x - b),$$

we take group multiplication to be

$$(a_1, b_1) \cdot (a_2, b_2) = (a_1 a_2, a_1 b_2 + b_1). \tag{14.1.3}$$

The inverse of (a, b) is

$$(a, b)^{-1} = \left(\frac{1}{a}, -\frac{b}{a} \right) \tag{14.1.4}$$

and the identity element is $(1, 0)$. Thus G is a two-dimensional Lie group. It is an example of a *semidirect product*. (See Exercise **9.3-1**.) As a set, the Lie algebra of G is $\mathfrak{g} = \mathbb{R}^2$; to compute the bracket on $\mathfrak{g}$ we shall first compute the adjoint representation. The inner automorphisms are given by

$$\begin{aligned} I_{(a,b)}(c, d) &= (a, b) \cdot (c, d) \cdot (a, b)^{-1} \\ &= (ac, ad + b) \cdot \left(\frac{1}{a}, -\frac{b}{a} \right) \\ &= (c, ad - bc + b), \end{aligned} \tag{14.1.5}$$

and so differentiating (14.1.5) with respect to (c, d) at the identity in the direction of $(u, v) \in \mathfrak{g}$, gives

$$\mathrm{Ad}_{(a,b)}(u, v) = (u, av - bu). \tag{14.1.6}$$

Differentiating (14.1.6) with respect to (a, b) in the direction (r, s) gives the Lie bracket

$$[(r, s), (u, v)] = (0, rv - su). \tag{14.1.7}$$

The adjoint orbit through (u, v) is $\{u\} \times \mathbb{R}$ if $(u, v) \neq (0, 0)$ and is $\{(0, 0)\}$ if $(u, v) = (0, 0)$. The *adjoint orbit* $\{u\} \times \mathbb{R}$ cannot be symplectic, as it is one dimensional. To compute the *coadjoint orbits*, denote elements of $\mathfrak{g}^*$ by $\begin{pmatrix} \alpha \\ \beta \end{pmatrix}$ and the pairing

$$\left\langle (u, v), \begin{pmatrix} \alpha \\ \beta \end{pmatrix} \right\rangle = \alpha u + \beta v \tag{14.1.8}$$

identifies $\mathfrak{g}^*$ with $\mathbb{R}^2$. Then

$$\begin{aligned} \left\langle \mathrm{Ad}^*_{(a,b)} \begin{pmatrix} \alpha \\ \beta \end{pmatrix}, (u, v) \right\rangle &= \left\langle \begin{pmatrix} \alpha \\ \beta \end{pmatrix}, \mathrm{Ad}_{(a,b)}(u, v) \right\rangle \\ &= \left\langle \begin{pmatrix} \alpha \\ \beta \end{pmatrix}, (u, av - bu) \right\rangle \\ &= \alpha u + \beta a v - \beta b u. \end{aligned} \tag{14.1.9}$$

Thus,

$$\mathrm{Ad}^*_{(a,b)} \begin{pmatrix} \alpha \\ \beta \end{pmatrix} = \begin{pmatrix} \alpha - \beta b \\ \beta a \end{pmatrix}. \tag{14.1.10}$$

If $\beta = 0$, the coadjoint orbit through $\begin{pmatrix} \alpha \\ \beta \end{pmatrix}$ is a single point. If $\beta \neq 0$, the orbit through $\begin{pmatrix} \alpha \\ \beta \end{pmatrix}$ is $\mathbb{R}^2$ minus the β-axis.

(c) Orbits in $\mathfrak{X}^*_{\mathrm{div}}$ Let $G = \mathrm{Diff}_{\mathrm{vol}}(\Omega)$, the group of volume-preserving diffeomorphisms of a region Ω in $\mathbb{R}^n$, with Lie algebra $\mathfrak{X}_{\mathrm{div}}(\Omega)$. In Example **(d)** of §10.2 we identified $\mathfrak{X}^*_{\mathrm{div}}(\Omega)$ with $\mathfrak{X}_{\mathrm{div}}(\Omega)$ by using the L^2-pairing on vector fields. Here we begin by finding a different representative of the dual $\mathfrak{X}^*_{\mathrm{div}}(\Omega)$, which is more convenient for explicitly determining the coadjoint action. Then we return to the identification above and will find the expression for the coadjoint action on $\mathfrak{X}_{\mathrm{div}}(\Omega)$; it will turn out to be more complicated.

The main technical ingredient used below is the Hodge decomposition theorem for manifolds with boundary. Here we state only the relevant facts to be used below. A k-form α is said to be ***tangent*** to $\partial\Omega$ if $i^*(*\alpha) = 0$. Let $\Omega_t^k(\Omega)$ denote all k-forms on M which are tangent to $\partial\Omega$. One of the Hodge decomposition theorems states that there is an L^2-orthogonal decomposition

$$\Omega^k(\Omega) = \mathbf{d}\Omega^{k-1}(\Omega) \oplus \{\alpha \in \Omega_t^k(\Omega) \mid \delta\alpha = 0\}.$$

This implies that the pairing

$$\langle\, , \rangle : \{\alpha \in \Omega_t^1(\Omega) \mid \delta\alpha = 0\} \times \mathfrak{X}_{\mathrm{div}}(\Omega) \to \mathbb{R}$$

given by

$$\langle M, X\rangle = \int_\Omega M_i X^i d^n x. \tag{14.1.11}$$

is weakly nondegenerate. Indeed, if

$$M \in \{\alpha \in \Omega_t^1(M) \mid \delta\alpha = 0\}$$

and $\langle M, X\rangle = 0$ for all $X \in \mathfrak{X}_{\mathrm{div}}(\Omega)$, then $\langle M, B\rangle = 0$ for all

$$B \in \{\Omega_t^1(\Omega) \mid \delta B = 0\}$$

because the index lowering operator $^\flat$ given by the metric on Ω induces an ismorphism between $\mathfrak{X}_{\mathrm{div}}(\Omega)$ and

$$\{\alpha \in \Omega_t^1(\Omega) \mid \delta B = 0\}.$$

Therefore, by the L^2-orthogonal decomposition quoted above, $M = \mathbf{d}f$ and hence $M = 0$. Similarly, if $X \in \mathfrak{X}_{\mathrm{div}}(\Omega)$ and $\langle M, X\rangle = 0$ for all $M \in \{\alpha \in \Omega_t^1(M) \mid \delta\alpha = 0\}$, then $\langle M, X^\flat\rangle = 0$ for all such M, and as before $X^\flat = \mathbf{d}f$, that is, $X = \nabla f$. But this implies $X = 0$ since $\mathfrak{X}_{\mathrm{div}}(\Omega)$ and gradients are L^2-orthogonal by the Stokes theorem. Therefore, we can identify

$$\mathfrak{X}_{\mathrm{div}}^*(\Omega) = \{M \in \Omega_t^1(\Omega) \mid \delta M = 0\}. \tag{14.1.12}$$

The coadjoint action of $\mathrm{Diff}_{\mathrm{vol}}(\Omega)$ on $\mathfrak{X}_{\mathrm{div}}^*(\Omega)$ is computed in the following way. Recall from Chapter **9** that $\mathrm{Ad}_\varphi(X) = \varphi_* X$ for $\varphi \in \mathrm{Diff}_{\mathrm{vol}}(\Omega)$ and $X \in \mathfrak{X}_{\mathrm{div}}(\Omega)$. Thus

$$\langle \mathrm{Ad}_{\varphi^{-1}}^* M, X\rangle = \langle M, \mathrm{Ad}_{\varphi^{-1}} X\rangle = \int_\Omega M \cdot \varphi^* X\, d^n x = \int_\Omega \varphi_* M \cdot X d^n x$$

by the change of variables formula. Therefore,

$$\mathrm{Ad}_{\varphi^{-1}}^* M = \varphi_* M \quad \text{and so} \quad \mathrm{Orb} M = \{\varphi_* M \mid \varphi \in \mathrm{Diff}_{\mathrm{vol}}(\Omega)\}. \tag{14.1.13}$$

Next, let us return to the identification of $\mathfrak{X}_{\mathrm{div}}(\Omega)$ with itself by the L^2-pairing on vector fields

$$\langle X, Y \rangle = \int_\Omega X \cdot Y d^n x. \tag{14.1.14}$$

The Helmholtz decomposition says that any vector field on Ω can be uniquely decomposed orthogonally in a sum of a gradient of a function and a divergence-free vector field tangent to $\partial\Omega$; this decomposition is equivalent to the Hodge decomposition on one-forms quoted before. This shows that (14.1.14) is a weakly nondegenerate pairing. For $\varphi \in \mathrm{Diff}_{\mathrm{vol}}(\Omega)$, denote by $(T\varphi)^\dagger$ the adjoint of $T\varphi : T\Omega \to T\Omega$ relative to the metric (14.1.14). By the change of variables formula,

$$\begin{aligned} \langle \mathrm{Ad}^*_{\varphi^{-1}} Y, X \rangle &= \langle Y, \mathrm{Ad}_{\varphi^{-1}} X \rangle = \int_\Omega Y \cdot \varphi^* X \, d^n x \\ &= \int_\Omega Y \cdot (T\varphi^{-1} \circ X \circ \varphi) \, d^n x \\ &= \int_\Omega ((T\varphi^{-1})^\dagger \circ Y \circ \varphi) \cdot X \, d^n x, \end{aligned}$$

that is,

$$\mathrm{Ad}^*_{\varphi^{-1}} Y = (T\varphi^{-1})^\dagger \circ Y \circ \varphi \tag{14.1.15}$$

and

$$\mathrm{Orb}\, Y = \{(T\varphi^{-1})^\dagger \circ Y \circ \varphi \mid \varphi \in \mathrm{Diff}_{\mathrm{vol}}(\Omega)\}. \tag{14.1.16}$$

This example shows that different pairings give rise to different formulae for the coadjoint action and that the choice of dual is dictated by the specific application one has in mind. For example, the pairing (14.1.14) was convenient for the Lie-Poisson bracket on $\mathfrak{X}_{\mathrm{div}}(\Omega)$ in Example **(d)** of §10.2. On the other hand, many computations involving the coadjoint action are simpler with the choice (14.1.12) of the dual corresponding to the pairing (14.1.11).

(d) Orbits in $\mathfrak{X}^*_{\mathrm{can}}$ Let $G = \mathrm{Diff}_{\mathrm{can}}(P)$ be the group of canonical transformations of a symplectic manifold P with $H^1(P) = 0$. Letting k be a function on P, and X_k the corresponding Hamiltonian vector field, and $\varphi \in G$, we have

$$\mathrm{Ad}_\varphi X_k = \varphi_* X_k = X_{k \circ \varphi^{-1}} \tag{14.1.17}$$

so identifying $\mathfrak{g}$ with $\mathcal{F}(P)$ modulo constants, or equivalently with functions on P with zero average, we get $\mathrm{Ad}_\varphi k = \varphi_* k = k \circ \varphi^{-1}$. On the dual space, which is identified with $\mathcal{F}(P)$ (modulo constants) via the L^2-pairing, a straightforward verification shows that

$$\mathrm{Ad}^*_{\varphi^{-1}} f = \varphi_* f = f \circ \varphi^{-1}. \tag{14.1.18}$$

One sometimes says that

$$\operatorname{Orb}(f) = \{f \circ \varphi^{-1} \mid \varphi \in \operatorname{Diff}_{\text{can}}(P)\}$$

consists of ***canonical rearrangements*** of f.

(e) Toda Orbit Another interesting example is the Toda orbit, which arises in the study of completely integrable systems. Let

$$\begin{aligned} \mathfrak{g} &= \text{Lie algebra of real } n \times n \text{ lower triangular matrices} \\ &\quad\ \text{of trace zero,} \\ G &= \text{lower triangular matrices with determinant one,} \end{aligned}$$

and identify

$$\mathfrak{g}^* = \text{the upper triangular matrices,}$$

using the pairing

$$\langle \xi, \mu \rangle = \operatorname{Trace}(\xi \mu),$$

where $\xi \in \mathfrak{g}$ and $\mu \in \mathfrak{g}^*$. Since $\operatorname{Ad}_A \xi = A \xi A^{-1}$, we get

$$\operatorname{Ad}^*_{A^{-1}} \mu = P(A \mu A^{-1}), \tag{14.1.19}$$

where $P : \mathfrak{sl}(n, \mathbb{R}) \to \mathfrak{g}^*$ is the projection sending any matrix to its upper triangular part. Now let

$$\mu = \begin{bmatrix} 0 & 1 & 0 & \cdots & 0 & 0 \\ 0 & 0 & 1 & \cdots & 0 & 0 \\ 0 & 0 & 0 & \cdots & 0 & 0 \\ \vdots & \vdots & \vdots & \ddots & \vdots & \vdots \\ 0 & 0 & 0 & \cdots & 0 & 1 \\ 0 & 0 & 0 & \cdots & 0 & 0 \end{bmatrix} \in \mathfrak{g}^*. \tag{14.1.20}$$

One finds that $\operatorname{Orb}(\mu) = \{P(A \mu A^{-1}) \mid A \in G\}$ consists of matrices of the form

$$L = \begin{bmatrix} b_1 & a_1 & 0 & 0 & \cdots & 0 & 0 \\ 0 & b_2 & a_2 & 0 & \cdots & 0 & 0 \\ 0 & 0 & b_3 & a_3 & \cdots & 0 & 0 \\ 0 & 0 & 0 & b_4 & \cdots & 0 & 0 \\ \vdots & \vdots & \vdots & \vdots & \ddots & \vdots & \vdots \\ 0 & 0 & 0 & 0 & \cdots & b_{n-1} & a_{n-1} \\ 0 & 0 & 0 & 0 & \cdots & 0 & b_n \end{bmatrix}, \tag{14.1.21}$$

where $\sum b_n = 0$. See Kostant [1980] and Symes [1982a,b] for further information.

(f) Coadjoint Orbits that Are Not Submanifolds The following example of a Lie group G, whose generic coadjoint orbits in $\mathfrak{g}^*$ are *not* submanifolds, is due to Kirillov [1976b], p. 293. Let α be irrational, define

$$G = \left\{ \left[\begin{array}{ccc} e^{it} & 0 & z \\ 0 & e^{i\alpha t} & w \\ 0 & 0 & 1 \end{array} \right] \middle| \, t \in \mathbb{R}, z, w \in \mathbb{C} \right\}, \tag{14.1.22}$$

and note the G is diffeomorphic to $\mathbb{R}^5$. As a group it is the semidirect product of

$$H = \left\{ \left[\begin{array}{cc} e^{it} & 0 \\ 0 & e^{i\alpha t} \end{array} \right] \middle| \, t \in \mathbb{R} \right\}$$

with $\mathbb{C}^2$, the action being by left multiplication of vectors in $\mathbb{C}^2$ by elements of H (see Exercise **9.3-1**). The Lie algebra $\mathfrak{g}$ of G is

$$\mathfrak{g} = \left\{ \left[\begin{array}{ccc} it & 0 & x \\ 0 & i\alpha t & y \\ 0 & 0 & 0 \end{array} \right] \middle| \, t \in \mathbb{R}, x, y \in \mathbb{C} \right\} \tag{14.1.23}$$

with the usual commutator bracket as Lie bracket. Identify $\mathfrak{g}^*$ with

$$\mathfrak{g}^* = \left\{ \left[\begin{array}{ccc} is & 0 & 0 \\ 0 & i\alpha s & 0 \\ a & b & 0 \end{array} \right] \middle| \, s \in \mathbb{R}, a, b \in \mathbb{C} \right\} \tag{14.1.24}$$

via the nondegenerate pairing in $\mathfrak{gl}(3, \mathbb{C})$ is given by

$$\langle A, B \rangle = \operatorname{Re} (\operatorname{trace}(AB)).$$

The adjoint action of

$$g = \left[\begin{array}{ccc} e^{it} & 0 & z \\ 0 & e^{i\alpha t} & w \\ 0 & 0 & 1 \end{array} \right] \quad \text{on} \quad \xi = \left[\begin{array}{ccc} is & 0 & x \\ 0 & i\alpha s & y \\ 0 & 0 & 0 \end{array} \right]$$

is given by

$$\operatorname{Ad}_g \xi = \left[\begin{array}{ccc} is & 0 & e^{it}x - isz \\ 0 & i\alpha s & e^{i\alpha t}y - i\alpha s w \\ 0 & 0 & 0 \end{array} \right]. \tag{14.1.25}$$

The coadjoint action of the same group element g on

$$\mu = \left[\begin{array}{ccc} iu & 0 & 0 \\ 0 & i\alpha u & 0 \\ a & b & 0 \end{array} \right]$$

is given by

$$\operatorname{Ad}^*_{g^{-1}}\mu = \begin{bmatrix} iu' & 0 & 0 \\ 0 & i\alpha u' & 0 \\ ae^{-it} & be^{-i\alpha t} & 0 \end{bmatrix}, \tag{14.1.26}$$

where

$$u' = u + \frac{1}{1+\alpha^2}\operatorname{Im}(ae^{-it}z + be^{-i\alpha t}\alpha w). \tag{14.1.27}$$

If $a, b \neq 0$, the orbit through μ is two dimensional; it is a cylindrical surface whose generator is the u'-axis and whose base is the curve in $\mathbb{C}^2$ given parametrically by $t \mapsto (ae^{-it}, be^{-i\alpha t})$. This curve, however, is the irrational flow on the torus with radii $|a|$ and $|b|$, that is, the cylindrical surface accumulates on itself and thus is not a submanifold of $\mathbb{R}^5$. We shall return to this example at the end of §14.6.

14.2 Tangent Vectors to Coadjoint Orbits

In general, orbits of a Lie group action, while manifolds in their own right, are not submanifolds of the ambient manifold; they are only injectively immersed manifolds. A notable exception occurs in the case of compact Lie groups: then all their orbits are embedded submanifolds. Coadjoint orbits are no exception to this global problem, as we saw in the preceding examples. We shall always regard them as injectively immersed submanifolds, diffeomorphic to G/G_μ, where $G_\mu = \{g \in G \mid \operatorname{Ad}^*_g \mu = \mu\}$ is the isotropy subgroup of the coadjoint action at a point μ in the orbit.

We now describe tangent vectors to coadjoint orbits. Let $\xi \in \mathfrak{g}$ and let $g(t)$ be a curve in G tangent to ξ at $t = 0$; for example, let $g(t) = \exp(t\xi)$. Let $\mathcal{O}$ be a coadjoint orbit, and $\mu \in \mathcal{O}$. If $\eta \in \mathfrak{g}$, then

$$\mu(t) = \operatorname{Ad}^*_{g(t)^{-1}}\mu \tag{14.2.1}$$

is a curve in $\mathcal{O}$ with $\mu(0) = \mu$. Differentiating the identity

$$\langle \mu(t), \eta \rangle = \langle \mu, \operatorname{Ad}_{g(t)^{-1}}\eta \rangle \tag{14.2.2}$$

with respect to t at $t = 0$, we get

$$\langle \mu'(0), \eta \rangle = -\langle \mu, \operatorname{ad}_\xi \eta \rangle = -\langle \operatorname{ad}^*_\xi \mu, \eta \rangle, \quad \text{and so} \quad \mu'(0) = -\operatorname{ad}^*_\xi \mu. \tag{14.2.3}$$

Thus,

$$T_\mu\mathcal{O} = \{\operatorname{ad}^*_\xi \mu \mid \xi \in \mathfrak{g}\}. \tag{14.2.4}$$

This calculation also proves that the infinitesimal generator of the coadjoint action is given by

$$\xi_{\mathfrak{g}^*}(\mu) = -\operatorname{ad}^*_\xi \mu. \tag{14.2.5}$$

The following characterization of the tangent space to coadjoint orbits is often useful. We let $\mathfrak{g}_\mu = \{\xi \in \mathfrak{g} \mid \operatorname{ad}^*_\xi \mu = 0\}$ be the coadjoint isotropy algebra of μ; it is the Lie algebra of the coadjoint isotropy group $G_\mu = \{g \in G \mid \operatorname{Ad}^*_g \mu = \mu\}$.

Proposition 14.2.1 *Let $\langle\, ,\rangle : \mathfrak{g}^* \times \mathfrak{g} \to \mathbb{R}$ be a weakly nondegenerate pairing and let $\mathcal{O}$ be the coadjoint orbit through $\mu \in \mathfrak{g}^*$. Let*

$$\mathfrak{g}_\mu^\circ := \{\nu \in \mathfrak{g}^* \mid \langle \nu, \eta\rangle = 0 \quad \textit{for all} \quad \eta \in \mathfrak{g}_\mu\}$$

be the ***annihilator*** *of $\mathfrak{g}_\mu$ in $\mathfrak{g}^*$. Then $T_\mu\mathcal{O} \subset \mathfrak{g}_\mu^\circ$. If $\mathfrak{g}$ is finite dimensional, then $T_\mu\mathcal{O} = \mathfrak{g}_\mu^\circ$. The same equality holds if $\mathfrak{g}$ and $\mathfrak{g}^*$ are Banach spaces, $T_\mu\mathcal{O}$ is closed in $\mathfrak{g}^*$, and the pairing is strongly nondegenerate.*

Proof For any $\xi \in \mathfrak{g}, \eta \in \mathfrak{g}_\mu$ we have

$$\langle \operatorname{ad}^*_\xi \mu, \eta\rangle = \langle \mu, [\xi, \eta]\rangle = -\langle \operatorname{ad}^*_\eta \mu, \xi\rangle = 0$$

which proves the inclusion $T_\mu\mathcal{O} \subset \mathfrak{g}_\mu^\circ$. If $\mathfrak{g}$ is finite dimensional, equality holds since $\dim T_\mu\mathcal{O} = \dim \mathfrak{g} - \dim \mathfrak{g}_\mu = \dim \mathfrak{g}_\mu^\circ$. If $\mathfrak{g}$ and $\mathfrak{g}^*$ are infinite-dimensional Banach spaces and $\langle\, ,\rangle : \mathfrak{g}^* \times \mathfrak{g} \to \mathbb{R}$ is a strong pairing, we can assume without loss of generality that it is the natural pairing between a Banach space and its dual. If $\mathfrak{g}_\mu^\circ \neq T_\mu\mathcal{O}$ pick $\nu \neq 0, \nu \in \mathfrak{g}_\mu^\circ, \nu \notin T_\mu\mathcal{O}$. By the Hahn-Banach theorem there is an $\eta \in \mathfrak{g}$ such that $\langle \nu, \eta\rangle = 1$ and $\langle \operatorname{ad}^*_\xi \mu, \eta\rangle = 0$ for all $\xi \in \mathfrak{g}$. The latter condition is equivalent to $\eta \in \mathfrak{g}_\mu$. On the other hand, since $\nu \in \mathfrak{g}_\mu^\circ$ we have $\langle \nu, \eta\rangle = 0$, which is a contradiction. ■

14.3 Examples of Tangent Vectors

(a) Rotation Group Identifying $(\mathfrak{so}(3), [\cdot\,,\cdot]) \cong (\mathbb{R}^3, \times)$ and $\mathfrak{so}(3)^* \cong \mathbb{R}^3$ via the natural pairing given by the Euclidean inner product, formula (14.2.5) reads as follows for $\mathbf{\Pi} \in \mathfrak{so}(3)^*$ and $\boldsymbol{\xi}, \boldsymbol{\eta} \in \mathfrak{so}(3)$,

$$\langle \boldsymbol{\xi}_{\mathfrak{so}(3)^*}(\mathbf{\Pi}), \boldsymbol{\eta}\rangle = -\mathbf{\Pi} \cdot (\boldsymbol{\xi} \times \boldsymbol{\eta}) = -(\mathbf{\Pi} \times \boldsymbol{\xi}) \cdot \boldsymbol{\eta} \qquad (14.3.1)$$

so that $\boldsymbol{\xi}_{\mathfrak{so}(3)^*}(\mathbf{\Pi}) = -\mathbf{\Pi} \times \boldsymbol{\xi} = \boldsymbol{\xi} \times \mathbf{\Pi}$. As expected, $\boldsymbol{\xi}_{\mathfrak{so}(3)^*}(\mathbf{\Pi}) \in T_\Pi \operatorname{Orb}(\mathbf{\Pi})$ is tangent to the sphere $\operatorname{Orb}(\mathbf{\Pi})$. Allowing $\boldsymbol{\xi}$ to vary in $\mathfrak{so}(3) \cong \mathbb{R}^3$, one obtains all of $T_\Pi \operatorname{Orb}(\mathbf{\Pi})$. ♦

(b) Affine Group on $\mathbb{R}$ Let $(u, v) \in \mathfrak{g}$ and consider the coadjoint orbit through the point $\begin{pmatrix} \alpha \\ \beta \end{pmatrix} \in \mathfrak{g}^*$. Then (14.2.5) reads

$$(u, v)_{\mathfrak{g}^*}\begin{pmatrix} \alpha \\ \beta \end{pmatrix} = \left\langle \begin{pmatrix} \alpha \\ \beta \end{pmatrix}, [\,\cdot\,, (u, v)]\right\rangle. \qquad (14.3.2)$$

But $\left\langle \begin{pmatrix} \alpha \\ \beta \end{pmatrix}, [(r,s),(u,v)] \right\rangle = \left\langle \begin{pmatrix} \alpha \\ \beta \end{pmatrix}, (0, rv - su) \right\rangle = rv\beta - su\beta$, and so

$$(u,v)_{\mathfrak{g}^*} \begin{pmatrix} \alpha \\ \beta \end{pmatrix} = \begin{pmatrix} v\beta \\ -u\beta \end{pmatrix}. \tag{14.3.3}$$

If $\beta \neq 0$, these vectors span $\mathfrak{g}^* = \mathbb{R}^2$ as they should. ♦

(c) The Group $\mathrm{Diff}_{\mathrm{vol}}$ For $G = \mathrm{Diff}_{\mathrm{vol}}$ and $M \in \mathfrak{X}^*_{\mathrm{div}}$, we get the tangent vectors to $\mathrm{Orb}(M)$ by differentiating (14.1.13) with respect to φ, yielding

$$T_M \,\mathrm{Orb}(M) = \{-\pounds_v M \mid v \text{ is divergence free and tangent to } \partial\Omega\}. \tag{14.3.4}$$

(d) The Group $\mathrm{Diff}_{\mathrm{can}}(P)$ For $G = \mathrm{Diff}_{\mathrm{can}}(P)$, we have

$$T_f \,\mathrm{Orb}(f) = \{-\{f,k\} \mid k \in \mathcal{F}(P)\}. \tag{14.3.5}$$

(e) The Toda Lattice The tangent space to the Toda orbit consists of matrices of the same form as L in (14.1.21) since those matrices form a linear space. The reader can check that (14.2.4) gives the same answer. ♦

14.4 The Symplectic Structure on Coadjoint Orbits

Theorem 14.4.1 (Coadjoint Orbit Theorem) *Let G be a Lie group and let $\mathcal{O} \subset \mathfrak{g}^*$ be a coadjoint orbit. Then $\mathcal{O}$ is a symplectic manifold. In fact, there are unique symplectic forms $\omega^{\pm}$ on $\mathcal{O}$ such that*

$$\omega^{\pm}(\mu)(\xi_{\mathfrak{g}^*}(\mu), \eta_{\mathfrak{g}^*}(\mu)) = \pm\langle \mu, [\xi, \eta]\rangle \tag{14.4.1}$$

for all $\mu \in \mathcal{O}$ and $\xi, \eta \in \mathfrak{g}$. We refer to $\omega^{\pm}$ as the ***coadjoint orbit symplectic structures*** *and, if there is danger of confusion, denote it $\omega^{\pm}_{\mathcal{O}}$.*

Proof We prove the result for ω^-, the argument for ω^+ being similar. First we show that formula (14.4.1) gives a well-defined form; that is, the right-hand side is independent of the particular $\xi \in \mathfrak{g}$ and $\eta \in \mathfrak{g}$ which define the tangent vectors $\xi_{\mathfrak{g}^*}(\mu)$ and $\eta_{\mathfrak{g}^*}(\mu)$. This follows by observing that $\xi_{\mathfrak{g}^*}(\mu) = \xi'_{\mathfrak{g}^*}(\mu)$ implies $-\langle \mu, [\xi, \eta]\rangle = -\langle \mu, [\xi', \eta]\rangle$ for all $\eta \in \mathfrak{g}$. Therefore, $\omega^-(\mu)(\xi_{\mathfrak{g}^*}(\mu), \eta_{\mathfrak{g}^*}(\mu)) = \omega^-(\xi'_{\mathfrak{g}^*}(\mu), \eta_{\mathfrak{g}^*}(\mu))$, so ω^- is well defined.

Second, we show that ω^- is nondegenerate. Since the pairing $\langle\, , \rangle$ is nondegenerate, $\omega^-(\mu)(\xi_{\mathfrak{g}^*}(\mu), \eta_{\mathfrak{g}^*}(\mu)) = 0$ for all $\eta_{\mathfrak{g}^*}(\mu)$ implies $-\langle \mu, [\xi, \eta]\rangle = 0$ for all η. This means that $0 = -\langle \mu, [\xi, \cdot]\rangle = \xi_{\mathfrak{g}^*}(\mu)$.

Finally, we show that ω^- is closed, that is $\mathbf{d}\omega^- = 0$. To do this we begin by defining, for each $\nu \in \mathfrak{g}^*$, the one-form ν_L on G by

$$\nu_L(g) = (T_g^* L_{g^{-1}})(\nu),$$

where $g \in G$. The one-form ν_L is readily checked to be left invariant; that is $L_g^* \nu_L = \nu_L$ for all $g \in G$. For $\xi \in \mathfrak{g}$, let ξ_L be the corresponding left invariant vector field on G, so $\nu_L(\xi_L)$ is a constant function on G (whose value at any point is $\langle \nu, \xi \rangle$). Choose $\nu \in \mathcal{O}$ and consider the surjective map $\varphi_\nu : G \to \mathcal{O}$ defined by $g \mapsto \mathrm{Ad}^*_{g^{-1}}(\nu)$ and the two-form $\sigma = \varphi_\nu^* \omega^-$ on G. We claim that

$$\sigma = \mathbf{d}\nu_L. \tag{14.4.2}$$

To prove this, notice that

$$(T_e \varphi_\nu)(\eta) = \eta_{\mathfrak{g}^*}(\nu) \tag{14.4.3}$$

so that the surjective map φ_ν is submersive at e. By definition of pull back, $\sigma(e)(\xi, \eta)$ equals

$$\begin{aligned}(\varphi_\nu^* \omega^-)(e)(\xi, \eta) &= \omega^-(\varphi_\nu(e))(T_e\varphi_\nu \cdot \xi, T_e\varphi_\nu \cdot \eta) \\ &= \omega^-(\nu)(\xi_{\mathfrak{g}^*}(\nu), \eta_{\mathfrak{g}^*}(\nu)) = -\langle \nu, [\xi, \eta] \rangle. \end{aligned} \tag{14.4.4}$$

Hence

$$\sigma(\xi_L, \eta_L)(e) = \sigma(e)(\xi, \eta) = -\langle \nu, [\xi, \eta] \rangle = -\langle \nu_L, [\xi_L, \eta_L] \rangle (e). \tag{14.4.5}$$

We shall need the relation $\sigma(\xi_L, \eta_L) = -\langle \nu_L, [\xi_L, \eta_L] \rangle$ at each point of G; to get it, we first prove two lemmas.

Lemma 14.4.2 $\mathrm{Ad}^*_{g^{-1}} : \mathcal{O} \to \mathcal{O}$ *preserves* ω^-, *i.e.,* $(\mathrm{Ad}^*_{g^{-1}})^* \omega^- = \omega^-$.

To prove this, we recall two identities from Chapter 9. First,

$$(\mathrm{Ad}_g \xi)_{\mathfrak{g}^*} = \mathrm{Ad}^*_{g^{-1}} \circ \xi_{\mathfrak{g}^*} \circ \mathrm{Ad}^*_g, \tag{14.4.6}$$

which is proved by letting ξ be tangent to a curve $h(\varepsilon)$ at $\varepsilon = 0$, recalling that

$$\mathrm{Ad}_g \xi = \left. \frac{d}{d\varepsilon} g h(\varepsilon) g^{-1} \right|_{\varepsilon=0} \tag{14.4.7}$$

and noting

$$\begin{aligned}(\mathrm{Ad}_g \xi)_{\mathfrak{g}^*}(\mu) &= \left. \frac{d}{d\varepsilon} \mathrm{Ad}^*_{(gh(\varepsilon)g^{-1})^{-1}} \mu \right|_{\varepsilon=0} \\ &= \left. \frac{d}{d\varepsilon} \mathrm{Ad}^*_{g^{-1}} \mathrm{Ad}^*_{h(\varepsilon)^{-1}} \mathrm{Ad}^*_g(\mu) \right|_{\varepsilon=0}. \end{aligned} \tag{14.4.8}$$

Second, we require the identity

$$\operatorname{Ad}_g[\xi, \eta] = [\operatorname{Ad}_g \xi, \operatorname{Ad}_g \eta], \tag{14.4.9}$$

which follows by differentiating the relation

$$I_g(I_h(k)) = I_g(h) I_g(k) I_g(h^{-1}) \tag{14.4.10}$$

with respect to h and k and evaluating at the identity.

Evaluating (14.4.6) at $\nu = \operatorname{Ad}^*_{g^{-1}} \mu$, we get

$$(\operatorname{Ad}_g \xi)_{\mathfrak{g}^*}(\nu) = \operatorname{Ad}^*_{g^{-1}} \cdot \xi_{\mathfrak{g}^*}(\mu) = T_\mu \operatorname{Ad}^*_{g^{-1}} \cdot \xi_{\mathfrak{g}^*}(\mu), \tag{14.4.11}$$

by linearity of $\operatorname{Ad}^*_{g^{-1}}$. Thus,

$$\begin{aligned}
&((\operatorname{Ad}^*_{g^{-1}})^* \omega^-)(\mu)(\xi_{\mathfrak{g}^*}(\mu), \eta_{\mathfrak{g}^*}(\mu)) \\
&\quad = \omega^-(\nu)(T_\mu \operatorname{Ad}^*_{\mathfrak{g}^{-1}} \cdot \xi_{\mathfrak{g}^*}(\mu), T_\mu \operatorname{Ad}^*_{\mathfrak{g}^{-1}} \cdot \eta_{\mathfrak{g}^*}(\mu)) \\
&\quad = \omega^-(\nu)((\operatorname{Ad}_g \xi)_{\mathfrak{g}^*}(\nu), (\operatorname{Ad}_g \eta)_{g^*}(\nu)) && \text{(by (14.4.11))} \\
&\quad = -\langle \nu, [\operatorname{Ad}_g \xi, \operatorname{Ad}_g \eta] \rangle && \text{(by definition of } \omega^-\text{))} \\
&\quad = -\langle \nu, \operatorname{Ad}_g[\xi, \eta] \rangle && \text{(by (14.4.9))} \\
&\quad = -\left\langle \operatorname{Ad}^*_g \nu, [\xi, \eta] \right\rangle = -\langle \mu, [\xi, \eta] \rangle \\
&\quad = \omega^-(\mu)(\xi_{\mathfrak{g}^*}(\mu), \eta_{\mathfrak{g}^*}(\mu)). \quad \blacktriangledown
\end{aligned} \tag{14.4.12}$$

Lemma 14.4.3 *σ is left invariant, i.e., $L^*_g \sigma = \sigma$ for all $g \in G$.*

Proof Using the equivariance identity $\varphi_\nu \circ L_g = \operatorname{Ad}^*_{g^{-1}} \circ \varphi_\nu$, we compute $L^*_g \sigma = L^*_g \varphi^*_\nu \omega^- = (\varphi_\nu \circ L_g)^* \omega^- = (\operatorname{Ad}^*_{g^{-1}} \circ \varphi_\nu)^* \omega^- = \varphi^*_\nu (\operatorname{Ad}^*_{g^{-1}})^* \omega^- = \varphi^*_\nu \omega^- = \sigma$. $\blacktriangledown$

Lemma 14.4.4 $\sigma(\xi_L, \eta_L) = -\langle \nu_L, [\xi_L, \eta_L] \rangle$.

Proof Both sides are left invariant and are equal at the identity by (14.4.5). $\blacktriangledown$

The exterior derivative $\mathbf{d}\alpha$ of a one-form α is given in terms of the Jacobi-Lie bracket by

$$(\mathbf{d}\alpha)(X, Y) = X[\alpha(Y)] - Y[\alpha(X)] - \alpha([X, Y]). \tag{14.4.13}$$

Since $\nu_L(\xi_L)$ is constant, $\eta_L[\nu_L(\xi_L)] = 0$ and $\xi_L[\nu_L(\eta_L)] = 0$, so Lemma **14.4.4** implies

$$\sigma(\xi_L, \eta_L) = (\mathbf{d}\nu_L)(\xi_L, \eta_L). \tag{14.4.14}$$

Lemma 14.4.5

$$\sigma = \mathbf{d}\nu_L. \tag{14.4.15}$$

Proof We shall prove that for any vector fields X and Y, $\sigma(X,Y) = (\mathbf{d}\nu_L)(X,Y)$. Indeed, since σ is left invariant,

$$\begin{aligned}
&\sigma(X,Y)(g)\\
&= (L^*_{g^{-1}}\sigma)(g)(X(g),Y(g)) = \sigma(e)(TL_{g^{-1}}\cdot X(g), TL_{g^{-1}}\cdot Y(g))\\
&= \sigma(e)(\xi,\eta) \quad (\text{where } \xi = TL_{g^{-1}}\cdot X(g) \text{ and } \eta = TL_{g^{-1}}\cdot Y(g))\\
&= \sigma(\xi_L,\eta_L)(e) = (\mathbf{d}\nu_L)(\xi_L,\eta_L)(e) \qquad (\text{by } (14.4.14))\\
&= (L^*_g\mathbf{d}\nu_L)(\xi_L,\eta_L)(e) \qquad (\text{since } \nu_L \text{ is left invariant})\\
&= (\mathbf{d}\nu_L)(g)(TL_g\cdot\xi_L(e), TL_g\cdot\eta_L(e))\\
&= (\mathbf{d}\nu_L)(g)(TL_g\cdot\xi, TL_g\cdot\eta) = (\mathbf{d}\nu_L)(g)(X(g),\ Y(g))\\
&= (\mathbf{d}\nu_L)(X,Y)(g). \quad \blacktriangledown
\end{aligned}$$

Since $\sigma = \mathbf{d}\nu_L$ by Lemma **14.4.5**, $\mathbf{d}\sigma = \mathbf{dd}\nu_L = 0$, and so $0 = \mathbf{d}\varphi_\nu^*\omega^- = \varphi_\nu^*\mathbf{d}\omega^-$. From $\varphi_\nu \circ L_g = \mathrm{Ad}^*_{g^{-1}} \circ \varphi_\nu$, it follows that submersivity of φ_ν at e is equivalent to submersivity of φ_ν at any $g \in G$, that is, φ_ν is a surjective submersion. Thus φ_ν^* is injective, and hence $\mathbf{d}\omega^- = 0$. ∎

Remark Any Lie group carries a natural connection associated to the left (or right) action. The calculation (14.4.13) is essentially the calculation of the curvature of this connection and, as such, is closely related to the *Maurer-Cartan equations* (see §**9.1**). ♦

Since coadjoint orbits are symplectic, we get the following:

Corollary 14.4.6 *Coadjoint orbits of finite-dimensional Lie groups are even dimensional.*

Corollary 14.4.7 *Let $G_\nu = \{g \in G \mid \mathrm{Ad}^*_{g^{-1}}\nu = \nu\}$ be the isotropy subgroup of the coadjoint action of $\nu \in \mathfrak{g}^*$. Then G_ν is a closed subgroup of G, and so the quotient G/G_ν is a smooth manifold with smooth projection $\pi : G \to G/G_\nu; g \mapsto g\cdot G_\nu$. We identify $G/G_\nu \cong \mathrm{Orb}(\nu)$ via the diffeomorphism $\rho : g\cdot G_\nu \in G/G_\nu \mapsto \mathrm{Ad}^*_{g^{-1}}(\nu) \in \mathrm{Orb}(\nu)$. Thus G/G_ν is symplectic, with symplectic form ω^- induced from $\mathbf{d}\nu_L$, that is,*

$$\mathbf{d}\nu_L = \pi^*\rho^*\omega^-$$

(respectively, $\mathbf{d}\nu_R = \pi^*\rho^*\omega^+$).

As we shall see in Example **(a)** of §14.5, ω^- is not exact in general, even though $\pi^*\rho^*\omega^-$ is.

Exercises

14.4-1 *Show that $J(\xi) = \langle \nu_L, \xi_L\rangle$ defines a momentum map.*

14.4-2 *Relate the calculations of this section to the Mauer-Cartan equations.*

14.4.3 *Give another proof that* $\mathbf{d}\omega^\pm = 0$ *by showing that* X_H *for* $\omega^\pm$ *coincides with that for the Lie-Poisson bracket and hence that Jacobi's identity holds.*

14.5 Examples of Symplectic Structures on Orbits

(a) **Rotation Group** Consider $\mathrm{Orb}(\mathbf{\Pi})$, the coadjoint orbit through $\mathbf{\Pi} \in \mathbb{R}^3$; then

$$\boldsymbol{\xi}_{\mathbb{R}^3}(\mathbf{\Pi}) = \boldsymbol{\xi} \times \mathbf{\Pi} \in T_\Pi(\mathrm{Orb}(\mathbf{\Pi})), \text{ and } \boldsymbol{\eta}_{\mathbb{R}^3}(\mathbf{\Pi}) = \boldsymbol{\eta} \times \mathbf{\Pi} \in T_\Pi(\mathrm{Orb}(\mathbf{\Pi})),$$

and so with the usual identification of $\mathfrak{so}(3)$ with $\mathbb{R}^3$, the (–) coadjoint orbit symplectic structure becomes

$$\omega^-(\boldsymbol{\xi}_{\mathbb{R}^3}(\mathbf{\Pi}), \boldsymbol{\eta}_{\mathbb{R}^3}(\mathbf{\Pi})) = -\mathbf{\Pi} \cdot (\boldsymbol{\xi} \times \boldsymbol{\eta}). \tag{14.5.1}$$

Recall that the oriented area of the (planar) parallelogram spanned by two vectors $\mathbf{v}, \mathbf{w} \in \mathbb{R}^3$, is given by $\mathbf{v} \times \mathbf{w}$ (the numerical area is $\|\mathbf{v} \times \mathbf{w}\|$). Thus, the oriented area spanned by $\boldsymbol{\xi}_{\mathbb{R}^3}(\mathbf{\Pi})$ and $\boldsymbol{\eta}_{\mathbb{R}^3}(\mathbf{\Pi})$ is $(\boldsymbol{\xi} \times \mathbf{\Pi}) \times (\boldsymbol{\eta} \times \mathbf{\Pi}) = [(\boldsymbol{\xi} \times \mathbf{\Pi}) \cdot \mathbf{\Pi}]\boldsymbol{\eta} - [(\boldsymbol{\xi} \times \mathbf{\Pi}) \cdot \boldsymbol{\eta}]\mathbf{\Pi} = \mathbf{\Pi}(\mathbf{\Pi} \cdot (\boldsymbol{\xi} \times \boldsymbol{\eta}))$.

The area element dA on a sphere in $\mathbb{R}^3$ assigns to each pair $(\mathbf{v}, \mathbf{w})$ of tangent vectors the number $dA(\mathbf{v}, \mathbf{w}) = \mathbf{n} \cdot (\mathbf{v} \times \mathbf{w})$, where $\mathbf{n}$ is the unit outward normal (this is the area of the parallelogram spanned by $\mathbf{v}$ and $\mathbf{w}$, taken "+" if $\mathbf{v}, \mathbf{w}, \mathbf{n}$ form a positively oriented basis and "–" otherwise). For a sphere of radius $\|\mathbf{\Pi}\|$ and tangent vectors $\mathbf{v} = \boldsymbol{\xi} \times \mathbf{\Pi}$ and $\mathbf{w} = \boldsymbol{\eta} \times \mathbf{\Pi}$, we have

$$\begin{aligned} dA(\boldsymbol{\xi} \times \mathbf{\Pi}, \boldsymbol{\eta} \times \mathbf{\Pi}) &= \frac{\mathbf{\Pi}}{\|\mathbf{\Pi}\|} \cdot ((\boldsymbol{\xi} \times \mathbf{\Pi}) \times (\boldsymbol{\eta} \times \mathbf{\Pi})) \\ &= \frac{\mathbf{\Pi}}{\|\mathbf{\Pi}\|} \cdot (((\boldsymbol{\xi} \times \mathbf{\Pi}) \cdot \mathbf{\Pi})\boldsymbol{\eta} - ((\boldsymbol{\xi} \times \mathbf{\Pi}) \cdot \boldsymbol{\eta})\mathbf{\Pi}) \\ &= \|\mathbf{\Pi}\| \mathbf{\Pi} \cdot (\boldsymbol{\xi} \times \boldsymbol{\eta}). \end{aligned} \tag{14.5.2}$$

Thus

$$\omega^- = -\frac{1}{\|\mathbf{\Pi}\|} dA. \tag{14.5.3}$$

The use of "dA" for the area element is, of course, a notational abuse since this two-form cannot be exact. Likewise,

$$\omega^+ = \frac{1}{\|\mathbf{\Pi}\|} dA. \tag{14.5.4}$$

Notice that $\omega^+/\|\mathbf{\Pi}\| = (dA)/\|\mathbf{\Pi}\|^2$ is the solid angle subtended by the area element dA. ♦

(b) Affine Group on $\mathbb{R}$ For $\beta \neq 0$, and $\mu = \begin{pmatrix} \alpha \\ \beta \end{pmatrix}$ on the open orbit $\mathcal{O}$, formula (14.4.1) gives

$$\begin{aligned} \omega^-(\mu)((r,s)_{\mathfrak{g}^*}(\mu),(u,v)_{\mathfrak{g}^*}(\mu) &= -\left\langle \begin{pmatrix} \alpha \\ \beta \end{pmatrix}, [(r,s),(u,v)] \right\rangle \\ &= \beta(rv - su), \end{aligned} \tag{14.5.5}$$

or in coordinates, $(q,p) \in \mathbb{R}^2$,

$$\omega^-(\mu) = \beta \mathbf{d}q \wedge \mathbf{d}p. \quad \blacklozenge \tag{14.5.6}$$

(c) The Group $\mathrm{Diff}_{\mathrm{vol}}$ For a coadjoint orbit of $G = \mathrm{Diff}_{\mathrm{vol}}(\Omega)$ the $(+)$ coadjoint orbit symplectic structure at a point M becomes

$$\omega^+(M)(-\pounds_v M, -\pounds_w M) = -\int_\Omega M \cdot [v,w]\, d^n x, \tag{14.5.7}$$

where $[v,w]$ is the Jacobi-Lie bracket. Note that we have indeed a minus sign on the right-hand side of (14.5.7) since $[v,w]$ is *minus* the left Lie algebra bracket. $\blacklozenge$

Exercises

14.5-1 The Group $\mathrm{Diff}_{\mathrm{can}}$ *For a coadjoint orbit for $G = \mathrm{Diff}_{\mathrm{can}}(P)$, show that the $(+)$ coadjoint orbit symplectic structure is*

$$\omega^+(f)(\{k,f\},\{h,f\}) = \int_P f\{k,h\}\, dq\, dp.$$

14.5-2 The Toda Lattice *For the Toda orbit, check that the orbit symplectic structure is*

$$\omega^+(f) = \sum_{i=1}^{n-1} \frac{1}{a_i} \mathbf{d}b_i \wedge \mathbf{d}a_i. \tag{14.5.8}$$

14.6 The Orbit Bracket via Restriction of the Lie-Poisson Bracket

Theorem 14.6.1 (Lie-Poisson-Coadjoint Orbit Compatibility) *The Lie-Poisson bracket and the coadjoint orbit symplectic structure are consistent in the following sense: for $F, H : \mathfrak{g}^* \to \mathbb{R}$ and $\mathcal{O}$ a coadjoint orbit in $\mathfrak{g}^*$,*

$$\{F,H\}_+|\mathcal{O} = \{F|\mathcal{O}, H|\mathcal{O}\}^+. \tag{14.6.1}$$

Here, the bracket $\{F, G\}_+$ *is the* $(+)$ *Lie-Poisson bracket, while the bracket on the right-hand side of* (14.6.1) *is the Poisson bracket defined by the* $(+)$ *coadjoint orbit symplectic structure on* $\mathcal{O}$. *Similarly,*

$$\{F, H\}_-|\mathcal{O} = \{F|\mathcal{O}, H|\mathcal{O}\}^-. \tag{14.6.2}$$

The following box summarizes the basic content of what the theorem says.

Two Approaches to the Lie-Poisson Bracket

Two different ways to produce the same Lie-Poisson bracket $\{F, H\}_-$(respectively, $\{F, H\}_+$) on $\mathfrak{g}^*$ are:

Extension Method

1. Take $F, H : \mathfrak{g}^* \to \mathbb{R}$;
2. extend F, H to $F_L, H_L : T^*G \to \mathbb{R}$ by left (respectively, right) invariance;
3. take the bracket $\{F_L, H_L\}$ with respect to the canonical symplectic structure on T^*G; and
4. restrict: $\{F_L, H_L\}|\mathfrak{g}^* = \{F, H\}_-$ (respectively, $\{F_R, H_R\}|\mathfrak{g}^* = \{F, H\}_+$).

Restriction Method

1. Take $F, H : \mathfrak{g}^* \to \mathbb{R}$;
2. form the restrictions $F|\mathcal{O}, H|\mathcal{O}$ to a coadjoint orbit; and
3. take the Poisson bracket $\{F|\mathcal{O}, H|\mathcal{O}\}^-$ with respect to the $-$ (respectively, $+$) orbit symplectic structure ω^- (respectively, ω^+) on the orbit $\mathcal{O}$: *for* $\mu \in \mathcal{O}$ *we have*

$$\{F|\mathcal{O}, H|\mathcal{O}\}^-(\mu) = \{F, H\}_-(\mu).$$

Proof of Theorem 14.6.1 Let $\mu \in \mathcal{O}$. By definition,

$$\{F, H\}_-(\mu) = -\left\langle \mu, \left[\frac{\delta F}{\delta \mu}, \frac{\delta H}{\delta \mu}\right]\right\rangle. \tag{14.6.3}$$

On the other hand,

$$\{F|\mathcal{O}, H|\mathcal{O}\}^-(\mu) = \omega^-(X_F, X_H)(\mu), \tag{14.6.4}$$

where X_F and X_H are the Hamiltonian vector fields on $\mathcal{O}$ generated by $F|\mathcal{O}$ and $H|\mathcal{O}$, and ω^- is the minus orbit symplectic form. Recall that the Hamiltonian vector field X_F on $\mathfrak{g}^*_-$ is given by

$$X_F(\mu) = \operatorname{ad}^*_\xi(\mu), \quad \text{where} \quad \xi = \frac{\delta F}{\delta \mu} \in \mathfrak{g}. \tag{14.6.5}$$

Motivated by this we prove the following:

Lemma 14.6.2 *Using the orbit symplectic form ω^-, for $\mu \in \mathcal{O}$ we have*

$$X_{F|\mathcal{O}}(\mu) = \operatorname{ad}^*_{\delta F/\delta\mu}(\mu). \tag{14.6.6}$$

Proof Let $\xi, \eta \in \mathfrak{g}$, so (14.4.1) gives

$$\omega^-(\mu)(\operatorname{ad}^*_\xi \mu, \operatorname{ad}^*_\eta \mu) = -\langle \mu, [\xi, \eta]\rangle = \langle \mu, \operatorname{ad}_\eta(\xi)\rangle = \langle \operatorname{ad}^*_\eta(\mu), \xi\rangle. \tag{14.6.7}$$

Letting $\xi = \delta F/\delta\mu$ and η be arbitrary, we get

$$\omega^-(\mu)(\operatorname{ad}^*_{\delta F/\delta\mu} \mu, \operatorname{ad}^*_\eta \mu) = \left\langle \operatorname{ad}^*_\eta \mu, \frac{\delta F}{\delta\mu}\right\rangle = \mathbf{d}F(\mu) \cdot \operatorname{ad}^*_\eta \mu. \tag{14.6.8}$$

Thus, $X_{F|\mathcal{O}}(\mu) = \operatorname{ad}^*_{\delta F/\delta\mu} \mu$, as required. ▼

To complete the proof of Theorem **14.6.1**, note that

$$\begin{aligned} \{F|\mathcal{O}, H|\mathcal{O}\}^-(\mu) &= \omega^-(\mu)(X_{F|\mathcal{O}}(\mu), X_{H|\mathcal{O}}(\mu)) \\ &= \omega^-(\mu)(\operatorname{ad}^*_{\delta F/\delta\mu}\mu, \operatorname{ad}^*_{\delta H/\delta\mu}\mu) \\ &= -\left\langle \mu, \left[\frac{\delta F}{\delta\mu}, \frac{\delta H}{\delta\mu}\right]\right\rangle = \{F, H\}_-(\mu). \quad \blacksquare \end{aligned}$$

Corollary 14.6.3

i *For $H \in \mathcal{F}(\mathfrak{g}^*)$, the trajectory of X_H starting at μ stays in* $\operatorname{Orb}(\mu)$.

ii *A function $C \in \mathcal{F}(\mathfrak{g}^*)$ is a Casimir iff $\delta C/\delta\mu \in \mathfrak{g}_\mu$ for all $\mu \in \mathfrak{g}^*$.*

iii *If $C \in \mathcal{F}(\mathfrak{g}^*)$ is* Ad^**-invariant (constant on orbits) then C is a Casimir. The converse is also true if all coadjoint orbits are connected.*

Proof i follows from the fact that $X_H(\nu)$ is tangent to the coadjoint orbit $\mathcal{O}$ for $\nu \in \mathcal{O}$, since $X_H(\nu) = \operatorname{ad}^*_{\delta H/\delta\mu}(\nu)$. **ii** follows from the definitions and formula (14.6.5) and **iii** follows from **ii** by writing out the condition of Ad^*-invariance as $C(\operatorname{Ad}^*_{g^{-1}} \mu) = C(\mu)$ and differentiating in g at $g = e$.

The converse is proved in the following way. If P is a Poisson manifold, $S \subset P$ is a symplectic leaf, and C is a Casimir function, then C is necessarily constant on S. Indeed, if C were not locally a constant on

S, then there would be a point $z \in S$ such that $\mathbf{d}C(z) \cdot v \neq 0$ for some $v \in T_z S$. But $T_z S$ is spanned by $X_k(z)$ for k varying over $\mathcal{F}(P)$ and hence $\mathbf{d}C(z) \cdot X_k(z) = \{C, k\}(z) = 0$. Therefore $\mathbf{d}C(z) \cdot v = 0$, a contradiction. Thus C is locally constant on S and hence constant, by connectedness of the leaf S. In particular, if all coadjoint orbits of a Lie group G in $\mathfrak{g}^*$ are connected, then a Casimir function C is constant on each orbit and hence Ad^*-invariant. ■

To illustrate **iii**, we note that for $G = SO(3)$, the function $C_\Phi(\mathbf{\Pi}) = \Phi\left(\frac{1}{2}\|\mathbf{\Pi}\|^2\right)$ is invariant under the coadjoint action $(\mathbf{A}, \mathbf{\Pi}) \mapsto \mathbf{A\Pi}$ and is therefore a Casimir function. Another example is given by $G = \mathrm{Diff}_{\mathrm{can}}(P)$, and the functional

$$C_\Phi(f) := \int_P \Phi(f)\, dq\, dp,$$

where $dq\, dp$ is the Liouville measure and Φ is any function of one variable. This is a Casimir since it is Ad^*-invariant by the change of variables formula.

In general, Ad^*-invariance of C is a stronger condition than C being a Casimir function. Indeed if C is Ad^*-invariant, differentiating the relation $C(\mathrm{Ad}^*_{g^{-1}}\mu) = C(\mu)$ relative to μ rather than g as we did in the proof of **iii**, we get

$$\frac{\delta C}{\delta(\mathrm{Ad}^*_{g^{-1}}\mu)} = \mathrm{Ad}_g \frac{\delta C}{\delta \mu} \tag{14.6.9}$$

for all $g \in G$. Taking $g \in G_\mu$, this relation becomes $\delta C/\delta\mu = \mathrm{Ad}_g(\delta C/\delta\mu)$, that is, $\delta C/\delta\mu$ belongs to the centralizer of G_μ in $\mathfrak{g}$, that is, to the set

$$\mathrm{Cent}(G_\mu, \mathfrak{g}) := \{\xi \in \mathfrak{g} \mid \mathrm{Ad}_g \xi = \xi \quad \text{for all} \quad g \in G_\mu\}.$$

Letting

$$\mathrm{Cent}(\mathfrak{g}_\mu, \mathfrak{g}) := \{\xi \in \mathfrak{g} \mid [\eta, \xi] = 0 \quad \text{for all} \quad \eta \in \mathfrak{g}_\mu\}$$

denote the centralizer of $\mathfrak{g}_\mu$ in $\mathfrak{g}$, we see by differentiating the relation defining $\mathrm{Cent}(G_\mu, \mathfrak{g})$ with respect to g at the identity, that $\mathrm{Cent}(G_\mu, \mathfrak{g}) \subset \mathrm{Cent}(\mathfrak{g}_\mu, \mathfrak{g})$. Thus, if C is Ad^*-invariant, then

$$\frac{\delta C}{\delta \mu} \in \mathfrak{g}_\mu \cap \mathrm{Cent}(\mathfrak{g}_\mu, \mathfrak{g}) = \mathrm{Cent}(\mathfrak{g}_\mu) = \text{ the center of } \mathfrak{g}_\mu.$$

Proposition 14.6.4 (Kostant[1979]) *If C is an Ad^*-invariant function on $\mathfrak{g}^*$, then $\delta C/\delta\mu$ lies in both $\mathrm{Cent}(G_\mu, \mathfrak{g})$ and in $\mathrm{Cent}(\mathfrak{g}_\mu)$. If C is a Casimir function, then $\delta C/\delta\mu$ lies in the center of $\mathfrak{g}_\mu$.*

Proof The preceding calculations prove the first statement. The last statement follows from the observation that if C is a Casimir function for G, it is also one for G_0, the connected component of the identity, and so from **14.6.3iii** it is G_0-Ad^*-invariant, so the first statement applies. ■

By the theorem of Duflo and Vergne [1969] (see Chapter 9), for generic $\mu \in \mathfrak{g}^*$, the coadjoint isotropy $\mathfrak{g}_\mu$ is abelian and therefore $\mathrm{Cent}(\mathfrak{g}_\mu) = \mathfrak{g}_\mu$ generically. The above corollary and proposition leave open, in principle, the possibility of non-Ad^*-invariant Casimir functions on $\mathfrak{g}^*$. This is not possible for Lie groups with connected coadjoint orbits, as we saw before. It is also not possible for semisimple Lie groups since any Casimir function is a functional of the basis of the ring of invariants. *If $C : \mathfrak{g}^* \to \mathbb{R}$ is a function such that $\delta C/\delta\mu \in \mathfrak{g}_\mu$ for all $\mu \in \mathfrak{g}^*$, but there is at least one $\nu \in \mathfrak{g}^*$ such that $\delta C/\delta\nu \notin \mathrm{Cent}(\mathfrak{g}_\nu)$, then C is a Casimir function that is not Ad^*-invariant.* This element $\nu \in \mathfrak{g}^*$ must be such that its coadjoint orbit is disconnected, it must be nongeneric, and $\mathfrak{g}$ must be non-semisimple. We know of no such example of a Casimir function.

On the other hand, the above statements provide easily verifiable criteria for the form or the nonexistence of Casimir functions on duals of Lie algebras. For example, if $\mathfrak{g}^*$ has open orbits whose union is dense, it cannot have Casimir functionals. Indeed, any Casimir would have to be constant on each orbit, and thus by continuity, on $\mathfrak{g}^*$. An example of such a Lie algebra is that of the affine group on the line discussed in Example **(b)** of §14.1. The same argument shows that Lie algebras with at least one dense orbit have no Casimir functionals.

Let us use Corollary **14.6.3** to determine all Casimir functions for the Lie algebra in Example **(f)** of §14.1. If

$$\mu = \begin{bmatrix} iu & 0 & 0 \\ 0 & i\alpha u & 0 \\ a & b & 0 \end{bmatrix} \in \mathfrak{g}^*, \quad \xi = \begin{bmatrix} is & 0 & x \\ 0 & i\alpha s & y \\ 0 & 0 & 0 \end{bmatrix} \in \mathfrak{g},$$

for $a, b, x, y \in \mathbb{C}$, $u, s \in \mathbb{R}$, then it is straightforward to check that

$$\mathrm{ad}_\xi^* \mu = \begin{bmatrix} iu'' & 0 & 0 \\ 0 & i\alpha u'' & 0 \\ -ixa & -i\alpha sb & 0 \end{bmatrix} \quad \text{for} \quad u'' = \frac{1}{1+\alpha^2}\mathrm{Im}(ax + \alpha bxy).$$

Thus if at least one of a, b is not zero, then

$$\mathfrak{g}_\mu = \left\{ \begin{bmatrix} 0 & 0 & x \\ 0 & 0 & y \\ 0 & 0 & 0 \end{bmatrix} \middle| \, \mathrm{Im}(ax + \alpha by) = 0 \right\},$$

whereas if $a = b = 0$, then $\mathfrak{g}_\mu = \mathfrak{g}$. For $C : \mathfrak{g}^* \to \mathbb{R}$ denote by

$$\frac{\delta C}{\delta \mu} = \begin{bmatrix} iC_u & 0 & C_a \\ 0 & i\alpha C_u & C_b \\ 0 & 0 & 0 \end{bmatrix},$$

where $C_u \in \mathbb{R}, C_a, C_b \in \mathbb{C}$ are the partial derivatives of C relative to the variables u, a, b. Thus, the condition $\delta C/\delta\mu \in \mathfrak{g}_\mu$ for all μ implies that $C_u = 0$, that is, C is independent of u and

$$\operatorname{Im}(aC_a + \alpha bC_b) = 0.$$

The same condition could have been obtained by lengthier direct calculations involving the Lie-Poisson bracket. Here are the highlights. The commutator bracket on $\mathfrak{g}$ is given by

$$\left[\begin{bmatrix} is & 0 & x \\ 0 & i\alpha s & y \\ 0 & 0 & 0 \end{bmatrix}, \begin{bmatrix} iu & 0 & z \\ 0 & i\alpha u & w \\ 0 & 0 & 0 \end{bmatrix}\right] = \begin{bmatrix} 0 & 0 & i(sz - ux) \\ 0 & 0 & i\alpha(sw - uy) \\ 0 & 0 & 0 \end{bmatrix},$$

so that for $\mu \in \mathfrak{g}^*$ parametrized by $u \in \mathbb{R}, a, b, \in \mathbb{C}$, we have

$$\begin{aligned} \{F, H\}_-(\mu) &= -\operatorname{Re}\operatorname{Tr}\left(\mu\left[\frac{\delta F}{\delta\mu}, \frac{\delta H}{\delta\mu}\right]\right) \\ &= \operatorname{Im}[a(F_u H_a - H_u F_a) + \alpha b(F_u H_b - H_u F_b)]. \end{aligned}$$

Taking $F_u = F_b = 0$ in $\{F, C\}_- = 0$, forces $C_u = 0$. Then the remaining condition reduces to $\operatorname{Im}(aC_a + \alpha bC_b) = 0$.

We solve this partial differential equation by the method of characteristics. Let $a = x + iy, b = u + iv$ so that $C_a = C_x + iC_y,\ C_b = C_u + iC_v$, and we get

$$xC_y + yC_x + \alpha vC_u + \alpha uC_v = 0.$$

The flow of the vector field with components $(y, x, \alpha v, \alpha u)$ is given by

$$\begin{aligned} F_t(x, y, u, v) &= \Big(x\cosh t + y\sinh t,\ x\sinh t + y\cosh t, \\ &\qquad u\cosh\alpha t + v\sinh\alpha t,\ u\sinh\alpha t + v\cosh\alpha t\Big) \end{aligned}$$

and thus any function $C = f(x^2 - y^2, u^2 - v^2)$ is constant on this flow. Therefore, these functions are all Casimir functionals for $\mathfrak{g}^*$.

One mathematical reason coadjoint orbits and the Lie-Poisson bracket are so important is that every Hamiltonian space is (a covering of) a coadjoint orbit. This is proved below.

If X and Y are topological spaces, a continuous surjective map $p : X \to Y$ is called a ***covering map*** if every point in Y has an open neighborhood U such that $p^{-1}(U)$ is a disjoint union of open sets in X, called the ***decks*** over U. Note that each deck is homeomorphic to U by p. If $p : M \to N$ is a surjective proper map of smooth manifolds which is also a local diffeomorphism, then it is a covering map. For example, $SU(2)$ (the spin group) forms a covering space of $SO(3)$ with two decks over each point and $SU(2)$ is simply connected while $SO(3)$ is not. (See Chapter 9.)

Transitive Hamiltonian actions have been characterized by Lie, Kostant, Kirillov, and Souriau in the following manner (see Kostant [1966]):

Theorem 14.6.5 (Kostant's Coadjoint Orbit Covering Theorem) *Let P be a Poisson manifold and let $\Phi : G \times P \to P$ be a left, transitive, Hamiltonian action with equivariant momentum map $\mathbf{J} : P \to \mathfrak{g}^*$. Then*

i $\mathbf{J} : P \to \mathfrak{g}^*_+$ *is a canonical submersion onto a coadjoint orbit of G in $\mathfrak{g}^*$.*

ii *If P is symplectic, $\mathbf{J}$ is a symplectic local diffeomorphism onto a coadjoint orbit endowed with the "+" orbit symplectic structure. If $\mathbf{J}$ is also proper, then it is a covering map.*

Proof i That $\mathbf{J}$ is a canonical map was proved in §12.5. Since Φ is transitive, choosing a $z_0 \in P$, any $z \in P$ can be written as $z = \Phi_g(z_0)$ for some $g \in G$. Thus, by equivariance

$$\begin{aligned} \mathbf{J}(P) = \{\mathbf{J}(z) \mid z \in P\} &= \{\mathbf{J}(\Phi_g(z_0)) \mid g \in G\} \\ &= \{\mathrm{Ad}^*_{g^{-1}}\mathbf{J}(z_0) \mid g \in G\} = \mathrm{Orb}(\mathbf{J}(z_0)). \end{aligned}$$

Again by equivariance, for $z \in P$ we have $T_z\mathbf{J}(\xi_P(z)) = -\mathrm{ad}^*_\xi \mathbf{J}(z)$, which has the form of a general tangent vector at $\mathbf{J}(z)$ to the orbit $\mathrm{Orb}(\mathbf{J}(z_0))$; thus, $\mathbf{J}$ is a submersion.

ii If P is symplectic with symplectic form Ω, $\mathbf{J}$ is a symplectic map if the orbit has the "+" symplectic form: $\omega^+(\mu)(\mathrm{ad}^*_\xi\mu, \mathrm{ad}^*_\eta\mu) = \langle \mu, [\xi, \eta]\rangle$. This is seen in the following way. Since $T_zP = \{\xi_P(z) \mid \xi \in \mathfrak{g}\}$ by transitivity of the action,

$$\begin{aligned} (\mathbf{J}^*\omega^+)(z)(\xi_P(z), \eta_P(z)) &= \omega^+(\mathbf{J}(z))(T_z\mathbf{J}(\xi_P(z)), T_z\mathbf{J}(\eta_P(z))) \\ &= \omega^+(\mathbf{J}(z))(\mathrm{ad}^*_\xi \mathbf{J}(z), \mathrm{ad}^*_\eta \mathbf{J}(z)) \\ &= \langle \mathbf{J}(z), [\xi, \eta]\rangle = J([\xi, \eta])(z) \\ &= \{J(\xi), J(\eta)\}(z) \qquad \text{(by equivariance)} \\ &= \Omega(z)(X_{J(\xi)}(z), X_{J(\eta)}(z)) \\ &= \Omega(z)(\xi_P(z), \eta_P(z)), \end{aligned}$$

which shows that $\mathbf{J}^*\omega^+ = \Omega$, that is, $\mathbf{J}$ is symplectic. Since any symplectic map is an immersion, $\mathbf{J}$ is a local diffeomorphism. If $\mathbf{J}$ is also proper, it is a symplectic covering map, as discussed above. ■

If $\mathbf{J}$ is proper and the symplectic manifold P is simply connected, the covering map in **ii** is a diffeomorphism; this follows from classical theorems about covering spaces (Spanier [1966]). It is clear that if Φ is not transitive, $\mathbf{J}(P)$ is a union of coadjoint orbits. See Guillemin and Sternberg [1984] and Grigore and Popp [1989] for more information.

Exercise

14.6-1 *Show that if C is a Casimir function on a Poisson manifold, then $\{F, K\}_C = C\{F, K\}$ is also a Poisson structure.*

14.7 The Special Linear Group on the Plane

In the Lie algebra $\mathfrak{sl}(2, \mathbb{R})$ of traceless real 2×2 matrices, introduce the basis

$$\mathbf{e} = \begin{bmatrix} 0 & 1 \\ 0 & 0 \end{bmatrix}, \quad \mathbf{f} = \begin{bmatrix} 0 & 0 \\ 1 & 0 \end{bmatrix}, \quad \mathbf{h} = \begin{bmatrix} 1 & 0 \\ 0 & -1 \end{bmatrix}.$$

Note that $[\mathbf{h}, \mathbf{e}] = 2\mathbf{e}$, $[\mathbf{h}, \mathbf{f}] = -2\mathbf{f}$, and $[\mathbf{e}, \mathbf{f}] = \mathbf{h}$. Identify $\mathfrak{sl}(2, \mathbb{R})$ with $\mathbb{R}^3$ via

$$x\mathbf{e} + y\mathbf{f} + z\mathbf{h} \in \mathfrak{sl}(2, \mathbb{R}) \mapsto (x, y, z) \in \mathbb{R}^3. \tag{14.7.1}$$

The nonzero structure constants are $c_{12}^3 = 1, c_{13}^1 = -2$, and $c_{23}^2 = 2$. We identify the dual space $\mathfrak{sl}(2, \mathbb{R})^*$ with $\mathbb{R}^3$ via the map

$$\alpha \in \mathfrak{sl}(2, \mathbb{R})^* \mapsto (a, b, c) \in \mathbb{R}^3, \tag{14.7.2}$$

where $(a, b, c) \in \mathbb{R}^3$ is uniquely determined by the equality

$$\langle \alpha, x\mathbf{e} + y\mathbf{f} + z\mathbf{h} \rangle = ax + by + cz, \quad \text{i.e.,} \quad \alpha(\mathbf{e}) = a, \alpha(\mathbf{f}) = b, \alpha(\mathbf{h}) = c. \tag{14.7.3}$$

One calculates that the $(\pm)$ Lie-Poisson bracket of $\mathfrak{sl}(2, \mathbb{R})^*$ induces the following Poisson brackets on $\mathbb{R}^3$: $\{c, a\} = 2a$ (respectively, $-2a$), $\{c, b\} = -2b$ (respectively, $+2b$), $\{a, b\} = c$ (respectively, $-c$), that is,

$$\begin{aligned} \{F, G\}_\pm(a, b, c) &= \mp 2a \left(\frac{\partial F}{\partial a} \frac{\partial G}{\partial c} - \frac{\partial F}{\partial c} \frac{\partial G}{\partial a} \right) \pm 2b \left(\frac{\partial F}{\partial b} \frac{\partial G}{\partial c} - \frac{\partial F}{\partial c} \frac{\partial G}{\partial b} \right) \\ &\quad \pm c \left(\frac{\partial F}{\partial a} \frac{\partial G}{\partial b} - \frac{\partial F}{\partial b} \frac{\partial G}{\partial a} \right). \end{aligned} \tag{14.7.4}$$

Any Casimir function of $\mathbb{R}^3$ endowed with the $(\pm)$ Lie-Poisson bracket of $\mathfrak{sl}(2, \mathbb{R})^*$ is of the form

$$C(a, b, c) = \Phi \left(ab + \frac{1}{4} c^2 \right) \tag{14.7.5}$$

for a C^1 function $\Phi : \mathbb{R} \to \mathbb{R}$. Thus the symplectic leaves are the sheets of the hyperboloids

$$C_0(a, b, c) := \frac{1}{2} \left(ab + \frac{1}{4} c^2 \right) = \text{constant} \neq 0, \tag{14.7.6}$$

the two nappes (without vertex) of the cone $ab + (1/4)c^2 = 0$, and the origin. The orbit symplectic structure on these hyperboloids is given by

$$\begin{aligned}
&\omega^-(a,b,c)(\mathrm{ad}^*_{(x,y,z)}(a,b,c), \mathrm{ad}^*_{(x',y',z')}(a,b,c)) \\
&\quad = -a(2zx' - 2xz') - b(2yz' - 2zy') - c(xy' - yx') \\
&\quad = -\frac{1}{\|\nabla C_0(a,b,c)\|} \text{ (area element of the hyperboloid)}. \quad (14.7.7)
\end{aligned}$$

To prove the last equality in (14.7.7), use the formulae

$$\begin{aligned}
\mathrm{ad}^*_{(x,y,z)}(a,b,c) &= (2az - cy, cx - 2bz, 2by - 2zx), \\
\mathrm{ad}^*_{(x,y,z)}(a,b,c) &\times ab^*_{(x',y',z')}(a,b,c) \\
&= (2bc(xy' - yx') + 4b^2(yz' - zy') + 4ab(zx' - xz'), \\
&\quad 2ac(xy' - yx') + 4ab(yz' - zy') + 4a^2(zx' - xz'), \\
&\quad c^2(xy' - yx') + 2bc(yz' - zy') + 2ac(zx' - xz')),
\end{aligned}$$

and the fact that $\nabla(ab + \frac{1}{4}c^2) = (b, a, \frac{1}{2}c)$ is normal to the hyperboloid to get, as in (14.5.3),

$$\begin{aligned}
&dA(a,b,c)(\mathrm{ad}^*_{(x,y,z)}(a,b,c), \mathrm{ad}^*_{(x',y',z')}(a,b,c)) \\
&= \frac{(b, a, \frac{1}{2}c)}{\|(b, a, \frac{1}{2}c)\|} \cdot (\mathrm{ad}^*_{(x,y,z)}(a,b,c) \times \mathrm{ad}^*_{(x',y',z')}(a,b,c)) \\
&= -\|\nabla C_0(a,b,c)\|)\omega^-(a,b,c)(\mathrm{ad}^*_{(x,y,z)}(a,b,c), \mathrm{ad}^*_{(x',y',z')}(a,b,c)).
\end{aligned}$$

14.8 The Euclidean Group of the Plane

We use the notation and terminology from Exercise **11.5-2**. Recall that $\mathfrak{se}(2)^*$ is isomorphic to $\mathbb{R}^3$ with the bracket.

$$\begin{aligned}
[(\omega, v_1, v_2), (\zeta, w_1, w_2)] &= (0, \zeta v_2 - \omega w_2, \omega w_1 - \zeta v_1) \\
&= (0, \omega \mathbb{J}^T \mathbf{w} - \zeta \mathbb{J}^T \mathbf{v}), \qquad (14.8.1)
\end{aligned}$$

where $\mathbf{v} = (v_1, v_2)$, $\mathbf{w} = (w_1, w_2)$ and

$$\mathbb{J} = \begin{bmatrix} 0 & 1 \\ -1 & 0 \end{bmatrix}, \quad \mathbb{J}^t = \mathbb{J}^{-1} = -\mathbb{J}. \qquad (14.8.2)$$

Thus $\mathfrak{se}(2)^*$ is identified with $\mathbb{R}^3$ via the dot product. Therefore, if $F : \mathfrak{se}(2)^* \cong \mathbb{R} \times \mathbb{R}^2 \to \mathbb{R}$, its functional derivative is

$$\frac{\delta F}{\delta(\mu, \alpha)} = \left(\frac{\partial F}{\partial \mu}, \nabla_\alpha F \right), \qquad (14.8.3)$$

where $(\mu, \alpha) \in \mathfrak{se}(2)^* \cong \mathbb{R} \times \mathbb{R}^2$ and $\nabla_\alpha F$ denotes the gradient of F with respect to α. The $(\pm)$ Lie-Poisson structure on $\mathfrak{se}(2)^*$ is given by

$$\{F, G\}_\pm(\mu, \alpha) = \pm\left(\frac{\partial F}{\partial \mu}\mathbb{J}\alpha \cdot \nabla_\alpha G - \frac{\partial G}{\partial \mu}\mathbb{J}\alpha \cdot \nabla_\alpha F\right). \tag{14.8.4}$$

One also checks that functions on $\mathfrak{se}(2)^*$, of the form

$$C(\mu, \alpha) = \Phi\left(\frac{1}{2}\|\alpha\|^2\right) \tag{14.8.5}$$

for a (smooth) function $\Phi : [0, \infty[\to \mathbb{R}$, are Casmir functions and that the symplectic leaves of $\mathfrak{se}(2)^*$ are the cylinders

$$\{(\mu, \alpha) \in \mathbb{R}^3 \mid \|\alpha\| = \text{constant} \neq 0\} \tag{14.8.6}$$

and the points on the μ-axis.

On the coadjoint orbit representing a cylinder about the μ-axis, the orbit symplectic structure is

$$\begin{aligned}
&\omega(\mu, \alpha)(\operatorname{ad}(\xi, \mathbf{u})^*(\mu, \alpha), \operatorname{ad}(\eta, \mathbf{v})^*(\mu, \alpha)) \\
&\quad = \pm(\xi\mathbb{J}\alpha \cdot \mathbf{v} - \eta\mathbb{J}\alpha \cdot \mathbf{u}) \\
&\quad = \pm(\text{area element } dA \text{ on the cylinder})/\|\alpha\|.
\end{aligned} \tag{14.8.7}$$

The last equality is proved in the following way. Since

$$\operatorname{ad}(\xi, \mathbf{u})^*(\mu, \alpha) = (-\mathbb{J}\alpha \cdot \mathbf{u}, \xi\mathbb{J}\alpha) \tag{14.8.8}$$

and the outward unit normal to the cylinder is $(0, \alpha)/\|\alpha\|$, the area element dA is given by

$$\begin{aligned}
&dA(\mu, \alpha)((-\mathbb{J}\alpha \cdot \mathbf{u}, \xi\mathbb{J}\alpha), (-\mathbb{J}\alpha \cdot \mathbf{v}, \eta\mathbb{J}\alpha)) \\
&\quad = \frac{(0, \alpha)}{\|\alpha\|} \cdot [((-\mathbb{J}\alpha \cdot \mathbf{u}, \xi\mathbb{J}\alpha) \times (-\mathbb{J}\alpha \cdot \mathbf{u}, \xi\mathbb{J}\alpha)] \\
&\quad = \|\alpha\|(\xi\mathbb{J}\alpha \cdot \mathbf{v} - \eta\mathbb{J}\alpha \cdot \mathbf{u}).
\end{aligned}$$

The Poisson structures of $\mathfrak{so}(3)^*, \mathfrak{sl}(2, \mathbb{R})^*$, and $\mathfrak{se}(2)^*$ fit together in a larger Poisson manifold. Weinstein [1983b] considers for every $\varepsilon \in \mathbb{R}$ the Lie algebra $\mathfrak{g}_\varepsilon$ with abstract basis $\mathbf{X}_1, \mathbf{X}_2, \mathbf{X}_3$ and relations

$$[\mathbf{X}_3, \mathbf{X}_1] = \mathbf{X}_2, \quad [\mathbf{X}_2, \mathbf{X}_3] = \mathbf{X}_1, \quad [\mathbf{X}_1, \mathbf{X}_2] = \varepsilon\mathbf{X}_3. \tag{14.8.9}$$

If $\varepsilon > 0$, the map

$$\mathbf{X}_1 \mapsto \sqrt{\varepsilon}(1, 0, 0)^\wedge, \quad \mathbf{X}_2 \mapsto \sqrt{\varepsilon}(0, 1, 0)^\wedge, \quad \mathbf{X}_3 \mapsto \sqrt{\varepsilon}(0, 0, 1)^\wedge, \tag{14.8.10}$$

defines an isomorphism of $\mathfrak{g}_\varepsilon$ with $\mathfrak{so}(3)$, while if $\varepsilon = 0$, the map

$$\mathbf{X}_1 \mapsto (0, 0, -1), \quad \mathbf{X}_2 \mapsto (0, -1, 0), \quad \mathbf{X}_3 \mapsto (-1, 0, 0), \tag{14.8.11}$$

defines an isomorphism of $\mathfrak{g}_0$ with $\mathfrak{se}(2)$, and if $\varepsilon < 0$, the map

$$\mathbf{X}_1 \mapsto \frac{\sqrt{-\varepsilon}}{2}\begin{bmatrix} 1 & 0 \\ 0 & -1 \end{bmatrix}, \quad \mathbf{X}_2 \mapsto \frac{\sqrt{-\varepsilon}}{2}\begin{bmatrix} 1 & 0 \\ 0 & 1 \end{bmatrix}, \quad \mathbf{X}_3 \mapsto \frac{1}{2}\begin{bmatrix} 0 & -1 \\ 1 & 0 \end{bmatrix}, \tag{14.8.12}$$

defines an isomorphism of $\mathfrak{g}_\varepsilon$ with $\mathfrak{sl}(2, \mathbb{R})$.

The (+) Lie-Poisson structure of $\mathfrak{g}_\varepsilon^*$ is given by the bracket relations

$$\{x_3, x_1\} = x_2, \quad \{x_2, x_3\} = x_1, \quad \{x_1, x_2\} = \varepsilon x_3, \tag{14.8.13}$$

for the coordinate functions $x_i \in \mathfrak{g}_\varepsilon^* = \mathbb{R}^3$, $\langle x_i, x_j \rangle = \delta_{ij}$.

In $\mathbb{R}^4$ with coordinate functions $(x_1, x_2, x_3, \varepsilon)$ consider the above bracket relations plus $\{\varepsilon, x_1\} = \{\varepsilon, x_2\} = \{\varepsilon, x_3\} = 0$. This defines a Poisson structure on $\mathbb{R}^4$ which is not of Lie-Poisson type. The leaves of this Poisson structure are all two dimensional in the space (x_1, x_2, x_3) and the Casimir functions are all functions of $x_1^2 + x_2^2 + \varepsilon x_3^2$ and ε. The inclusion of $\mathfrak{g}_\varepsilon^*$ in $\mathbb{R}^4$ with the above Poisson structure is a canonical map. The leaves of $\mathbb{R}^4$ with the above Poisson structure as ε passes through zero is given in Figure 14.8.1.

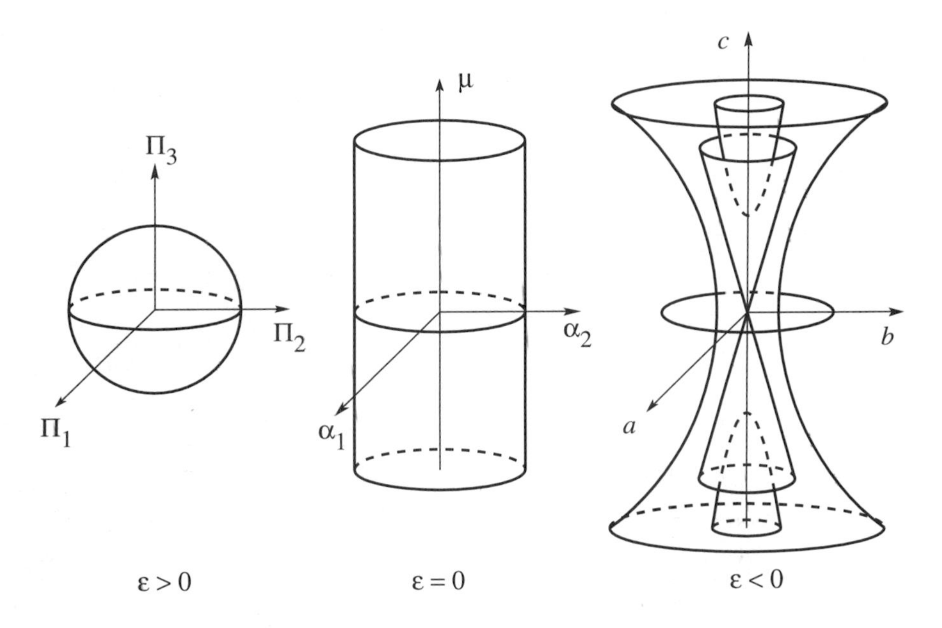

FIGURE 14.8.1. The orbit structure for $\mathfrak{so}(3)^*, \mathfrak{se}(3)^*$, and $\mathfrak{sl}(2, \mathbb{R})^*$.

14.9 The Euclidean Group of Three-Space

The Euclidean Group, its Lie Algebra and its Dual An element of $SE(3)$ is a pair $(\mathbf{A}, \mathbf{a})$ where $\mathbf{A} \in SO(3)$ and $\mathbf{a} \in \mathbb{R}^3$; the action of $SE(3)$ on $\mathbb{R}^3$ is the rotation $\mathbf{A}$ followed by translation by the vector $\mathbf{a}$ and has the expression

$$(\mathbf{A}, \mathbf{a}) \cdot \mathbf{x} = \mathbf{A}\mathbf{x} + \mathbf{a}. \tag{14.9.1}$$

Using this formula, one sees that multiplication and inversion in $SE(3)$ are given by

$$(\mathbf{A}, \mathbf{a})(\mathbf{B}, \mathbf{b}) = (\mathbf{A}\mathbf{B}, \mathbf{A}\mathbf{b} + \mathbf{a}) \tag{14.9.2}$$

and

$$(\mathbf{A}, \mathbf{a})^{-1} = (\mathbf{A}^{-1}, -\mathbf{A}^{-1}\mathbf{a}), \tag{14.9.3}$$

for $\mathbf{A}, \mathbf{B} \in SO(3)$ and $\mathbf{a}, \mathbf{b} \in \mathbb{R}^3$. The identity element is $(\mathbf{I}, \mathbf{0})$. Note that $SE(3)$ embeds into $SL(4; \mathbb{R})$ via the map

$$(\mathbf{A}, \mathbf{a}) \mapsto \begin{bmatrix} \mathbf{A} & \mathbf{a} \\ 0 & 1 \end{bmatrix}; \tag{14.9.4}$$

thus one can operate with $SE(3)$ as one would with matrix Lie groups by using this embedding. In particular, the Lie algebra $\mathfrak{se}(3)$ of $SE(3)$ is isomorphic to a Lie subalgebra of $\mathfrak{sl}(4; \mathbb{R})$ with elements of the form

$$\begin{bmatrix} \hat{\mathbf{x}} & \mathbf{y} \\ 0 & 0 \end{bmatrix}, \quad \text{where} \quad \mathbf{x}, \mathbf{y} \in \mathbb{R}^3, \tag{14.9.5}$$

and a Lie algebra bracket equal to the commutator bracket of matrices. This shows that the Lie bracket operation on $\mathfrak{se}(3)$ is given by

$$[(\mathbf{x}, \mathbf{y}), (\mathbf{x}', \mathbf{y}')] = (\mathbf{x} \times \mathbf{x}', \mathbf{x} \times \mathbf{y}' - \mathbf{x}' \times \mathbf{y}). \tag{14.9.6}$$

Since

$$\begin{bmatrix} \mathbf{A} & \mathbf{a} \\ 0 & 1 \end{bmatrix}^{-1} = \begin{bmatrix} \mathbf{A}^{-1} & -\mathbf{A}^{-1}\mathbf{a} \\ 0 & 1 \end{bmatrix}$$

and

$$\begin{bmatrix} \mathbf{A} & \mathbf{a} \\ 0 & 1 \end{bmatrix} \begin{bmatrix} \hat{\mathbf{x}} & \mathbf{y} \\ 0 & 0 \end{bmatrix} \begin{bmatrix} \mathbf{A}^{-1} & -\mathbf{A}^{-1}\mathbf{a} \\ 0 & 1 \end{bmatrix} = \begin{bmatrix} \mathbf{A}\hat{\mathbf{x}}\mathbf{A}^{-1} & -\mathbf{A}\hat{\mathbf{x}}\mathbf{A}^{-1}\mathbf{a} + \mathbf{A}\mathbf{y} \\ 0 & 0 \end{bmatrix},$$

we see that the adjoint action of $SE(3)$ on $\mathfrak{se}(3)$ has the expression

$$\mathrm{Ad}_{(\mathbf{A}, \mathbf{a})}(\mathbf{x}, \mathbf{y}) = (\mathbf{A}\mathbf{x}, \mathbf{A}\mathbf{y} - \mathbf{A}\mathbf{x} \times \mathbf{a}). \tag{14.9.7}$$

The (6×6)-matrix of $\mathrm{Ad}_{(\mathbf{A}, \mathbf{a})}$ is given by

$$\begin{bmatrix} \mathbf{A} & 0 \\ \hat{\mathbf{a}}\mathbf{A} & \mathbf{A} \end{bmatrix}. \tag{14.9.8}$$

Identifying the dual of $\mathfrak{se}(3)$ with $\mathbb{R}^3 \times \mathbb{R}^3$ by the dot product in every factor, it follows that the matrix of $\mathrm{Ad}^*_{(A,a)^{-1}}$ is given by the inverse transpose of the (6×6)-matrix (14.9.8), that is, it equals

$$\begin{bmatrix} \mathbf{A} & \hat{\mathbf{a}}\mathbf{A} \\ 0 & \mathbf{A} \end{bmatrix}. \tag{14.9.9}$$

Thus, the coadjoint action of $SE(3)$ on $\mathfrak{se}(3)^* = \mathbb{R}^3 \times \mathbb{R}^3$ has the expression

$$\mathrm{Ad}^*_{(A,a)^{-1}}(\mathbf{u}, \mathbf{v}) = (\mathbf{A}\mathbf{u} + \mathbf{a} \times \mathbf{A}\mathbf{v}, \mathbf{A}\mathbf{v}). \tag{14.9.10}$$

(This Lie algebra is a semidirect product and all formulae derived here "by hand" are special cases of general ones that may be found in works on semidirect products; see, for example, Marsden, Ratiu, and Weinstein [1984a,b].)

Coadjoint Orbits in $\mathfrak{se}(3)^*$ Let $\{\mathbf{e}_1, \mathbf{e}_2, \mathbf{e}_3, \mathbf{f}_1, \mathbf{f}_2, \mathbf{f}_3\}$ be an orthonormal basis of $\mathfrak{se}(3) = \mathbb{R}^3 \times \mathbb{R}^3$ such that $\mathbf{e}_i = \mathbf{f}_i, i = 1, 2, 3$. The dual basis of $\mathfrak{se}(3)^*$ via the dot product is again $\{\mathbf{e}_1, \mathbf{e}_2, \mathbf{e}_3, \mathbf{f}_1, \mathbf{f}_2, \mathbf{f}_3\}$. Let $\mathbf{e}$ and $\mathbf{f}$ denote arbitrary vectors satisfying $\mathbf{e} \in \mathrm{span}\{\mathbf{e}_1, \mathbf{e}_2, \mathbf{e}_3\}$ and $\mathbf{f} \in \mathrm{span}\{\mathbf{f}_1, \mathbf{f}_2, \mathbf{f}_3\}$. For the coadjoint action the only zero-dimensional orbit is the origin. Since $\mathfrak{se}(3)$ is six dimensional, there can also be two- and four-dimensional coadjoint orbits. These in fact occur and fall into three types.

Type I The orbit through $(\mathbf{e}, \mathbf{0})$ equals

$$SE(3) \cdot (\mathbf{e}, \mathbf{0}) = \{(\mathbf{A}\mathbf{e}, \mathbf{0}) \mid \mathbf{A} \in SO(3)\} = S^2_{\|\mathbf{e}\|}, \tag{14.9.11}$$

the two-sphere of radius $\|\mathbf{e}\|$.

Type II The orbit through $(\mathbf{0}, \mathbf{f})$

$$\begin{aligned} & SE(3) \cdot (\mathbf{0}, \mathbf{f}) \\ = \;& \{(\mathbf{a} \times \mathbf{A}\mathbf{f}, \mathbf{A}\mathbf{f}) | \mathbf{A} \in SO(3), \mathbf{a} \in \mathbb{R}^3\} \\ = \;& \{(\mathbf{u}, \mathbf{A}\mathbf{f}) | \mathbf{A} \in SO(3), \mathbf{u} \perp \mathbf{A}\mathbf{f}\} = TS^2_{\|\mathbf{f}\|}, \end{aligned} \tag{14.9.12}$$

the tangent bundle of the two-sphere of radius $\|\mathbf{f}\|$; note the vector part is in the first slot.

Type III The orbit through $(\mathbf{e}, \mathbf{f})$, where $\mathbf{e}, \mathbf{f} \neq \mathbf{0}$, equals

$$SE(3) \cdot (\mathbf{e}, \mathbf{f}) = \{(\mathbf{A}\mathbf{e} + \mathbf{a} \times \mathbf{A}\mathbf{f}, \mathbf{A}\mathbf{f}) \mid \mathbf{A} \in SO(3), \mathbf{a} \in \mathbb{R}^3\}. \tag{14.9.13}$$

We will prove below that this orbit is diffeomorphic to $TS^2_{\|\mathbf{f}\|}$. Consider the smooth map

$$\varphi : (\mathbf{A}, \mathbf{a}) \in SE(3) \mapsto \left(\mathbf{A}\mathbf{e} + \mathbf{a} \times \mathbf{A}\mathbf{f} - \frac{\mathbf{e} \cdot \mathbf{f}}{\|\mathbf{f}\|^2} \mathbf{A}\mathbf{f}, \mathbf{A}\mathbf{f} \right) \in TS^2_{\|\mathbf{f}\|} \tag{14.9.14}$$

which is right invariant under the isotropy group

$$SE(3)_{(\mathrm{e},\mathrm{f})} = \{(\mathbf{B}, \mathbf{b}) \mid \mathbf{Be} + \mathbf{b} \times \mathbf{f} = \mathbf{e}, \;\; \mathbf{Bf} = \mathbf{f}\} \tag{14.9.15}$$

(see (14.9.10)), that is,

$$\varphi((\mathbf{A}, \mathbf{a})(\mathbf{B}, \mathbf{b})) = \varphi(\mathbf{A}, \mathbf{a})$$

for all $(\mathbf{A}, \mathbf{a}) \in SE(3)$ and $(\mathbf{B}, \mathbf{b}) \in SE(3)_{(\mathrm{e},\mathrm{f})}$. Thus φ induces a smooth map $\bar{\varphi} : SE(3)/SE(3)_{(\mathrm{e},\mathrm{f})} \to TS^2_{\|\mathrm{f}\|}$. The map $\bar{\varphi}$ is injective, for if $\varphi(\mathbf{A}, \mathbf{a}) = \varphi(\mathbf{A}', \mathbf{a}')$, then

$$(\mathbf{A}, \mathbf{a})^{-1}(\mathbf{A}', \mathbf{a}') = (\mathbf{A}^{-1}\mathbf{A}', \mathbf{A}^{-1}(\mathbf{a}' - \mathbf{a})) \in SE(3)_{(\mathrm{e},\mathrm{f})}$$

as is easily checked. To see that φ (and hence $\bar{\varphi}$) is surjective, let $(\mathbf{u}, \mathbf{v}) \in TS^2_{\|\mathrm{f}\|}$, that is, $\|\mathbf{v}\| = \|\mathbf{f}\|$ and $\mathbf{u} \cdot \mathbf{v} = 0$. Then choose an $\mathbf{A} \in SO(3)$ such that $\mathbf{Af} = \mathbf{v}$ and let $\mathbf{a} = [\mathbf{v} \times (\mathbf{u} - \mathbf{Ae})]/\|\mathbf{f}\|^2$. It is then straightforward to check that $\varphi(\mathbf{A}, \mathbf{a}) = (\mathbf{u}, \mathbf{v})$ by (14.9.14). Thus $\bar{\varphi}$ is a bijective map. Since the derivative of φ at $(\mathbf{A}, \mathbf{a})$ in the direction $T_{(\mathrm{I},0)}L_{(\mathrm{A},\mathrm{a})}(\hat{\mathbf{x}}, \mathbf{y}) = (\mathbf{A}\hat{\mathbf{x}}, \mathbf{Ay})$ equals

$$\begin{aligned} T_{(\mathrm{A},\mathrm{a})}\varphi(\mathbf{A}\hat{\mathbf{x}}, \mathbf{Ay}) &= \left.\frac{d}{dt}\right|_{t=0} \varphi(\mathbf{A}e^{t\hat{\mathbf{x}}}, \mathbf{a} + t\mathbf{Ay}) \\ &= (\mathbf{A}(\mathbf{x} \times \mathbf{e} + \mathbf{y} \times \mathbf{f}) + \mathbf{a} \times \mathbf{A}(\mathbf{x} \times \mathbf{f}) \\ &\quad - \frac{\mathbf{e} \cdot \mathbf{f}}{\|\mathbf{f}\|^2}\mathbf{A}(\mathbf{x} \times \mathbf{f}), \mathbf{A}(\mathbf{x} \times \mathbf{f})) \end{aligned} \tag{14.9.16}$$

we see that its kernel consists of left translates by $(\mathbf{A}, \mathbf{a})$ of

$$\{(\mathbf{x}, \mathbf{y}) \in \mathfrak{se}(3) \mid \mathbf{x} \times \mathbf{e} + \mathbf{y} \times \mathbf{f} = \mathbf{0}, \mathbf{x} \times \mathbf{f} = 0\}. \tag{14.9.17}$$

However, taking the derivatives of the defining relations in (14.9.15) at $(\mathbf{B}, \mathbf{b}) = (\mathbf{I}, \mathbf{0})$ we see that (14.9.17) coincides with $\mathfrak{se}(3)_{(\mathrm{e},\mathrm{f})}$. This shows that $\overline{\varphi}$ is an immersion and hence, since $\dim(SE(3)/SE(3)_{(\mathrm{e},\mathrm{f})}) = \dim TS^2_{\|\mathrm{f}\|} = 4$, it follows that $\overline{\varphi}$ is a local diffeomorphism. Therefore φ is a diffeomorphism.

To compute the tangent spaces to these orbits, we use Proposition **14.2.1** which states that the annihilator of the coadjoint isotropy subalgebra at μ equals $T_\mu\mathcal{O}$. The coadjoint action of the Lie algebra $\mathfrak{se}(3)$ on its dual $\mathfrak{se}(3)^*$ is computed to be

$$\mathrm{ad}^*_{(\mathrm{x},\mathrm{y})}(\mathbf{u}, \mathbf{v}) = (\mathbf{u} \times \mathbf{x} + \mathbf{v} \times \mathbf{y}, \mathbf{v} \times \mathbf{x}). \tag{14.9.18}$$

Thus the isotropy subalgebra $\mathfrak{se}(3)_{(\mathrm{u},\mathrm{v})}$ is given again by (14.9.17), that is, it equals $\{(\mathbf{x}, \mathbf{y}) \in \mathfrak{se}(3) \mid \mathbf{u} \times \mathbf{x} + \mathbf{v} \times \mathbf{y} = \mathbf{0}, \mathbf{v} \times \mathbf{x} = \mathbf{0}\}$. Let $\mathcal{O}$ denote a nonzero coadjoint orbit in $\mathfrak{se}(3)^*$. Then the tangent space at a point in $\mathcal{O}$ is given as follows for each of the three types of orbits:

Type I Since

$$\mathfrak{se}(3)_{(\mathrm{e},0)} = \{(\mathbf{x}, \mathbf{y}) \in \mathfrak{se}(3) \mid \mathbf{e} \times \mathbf{x} = \mathbf{0}\} = \mathrm{span}(\mathbf{e}) \times \mathbb{R}^3, \tag{14.9.19}$$

it follows that the tangent space to $\mathcal{O}$ at $(\mathbf{e}, \mathbf{0})$ is the tangent space to the sphere of radius $\|\mathbf{e}\|$ at the point $\mathbf{e}$ in the first factor.

Type II Since

$$\begin{aligned} \mathfrak{se}(3)_{(0,\mathrm{f})} &= \{(\mathbf{x}, \mathbf{y}) \in \mathfrak{se}(3) \mid \mathbf{f} \times \mathbf{y} = \mathbf{0}, \mathbf{f} \times \mathbf{x} = \mathbf{0}\} \\ &= \mathrm{span}(\mathbf{f}) \times \mathrm{span}(\mathbf{f}), \end{aligned} \tag{14.9.20}$$

it follows that the tangent space to $\mathcal{O}$ at $(\mathbf{0}, \mathbf{f})$ equals $\mathbf{f}^\perp \times \mathbf{f}^\perp$, where $\mathbf{f}^\perp$ denotes the plane perpendicular to $\mathbf{f}$.

Type III Since

$$\begin{aligned} \mathfrak{se}(3)_{(\mathrm{e},\mathrm{f})} &= \{(\mathbf{x}, \mathbf{y}) \in \mathfrak{se}(3) \mid \mathbf{e} \times \mathbf{x} + \mathbf{f} \times \mathbf{y} = \mathbf{0} \text{ and } \mathbf{f} \times \mathbf{x} = \mathbf{0}\} \\ &= \{(c_1\mathbf{f}, c_1\mathbf{e} + c_2\mathbf{f}) \mid c_1, c_2 \in \mathbb{R}\}, \end{aligned} \tag{14.9.21}$$

the tangent space at $(\mathbf{e}, \mathbf{f})$ to $\mathcal{O}$ is the orthogonal complement of the space spanned by $(\mathbf{f}, \mathbf{e})$ and $(\mathbf{0}, \mathbf{f})$, that is, it equals

$$\{(\mathbf{u}, \mathbf{v}) \mid \mathbf{u} \cdot \mathbf{f} + \mathbf{v} \cdot \mathbf{e} = \mathbf{0} \quad \text{and} \quad \mathbf{v} \cdot \mathbf{f} = \mathbf{0}\}.$$

The Symplectic Form on Orbits Let $\mathcal{O}$ denote a nonzero orbit of $\mathfrak{se}(3)^*$. We consider the different oribt types separately, as above.

Type I If $\mathcal{O}$ contains a point of the form $(\mathbf{e}, \mathbf{0})$, the orbit $\mathcal{O}$ equals $S^2_{\|\mathrm{e}\|} \times \{\mathbf{0}\}$. The minus orbit symplectic form is

$$\omega^-(\mathbf{e}, \mathbf{0})(\mathrm{ad}^*_{(\mathrm{x},\mathrm{y})}(\mathbf{e}, \mathbf{0}), \mathrm{ad}^*_{(\mathrm{a},\mathrm{b})}(\mathbf{e}, \mathbf{0})) = -\mathbf{e} \cdot (\mathbf{x} \times \mathbf{x}'). \tag{14.9.22}$$

Thus the symplectic form on $\mathcal{O}$ at $(\mathbf{e}, \mathbf{0})$ is $-1/\|\mathbf{e}\|$ times the area element of the sphere of radius $\|\mathbf{e}\|$ (see (14.5.1) and (14.5.3)).

Type II If $\mathcal{O}$ contains a point of the form $(\mathbf{0}, \mathbf{f})$, then $\mathcal{O}$ equals $TS^2_{\|\mathrm{f}\|}$. Let $(\mathbf{u}, \mathbf{v}) \in \mathcal{O}$, that is, $\|\mathbf{v}\| = \|\mathbf{f}\|$ and $\mathbf{u} \perp \mathbf{v}$. The symplectic form in this case is

$$\begin{aligned} &\omega^-(\mathbf{u}, \mathbf{v})(\mathrm{ad}^*_{(\mathrm{x},\mathrm{y})}(\mathbf{u}, \mathbf{v}), \mathrm{ad}^*_{(\mathrm{a},\mathrm{b})}(\mathbf{u}, \mathbf{v})) \\ &\qquad = -\mathbf{u} \cdot (\mathbf{x} \times \mathbf{x}') - \mathbf{v} \cdot (\mathbf{x} \times \mathbf{y}' - \mathbf{x}' \times \mathbf{y}). \end{aligned} \tag{14.9.23}$$

We shall prove below that this form is exact, namely, $\omega^- = -\mathbf{d}\theta$, where

$$\theta(\mathbf{u}, \mathbf{v}) \cdot \mathrm{ad}^*_{(\mathrm{x},\mathrm{y})}(\mathbf{u}, \mathbf{v}) = \mathbf{u} \cdot \mathbf{x}. \tag{14.9.24}$$

First, note that θ is indeed well defined, for if

$$\mathrm{ad}^*_{(\mathbf{x},\mathbf{y})}(\mathbf{u},\mathbf{v}) = \mathrm{ad}^*_{(\mathbf{x}',\mathbf{y}')}(\mathbf{u},\mathbf{v}),$$

by (14.9.18) we have $(\mathbf{x}-\mathbf{x}')\times\mathbf{v}=0$, that is, $\mathbf{x}-\mathbf{x}'=c\mathbf{v}$ for some constant $c\in\mathbb{R}$, and since $\mathbf{u}\perp\mathbf{v}$, we conclude from here that $\mathbf{u}\cdot\mathbf{x}=\mathbf{u}\cdot\mathbf{x}'$. Second, in order to compute $\mathbf{d}\theta$, we shall use the formula

$$\mathbf{d}\theta(X,Y) = X[\theta(Y)] - Y[\theta(X)] - \theta([X,Y])$$

for any vector fields X, Y on $\mathcal{O}$. Third, we shall choose X and Y as follows:

$$\begin{aligned} X(\mathbf{u},\mathbf{v}) &= (\mathbf{x},\mathbf{y})_{\mathfrak{se}(3)^*}(\mathbf{u},\mathbf{v}) = -\mathrm{ad}^*_{(\mathbf{x},\mathbf{y})}(\mathbf{u},\mathbf{v}), \\ Y(\mathbf{u},\mathbf{v}) &= (\mathbf{x}',\mathbf{y}')_{\mathfrak{se}(3)^*}(\mathbf{u},\mathbf{v}) = -\mathrm{ad}^*_{(\mathbf{x}',\mathbf{y}')}(\mathbf{u},\mathbf{v}), \end{aligned}$$

for fixed $\mathbf{x},\mathbf{y},\mathbf{x}',\mathbf{y}'\in\mathbb{R}^3$. Fourth, to compute $X[\theta(Y)](\mathbf{u},\mathbf{v})$, consider the path

$$(\mathbf{u}(\epsilon),\mathbf{v}(\epsilon)) = (e^{-\epsilon\hat{\mathbf{x}}}\mathbf{u} + \epsilon(\mathbf{v}\times\mathbf{y}), e^{-\epsilon\hat{\mathbf{x}}}\mathbf{v}),$$

which satisfies $(\mathbf{u}(0),\mathbf{v}(0)) = (\mathbf{u},\mathbf{v})$ and

$$(\mathbf{u}'(0),\mathbf{v}'(0)) = (\mathbf{u}\times\mathbf{x}+\mathbf{v}\times\mathbf{y}, \mathbf{v}\times\mathbf{x}) = \mathrm{ad}^*_{(\mathbf{x},\mathbf{y})}(\mathbf{u},\mathbf{v}).$$

Then

$$\begin{aligned} X[\theta(Y)](\mathbf{u},\mathbf{v}) &= \left.\frac{d}{d\epsilon}\right|_{\epsilon=0} \theta(Y)(\mathbf{u}(\epsilon),\mathbf{v}(\epsilon)) \\ &= \left.\frac{d}{d\epsilon}\right|_{\epsilon=0} \mathbf{u}(\epsilon)\cdot\mathbf{x}' = (\mathbf{u}\times\mathbf{x}+\mathbf{v}\times\mathbf{y})\cdot\mathbf{x}'. \end{aligned}$$

Similarly, $Y[\theta(X)](\mathbf{u},\mathbf{v}) = (\mathbf{u}\times\mathbf{x}'+\mathbf{v}\times\mathbf{y}')\cdot\mathbf{x}$. Finally,

$$\begin{aligned} [X,Y](\mathbf{u},\mathbf{v}) &= [(\mathbf{x},\mathbf{y})_{\mathfrak{se}(3)^*}, (\mathbf{x}',\mathbf{y}')_{\mathfrak{se}(3)^*}](\mathbf{u},\mathbf{v}) \\ &= -[(\mathbf{x},\mathbf{y}),(\mathbf{x}',\mathbf{y}')]_{\mathfrak{se}(3)^*}(\mathbf{u},\mathbf{v}) \\ &= -(\mathbf{x}\times\mathbf{x}', \mathbf{x}\times\mathbf{y}'-\mathbf{x}'\times\mathbf{y})_{\mathfrak{se}(3)^*}(\mathbf{u},\mathbf{v}) \\ &= \mathrm{ad}^*_{(\mathbf{x}\times\mathbf{x}', \mathbf{x}\times\mathbf{y}'-\mathbf{x}'\times\mathbf{y})}(\mathbf{u},\mathbf{v}). \end{aligned}$$

Therefore

$$\begin{aligned} -\mathbf{d}\theta(\mathbf{u},\mathbf{v})&(\mathrm{ad}^*_{(\mathbf{x},\mathbf{y})}(\mathbf{u},\mathbf{v}), \mathrm{ad}^*_{(\mathbf{x}',\mathbf{y}')}(\mathbf{u},\mathbf{v})) = -X[\theta(Y)](\mathbf{u},\mathbf{v}) \\ &\quad + Y[\theta(X)](\mathbf{u},\mathbf{v}) + \theta([X,Y])(\mathbf{u},\mathbf{v}) \\ &= -(\mathbf{u}\times\mathbf{x}+\mathbf{v}\times\mathbf{y})\cdot\mathbf{x}' + (\mathbf{u}\times\mathbf{x}'+\mathbf{v}\times\mathbf{y}')\cdot\mathbf{x} + \mathbf{u}\cdot(\mathbf{x}\times\mathbf{x}') \\ &= -\mathbf{u}\cdot(\mathbf{x}\times\mathbf{x}') - \mathbf{v}\cdot(\mathbf{x}\times\mathbf{y}'-\mathbf{x}'\times\mathbf{y}), \end{aligned}$$

which coincides with (14.9.23).

The form θ given by (14.9.24) is the canonical symplectic structure when we identify $TS^2_{\|\mathbf{f}\|}$ with $T^*S^2_{\|\mathbf{f}\|}$ using the Euclidean metric.

Type III If $\mathcal{O}$ contains $(\mathbf{e}, \mathbf{f})$ where $\mathbf{e} \neq \mathbf{0}$ and $\mathbf{f} \neq \mathbf{0}$, then $\mathcal{O}$ is diffeomorphic to $T^*S^2_{\|\mathbf{f}\|}$ in the following way. The map $\varphi : SE(3) \to T^*S^2_{\|\mathbf{f}\|}$ given by (14.9.14) induces a diffeomorphism $\overline{\varphi} : SE(3)/SE(3)_{(\mathbf{e},\mathbf{f})} \to T^*S^2_{\|\mathbf{f}\|}$. However, the orbit $\mathcal{O}$ through $(\mathbf{e}, \mathbf{f})$ is diffeomorphic to $SE(3)/SE(3)_{(\mathbf{e},\mathbf{f})}$ by the diffeomorphism

$$(\mathbf{A}, \mathbf{a}) \mapsto \mathrm{Ad}^*_{(\mathbf{A},\mathbf{a})^{-1}}(\mathbf{e}, \mathbf{f}). \tag{14.9.25}$$

Therefore the diffeomorphism $\Phi : \mathcal{O} \to T^*S^2_{\|\mathbf{f}\|}$ is given by

$$\begin{aligned} \Phi(\mathrm{Ad}^*_{(\mathbf{A},\mathbf{a})^{-1}}(\mathbf{e}, \mathbf{f})) &= \Phi(\mathbf{A}\mathbf{e} + \mathbf{a} \times \mathbf{A}\mathbf{f}, \mathbf{A}\mathbf{f}) \\ &= \left(\mathbf{A}\mathbf{e} + \mathbf{a} \times \mathbf{A}\mathbf{f} - \frac{\mathbf{e}\cdot\mathbf{f}}{\|\mathbf{f}\|^2}\mathbf{A}\mathbf{f}, \mathbf{A}\mathbf{f}\right). \end{aligned} \tag{14.9.26}$$

If $(\overline{\mathbf{u}}, \overline{\mathbf{v}}) \in \mathcal{O}$, the orbit symplectic structure is given by formula (14.9.23), where $\overline{\mathbf{u}} = \mathbf{A}\mathbf{e} + \mathbf{a} \times \mathbf{A}\mathbf{f}, \overline{\mathbf{v}} = \mathbf{A}\mathbf{f}$ for some $\mathbf{A} \in SO(3), \mathbf{a} \in \mathbb{R}^3$. Let

$$\begin{aligned} \mathbf{u} &= \mathbf{A}\mathbf{e} + \mathbf{a} \times \mathbf{A}\mathbf{f} - \frac{\mathbf{e}\cdot\mathbf{f}}{\|\mathbf{f}\|^2}\mathbf{A}\mathbf{f} = \overline{\mathbf{u}} - \frac{\mathbf{e}\cdot\mathbf{f}}{\|\mathbf{f}\|^2}\overline{\mathbf{v}}, \\ \mathbf{v} &= \mathbf{A}\mathbf{f} = \overline{\mathbf{v}}, \end{aligned} \tag{14.9.27}$$

the pair of vectors $(\mathbf{u}, \mathbf{v})$ representing an element of TS^2. Note that $\|\mathbf{v}\| = \|\mathbf{f}\|$ and $\mathbf{u}\cdot\mathbf{v} = 0$. Then a tangent vector to $TS^2_{\|\mathbf{f}\|}$ at $(\mathbf{u}, \mathbf{v})$ can be represented as $\mathrm{ad}^*_{(\mathbf{x},\mathbf{y})}(\mathbf{u}, \mathbf{v}) = (\mathbf{u} \times \mathbf{x} + \mathbf{v} \times \mathbf{y}, \mathbf{v} \times \mathbf{x})$ so that by (14.9.26) we get

$$\begin{aligned} T_{(\mathbf{u},\mathbf{v})}\Phi^{-1}(\mathrm{ad}^*_{(\mathbf{x},\mathbf{y})}(\mathbf{u}, \mathbf{v})) &= \left.\frac{d}{d\epsilon}\right|_{\epsilon=0} \Phi^{-1}(e^{-\epsilon\hat{\mathbf{x}}}\mathbf{u} + \epsilon(\mathbf{v} \times \mathbf{y}), e^{\epsilon\hat{\mathbf{x}}}\mathbf{v}) \\ &= \left.\frac{d}{d\epsilon}\right|_{\epsilon=0} \left(e^{-\epsilon\hat{\mathbf{x}}}\mathbf{u} + \epsilon(\mathbf{v} \times \mathbf{y}) + \frac{\mathbf{e}\cdot\mathbf{f}}{\|f\|^2}e^{-\epsilon\hat{\mathbf{x}}}\mathbf{v}, e^{-\epsilon\hat{\mathbf{x}}}\mathbf{v}\right) \\ &= \left(\mathbf{u} \times \mathbf{x} + \mathbf{v} \times \mathbf{y} + \frac{\mathbf{e}\cdot\mathbf{f}}{\|f\|^2}(\mathbf{v} \times \mathbf{x}), \mathbf{v} \times \mathbf{x}\right) \\ &= (\overline{\mathbf{u}} \times \mathbf{x} + \overline{\mathbf{v}} \times \mathbf{y}, \overline{\mathbf{v}} \times \mathbf{x}) \\ &= \mathrm{ad}^*_{(\mathbf{x},\mathbf{y})}(\overline{\mathbf{u}}, \overline{\mathbf{v}}). \end{aligned}$$

Therefore, the push-forward of the orbit symplectic form ω^- to $TS^2_{\|\mathbf{f}\|}$ is

$$\begin{aligned} &(\Phi_*\omega^-)(\mathbf{u}, \mathbf{v})(\mathrm{ad}^*_{(\mathbf{x},\mathbf{y})}(\mathbf{u}, \mathbf{v}), \mathrm{ad}^*_{(\mathbf{x}',\mathbf{y}')}(\mathbf{u}, \mathbf{v})) \\ &\quad= \omega^-(\overline{\mathbf{u}}, \overline{\mathbf{v}})(T_{(\mathbf{u},\mathbf{v})}\Phi^{-1}(\mathrm{ad}^*_{(\mathbf{x},\mathbf{y})}(\mathbf{u}, \mathbf{v})), T_{(\mathbf{u},\mathbf{v})}\Phi^{-1}(\mathrm{ad}^*_{(\mathbf{x}',\mathbf{y}')}(\mathbf{u}, \mathbf{v})) \\ &\quad= \omega^-(\overline{\mathbf{u}}, \overline{\mathbf{v}})(\mathrm{ad}^*_{(\mathbf{x},\mathbf{y})}(\overline{\mathbf{u}}, \overline{\mathbf{v}}), \mathrm{ad}^*_{(\mathbf{x}',\mathbf{y}')}(\overline{\mathbf{u}}, \overline{\mathbf{v}})) \\ &\quad= -\overline{\mathbf{u}}\cdot(\mathbf{x} \times \mathbf{x}') - \overline{\mathbf{v}}\cdot(\mathbf{x} \times \mathbf{y}' - \mathbf{x}' \times \mathbf{y}) \\ &\quad= -\mathbf{u}\cdot(\mathbf{x} \times \mathbf{x}') - \mathbf{v}\cdot(\mathbf{x} \times \mathbf{y}' - \mathbf{x}' \times \mathbf{y}) - \frac{\mathbf{e}\cdot\mathbf{f}}{\|\mathbf{f}\|^2}\mathbf{v}\cdot(\mathbf{x} \times \mathbf{x}'). \end{aligned} \tag{14.9.28}$$

The first two terms represent the canonical symplectic structure on $TS^2_{\|\mathbf{f}\|}$ (identified via the Euclidean metric with $T^*S^2_{\|\mathbf{f}\|}$), as we have seen in the analysis of type II orbits. The third term is the following two-form on $TS^2_{\|\mathbf{f}\|}$

$$\beta(\mathbf{u},\mathbf{v})\left(\mathrm{ad}^*_{(\mathbf{x},\mathbf{y})}(\mathbf{u},\mathbf{v}), \mathrm{ad}^*_{(\mathbf{x}',\mathbf{y}')}(\mathbf{u},\mathbf{v})\right) = -\frac{\mathbf{e}\cdot\mathbf{f}}{\|\mathbf{f}\|^2}\mathbf{v}\cdot(\mathbf{x}\times\mathbf{x}'). \qquad (14.9.29)$$

As in the case of θ for type II orbits, it is easily seen that (14.9.28) correctly defines a two-form on $TS^2_{\|\mathbf{f}\|}$. It is necessarily closed since it is the difference between $\Phi_*\omega^-$ and the canonical two-form on $TS^2_{\|\mathbf{f}\|}$. The two-form β is a magnetic term in the sense of §**6.6**.

We remark that the semidirect product theory of Marsden, Ratiu, and Weinstein [1984a,b], combined with cotangent bundle reduction theory, (see, for example, Marsden [1992]) can be used to give an alternative approach to the computation of the orbit symplectic forms.

Exercises

14.9-1 *Let K be a quadratic form on $\mathbb{R}^3$ and let $\mathbf{K}$ be the associated symmetric* (3×3)*-matrix. Let*

$$\{F,L\}_K = -\nabla K\cdot(\nabla F\times\nabla L).$$

Show that this is the Lie-Poisson bracket for the Lie algebra structure

$$[\mathbf{u},\mathbf{v}]_K = \mathbf{K}(\mathbf{u}\times\mathbf{v}).$$

What is the underlying Lie group?

14.9-2 *Determine the coadjoint orbits for the Lie algebra in the preceding exercise and calculate the orbit symplectic structure. Specialize to the case* $SO(2,1)$.

14.9-3 *Classify the coadjoint orbits of* $SU(1,1)$*, namely, the group of complex* (2×2) *matrices of determinant one, of the form*

$$g = \begin{pmatrix} a & b \\ \bar{a} & \bar{b} \end{pmatrix}.$$

15
The Free Rigid Body

As an application of the theory developed so far, we discuss the motion of a free rigid body about a fixed point. We begin with a discussion of the kinematics of rigid body motion. Our description of the kinematics of rigid bodies follows some of the notations and conventions of continuum mechanics, as in Marsden and Hughes [1983].

15.1 Material, Spatial, and Body Coordinates

Consider a rigid body, free to move in $\mathbb{R}^3$. A ***reference configuration*** $\mathcal{B}$ of the body is the closure of an open set in $\mathbb{R}^3$ with a piecewise smooth boundary. Points in $\mathcal{B}$, denoted $X = (X^1, X^2, X^3) \in \mathcal{B}$ relative to an orthonormal basis $(\mathbf{E}_1, \mathbf{E}_2, \mathbf{E}_3)$ are called ***material points*** and $X^i, i = 1, 2, 3$, are called ***material coordinates.*** A ***configuration*** of $\mathcal{B}$ is a mapping $\varphi : \mathcal{B} \to \mathbb{R}^3$ which is, for our purposes, C^1, orientation preserving, and invertible on its image. Points in the image of φ are called ***spatial points*** and denoted by lowercase letters. Let $(\mathbf{e}_1, \mathbf{e}_2, \mathbf{e}_3)$ be a right-handed orthonormal basis of $\mathbb{R}^3$. Coordinates for spatial points, such as $x = (x^1, x^2, x^3) \in \mathbb{R}^3, i = 1, 2, 3$, relative to the basis $(\mathbf{e}_1, \mathbf{e}_2, \mathbf{e}_3)$ are called ***spatial coordinates.*** Dually, one can consider material quantities such as maps defined on $\mathcal{B}$, say $Z : \mathcal{B} \to \mathbb{R}$. Then we can form spatial quantities by composition: $z_t = Z_t \circ \varphi_t^{-1}$. Spatial quantities are also called ***Eulerian quantities*** and material quantities are often called ***Lagrangian quantities.***

A ***motion*** of $\mathcal{B}$ is a time-dependent family of configurations, written $x =$

$\varphi(X, t) = \varphi_t(X)$ or simply $x(X, t)$ or $x_t(X)$. Spatial quantities are functions of x, and are typically written as lowercase letters. By composition with φ_t, spatial quantities become functions of the material points X.

Rigidity of the body means that the distances between points of the body are fixed as the body moves. We shall assume that no external forces act on the body and that the center of mass is fixed at the origin (see Exercise **15.1-1**). Since any isometry of $\mathbb{R}^3$ that leaves the origin fixed is a rotation (a 1932 theorem of Mazur and Ulam), we can write

$$x(X, t) = \mathbf{R}(t)X, \quad \text{i.e.,} \quad x^i = \mathbf{R}^i_j(t)X^j, \quad i, j = 1, 2, 3, \text{ sum on } j,$$

where x^i are the components of x relative to the basis $\mathbf{e}_1, \mathbf{e}_2, \mathbf{e}_3$ fixed in space, and $[\mathbf{R}^i_j]$ is the matrix of $\mathbf{R}$ relative to the basis $(\mathbf{E}_1, \mathbf{E}_2, \mathbf{E}_3)$ and $(\mathbf{e}_1, \mathbf{e}_2, \mathbf{e}_3)$. The motion is assumed to be continuous and $\mathbf{R}(0)$ is the identity, so $\det(\mathbf{R}(t)) = 1$ and thus $\mathbf{R}(t) \in SO(3)$, the proper orthogonal group. Thus, *the configuration space for the rotational motion of a rigid body may be identified with $SO(3)$. Consequently, the velocity phase space of the free rigid body is $TSO(3)$ and the momentum phase space is the cotangent bundle $T^*SO(3)$. Euler angles*, discussed shortly, are the traditional way to parametrize $SO(3)$.

In addition to the material and spatial coordinates, there is a third set, the *convected* or *body coordinates.* These are the coordinates associated with the moving basis, and the description of the rigid body motion in these coordinates, due to Euler, becomes very simple. As before, let $\mathbf{E}_1, \mathbf{E}_2, \mathbf{E}_3$ be an orthonormal basis fixed in the reference configuration. Let the time-dependent basis $\boldsymbol{\xi}_1, \boldsymbol{\xi}_2, \boldsymbol{\xi}_3$ be defined by $\boldsymbol{\xi}_i = \mathbf{R}(t)\mathbf{E}_i$, $i = 1, 2, 3$, so $\boldsymbol{\xi}_1, \boldsymbol{\xi}_2, \boldsymbol{\xi}_3$ move attached to the body. The ***body coordinates*** of a vector in $\mathbb{R}^3$ are its components relative to $\boldsymbol{\xi}_i$. For the rigid body anchored at the origin and rotating in space, $(\mathbf{e}_1, \mathbf{e}_2, \mathbf{e}_3)$ is thought of as a basis fixed in space, whereas $(\boldsymbol{\xi}_1, \boldsymbol{\xi}_2, \boldsymbol{\xi}_3)$ is a basis fixed in the body and moving with it. For this reason $(\mathbf{e}_1, \mathbf{e}_2, \mathbf{e}_3)$ is called the ***spatial coordinate system*** and $(\boldsymbol{\xi}_1, \boldsymbol{\xi}_2, \boldsymbol{\xi}_3)$ the ***body coordinate system***.

Exercise

15.1-1 *Start with $SE(3)$ as the configuration space for the rigid body and "reduce out" (see §10.7, the Euler-Poincaré, and Lie-Poisson reduction theorems) translations to arrive at $SO(3)$ as the configuration space.*

15.2 The Lagrangian of the Free Rigid Body

If $X \in \mathcal{B}$ is a material point of the body, the corresponding trajectory followed by X in space is $x(t) = \mathbf{R}(t)X$, where $\mathbf{R}(t) \in SO(3)$. The ***material*** or ***Lagrangian velocity*** $V(X,t)$ is defined by

$$V(X,t) = \frac{\partial x(X,t)}{\partial t} = \dot{\mathbf{R}}(t)X, \tag{15.2.1}$$

while the ***spatial*** or ***Eulerian velocity*** $v(x,t)$ is defined by

$$v(x,t) = V(X,t) = \dot{\mathbf{R}}(t)\mathbf{R}(t)^{-1}x, \tag{15.2.2}$$

and the ***body*** or ***convective velocity*** $\mathcal{V}(X,t)$ is defined by taking the velocity regarding X as time-dependent and x fixed, that is, $X(x,t) = \mathbf{R}(t)^{-1}x$:

$$\begin{aligned} \mathcal{V}(X,t) &= -\frac{\partial X(x,t)}{\partial t} = \mathbf{R}(t)^{-1}\dot{\mathbf{R}}(t)\mathbf{R}(t)^{-1}x \\ &= \mathbf{R}(t)^{-1}\dot{\mathbf{R}}(t)X \\ &= \mathbf{R}(t)^{-1}V(X,t) \\ &= \mathbf{R}(t)^{-1}v(x,t). \end{aligned} \tag{15.2.3}$$

Assume that the mass distribution of the body is described by a compactly supported density measure $\rho_0 d^3X$ in the reference configuration, which is zero at points outside the body. The Lagrangian, taken to be the kinetic energy, is given by any of the following expressions that are related to one another by a change of variables and the identities $\|\mathcal{V}\| = \|V\| = \|v\|$:

$$\begin{aligned} L &= \frac{1}{2}\int_{\mathcal{B}} \rho_0(X)\|V(X,t)\|^2\, d^3X & \text{(material)} \quad (15.2.4) \\ &= \frac{1}{2}\int_{\mathrm{R}(t)\mathcal{B}} \rho_0(\mathbf{R}(t)^{-1}x)\|v(x,t)\|^2\, d^3x & \text{(spatial)} \quad (15.2.5) \\ &= \frac{1}{2}\int_{\mathcal{B}} \rho_0(X)\|\mathcal{V}(X,t)\|^2\, d^3X & \text{(convective or body).} \quad (15.2.6) \end{aligned}$$

Differentiating $\mathbf{R}(t)^T\mathbf{R}(t) =$ Identity and $\mathbf{R}(t)\mathbf{R}(t)^T =$ Identity with respect to t, it follows that both $\mathbf{R}(t)^{-1}\dot{\mathbf{R}}(t)$ and $\dot{\mathbf{R}}(t)\mathbf{R}(t)^{-1}$ are skew-symmetric. Moreover, by (15.2.2), (15.2.3), and the classical definition $\mathbf{v} = \boldsymbol{\omega} \times \mathbf{r} = \hat{\boldsymbol{\omega}}\mathbf{r}$ of angular velocity, it follows that the vectors $\boldsymbol{\omega}(t)$ and $\boldsymbol{\Omega}(t)$ in $\mathbb{R}^3$ defined by

$$\hat{\boldsymbol{\omega}}(t) = \dot{\mathbf{R}}(t)\mathbf{R}(t)^{-1} \tag{15.2.7}$$

and

$$\hat{\boldsymbol{\Omega}}(t) = \mathbf{R}(t)^{-1}\dot{\mathbf{R}}(t) \tag{15.2.8}$$

represent the ***spatial*** and ***convective angular velocities*** of the body. Note that $\boldsymbol{\omega}(t) = \mathbf{R}(t)\boldsymbol{\Omega}(t)$, or as matrices,

$$\hat{\boldsymbol{\omega}} = \mathrm{Ad}_{\mathrm{R}}\hat{\boldsymbol{\Omega}} = \mathbf{R}\hat{\boldsymbol{\Omega}}\mathbf{R}^{-1}.$$

Let us show that $L : TSO(3) \to \mathbb{R}$ given by (15.2.4) is left-invariant. Indeed, if $\mathbf{B} \in SO(3)$, left translation by $\mathbf{B}$ is $L_{\mathrm{B}}\mathbf{R} = \mathbf{BR}$ and $TL_{\mathrm{B}}(\mathbf{R}, \dot{\mathbf{R}}) = (\mathbf{BR}, \mathbf{B}\dot{\mathbf{R}})$, so

$$\begin{aligned} L(TL_{\mathrm{B}}(\mathbf{R}, \dot{\mathbf{R}})) &= \frac{1}{2}\int_{\mathcal{B}} \rho_0(X)\|\mathbf{B}\dot{\mathbf{R}}(X)\|^2\, d^3X \\ &= \frac{1}{2}\int_{\mathcal{B}} \rho_0(X)\|\dot{\mathbf{R}}(X)\|^2\, d^3X = L(\mathbf{R}, \dot{\mathbf{R}}) \quad (15.2.9) \end{aligned}$$

since $\mathbf{R}$ is orthogonal.

By Lie-Poisson reduction of dynamics (Chapter **13**), the corresponding Hamiltonian system on $T^*SO(3)$, which is necessarily also left invariant, induces a Lie-Poisson system on $\mathfrak{so}(3)^*$ and this system leaves invariant the coadjoint orbits $\|\mathbf{\Pi}\| = \text{constant}$. Alternatively, by Euler-Poincaré reduction of dynamics, we get a system of equations in terms of body angular velocity on $so(3)$.

Reconstruction of the dynamics on $TSO(3)$ is simply this: given $\hat{\mathbf{\Omega}}(t)$, determine $\mathbf{R}(t) \in SO(3)$ from (15.2.8):

$$\dot{\mathbf{R}}(t) = \mathbf{R}(t)\hat{\mathbf{\Omega}}(t), \tag{15.2.10}$$

which is a time-dependent linear equation for $\mathbf{R}(t)$.

15.3 The Lagrangian and Hamiltonian for the Rigid Body in Body Representation

From (15.2.6), (15.2.3), and (15.2.8) of the previous section, the rigid body Lagrangian is

$$L = \frac{1}{2}\int_{\mathcal{B}} \rho_0(X)\|\mathbf{\Omega} \times X\|^2\, d^3X. \tag{15.3.1}$$

Introducing the new inner product

$$\langle\langle \mathbf{a}, \mathbf{b} \rangle\rangle := \int_{\mathcal{B}} \rho_0(X)(\mathbf{a} \times X) \cdot (\mathbf{b} \times X)\, d^3X,$$

which is determined by the density $\rho_0(X)$ of the body, (15.3.1) becomes

$$L(\mathbf{\Omega}) = \frac{1}{2}\langle\langle \mathbf{\Omega}, \mathbf{\Omega} \rangle\rangle. \tag{15.3.2}$$

Define the linear isomorphism $\mathbf{I} : \mathbb{R}^3 \to \mathbb{R}^3$ by $\mathbf{Ia} \cdot \mathbf{b} = \langle\langle \mathbf{a}, \mathbf{b} \rangle\rangle$ for all $\mathbf{a}, \mathbf{b} \in \mathbb{R}^3$; this is possible and uniquely determines $\mathbf{I}$, since both the dot product and $\langle\langle\, , \rangle\rangle$ are nondegenerate bilinear forms (assuming the rigid body is not concentrated on a line). It is clear that $\mathbf{I}$ is symmetric with respect to the dot product and is positive-definite. Let $(\mathbf{E}_1, \mathbf{E}_2, \mathbf{E}_3)$ be an orthonormal basis for material coordinates. The matrix of $\mathbf{I}$ is

$$\mathbf{I}_{ij} = \mathbf{E}_i \cdot \mathbf{I}\mathbf{E}_j = \langle\langle \mathbf{E}_i, \mathbf{E}_j \rangle\rangle = \begin{cases} -\displaystyle\int_{\mathcal{B}} \rho_0(X) X^i X^j \, d^3X, & i \neq j, \\ \displaystyle\int_{\mathcal{B}} \rho_0(X)(\|X\|^2 - (X^i)^2) \, d^3X, & i = j, \end{cases}$$

which are the classical expressions of the matrix of the ***inertia tensor***. Since $\mathbf{I}$ is symmetric, it can be diagonalized; an orthonormal basis in which it is diagonal is a ***principal axis body frame*** and the diagonal elements I_1, I_2, I_3 are the ***principal moments of inertia*** of the rigid body. In what follows we work in a principal axis reference and body frame, $(\mathbf{E}_1, \mathbf{E}_2, \mathbf{E}_3)$.

Since $\mathfrak{so}(3)^*$ and $\mathbb{R}^3$ are identified by the dot product (not by $\langle\langle\, , \rangle\rangle$), the linear functional $\langle\langle \mathbf{\Omega}, \cdot \rangle\rangle$—the Legendre transformation of $\mathbf{\Omega}$—on $\mathfrak{so}(3) \cong \mathbb{R}^3$ is identified with $\mathbf{I\Omega} := \mathbf{\Pi} \in \mathfrak{so}(3)^* \cong \mathbb{R}^3$ because $\mathbf{\Pi} \cdot \mathbf{a} = \langle\langle \mathbf{\Omega}, \mathbf{a} \rangle\rangle$ for all $\mathbf{a} \in \mathbb{R}^3$. With $\mathbf{I} = \operatorname{diag}(I_1, I_2, I_3)$, (15.3.2) defines a function

$$K(\mathbf{\Pi}) = \frac{1}{2}\left(\frac{\Pi_1^2}{I_1} + \frac{\Pi_2^2}{I_2} + \frac{\Pi_3^2}{I_3}\right) \tag{15.3.3}$$

that represents the expression for the kinetic energy on $\mathfrak{so}(3)^*$; note that $\mathbf{\Pi} = \mathbf{I\Omega}$ is the ***angular momentum in the body frame***. Indeed, for any $\mathbf{a} \in \mathbb{R}^3$, the identity $(X \times (\mathbf{\Omega} \times X)) \cdot \mathbf{a} = (\mathbf{\Omega} \times X) \cdot (\mathbf{a} \times X)$ and the classical expression of the angular momentum in the body frame, namely,

$$\int_{\mathcal{B}} (X \times \mathcal{V}) \rho_0(X) \, d^3X \tag{15.3.4}$$

gives

$$\begin{aligned} \left(\int_{\mathcal{B}} (X \times \mathcal{V}) \rho_0(X) \, d^3X \right) \cdot \mathbf{a} &= \int_{\mathcal{B}} (X \times (\mathbf{\Omega} \times X)) \cdot \mathbf{a} \rho_0(X) \, d^3X \\ &= \int_{\mathcal{B}} (\mathbf{\Omega} \times X) \cdot (\mathbf{a} \times X) \rho_0(X) \, d^3X \\ &= \langle\langle \mathbf{\Omega}, \mathbf{a} \rangle\rangle = \mathbf{I\Omega} \cdot \mathbf{a} = \mathbf{\Pi} \cdot \mathbf{a}, \end{aligned}$$

that is, the expression (15.3.4) equals $\mathbf{\Pi}$.

The ***angular momentum in space*** has the expression

$$\boldsymbol{\pi} = \int_{\mathrm{R}(\mathcal{B})} (x \times v) \rho(x) \, d^3x, \tag{15.3.5}$$

where $\rho(x) = \rho_0(X)$ is the ***spatial mass density*** and $v = \boldsymbol{\omega} \times x$ is the spatial velocity (see (15.2.2) and (15.2.7)). For any $\mathbf{a} \in \mathbb{R}^3$,

$$\begin{aligned} \boldsymbol{\pi} \cdot \mathbf{a} &= \int_{\mathrm{R}(\mathcal{B})} (x \times (\boldsymbol{\omega} \times x)) \cdot \mathbf{a} \rho(x) \, d^3X \\ &= \int_{\mathrm{R}(\mathcal{B})} (\boldsymbol{\omega} \times x) \cdot (\mathbf{a} \times x) \rho(x) \, d^3X. \end{aligned} \tag{15.3.6}$$

Changing variables $x = \mathbf{R}X$, (15.3.6) becomes

$$\begin{aligned}
&\int_{\mathcal{B}} (\boldsymbol{\omega} \times \mathbf{R}X) \cdot (\mathbf{a} \times \mathbf{R}X) \rho_0(X)\, d^3X \\
&\quad = \int_{\mathcal{B}} (\mathbf{R}^T\boldsymbol{\omega} \times X) \cdot (\mathbf{R}^T\mathbf{a} \times X) \rho_0(X)\, d^3X \\
&\quad = \langle\!\langle \boldsymbol{\Omega}, \mathbf{R}^T\mathbf{a} \rangle\!\rangle = \boldsymbol{\Pi} \cdot \mathbf{R}^T\mathbf{a} = \mathbf{R}\boldsymbol{\Pi} \cdot \mathbf{a},
\end{aligned}$$

that is,

$$\boldsymbol{\pi} = \mathbf{R}\boldsymbol{\Pi}. \tag{15.3.7}$$

Since L given by (15.3.2) is left invariant on $TSO(3)$, the function K defined on $\mathfrak{so}(3)^*$ by (15.3.3) defines the Lie-Poisson equations of motion on $\mathfrak{so}(3)^*$ relative to the bracket

$$\{F, H\}(\boldsymbol{\Pi}) = -\boldsymbol{\Pi} \cdot (\nabla F(\boldsymbol{\Pi}) \times \nabla H(\boldsymbol{\Pi})). \tag{15.3.8}$$

Since $\nabla K(\boldsymbol{\Pi}) = \mathbf{I}^{-1}\boldsymbol{\Pi}$, we get from (15.3.8) the rigid body equations

$$\dot{\boldsymbol{\Pi}} = -\nabla K(\boldsymbol{\Pi}) \times \boldsymbol{\Pi} = \boldsymbol{\Pi} \times \mathbf{I}^{-1}\boldsymbol{\Pi}, \tag{15.3.9}$$

that is, they are the standard ***Euler equations***:

$$\begin{aligned}
\dot{\Pi}_1 &= \frac{I_2 - I_3}{I_2 I_3}\Pi_2\Pi_3, \\
\dot{\Pi}_2 &= \frac{I_3 - I_1}{I_1 I_3}\Pi_1\Pi_3, \\
\dot{\Pi}_3 &= \frac{I_1 - I_2}{I_1 I_2}\Pi_1\Pi_2.
\end{aligned} \tag{15.3.10}$$

The fact that these equations preserve coadjoint orbits amounts, in this case, to the easily verified fact that

$$\Pi^2 := \|\boldsymbol{\Pi}\|^2 \tag{15.3.11}$$

is a constant of the motion. In terms of coadjoint orbits, these equations are Hamiltonian on each sphere in $\mathbb{R}^3$ with Hamiltonian function K. The functions

$$C_\Phi(\boldsymbol{\Pi}) = \Phi\left(\frac{1}{2}\|\boldsymbol{\Pi}\|^2\right), \tag{15.3.12}$$

for any $\Phi : \mathbb{R} \to \mathbb{R}$, are easily seen to be Casimir functions.

The conserved momentum resulting from left invariance is the ***spatial angular momentum***:

$$\boldsymbol{\pi} = \mathbf{R}\boldsymbol{\Pi}. \tag{15.3.13}$$

Using left invariance, or a direct calculation, one finds that $\boldsymbol{\pi}$ is constant in time. Indeed,

$$\begin{aligned}
\dot{\boldsymbol{\pi}} &= (\mathbf{R}\boldsymbol{\Pi})^{\cdot} = \dot{\mathbf{R}}\boldsymbol{\Pi} + \mathbf{R}\dot{\boldsymbol{\Pi}} = \boldsymbol{\omega} \times \mathbf{R}\boldsymbol{\Pi} + \mathbf{R}\dot{\boldsymbol{\Pi}} \\
&= \mathbf{R}\boldsymbol{\Omega} \times \mathbf{R}\boldsymbol{\Pi} + \mathbf{R}\dot{\boldsymbol{\Pi}} = \mathbf{R}(-\boldsymbol{\Pi} \times \mathbf{I}^{-1}\boldsymbol{\Pi} + \dot{\boldsymbol{\Pi}}) = 0.
\end{aligned}$$

The flow lines are given by intersecting the ellipsoids K = constant with the coadjoint orbits which are two-spheres. For distinct moments of inertia $I_1 > I_2 > I_3$, the flow on the sphere has saddle points at $(0, \pm\Pi, 0)$ and centers at $(\pm\Pi, 0, 0)$, $(0, 0, \pm\Pi)$. The saddles are connected by four heteroclinic orbits, as indicated in Figure 15.3.1.

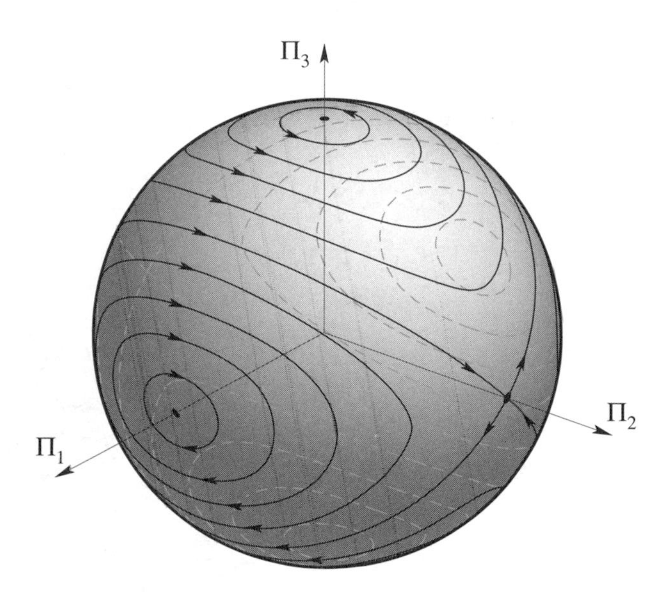

FIGURE 15.3.1. Rigid body flow on the angular momentum spheres.

In §15.10 we prove:

Rigid Body Stability Theorem *In the motion of a free rigid body, rotation around the long and short axes are (Liapunov) stable and rotation about the middle axis is unstable.*

Even though we completely solved the rigid body equations in body representation, the actual configuration of the body, that is, its attitude in space, has not been determined yet. This will be done in §15.8. Also, one has to be careful about the meaning of stability in space versus material versus body representation.

Euler's equations are very general. The n-dimensional case has been treated by Mishchenko and Fomenko [1976, 1978a], Adler and van Moerbeke [1980a,b], and Ratiu [1980, 1981, 1982] in connection with Lie algebras and algebraic geometry. The Russian school has generalized these equations further to a large class of Lie algebras and proved their complete integrability in a long series of papers starting in 1978; see the treatise of Fomenko and Trofimov [1989] and references therein.

15.4 Kinematics on Lie Groups

We now generalize the notation used for the rigid body to any Lie group. This abstraction unifies ideas common to rigid bodies, fluids, and plasmas in a consistent way. If G is a Lie group, and $H : T^*G \to \mathbb{R}$ is a Hamiltonian, we say it is described in the ***material picture***. If $\alpha \in T_g^*G$, its ***spatial representation*** is defined by

$$\alpha^S = T_e^* R_g(\alpha), \tag{15.4.1}$$

while its ***body representation*** is

$$\alpha^B = T_e^* L_g(\alpha). \tag{15.4.2}$$

Similar notation is used for TG; if $V \in T_gG$, we get

$$V^S = T_g R_{g^{-1}}(V) \tag{15.4.3}$$

and

$$V^B = T_g L_{g^{-1}}(V). \tag{15.4.4}$$

Thus, we get body and space isomorphisms as follows:

$$\text{(Body)}\ \ G \times \mathfrak{g}^* \xleftarrow{\text{LeftTranslate}} T^*G \xrightarrow{\text{RightTranslate}} G \times \mathfrak{g}^* \ \ \text{(Space)}.$$

Thus

$$\alpha^S = \operatorname{Ad}^*_{g^{-1}} \alpha^B \tag{15.4.5}$$

and

$$V^S = \operatorname{Ad}_g {}^{‘}V^B. \tag{15.4.6}$$

Part of the general theory of Chapter 13 says that if H is left (respectively, right) invariant on T^*G, it induces a Lie-Poisson system on $\mathfrak{g}^*_-$ (respectively, $\mathfrak{g}^*_+$).

Exercise

15.4-1 (Cayley-Klein Parameters) *Recall that the Lie algebras of $SO(3)$ and $SU(2)$ are the same. Recall also that $SU(2)$ acts symplectically on $\mathbb{C}^2$ by multiplication of (complex) matrices. Use this to produce a momentum map $\mathbf{J} : \mathbb{C}^2 \to su(2)^* \cong \mathbb{R}^3$.*

- **i** *Write down $\mathbf{J}$ explicitly.*
- **ii** *Verify by hand that $\mathbf{J}$ is a Poisson map.*
- **iii** *If H is the rigid body Hamiltonian, compute $H_{CK} = H \circ J$.*
- **iv** *Write down Hamilton's equations for H_{CK} and discuss the collective Hamiltonian theorem in this context.*
- **v** *Find this material, and relate it to the present context in one of the standard books (Whittaker, Pars, Hamel, or Goldstein, for example).*

15.5 Poinsot's Theorem

Recall from §**15.3** that the spatial angular momentum vector $\boldsymbol{\pi}$ is constant under the flow of the free rigid body. Also, if $\boldsymbol{\omega}$ is the angular velocity in space, then

$$\boldsymbol{\omega} \cdot \boldsymbol{\pi} = \boldsymbol{\Omega} \cdot \boldsymbol{\Pi} = 2K \tag{15.5.1}$$

is a constant. From this, it follows that $\boldsymbol{\omega}$ moves in an (affine) plane perpendicular to the fixed vector $\boldsymbol{\pi}$, called the ***invariable plane***. The distance from the origin to this plane is $2K/\|\boldsymbol{\pi}\|$. The ***ellipsoid of inertia in the body*** is defined by

$$\mathfrak{E} = \{\boldsymbol{\Omega} \in \mathbb{R}^3 \mid \boldsymbol{\Omega} \cdot \mathbf{I}\boldsymbol{\Omega} = 2K\}.$$

The ***ellipsoid of inertia in space*** is

$$\mathbf{R}(\mathfrak{E}) = \{\mathbf{u} \in \mathbb{R}^3 \mid \mathbf{u} \cdot \mathbf{RIR}^{-1}\mathbf{u} = 2K\},$$

where $\mathbf{R} = \mathbf{R}(t) \in SO(3)$ denotes the configuration of the body at time t.

Theorem 15.5.1 (Poinsot's Theorem) *The moment of inertia ellipsoid in space rolls without slipping on the invariable plane.*

Proof First, we determine the planes perpendicular to the fixed vector $\boldsymbol{\pi}$ and tangent to $\mathbf{R}(\mathfrak{E})$. See Figure 15.5.1. At the point of tangency $\mathbf{u}$, the vector $\mathbf{2RIR}^{-1}\mathbf{u}$ (the gradient of the expression defining $\mathbf{R}(\mathfrak{E})$) is proportional to $\boldsymbol{\pi}$, that is, there is an $a \in \mathbb{R}$ such that $\mathbf{RIR}^{-1}\mathbf{u} = a\boldsymbol{\pi}$, or

$$\mathbf{u} = a\mathbf{RI}^{-1}\mathbf{R}^{-1}\boldsymbol{\pi} = a\mathbf{RI}^{-1}\boldsymbol{\Pi} = a\mathbf{R}\boldsymbol{\Omega} = a\boldsymbol{\omega}$$

by (15.3.7), the definition of $\boldsymbol{\Pi}$, and the relation $\boldsymbol{\omega} = \mathbf{R}\boldsymbol{\Omega}$. However, this point $\mathbf{u} = a\boldsymbol{\omega}$ must belong to $\mathbf{R}(\mathfrak{E})$ so that using the same relations again, we get

$$2K = a^2\boldsymbol{\omega} \cdot \mathbf{RIR}^{-1}\boldsymbol{\omega} = a^2\boldsymbol{\Omega} \cdot \mathbf{I}\boldsymbol{\Omega} = 2a^2K,$$

whence $a = \pm 1$, that is, there are exactly two planes perpendicular to $\boldsymbol{\pi}$ and tangent at $\pm\boldsymbol{\omega}$ to $\mathbf{R}(\mathfrak{E})$.

Second, we show that the plane tangent to $\mathbf{R}(\mathfrak{E})$ at $\boldsymbol{\omega}$ is the invariable plane. Indeed, since the equation of this plane is $\mathbf{u} \cdot \boldsymbol{\pi} = C$ for some constant C and $\boldsymbol{\omega}$ is in the plane, it follows that $C = \boldsymbol{\omega} \cdot \boldsymbol{\pi} = 2K$, that is, the equation of the plane is $\mathbf{u} \cdot \boldsymbol{\pi} = 2K$, which is the invariable plane.

Third, since the point of tangency is $\boldsymbol{\omega}$, which is the instantaneous axis of rotation, its velocity is zero, that is, the rolling of the inertia ellipsoid on the invariable plane takes place without slipping. ■

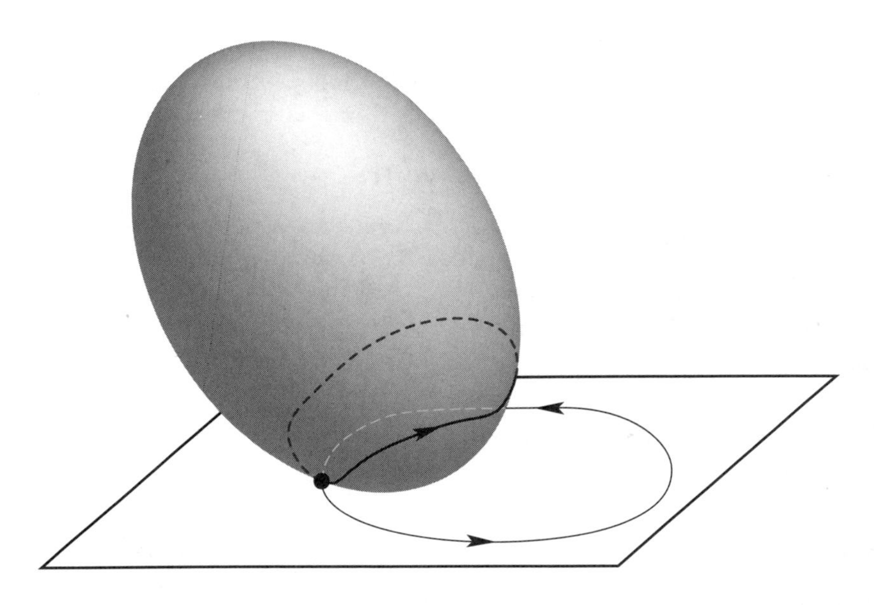

FIGURE 15.5.1. The geometry of Poinsot's theorem.

15.6 Euler Angles

In what follows, we adopt the conventions of Arnold [1989], Cabannes [1962], Goldstein [1980], and Hamel [1949]; these are different from the ones used by the British school (Whittaker [1927] and Pars [1965]).

Let (x^1, x^2, x^3) and (χ^1, χ^2, χ^3) denote the components of a vector written in the basis $(\mathbf{e}_1, \mathbf{e}_2, \mathbf{e}_3)$ and $(\boldsymbol{\xi}_1, \boldsymbol{\xi}_2, \boldsymbol{\xi}_3)$, respectively. We pass from the basis $(\mathbf{e}_1, \mathbf{e}_2, \mathbf{e}_3)$ to the basis $(\boldsymbol{\xi}_1, \boldsymbol{\xi}_2, \boldsymbol{\xi}_3)$ by means of three consecutive counterclockwise rotations (see Figure 15.6.1). First rotate $(\mathbf{e}_1, \mathbf{e}_2, \mathbf{e}_3)$ by an angle φ around $\mathbf{e}_3$ and denote the resulting basis and coordinates by $(\mathbf{e}'_1, \mathbf{e}'_2, \mathbf{e}'_3)$ and (x'_1, x'_2, x'_3), respectively. The new coordinates (x'^1, x'^2, x'^3) are expressed in terms of the old coordinates (x^1, x^2, x^3) of the *same point* by

$$\begin{bmatrix} x'^1 \\ x'^2 \\ x'^3 \end{bmatrix} = \begin{bmatrix} \cos\varphi & \sin\varphi & 0 \\ -\sin\varphi & \cos\varphi & 0 \\ 0 & 0 & 1 \end{bmatrix} \begin{bmatrix} x^1 \\ x^2 \\ x^3 \end{bmatrix}. \tag{15.6.1}$$

Denote the change of basis matrix (15.6.1) in $\mathbb{R}^3$ by $\mathbf{R}_1$. Second, rotate $(\mathbf{e}'_1, \mathbf{e}'_2, \mathbf{e}'_3)$ by the angle θ around $\mathbf{e}'_1$ and denote the resulting basis and coordinate system by $(\mathbf{e}''_1, \mathbf{e}''_2, \mathbf{e}''_3)$ and (x''^1, x''^2, x''^3), respectively. The new coordinates (x''^1, x''^2, x''^3) are expressed in terms of the old coordinates

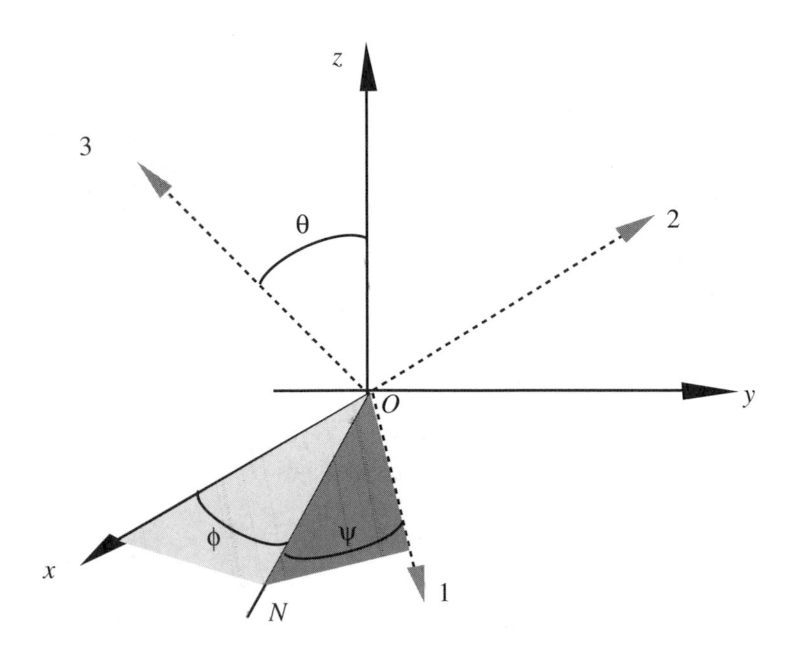

FIGURE 15.6.1. Euler angles.

(x'^1, x'^2, x'^3) by

$$\begin{bmatrix} x''^1 \\ x''^2 \\ x''^3 \end{bmatrix} = \begin{bmatrix} 1 & 0 & 0 \\ 0 & \cos\theta & \sin\theta \\ 0 & -\sin\theta & \cos\theta \end{bmatrix} \begin{bmatrix} x'^1 \\ x'^2 \\ x'^3 \end{bmatrix}. \tag{15.6.2}$$

Denote the change of basis matrix in (15.6.2) by $\mathbf{R}_2$. The $\mathbf{e}'_1$-axis, that is, the intersection of the $(\mathbf{e}_1, \mathbf{e}_2)$-plane with the $(\mathbf{e}''_1, \mathbf{e}''_2)$-plane is called the ***line of nodes*** and is denoted by ON. Finally, rotate by the angle ψ around $\mathbf{e}''_3$. The resulting basis is $(\boldsymbol{\xi}_1, \boldsymbol{\xi}_2, \boldsymbol{\xi}_3)$ and the new coordinates (χ^1, χ^2, χ^3) are expressed in terms of the old coordinates (x''^1, x''^2, x''^3) by

$$\begin{bmatrix} \chi^1 \\ \chi^2 \\ \chi^3 \end{bmatrix} = \begin{bmatrix} \cos\psi & \sin\psi & 0 \\ -\sin\psi & \cos\psi & 0 \\ 0 & 0 & 1 \end{bmatrix} \begin{bmatrix} x''^1 \\ x''^2 \\ x''^3 \end{bmatrix}. \tag{15.6.3}$$

Let $\mathbf{R}_3$ denote the change of basis matrix in (15.6.3). The rotation $\mathbf{R}$ sending (x^1, x^2, x^3) to (χ^1, χ^2, χ^3) is described by the matrix $\mathbf{P} = \mathbf{R}_3\mathbf{R}_2\mathbf{R}_1$ given by

$$\begin{bmatrix} \cos\psi\cos\varphi - \cos\theta\sin\varphi\sin\psi & \cos\psi\sin\varphi + \cos\theta\cos\varphi\sin\psi & \sin\theta\sin\psi \\ -\sin\psi\cos\varphi - \cos\theta\sin\varphi\cos\psi & -\sin\psi\sin\varphi + \cos\theta\cos\varphi\cos\psi & \sin\theta\cos\psi \\ \sin\theta\sin\varphi & -\sin\theta\cos\varphi & \cos\theta \end{bmatrix}. \tag{15.6.4}$$

Thus, $\chi = \mathbf{P}x$; equivalently, since the *same* point is expressed in two ways as $\sum_{i=1}^{3} \chi^i \boldsymbol{\xi}_i = \sum_{j=1}^{3} x^i \mathbf{e}_j$, we get

$$\sum_{j=1}^{3} x^j \mathbf{e}_j = \sum_{i=1}^{3} \chi^i \boldsymbol{\xi}_i = \sum_{i=1}^{3} \left(\sum_{j=1}^{3} P_{ij} x^j \right) \boldsymbol{\xi}_i = \sum_{j=1}^{3} x^j \sum_{i=1}^{3} P_{ij} \boldsymbol{\xi}_i,$$

that is,

$$\mathbf{e}_j = \sum_{i=1}^{3} P_{ij} \boldsymbol{\xi}_i, \tag{15.6.5}$$

and hence $\mathbf{P}$ is the change of basis matrix between the rotated basis $(\boldsymbol{\xi}_1, \boldsymbol{\xi}_2, \boldsymbol{\xi}_3)$, and the fixed spatial basis $(\mathbf{e}_1, \mathbf{e}_2, \mathbf{e}_3)$. On the other hand, (15.6.5) represents the matrix expression of the rotation $\mathbf{R}^T$ sending $\boldsymbol{\xi}_j$ to $\mathbf{e}_j$, that is, the matrix $[\mathbf{R}]_\xi$ of $\mathbf{R}$ in the basis $(\boldsymbol{\xi}_1, \boldsymbol{\xi}_2, \boldsymbol{\xi}_3)$ is P^T:

$$[\mathbf{R}]_\xi = \mathbf{P}^T, \quad \text{i.e.,} \quad \mathbf{R}\boldsymbol{\xi}_i = \sum_{i=1}^{3} P_{ij} \boldsymbol{\xi}_j. \tag{15.6.6}$$

Consequently, the matrix $[\mathbf{R}]_{\mathrm{e}}$ of $\mathbf{R}$ in the basis $(\mathbf{e}_1, \mathbf{e}_2, \mathbf{e}_3)$ is given by P:

$$[\mathbf{R}]_{\mathrm{e}} = \mathbf{P}, \quad \text{i.e.,} \quad \mathbf{R}\mathbf{e}_j = \sum_{i=1}^{3} P_{ij} \mathbf{e}_i. \tag{15.6.7}$$

It is straightforward to check that if $0 \leq \varphi < 2\pi$, $0 \leq \psi < 2\pi, 0 \leq \theta < \pi$, there is a bijective map between the (φ, ψ, θ) variables and $SO(3)$. However, this bijective map does not define a chart, since its differential vanishes, for example, at $\varphi = \psi = \theta = 0$. The differential is nonzero for $0 < \varphi < 2\pi, 0 < \psi < 2\pi, 0 < \theta < \pi$, and on this domain, the Euler angles do form a chart.

15.7 The Hamiltonian of the Free Rigid Body in the Material Description via Euler Angles

To express the kinetic energy in terms of Euler angles, we choose the basis $\mathbf{E}_1, \mathbf{E}_2, \mathbf{E}_3$ of $\mathbb{R}^3$ in the reference configuration to equal the basis $(\mathbf{e}_1, \mathbf{e}_2, \mathbf{e}_3)$ of $\mathbb{R}^3$ in the spatial coordinate system. Thus, the matrix representation of $\mathbf{R}(t)$ in the basis $\boldsymbol{\xi}_1, \boldsymbol{\xi}_2, \boldsymbol{\xi}_3$ equals $\mathbf{P}^T$, where $\mathbf{P}$ is given by (15.6.4). In this way, $\boldsymbol{\omega}$ and $\boldsymbol{\Omega}$ have the following expressions in the basis $\boldsymbol{\xi}_1, \boldsymbol{\xi}_2, \boldsymbol{\xi}_3$:

$$\boldsymbol{\omega} = \begin{bmatrix} \dot{\theta}\cos\varphi + \dot{\psi}\sin\varphi\sin\theta \\ \dot{\theta}\sin\varphi - \dot{\psi}\cos\varphi\sin\theta \\ \dot{\varphi} + \dot{\psi}\cos\theta \end{bmatrix}, \quad \boldsymbol{\Omega} = \begin{bmatrix} \dot{\theta}\cos\psi + \dot{\varphi}\sin\psi\sin\theta \\ -\dot{\theta}\sin\psi + \dot{\varphi}\cos\psi\sin\theta \\ \dot{\varphi}\cos\theta + \dot{\psi} \end{bmatrix}. \tag{15.7.1}$$

By definition of $\mathbf{\Pi}$, it follows that

$$\mathbf{\Pi} = \begin{bmatrix} I_1(\dot\varphi \sin\theta \sin\psi + \dot\theta \cos\psi) \\ I_2(\dot\varphi \sin\theta \cos\psi - \dot\theta \sin\psi) \\ I_3(\dot\varphi \cos\theta + \dot\psi) \end{bmatrix}. \tag{15.7.2}$$

This expresses $\mathbf{\Pi}$ in terms of coordinates on $T(SO(3))$. Since $T(SO(3))$ and $T^*(SO(3))$ are to be identified by the metric defined as the left invariant metric given at the identity by $\langle\langle\, , \rangle\rangle$, the variables $(p_\varphi, p_\psi, p_\theta)$ canonically conjugate to $(\varphi, \psi.\theta)$ are given by the Legendre transformation $p_\varphi = \partial K/\partial\dot\varphi, p_\psi = \partial K/\partial\dot\psi, p_\theta = \partial K/\partial\dot\theta$, where the expression of the kinetic energy on $T(SO(3))$ is obtained by plugging (15.7.2) into (15.3.3). We get

$$\begin{aligned} p_\varphi &= I_1(\dot\varphi \sin\theta \sin\psi + \dot\theta\cos\psi)\sin\theta\sin\psi \\ &\quad + I_2(\dot\varphi\sin\theta\cos\varphi - \dot\theta\sin\psi)\sin\theta\cos\psi + I_3(\dot\varphi\cos\theta + \dot\psi)\cos\theta, \\ p_\psi &= I_3(\dot\varphi\cos\theta + \dot\psi), \\ p_\theta &= I_1(\dot\varphi\sin\theta\sin\psi + \dot\theta\cos\psi)\cos\psi \\ &\quad - I_2(\dot\varphi\sin\theta\cos\psi - \dot\theta\sin\psi)\sin\psi, \end{aligned} \tag{15.7.3}$$

whence by (15.7.2)

$$\mathbf{\Pi} = \begin{bmatrix} ((p_\varphi - p_\psi\cos\theta)\sin\psi + p_\theta\sin\theta\cos\psi)/\sin\theta \\ ((p_\varphi - p_\psi cos\theta)\cos\psi - p_\theta\sin\theta\sin\psi)/\sin\theta \\ p_\psi \end{bmatrix}, \tag{15.7.4}$$

and so by (15.3.3) we get the coordinate expression of the kinetic energy in the material picture to be

$$\begin{aligned} &K(\varphi, \psi, \theta, p_\varphi, p_\psi, p_\theta) \\ &\quad = \frac{1}{2}\left\{ \frac{[(p_\varphi - p_\psi\cos\theta)\sin\psi + p_\theta\sin\theta\cos\psi]^2}{I_1\sin^2\theta} \right. \\ &\quad\quad \left. + \frac{[(p_\varphi - p_\psi\cos\theta)\cos\psi - p_\theta\sin\theta\sin\psi]^2}{I_1\sin^2\theta} + \frac{p_\psi^2}{I_3} \right\}. \end{aligned} \tag{15.7.5}$$

This expression for the kinetic energy has an invariant expression on the cotangent bundle $T^*(SO(3))$. In fact,

$$K(\alpha_R) = \frac{1}{2}\langle\langle \mathbf{\Omega}, \mathbf{\Omega} \rangle\rangle = \frac{1}{4}\mathrm{Tr}(\mathbf{I}\mathbf{R}^{-1}\dot{\mathbf{R}}\mathbf{R}^{-1}\dot{\mathbf{R}}), \tag{15.7.6}$$

where $\alpha_R \in T_R^*(SO(3))$ is defined by $\langle \alpha, \mathbf{R}\hat{\mathbf{v}} \rangle = \langle\langle \mathbf{\Omega}, \mathbf{v} \rangle\rangle$ for all $\mathbf{v} \in \mathbb{R}^3$.

The equation of motion (15.3.9) can also be derived "by hand" without appeal to Lie-Poisson or Euler-Poincaré reduction as follows. Hamilton's canonical equations

$$\dot\varphi = \frac{\partial K}{\partial p_\varphi}, \quad \dot\psi = \frac{\partial K}{\partial p_\psi}, \quad \dot\theta = \frac{\partial K}{\partial p_\theta},$$

$$\dot{p}_\varphi = -\frac{\partial K}{\partial \varphi}, \quad \dot{p}_\psi = -\frac{\partial K}{\partial \psi}, \quad \dot{p}_\theta = -\frac{\partial K}{\partial \theta},$$

in a chart given by the Euler angles, become after direct substitution and a somewhat lengthy calculation,

$$\dot{\mathbf{\Pi}} = \mathbf{\Pi} \times \mathbf{\Omega}.$$

For $F, G : T^*(SO(3)) \to \mathbb{R}$, that is, F, G are functions of $(\varphi, \psi, \theta, p_\varphi, p_\psi, p_\theta)$ in a chart given by Euler angles, the standard canonical Poisson bracket is

$$\begin{aligned} \{F, G\} &= \frac{\partial F}{\partial \varphi}\frac{\partial G}{\partial p_\varphi} - \frac{\partial F}{\partial p_\varphi}\frac{\partial G}{\partial \varphi} + \frac{\partial F}{\partial \psi}\frac{\partial G}{\partial p_\psi} \\ &\quad - \frac{\partial F}{\partial p_\psi}\frac{\partial G}{\partial \psi} + \frac{\partial F}{\partial \theta}\frac{\partial G}{\partial p_\theta} - \frac{\partial F}{\partial p_\theta}\frac{\partial G}{\partial \theta}. \end{aligned} \tag{15.7.7}$$

A computation shows that after the substitution $(\varphi, \psi, \theta, p_\varphi, p_\psi, p_\theta) \mapsto (\Pi_1, \Pi_2, \Pi_3)$, this becomes

$$\{F, G\}(\mathbf{\Pi}) = -\mathbf{\Pi} \cdot (\nabla F(\mathbf{\Pi}) \times \nabla G(\mathbf{\Pi})) \tag{15.7.8}$$

which is the (−) Lie-Poisson bracket. This provides a direct check on the Lie-Poisson reduction theorem in Chapter **13**. Thus (15.7.4) defines a canonical map between Poisson manifolds. The apparently "miraculous" groupings and cancellations of terms that occur in this calculation should make the reader appreciate the general theory.

Exercise

15.7-1 *Verify that* (15.7.8) *holds by a* ***direct*** *calculation using substitution and the chain rule.*

15.8 The Analytical Solution of the Free Rigid Body Problem

We now give the analytical solution of the Euler equations. These formulae are useful when, for example, one is dealing with perturbations leading chaos via the Poincaré-Melnikov method, as in Ziglin [1980a,b], Holmes and Marsden [1983], and Koiller [1985]. For the last part of this section, the reader is assumed to be familiar with Jacobi's elementary elliptic functions; see, for example, Lawden [1989]. Let us make the following simplifying notations

$$a_1 = \frac{I_2 - I_3}{I_2 I_3} \geq 0, \quad a_2 = \frac{I_3 - I_1}{I_1 I_3} \leq 0, \quad \text{and} \quad a_3 = \frac{I_1 - I_2}{I_1 I_2} \geq 0,$$

where we assume $I_1 \geq I_2 \geq I_3 > 0$. Then Euler's equations $\dot{\mathbf{\Pi}} = \mathbf{\Pi} \times \mathbf{I}^{-1}\mathbf{\Pi}$ can be written as

$$\begin{aligned} \dot{\Pi}_1 &= a_1 \Pi_2 \Pi_3, \\ \dot{\Pi}_2 &= a_2 \Pi_3 \Pi_1, \\ \dot{\Pi}_3 &= a_3 \Pi_1 \Pi_2. \end{aligned} \tag{15.8.1}$$

For the analysis that follows it is important to recall that *the angular momentum in space is fixed* and that the instantaneous axis of rotation of the body in body coordinates is given by the angular velocity vector $\mathbf{\Omega}$.

Case 1: $I_1 = I_2 = I_3$. Then $a_1 = a_2 = a_3 = 0$ and we conclude that $\mathbf{\Pi}$, and thus $\mathbf{\Omega}$ are both constant. Hence the body rotates with constant angular velocity about a fixed axis. In Figure 15.3.1, all points on the sphere become fixed points.

Case 2: $I_1 = I_2 > I_3$. Then $a_3 = 0$ and $a_2 = -a_1$. Since $a_3 = 0$ it follows from (15.8.1) that $\Pi_3 =$ constant, and thus denoting $\lambda = -a_1 \Pi_3$ we get $a_2 \Pi_3 = \lambda$. Thus, (15.8.1) become

$$\begin{aligned} \dot{\Pi}_1 + \lambda \Pi_2 &= 0, \\ \dot{\Pi}_2 - \lambda \Pi_1 &= 0, \end{aligned}$$

which has solution for initial data given at time $t = 0$ given by

$$\begin{aligned} \Pi_1 &= \Pi_1(0) \cos \lambda t - \Pi_2(0) \sin \lambda t, \\ \Pi_2 &= \Pi_2(0) \cos \lambda t + \Pi_1(0) \sin \lambda t. \end{aligned}$$

These formulae say that the axis of symmetry OZ of the body rotates *relative to the body* with angular velocity λ. It is straightforward to check that OZ, $\mathbf{\Omega}$, and $\mathbf{\Pi}$ are in the same plane and that $\mathbf{\Pi}$ and $\mathbf{\Omega}$ make constant angles with OZ and thus among themselves. In addition, since $I_1 = I_2$, we have

$$\begin{aligned} \|\mathbf{\Omega}\|^2 &= \frac{\Pi_1^2}{I_1^2} + \frac{\Pi_2^2}{I_2^2} + \frac{\Pi_3^2}{I_3^2} \\ &= \left(\frac{\Pi_1^2}{I_1} + \frac{\Pi_2^2}{I_2} + \frac{\Pi_3^2}{I_3} \right) \frac{1}{I_1} - \frac{\Pi_3^2}{I_3} \left(\frac{1}{I_1} - \frac{1}{I_3} \right) \\ &= \frac{2K}{I_1} - \frac{a_2 \Pi_3^2}{I_3} = \text{constant}. \end{aligned}$$

Therefore the corresponding spatial objects Oz (the symmetry axis of the inertia ellipsoid in space), $\boldsymbol{\omega}$, and $\boldsymbol{\pi}$ enjoy the same properties and hence the axis of rotation in the body (given by $\mathbf{\Omega}$) makes a constant angle with the angular momentum vector that is fixed in space, and thus the axis of rotation describes a right circular cone of constant angle in space. At the same time, the axis of rotation in the body (given by $\mathbf{\Omega}$) makes a constant angle with Oz, thus tracing a second cone in the body. See Figure 15.8.1.

Consequently, the motion can be described by the rolling of a cone of constant angle in the body on a second cone of constant angle fixed in space. Whether the cone in the body rolls outside or inside the cone in space is determined by the sign of λ. Since $Oz, \boldsymbol{\omega}$, and $\boldsymbol{\pi}$ remain coplanar during the motion, $\boldsymbol{\omega}$ and Oz rotate about the fixed vector $\boldsymbol{\pi}$ with the same angular velocity, namely, the component of $\boldsymbol{\omega}$ along $\boldsymbol{\pi}$ in the decomposition of $\boldsymbol{\omega}$ relative to $\boldsymbol{\pi}$ and the Oz-axis. This angular velocity is called the ***angular velocity of precession***. Let $\mathbf{e}$ denote the unit vector along Oz and write $\boldsymbol{\omega} = \alpha\boldsymbol{\pi} + \beta\mathbf{e}$. Therefore

$$\begin{aligned} 2K &= \boldsymbol{\omega}\cdot\boldsymbol{\pi} = \alpha\|\boldsymbol{\pi}\|^2 + \beta\mathbf{e}\cdot\boldsymbol{\pi} = \alpha\|\boldsymbol{\pi}\|^2 + \beta\Pi_3, \\ \frac{\Pi_3}{I_3} &= \boldsymbol{\Omega}^3 = \boldsymbol{\omega}\cdot\mathbf{e} = \alpha\boldsymbol{\pi}\cdot\mathbf{e} + \beta = \alpha\Pi_3 + \beta, \end{aligned}$$

and

$$\beta = -a_2\Pi_3,$$

so that $\alpha = 1/I_1$ and $\beta = -a_2\Pi_3$. Therefore the *angular velocity of precession equals* $\mathbf{\Pi}_S/I_1$.

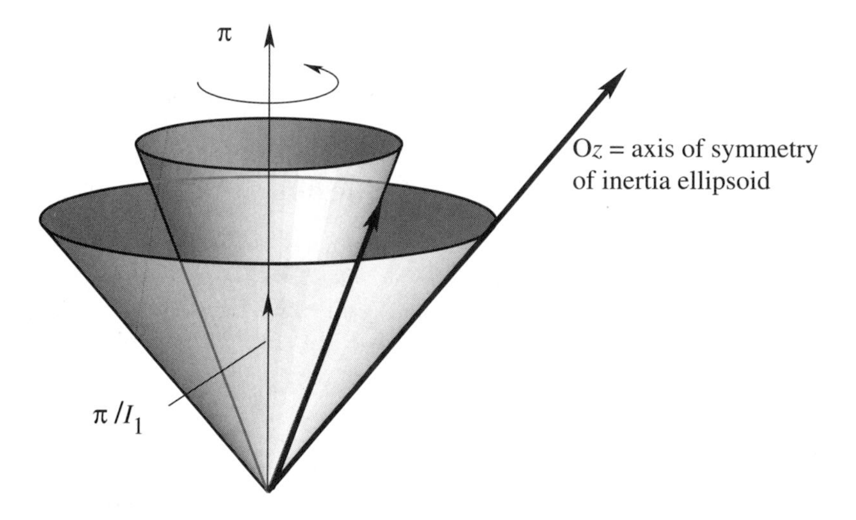

FIGURE 15.8.1. The geometry for integrating Euler's equations.

On the $\mathbf{\Pi}$-sphere, the dynamics reduce to two fixed points surrounded by oppositely oriented periodic lines of latitude and separated by an equator of fixed points. A similar analysis applies if $I_1 > I_2 = I_3$.

Case 3: $I_1 > I_2 > I_3$. The two integrals of energy and angular momentum

$$\frac{\Pi_1^2}{I_1} + \frac{\Pi_2^2}{I_2} + \frac{\Pi_3^3}{I_3} = 2h = ab^2, \tag{15.8.2}$$

$$\Pi_1^2 + \Pi_2^2 + \Pi_3^2 = \|\mathbf{\Pi}\|^2 = a^2b^2, \tag{15.8.3}$$

where $a = \|\mathbf{\Pi}\|^2/2h, b = 2h/\|\mathbf{\Pi}\|$ are positive constants, enable us to express Π_1 and Π_3 in terms of Π_2 as

$$\Pi_1^2 = \frac{I_1(I_2 - I_3)}{I_2(I_1 - I_3)}(\alpha^2 - \Pi_2^2) \tag{15.8.4}$$

and

$$\Pi_3^2 = \frac{I_3(I_1 - I_2)}{I_2(I_1 - I_3)}(\beta^2 - \Pi_2^2), \tag{15.8.5}$$

where α and β are positive constants given by

$$\alpha^2 = \frac{aI_2(a - I_3)b^2}{I_2 - I_3} \quad \text{and} \quad \beta^2 = \frac{aI_2(I_1 - a)b^2}{I_1 - I_2}. \tag{15.8.6}$$

By the definition of a, note that $I_1 \geq a \geq I_3$. The endpoints of the interval $[I_1, I_3]$ are easy to deal with. If $a = I_1$, then $\Pi_2 = \Pi_3 = 0$ and the motion is a steady rotation about the $\mathbf{\Pi}$-axis with body angular velocity $\pm b$. Similarly, if $a = I_3$, then $\Pi_1 = \Pi_2 = 0$. So we can assume that $I_1 > a > I_3$. With these expressions, the square of (15.8.1) becomes

$$(\dot{\Pi}_2)^2 = a_1 a_3 (\alpha^2 - \Pi_2^2)(\beta^2 - \Pi_2^2) \tag{15.8.7}$$

that is,

$$t = \int_{\Pi_2(0)}^{\Pi_2} \frac{du}{\sqrt{a_1 a_3 (\alpha^2 - u^2)(\beta^2 - u^2)}} \tag{15.8.8}$$

which shows that Π_2, and hence Π_1, Π_3 are elliptic functions of time.

In case the quartic under the square root has double roots, that is, $\alpha = \beta$, (15.8.8) can be integrated explicitly by means of elementary functions. By (15.8.6) if follows that

$$\beta^2 - \alpha^2 = \frac{ab^2 I_2(I_1 - I_3)(I_2 - a)}{(I_1 - I_2)(I_2 - I_3)}.$$

Thus $\alpha = \beta$ if and only if $a = I_2$ which in turn forces $\alpha = \beta = ab = \|\mathbf{\Pi}\|$ and $\|\mathbf{\Pi}\|^2 = 2hI_2$. Thus (15.8.7) becomes

$$(\dot{\Pi}_2)^2 = a_1 a_3 (\|\mathbf{\Pi}\|^2 - \Pi_2^2)^2. \tag{15.8.9}$$

If $\|\mathbf{\Pi}\|^2 = 2hI_2$ is satisfied, the intersection of the sphere of constant angular momentum $\|\mathbf{\Pi}\|$ with the elliptical energy surface corresponding to the value $2h$ consists of two great circles on the sphere going through the Π_2-axis in the planes

$$\Pi_3 = \pm \Pi_1 \sqrt{\frac{a_3}{a_1}}.$$

In other words, the solution of (15.8.9) consists of four heteroclinic orbits and the values $\Pi_2 = \pm\|\mathbf{\Pi}\|$. Equation (15.8.9) is solved by putting $\Pi_2 =$

$\|\mathbf{\Pi}\| \tanh \theta$. Setting $\Pi_2(0) = 0$ for simplicity we get the four heteroclinic orbits

$$\begin{aligned} \Pi_1^+(t) &= \pm\|\mathbf{\Pi}\|\sqrt{\frac{a_1}{-a_2}}\operatorname{sech}(-\sqrt{a_1 a_3}\,\|\mathbf{\Pi}\|t), \\ \Pi_2^+(t) &= \pm\|\mathbf{\Pi}\|\tanh(-\sqrt{a_1 a_3}\,\|\mathbf{\Pi}\,\|t), \\ \Pi_3^+(t) &= \pm\|\mathbf{\Pi}\|\sqrt{\frac{a_3}{-a_2}}\operatorname{sech}(-\sqrt{a_1 a_3}\,\|\mathbf{\Pi}\|t), \end{aligned} \tag{15.8.10}$$

when

$$\Pi_3 = \Pi_1\sqrt{\frac{a_3}{a_1}}$$

and

$$\Pi_1^-(t) = \Pi_1^+(-t),\;\; \Pi_2^-(t) = \Pi_2^+(-t),\;\; \Pi_3^-(t) = \Pi_3^+(-t),$$

when

$$\Pi_3 = -\Pi_1\sqrt{\frac{a_3}{a_1}}.$$

If $\alpha \neq \beta$, then $a \neq I_2$, and the integration is performed with the aid of Jacobi's elliptic functions (see Whittaker and Watson [1940], Chapter 22, or Lawden [1989]). For example, the elliptic function $\operatorname{sn} u$ with modulus k is given by

$$\operatorname{sn} u = u - \frac{1}{3!}(1+k^2)u^3 + \frac{1}{5!}(1+14k^2+k^4)u^5 - \dots$$

and its inverse is

$$\operatorname{sn}^{-1} x = \int_0^x \frac{1}{\sqrt{(1-t^2)(1-k^2t^2)}}\,dt, \quad 0 \le x \le 1.$$

Assuming $I_1 > I_2 > a > I_3$ or, equivalently, $\alpha < \beta$, the substitution of the elliptic function $\Pi_2 = \alpha \operatorname{sn} u$ in (15.8.8) with the modulus

$$k = \alpha/\beta = \left[\frac{(I_1 - I_2)(a - I_3)}{(I_1 - a)(I_2 - I_3)}\right]^{1/2},$$

gives $\dot{u}^2 = ab^2(I_1 - a)(I_2 - I_3)/I_1 I_2 I_3 = \mu^2$. We will need the identities

$$\operatorname{cn}^2 u = 1 - \operatorname{sn}^2 u, \quad \operatorname{dn}^2 u = 1 - k^2 \operatorname{sn}^2 u, \quad \text{and} \quad \frac{d}{dx}\operatorname{sn} x = \operatorname{cn} x \operatorname{dn} x.$$

With initial condition $\Pi_2(0) = 0$, this gives

$$\Pi_2 = \alpha \operatorname{sn}(\mu t). \tag{15.8.11}$$

Thus Π_2 varies between α and $-\alpha$. Choosing the time direction appropriately, we can assume without loss of generality that $\dot{\Pi}_2(0) > 0$. Note that Π_1 vanishes when Π_2 equals $\pm\alpha$ by (15.8.4), but that Π_3^2 attains its maximal value

$$\frac{I_3(I_1 - I_2)}{I_2(I_1 - I_3)}(\beta^2 - \alpha^2) = \frac{I_3(I_2 - a)ab^2}{(I_2 - I_3)} \tag{15.8.12}$$

by (15.8.5). The minimal value of Π_3^2 occurs when $\Pi_2 = 0$, that is, it is

$$\frac{I_3(I_1 - I_2)}{I_2(I_1 - I_3)}\beta^2 = \frac{I_3(I_1 - a)ab^2}{(I_1 - I_3)} =: \delta^2, \tag{15.8.13}$$

again by (15.8.5). Thus the sign of Π_3 is constant throughout the motion. Let us assume it is positive. This hypothesis together with $\dot{\Pi}_2(0) > 0$ and $a_2 < 0$ imply that $\Pi_1(0) < 0$.

Solving for Π_1 and Π_3 from (15.8.2) and (15.8.3) and remembering that $\Pi_1(0) < 0$ gives $\Pi_1(t) = -\gamma\,\mathrm{cn}(\mu t), \Pi_3(t) = \delta\,\mathrm{dn}(\mu t)$, where δ is given by (15.8.13) and

$$\gamma^2 = \frac{I_1(a - I_3)ab^2}{(I_1 - I_3)}. \tag{15.8.14}$$

Note that $\beta > \alpha > \gamma$ and, as usual, the values of γ and δ are taken to be positive. The solution of the Euler equations is therefore

$$\Pi_1(t) = -\gamma\,\mathrm{cn}(\mu t), \quad \Pi_2(t) = \alpha\,\mathrm{sn}(\mu t), \quad \Pi_3(t) = \delta\,\mathrm{dn}(\mu t), \tag{15.8.15}$$

with α, γ, δ given by (15.8.6), (15.8.13), (15.8.14). If κ denotes the period invariant of Jacobi's elliptic functions then Π_1 and Π_2 have period $4\kappa/\mu$ whereas Π_3 has period $2\kappa/\mu$.

Exercise

15.8-1 *Continue this integration process and find formulas for the attitude matrix $A(t)$ as functions of time with $A(0) =$* Identity *and with given body angular momentum (or velocity).*

15.9 Rigid Body Stability

Following the energy-momentum-Casimir method step by step (see the Introduction), we begin with the equations

$$\dot{\mathbf{\Pi}} = \frac{d\mathbf{\Pi}}{dt} = \mathbf{\Pi} \times \mathbf{\Omega}, \tag{15.9.1}$$

where $\mathbf{\Pi}, \mathbf{\Omega} \in \mathbb{R}^3, \mathbf{\Omega}$ is the angular velocity, and $\mathbf{\Pi}$ is the angular momentum, both viewed in the body; the relation between $\mathbf{\Pi}$ and $\mathbf{\Omega}$ is given by $\Pi_j = I_j\Omega^j$, $j = 1, 2, 3$, where $I = (I_1, I_2, I_3)$ is the diagonalized moment of inertia tensor, $I_1, I_2, I_3 > 0$. This system is Hamiltonian in the Lie-Poisson structure of $\mathbb{R}^3$ given by (15.3.8) and relative to the kinetic energy Hamiltonian

$$H(\mathbf{\Pi}) = \frac{1}{2}\mathbf{\Pi} \cdot \mathbf{\Omega} = \frac{1}{2}\sum_{i=0}^{3} \frac{\Pi_i^2}{I_i}. \tag{15.9.2}$$

Recall from (15.3.12) that for a smooth function $\Phi : \mathbb{R} \to \mathbb{R}$,

$$C_\Phi(\mathbf{\Pi}) = \Phi\left(\frac{1}{2}\|\mathbf{\Pi}\|^2\right) \tag{15.9.3}$$

is a Casimir function.

1 First Variation

We find a Casimir function C_Φ such that $H_{C_\Phi} := H + C_\Phi$ has a critical point at a given equilibrium point of (15.9.1). Such points occur when $\mathbf{\Pi}$ is parallel to $\mathbf{\Omega}$. We can assume without loss of generality, that $\mathbf{\Pi}$ and $\mathbf{\Omega}$ point in the Ox-direction. After normalizing if necessary, we can assume that the equilibrium solution is $\mathbf{\Pi}_e = (1, 0, 0)$. The derivative of

$$H_{C_\Phi}(\mathbf{\Pi}) = \frac{1}{2}\sum_{i=0}^{3} \frac{\Pi_i^2}{I_i} + \Phi\left(\frac{1}{2}\|\mathbf{\Pi}\|^2\right)$$

is

$$\mathbf{D}H_{C_\Phi}(\mathbf{\Pi}) \cdot \delta\mathbf{\Pi} = \left(\mathbf{\Omega} + \Phi'\left(\frac{1}{2}\|\mathbf{\Pi}\|^2\right)\mathbf{\Pi}\right) \cdot \delta\mathbf{\Pi}. \tag{15.9.4}$$

This equals zero at $\mathbf{\Pi}_e = (1, 0, 0)$, provided that

$$\Phi'\left(\frac{1}{2}\right) = -\frac{1}{I_1}. \tag{15.9.5}$$

2 Second Variation

Using (15.9.4), the second derivative of H_{C_Φ} at the equilibrium $\mathbf{\Pi}_e = (1, 0, 0)$ is

$$\begin{aligned}
&\mathbf{D}^2 H_{C_\Phi}(\mathbf{\Pi}_e) \cdot (\delta\mathbf{\Pi}, \delta\mathbf{\Pi}) \\
&= \delta\mathbf{\Omega} \cdot \delta\mathbf{\Pi} + \Phi'\left(\frac{1}{2}\|\mathbf{\Pi}_e\|^2\right)\|\delta\mathbf{\Pi}\|^2 + (\mathbf{\Pi}_e \cdot \delta\mathbf{\Pi})^2 \Phi''\left(\frac{1}{2}\|\mathbf{\Pi}_e\|^2\right) \\
&= \sum_{i=0}^{3} \frac{(\delta\Pi_i)^2}{I_i} - \frac{\|\delta\mathbf{\Pi}\|^2}{I_1} + \Phi''\left(\frac{1}{2}\right)(\delta\Pi_1)^2 \\
&= \left(\frac{1}{I_2} - \frac{1}{I_1}\right)(\delta\Pi_2)^2 + \left(\frac{1}{I_3} - \frac{1}{I_1}\right)(\delta\Pi_3)^2 + \Phi''\left(\frac{1}{2}\right)(\delta\Pi_1)^2.
\end{aligned} \tag{15.9.6}$$

3 Definiteness
This quadratic form is positive-definite if and only if

$$\Phi''\left(\frac{1}{2}\right) > 0 \tag{15.9.7}$$

and

$$I_1 > I_2, \quad I_1 > I_3. \tag{15.9.8}$$

Consequently,

$$\Phi(x) = -\frac{1}{I_1}x + \left(x - \frac{1}{2}\right)^2$$

satisfies (15.9.5) and makes the second derivative of H_{C_Φ} at $(1,0,0)$ positive-definite, so *stationary rotation around the longest axis is (Liapunov) stable.*

The quadratic form is negative-definite provided

$$\Phi''\left(\frac{1}{2}\right) < 0 \tag{15.9.9}$$

and

$$I_1 < I_2, \quad I_1 < I_3. \tag{15.9.10}$$

It is obvious that we may find a function Φ satisfying the requirements (15.9.5) and (15.9.9); for example, $\Phi(x) = -(1/I_1)x - \left(x - \frac{1}{2}\right)^2$. This proves that *rotation around the short axis is (Liapunov) stable.*

Finally, the quadratic form (15.9.6) is indefinite if

$$I_1 > I_2, \quad I_3 > I_1, \tag{15.9.11}$$

or the other way around. We cannot show by this method that rotation around the middle axis is *unstable*. We shall prove, by using a spectral analysis, that rotation about the middle axis is, in fact, unstable. Linearizing (15.9.1) at $\mathbf{\Pi}_e = (1,0,0)$ yields the linear constant coefficient system

$$\begin{aligned}
(\delta\dot{\Pi}) &= \delta\Pi \times \mathbf{\Omega}_e + \Pi_e \times \delta\mathbf{\Omega} \\
&= \left(0, \frac{I_3 - I_1}{I_3 I_1}\delta\Pi_3, \frac{I_1 - I_2}{I_1 I_2}\delta\Pi_2\right) \\
&= \begin{bmatrix} 0 & 0 & 0 \\ 0 & 0 & \dfrac{I_3 - I_1}{I_3 I_1} \\ 0 & \dfrac{I_1 - I_2}{I_1 I_2} & 0 \end{bmatrix} \delta\Pi.
\end{aligned} \tag{15.9.12}$$

On the tangent space at $\mathbf{\Pi}_e$ to the sphere of radius $\|\mathbf{\Pi}_e\| = 1$, the linear operator given by this linearized vector field has a matrix given by the lower right (2×2)-block whose eigenvalues are

$$\pm\frac{1}{I_1\sqrt{I_2 I_3}}\sqrt{(I_1 - I_2)(I_3 - I_1)}.$$

Both of them are real by (15.9.11) and one is strictly positive. Thus $\mathbf{\Pi}_e$ is spectrally unstable and thus is unstable.

We summarize the results in the following theorem.

Theorem 15.9.1 (Rigid Body Stability Theorem) *In the motion of a free rigid body, rotation around the long and short axes is (Liapunov) stable and around the middle axis is unstable.*

It is important to keep the Casimir functions as general as possible, because otherwise (15.9.5) and (15.9.9) could be contradictory. Had we simply chosen

$$\Phi(x) = -\frac{1}{I_1}x + \left(x - \frac{1}{2}\right)^2,$$

(15.9.5) would be verified, but (15.9.9) would not. It is only the choice of *two different* Casimirs that enables us to prove the two stability results, even though the level surfaces of these Casimirs are the same.

Remarks

1. As we have seen, rotations about the intermediate axis are unstable and this is even for the linearized equations. The unstable homoclinic orbit that connect the two unstable points have interesting features. Not only are they interesting because of the chaotic solutions via the Poincaré-Melnikov method that can be obtained in various perturbed systems (see Holmes and Marsden [1983], Wiggins [1988], and references therein), but already, the orbit itself is interesting since a rigid body tossed about its middle axis will undergo an interesting half twist when the opposite saddle point is reached, even though the rotation axis has returned to where it was. The reader can easily perform the experiment; see Ashbaugh, Chicone, and Cushman [1990] and Montgomery [1991a] for more information.

2. The same stability theorem can also be proved by working with the second derivative along a coadjoint orbit in $\mathbb{R}^3$; that is, a two-sphere; see Arnold [1966a]. This coadjoint orbit method also *suggests* instability of rotation around the intermediate axis.

3. Dynamic stability on the Π-sphere has been shown. What about the stability of the dynamically rigid body we "see"? This can be deduced from what we have done. Probably the best approach though is to use the relation between the reduced and unreduced dynamics; see Simo, Lewis, and Marsden [1991] and Lewis [1992] for more information.

4. When the body angular momentum undergoes a periodic motion, the actual motion of the rigid body in space is not periodic. In the introduction we described the associated geometric phase.

5. See Lewis and Simo [1990] and Simo, Lewis, and Marsden [1991] for related work on deformable elastic bodies (pseudo-rigid bodies).

Exercises

15.9-1 *Let* $\mathbf{B}$ *be a given fixed vector in* $\mathbb{R}^3$ *and let* M *evolve by* $\dot{M} = M \times B$. *Show that this evolution is Hamiltonian. Determine the equilibria and their stability.*

15.9-2 *Consider the following modification of the Euler equations:*

$$\dot{\mathbf{\Pi}} = \mathbf{\Pi} \times \Omega + \alpha \mathbf{\Pi} \times (\mathbf{\Pi} \times \Omega),$$

where α *is a positive constant. Show that,*

(a) *The spheres* $\|\mathbf{\Pi}\|^2$ *are preserved.*

(b) *Energy is strictly decreasing except at equilibria.*

(c) *The equations can be written in the form*

$$\dot{F} = \{F, H\}_{\text{rb}} + \{F, H\}_{\text{sym}},$$

where the first bracket is the usual rigid body bracket and the second is the symmetric bracket

$$\{F, K\}_{\text{sym}} = \alpha(\Pi \times \nabla F) \cdot (\Pi \times \nabla K).$$

15.10 Heavy Top Stability

The heavy top equations are

$$\frac{d\mathbf{\Pi}}{dt} = \mathbf{\Pi} \times \mathbf{\Omega} + Mgl\mathbf{\Gamma} \times \boldsymbol{\chi}, \tag{15.10.1}$$

$$\frac{d\mathbf{\Gamma}}{dt} = \mathbf{\Gamma} \times \mathbf{\Omega}, \tag{15.10.2}$$

where $\mathbf{\Pi}, \mathbf{\Gamma}, \boldsymbol{\chi} \in \mathbb{R}^3$. Here $\mathbf{\Pi}$ and $\mathbf{\Omega}$ are the angular momentum and angular velocity in the body, $\Pi_i = I_i \Omega^i, I_i > 0,\ i = 1, 2, 3$, with $I = (I_1, I_2, I_3)$ the moment of inertia tensor. The vector $\mathbf{\Gamma}$ represents the motion of the unit vector along the Oz-axis as seen from the body, and the constant vector $\boldsymbol{\chi}$ is the unit vector along the line segment of length l connecting the fixed point to the center mass of the body; M is the total mass of the body, and g is the strength of the gravitational acceleration, which is along Oz pointing down.

This system is Hamiltonian in the Lie-Poisson structure of $\mathbb{R}^3 \times \mathbb{R}^3$ given in the Introduction relative to the heavy top Hamiltonian

$$H(\mathbf{\Pi}, \mathbf{\Gamma}) = \frac{1}{2}\mathbf{\Pi} \cdot \mathbf{\Omega} + Mgl\mathbf{\Gamma} \cdot \boldsymbol{\chi}. \tag{15.10.3}$$

The Poisson structure (with $\|T\| = 1$ imposed) foreshadows that of $T^*SO(3)/S^1$, where S^1 acts by rotation about the axis of gravity. The fact that one gets the Lie-Poisson bracket for a semi-direct product Lie algebra is a special case of the general theory of reduction and semi-direct products (Marsden, Ratiu and Weinstein [1984a,b])

The functions $\mathbf{\Pi} \cdot \mathbf{\Gamma}$ and $\|\mathbf{\Gamma}\|^2$ are Casimir functions, as is

$$C(\mathbf{\Pi}, \mathbf{\Gamma}) = \Phi(\mathbf{\Pi} \cdot \mathbf{\Gamma}, \|\mathbf{\Gamma}\|^2), \tag{15.10.4}$$

where Φ is any smooth function from $\mathbb{R}^2$ to $\mathbb{R}$.

We shall be concerned here with the Lagrange top. This is a heavy top for which $I_1 = I_2$, that is, it is symmetric, and the center of mass lies on the axis of symmetry in the body, that is, $\boldsymbol{\chi} = (0, 0, 1)$. This assumption simplifies the equations of motion (15.10.1) to

$$\begin{aligned}
\dot{\Pi}_1 &= \frac{I_2 - I_3}{I_2 I_3}\Pi_2\Pi_3 + Mgl\Gamma_2, \\
\dot{\Pi}_2 &= \frac{I_3 - I_1}{I_1 I_3}\Pi_1\Pi_3 - Mgl\Gamma_1, \\
\dot{\Pi}_3 &= \frac{I_1 - I_2}{I_1 I_2}\Pi_1\Pi_2.
\end{aligned}$$

Since $I_1 = I_2$, we have $\dot{\Pi}_3 = 0$; thus Π_3 and hence any function $\varphi(\Pi_3)$ of Π_3 is conserved.

1 First Variation

We shall study the equilibrium solution

$$\mathbf{\Pi}_e = (0, 0, \Pi_3^0), \quad \mathbf{\Gamma}_e = (0, 0, 1),$$

where $\Pi_3^0 \neq 0$, which represents the spinning of a symmetric top in its upright position. To begin, we consider conserved quantities of the form $H_{\Phi,\varphi} = H + \Phi(\mathbf{\Pi} \cdot \mathbf{\Gamma}, \|\mathbf{\Gamma}\|^2) + \varphi(\Pi_3)$ and which have a critical point at the equilibrium. The first derivative of $H_{\Phi,\varphi}$ is given by

$$\begin{aligned}
\mathbf{D}H_{\Phi,\varphi}(\mathbf{\Pi}, \mathbf{\Gamma}) \cdot (\delta\mathbf{\Pi}, \delta\Gamma) &= (\mathbf{\Omega} + \dot{\Phi}(\mathbf{\Pi} \cdot \mathbf{\Gamma}, \|\mathbf{\Gamma}\|^2)\mathbf{\Gamma}) \cdot \delta\mathbf{\Pi} \\
&\quad + [Mgl\boldsymbol{\chi} + \dot{\Phi}(\mathbf{\Pi} \cdot \mathbf{\Gamma}, \|\mathbf{\Gamma}\|^2)\mathbf{\Pi} \\
&\quad + 2\Phi'(\mathbf{\Pi} \cdot \mathbf{\Gamma}, \|\mathbf{\Gamma}\|^2)\mathbf{\Gamma}] \cdot \delta\mathbf{\Gamma} + \varphi'(\Pi_3)\delta\Pi_3,
\end{aligned}$$

where $\dot{\Phi} = \partial\Phi/\partial(\mathbf{\Pi} \cdot \mathbf{\Gamma})$ and $\Phi' = \partial\Phi/\partial(\|\mathbf{\Gamma}\|^2)$. At the equilibrium solution $(\mathbf{\Pi}_e, \mathbf{\Gamma}_e)$ the first derivative of $H_{\Phi,\varphi}$ vanishes, provided that

$$\frac{\Pi_3^0}{I_3} + \dot{\Phi}(\Pi_3^0, 1) + \varphi'(\Pi_3^0) = 0$$

and that

$$Mgl + \dot{\Phi}(\Pi_3^0, 1)\Pi_3^0 + 2\Phi'(\Pi_3^0, 1) = 0;$$

the remaining equations, involving indices 1 and 2, are trivially verified. Solving for $\dot{\Phi}(\Pi_3^0, 1)$ and $\Phi'(\Pi_3^0, 1)$ we get the conditions

$$\dot{\Phi}(\Pi_3^0, 1) = -\left(\frac{1}{I_3} + \frac{\varphi'(\Pi_3^0)}{\Pi_3^0}\right)\Pi_3^0, \tag{15.10.5}$$

$$\Phi'(\Pi_3^0, 1) = \frac{1}{2}\left(\frac{1}{I_3} + \frac{\varphi'(\Pi_3^0)}{\Pi_3^0}\right)(\Pi_3^0)^2 - \frac{1}{2}Mgl. \tag{15.10.6}$$

2 Second Variation

We shall check for definiteness of the second variation of $H_{\Phi,\varphi}$ at the equilibrium point $(\mathbf{\Pi}_e, \mathbf{\Gamma}_e)$. To simplify the notation we shall set

$$a = \varphi''(\Pi_3^0), \qquad b = 4\Phi''(\Pi_3^0, 1), \qquad c = \ddot{\Phi}(\Pi_3^0, 1), \qquad d = 2\dot{\Phi}'(\Pi_3^0, 1).$$

With this notation, (15.10.5) and (15.10.6), we find that the matrix of the second derivative at (Π_e, Γ_e) is

$$\begin{bmatrix} 1/I_1 & 0 & 0 & \dot{\Phi}(\Pi_3^0,1) & 0 & 0 \\ 0 & 1/I_1 & 0 & 0 & \dot{\Phi}(\Pi_3^0,1) & 0 \\ 0 & 0 & (1/I_3)+a+c & 0 & 0 & a_{36} \\ \dot{\Phi}(\Pi_3^0,1) & 0 & 0 & 2\Phi'(\Pi_3^0,1) & 0 & 0 \\ 0 & \dot{\Phi}(\Pi_3^0,1) & 0 & 0 & 2\Phi'(\Pi_3^0,1) & 0 \\ 0 & 0 & a_{36} & 0 & 0 & a_{66} \end{bmatrix}, \tag{15.10.7}$$

where $a_{36} = \dot{\Phi}(\Pi_3^0, 1) + \Pi_3^0 c + d$ and $a_{66} = 2\Phi'(\Pi_3^0, 1) + b + (\Pi_3^0)^2 c + \Pi_3^0 d$.

3 Definiteness

The computations for this part will be done using the following formula from linear algebra. If

$$M = \begin{bmatrix} A & B \\ C & D \end{bmatrix}$$

is a $(p+q) \times (p+q)$ matrix and if the $(p \times p)$-matrix A is invertible, then

$$\det M = \det A \det(D - CA^{-1}B).$$

If the quadratic form given by (15.10.7) is definite, it must be positive-definite since the $(1,1)$-entry is positive. Recalling that $I_1 = I_2$, the six principal determinants have the following values:

$$\frac{1}{I_1}, \quad \frac{1}{I_1^2}, \quad \frac{1}{I_1^2}\left(\frac{1}{I_3} + a + c\right),$$

$$\frac{1}{I_1}\left(\frac{1}{I_3} + a + c\right)\left(\frac{2}{I_1}\Phi'(\Pi_3^0, 1) - \dot{\Phi}(\Pi_3^0, 1)^2\right),$$

$$\left(\frac{2}{I_1}\Phi'(\Pi_3^0,1)-\dot{\Phi}(\Pi_3^0,1)^2\right)^2\left(\frac{1}{I_3}+a+c\right),$$

and

$$\left(\frac{2}{I_1}\Phi'(\Pi_3^0,1)-\dot{\Phi}(\Pi_3^0,1)^2\right)^2(\dot{\Phi}(\Pi_3^0,1)+\Pi_3^0c+d)^2(2\Phi'(\Pi_3^0,1)+b+(\Pi_3^0)^2c+\Pi_3^0d).$$

Consequently, the quadratic form given by (15.10.7) is positive-definite, if and only if

$$\frac{1}{I_3}+a+c>0, \tag{15.10.8}$$

$$\frac{2}{I_1}\Phi'(\Pi_3^0,1)-\dot{\Phi}(\Pi_3^0,1)^2>0, \tag{15.10.9}$$

and

$$2\Phi'(\Pi_3^0,1)+b+(\Pi_3^0)^2c+\Pi_3^0d>0. \tag{15.10.10}$$

Conditions (15.10.8) and (15.10.10) can always be satisfied if we choose the numbers a, b, c, and d appropriately; for example, $a=c=d=0$ and b sufficiently large and positive. Thus, the determining condition for stability is (15.10.9). By (15.10.5) and (15.10.6), this becomes

$$\frac{1}{I_1}\left[\left(\frac{1}{I_3}+\frac{\varphi'(\Pi_3^0)}{\Pi_3^0}\right)(\Pi_3^0)^2-Mgl\right]-\left(\frac{1}{I_3}+\frac{\varphi'(\Pi_3^0)}{\Pi_3^0}\right)^2(\Pi_3^0)^2>0. \tag{15.10.11}$$

We can choose $\varphi'(\Pi_3^0)$ so that

$$\frac{1}{I_3}+\frac{\varphi'(\Pi_3^0)}{\Pi_3^0}=e$$

has any value we wish. The left side of (15.10.11) is a quadratic polynomial in e, whose leading coefficient is negative. In order for this to be positive for some e, it is necessary and sufficient for the discriminant

$$\frac{(\Pi_3^0)^4}{I_1^2}-\frac{4(\Pi_3^0)^2Mgl}{I_1}$$

to be positive; that is,

$$(\Pi_3^0)^2>4MglI_1$$

which is the classical stability condition for a fast top. We have proved the first part of the following:

Theorem 15.10.1 (Heavy Top Stability Theorem) *An upright spinning Lagrange top is stable provided that the angular velocity is strictly larger than $2\sqrt{MglI_1}/I_3$. It is unstable if the angular velocity is smaller than this value.*

The second part of the theorem is proved, as in §15.9, by a spectral analysis of the linearized equations, namely

$$(\delta\dot{\mathbf{\Pi}}) = \delta\mathbf{\Pi} \times \mathbf{\Omega} + \mathbf{\Pi}_e \times \delta\mathbf{\Omega} + Mgl\delta\mathbf{\Gamma} \times \boldsymbol{\chi}, \tag{15.10.12}$$

$$(\delta\dot{\mathbf{\Gamma}}) = \delta\mathbf{\Gamma} \times \mathbf{\Omega} + \mathbf{\Gamma}_e \times \delta\mathbf{\Omega}, \tag{15.10.13}$$

on the tangent space to the coadjoint orbit in $\mathfrak{se}(3)^*$ through $(\mathbf{\Pi}_e, \mathbf{\Gamma}_e)$ given by

$$\{(\delta\mathbf{\Pi}, \delta\mathbf{\Gamma}) \in \mathbb{R}^3 \times \mathbb{R}^3 \mid \delta\mathbf{\Pi} \cdot \mathbf{\Gamma}_e + \mathbf{\Pi}_e \cdot \delta\mathbf{\Gamma} = 0 \quad \text{and} \quad \delta\mathbf{\Gamma} \cdot \mathbf{\Gamma}_e = 0\}$$
$$\cong \{(\delta\mathbf{\Pi}_1, \delta\mathbf{\Pi}_2, \delta\mathbf{\Gamma}_1, \delta\mathbf{\Gamma}_2)\} = \mathbb{R}^4. \tag{15.10.14}$$

The matrix of the linearized system of equations on this space is computed to be

$$\begin{bmatrix} 0 & \dfrac{\Pi_3^0}{I_3}\dfrac{I_1 - I_3}{I_1} & 0 & Mgl \\ -\dfrac{\Pi_3^0}{I_3}\dfrac{I_1 - I_3}{I_1} & 0 & -Mgl & 0 \\ 0 & -\dfrac{1}{I_1} & 0 & \dfrac{\Pi_3^0}{I_3} \\ \dfrac{1}{I_1} & 0 & -\dfrac{\Pi_3^0}{I_3} & 0 \end{bmatrix}. \tag{15.10.15}$$

The matrix (15.10.15) has characteristic polynomial

$$\lambda^4 + \frac{1}{I_1^2}\left[(I_1^2 + (I_1 - I_3)^2)\left(\frac{\Pi_3^0}{I_3}\right)^2 - 2MglI_1\right]\lambda^2$$
$$+\frac{1}{I_1^2}\left[(I_1 - I_3)\left(\frac{\Pi_3^0}{I_3}\right)^2 + Mgl\right]^2, \tag{15.10.16}$$

whose discriminant as a quadratic polynomial in λ^2 is

$$\frac{1}{I_1^4}(2I_1 - I_3)^2\left(\frac{\Pi_3^0}{I_3}\right)^2\left(I_3^2\left(\frac{\Pi_3^0}{I_3}\right)^2 - 4MglI_1\right).$$

This discriminant is negative if and only if

$$\Pi_3^0 < 2\sqrt{MglI_1}.$$

Under this condition the four roots of the characteristic polynomial are all distinct and equal to $\lambda_0, \bar{\lambda}_0, -\lambda_0, -\bar{\lambda}_0$ for some $\lambda_0 \in \mathbb{C}$, where $\operatorname{Re}\lambda_0 \neq 0$ and $\operatorname{Im}\lambda_0 \neq 0$. Thus at least two of these roots have real part strictly larger than zero thereby showing that $(\mathbf{\Pi}_e, \mathbf{\Gamma}_e)$ is spectrally unstable and hence unstable.

When $I_2 = I_1 + \epsilon$ for small ϵ, the conserved quantity $\varphi(\Pi_3)$ is no longer available. In this case, a sufficiently fast top is still linearly stable, and nonlinear stability can be assessed by KAM theory. Other regions of phase space are known to possess chaotic dynamics in this case (Holmes and Marsden [1983]). For more information on stability and bifurcation in the heavy top, we refer to Lewis, Ratiu, Simo, and Marsden [1992].

Exercise

15.10-1(a) *Show that* $\tilde{H}(\mathbf{\Pi}, \mathbf{\Gamma}) = H(\mathbf{\Pi}, \mathbf{\Gamma}) + \|\mathbf{\Gamma}\|^2/2$, *where* H *is given by* (15.10.3), *generates the same equations of motion* (15.10.1) *and* (15.10.2).

(b) *Taking the Legendre transform of* $\tilde{H}$, *show that the equations can be written in Euler-Poincaré form.*

15.11 The Rigid Body and the Pendulum

This section, following Holm and Marsden [1991], shows how the rigid body and the pendulum are linked.

Euler's equations are expressible in vector form as

$$\frac{d}{dt}\mathbf{\Pi} = \nabla H \times \nabla L, \tag{15.11.1}$$

where H is the energy,

$$H = \frac{\Pi_1^2}{2I_1} + \frac{\Pi_2^2}{2I_2} + \frac{\Pi_3^2}{2I_3}, \tag{15.11.2}$$

$$\nabla H = \left(\frac{\partial H}{\partial \Pi_1}, \frac{\partial H}{\partial \Pi_2}, \frac{\partial H}{\partial \Pi_3} \right) = \left(\frac{\Pi_1}{I_1}, \frac{\Pi_2}{I_2}, \frac{\Pi_3}{I_3} \right), \tag{15.11.3}$$

is the gradient of H and L is the square of the body angular momentum,

$$L = \frac{1}{2}\left(\Pi_1^2 + \Pi_2^2 + \Pi_3^2 \right). \tag{15.11.4}$$

Since both H and L are conserved, the rigid body motion itself takes place, as we know, along the intersections of the level surfaces of the energy (ellipsoids) and the angular momentum (spheres) in $\mathbb{R}^3$. The centers of the energy ellipsoids and the angular momentum spheres coincide. This, along with the $(\mathbb{Z}_2)^3$ symmetry of the energy ellipsoids, implies that the two sets of level surfaces in $\mathbb{R}^3$ develop collinear gradients (for example, tangencies) at pairs of points which are diametrically opposite on an angular momentum sphere. At these points, collinearity of the gradients of H and L implies stationary rotations, that is, equilibria.

Euler's equations for the rigid body may also be written as

$$\frac{d}{dt}\mathbf{\Pi} = \nabla K \times \nabla N, \tag{15.11.5}$$

where K and N are linear combinations of energy and angular momentum of the form

$$\begin{pmatrix} K \\ N \end{pmatrix} = \begin{bmatrix} a & b \\ c & d \end{bmatrix} \begin{pmatrix} H \\ L \end{pmatrix}, \tag{15.11.6}$$

with real constants a, b, c, and d satisfying $ad - bc = 1$. Thus, the equations of rigid body motion are unchanged (so the trajectories of the motion in $\mathbb{R}^3$ remain unchanged) when the energy H and angular momentum L are replaced by the $SL(2, \mathbb{R})$-linear combinations K and N. Notice that K will be a quadratic form and that it can occur with *any signature.* The diagonal form is $(0, 0, 0)$ iff $I_1 = I_2 = I_3$ and $(*, 0, 0)$, iff $I_2 = I_3$. If all moments of inertia I_1, I_2, I_3 are distinct and can always arrange a and b such that the diagonal form of K is $(0, *, *)$.

Assuming $I_1 < I_2 < I_3$, one may choose to eliminate one of the terms in each of H and L, by choosing the linear combination given by

$$\begin{bmatrix} a & b \\ c & d \end{bmatrix} = \begin{bmatrix} 1 & -1/I_3 \\ -c & c/I_1 \end{bmatrix} \quad \text{with} \quad c = \frac{1}{\left(\frac{1}{I_1} - \frac{1}{I_3}\right)} > 0, \tag{15.11.7}$$

so that

$$K = \frac{1}{2}\left(\frac{1}{I_1} - \frac{1}{I_3}\right)\Pi_1^2 + \frac{1}{2}\left(\frac{1}{I_2} - \frac{1}{I_3}\right)\Pi_2^2, \tag{15.11.8}$$

and

$$N = \frac{c}{2}\left(\frac{1}{I_1} - \frac{1}{I_2}\right)\Pi_2^2 + \frac{1}{2}\Pi_3^2. \tag{15.11.9}$$

With this choice, the orbits for Euler's equations for rigid body dynamics are realized as motion along the intersections of two, orthogonally oriented, *elliptic cylinders,* one elliptic cylinder being a level surface of K, with its translation axis along Π_3 (where $K = 0$), and the other a level surface of N, with its translation axis along Π_1 (where $N = 0$).

For a general choice of K and N, equilibria occur at points where the gradients of K and N are collinear. This can occur at points where the level sets are tangent (and the gradients both are nonzero), or at points where one of the gradients vanishes. In the elliptic cylinder case above, these two cases are points where the elliptic cylinders are tangent, and at points where the axis of one cylinder punctures normally through the surface of the other. The elliptic cylinders are tangent at one $\mathbb{Z}_2$-symmetric pair of points along the Π_2 axis, and the elliptic cylinders have normal axial punctures at two other $\mathbb{Z}_2$-symmetric pairs of points along the Π_1 and Π_3 axes.

Let us pursue the elliptic cylinders point of view further. We now change variables in the rigid body equations within a level surface of K. To simplify notation, we first define the three positive constants $k_i^2, i = 1, 2, 3$, by setting

$$K = \frac{\Pi_1^2}{2k_1^2} + \frac{\Pi_2^2}{2k_2^2} \quad \text{and} \quad N = \frac{\Pi_2^2}{2k_3^2} + \frac{1}{2}\Pi_3^2. \tag{15.11.10}$$

For

$$\frac{1}{k_1^2} = \frac{1}{I_1} - \frac{1}{I_3}, \frac{1}{k_2^2} = \frac{1}{I_2} - \frac{1}{I_3}, \frac{1}{k_3^2} = c\left(\frac{1}{I_1} - \frac{1}{I_2}\right) \tag{15.11.11}$$

for c given by (15.11.7). On the surface $K =$ constant, and setting $r = \sqrt{2K}$ $=$ constant, define new variables θ and p by

$$\Pi_1 = k_1 r \cos\theta, \quad \Pi_2 = k_2 r \sin\theta, \quad \Pi_3 = p. \tag{15.11.12}$$

In terms of these variables, the constants of the motion become

$$K = \frac{1}{2}r^2 \quad \text{and} \quad N = \frac{1}{2}p^2 + \left(\frac{k_2^2}{2k_3^2}r^2\right)\sin^2\theta. \tag{15.11.13}$$

As we shall show below, using a Poisson structure relevant to the equations of motion in the form $\frac{d}{dt}\mathbf{\Pi} = \nabla K \times \nabla N$, the variables θ and p are, up to a scale factor, canonically conjugate, that is, the Poisson bracket of two functions of θ and p are given in standard canonical form (up to a scale factor) as follows:

$$\{F, G\}_{\text{EllipCyl}} = \frac{1}{k_1 k_2}\left(\frac{\partial F}{\partial p}\frac{\partial G}{\partial \theta} - \frac{\partial F}{\partial \theta}\frac{\partial G}{\partial p}\right). \tag{15.11.14}$$

In particular,

$$\{p, \theta\}_{\text{EllipCyl}} = \frac{1}{k_1 k_2}. \tag{15.11.15}$$

The quantity N is the Hamiltonian in these variables—note that N has the form of kinetic plus potential energy—and the equations of motion express themselves in Hamiltonian form in terms of the canonical Poisson bracket. Namely,

$$\frac{d}{dt}\theta = \{N, \theta\}_{\text{EllipCyl}} = \frac{1}{k_1 k_2}\frac{\partial N}{\partial p} = \frac{1}{k_1 k_2}p, \tag{15.11.16}$$

$$\frac{d}{dt}p = \{N, p\}_{\text{EllipCyl}} = \frac{-1}{k_1 k_2}\frac{\partial N}{\partial \theta} = \frac{-1}{k_1 k_2}\frac{k_2^2}{k_3^2}r^2 \sin\theta\cos\theta. \tag{15.11.17}$$

Combining these equations of motion gives

$$\frac{d^2}{dt^2}\theta = \frac{-r^2}{2k_1^2 k_3^2}\sin 2\theta, \tag{15.11.18}$$

or, in terms of the original rigid body parameters,

$$\frac{d^2}{dt^2}\theta = -K\left(\frac{1}{I_1} - \frac{1}{I_2}\right)\sin 2\theta. \tag{15.11.19}$$

Thus, we have proved

Proposition 15.11.1 *Rigid body motion reduces to pendulum motion on level surfaces of K.*

Another way of saying this is as follows: regard rigid body angular momentum space as the union of the level surfaces of K, so the dynamics of the rigid body is recovered by looking at the dynamics on each of these level surfaces. On each level surface, the dynamics is equivalent to a simple pendulum. In this sense, we have proved:

Corollary 15.11.2 *The dynamics of a rigid body in three-dimensional body angular momentum space is a union of two-dimensional simple pendula phase portraits.*

By restricting to a nonzero level surface of K, the pair of rigid body equilibria along the Π_3 axis are excluded. (This pair of equilibria can be included by permuting the indices of the moments of inertia.) The other two pairs of equilibria, along the Π_1 and Π_2 axes, lie in the $p = 0$ plane at $\theta = 0, \pi/2, \pi$, and $3\pi/2$. Since K is positive, the stability of each equilibrium point is determined by the relative sizes of the principal moments of inertia, which affect the overall sign of the right-hand side of the pendulum equation. The well-known results about stability of equilibrium rotations along the least and greatest principal axes, and instability around the intermediate axis, are immediately recovered from this overall sign, combined with the stability properties of the pendulum equilibria. For $K > 0$ and $I_1 < I_2 < I_3$, this overall sign is negative, so the equilibria at $\theta = 0$ and π (along the Π_1 axis) are stable, while those at $\theta = \pi/2$ and $3\pi/2$ (along the Π_2 axis) are unstable. The factor of 2 in the argument of the sine in the pendulum equation is explained by the $\mathbb{Z}_2$ symmetry of the level surfaces of K (or, just as well, by their invariance under $\theta \mapsto \theta + \pi$). Under this discrete symmetry operation, the equilibria at $\theta = 0$ and $\pi/2$ exchange with their counterparts at $\theta = \pi$ and $3\pi/2$, respectively, while the elliptical level surface of K is left invariant. By construction, the Hamiltonian N in the reduced variables θ and p is also invariant under this discrete symmetry.

Let us return to the derivation of the Poisson bracket (15.11.4) on the level surface $K =$ constant. Recall that the *rigid body Poisson bracket* on two functions F_1 and F_2 of $\mathbf{\Pi}$ is given by the *minus Lie-Poisson bracket* for $\mathfrak{so}(3)^*$:

$$\{F_1, F_2\} = -\mathbf{\Pi} \cdot (\nabla F_1 \times \nabla F_2). \tag{15.11.20}$$

If Euler's equations are rewritten as

$$\frac{d}{dt}\mathbf{\Pi} = \nabla K \times \nabla N,$$

where K and N are given as above by an $SL(2, \mathbb{R})$ matrix

$$\begin{pmatrix} K \\ N \end{pmatrix} = \begin{bmatrix} a & b \\ c & d \end{bmatrix} \begin{pmatrix} H \\ L \end{pmatrix},$$

one checks that the equations are Hamiltonian with energy N and the Poisson bracket

$$\{F_1, F_2\}_K = -\nabla K \cdot (\nabla F_1 \times \nabla F_2). \tag{15.11.21}$$

As we saw in Exercise **1.3-2**, this defines a Poisson structure and K is a Casimir function for this bracket. One can now directly verify the formula $\{F, G\}_{\text{EllipCyl}}$ for the Poisson bracket on level sets of the function K in the elliptic cylinder case by a straightforward calculation.

The rigid body can, correspondingly, be regarded as a left invariant system on the group $O(K)$ or $SE(2)$. The special case of $SE(2)$ is the one in which the orbits are cotangent bundles. The fact that one gets a cotangent bundle in this situation is a special case of the cotangent bundle reduction theorem (Volume II) using the semidirect product reduction theorem; see Marsden, Ratiu, and Weinstein [1984a,b]. For the Euclidean group it says that the coadjoint orbits of the Euclidean group of the plane are given by reducing the cotangent bundle of the rotation group of the plane by the trivial group, giving the cotangent bundle of a circle with its canonical symplectic structure up to a factor. This is the abstract explanation of why, in the elliptic cylinder case above, the variables θ and p were, up to a factor, canonically conjugate. This general theory is also consistent with the fact that the Hamiltonian N is of the form kinetic plus potential energy. In fact, in the cotangent bundle reduction theorem, one always gets a Hamiltonian of this form, with the potential being changed by the addition of an amendment to give the *amended potential.* In the case of the pendulum equation, the original Hamiltonian is purely kinetic energy and so the potential term in N, namely $(k_2^2 r^2/2k_3^2)\sin^2\theta$, is entirely amendment. See Volume II for the general theory.

Putting the above discussion together with Exercises **14.9-1** and **14.9-2**, one gets

Theorem 15.11.3 *Euler's equations for a free rigid body are Lie-Poisson with the Hamiltonian N for the Lie algebra $\mathbb{R}^3_K$ where the underlying Lie group is the orthogonal group of K if the quadratic form is nondegenerate, and is the Euclidean group of the plane if K has signature $(+,+,0)$. In particular, all the groups $SO(3), SO(2,1)$, and $SE(2)$ occur as the parameters a, b, c, and d are varied. (If the body is a Lagrange body, then the Heisenberg group occurs as well.)*

The same richness of Hamiltonian structure was found in the Maxwell-Bloch system in David and Holm [1992] (see also David, Holm, and Tratnick [1990]). As in the case of the rigid body, the $\mathbb{R}^3$ motion for the Maxwell-Bloch system may also be realized as motion along the intersections of two orthogonally oriented cylinders. However, in this case, one cylinder is parabolic in cross section, while the other is circular. Upon passing to parabolic cylindrical coordinates, the Maxwell-Bloch system reduces to the ideal Duffing equation, while in circular cylindrical coordinates, the pendulum equation results. The $SL(2,\mathbb{R})$ matrix transformation in the Maxwell-Bloch case provides a parametrized array of (offset) ellipsoids, hyperboloids, and cylinders, along whose intersections the $\mathbb{R}^3$ motion takes place.

References

Abarbanel, H.D.I. and D.D. Holm [1987] Nonlinear stability analysis of inviscid flows in three dimensions: incompressible fluids and barotropic fluids. *Phys. Fluids* **30**, 3369–3382.

Abarbanel, H.D.I., D.D. Holm, J.E. Marsden, and T.S. Ratiu [1986] Nonlinear stability analysis of stratified fluid equilibria. *Phil. Trans. Roy. Soc. London A* **318**, 349–409; also Richardson number criterion for the nonlinear stability of three-dimensional stratified flow. *Phys. Rev. Lett.* **52** [1984], 2552–2555.

Abraham, R. and J.E. Marsden [1978] *Foundations of Mechanics.* Second Edition, Addison-Wesley.

Abraham, R., J.E. Marsden, and T.S. Ratiu [1988] *Manifolds, Tensor Analysis, and Applications.* Second Edition, Applied Mathematical Sciences **75**, Springer-Verlag.

Adams, M.R., J. Harnad, and E. Previato [1988] Isospectral Hamiltonian flows in finite and infinite dimensions I. Generalized Moser systems and moment maps into loop algebras. *Comm. Math. Phys.* **117**, 451–500.

Adams, M.R., T.S. Ratiu, and R. Schmid [1986a] A Lie group structure for pseudodifferential operators. *Math. Ann.* **273**, 529–551.

Adams, M.R., T.S. Ratiu, and R. Schmid [1986b] A Lie group structure for Fourier integral operators. *Math. Ann.* **276**, 19–41.

Adler, M. and van Moerbeke, P. [1980a] Completely integrable systems, Euclidean Lie algebras and curves. *Adv. in Math.* **38**, 267–317.

Adler, M. and van Moerbeke, P. [1980b] Linearization of Hamiltonian systems, Jacobi varieties and representation theory. *Adv. in Math.* **38**, 318–379.

Aeyels, D. and M. Szafranski [1988] Comments on the stabilizability of the angular velocity of a rigid body. *Systems Control Lett.* **10**, 35–39.

Aharonov, Y. and J. Anandan [1987] Phase change during acyclic quantum evolution. *Phys. Rev. Lett.* **58**, 1593–1596.

Alber, M. and J.E. Marsden [1992] On geometric phases for soliton equations. *Comm. Math. Phys.* **149**, 217–240.

Altmann, S.L. [1986] *Rotations, Quaternions, and Double Groups.* Oxford University Press.

Anandan, J. [1988] Geometric angles in quantum and classical physics. *Phys. Lett.* **A 129**, 201–207.

Arens, R. [1970] A quantum dynamical, relativistically invariant rigid body system. *Trans. Amer. Math. Soc.* **147**, 153–201.

Arms, J.M. [1981] The structure of the solution set for the Yang-Mills equations. *Math. Proc. Camb. Philos. Soc.* **90**, 361–372.

Arms, J.M., R.H. Cushman, and M. Gotay [1991] A universal reduction procedure for hamiltonian group actions. *The Geometry of Hamiltonian systems,* T. Ratiu, ed., MSRI Series, **22** Springer-Verlag, 33–52.

Arms, J.M., A. Fischer, and J.E. Marsden [1975] Une approche symplectique pour des théorémes de décomposition en géométrie ou relativité générale. *C. R. Acad. Sci. Paris* **281**, 517–520.

Arms, J.M., J.E. Marsden, and V. Moncrief [1981] Symmetry and bifurcations of momentum mappings. *Comm. Math. Phys.* **78**, 455–478.

Arms, J.M., J.E. Marsden, and V. Moncrief [1982] The structure of the space solutions of Einstein's equations: II Several Killings fields and the Einstein-Yang-Mills equations. *Ann. of Phys.* **144**, 81–106.

Arnold, V.I. [1964] Instability of dynamical systems with several degrees of freedom. *Dokl. Akad. Nauk SSSR* **156**, 9–12.

Arnold, V.I. [1965a] Sur une propriété topologique des applications globelement canoniques de la mécanique classique. *C.R. Acad. Sci. Paris* **26**, 3719–3722.

Arnold, V.I. [1965b] Conditions for nonlinear stability of the stationary plane curvilinear flows of an ideal fluid. *Dokl. Mat. Nauk SSSR* **162**, 773–777.

Arnold, V.I. [1965c] Variational principle for three-dimensional steady-state flows of an ideal fluid. *J. Appl. Math. Mech.* **29**, 1002–1008.

Arnold, V.I. [1966a] Sur la géometrie differentielle des groupes de Lie de dimenson infinie et ses applications à l'hydrodynamique des fluids parfaits. *Ann. Inst. Fourier, Grenoble* **16**, 319–361.

Arnold, V.I. [1966b] On an a priori estimate in the theory of hydrodynamical stability. *Izv. Vyssh. Uchebn. Zaved. Mat. Nauk* **54**, 3–5; English Translation: *Amer. Math. Soc. Transl.* **79** [1969], 267–269.

Arnold, V.I. [1966c] Sur un principe variationnel pour les découlements stationaires des liquides parfaits et ses applications aux problemes de stabilité non linéaires. *J. Mécanique* **5**, 29–43.

Arnold, V.I. [1967] Characteristic class entering in conditions of quantization. *Funct. Anal. Appl.* **1**, 1–13.

Arnold, V.I. [1968] Singularities of differential mappings. *Russian Math. Surveys* **23**, 1–43.

Arnold, V.I. [1969] Hamiltonian character of the Euler equations of the dynamics of solids and of an ideal fluid. *Uspekhi Mat. Nauk* **24**, 225–226.

Arnold, V.I. [1972] Note on the behavior of flows of a three dimensional ideal fluid under a small perturbation of the initial velocity field. *Appl. Math. Mech.* **36**, 255–262.

Arnold, V.I. [1984] *Catastrophe Theory.* Springer-Verlag.

Arnold, V.I. [1988] *Dynamical Systems III.* Encyclopedia of Mathematics **3**, Springer-Verlag.

Arnold, V.I. [1989] *Mathematical Methods of Classical Mechanics.* Second Edition, Graduate Texts in Mathematics **60**, Springer-Verlag.

Arnold, V.I. and B. Khesin [1992] Topological methods in hydrodynamics. *Ann. Rev. Fluid Mech.* **24**, 145–166.

Arnold, V.I., V.V. Kozlov, and A.I. Neishtadt [1988] Mathematical aspects of classical and celestial mechanics, in: *Dynamical Systems III,* V.I. Arnold, ed. Springer-Verlag.

Ashbaugh, M.S., C.C. Chicone, and R.H. Cushman [1990] The twisting tennis racket. *Dyn. Diff. Eqns.* **3**, 67–85.

Atiyah, M. [1982] Convexity and commuting Hamiltonians. *Bull. London Math. Soc.* **14**, 1–15.

Atiyah, M. [1983] Angular momentum, convex polyhedra and algebraic geometry. *Proc. Edinburgh Math. Soc.* **26**, 121–138.

Atiyah, M. and R. Bott [1984] The moment map and equivariant cohomology. *Topology* **23**, 1–28.

Audin, M. [1991] *The Topology of Torus Actions on Symplectic Manifolds.* Progress in Math **93**, Birkhäuser.

Austin, M. and P.S. Krishnaprasad [1993] Almost Poisson Integration of Rigid Body Systems. *J. Comp. Phys.* **106**.

Baider, A., R.C. Churchill, and D.L. Rod [1990] Monodromy and nonintegrability in complex Hamiltonian systems. *J. Dyn. Diff. Eqns.* **2**, 451–481.

Baillieul, J. [1987] Equilibrium mechanics of rotating systems. *Proc. CDC* **26**, 1429–1434.

Baillieul, J. and M. Levi [1987] Rotational elastic dynamics. *Physica D* **27**, 43–62.

Baillieul, J. and M. Levi [1991] Constrained relative motions in rotational mechanics. *Arch. Rat. Mech. Anal.* **115**, 101–135.

Ball, J.M. and J.E. Marsden [1984] Quasiconvexity at the boundary, positivity of the second variation and elastic stability. *Arch. Rat. Mech. Anal.* **86**, 251–277.

Bao, D., J.E. Marsden, and R. Walton [1984] The Hamiltonian structure of general relativistic perfect fluids. *Comm. Math. Phys.* **99**, 319–345.

Bao, D. and V.P. Nair [1985] A note on the covariant anomaly as an equivariant momentum mapping. *Comm. Math. Phys.* **101**, 437–448.

Bates, L. and J. Sniatycki [1993] Nonholonomic reduction. *Rep. Math. Phys.*

Batt, J. and G. Rein [1991] A rigorous stability result for the Vlasov-Poisson system in three dimensions. *Ann. Mat. Pura Appl.* (to appear).

Benjamin, T.B. [1972] The stability of solitary waves. *Proc. Roy. Soc. London* **328A**,153–183.

Benjamin, T.B. [1984] Impulse, flow force and variational principles. *IMA J. Appl. Math.* **32**, 3–68.

Benjamin, T.B. and P.J. Olver [1982] Hamiltonian structure, symmetrics and conservation laws for water waves. *J. Fluid Mech.* **125**, 137–185.

Berezin, F.A. [1967] Some remarks about the associated envelope of a Lie algebra. *Funct. Anal. Appl.* **1**, 91–102.

Bernstein, B. [1958] Waves in a plasma in a magnetic field. *Phys. Rev.* **109**, 10–21.

Berry, M. [1984] Quantal phase factors accompanying adiabatic changes. *Proc. Roy. Soc. London A* **392**, 45–57.

Berry, M. [1985] Classical adiabatic angles and quantal adiabatic phase. *J. Phys. A. Math. Gen.* **18**, 15–27.

Berry, M. [1990] Anticipations of the geometric phase. *Physics Today*, December, 1990, 34–40.

Berry, M. and J. Hannay [1988] Classical non-adiabatic angles. *J. Phys. A. Math. Gen.* **21**, 325–333.

Besse, A.L. [1987] *Einstein Manifolds.* Springer-Verlag.

Bhaskara, K.H. and K. Viswanath [1988] *Poisson Algebras and Poisson Manifolds.* Longman (UK) and Wiley (US).

Bialynicki-Birula, I., J.C. Hubbard, and L.A. Turski [1984] Gauge-independent canonical formulation of relativistic plasma theory. *Physica A* **128**, 509–519.

Birnir, B. [1986] Chaotic perturbations of KdV. *Physica D* **19**, 238–254.

Birnir, B. and R. Grauer [1994] An explicit description of the global attractor of the damped and driven Sine-Gordon equation. *Comm. Math. Phys.* (to appear).

Bloch, A.M., R.W. Brockett, and T.S. Ratiu [1990] A new formulation of the generalized Toda Lattice equations and their fixed point analysis via the momentum map. *Bull. Amer. Math. Soc.* **23**, 477–485.

Bloch, A.M., R.W. Brockett, and T.S. Ratiu [1992] Completely integrable gradient flows. *Comm. Math. Phys.* **147**, 57–74.

Bloch, A.M., H. Flaschka, and T.S. Ratiu [1990] A convexity theorem for isospectral manifolds of Jacobi matrices in a compact Lie algebra. *Duke Math. J.* **61**, 41–65.

Bloch, A.M., P.S. Krishnaprasad, J.E. Marsden, and R. Murray [1994] Nonholonomic mechanical systems with symmetry. (In preparation.)

Bloch, A.M., P.S. Krishnaprasad, J.E. Marsden, and T.S. Ratiu [1991] Asymptotic stability, instability, and stabilization of relative equilibria. *Proc. ACC., Boston IEEE*, 1120–1125.

Bloch, A.M., P.S. Krishnaprasad, J.E. Marsden, and T.S. Ratiu [1994] Dissipation Induced Instabilities, *Ann. Inst. H. Poincaré, Analyse Nonlineare* **11**, 37–90.

Bloch, A.M., P.S. Krishnaprasad, J.E. Marsden, and T.S. Ratiu [1994b] The Euler-Poincaré equations and double bracket dissipation. (Preprint.)

Bloch, A.M., P.S. Krishnaprasad, J.E. Marsden, and G. Sánchez de Alvarez [1992] Stabilization of rigid body dynamics by internal and external torques. *Automatica* **28**, 745–756.

Bloch, A.M. and J.E. Marsden [1989] Controlling homoclinic orbits. *Theoretical and Computational Fluid Mechanics* **1**, 179–190.

Bloch, A.M. and J.E. Marsden [1990] Stabilization of rigid body dynamics by the energy-Casimir method. *Systems Control Lett.* **14**, 341–346.

Bobenko, A.I., A.G. Reyman, and M.A. Semenov-Tian-Shansky [1989] The Kowalewski top 99 years later: A Lax pair, generalizations and explicit solutions. *Comm. Math. Phys.* **122**, 321–354.

Bogoyavlensky, O.I. [1985] *Methods in the Qualitative Theory of Dynamical Systems in Astrophysics and Gas Dynamics.* Springer-Verlag.

Bolza, O. [1973] *Lectures on the Calculus of Variations.* Chicago University Press (1904). Reprinted by Chelsea, (1973).

Bona, J. [1975] On the stability theory of solitary waves. *Proc. Roy. Soc. London* **344A**, 363–374.

Born, M. and L. Infeld [1935] On the quantization of the new field theory. *Proc. Roy. Soc. London A* **150**, 141.

Bortolotti, F. [1926] *Rend. R. Naz. Lincei* **6a**, 552.

Bourbaki, N. [1971] *Variétés differentielles et analytiqes. Fascicule de résultats.* **33**. Hermann.

Bourguignon, J.P. and H. Brezis [1974] Remarks on the Euler equation. *J. Funct. Anal.* **15**, 341–363.

Boya, L.J., J.F. Carinena, and J.M. Gracia-Bondia [1991] Symplectic structure of the Aharonov-Anandan geometric phase. *Phys. Lett. A* **161**, 30–34.

Bretherton, F.P. [1970] A note on Hamilton's principle for perfect fluids. *J. Fluid Mech.* **44**, 19-31.

Bridges, T. [1990] Bifurcation of periodic solutions near a collision of eigenvalues of opposite signature. *Math. Proc. Camb. Philos. Soc.* **108**, 575-601.

Brizard, A. [1992] Hermitian structure for linearised ideal MHD equations with equilibrium flow. *Phys. Lett. A* **168**, 357–362

Brockett, R.W. [1973] Lie algebras and Lie groups in control theory. *Geometric Methods in Systems Theory*, Proc. NATO Advanced Study Institute, R.W. Brockett and D.Q. Mayne (eds.), Reidel, 43–82.

Brockett, R.W. [1976] Nonlinear systems and differential geometry. *Proc. IEEE* **64**, No 1, 61–72.

Brockett, R.W. [1981] Control theory and singular Riemannian geometry. *New Directions in Applied Mathematics*, P.J. Hilton and G.S. Young (eds.), Springer-Verlag.

Brockett, R.W. [1983] Asymptotic stability and feedback stabilization. *Differential Geometric Control Theory*, R.W. Brockett, R.S. Millman, and H. Sussman (eds.), Birkhauser.

Brockett, R.W. [1987] On the control of vibratory actuators. *Proc. 1987 IEEE Conf. Decision and Control*, 1418–1422.

Brockett, R.W. [1989] On the rectification of vibratory motion. *Sensors and Actuators* **20**, 91–96.

Broer, H., S.N. Chow, Y.Kim, and G. Vegter [1993] A normally elliptic Hamiltonian bifurcation. *Z. Angew Math. Phys.* **44**, 389–432.

Burov, A.A. [1986] On the non-existence of a supplementary integral in the problem of a heavy two-link pendulum. *PMM USSR* **50**, 123–125.

Busse, F.H. [1984] Oscillations of a rotating liquid drop. *J. Fluid Mech.* **142**, 1-8.

Cabannes, H. [1962] *Cours de Mécanique Générale.* Dunod.

Calabi, E. [1970] On the group of automorphisms of a symplectic manifold. In *Problems in Analysis*, Princeton University Press, 1–26.

Camassa, R. and D.D. Holm [1992] Dispersive barotropic equations for stratified mesoscale ocean dynamics. *Physica D* **60**, 1–15.

Carinena, J.F., E. Martinez, and J. Fernandez-Nunez [1992] Noether's theorem in time-dependent Lagrangian mechanics. *Rep. Math. Phys.* **31**, 189-203.

Cartan, E. [1922] Sur les petites oscillations d'une masse fluide. *Bull. Sci. Math.* **46**, 317–352 and 356–369.

Cartan, E. [1923] Sur les variétés a connexion affine et théorie de relativité generalizée. *Ann. Ecole Norm. Sup.* **40**, 325–412; **41**, 1–25.

Cartan, E. [1928a] Sur la représentation géométrique des systèmes matèriels non holonomes, *Atti. Cong. Int. Matem.* **4**, 253–261.

Cartan, E. [1928b] Sur la stabilité ordinaire des ellipsoides de Jacobi. *Proc. Int. Math. Cong. Toronto* **2**, 9–17.

Casimir, H.B.G. [1931] *Rotation of a Rigid Body in Quantum Mechanics.* Thesis, J.B. Wolters' Uitgevers-Maatschappij, N.V. Groningen, den Haag, Batavia.

Cendra, H. and J.E. Marsden [1987] Lin constraints, Clebsch potentials and variational principles. *Physica D* **27**, 63–89.

Cendra, H., A. Ibort, and J.E. Marsden [1987] Variational principal fiber bundles: a geometric theory of Clebsch potentials and Lin constraints. *J. Geom. Phys.* **4**, 183–206.

Chandrasekhar, S. [1961] *Hydrodynamic and Hydromagnetic Instabilities.* Oxford University Press.

Chandrasekhar, S. [1977] *Ellipsoidal Figures of Equilibrium.* Dover.

Channell, P. [1983] Symplectic integration algorithms. *Los Alamos National Laboratory Report AT-6:ATN-83-9*.

Channell, P. and C. Scovel [1990] Symplectic integration of Hamiltonian systems. *Nonlinearity* **3**, 231–259.

Chen, F.F. [1974] *Introduction to Plasma Physics*. Plenum.

Chern, S.J. and J.E. Marsden [1990] A note on symmetry and stability for fluid flows. *Geo. Astro. Fluid. Dyn.* **51**, 1–4.

Chernoff, P.R. and J.E. Marsden [1974] *Properties of Infinite Dimensional Hamiltonian systems*. Springer Lect. Notes in Math. **425**.

Chetaev, N.G. [1961] *The Stability of Motion*. Pergamon.

Chetaev, N.G. [1989] *Theoretical Mechanics*. Springer-Verlag.

Chirikov, B.V. [1979] A universal instability of many dimensional oscillator systems. *Phys. Rep.* **52**, 263–379.

Chorin, A.J, T.J.R. Hughes, J.E. Marsden, and M. McCracken [1978] Product formulas and numerical algorithms. *Comm. Pure Appl. Math.* **31**, 205–256.

Chorin, A.J. and J.E. Marsden [1993] *A Mathematical Introduction to Fluid Mechanics*. Third Edition, Texts in Applied Mathematical Sciences **4**, Springer-Verlag.

Chow, S.N. and J.K. Hale [1982] *Methods of Bifurcation Theory*. Springer-Verlag.

Chow, S.N., J.K. Hale, and J. Mallet-Paret [1980] An example of bifurcation to homoclinic orbits. *J. Diff. Eqns.* **37**, 351–373.

Clebsch, A. [1857] Über eine allgemeine Transformation der hydrodynamischen Gleichungen. *Z. Reine Angew. Math.* **54**, 293–312.

Clebsch, A. [1859] Über die Integration der hydrodynamischen Gleichungen. *Z. Reine Angew. Math.* **56**, 1–10.

Clemmow, P.C. and J.P. Dougherty [1959] *Electrodynamics of Particles and Plasmas*. Addison-Wesley.

Conn, J.F. [1984] Normal forms for Poisson structures. *Ann. of Math.* **119**, 576–601.

Conn, J.F. [1985] Normal forms for Poisson structures. *Ann. of Math.* **121**, 565–593.

Cordani, B. [1986] Kepler problem with a magnetic monopole. *J. Math. Phys.* **27**, 2920–2921.

Corson, E.M. [1953] *Introduction to Tensors, Spinors and Relativistic Wave Equations*. Hafner.

Crouch, P.E. [1986] Spacecraft attitude control and stabilization: application of geometric control to rigid body models. *IEEE Trans. Auto. Cont.* **29**, 321–331.

Cushman, R. and D. Rod [1982] Reduction of the semi-simple 1:1 resonance. *Physica D* **6**, 105–112.

Cushman, R. and R. Sjamaar [1991] On singular reduction of Hamiltonian spaces. *Symplectic Geometry and Mathematical Physics*, ed. by P. Donato, C. Duval, J. Elhadad, and G.M. Tuynman, Birkhaüser,114–128.

Dashen, R.F. and D.H. Sharp [1968] Currents as coordinates for hadrons. *Phys. Rev.* **165**, 1857–1866.

David, D. and D.D. Holm [1992] Multiple Lie-Poisson structures. reductions, and geometric phases for the Maxwell-Bloch travelling wave equations, *J. Nonlinear Sci.* **2**, 241–262.

David, D., D.D. Holm, and M. Tratnik [1990] Hamiltonian chaos in nonlinear optical polarization dynamics. *Phys. Rep.* **187**, 281–370.

Davidson, R.C. [1972] *Methods in Nonlinear Plasma Theory*. Academic Press.

De Leon, M. M. H. Mello, and P.R. Rodrigues [1992] Reduction of nondegenerate nonautonomous Lagrangians. *Cont. Math. AMS* **132**, 275-306.

Deift, P.A. and L.C. Li [1989] Generalized affine lie algebras and the solution of a class of flows associated with the QR eigenvalue algorithm. *Comm. Pure Appl. Math.* **42**, 963–991.

Dellnitz, M., J.E. Marsden, I. Melbourne, and J. Scheurle [1992] Generic bifurcations of pendula. *Int. Series on Num. Math.* **104**, 111-122. ed. by G. Allgower, K. Böhmer, and M. Golubitsky, Birkhaüser.

Dellnitz, M., I. Melbourne, and J.E. Marsden [1992] Generic bifurcation of Hamiltonian vector fields with symmetry. *Nonlinearity* **5**, 979–996.

Delshams, A. and T.M. Seara [1991] An asymptotic expression for the splitting of separatrices of the rapidly forced pendulum. *Comm. Math. Phys.* **150**, 433–463.

Delzant, T. [1988] Hamiltoniens périodiques et images convexes de l'application moment. *Bull. Soc. Math. France,* **116**, 315–339.

Delzant, T. [1990] Classification des actions hamiltoniennes complètement intégrables de rang deux. *Ann. Global Anal. Geom.* **8**, 87–112.

Deprit, A. [1983] Elimination of the nodes in problems of N bodies. *Celestial Mech.* **30**, 181–195.

deVogelaére, R. [1956] Methods of integration which preserve the contact transformation property of the Hamiltonian equations. Department of Mathematics, University of Notre Dame Report **4**.

Dirac, P.A.M. [1930] *The Principles of Quantum Mechanics.* Oxford University Press.

Dirac, P.A.M. [1950] Generalized Hamiltonian mechanics. *Canad. J. Math.* **2**, 129–148.

Dirac, P.A.M. [1964] *Lectures on Quantum Mechanics.* Belfer Graduate School of Science, Monograph Series **2**, Yeshiva University.

Duflo, M. and M. Vergne [1969] Une proprieté de la représentation coadjointe d'une algébre de Lie. *C.R. Acad. Sci. Paris* **268**, 583–585.

Duistermaat, H. [1983] *Bifurcations of periodic solutions near equilibrium points of Hamiltonian systems.* Springer Lect. Notes in Math. **1057**, 57–104.

Duistermaat, H. [1984] Non-integrability of 1:2:2 resonance. *Ergodic Theory Dynamical Systems* **4**, 553.

Duistermaat, J.J. [1980] On global action angle coordinates. *Comm. Pure Appl. Math.* **33**, 687–706.

Duistermaat, J.J. and G.J. Heckman [1982] On the variation in the cohomology of the symplectic form of the reduced phase space. *Inv. Math* **69**, 259–269, **72**, 153–158

Dzyaloshinskii, I.E. and G.E. Volovick [1980] Poisson brackets in condensed matter physics. *Ann. of Phys.* **125**, 67–97.

Ebin, D.G. [1970] On the space of Riemannian metrics. *Symp. Pure Math., Am. Math. Soc.* **15**, 11–40.

Ebin, D.G. and J.E. Marsden [1970] Groups of diffeomorphisms and the motion of an incompressible fluid. *Ann. Math.* **92**, 102–163.

Eckard, C. [1960] Variational principles of hydrodynamics. *Phys. Fluids* **3**, 421–427.

Emmrich, C. and H. Römer [1990] Orbifolds as configuration spaces of systems with gauge symmetries. *Comm. Math. Phys.* **129**, 69–94.

Enos, M.J. [1993] On an optimal control problem on $SO(3) \times SO(3)$ and the falling cat. *Fields Inst. Comm.* **1**, 75–112.

Ercolani, N., M.G. Forest, and D.W. McLaughlin [1990] Geometry of the modulational instability, III. Homoclinic orbits for the periodic sine-Gordon equation. *Physica D* **43**, 349.

Ercolani, N., M.G. Forest, D.W. McLaughlin, and R. Montgomery [1987] Hamiltonian structure of modulation equation for the sine-Gordon equation. *Duke Math. J.* **55**, 949-983.

Feng, K. [1986] Difference schemes for Hamiltonian formalism and symplectic geometry. *J. Comp. Math.* **4**, 279–289.

Feng, K. and Z. Ge [1988] On approximations of Hamiltonian systems. *J. Comp. Math.* **6**, 88–97.

Feng, Z. and S. Wiggins [1992], On the existence of chaos in a class of two degree of freedom, damped, parametrically forced mechanical systems with broken $O(2)$ symmetry. *ZAMP*, **44**, 201–248.

Finn, J.M. and G. Sun [1987] Nonlinear Stability and the energy-casimir method. *Comm. on Plasma Phys. and Controlled Fusion* **XI**, 7–25.

Fischer, A.E. and J.E. Marsden [1972] The Einstein equations of evolution — a geometric approach. *J. Math. Phys.* **13**, 546–68.

Fischer, A.E. and J.E. Marsden [1979] Topics in the dynamics of general relativity. *Isolated Gravitating Systems in General Relativity*, J. Ehlers (ed.), Italian Physical Society, 322–395.

Fischer, A.E., J.E. Marsden, and V. Moncrief [1980] The structure of the space of solutions of Einstein's equations, I: One Killing field. *Ann. Inst. H. Poincaré* **33**, 147–194.

Flaschka, H. [1976], The Toda lattice. *Phys. Rev. B* **9**, 1924-1925.

Flaschka, H., A. Newell, and T.S. Ratiu [1983a] Kac-Moody Lie algebras and soliton equations II. Lax equations associated with $A_1^{(1)}$. *Physica D* **9**, 300–323.

Flaschka, H., A. Newell, and T.S. Ratiu [1983b] Kac-Moody Lie algebras and soliton equations III. Stationary equations associated with $A_1^{(1)}$. *Physica D* **9**, 324–332.

Fomenko, A.T [1988a] *Symplectic Geometry.* Gordon and Breach.

Fomenko, A.T [1988b] *Integrability and Nonintegrability in Geometry and Mechanics.* Kluwer Academic.

Fomenko, A.T and V.V. Trofimov [1989] *Integrable Systems on Lie Algebras and Symmetric Spaces.* Gordon and Breach.

Fong, U. and K.R. Meyer [1975] Algebras of integrals. *Rev. Colombiana Mat.* **9**, 75–90.

Fontich, E. and C. Simo [1990] The splitting of separatrices for analytic diffeomorphisms. *Erg. Thy. Dyn. Syst.* **10**, 295–318.

Fowler, T.K. [1963] Liapunov's stability criteria for plasmas. *J. Math. Phys.* **4**, 559–569.

Friedlander, S. and M.M. Vishik [1990] Nonlinear stability for stratified magnetohydrodynamics. *Geophys. Astrophys. Fluid Dyn.* **55**, 19–45.

Galin, D.M. [1982] Versal deformations of linear Hamiltonian systems. *AMS Transl.* **118**, 1–12 (1975 *Trudy Sem. Petrovsk.* **1**, 63–74).

Gallavotti, G. [1983] *The Elements of Mechanics.* Springer-Verlag.

Gantmacher, F.R. [1959] *Theory of Matrices.* Chelsea.

Gardner, C.S. [1971] Korteweg-de Vries equation and generalizations IV. The Korteweg-de Vries equation as a Hamiltonian system. *J. Math. Phys.* **12**, 1548–1551.

Ge, Z. [1990] Generating functions, Hamilton-Jacobi equation and symplectic groupoids over Poisson manifolds. *Indiana Univ. Math. J.* **39**, 859–876.

Ge, Z. [1991a] Equivariant symplectic difference schemes and generating functions. *Physica D* **49**, 376–386.

Ge, Z. [1991b] A constrained variational problem and the space of horizontal paths. *Pacific J. Math.* **149**, 61–94.

Ge, Z. and J.E. Marsden [1988] Lie-Poisson integrators and Lie-Poisson Hamilton-Jacobi theory. *Phys. Lett. A* **133**, 134–139.

Gelfand, I.M. and I.Y. Dorfman [1979] Hamiltonian operators and the algebraic structures connected with them. *Funct. Anal. Appl.* **13**, 13–30.

Gelfand, I.M. and S.V. Fomin [1963] *Calculus of Variations.* Prentice-Hall.

Gibbons, J. [1981] Collisionless Boltzmann equations and integrable moment equations. *Physica A* **3**, 503–511.

Gibbons, J., D.D. Holm, and B.A. Kuperschmidt [1982] Gauge-invariance Poisson brackets for chromohydrodynamics. *Phys. Lett.* **90A**.

Godbillon, C. [1969] *Géometrie Différentielle et Mécanique Analytique.* Hermann.

Goldin, G.A. [1971] Nonrelativistic current algebras as unitary representations of groups. *J. Math. Phys.* **12**, 462–487.

Goldman, W.M. and J.J. Millson [1990] Differential graded Lie algebras and singularities of level sets of momentum mappings. *Comm. Math. Phys.* **131**, 495–515.

Goldstein, H. [1980] *Classical Mechanics.* Second Edition, Addison-Wesley.

Golin, S., A. Knauf, and S. Marmi [1989] The Hannay angles: geometry, adiabaticity, and an example. *Comm. Math. Phys.* **123**, 95–122.

Golin, S. and S. Marmi [1990] A class of systems with measurable Hannay angles. *Nonlinearity* **3**, 507–518.

Golubitsky, M., M. Krupa, and C. Lim [1991] Time reversibility and particle sedimentation. *SIAM J. Appl. Math.* **51**, 49–72.

Golubitsky, M., J.E. Marsden, I. Stewart, and M. Dellnitz [1994] The constrained Liapunov Schmidt procedure and periodic orbits. *Fields Inst. Comm.* (to appear).

Golubitsky, M., and D. Schaeffer [1985] *Singularities and Groups in Bifurcation Theory.* Vol. 1, Applied Mathematical Sciences **69**, Springer-Verlag.

Golubitsky, M. and I. Stewart [1987] Generic bifurcation of Hamiltonian systems with symmetry. *Physica D* **24**, 391–405.

Golubitsky, M., I. Stewart, and D. Schaeffer [1988] *Singularities and Groups in Bifurcation Theory.* Vol. 2, Applied Mathematical Sciences **69**, Springer-Verlag.

Goodman, L.E. and A.R. Robinson [1958] Effects of finite rotations on gyroscopic sensing devices. *J. of Appl. Mech* **28**, 210–213. (See also *Trans. ASME* **80**, 210–213.)

Gotay, M.J. [1988] A multisymplectic approach to the KdV equation: in *Differential Geometric Methods in Theoretical Physics.* Kluwer, 295–305.

Gotay, M.J., J.A. Isenberg, J.E. Marsden, and R. Montgomery [1994] *Momentum Maps and Classical Relativistic Fields.* (In preparation.)

Gotay, M.J., R. Lashof, J. Sniatycki, and A. Weinstein [1980] Closed forms on symplectic fiber bundles. *Comm. Math. Helv.* **58**, 617–621.

Gotay, M.J. and J.E. Marsden [1992] Stress-energy-momentum tensors and the Belifante-Resenfeld formula. *Cont. Math. AMS* **132**, 367–392.

Gotay, M.J., J.M. Nester, and G. Hinds [1979] Presymplectic manifolds and the Dirac-Bergmann theory of constraints. *J. Math. Phys.* **19**, 2388–2399.

Gozzi, E. and W.D. Thacker [1987] Classical adiabatic holonomy in a Grassmannian system. *Phys. Rev. D* **35**, 2388–2396.

Greenspan, B.D. and P.J. Holmes [1983] Repeated resonance and homoclinic bifurcations in a periodically forced family of oscillators. *SIAM J. Math. Anal.* **15**, 69–97.

Griffa, A. [1984] Canonical transformations and variational principles for fluid dynamics. *Physica A* **127**, 265–281.

Grigore, D.R. and O.T. Popp [1989] The complete classification of generalized homogeneous symplectic manifolds. *J. Math. Phys.* **30**, 2476–2483.

Grossman, R., P.S. Krishnaprasad, and J.E. Marsden [1988] The dynamics of two coupled rigid bodies. *Dynamical Systems Approaches to Nonlinear Problems in Systems and Circuits*, Salam and Levi (eds.). SIAM, 373–378.

Gruendler, J. [1985] The existence of homoclinic orbits and the methods of Melnikov for systems. *SIAM J. Math. Anal.* **16**, 907–940.

Guckenheimer, J. and P. Holmes [1983] *Nonlinear Oscillations, Dynamical Systems and Bifurcations of Vector Fields.* Applied Mathematical Sciences **43**, Springer-Verlag.

Guckenheimer, J. and P. Holmes [1988] Structurally stable heteroclinic cycles. *Math. Proc. Camb. Philos. Soc.* **103**, 189–192.

Guckenheimer, J. and A. Mahalov [1992a], Resonant triad interactions in symmetric systems. *Physica D* **54**, 267–310.

Guckenheimer, J. and A. Mahalov [1992b] Instability induced by symmetry reduction. *Phys. Rev. Lett.* **68**, 2257–2260.

Guichardet, A. [1984] On rotation and vibration motions of molecules. *Ann. Inst. H. Poincaré* **40**, 329–342.

Guillemin, V. and A. Pollack [1974] *Differential Topology.* Prentice-Hall.

Guillemin, V. and E. Prato [1990] Heckman, Kostant, and Steinberg formulas for symplectic manifolds. *Adv. in Math.* **82**, 160–179.

Guillemin, V. and S. Sternberg [1977] *Geometric Asymptotics.* Amer. Math. Soc. Surveys, **14**. (Revised edition, 1990.)

Guillemin, V. and S. Sternberg [1980] The moment map and collective motion. *Ann. of Phys.* **1278**, 220–253.

Guillemin, V. and S. Sternberg [1982] Convexity properties of the moment map. *Inv. Math.* **67**, 491-513, **77** (1984) 533–546.

Guillemin, V. and S. Sternberg [1983], On the method of Symes for integrating systems of the Toda type. *Lett. Math. Phys.* **7**, 113-115.

Guillemin, V. and S. Sternberg [1984] *Symplectic Techniques in Physics.* Cambridge University Press.

Guillemin, V. and A. Uribe [1987] Reduction, the trace formula, and semiclassical asymptotics. *Proc. Nat. Acad. Sci.* **84**, 7799–7801.

Hahn, W. [1967] *Stability of Motion.* Springer-Verlag.

Hale, J.K. [1963] *Oscillations in Nonlinear Systems.* McGraw-Hill.

Haller, G. and S. Wiggins [1993] Orbit homoclinic to resonances: the Hamiltonian case. *Physica D* **66**, 293–346.

Hamel, G. [1904] Die Lagrange-Eulerschen Gleichungen der Mechanik. *Z. Mathematik u. Physik* **50**, 1–57.

Hamel, G. [1949] *Theoretische Mechanik*. Springer-Verlag.

Hamilton, W.R. [1834] On a general method in dynamics . *Phil. Trans. Roy. Soc. Lon.* **Part II**, 247–308, **Part I for 1835**, 95–144.

Hannay, J. [1985] Angle variable holonomy in adiabatic excursion of an itegrable Hamiltonian. *J. Phys. A: Math. Gen.* **18**, 221–230.

Hanson, A., T. Regge, and C. Teitelboim [1976] Constrained Hamiltonian systems. *Accademia Nazionale Dei Lincei, Rome*, 1–135.

Helgason, S. [1978] *Differential Geometry, Lie Groups and Symmetric Spaces*, Academic Press.

Henon, M. [1982] Vlasov Equation? *Astron. Astrophys.* **114**, 211–212.

Herivel, J.W. [1955] The derivation of the equation of motion of an ideal fluid by Hamilton's principle. *Proc. Camb. Phil. Soc.* **51**, 344-349.

Hermann, R. [1968] *Differential Geometry and the Calculus of Variations.* Math. Science Press.

Hermann, R. [1973] *Geometry, Physics, and Systems.* Marcel Dekker.

Hirsch, M. and S. Smale [1974] *Differential Equations, Dynamical Systems and Linear Algebra.* Academic Press.

Hirschfelder, J.O. and J.S. Dahler [1956] The kinetic energy of relative motion. *Proc. Nat. Acad. Sci.* **42**, 363–365.

Holm, D.D. and B.A. Kuperschmidt [1983] Poisson brackets and clebsch representations for magnetohydrodynamics, multifluid plasmas, and elasticity. *Physica D* **6**, 347–363.

Holm, D.D., B.A. Kupershmidt, and C.D. Levermore [1985] Hamiltonian differencing of fluid dynamics. *Adv. in Appl. Math.* **6**, 52–84.

Holm, D.D. and J.E. Marsden [1991] The rotor and the pendulum. *Symplectic Geometry and Mathematical Physics*, P. Donato, C. Duval, J. Elhadad, and G.M. Tuynman (eds.), Birkhaüser, pp. 189-203.

Holm, D.D., J.E. Marsden, and T.S. Ratiu [1986] The Hamiltonian structure of continuum mechanics in material, spatial and convective representations. *Séminaire de Mathématiques Supérieurs, Les Presses de l'Univ. de Montréal* **100**, 11–122.

Holm, D.D., J.E. Marsden, T.S. Ratiu, and A. Weinstein [1985] Nonlinear stability of fluid and plasma equilibria. *Phys. Rep.* **123**, 1–116.

Holmes, P.J. (ed) [1980a] *New Approaches to Nonlinear Problems in Dynamics.* SIAM.

Holmes, P.J. [1980b] Averaging and chaotic motions in forced oscillations. *SIAM J. Appl. Math.* **38**, 68–80 and **40**, 167–168.

Holmes, P.J. and J.E. Marsden [1981] A partial differential equation with infinitely many periodic orbits: chaotic oscillations of a forced beam. *Arch. Rat. Mech. Anal.* **76**, 135-166.

Holmes, P.J. and J.E. Marsden [1982a] Horseshoes in perturbations of Hamiltonian systems with two degrees of freedom. *Comm. Math. Phys.* **82**, 523–544.

Holmes, P.J. and J.E. Marsden [1982b] Melnikov's method and Arnold diffusion for perturbations of integrable Hamiltonian systems. *J. Math. Phys.* **23**, 669–675.

Holmes, P.J. and J.E. Marsden [1983] Horseshoes and Arnold diffusion for Hamiltonian systems on Lie groups. *Indiana Univ. Math. J.* **32**, 273–310.

Holmes, P.J., J.E. Marsden, and J. Scheurle [1988] Exponentially small splittings of separatrices with applications to KAM theory and degenerate bifurcations. *Contemp. Math.* **81**, 213–244.

Horn, A. [1954] Doubly stochastic matrices and the diagonal of a rotation matrix. *Amer. J. Math.* **76**, 620–630.

Howard, J.E. and R.S. MacKay [1987a] Linear stability of symplectic maps. *J. Math. Phys.* **28**, 1036–1051.

Howard, J.E. and R.S. MacKay [1987b] Calculation of linear stability boundaries for equilibria of Hamiltonian systems. *Phys. Lett. A* **122**, 331–334.

Hughes, T.J.R., T. Kato, and J.E. Marsden [1977] Well-posed quasi-linear second-order hyperbolic systems with applications to nonlinear elastodynamics and general relativity. *Arch. Rat. Mech. Anal.* **90**, 545–561.

Iacob, A. [1971] Invariant manifolds in the motion of a rigid body about a fixed point. *Rev. Roumaine Math. Pures Appl.* **16**, 1497–1521.

Ichimaru, S. [1973] *Basic Principles of Plasma Physics.* Addison-Wesley.

Isenberg, J. and J.E. Marsden [1982] A slice theorem for the space of solutions of Einstein's equations. *Phys. Rep.* **89**, 179–222.

Ishlinskii, A. [1952] *Mechanics of Special Gyroscopic Systems (in Russian).* National Acad. Ukrainian SSR, Kiev.

Ishlinskii, A. [1963] *Mechanics of Gyroscopic Systems (in English).* Israel Program for Scientific Translations, Jerusalem, 1965, (also available as a NASA technical translation).

Ishlinskii, A. [1976] *Orientation, Gyroscopes and Inertial Navigation (in Russian).* Nauka, Moscow.

Iwai, T. [1982] The symmetry group of the harmonic oscillator and its reduction. *J. Math. Phys.* **23**, 1088–1092.

Iwai, T. [1985] On reduction of two degrees of freedom Hamiltonian systems by an S^1 action, and $SO_0(1,2)$ as a dynamical group. *J. Math. Phys.* **26**, 885-893.

Iwai, T. [1987a] A gauge theory for the quantum planar three-body system. *J. Math. Phys.* **28**, 1315–1326.

Iwai, T. [1987b] A geometric setting for internal motions of the quantum three-body system. *J. Math. Phys.* **28**, 1315–1326.

Iwai, T. [1987c] A geometric setting for classical molecular dynamics. *Ann. Inst. Henri Poincaré, Phys. Théor.* **47**, 199–219.

Iwai, T. [1990a] On the Guichardet/Berry connection. *Phys. Lett. A* **149**, 341–344.

Iwai, T. [1990b] The geometry of the $SU(2)$ Kepler problem. *J. Geom. Phys.* **7**, 507–535.

Iwiínski, Z.R. and L.A. Turski [1976] Canonical theories of systems interacting electromagnetically. *Letters in Applied and Engineering Sciences* **4**, 179–191.

Jacobi, C.G.K. [1837] Note sur l'intégration des équations différentielles de la dynamique. *C.R. Acad. Sci., Paris* **5**, 61.

Jacobi, C.G.K. [1843] *J. Math.* **26**, 115.

Jacobi, C.G.K. [1866] *Vorlesungen über Dynamik.* (Based on lectures given in 1842-3) Verlag G. Reimer; Reprinted by Chelsea, 1969.

Jacobson, N. [1962] *Lie Algebras.* Interscience, reprinted by Dover.

Jeans, J. [1919] *Problems of Cosmogony and Stellar Dynamics.* Cambridge University Press.

Jellinek, J. and D.H. Li [1989] Separation of the energy of overall rotations in an N-body system. *Phys. Rev. Lett.* **62**, 241–244.

Jepson, D.W. and J.O. Hirschfelder [1958] Set of coordinate systems which diagonalize the kinetic energy of relative motion. *Proc. Nat. Acad. Sci.* **45**, 249–256.

Jost, R. [1964] Poisson brackets (An unpedagogical lecture). *Rev. Mod. Phys.* **36**, 572–579.

Kammer, D.C. and G.L. Gray [1992] A nonlinear control design for energy sink simulation in the Euler-Poinsot problem. *J. Astr. Sci.* **41**, 53–72.

Kane, T.R. and M. Scher [1969] A dynamical explanation of the falling cat phenomenon. *Int. J. Solids Structures* **5**, 663–670.

Karasev, M.V. and V.P. Maslov [1993] *Nonlinear Poisson Brackets. Geometry and Quantization.* Transl. of Math. Monographs. **119**. Amer. Math. Soc.

Kato, T. [1950] On the adiabatic theorem of quantum mechanics. *J. Phys. Soc. Japan* **5**, 435–439.

Kato, T. [1967] On classical solutions of the two-dimensional non-stationary Euler equation. *Arch. Rat. Mech. Anal.* **25**, 188–200.

Kato, T. [1972] Nonstationary flows of viscous and ideal fluids in $\mathbb{R}^3$. *J. Funct. Anal.* **9**, 296–305.

Kato, T. [1975] On the initial value problem for quasi-linear symmetric hyperbolic systems. *Arch. Rat. Mech. Anal.* **58** , 181–206.

Kato, T. [1984] *Perturbation Theory for Linear Operators.* Springer-Verlag.

Kato, T. [1985] *Abstract Differential Equations and Nonlinear Mixed Problems.* Lezioni Fermiane, Scuola Normale Superiore, Accademia Nazionale dei Lincei.

Katz, S. [1961] Lagrangian density for an inviscid, perfect, compressible plasma. *Phys. Fluids* **4**, 345–348.

Kaufman, A. [1982] Elementary derivation of Poisson structures for fluid dynamics and electrodynamics. *Phys. Fluids* **25**, 1993–1994.

Kaufman, A. and R.L. Dewar [1984] Canonical derivation of the Vlasov-Coulomb noncanonical Poisson structure. *Contemp. Math.* **28**, 51–54.

Kazhdan, D., B. Kostant, and S. Sternberg [1978] Hamiltonian group actions and dynamical systems of Calogero type. *Comm. Pure Appl. Math.* **31**, 481–508.

Khesin, B.A. [1992] Ergodic interpretation of integral hydrodynamic invariants. *J. Geom. Phys.* (to appear).

Khesin, B.A. and Y. Chekanov [1989] Invariants of the Euler equations for ideal or barotropic hydrodynamics and superconductivity in D dimensions. *Physica D* **40**, 119–131.

Kijowski, J. and W. Tulczyjew [1979] *A Symplectic Framework for Field Theories.* Springer Lect. Notes in Phys. **107**.

Kirillov, A.A. [1962] Unitary representations of nilpotent Lie groups. *Russian Math. Surveys* **17**, 53–104.

Kirillov, A.A. [1976a] Local Lie Algebras. *Russian Math. Surveys* **31**, 55–75.

Kirillov, A.A. [1976b] *Elements of the Theory of Representations.* Grundlehren Math. Wiss., Springer-Verlag.

Kirwan, F.C. [1984] *Cohomology Quotients in Symplectic and Algebraic Geometry.* Princeton Math. Notes **31**, Princeton University Press.

Kirwan, F.C. [1985] Partial desingularization of quotients of nonsingular varieties and their Betti numbers. *Ann. of Math.* **122**, 41–85.

Kirwan, F.C. [1988] The topology of reduced phase spaces of the motion of vortices on a sphere. *Physica D* **30**, 99–123.

Klein, F. [1897] *The Mathematical Theory of the Top.* Scribner.

Klein, M. [1970] *Paul Ehrenfest.* North-Holland.

Klingenberg, W. [1978] *Lectures on Closed Geodesics.* Grundlehren Math. Wiss. **230**, Springer-Verlag.

Knobloch, E. and J.D. Gibbon [1991] Coupled NLS equations for counterpropagating waves in systems with reflection symmetry. *Phys. Lett. A* **154**, 353–356.

Knobloch, E., Mahalov, and J.E. Marsden [1994] Normal forms for three-dimensional parametric instabilities in ideal hydrodynamics. *Physica D* **73**, 49–81.

Knobloch, E. and M. Silber [1992], Hopf bifurcation with $Z_4 \times T^2$ symmetry: in *Bifurcation and Symmetry*, K. Boehmer (ed.). Birkhäuser.

Kobayashi and Nomizu [1963] *Foundations of Differential Geometry.* Wiley

Kocak, H., F. Bisshopp, T. Banchoff, and D. Laidlaw [1986] Topology and Mechanics with Computer Graphics. *Adv. in Appl. Math.* **7**, 282–308.

Koiller, J. [1985] On Aref's vortex motions with a symmetry center. *Physica D* **16**, 27–61.

Koiller, J. [1992] Reduction of some classical nonholonomic systems with symmetry. *Arch. Rat. Mech. Anal.* **118**, 113–148.

Koiller, J., J.M. Balthazar, and T. Yokoyama [1987] Relaxation-Chaos phenomena in celestial mechanics. *Physica D* **26**, 85–122.

Koiller, J., I.D. Soares, and J.R.T. Melo Neto [1985] Homoclinic phenomena in gravitational collapse. *Phys. Lett.* **110A**, 260–264.

Kopell, N. and R.B. Washburn, Jr. [1982] Chaotic motions in the two degree-of-freedom swing equations. *IEEE Trans. Circuits and Systems*, **29**, 738–746.

Korteweg, D.J. and G. de Vries [1895] On the change of form of long waves advancing in a rectangular canal and on a new type of long stationary wave. *Phil. Mag.* **39**, 422–433.

Kosmann-Schwarzbach, Y. and F. Magri [1990] Poisson-Nijenhuis structures. *Ann. Inst. H. Poincaré* **53**, 35–81.

Kostant, B. [1966] Orbits, symplectic structures and representation theory. *Proc. US-Japan Seminar on Diff. Geom., Kyoto. Nippon Hyronsha, Tokyo* 77.

Kostant, B. [1970] *Quantization and unitary representations.* Springer Lect. Notes in Math. **570**, 177–306.

Kostant, B. [1973], On Convexity, the Weyl group and the Iwasawa decomposition. *Ann. Sci. École Norm. Sup.* **6**, 413-455.

Kostant, B. [1978] On Whittaker vectors and representation theory. *Inv. Math.* **48**, 101–184.

Kostant, B. [1979] The solution to a generalized Toda lattice and representation theory. *Adv. in Math.* **34**, 195–338.

Koszul, J.L. [1985] Crochet de Schouten-Nijenhuis et cohomologie, *É. Cartan et les mathématiques d'Aujourdhui, Astérisque hors série, Soc. Math. France*, 257–271.

Kovačič, G. and S. Wiggins [1992] Orbits homoclinic to resonances, with an application to chaos in the damped and forced sine-Gordon equation. *Physica D* **57**, 185.

Krall, N.A. and A.W. Trivelpiece [1973] *Principles of Plasma Physics.* McGraw-Hill.

Krein, M.G. [1950] A generalization of several investigations of A.M. Liapunov on linear differential equations with periodic coefficients. *Dokl. Akad. Nauk. SSSR* **73**, 445–448.

Krishnaprasad, P.S. [1985] Lie-Poisson structures, dual-spin spacecraft and asymptotic stability. *Nonlinear Anal. TMA* **9**, 1011–1035.

Krishnaprasad, P.S. [1989] Eulerian many-body problems. *Cont. Math. AMS* **97**, 187–208.

Krishnaprasad, P.S. and J.E. Marsden [1987] Hamiltonian structure and stability for rigid bodies with flexible attachments. *Arch. Rat. Mech. Anal.* **98**, 137–158.

Krupa, M. [1990] Bifurcations of relative equilibria. *SIAM J. Math. Anal.* **21**, 1453–1486.

Kruse, H.P. [1993] The dynamics of a liquid drop between two plates. Thesis, University of Hamburg.

Kruse, H.P., J.E. Marsden, and J. Scheurle [1993] On uniformly rotating field drops trapped between two parallel plates. *Lect. in Appl. Math. AMS* **29**, 307–317.

Kummer, M. [1975] An interaction of three resonant modes in a nonlinear lattice. *J. Math. Anal. App.* **52**, 64.

Kummer, M. [1979] On resonant classical Hamiltonians with two degrees of freedom near an equilibrium point. *Stochastic Behavior in Classical and Quantum Hamiltonian Systems.* Springer Lect. Notes in Phys. **93**.

Kummer, M. [1981] On the construction of the reduced phase space of a Hamiltonian system with symmetry. *Indiana Univ. Math. J.* **30**, 281–291.

Kummer, M. [1986] On resonant Hamiltonian systems with finitely many degrees of freedom. In *Local and Global Methods of Nonlinear Dynamics,* Lect. Notes in Phys. **252**, Springer-Verlag.

Kummer, M. [1990] On resonant classical hamiltonians with n frequencies. *J. Diff. Eqns.* **83**, 220–243.

Kunzle, H.P. [1969] Degenerate Lagrangian systems. *Ann. Inst. H. Poincaré* **11**, 393–414.

Kunzle, H.P. [1972] Canonical dynamics of spinning particles in gravitational and electromagnetic fields. *J. Math. Phys.* **13**, 739–744.

Kuperschmidt, B.A. and T. Ratiu [1983] Canonical maps between semidirect products with applications to elasticity and superfluids *Comm. Math. Phys.* **90**, 235–250.

Lagrange, J.L. [1788] *Mécanique Analytique.* Chez la Veuve Desaint.

Lanczos, C. [1949] *The Variational Principles of Mechanics.* University of Toronto Press.

Larsson, J. [1992] An action principle for the Vlasov equation and associated Lie perturbation equations. *J. Plasma Phys.* **48**, 13–35; **49**, 255-270.

Laub, A.J. and K.R. Meyer [1974] Canonical forms for symplectic and Hamiltonian matrices. *Celestial Mech.* **9**, 213–238.

Lawden, D.F. [1989] *Elliptic Functions and Applications.* Applied Mathematical Sciences **80**, Springer-Verlag.

Lerman, E. [1989] On the centralizer of invariant functions on a hamiltonian G-space. *J. Diff. Geom.* **30**, 805–815.

Levi, M. [1989] Morse theory for a model space structure. *Cont. Math. AMS* **97**, 209–216

Levi, M. [1993] Geometric phases in the motion of rigid bodies, *Arch. Rat. Mech. Anal.* **122**, 213–229.

Lewis, D. [1989] Nonlinear stability of a rotating planar liquid drop. *Arch. Rat. Mech. Anal.* **106**, 287–333.

Lewis, D [1992] Lagrangian block diagonalization. *Dyn. Diff. Eqns.* **4**, 1–42.

Lewis, D., J.E. Marsden, R. Montgomery, and T.S. Ratiu [1986] The Hamiltonian structure for dynamic free boundary problems. *Physica D* **18**, 391–404.

Lewis, D., J.E. Marsden, and T.S. Ratiu [1987] Stability and bifurcation of a rotating liquid drop. *J. Math. Phys.* **28**, 2508–2515.

Lewis, D., T.S. Ratiu, J.C. Simo, and J.E. Marsden [1992] The heavy top, a geometric treatment. *Nonlinearity* **5**, 1–48.

Lewis, D. and J.C. Simo [1990] Nonlinear stability of rotating pseudo-rigid bodies. *Proc. Roy. Soc. London A* **427**, 281–319.

Li, C.W. and Qin, M.Z. [1988] A symplectic difference scheme for the infinite dimensional Hamilton system. *J. Comp. Math.* **6**, 164–174.

Liapunov, A.M. [1892] *Problème Générale de la Stabilité du Mouvement.* Kharkov. French translation in *Ann. Fac. Sci. Univ.* Toulouse, **9**, 1907; reproduced in *Ann. Math. Studies*, **17**, Princeton University Press, 1949.

Liapunov, A.M. [1897] Sur l'instabilité de l'équilibre dans certains cas où la fonction de forces n'est pas maximum. *J. Math. Appl.* **3**, 81–84.

Libermann, P. and C.M. Marle [1987] *Symplectic Geometry and Analytical Mechanics.* Kluwer Academic.

Lichnerowicz, A. [1951] Sur les variétés symplectiques. *C.R. Acad. Sci. Paris* **233**, 723–726.

Lichnerowicz, A. [1973] Algébre de Lie des automorphismes infinitésimaux d'une structure de contact. *J. Math. Pures Appl.* **52**, 473–508.

Lichnerowicz, A. [1975a] Variété symplectique et dynamique associée à une sous-variété. *C.R. Acad. Sci. Paris, Sér. A* **280**, 523–527.

Lichnerowicz, A. [1975b] Structures de contact et formalisme Hamiltonien invariant. *C.R. Acad. Sci. Paris, Sér. A.* **281**, 171–175.

Lichnerowicz, A. [1976] Variétés symplectiques, variétés canoniques, et systèmes dynamiques, in *Topics in Differential Geometry*, H. Rund and W. Forbes (eds.). Academic Press.

Lichnerowicz, A. [1977] Les variétés de Poisson et leurs algèbres de Lie associées. *J. Diff. Geom.* **12**, 253–300.

Lichnerowicz, A. [1978] Deformation theory and quantization. *Group Theoretical Methods in Physics,* Springer Lect. Notes in Phys. **94**, 280–289.

Lichtenberg, A.J. and M.A. Liebermann [1983] *Regular and Stochastic Motion.* Applied Mathematical Sciences **38**. Springer-Verlag, 2nd edition [1991].

Lie, S. [1890] *Theorie der Transformationsgruppen. Zweiter Abschnitt.* Teubner.

Lin, C.C. [1963] Hydrodynamics of helium II. *Proc. Int. Sch. Phys.* **21**, 93–146.

Littlejohn, R.G. [1983] Variational principles of guiding center motion. *J. Plasma Physics* **29**, 111-125.

Littlejohn, R.G. [1984] Geometry and guiding center motion. *Cont. Math. AMS* **28**, 151-168.

Littlejohn, R.G. [1988] Cyclic evolution in quantum mechanics and the phases of Bohr-Sommerfeld and Maslov. *Phys. Rev. Lett.* **61**, 2159–2162.

Love, A.E.H [1944] *A Treatise on the Mathematical Theory of Elasticity.* Dover.

Low, F.E. [1958] A Lagrangian formulation of the Boltzmann-Vlasov equation for plasmas. *Proc. Roy. Soc. London A* **248**, 282–287.

Lu, J.H. and T.S. Ratiu [1991] On Kostant's convexity theorem. *J. AMS* **4**, 349–364.

Lundgren, T.S. [1963] Hamilton's variational principle for a perfectly conducing plasma continuum. *Phys. Fluids* **6**, 898–904.

MacKay, R.S. [1991] Movement of eigenvalues of Hamiltonian equilibria under non-Hamiltonian perturbation. *Phys. Lett. A* **155**, 266–268.

MacKay, R.S. and J.D. Meiss (Eds.) [1987] *Hamiltonian Dynamical Systems.* Adam Higler, IOP Publishing.

Maddocks, J. [1991] On the stability of relative equilibria. *IMA J. Appl. Math.* **46**, 71–99.

Maddocks, J. and R.L. Sachs [1992] On the stability of KdV multi-solitons. (Preprint.)

Manin, Y.I. [1979] Algebraic aspects of nonlinear differential equations. *J. Soviet Math.* **11**, 1-122.

Marle, C.M. [1976] Symplectic manifolds, dynamical groups and Hamiltonian mechanics; in *Differential Geometry and Relativity*, M. Cahen and M. Flato (eds.), Reidel.

Marsden, J.E. [1981] *Lectures on Geometric Methods in Mathematical Physics.* SIAM.

Marsden, J.E. [1982] A group theoretic approach to the equations of plasma physics. *Canad. Math. Bull.* **25**, 129–142.

Marsden, J.E. [1987] *Appendix to* Golubitsky and Stewart [1987].

Marsden, J.E. [1992], *Lectures on Mechanics.* London Mathematical Society Lecture Note Series, **174**, Cambridge University Press.

Marsden, J.E., D.G. Ebin, and A. Fischer [1972] Diffeomorphism groups, hydrodynamics and relativity. *Proceedings of the 13th Biennial Seminar on Canadian Mathematics Congress*, pp. 135–279.

Marsden, J.E. and T.J.R. Hughes [1983] *Mathematical Foundations of Elasticity.* Prentice-Hall. Dover edition [1994].

Marsden, J.E. and M. McCracken [1976] *The Hopf Bifurcation and its Applications*. Springer Applied Mathematics Series **19**.

Marsden, J.E., R. Montgomery, P. Morrison, and W.B. Thompson [1986] Covariant Poisson brackets for classical fields. *Ann. of Phys.* **169**, 29–48.

Marsden, J.E., R. Montgomery, and T. Ratiu [1989] Cartan-Hannay-Berry phases and symmetry. *Cont. Math. AMS* **97**, 279-295.

Marsden, J.E., R. Montgomery, and T.S. Ratiu [1990] *Reduction, Symmetry, and Phases in Mechanics.* Memoirs AMS **436**.

Marsden, J.E. and P.J. Morrison [1984] Noncanonical Hamiltonian field theory and reduced MHD. *Cont. Math. AMS* **28**, 133–150.

Marsden, J.E., P.J. Morrison, and A. Weinstein [1984] The Hamiltonian structure of the BBGKY hierarchy equations. *Cont. Math. AMS* **28**, 115–124.

Marsden, J.E., O.M. O'Reilly, F.J. Wicklin, and B.W. Zombro [1991] Symmetry, stability, geometric phases, and mechanical integrators. *Nonlinear Science Today* **1**, 4–11; **1**, 14–21.

Marsden, J.E. and T.S. Ratiu [1986] Reduction of Poisson manifolds. *Lett. Math. Phys.* **11**, 161–170.

Marsden, J.E., T.S. Ratiu, and G. Raugel [1991] Symplectic connections and the linearization of Hamiltonian systems. *Proc. Roy. Soc. Edinburgh A* **117**, 329-380

Marsden, J.E., T.S. Ratiu, and A. Weinstein [1984a] Semi-direct products and reduction in mechanics. *Trans. Amer. Math. Soc.* **281**, 147–177.

Marsden, J.E., T.S. Ratiu, and A. Weinstein [1984b] Reduction and Hamiltonian structures on duals of semidirect product Lie Algebras. *Cont. Math. AMS* **28**, 55–100.

Marsden, J.E. and J. Scheurle [1993a] Lagrangian reduction and the double spherical pendulum. *ZAMP* **44**, 17–43.

Marsden, J.E. and J. Scheurle [1993b] The reduced Euler-Lagrange equations. *Fields Institute Comm.* **1**, 139–164.

Marsden, J.E. and A.J. Tromba [1988] *Vector Calculus.* Freeman.

Marsden, J.E. and A. Weinstein [1974] Reduction of symplectic manifolds with symmetry. *Rep. Math. Phys.* **5**, 121–130.

Marsden, J.E. and A. Weinstein [1982] The Hamiltonian structure of the Maxwell-Vlasov equations. *Physica D* **4**, 394–406.

Marsden, J.E. and A. Weinstein [1983] Coadjoint orbits, vortices and Clebsch variables for incompressible fluids. *Physica D* **7**, 305–323.

Marsden, J.E., A. Weinstein, T.S. Ratiu, R. Schmid, and R.G. Spencer [1983] Hamiltonian systems with symmetry, coadjoint orbits and plasma physics. Proc. IUTAM-IS1MM Symposium on *Modern Developments in Analytical Mechanics*, Torino 1982, *Atti della Acad. della Sc. di Torino* **117**, 289–340.

Martin, J.L. [1959] Generalized classical dynamics and the "classical analogue" of a Fermi oscillation. *Proc. Roy. Soc. London A* **251**, 536.

Maslov, V.P. [1965] *Theory of Perturbations and Asymptotic Methods.* Moscow State University.

Mazer, A. and T.S. Ratiu [1989] Hamiltonian formulation of adiabatic free boundary Euler flows. *J. Geom. Phys.* **6**, 271–291.

McLaughlin, D.W., E.A. Overman, S. Wiggins, and C. Xion [1993] Homoclinic orbits in a four-dimensional model of a perturbed NLS equation: a geometric singular perturbation study. (Preprint.)

Melbourne, I., P. Chossat, and M. Golubitsky [1989] Heteroclinic cycles involving periodic solutions in mode interactions with $O(2)$ symmetry. *Proc. Roy. Soc. Edinburgh* **133A**, 315–345.

Melnikov, V.K. [1963] On the stability of the center for time periodic perturbations. *Trans. Moscow Math. Soc.* **12**, 1–57.

Meyer, K.R. [1973] Symmetries and integrals in mechanics. In *Dynamical Systems*, M. Peixoto (ed.). Academic Press, pp. 259–273.

Meyer, K.R. [1981] Hamiltonian systems with a discrete symmetry. *J. Diff. Eqns.* **41**, 228–238.

Meyer, K.R. and R. Hall [1992] *Hamiltonian Mechanics and the n-body Problem.* Applied Mathematical Sciences **90**. Springer-Verlag.

Meyer, K.R. and D.G. Saari (eds.) [1988] *Hamiltonian Dynamical Systems.* Cont. Math. AMS, **81**.

Mielke, A. [1992] *Hamiltonian and Lagrangian Flows on Center Manifolds, with Applications to Elliptic Variational Problems.* Springer Lect. Notes in Math., **1489**.

Mikhailov, G.K. and V.Z. Parton [1990] *Stability and Analytical Mechanics. Applied Mechanics, Soviet Reviews.* **1**, Hemisphere.

Miller, S.C. and R.H. Good [1953] A WKB-type approximation to the Schrödinger equation. *Phys. Rev.* **91**, 174–179.

Milnor, J. [1963] *Morse Theory*. Princeton University Press.

Milnor, J. [1965] *Topology from the Differential Viewpoint*. University of Virginia Press.

Mishchenko, A.S. and Fomenko, A.T. [1976] On the integration of the Euler equations on semisimple Lie algebras. *Sov. Math. Dokl.* **17**, 1591–1593.

Mishchenko, A.S. and Fomenko, A.T. [1978a] Euler equations on finite dimensional Lie groups. *Math. USSR, Izvestija* **12**, 371–389.

Mishchenko, A.S. and Fomenko, A.T. [1978b] Generalized Liouville method of integration of Hamiltonian systems. *Funct. Anal. Appl.* **12**, 113–121.

Mishchenko, A.S. and Fomenko, A.T. [1979] *Symplectic Lie group actions*. Springer Lecture Notes in Mathematics **763**, 504–539.

Misner, C., K. Thorne, and J.A. Wheeler [1973] *Gravitation*. W.H. Freeman, San Francisco.

Mobbs, S.D. [1982] Variational principles for perfect and dissipative fluid flows. *Proc. Roy. Soc. London A* **381**, 457–468.

Montaldi, J.A., R.M. Roberts, and I.N. Stewart [1988] Periodic solutions near equilibria of symmetric Hamiltonian systems. *Phil. Trans. Roy. Soc. London A* **325**, 237–293.

Montaldi, J.A., R.M. Roberts, and I.N. Stewart [1990] Existence of nonlinear normal modes of symmetric Hamiltonian systems. *Nonlinearity* **3**, 695–730, 731–772.

Montgomery, R. [1984] Canonical formulations of a particle in a Yang-Mills field. *Lett. Math. Phys.* **8**, 59–67.

Montgomery, R. [1985] Analytic proof of chaos in the Leggett equations for superfluid ^{3}He. *J. Low Temp. Phys.* **58**, 417–453.

Montgomery, R. [1988] The connection whose holonomy is the classical adiabatic angles of Hannay and Berry and its generalization to the non-integrable case. *Comm. Math. Phys.* **120**, 269–294.

Montgomery, R. [1990] Isoholonomic problems and some applications. *Comm. Math. Phys.* **128**, 565–592.

Montgomery, R. [1991a] How much does a rigid body rotate? A Berry's phase from the eighteenth century. *Amer. J. Phys.* **59**, 394–398.

Montgomery, R. [1991b] *The Geometry of Hamiltonian Systems*, T. Ratiu ed., MSRI Series, **22**. Springer-Verlag.

Montgomery, R., J.E. Marsden, and T.S. Ratiu [1984] Gauged Lie-Poisson structures. *Cont. Math. AMS* **28**, 101–114.

Morozov, V.M., V.N. Rubanovskii, V.V. Rumiantsev, and V.A. Samsonov [1973] On the bifurcation and stability of the steady state motions of complex mechanical systems. *PMM* **37**, 387–399.

Morrison, P.J. [1980] The Maxwell-Vlasov equations as a continuous Hamiltonian system. *Phys. Lett. A* **80**, 383–386.

Morrison, P.J. [1982] Poisson brackets for fluids and plasmas, in *Mathematical Methods in Hydrodynamics and Integrability in Related Dynamical Systems*. M. Tabor and Y.M. Treve (eds.) AIP Conf. Proc. **88**.

Morrison, P.J. [1986] A paradigm for joined Hamiltonian and dissipative systems. *Physica D* **18**, 410–419.

Morrison, P.J. [1987] Variational principle and stability of nonmonotone Vlasov-Poisson equilibria. *Z. Naturforsch.* **42a**, 1115–1123.

Morrison, P.J. [1994] Hamiltonian description of the ideal fluid. *Univ. Texas IFSR Preprint* #640-Review.

Morrison, P.J. and S. Eliezer [1986] Spontaneous symmetry breaking and neutral stability on the noncanonical Hamiltonian formalism. *Phys. Rev. A* **33**, 4205.

Morrison, P.J. and D. Pfirsch [1990] The free energy of Maxwell-Vlasov equilibria. *Phys. Fluids B* **2**, 1105–1113.

Morrison, P.J. and D. Pfirsch [1992] Dielectric energy versus plasma energy, and Hamiltonian action-angle variables for the Vlasov equation. *Phys. Fluids B* **4**, 3038–3057.

Morrison, P.J. and J.M. Greene [1980] Noncanonical Hamiltonian density formulation of hydrodynamics and ideal magnetohydrodynamics. *Phys. Rev. Lett.* **45**, 790–794, errata **48** (1982), 569.

Morrison, P.J. and R.D. Hazeltine [1984] Hamiltonian formulation of reduced magnetohydrodynamics. *Phys. Fluids* **27**, 886–897.

Moser, J. [1958] New aspects in the theory of stability of Hamiltonian systems. *Comm. Pure Appl. Math.* **XI**, 81–114.

Moser, J. [1965] On the volume elements on a manifold. *Trans. Amer. Math. Soc.* **120**, 286–294.

Moser, J. [1973] *Stable and Random Motions in Dynamical Systems with Special Emphasis on Celestial Mechanics.* Princeton University Press.

Moser, J. [1974], *Finitely Many Mass Points on the Line Under the Influence of an Exponential Potential.* Springer Lect. Notes in Phys. **38**, 417-497.

Moser, J. [1975] Three integrable Hamiltonian systems connected with isospectral deformations. *Adv. in Math.* **16**, 197–220.

Moser, J. [1976] Periodic orbits near equilibrium and a theorem by Alan Weinstein. *Comm. Pure Appl. Math.* **29**, 727–747.

Moser, J. [1980] Various aspects of integrable Hamiltonian systems. *Dynamical Systems,* Progress in Math. **8**, Birkhäuser.

Murray, R.M. and S.S. Sastry [1993] Nonholonomic motion planning: steering using sinusoids. *IEEE Trans. on Automatic Control* **38**, 700–716.

Naimark, Ju. I. and N.A. Fufaev [1972] *Dynamics of Nonholonomic Systems.* Translations of Mathematical Monographs, Amer. Math. Soc., vol. **33**.

Nambu, Y. [1973] Generalized Hamiltonian dynamics. *Phys. Rev. D* **7**, 2405–2412.

Nekhoroshev, N.M. [1971a] Behavior of Hamiltonian systems close to integrable. *Funct. Anal. Appl.* **5**, 338–339.

Nekhoroshev, N.M. [1971b] Action angle variables and their generalizations. *Trans. Moscow Math. Soc.* **26**, 180–198.

Nekhoroshev, N.M. [1977] An exponential estimate of the time of stability of nearly integrable Hamiltonian systems. *Russ. Math. Surveys* **32**, 1–65.

Neishtadt, A. [1984] The separation of motions in systems with rapidly rotating phase. *P.M.M. USSR* **48**, 133–139.

Nelson, E. [1959] Analytic vectors. *Ann. Math.* **70**, 572–615.

Newcomb, W.A. [1958] Appendix in Bernstein [1958].

Newcomb, W.A. [1962] Lagrangian and Hamiltonian methods in Magnetohydrodynamics. *Nuc. Fusion* **Suppl., part 2**, 451–463.

Newell, A.C. [1985] *Solitons in Mathematics and Physics*. SIAM.

Nijenhuis, A. [1953] On the holonomy ogroup of linear connections. *Indag. Math.* **15**, 233-249, **16** (1954), 17–25.

Nijenhuis, A. [1955] Jacobi-type identities for bilinear differential con comitants of certain tensor fields. *Indag. Math.* **17**, 390-403.

Nill, F. [1983] An effective potential for classical Yang-Mills fields as outline for bifurcation on gauge orbit space. *Ann. Phys.* **149**, 179–202.

Nirenberg, L. [1959] On elliptic partial differential equations. *Ann. Scuola. Norm. Sup. Pisa* **13**(3), 115–162.

Noether, E. [1918] Invariante Variationsprobleme. *Kgl. Ges. Wiss. Nachr. Göttingen. Math. Physik.* **2**, 235–257.

Oh, Y.G. [1987] A stability criterion for Hamiltonian systems with symmetry. *J. Geom. Phys.* **4**, 163–182.

Oh, Y.G., N. Sreenath, P.S. Krishnaprasad, and J.E. Marsden [1989] The dynamics of coupled planar rigid bodies part 2: bifurcations, periodic solutions, and chaos. *Dynamics Diff. Eqns.* **1**, 269–298.

Olver, P.J. [1980] On the Hamiltonian structure of evolution equations. *Math. Proc. Camb. Philps. Soc.* **88**, 71–88.

Olver, P.J. [1984] Hamiltonian perturbation theory and water waves. *Cont. Math. AMS* **28**, 231–250.

Olver, P.J. [1986] *Applications of Lie Groups to Differential Equations.* Graduate Texts in Mathematics **107**, Springer-Verlag.

Olver, P.J. [1988] Darboux' theorem for Hamiltonian differential operators. *J. Diff. Eqns.* **71**, 10–33.

Otto, M. [1987] A reduction scheme for phase spaces with almost Kähler symmetry regularity results for momentum level sets. *J. Geom. Phys.* **4**, 101–118.

Ovsienko, V.Y. and B.A. Khesin [1987] Korteweg-de Vries superequations as an Euler equation. *Funct. Anal. Appl.* **21**, 329–331.

Palais, R.S. [1968] *Foundations of Global Non-Linear Analysis.* Benjamin.

Paneitz, S.M. [1981] Unitarization of symplectics and stability for causal differential equations in Hilbert space. *J. Funct. Anal.* **41**, 315–326.

Pars, L.A. [1965] *A Treatise on Analytical Dynamics.* Wiley.

Patrick, G. [1989] The dynamics of two coupled rigid bodies in three space. *Cont. Math. AMS* **97**, 315–336.

Pauli, W. [1933] *General Principles of Quantum Mechanics.* Reprinted in English translation by Springer-Verlag (1981).

Pauli, W. [1953] On the Hamiltonian structure of non-local field theories. *Il Nuovo Cimento* **10**, 648–667.

Penrose, O. [1960] Electrostatic instabilities of a uniform non-Maxwellian plasma. *Phys. Fluids* **3**, 258–265.

Percival, I. and D. Richards [1982] *Introduction to Dynamics.* Cambridge Univ. Press.

Perelomov, A.M. [1990] *Integrable Systems of Classical Mechanics and Lie Algebras.* Birkhäuser.

Poincaré, H. [1885] Sur l'équilibre d'une masse fluide animée d'un mouvement de rotation. *Acta Math.* **7**, 259.

Poincaré, H. [1890] Sur la problème des trois corps et les équations de la dynamique. *Acta Math.* **13**, 1–271.

Poincaré, H. [1892] Les formes d'équilibre d'une masse fluide en rotation. *Revue Générale des Sciences* **3**, 809–815.

Poincaré, H. [1901a] Sur la stabilité de l'équilibre des figures piriformes affectées par une masse fluide en rotation. *Philos. Trans. A* **198**, 333–373.

Poincaré, H. [1901b] Sur une forme nouvelle des équations de la mécanique. *C.R. Acad. Sci.* **132**, 369–371.

Poincaré, H. [1910] Sur la precession des corps deformables. *Bull Astron* **27**, 321–356.

Potier-Ferry, M. [1982] On the mathematical foundations of elastic stability theory. *Arch. Rat. Mech. Anal.* **78**, 55–72.

Pressley, A. and Segal G. [1986] *Loop Groups.* Oxford Univ. Press.

Pullin, D.I. and P.G. Saffman [1991] Long time symplectic integration: the example of four-vortex motion. *Proc. Roy. Soc. London A* **432**, 481-494.

Puta, M. [1993] *Hamiltonian Mechanical Systems and Geometric Quantization.* Kluwer.

Rais, M. [1972] Orbites de la représentation coadjointe d'un groupe de Lie, *Représentations des Groupes de Lie Résolubles.* P. Bernat, N. Conze, M. Duflo, M. Lévy-Nahas, M. Rais, P. Renoreard, M. Vergne, eds. *Monographies de la Société Mathématique de France, Dunod, Paris* **4**, 15–27.

Ratiu, T.S. [1980], *Thesis.* University of California at Berkeley.

Ratiu, T.S. [1981] Euler-Poisson equations on Lie algebras and the N-dimensional heavy rigid body. *Proc. Nat. Acad. Sci. USA* **78**, 1327–1328.

Ratiu, T.S. [1982] Euler-Poisson equations on Lie algebras and the N-dimensional heavy rigid body. *Amer. J. Math.* **104**, 409–448, 1337.

Rayleigh, J.W.S. [1880] On the stability or instability of certain fluid motions. *Proc. London. Math. Soc.* **11**, 57–70.

Rayleigh, Lord [1916] On the dynamics of revolving fluids. *Proc. Roy. Soc. London A* **93**, 148–154.

Reeb, G. [1949] Sur les solutions périodiques de certains systèmes différentiels canoniques. *C.R. Acad. Sci. Paris* **228**, 1196–1198.

Reeb, G. [1952] Variétés symplectiques, variétés presque-complexes et systèmes dynamiques. *C.R. Acad. Sci. Paris* **235**, 776–778.

Reed, M. and B. Simon [1974] *Methods on Modern Mathematical Physics.* Vol. 1: *Functional Analysis.* Vol. 2: *Self-adjointness and Fourier Analysis.* Academic Press.

Reyman, A.G. and M.A. Semenov-Tian-Shansky [1990] Group theoretical methods in the theory of integrable systems. *Encyclopedia of Mathematical Sciences* **16**, Springer-Verlag.

Riemann, B. [1860] Untersuchungen über die Bewegung eines flüssigen gleich-artigen Ellipsoides. *Abh. d. Königl. Gesell. der Wiss. zu Göttingen* **9**, 3–36.

Riemann, B. [1861] Ein Beitrag zu den Untersuchungen über die Bewegung eines flüssigen gleichartigen Ellipsoides. *Abh. d. Königl. Gesell. der Wiss. zu Göttingen.*

Robbins, J.M. and M.V. Berry [1992] The geometric phase for chaotic systems. *Proc. Roy. Soc. London A* **436**, 631–661.

Robinson, C. [1970] Generic properties of conservative systems, I, II. *Amer. J. Math.* **92**, 562–603.

Robinson, C. [1975] Fixing the center of mass in the n-body problem by means of a group action. *Colloq. Intern. CNRS* **237**.

Robinson, C. [1988] Horseshoes for autonomous Hamiltonian systems using the Melnikov integral. *Ergodic Theory Dynamical Systems* **8***, 395–409.

Rodrigues, [1840] Title. *J. de Math.* **5**, 380.

Rosenbluth, M.N. [1964] Topics in microinstabilities. *Adv. Plasma Phys.* **137**, 248.

Routh, E.J. [1877] *Stability of a given state of motion.* Macmillan. Reprinted in *Stability of Motion*, A.T. Fuller (ed.), Halsted Press 1975.

Routh, E.J. [1884] *Advanced Rigid Dynamics.* Macmillian.

Rumjantsev, V.V. [1982] On stability problem of a top. *Rend. Sem. Mat. Univ. Padova* **68**, 119–128.

Ruth, R. [1983] A canonical integration techniques, *IEEE Trans. Nucl. Sci.* **30**, 2669–2671.

Rytov, S.M. [1938] Sur la transition de l'optique ondulatoire à l'optique géométrique. *Dokl. Akad. Nauk SSSR* **18**, 263–267.

Salam, F.M.A., J.E. Marsden, and P.P. Varaiya [1983] Arnold diffusion in the swing equations of a power system. *IEEE Trans. CAS* **30**, 697–708, **31** 673–688.

Salam, F.A. and S. Sastry [1985] Complete Dynamics of the forced Josephson junction; regions of chaos. *IEEE Trans. CAS* **32** 784–796.

Salmon, R. [1988] Hamiltonian fluid mechanics. *Ann. Rev. Fluid Mech.* **20**, 225–256.

Sánchez de Alvarez, G. [1986] *Thesis.* University of California at Berkeley.

Sánchez de Alvarez, G. [1989] Controllability of Poisson control systems with symmetry. *Cont. Math. AMS* **97**, 399–412.

Sanders, J.A. [1982] Melnikov's method and averaging. *Celestial Mech.* **28**, 171–181.

Sanders, J.A. and F. Verhulst [1985] *Averaging Methods in Nonlinear Dynamical Systems.* Applied Mathematical Sciences **59**, Springer-Verlag.

Sattinger, D.H. and D. L. Weaver [1986], *Lie Groups and Lie Algebras in Physics, Geometry, and Mechanics.* Applied Mathematical Sciences **61**, Springer-Verlag.

Satzer, W.J. [1977] Canonical reduction of mechanical systems invariant under abelian group actions with an application to celestial mechanics. *Indiana Univ. Math. J.* **26**, 951–976.

Scheurle, J. [1989] Chaos in a rapidly forced pendulum equation. *Cont. Math. AMS* **97**, 411–419.

Scheurle, J., J.E. Marsden, and P.J. Holmes [1991] Exponentially small estimates for separatrix splittings. *Proc. Conf. Beyond all Orders,* H. Segur and S. Tanveer (eds.), Birkhäuser.

Schouten, J.A. [1940] *Ricci Calculus* (2nd Edition 1954). Springer-Verlag.

Schur, I. [1923] Über eine Klasse von Mittelbildungen mit Anwendungen auf Determinantentheorie. *Sitzungsberichte der Berliner Math. Gessellshaft* **22**, 9–20.

Scovel, C. [1991] Symplectic numerical integration of Hamiltonian systems. *Geometry of Hamiltonian systems*, ed. T. Ratiu, MSRI Series **22**, Springer-Verlag, pp. 463–496.

Segal, G. [1991] The geometry of the KdV equation. *Int. J. Mod. Phys.A* **6**, 2859–2869.

Segal, I. [1962] Nonlinear semigroups. *Ann. Math.* **78**, 339–364.

Seliger, R.L. and G.B. Whitham [1968] Variational principles in continuum mechanics. *Proc. Roy. Soc. London* **305**, 1–25.

Serrin, J. [1959] Mathematical principles of classical fluid mechanics. *Handbuch der Physik* **VIII-I**, 125–263. Springer-Verlag.

Shahshahani, S. [1972] Dissipative systems on manifolds. *Inv. Math.* **16**, 177–190.

Shapere, A. and F. Wilczek [1987], Self-propulsion at low Reynolds number. *Phys. Rev. Lett.* **58**, 2051-2054.

Shapere, A. and F. Wilczeck [1989] Geometry of self-propulsion at low Reynolds number. *J. Fluid Mech.* **198**, 557–585.

Shinbrot, T., C. Grebogi, J. Wisdom, and J.A. Yorke [1992] Chaos in a double pendulum. *Amer. J. Phys.* **60**, 491–499.

Simo, J.C. and D.D. Fox [1989] On a stress resultant, geometrically exact shell model. Part I: Formulation and optimal parametrization. *Comp. Meth. Appl. Mech. Engr.* **72**, 267–304.

Simo, J.C., D.D. Fox, and M.S. Rifai [1990], On a stress resultant geometrically exact shell model. Part III: Computational aspects of the nonlinear theory. *Comput. Methods Applied Mech. and Engr.* **79**, 21-70.

Simo, J.C. and D. Lewis [1994] Conserving algorithms for the dynamics of Hamiltonian systems on Lie groups. *J. Nonlinear Sci.*, to appear.

Simo, J.C., D.R. Lewis, and J.E. Marsden [1991] Stability of relative equilibria I: The reduced energy momentum method. *Arch. Rat. Mech. Anal.* **115**, 15-59.

Simo, J.C. and J.E. Marsden [1984] On the rotated stress tensor and a material version of the Doyle Ericksen formula. *Arch. Rat. Mech. Anal.* **86**, 213–231.

Simo, J.C., J.E. Marsden, and P.S. Krishnaprasad [1988] The Hamiltonian structure of nonlinear elasticity: The material, spatial, and convective representations of solids, rods, and plates. *Arch. Rat. Mech. Anal.* **104**, 125–183.

Simo, J.C., T.A. Posbergh, and J.E. Marsden [1990] Stability of coupled rigid body and geometrically exact rods: block diagonalization and the energy-momentum method. *Phys. Rep.* **193**, 280–360.

Simo, J.C., T.A. Posbergh, and J.E. Marsden [1991] Stability of relative equilibria II: Three dimensional elasticity. *Arch. Rat. Mech. Anal.* **115**, 61–100.

Simo, J.C., M.S. Rifai, and D.D Fox [1992], On a stress resultant geometrically exact shell models. Part VI: Conserving algorithms for nonlinear dynamics. *Comp. Meth. Appl. Mech. Engng* **34**, 117–164.

Simo, J.C. and N. Tarnow [1992] The discrete energy momentum method. Conserving algorithms for nonlinear elastodynamics. *ZAMP* **43**, 757–792.

Simo, J.C., N. Tarnow, and K.K. Wong [1992] Exact energy-momentum conserving algorithms and symplectic schemes for nonlinear dynamics. *Comput. Methods Appl. Mech. Engr.* **1**, 63-116.

Simo, J.C. and L. VuQuoc [1985] Three-dimensional finite strain rod model. Part II. Computational Aspects. *Comput. Methods Appl. Mech. Engr.* **58**, 79–116.

Simo, J.C. and L. VuQuoc [1988a] On the dynamics in space of rods undergoing large overall motions-a geometrically exact approach. *Comput. Methods Appl. Mech. Engr.* **66**, 125–161.

Simo, J.C. and L. VuQuoc [1988b] The role of nonlinear theories in the dynamics of fast rotating flexible structures. *J. Sound Vibration* **119**, 487–508.

Simo, J.C. and K.K. Wong [1989] Unconditionally stable algorithms for the orthogonal group that exactly preserve energy and momentum. *Int. J. Num. Meth. Engr.* **31**, 19–52.

Simon, B. [1983] Holonomy, the Quantum Adiabatic Theorem, and Berry's Phase. *Phys. Rev. Letters* **51**, 2167–2170.

Sjamaar, R. and E. Lerman [1991] Stratified symplectic spaces and reduction. *Ann. of Math.* **134**, 375–422.

Slawianowski, J.J. [1971] Quantum relations remaining valid on the classical level. *Rep. Math. Phys.* **2**, 11–34.

Slebodzinski, W. [1931] Sur les équations de Hamilton. *Bull. Acad. Roy. de Belg.* **17**, 864–870.

Slebodzinski, W. [1970] *Exterior Forms and Their Applications.* Polish Scientific.

Slemrod, M. and J.E. Marsden [1985] Temporal and spatial chaos in a van der Waals fluid due to periodic thermal fluctuations. *Adv. Appl. Math.* **6**, 135–158.

Smale, S. [1964] Morse theory and a nonlinear generalization of the Dirichlet problem. *Ann. Math.* **80**, 382–396.

Smale, S. [1967] Differentiable dynamical systems. *Bull. Amer. Math. Soc.* **73**, 747–817. (Reprinted in *The Mathematics of Time.* Springer-Verlag, by S. Smale [1980].)

Smale, S. [1970] Topology and Mechanics. *Inv. Math.* **10**, 305–331, **11**, 45–64.

Sniatycki, J. [1974] Dirac brackets in geometric dynamics. *Ann. Inst. H. Poincaré* **20**, 365–372.

Sniatycki, J. and W. Tulczyjew [1971] Canonical dynamics of relativistic charged particles. *Ann. Inst. H. Poincaré* **15**, 177–187.

Sontag, E.D. and H.J. Sussman [1988] Further comments on the stabilization of the angular velocity of a rigid body. *Systems Control Lett.* **12**, 213–217.

Souriau, J.M. [1970, ©1969] *Structure des Systèmes Dynamiques.* Dunod, Paris.

Spanier, E.H. [1966] *Algebraic Topology.* McGraw-Hill (Reprinted by Springer-Verlag.)

Spencer, R.G. and A.N. Kaufman [1982] Hamiltonian structure of two-fluid plasma dynamics. *Phys. Rev. A* **25**, 2437–2439.

Spivak, M. [1976] *A Comprehensive Introduction to Differential Geometry.* Publish or Perish.

Sreenath, N., Y.G. Oh, P.S. Krishnaprasad, and J.E. Marsden [1988] The dynamics of coupled planar rigid bodies. Part 1: Reduction, equilibria and stability. *Dyn. Stab. Systems* **3**, 25–49.

Stefan, P. [1974] Accessible sets, orbits and foliations with singularities, *Proc. Lond. Math. Soc.* **29**, 699–713.

Sternberg, S. [1963] *Lectures on Differential Geometry.* Prentice-Hall. (Reprinted by Chelsea.)

Sternberg, S. [1969] *Celestial Mechanics,* Vols. I, II. Benjamin-Cummings.

Sternberg, S. [1975] Symplectic homogeneous spaces. *Trans. Amer. Math. Soc.* **212**, 113–130.

Sternberg, S. [1977] Minimal coupling and the symplectic mechanics of a classical particle in the presence of a Yang–Mills field. *Proc. Nat. Acad. Sci.* **74**, 5253–5254.

Su, C.A. [1961] Variational principles in plasma dynamics. *Phys. Fluids* **4**, 1376–1378.

Sudarshan, E.C.G. and N. Mukunda [1974] *Classical Mechanics: A Modern Perspective.* Wiley, 1974; Second Edition, Krieber, 1983.

Sussman, H. [1973] Orbits of families of vector fields and integrability of distributions. *Trans. Amer. Math. Soc.* **180**, 171–188.

Symes, W.W. [1980] Hamiltonian group actions and integrable systems. *Physica D* **1**, 339–374.

Symes, W.W. [1982a], Systems of Toda type, inverse spectral problems and representation theory. *Inv. Math.* **59**, 13-51.

Symes, W.W. [1982b], The QR algorithm and scattering for the nonperiodic Toda lattice. *Physica D* **4**, 275-280.

Szeri, A.J. and P.J. Holmes [1988] Nonlinear stability of axisymmetric swirling flow. *Phil. Trans. Roy. Soc. London A* **326**, 327–354.

Temam, R. [1975] On the Euler equations of incompressible perfect fluids. *J. Funct. Anal.* **20**, 32–43.

Thirring, W.E. [1978] *A Course in Mathematical Physics.* Springer-Verlag.

Thomson, W. (Lord Kelvin) and P.G. Tait [1879] *Treatise on Natural Philosophy.* Cambridge University Press.

Toda, M. [1975], Studies of a non-linear lattice. *Phys. Rep. Phys. Lett.* **8**, 1–125.

Tulczyjew, W.M. [1977] The Legendre transformation. *Ann. Inst. Poincaré* **27**, 101–114.

Vaisman, I. [1987] *Symplectic Geometry and Secondary Characteristic Classes.* Progress in Mathematics **72**, Birkhäuser.

Vaisman, I. [1994] *Lectures on the Geometry of Poisson Manifolds.* Progress in Mathematics **118**, Birkhäuser.

van der Meer, J.C. [1985] *The Hamiltonian Hopf Bifurcation.* Springer Lect. Notes in Math. **1160**.

van der Meer, J.C. [1990] Hamiltonian Hopf bifurcation with symmetry. *Nonlinearity* **3**, 1041–1056.

van der Schaft, A.J. [1982] Hamiltonian dynamics with external forces and observations. *Math. Systems Theory* **15**, 145–168.

van der Schaft, A.J. [1986] Stabilization of Hamiltonian systems. *Nonlinear Amer. TMA* **10**, 1021–1035.

van der Schaft, A.J. and D.E. Crouch [1987] Hamiltonian and self-adjoint control systems. *Systems Control Lett.* **8**, 289–295.

van Kampen, N.G. and B.U. Felderhof [1967] *Theoretical Methods in Plasma Physics.* North-Holland.

van Saarloos, W. [1981] A canonical transformation relating the Lagrangian and Eulerian descriptions of ideal hydrodynamics. *Physica A* **108**, 557–566.

Varadarajan. V.S. [1974] *Lie Groups, Lie Algebras and Their Representations.* Prentice Hall. (Reprinted in Graduate Texts in Mathematics, Springer-Verlag.)

Vershik, A.M. and Faddeev [1981] Lagrangian mechanics in invariant form. *Sel. Math. Sov.* **1**, 339–350.

Vershik, A.M. and V. Ya Gershkovich [1988] Non-holonomic Riemannian manifolds. Encyclopedia of Math. *Dynamical Systems* **7**, Springer-Verlag.

Vinogradov, A.M. and I.S. Krasilshchik [1975] What is the Hamiltonian formalism? *Russ. Math. Surveys* **30**, 177-202.

Vinogradov, A.M. and B.A. Kuperschmidt [1977] The structures of Hamiltonian mechanics. *Russ. Math. Surveys* **32**, 177–243.

Vladimirskii, V.V. [1941] Über die Drehung der Polarisationsebene im gekrümmten Lichtstrahl *Dokl. Akad. Nauk USSR* **21**, 222–225.

Wan, Y.H. [1986] The stability of rotating vortex patches. *Comm. Math. Phys.* **107**, 1–20.

Wan, Y.H. [1988a] Instability of vortex streets with small cores. *Phys. Lett. A* **127**, 27–32.

Wan, Y.H. [1988b] Desingularizations of systems of point vortices. *Physica D* **32**, 277–295.

Wan, Y.H. [1988c] Variational principles for Hill's spherical vortex and nearly spherical vortices. *Trans. Amer. Math. Soc.* **308**, 299–312.

Wan, Y.H. and M. Pulvirente [1984] Nonlinear stability of circular vortex patches. *Comm. Math. Phys.* **99**, 435–450.

Wang, L.S. and P.S. Krishnaprasad [1992] Gyroscopic control and stabilization. *J. Nonlinear Sci.* **2**, 367–415.

Wang, L.S., P.S. Krishnaprasad, and J.H. Maddocks [1991] Hamiltonian dynamics of a rigid body in a central gravitational field. *Cel. Mech. Dyn. Astr.* **50**, 349–386.

Weber, R.W. [1986] Hamiltonian systems with constraints and their meaning in mechanics. *Arch. Rat. Mech. Anal.* **91**, 309–335.

Weinstein, A. [1971] Symplectic manifolds and their Lagrangian submanifolds. *Adv. in Math.* **6**, 329–346 (see also *Bull. Amer. Math. Soc.* **75** (1969), 1040–1041).

Weinstein, A. [1973] Normal modes for nonlinear Hamiltonian systems. *Inv. Math.* **20**, 47–57.

Weinstein, A. [1977] *Lectures on Symplectic Manifolds.* CBMS Regional Conf. Ser. in Math. **29**, Amer. Math. Soc.

Weinstein, A. [1978a] A universal phase space for particles in Yang-Mills fields. *Lett. Math. Phys.* **2**, 417–420.

Weinstein, A. [1978b] Bifurcations and Hamilton's principle. *Math. Z.* **159**, 235–248.

Weinstein, A. [1981] Neighborhood classification of isotropic embeddings. *J. Diff. Geom.* **16**, 125–128.

Weinstein, A. [1983a] Sophus Lie and symplectic geometry. *Exposition Math.* **1**, 95–96.

Weinstein, A. [1983b] The local structure of Poisson manifolds. *J. Diff. Geom.* **18**, 523–557

Weinstein, A. [1984] Stability of Poisson-Hamilton equilibria. *Cont. Math. AMS* **28**, 3–14.

Weinstein, A. [1990] Connections of Berry and Hannay type for moving Lagrangian submanifolds. *Adv. in Math.* **82** 133–159.

Weinstein, A. [1994] Another look at Lagranges equations. *Fields Inst. Comm.*, to appear.

Whittaker, E. T. [1927] *A Treatise on the Analytical Dynamics of Particles and Rigid Bodies.* Cambridge University Press.

Whittaker, E.T. and G.N. Watson [1940] *A Course of Modern Analysis, 4th ed.* Cambridge University Press.

Wiggins, S. [1988] *Global Bifurcations and Chaos.* Texts in Applied Mathematical Sciences **73**, Springer-Verlag.

Wiggins, S. [1990] *Introduction to Applied Nonlinear Dynamical Systems and Chaos.* Texts in Applied Mathematical Sciences **2**, Springer-Verlag.

Wiggins, S. [1992] *Chaotic Transport in Dynamical Systems.* Interdisciplinary Mathematical Sciences, Springer-Verlag.

Wiggins, S. [1993] *Global Dynamics, Phase Space Transport, Orbits Homoclinic to Resonances and Applications.* Fields Institute Monographs. **1**, Amer. Math. Soc.

Wilczek, F. and A. Shapere [1989] Geometry of self-propulsion at low Reynold's number. Efficiencies of self-propulsion at low Reynold's number, *J. Fluid Mech.* **198**, 587–599.

Williamson, J. [1936] On an algebraic problem concerning the normal forms of linear dynamical systems, *Amer. J. Math.* **58**, 141–163. **59**, 599-617.

Wintner, A [1941] *The Analytical Foundations of Celestial Mechanics.* Princeton University Press.

Wisdom, J., S.J. Peale, and F. Mignard [1984] The chaotic rotation of Hyperion. *Icarus* **58**, 137–152.

Wong, S.K. [1970] Field and particle equations for the classical Yang-Mills field and particles with isotopic spin. *Il Nuovo Cimento* **65**, 689–694.

Woodhouse, N.M.J. [1992] *Geometric Quantization.* Clarendon Press, Oxford University Press, 1980, Second Edition, 1992.

Xiao, L. and M.E. Kellman [1989] Unified semiclassical dynamics for molecular resonance spectra. *J. Chem. Phys.* **90**, 6086–6097.

Yang, C.N. [1985] Fiber bundles and the physics of the magnetic monopole. *The Chern Symposium.* Springer-Verlag, pp. 247–254.

Yang, R. and P.S. Krishnaprasad [1990] On the dynamics of floating four bar linkages. *Proc. 28th IEEE Conf. on Decision and Control.*

Zakharov, V.E. [1971] Hamiltonian formalism for hydrodynamic plasma models. *Sov. Phys. JETP* **33**, 927–932.

Zakharov, V.E. [1974] The Hamiltonian formalism for waves in nonlinear media with dispersion. *Izvestia Vuzov, Radiofizika* **17**.

Zakharov, V.E. and L.D. Faddeev [1972] Korteweg-deVries equation: a completely integrable Hamiltonian system. *Funct. Anal. Appl.* **5**, 280–287.

Zakharov, V.E. and E.A. Kuznetsov [1971] Variational principle and canonical variables in magnetohydrodynamics. *Sov. Phys. Dokl.* **15**, 913–914.

Zakharov, V.E. and E.A. Kuznetsov [1974] Three-dimensional solitons. *Sov. Phys. JETP* **39**, 285–286.

Zakharov, V.E. and E.A. Kuznetsov [1984] Hamiltonian formalism for systems of hydrodynamic type. *Math. Phys. Rev.* **4**, 167–220.

Ziglin, S.L. [1980a] Decomposition of separatrices, branching of solutions and nonexistena of an integral in the dynamics of a rigid body. *Trans. Moscow Math. Soc.* **41**, 287.

Ziglin, S.L. [1980b] Nonintegrability of a problem on the motion of four point vortices. *Sov. Math. Dokl.* **21**, 296–299.

Ziglin, S.L. [1981] Branching of solutions and nonexistence of integrals in Hamiltonian systems. *Dokl. Akad. Nauk SSSR* **257**, 26–29; *Funct. Anal. Appl.* **16**, 30–41, **17**, 8–23.

Zombro, B. and P. Holmes [1993] Reduction, stability instability and bifurcation in rotationally symmetric Hamiltonian systems. *Dyn. Stab. Systems* **8** 41–71.

Index